T0189915

Lecture Notes in Computer Science 12534

More information about this subseries at http://www.springer.com/series/7407

Haiqin Yang · Kitsuchart Pasupa ·
Andrew Chi-Sing Leung ·
James T. Kwok · Jonathan H. Chan ·
Irwin King (Eds.)

Neural Information Processing

27th International Conference, ICONIP 2020 Bangkok, Thailand, November 23–27, 2020 Proceedings, Part III

Springer

Editors
Haiqin Yang
Department of AI
Ping An Life
Shenzhen, China

Andrew Chi-Sing Leung
City University of Hong Kong
Kowloon, Hong Kong

Jonathan H. Chan
School of Information Technology
King Mongkut's University
of Technology Thonburi
Bangkok, Thailand

Kitsuchart Pasupa
Faculty of Information Technology
King Mongkut's Institute
of Technology Ladkrabang
Bangkok, Thailand

James T. Kwok
Department of Computer Science
and Engineering
Hong Kong University of Science
and Technology
Hong Kong, Hong Kong

Irwin King
The Chinese University of Hong Kong
New Territories, Hong Kong

ISSN 0302-9743 ISSN 1611-3349 (electronic)
Lecture Notes in Computer Science
ISBN 978-3-030-63835-1 ISBN 978-3-030-63836-8 (eBook)
https://doi.org/10.1007/978-3-030-63836-8

LNCS Sublibrary: SL1 – Theoretical Computer Science and General Issues

This Springer imprint is published by the registered company Springer Nature Switzerland AG
The registered company address is: Gewerbestrasse 11, 6330 Cham, Switzerland

Preface

This book is a part of the five-volume proceedings of the 27th International Conference on Neural Information Processing (ICONIP 2020), held during November 18–22, 2020. The conference aims to provide a leading international forum for researchers, scientists, and industry professionals who are working in neuroscience, neural networks, deep learning, and related fields to share their new ideas, progresses, and achievements. Due to the outbreak of COVID-19, this year's conference, which was supposed to be held in Bangkok, Thailand, was organized as fully virtual conference.

The research program of this year's edition consists of four main categories, Theory and Algorithms, Computational and Cognitive Neurosciences, Human-Centered Computing, and Applications, for refereed research papers with nine special sessions and one workshop. The research tracks attracted submissions from 1,083 distinct authors from 44 countries. All the submissions were rigorously reviewed by the conference Program Committee (PC) comprising 84 senior PC members and 367 PC members. A total of 1,351 reviews were provided, with each submission receiving at least 2 reviews, and some papers receiving 3 or more reviews. This year, we also provided rebuttals for authors to address the errors that exist in the review comments. Meta-reviews were provided with consideration of both authors' rebuttal and reviewers' comments. Finally, we accepted 187 (30.25%) of the 618 full papers that were sent out for review in three volumes of Springer's series of *Lecture Notes in Computer Science* (LNCS) and 189 (30.58%) of the 618 in two volumes of Springer's series of *Communications in Computer and Information Science* (CCIS).

We would like to take this opportunity to thank all the authors for submitting their papers to our conference, and the senior PC members, PC members, as well as all the Organizing Committee members for their hard work. We hope you enjoyed the research program at the conference.

November 2020

Haiqin Yang
Kitsuchart Pasupa

Organization

Honorary Chairs

Jonathan Chan	King Mongkut's University of Technology Thonburi, Thailand
Irwin King	Chinese University of Hong Kong, Hong Kong

General Chairs

Andrew Chi-Sing Leung	City University of Hong Kong, Hong Kong
James T. Kwok	Hong Kong University of Science and Technology, Hong Kong

Program Chairs

Haiqin Yang	Ping An Life, China
Kitsuchart Pasupa	King Mongkut's Institute of Technology Ladkrabang, Thailand

Local Arrangements Chair

Vithida Chongsuphajaisiddhi	King Mongkut University of Technology Thonburi, Thailand

Finance Chairs

Vajirasak Vanijja	King Mongkut's University of Technology Thonburi, Thailand
Seiichi Ozawa	Kobe University, Japan

Special Sessions Chairs

Kaizhu Huang	Xi'an Jiaotong-Liverpool University, China
Raymond Chi-Wing Wong	Hong Kong University of Science and Technology, Hong Kong

Tutorial Chairs

Zenglin Xu	Harbin Institute of Technology, China
Jing Li	Hong Kong Polytechnic University, Hong Kong

Proceedings Chairs

Xinyi Le	Shanghai Jiao Tong University, China
Jinchang Ren	University of Strathclyde, UK

Publicity Chairs

Zeng-Guang Hou	Chinese Academy of Sciences, China
Ricky Ka-Chun Wong	City University of Hong Kong, Hong Kong

Senior Program Committee

Sabri Arik	Istanbul University, Turkey
Davide Bacciu	University of Pisa, Italy
Yi Cai	South China University of Technology, China
Zehong Cao	University of Tasmania, Australia
Jonathan Chan	King Mongkut's University of Technology Thonburi, Thailand
Yi-Ping Phoebe Chen	La Trobe University, Australia
Xiaojun Chen	Shenzhen University, China
Wei Neng Chen	South China University of Technology, China
Yiran Chen	Duke University, USA
Yiu-ming Cheung	Hong Kong Baptist University, Hong Kong
Sonya Coleman	Ulster University, UK
Daoyi Dong	University of New South Wales, Australia
Leonardo Franco	University of Malaga, Spain
Jun Fu	Northeastern University, China
Xin Geng	Southeast University, China
Ping Guo	Beijing Normal University, China
Pedro Antonio Gutiérrez	Universidad de Córdoba, Spain
Wei He	University of Science and Technology Beijing, China
Akira Hirose	The University of Tokyo, Japan
Zengguang Hou	Chinese Academy of Sciences, China
Kaizhu Huang	Xi'an Jiaotong-Liverpool University, China
Kazushi Ikeda	Nara Institute of Science and Technology, Japan
Gwanggil Jeon	Incheon National University, South Korea
Min Jiang	Xiamen University, China
Abbas Khosravi	Deakin University, Australia
Wai Lam	Chinese University of Hong Kong, Hong Kong
Chi Sing Leung	City University of Hong Kong, Hong Kong
Kan Li	Beijing Institute of Technology, China
Xi Li	Zhejiang University, China
Jing Li	Hong Kong Polytechnic University, Hong Kong
Shuai Li	University of Cambridge, UK
Zhiyong Liu	Chinese Academy of Sciences, China
Zhigang Liu	Southwest Jiaotong University, China

Wei Liu	Tencent, China
Jun Liu	Xi'an Jiaotong University, China
Jiamou Liu	The University of Auckland, New Zealand
Lingjia Liu	Virginia Tech, USA
Jose A. Lozano	UPV/EHU, Spain
Bao-liang Lu	Shanghai Jiao Tong University, China
Jiancheng Lv	Sichuan University, China
Marley M. B. R. Vellasco	PUC of Rio de Janeiro, Brazil
Hiroshi Mamitsuka	Kyoto University, Japan
Leandro Minku	University of Birmingham, UK
Chaoxu Mu	Tianjin University, China
Wolfgang Nejdl	L3S Research Center, Germany
Quoc Viet Hung Nguyen	Griffith University, Australia
Takashi Omori	Tamagawa University, Japan
Seiichi Ozawa	Kobe University, Japan
Weike Pan	Shenzhen University, China
Jessie Ju Hyun Park	Yeungnam University, Japan
Kitsuchart Pasupa	King Mongkut's Institute of Technology Ladkrabang, Thailand
Abdul Rauf	Research Institute of Sweden, Sweden
Imran Razzak	Deakin University, Australia
Jinchang Ren	University of Strathclyde, UK
Hayaru Shouno	The University of Electro-Communications, Japan
Ponnuthurai Suganthan	Nanyang Technological University, Singapore
Yang Tang	East China University of Science and Technology, China
Jiliang Tang	Michigan State University, USA
Ivor Tsang	University of Technology Sydney, Australia
Peerapon Vateekul	Chulalongkorn University, Thailand
Brijesh Verma	Central Queensland University, Australia
Li-Po Wang	Nanyang Technological University, Singapore
Kok Wai Wong	Murdoch University, Australia
Ka-Chun Wong	City University of Hong Kong, Hong Kong
Raymond Chi-Wing Wong	Hong Kong University of Science and Technology, Hong Kong
Long Phil Xia	Peking University, Shenzhen Graduate School, China
Xin Xin	Beijing Institute of Technology, China
Guandong Xu	University of Technology Sydney, Australia
Bo Xu	Chinese Academy of Sciences, China
Zenglin Xu	Harbin Institute of Technology, China
Rui Yan	Peking University, China
Xiaoran Yan	Indiana University Bloomington, USA
Haiqin Yang	Ping An Life, China
Qinmin Yang	Zhejiang University, China
Zhirong Yang	Norwegian University of Science and Technology, Norway

De-Nian Yang	Academia Sinica, Taiwan
Zhigang Zeng	Huazhong University of Science and Technology, China
Jialin Zhang	Chinese Academy of Sciences, China
Min Ling Zhang	Southeast University, China
Kun Zhang	Carnegie Mellon University, USA
Yongfeng Zhang	Rutgers University, USA
Dongbin Zhao	Chinese Academy of Sciences, China
Yicong Zhou	University of Macau, Macau
Jianke Zhu	Zhejiang University, China

Program Committee

Muideen Adegoke	City University of Hong Kong, Hong Kong
Sheraz Ahmed	German Research Center for Artificial Intelligence, Germany
Shotaro Akaho	National Institute of Advanced Industrial Science and Technology, Japan
Sheeraz Akram	University of Pittsburgh, USA
Abdulrazak Alhababi	Universiti Malaysia Sarawak, Malaysia
Muhamad Erza Aminanto	University of Indonesia, Indonesia
Marco Anisetti	University of Milan, Italy
Sajid Anwar	Institute of Management Sciences, Pakistan
Muhammad Awais	COMSATS University Islamabad, Pakistan
Affan Baba	University of Technology Sydney, Australia
Boris Bacic	Auckland University of Technology, New Zealand
Mubasher Baig	National University of Computer and Emerging Sciences, Pakistan
Tao Ban	National Information Security Research Center, Japan
Sang Woo Ban	Dongguk University, South Korea
Kasun Bandara	Monash University, Australia
David Bong	Universiti Malaysia Sarawak, Malaysia
George Cabral	Rural Federal University of Pernambuco, Brazil
Anne Canuto	Federal University of Rio Grande do Norte, Brazil
Zehong Cao	University of Tasmania, Australia
Jonathan Chan	King Mongkut's University of Technology Thonburi, Thailand
Guoqing Chao	Singapore Management University, Singapore
Hongxu Chen	University of Technology Sydney, Australia
Ziran Chen	Bohai University, China
Xiaofeng Chen	Chongqing Jiaotong University, China
Xu Chen	Shanghai Jiao Tong University, China
He Chen	Hebei University of Technology, China
Junjie Chen	Inner Mongolia University, China
Mulin Chen	Northwestern Polytechnical University, China
Junying Chen	South China University of Technology, China

Chuan Chen	Sun Yat-sen University, China
Liang Chen	Sun Yat-sen University, China
Zhuangbin Chen	Chinese University of Hong Kong, Hong Kong
Junyi Chen	City University of Hong Kong, Hong Kong
Xingjian Chen	City University of Hong Kong, Hong Kong
Lisi Chen	Hong Kong Baptist University, Hong Kong
Fan Chen	Duke University, USA
Xiang Chen	George Mason University, USA
Long Cheng	Chinese Academy of Sciences, China
Aneesh Chivukula	University of Technology Sydney, Australia
Sung Bae Cho	Yonsei University, South Korea
Sonya Coleman	Ulster University, UK
Fengyu Cong	Dalian University of Technology, China
Jose Alfredo Ferreira Costa	Federal University of Rio Grande do Norte, Brazil
Ruxandra Liana Costea	Polytechnic University of Bucharest, Romania
Jean-Francois Couchot	University of Franche-Comté, France
Raphaël Couturier	University Bourgogne Franche-Comté, France
Zhenyu Cui	University of the Chinese Academy of Sciences, China
Debasmit Das	Qualcomm, USA
Justin Dauwels	Nanyang Technological University, Singapore
Xiaodan Deng	Beijing Normal University, China
Zhaohong Deng	Jiangnan University, China
Mingcong Deng	Tokyo University, Japan
Nat Dilokthanakul	Vidyasirimedhi Institute of Science and Technology, Thailand
Hai Dong	RMIT University, Australia
Qiulei Dong	Chinese Academy of Sciences, China
Shichao Dong	Shenzhen Zhiyan Technology Co., Ltd., China
Kenji Doya	Okinawa Institute of Science and Technology, Japan
Yiqun Duan	University of Sydney, Australia
Aritra Dutta	King Abdullah University of Science and Technology, Saudi Arabia
Mark Elshaw	Coventry University, UK
Issam Falih	Paris 13 University, France
Ozlem Faydasicok	Istanbul University, Turkey
Zunlei Feng	Zhejiang University, China
Leonardo Franco	University of Malaga, Spain
Fulvio Frati	Università degli Studi di Milano, Italy
Chun Che Fung	Murdoch University, Australia
Wai-Keung Fung	Robert Gordon University, UK
Claudio Gallicchio	University of Pisa, Italy
Yongsheng Gao	Griffith University, Australia
Cuiyun Gao	Harbin Institute of Technology, China
Hejia Gao	University of Science and Technology Beijing, China
Yunjun Gao	Zhejiang University, China

Xin Gao	King Abdullah University of Science and Technology, Saudi Arabia
Yuan Gao	Uppsala University, Sweden
Yuejiao Gong	South China University of Technology, China
Xiaotong Gu	University of Tasmania, Australia
Shenshen Gu	Shanghai University, China
Cheng Guo	Chinese Academy of Sciences, China
Zhishan Guo	University of Central Florida, USA
Akshansh Gupta	Central Electronics Engineering Research Institute, India
Pedro Antonio Gutiérrez	University of Córdoba, Spain
Christophe Guyeux	University Bourgogne Franche-Comté, France
Masafumi Hagiwara	Keio University, Japan
Ali Haidar	University of New South Wales, Australia
Ibrahim Hameed	Norwegian University of Science and Technology, Norway
Yiyan Han	Huazhong University of Science and Technology, China
Zhiwei Han	Southwest Jiaotong University, China
Xiaoyun Han	Sun Yat-sen University, China
Cheol Han	Korea University, South Korea
Takako Hashimoto	Chiba University of Commerce, Japan
Kun He	Shenzhen University, China
Xing He	Southwest University, China
Xiuyu He	University of Science and Technology Beijing, China
Wei He	University of Science and Technology Beijing, China
Katsuhiro Honda	Osaka Prefecture University, Japan
Yao Hu	Alibaba Group, China
Binbin Hu	Ant Group, China
Jin Hu	Chongqing Jiaotong University, China
Jinglu Hu	Waseda University, Japan
Shuyue Hu	National University of Singapore, Singapore
Qingbao Huang	Guangxi University, China
He Huang	Soochow University, China
Kaizhu Huang	Xi'an Jiaotong-Liverpool University, China
Chih-chieh Hung	National Chung Hsing University, Taiwan
Mohamed Ibn Khedher	IRT SystemX, France
Kazushi Ikeda	Nara Institute of Science and Technology, Japan
Teijiro Isokawa	University of Hyogo, Japan
Fuad Jamour	University of California, Riverside, USA
Jin-Tsong Jeng	National Formosa University, Taiwan
Sungmoon Jeong	Kyungpook National University, South Korea
Yizhang Jiang	Jiangnan University, China
Wenhao Jiang	Tencent, China
Yilun Jin	Hong Kong University of Science and Technology, Hong Kong

Wei Jin	Michigan State University, USA
Hamid Karimi	Michigan State University, USA
Dermot Kerr	Ulster University, UK
Tariq Khan	Deakin University, Australia
Rhee Man Kil	Korea Advanced Institute of Science and Technology, South Korea
Sangwook Kim	Kobe University, Japan
Sangwook Kim	Kobe University, Japan
DaeEun Kim	Yonsei University, South Korea
Jin Kyu Kim	Facebook, Inc., USA
Mutsumi Kimura	Ryukoku University, Japan
Yasuharu Koike	Tokyo Institute of Technology, Japan
Ven Jyn Kok	National University of Malaysia, Malaysia
Aneesh Krishna	Curtin University, Australia
Shuichi Kurogi	Kyushu Institute of Technology, Japan
Yoshimitsu Kuroki	National Institute of Technology, Kurume College, Japan
Susumu Kuroyanagi	Nagoya Institute of Technology, Japan
Weng Kin Lai	Tunku Abdul Rahman University College, Malaysia
Wai Lam	Chinese University of Hong Kong, Hong Kong
Kittichai Lavangnananda	King Mongkut's University of Technology Thonburi, Thailand
Xinyi Le	Shanghai Jiao Tong University, China
Teerapong Leelanupab	King Mongkut's Institute of Technology Ladkrabang, Thailand
Man Fai Leung	City University of Hong Kong, Hong Kong
Gang Li	Deakin University, Australia
Qian Li	University of Technology Sydney, Australia
Jing Li	University of Technology Sydney, Australia
JiaHe Li	Beijing Institute of Technology, China
Jian Li	Huawei Noah's Ark Lab, China
Xiangtao Li	Jilin University, China
Tao Li	Peking University, China
Chengdong Li	Shandong Jianzhu University, China
Na Li	Tencent, China
Baoquan Li	Tianjin Polytechnic University, China
Yiming Li	Tsinghua University, China
Yuankai Li	University of Science and Technology of China, China
Yang Li	Zhejiang University, China
Mengmeng Li	Zhengzhou University, China
Yaxin Li	Michigan State University, USA
Xiao Liang	Nankai University, China
Hualou Liang	Drexel University, USA
Hao Liao	Shenzhen University, China
Ming Liao	Chinese University of Hong Kong, Hong Kong
Alan Liew	Griffith University, Australia

Kazuteru Miyazaki	National Institution for Academic Degrees and Quality Enhancement of Higher Education, Japan
Bonaventure Molokwu	University of Windsor, Canada
Hiromu Monai	Ochanomizu University, Japan
J. Manuel Moreno	Universitat Politècnica de Catalunya, Spain
Francisco J. Moreno-Barea	University of Malaga, Spain
Chen Mou	Nanjing University of Aeronautics and Astronautics, China
Ahmed Muqeem Sheri	National University of Sciences and Technology, Pakistan
Usman Naseem	University of Technology Sydney, Australia
Mehdi Neshat	The University of Adelaide, Australia
Quoc Viet Hung Nguyen	Griffith University, Australia
Thanh Toan Nguyen	Griffith University, Australia
Dang Nguyen	University of Canberra, Australia
Thanh Tam Nguyen	Ecole Polytechnique Federale de Lausanne, France
Giang Nguyen	Korea Advanced Institute of Science and Technology, South Korea
Haruhiko Nishimura	University of Hyogo, Japan
Stavros Ntalampiras	University of Milan, Italy
Anupiya Nugaliyadde	Murdoch University, Australia
Toshiaki Omori	Kobe University, Japan
Yuangang Pan	University of Technology Sydney, Australia
Weike Pan	Shenzhen University, China
Teerapong Panboonyuen	Chulalongkorn University, Thailand
Paul S. Pang	Federal University Australia, Australia
Lie Meng Pang	Southern University of Science and Technology, China
Hyeyoung Park	Kyungpook National University, South Korea
Kitsuchart Pasupa	King Mongkut's Institute of Technology Ladkrabang, Thailand
Yong Peng	Hangzhou Dianzi University, China
Olutomilayo Petinrin	City University of Hong Kong, Hong Kong
Geong Sen Poh	National University of Singapore, Singapore
Mahardhika Pratama	Nanyang Technological University, Singapore
Emanuele Principi	Università Politecnica delle Marche, Italy
Yiyan Qi	Xi'an Jiaotong University, China
Saifur Rahaman	International Islamic University Chittagong, Bangladesh
Muhammad Ramzan	Saudi Electronic University, Saudi Arabia
Yazhou Ren	University of Electronic Science and Technology of China, China
Pengjie Ren	University of Amsterdam, The Netherlands
Colin Samplawski	University of Massachusetts Amherst, USA
Yu Sang	Liaoning Technical University, China
Gerald Schaefer	Loughborough University, UK

Rafal Scherer	Czestochowa University of Technology, Poland
Xiaohan Shan	Chinese Academy of Sciences, China
Hong Shang	Tencent, China
Nabin Sharma	University of Technology Sydney, Australia
Zheyang Shen	Aalto University, Finland
Yin Sheng	Huazhong University of Science and Technology, China
Jin Shi	Nanjing University, China
Wen Shi	South China University of Technology, China
Zhanglei Shi	City University of Hong Kong, Hong Kong
Tomohiro Shibata	Kyushu Institute of Technology, Japan
Hayaru Shouno	The University of Electro-Communications, Japan
Chiranjibi Sitaula	Deakin University, Australia
An Song	South China University of Technology, China
mofei Song	Southeast University, China
Liyan Song	Southern University of Science and Technology, China
Linqi Song	City University of Hong Kong, Hong Kong
Yuxin Su	Chinese University of Hong Kong, Hong Kong
Jérémie Sublime	Institut supérieur d'électronique de Paris, France
Tahira Sultana	UTM Malaysia, Malaysia
Xiaoxuan Sun	Beijing Normal University, China
Qiyu Sun	East China University of Science and Technology, China
Ning Sun	Nankai University, China
Fuchun Sun	Tsinghua University, China
Norikazu Takahashi	Okayama University, Japan
Hiu-Hin Tam	City University of Hong Kong, Hong Kong
Hakaru Tamukoh	Kyushu Institute of Technology, Japan
Xiaoyang Tan	Nanjing University of Aeronautics and Astronautics, China
Ying Tan	Peking University, China
Shing Chiang Tan	Multimedia University, Malaysia
Choo Jun Tan	Wawasan Open University, Malaysia
Gouhei Tanaka	The University of Tokyo, Japan
Yang Tang	East China University of Science and Technology, China
Xiao-Yu Tang	Zhejiang University, China
M. Tanveer	Indian Institutes of Technology, India
Kai Meng Tay	Universiti Malaysia Sarawak, Malaysia
Chee Siong Teh	Universiti Malaysia Sarawak, Malaysia
Ya-Wen Teng	Academia Sinica, Taiwan
Andrew Beng Jin Teoh	Yonsei University, South Korea
Arit Thammano	King Mongkut's Institute of Technology Ladkrabang, Thailand
Eiji Uchino	Yamaguchi University, Japan

Nhi N.Y. Vo	University of Technology Sydney, Australia
Hiroaki Wagatsuma	Kyushu Institute of Technology, Japan
Nobuhiko Wagatsuma	Tokyo Denki University, Japan
Yuanyu Wan	Nanjing University, China
Feng Wan	University of Macau, Macau
Dianhui Wang	La Trobe University, Australia
Lei Wang	Beihang University, China
Meng Wang	Beijing Institute of Technology, China
Sheng Wang	Henan University, China
Meng Wang	Southeast University, China
Chang-Dong Wang	Sun Yat-sen University, China
Qiufeng Wang	Xi'an Jiaotong-Liverpool University, China
Zhenhua Wang	Zhejiang University of Technology, China
Yue Wang	Chinese University of Hong Kong, Hong Kong
Jiasen Wang	City University of Hong Kong, Hong Kong
Jin Wang	Hanyang University, South Korea
Wentao Wang	Michigan State University, USA
Yiqi Wang	Michigan State University, USA
Peerasak Wangsom	CAT Telecom PCL, Thailand
Bunthit Watanapa	King Mongkut's University of Technology Thonburi, Thailand
Qinglai Wei	Chinese Academy of Sciences, China
Yimin Wen	Guilin University of Electronic Technology, China
Guanghui Wen	Southeast University, China
Ka-Chun Wong	City University of Hong Kong, Hong Kong
Kuntpong Woraratpanya	King Mongkut's Institute of Technology Ladkrabang, Thailand
Dongrui Wu	Huazhong University of Science and Technology, China
Qiujie Wu	Huazhong University of Science and Technology, China
Zhengguang Wu	Zhejiang University, China
Weibin Wu	Chinese University of Hong Kong, Hong Kong
Long Phil Xia	Peking University, Shenzhen Graduate School, China
Tao Xiang	Chongqing University, China
Jiaming Xu	Chinese Academy of Sciences, China
Bin Xu	Northwestern Polytechnical University, China
Qing Xu	Tianjin University, China
Xingchen Xu	Fermilab, USA
Hui Xue	Southeast University, China
Nobuhiko Yamaguchi	Saga University, Japan
Toshiyuki Yamane	IBM Research, Japan
Xiaoran Yan	Indiana University, USA
Shankai Yan	National Institutes of Health, USA
Jinfu Yang	Beijing University of Technology, China
Xu Yang	Chinese Academy of Sciences, China

Feidiao Yang	Chinese Academy of Sciences, China
Minghao Yang	Chinese Academy of Sciences, China
Jianyi Yang	Nankai University, China
Haiqin Yang	Ping An Life, China
Xiaomin Yang	Sichuan University, China
Shaofu Yang	Southeast University, China
Yinghua Yao	University of Technology Sydney, Australia
Jisung Yoon	Indiana University, USA
Junichiro Yoshimoto	Nara Institute of Science and Technology, Japan
Qi Yu	University of New South Wales, Australia
Zhaoyuan Yu	Nanjing Normal University, China
Wen Yu	CINVESTAV-IPN, Mexico
Chun Yuan	Tsinghua University, China
Xiaodong Yue	Shanghai University, China
Li Yun	Nanjing University of Posts and Telecommunications, China
Jichuan Zeng	Chinese University of Hong Kong, Hong Kong
Yilei Zhang	Anhui Normal University, China
Yi Zhang	Beijing Institute of Technology, China
Xin-Yue Zhang	Chinese Academy of Sciences, China
Dehua Zhang	Chinese Academy of Sciences, China
Lei Zhang	Chongqing University, China
Jia Zhang	Microsoft Research, China
Liqing Zhang	Shanghai Jiao Tong University, China
Yu Zhang	Southeast University, China
Liang Zhang	Tencent, China
Tianlin Zhang	University of Chinese Academy of Sciences, China
Rui Zhang	Xi'an Jiaotong-Liverpool University, China
Jialiang Zhang	Zhejiang University, China
Ziqi Zhang	Zhejiang University, China
Jiani Zhang	Chinese University of Hong Kong, Hong Kong
Shixiong Zhang	City University of Hong Kong, Hong Kong
Jin Zhang	Norwegian University of Science and Technology, Norway
Jie Zhang	Newcastle University, UK
Kun Zhang	Carnegie Mellon University, USA
Yao Zhang	Tianjin University, China
Yu Zhang	University of Science and Technology Beijing, China
Zhijia Zhao	Guangzhou University, China
Shenglin Zhao	Tencent, China
Qiangfu Zhao	University of Aizu, Japan
Xiangyu Zhao	Michigan State University, USA
Xianglin Zheng	University of Tasmania, Australia
Nenggan Zheng	Zhejiang University, China
Wei-Long Zheng	Harvard Medical School, USA
Guoqiang Zhong	Ocean University of China, China

Contents – Part III

Biomedical Information

Classification of Neuroblastoma Histopathological Images Using Machine Learning 3
Adhish Panta, Matloob Khushi, Usman Naseem, Paul Kennedy, and Daniel Catchpoole

Data Mining ENCODE Data Predicts a Significant Role of SINA3 in Human Liver Cancer 15
Matloob Khushi, Usman Naseem, Jonathan Du, Anis Khan, and Simon K. Poon

Diabetic Retinopathy Detection Using Multi-layer Neural Networks and Split Attention with Focal Loss 26
Usman Naseem, Matloob Khushi, Shah Khalid Khan, Nazar Waheed, Adnan Mir, Atika Qazi, Bandar Alshammari, and Simon K. Poon

Enhancer-DSNet: A Supervisedly Prepared Enriched Sequence Representation for the Identification of Enhancers and Their Strength 38
Muhammad Nabeel Asim, Muhammad Ali Ibrahim, Muhammad Imran Malik, Andreas Dengel, and Sheraz Ahmed

Machine Learned Pulse Transit Time (MLPTT) Measurements from Photoplethysmography 49
Philip Mehrgardt, Matloob Khushi, Anusha Withana, and Simon Poon

Weight Aware Feature Enriched Biomedical Lexical Answer Type Prediction 63
Keqin Peng, Wenge Rong, Chen Li, Jiahao Hu, and Zhang Xiong

Neural Data Analysis

Decoding Olfactory Cognition: EEG Functional Modularity Analysis Reveals Differences in Perception of Positively-Valenced Stimuli 79
Nida Itrat Abbasi, Sony Saint-Auret, Junji Hamano, Anumita Chaudhury, Anastasios Bezerianos, Nitish V. Thakor, and Andrei Dragomir

Identifying Motor Imagery-Related Electroencephalogram Features During Motor Execution 90
Yuki Kokai, Isao Nambu, and Yasuhiro Wada

Inter and Intra Individual Variations of Cortical Functional Boundaries Depending on Brain States 98
Zhen Zhang, Junhai Xu, Luqi Cheng, Cheng Chen, and Lingzhong Fan

Phase Synchronization Indices for Classification of Action Intention Understanding Based on EEG Signals 110
Xingliang Xiong, Xuesong Lu, Lingyun Gu, Hongfang Han, Zhongxian Hong, and Haixian Wang

The Evaluation of Brain Age Prediction by Different Functional Brain Network Construction Methods. 122
Hongfang Han, Xingliang Xiong, Jianfeng Yan, Haixian Wang, and Mengting Wei

Transfer Dataset in Image Segmentation Use Case 135
Anna Wróblewska, Sylwia Sysko-Romańczuk, and Karol Prusinowski

Neural Network Models

A Gaussian Process-Based Incremental Neural Network for Online Regression 149
Xiaoyu Wang, Lucian Gheorghe, and Jun-ichi Imura

Analysis on the Boltzmann Machine with Random Input Drifts in Activation Function 162
Wenhao Lu, Chi-Sing Leung, and John Sum

Are Deep Neural Architectures Losing Information? Invertibility is Indispensable 172
Yang Liu, Zhenyue Qin, Saeed Anwar, Sabrina Caldwell, and Tom Gedeon

Automatic Dropout for Deep Neural Networks 185
Veena Dodballapur, Rajanish Calisa, Yang Song, and Weidong Cai

Bayesian Randomly Wired Neural Network with Variational Inference for Image Recognition 197
Pegah Tabarisaadi, Abbas Khosravi, and Saeid Nahavandi

Brain-Inspired Framework for Image Classification with a New Unsupervised Matching Pursuit Encoding. 208
Shiming Song, Chenxiang Ma, and Qiang Yu

Estimating Conditional Density of Missing Values Using Deep Gaussian Mixture Model 220
Marcin Przewięźlikowski, Marek Śmieja, and Łukasz Struski

Environmentally-Friendly Metrics for Evaluating the Performance of Deep Learning Models and Systems 232
Sorin Liviu Jurj, Flavius Opritoiu, and Mircea Vladutiu

Hybrid Deep Shallow Network for Assessment of Depression Using Electroencephalogram Signals. 245
Abdul Qayyum, Imran Razzak, and Wajid Mumtaz

Iterative Imputation of Missing Data Using Auto-Encoder Dynamics. 258
Marek Śmieja, Maciej Kołomycki, Łukasz Struski, Mateusz Juda, and Mário A. T. Figueiredo

Multi-objective Evolution for Deep Neural Network Architecture Search 270
Petra Vidnerová and Roman Neruda

Neural Architecture Search for Extreme Multi-label Text Classification 282
Loïc Pauletto, Massih-Reza Amini, Rohit Babbar, and Nicolas Winckler

Non-linear ICA Based on Cramer-Wold Metric. 294
Przemysław Spurek, Aleksandra Nowak, Jacek Tabor, Łukasz Maziarka, and Stanisław Jastrzębski

Oblique Random Forests on Residual Network Features. 306
Wen Xin Cheng, P. N. Suganthan, and Rakesh Katuwal

P2ExNet: Patch-Based Prototype Explanation Network 318
Dominique Mercier, Andreas Dengel, and Sheraz Ahmed

Prediction of Taxi Demand Based on CNN-BiLSTM-Attention Neural Network 331
Xudong Guo

Pruning Long Short Term Memory Networks and Convolutional Neural Networks for Music Emotion Recognition 343
Madeline Brewer and Jessica Sharmin Rahman

Unsupervised Multi-layer Spiking Convolutional Neural Network Using Layer-Wise Sparse Coding 353
Regina Esi Turkson, Hong Qu, Yuchen Wang, and Moses J. Eghan

VAEPP: Variational Autoencoder with a Pull-Back Prior 366
Wenxiao Chen, Wenda Liu, Zhenting Cai, Haowen Xu, and Dan Pei

Why Do Deep Neural Networks with Skip Connections and Concatenated Hidden Representations Work? 380
Oyebade K. Oyedotun and Djamila Aouada

Recommender Systems

AMBR: Boosting the Performance of Personalized Recommendation via Learning from Multi-behavior Data 395
Chen Wang, Shilu Lin, Zhicong Zhong, Yipeng Zhou, and Di Wu

Asymmetric Pairwise Preference Learning for Heterogeneous One-Class Collaborative Filtering 407
Yongxin Ni, Zhuoxin Zhan, Weike Pan, and Zhong Ming

DPR-Geo: A POI Recommendation Model Using Deep Neural Network and Geographical Influence 420
Jun Zeng, Haoran Tang, and Junhao Wen

Feature Aware and Bilinear Feature Equal Interaction Network for Click-Through Rate Prediction. 432
Lang Luo, Yufei Chen, Xianhui Liu, and Qiujun Deng

GFEN: Graph Feature Extract Network for Click-Through Rate Prediction. . . 444
Mei Yu, Chengchang Zhen, Ruiguo Yu, Xuewei Li, Tianyi Xu, Mankun Zhao, Hongwei Liu, Jian Yu, and Xuyuan Dong

JUST-BPR: Identify Implicit Friends with Jump and Stay for Social Recommendation 455
Runsheng Wang, Min Gao, Junwei Zhang, and Quanwu Zhao

Leveraging Knowledge Context Information to Enhance Personalized Recommendation 467
Jiong Wang, Yingshuai Kou, Yifei Zhang, Neng Gao, and ChenYang Tu

LHRM: A LBS Based Heterogeneous Relations Model for User Cold Start Recommendation in Online Travel Platform 479
Ziyi Wang, Wendong Xiao, Yu Li, Zulong Chen, and Zhi Jiang

Match4Rec: A Novel Recommendation Algorithm Based on Bidirectional Encoder Representation with the Matching Task 491
Lingxiao Zhang, Jiangpeng Yan, Yujiu Yang, and Li Xiu

Multi-level Feature Extraction in Time-Weighted Graphical Session-Based Recommendation 504
Mei Yu, Suiwu Li, Ruiguo Yu, Xuewei Li, Tianyi Xu, Mankun Zhao, Hongwei Liu, and Jian Yu

Time Series Analysis

3ETS+RD-LSTM: A New Hybrid Model for Electrical Energy Consumption Forecasting 519
Grzegorz Dudek, Paweł Pełka, and Slawek Smyl

A Deep Time Series Forecasting Method Integrated with Local-Context Sensitive Features 532
Tianyi Chen, Canghong Jin, Tengran Dong, and Dongkai Chen

Benchmarking Adversarial Attacks and Defenses for Time-Series Data 544
Shoaib Ahmed Siddiqui, Andreas Dengel, and Sheraz Ahmed

Correlation-Aware Change-Point Detection via Graph Neural Networks 555
Ruohong Zhang, Yu Hao, Donghan Yu, Wei-Cheng Chang, Guokun Lai, and Yiming Yang

DPAST-RNN: A Dual-Phase Attention-Based Recurrent Neural Network Using Spatiotemporal LSTMs for Time Series Prediction 568
Shajia Shan, Ziyu Shen, Bin Xia, Zheng Liu, and Yun Li

ForecastNet: A Time-Variant Deep Feed-Forward Neural Network Architecture for Multi-step-Ahead Time-Series Forecasting 579
Joel Janek Dabrowski, YiFan Zhang, and Ashfaqur Rahman

Memetic Genetic Algorithms for Time Series Compression by Piecewise Linear Approximation 592
Tobias Friedrich, Martin S. Krejca, J. A. Gregor Lagodzinski, Manuel Rizzo, and Arthur Zahn

Sensor Drift Compensation Using Robust Classification Method 605
Guopei Wu, Junxiu Liu, Yuling Luo, and Senhui Qiu

SpringNet: Transformer and Spring DTW for Time Series Forecasting 616
Yang Lin, Irena Koprinska, and Mashud Rana

U-Sleep: A Deep Neural Network for Automated Detection of Sleep Arousals Using Multiple PSGs 629
Shenglan Yang, Bijue Jia, Yao Chen, Zhan ao Huang, Xiaoming Huang, and Jiancheng Lv

Author Index 641

Biomedical Information

Classification of Neuroblastoma Histopathological Images Using Machine Learning

Adhish Panta[1], Matloob Khushi[1(✉)], Usman Naseem[1], Paul Kennedy[3], and Daniel Catchpoole[2,3(✉)]

[1] School of Computer Science, The University of Sydney, Camperdown, NSW 2006, Australia
matloob.khushi@sydney.edu.au

[2] Children's Cancer Research Unit, The Children's Hospital at Westmead, Westmead, NSW, Australia
daniel.catchpoole@health.nsw.gov.au

[3] School of Computer Science, UTS, Sydney, Australia

Abstract. Neuroblastoma is the most common cancer in young children accounting for over 15% of deaths in children due to cancer. Identification of the class of neuroblastoma is dependent on histopathological classification performed by pathologists which are considered the gold standard. However, due to the heterogeneous nature of neuroblast tumours, the human eye can miss critical visual features in histopathology. Hence, the use of computer-based models can assist pathologists in classification through mathematical analysis. There is no publicly available dataset containing neuroblastoma histopathological images. So, this study uses dataset gathered from The Tumour Bank at Kids Research at The Children's Hospital at Westmead, which has been used in previous research. Previous work on this dataset has shown maximum accuracy of 84%. One main issue that previous research fails to address is the class imbalance problem that exists in the dataset as one class represents over 50% of the samples. This study explores a range of feature extraction and data undersampling and over-sampling techniques to improve classification accuracy. Using these methods, this study was able to achieve accuracy of over 90% in the dataset. Moreover, significant improvements observed in this study were in the minority classes where previous work failed to achieve high level of classification accuracy. In doing so, this study shows importance of effective management of available data for any application of machine learning.

1 Introduction

Neuroblastoma is the most common cancer diagnosed in children in the first year of life and accounts for nearly 15% of deaths in children due to cancer [1,2]. Neuroblast tumours evolve from immature neuroblasts in the sympathetic nervous system during the embryonic, fetal or postnatal stage in children. The disease

H. Yang et al. (Eds.): ICONIP 2020, LNCS 12534, pp. 3–14, 2020.
https://doi.org/10.1007/978-3-030-63836-8_1

spreads typically through bone, bone marrow and the liver, and the tumour can be spotted as mass lesions in areas like neck, chest, abdomen and pelvis [1,2]. Cellular heterogeneity is one of the distinctive features of neuroblastoma [3]. As a result of this feature, neuroblastic tumours show unexpected clinical behaviour, which includes spontaneous regression and aggressive progression. As such, it is common to see genetic materials achieve gains and loss rapidly when neuroblastoma is evolving.

Pathologists use the Shimada system to classify whether a tumour is favourable or unfavourable, which is considered the gold standard in neuroblastoma classification [3]. The Shimada system considers three key factors, which are: age of the patient, the category of the Neuroblast tumour and the Mitosis-Karyorrhexis index (MKI) [3]. To categorise neuroblast tumours, pathologists examine thin tissues using optical microscopes in different magnifications. While identifying the category of tumours, pathologists use several morphological features such as the presence of neuropil, cellularity, nuclear size and shape [3]. However, due to the complex and heterogeneous nature of neuroblastoma, pathologists can get misleading results. The use of machine learning techniques for feature extraction can reveal information and relationships not visible to the human eye. Moreover, the use of Computer-Aided Design (CAD) systems in the health sector offer benefits such as improvement of the overall speed and quality of the diagnosis process by eliminating human fatigue, acting as a tool for the second opinion and assisting with the shortage of medical experts [3]. These benefits are especially significant for neuroblastoma as the age of a patient at the time of diagnosis is vital for the prognosis outcome.

There is a lack of significant exploration of the use of machine learning techniques in for classification of Nuroblastoma types. One of the recent work was done by S. Gheisari et al. [5–7] who classified neuroblastoma histopathological images into five categories which were: undifferentiated neuroblastoma, ganglioneuroblastoma, ganglioneuroma, poorly-differentiated neuroblastoma, and differentiating neuroblastoma. A range of low level and high-level feature extraction techniques were used in previous research to achieve accuracy of around 84%. From a medical perspective, more accurate results would be desirable to increase confidence and improve the chances of computer-based systems being used to assist experts. Furthermore, [5–7] also identified that there was a high degree of misclassification between poorly-differentiated and differentiating neuroblastoma classes. From a biological perspective, these misclassifications are significant as they can result in patients being overtreated or undertreated. Hence there is keen interest in improving existing methods.

This paper aims to improve the previous work performed by S. Gheisari et al. [5–7], where neuroblastoma images were successfully classified into five categories. The dataset used for this work is the same as used in [5–7] which was gathered from The Tumour Bank at Kids Research at The Children's Hospital at Westmead. Through the exploration of previously used feature extraction methods and existing data optimisation techniques, this study aims to improve the overall accuracy metrics achieved in previous neuroblastoma research. The contributions made by this study is summarised below:

1. Explored the impact of different feature extraction and machine learning techniques on the performance for classification of neuroblastoma histopathological images.
2. Evaluated the effects of combining outcomes from different feature extraction techniques on performance metrics.
3. Explored the effects of data engineering techniques such as resampling the given dataset on the overall performance.
4. Improved overall accuracy metrics through the use of these techniques

The rest of the paper is as follows: Sect. 2 discusses the methodology used. Section 3 shows the results. Section 4 presents the discussion and Sect. 5 provides a conclusion.

2 Methodology

To effectively conduct the intended research, this work follows a structure commonly used for the application of machine learning in the medical informatics domain. The first step involves relevant data needs to be collected, or existing dataset needs to be selected to conduct the experiments. Collecting medical image is a complex process which involves taking tissue samples from high resolution microscopes. It would also require expert medical professionals to categorize the data so that models can be trained. So, the collection of raw data was considered out of scope, and an existing neuroblastoma dataset was used to conduct the experiments. The next step involves feature engineering, where the aim is to extract relevant features from available image data. Multiple feature and data engineering techniques are explored in work conducted. Then machine learning classification algorithms are implemented for classification. In this study, SVM classification is used for uniform comparison as a greater focus on the feature extraction and data engineering methods. Finally, the designed models are tested and evaluated using 5-fold cross-validation, accuracy, precision, recall and F-1 score.

2.1 Dataset

The dataset used for this study is the same as used by S. Gheisari et al. in [5–7]. The dataset was gathered from The Tumour Bank at Kids Research at The Children's Hospital at Westmead and is the most comprehensive available dataset for research in neuroblastoma. There is no publicly available dataset for neuroblastoma research, so the authors granted access to this dataset. The dataset contains 1043 images gathered from stained tissue biopsy slides of 125 patients. The tissue slides were scanned by using a software called Imagescope under 40x magnification. Each image was cropped to include 300 by 300 pixels to provide a balance between achieving a reasonable computational time and preserving critical information in each image. It was also ensured that each cropped image contained areas that best represent each category of neuroblastoma. This is a slight limitation of the dataset because it does not directly mimic a real-world scenario where data can have noise points. Expert pathologists classified images in the dataset according to the Shimada System where the categories included: poorly differentiated neuroblastoma, differentiating neuroblastoma, undifferentiating neuroblastoma, ganglioneuroma and ganglioneuroblastoma.

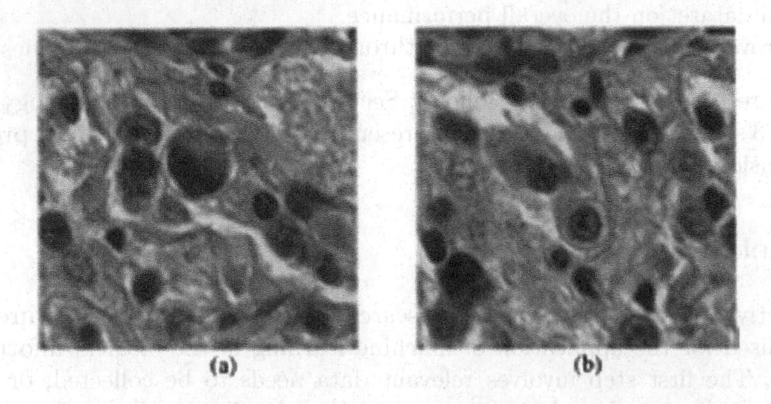

Fig. 1. Example of intra-class variance in neuroblastoma

As discussed previously, neuroblastoma has a high degree of intra-class variance. The extent of intra-class variance in the available dataset can be seen in Fig. 1. Both (a) and (b) in the figure belong to the same class of neuroblastoma (differentiating neuroblastoma). However, as the blue circled section in the image indicates, they have neuroblast cells of differing sizes. This is a prominent feature of neuroblast tumours and has been extensively captured in the dataset. The overall dataset used for the experimental setup can be summarised in Table 1. The table shows data distribution for each class and the number of patients from whom the images were gathered. It can be seen that overall, there are 1043 data samples from 125 patients. It can also be seen that there is a high degree of imbalance in the used dataset. The most common class is poorly-differentiated neuroblastoma as it represents over 50% of the data with 571 samples. For the least common class (ganglioneuroblastoma), there are only 46 samples available which represents less than 5% of the dataset and is gathered from 8 out of 125 patients.

2.2 Experimental Setup

The study explored a range of experimental approaches around feature extraction and data undersampling and oversampling to determine the effects they have in the given dataset. These experimental setups are explained below. To evaluate the setups, k-fold stratified k-fold cross validation was used. A stratified k-fold cross validation ensures that for each split in k, the original distribution of dataset is maintained. Accuracy, precision, recall and F1-score were used as

Table 1. Dataset used in the study

Neuroblastoma tumour class	No. cropped images	No. patients
Poorly differentiated	571	77
Differentiating	187	12
Undifferentiated	155	10
Ganglioneuroma	84	28
Ganglioneuroblastoma	46	8
Total	1043	125

metrics used for comparison of the different approaches and are presented in the results section.

Setup 1 – Scale Invariant Feature Transform (SIFT) + Bag of Visual Words (BOVW) + SVM

SIFT: SIFT feature extractor was introduced by [8] and is used to extract distinctive features that are invariant to scale rotation and illumination. The SIFT method finds the keypoints in an image by executing four key steps: detect scale-space extrema, localise keypoints, assignment orientation and descriptor representation. For any given image, the SIFT algorithm returns two key components. A set of keypoints and a descriptor for each keypoint. SIFT has three key parameters which are the width of the gaussian for scale-space extrema detection, contrast threshold for the elimination of low contrast keypoints and edge threshold for the elimination of edges. These parameters have been tuned according to the recommendations from [7]. The width of the gaussian was set to 1.7; the contrast threshold was set to 0.04 and edge threshold was set to 11.

BOVW: SIFT extracts thousands of feature points from each image, each of which is described by 128 element vectors. BOVW is a commonly used technique to encode features in image processing that is adapted from the Bag of Words algorithm used in Natural Language Processing (NLP) [9]. The BOVW algorithm takes the SIFT features extracted in the previous step and performs clustering over the data. Each cluster identified in this process acts as a visual vocabulary which describes the image. Once the visual vocabulary is established, for each image, a frequency histogram is created to count the occurrence of each feature. Clustering for BOVW is implemented using the k-means clustering algorithm. The number of clusters defines the size of the codebook, and according to the parameter tuning of [7], the cluster size has been set to 500. A visual representation of BOWV can been seen in Fig. 2.

Classification: For classification, Support Vector Machines (SVM) was used. Other common classifiers such as K-Nearest Kenghbours and Naive Bayes were not considered because [7] had already established that SVM was the optimal classifier for the given dataset. As the study focused on feature extraction and data resampling techniques, exploration of classification algorithms was

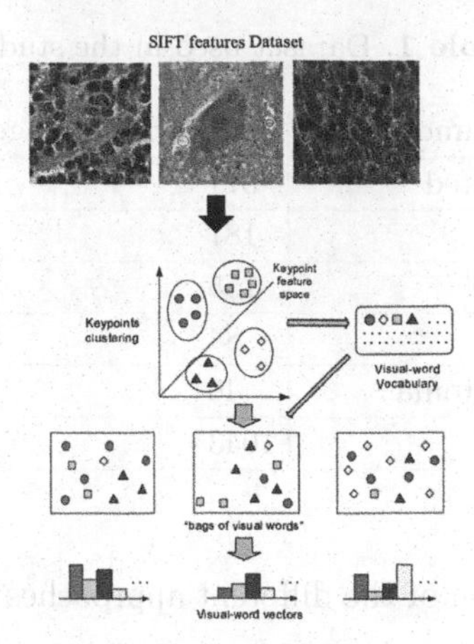

Fig. 2. Illustration of BOVW

considered out of scope for this research. So, for all the experimental setups discussed in this study, the SVM model was used. The SVM model used in the study, was setup with a RBF kernel with the kernel parameter set to 0.004. The SVM kernel and parameters were kept consistent for all other experimental setups used in the study.

Setup 2: SURF + Bag of Visual Words (BOVW) + SVM

The second experimental setup is similar to the first with the only alteration made to the feature extraction technique used. SURF is used as a feature extractor rather than SIFT in this setup. SIFT is quite effective but is generally more computationally expensive. SURF was developed to improve the speed of SIFT and can be up to three times faster while still providing features invariant to scale, illumination, blur and rotation. H. Bay et al. [10] has also shown that the SURF feature extraction method can outperform SIFT on multiple occasions. SURF has one key parameter – hessian threshold, which is used for keypoint detection. For this study, this parameter was set to 600 as used in [10]. However, [10,11] indicate that SURF can outperform SIFT in multiple scenarios. Also, each feature in SURF is described by a 64-dimensional vector as opposed to 128 vector descriptors of SIFT. This means that the BOVW will need parameter tuning. So, the cluster size of 100, 200 and 300 was considered for construction of codebook in BOVW. For classification, SVM classifier with RBF kernel was used.

Setup 3: Combining SIFT and SURF Features

The third experimental setup involves combining SURF and SITF feature extractors. A similar approach was previously implemented by L. Lenc and P. Král [12] for facial recognition, where results showed that combining SIFT and SURF can outperform state of the art in facial recognition. However, this approach has not been explored in the feature extraction of neuroblastoma tumours. It is expected using this approach of combining keypoint locations and descriptors from two algorithms will provide features that are more robust as compared to using them individually. The process of combining features from SIFT and SURF is as follows: (1) Extract interest points using SIFT extractor; (2) Extract interest points using SURF extractor; (3) Gather descriptors from SIFT extractor; (4) Gather descriptors from SURF extractor (5) Concatenate both descriptors to create a combined descriptor.

Setup 4: Subsampled Features

The experimental setups explored in the previous sections only consider different feature extraction approaches. However, the class imbalance issue remained unanswered. This issue was also not addressed in previous work on neuroblastoma classification by S. Gheisari et al. in [5–7].

Standard techniques used to manage the class imbalance problem are data sampling methods such as oversampling and undersampling [13,14]. Oversampling through image augmentation is a common approach used in research but is not suitable for this work because SIFT and SURF are robust to scale, blur, rotation and illumination. Hence this setups looked at undersampling to the representation numbers of the majority classes. While this method might appear suitable for the given problem, it has one severe limitation. As discussed in [15] merely removing data samples from the majority class can result in loss of vital information. Thus, the approach of removing image samples from the available dataset would not be a suitable approach for the given problem. Instead, an approach similar to that suggested in [13,15] appeared more feasible. So, features extracted from the images will be undersampled. This method will ensure that the most relevant features from each class is preserved.

As performed in previous experimental setups, a feature extraction method is first used to extract robust features from all the images in the training set. For this setup SIFT extractor with preciously set parameters (Gaussian width = 1.7, contrast threshold = 0.04 and edge threshold = 11) is used. After computing the feature points and corresponding descriptors, these descriptors are ordered based on the class they represent. For each class, the set of descriptors are then sorted based on relevance. Once vectors are organised by class and sorted, undersampling is performed to reduce the number of feature vectors in the majority classes. To do this, the number of feature vectors in the lowest class is taken as a limit for the number of features for each class. In the case of the dataset used, the minimum number of features would always be gathered from the ganglioneuroblastoma class. After the undersampling step, as before BOVW and SVM is used. For the BOVW, cluster sizes of 100, 200, and 300 were tested for parameter tuning.

Setup 5: Resampling for classification

The resampling method proposed in the previous setup only addresses the class imbalance for BOVW where clustering was used. It fails to address the class imbalance issue for classification for SVM. Thus, this experimental setup aims to address this imbalance by exploring oversampling and undersampling techniques before performing classification.

To oversample the minority classes for classification, a method known Synthetic Minority Oversampling Technique (SMOTE) is used in this experiment. This method was introduced by N. V. Chawla et al. [13], where additional samples for the minority class is generated synthetically by observing existing examples. SMOTE generates new data for the minority class by considering the nearest neighbours of existing samples. Another technique that can be applied for resampling is known as near-miss. This method is an undersampling approach where instances of the majority class are removed if they are too close to each other. In this method, first distances between all samples in the majority class and the minority class are calculated.

A combined approach of using SMOTE to oversample the minority classes and then using near-miss to undersample on the majority class (poorly differentiated neuroblastoma) is also explored in this experimental setup. First SIFT is used for feature extraction. Then the undersampling method, as discussed in setup 4, is used. The undersampled feature points are then fed through BOVW for feature encoding. After this, oversampling and undersampling methods are used for the available training data and then finally SVM classifier is used for classification.

3 Results

This section presents the results achieved with the multiple experimental setups used. Table 2 summaries the key metrics of accuracy, precision, recall and F-1 score achieved from different experimental methods discussed in the previous section. The results table include a summary of the achieved metrics using a stratified 5-fold cross validation. For the sake of comparison, the train-test split in dataset remained consistent throughout the different experimental setups. The results presented in table shows that the final setup generated the best outcome for accuracy, precision, recall F-1 scores. The final setup used the selective undersampling approach for SIFT and SMOTE for classification. The achieved result, provides evidence that the proposed data manipulation techniques in this paper provided significant improvements in the neuroblastoma dataset. The dramatic improvements in the overall metrics become even more evident when Fig. 3 is observed. This figure shows comparison between results achieved in previous study and the best results achieved in this study (setup 5) in a class level breakdown. It can be seen that overall, the proposed data sampling approaches provide significant improvements in classification of minority classes such as ganglioneuroblastoma and ganglioneuroma. It also addresses some issues identified in previous work such as the tendency of human experts and machine learning models to misclassify between differentiating and poorly differentiated classes. Overall, the results demonstrate the improvements achieved by this study as compared to previous work on the same dataset.

Table 2. Results achieved by the experimental setups

Method	Accuracy	Precision	Recall	F-1 Score
Setup 1: Sift for feature extraction	0.8304	0.8573	0.6978	0.7483
Setup 2: SURF + BOVW + SVM	0.7969	0.8081	0.6619	0.7035
Setup 3: Combining SIFT and SURF	0.8477	0.8619	0.7373	0.7801
Setup 4: Subsampled SIFT features	0.8822	0.9151	0.8192	0.8530
Setup 5: Resampling for classification	**0.9003**	**0.9060**	**0.8730**	**0.8868**

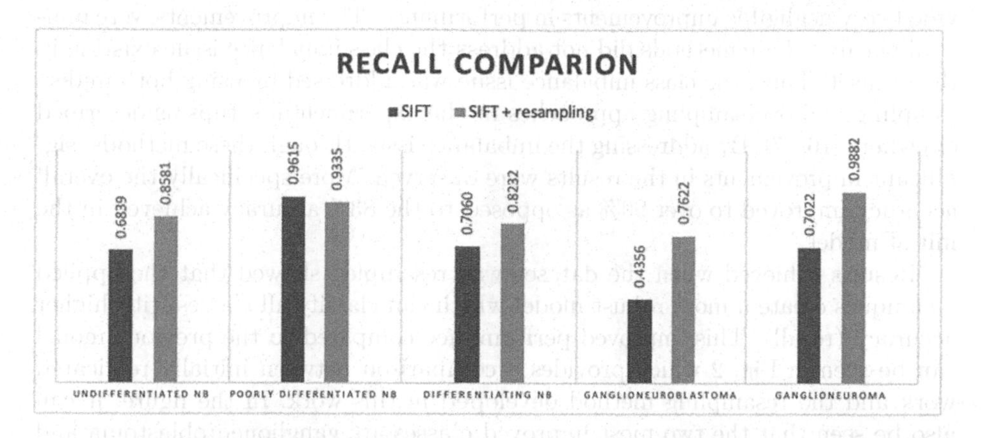

Fig. 3. Class level breakdown and comparison between results achieved with and without sampling approaches

4 Discussion

The work was conducted with the overarching aim of improving classification accuracy in neuroblastoma histopathological images in a given dataset. Due to the complex texture of neuro-blast tumours, machine learning-based classification can be considered beneficial as they can extract features based on mathematical feature extraction techniques. Experimental setups were designed to address the aims of exploring different feature extraction techniques and analaysing the impacts of data engineering techniques such as oversampling and undersampling.

Both SIFT and SURF extract features that are robust to these properties, so their performance were evaluated. Comparing, the results achieved for SIFT and SURF, it was evident that SURF performed poorly compared to SIFT for all the classes. While both SIFT and SURF extract features that are robust to features like, scale and rotation, the mathematical approach taken in these approaches are significantly different. While SIFT uses convolutions of Difference of Gaussian to determine the scale space, SURF uses Laplacian of Gaussian approach with boxed filters to approximate the Difference of Gaussians for scale space detection. As a result of this, features extracted by SIFT are more robust to scale variations compared to SURF. Thus, SURF performed poorly because there is high intra-class scale variance in neuroblast tumours which SIFT was more accurately able to capture.

A combined SIFT and SURF feature extraction method was used to evaluate the effects of combining different feature extraction techniques. Results showed that this approach provided only a slight improvement in overall classification performance. The approach only achieved an improvement of 1.5% as the accuracy of 84.5% was recorded. The only class affected was the undifferentiated neuroblastoma class which saw an improvement of around 7%.

The results achieved by SURF and a combination of SIFT and SURF provided very negligible improvements in performance. The improvements were minimal because these methods did not address the class imbalance issue existing in the dataset. Thus, the class imbalance issue was addressed by using both undersampling and oversampling approaches in the experimental setups as described elsewhere [16, 17]. By addressing the imbalance issue through these methods, significant improvements in the results were observed. More specifically the overall accuracy improved to over 90% as opposed to the 83% accuracy achieved in the initial model.

Results achieved when the dataset was resampled showed that the applied techniques create a more robust model which can classify all classes with higher accuracy (recall). This improved performance compared to the previous model can be seen in Fig. 2 which provides a comparison between initially replicated work and the resampling method developed in this work. In the figure, it can also be seen that the two most improved classes are ganglioneuroblastoma and ganglioneuroma classes where recall score increased by over 30%. Not surprisingly, these two categories were also the two least common class in the dataset. The use of sampling techniques did reduce the accuracy of the poorly differentiated class by 3%, but this is a worthy tradeoff as it is preferable to have higher accuracy overall classes than for just one class. Differentiating neuroblastoma was also identified as a problematic class in previous research because expert pathologists struggled with classification between poorly differentiated and differentiating class. However, the current approach was able to improve the recall of differentiating neuroblastoma class by over 12%. There is also significant medical significance of these improvements in addition to the statistical significance. Treatment plans for patients are dependent on the category of neuroblastoma they have. So, a more accurate model translates to lower possibilities of patients receiving inadequate treatment.

While the study improves on previous work there are few limitations that can be addressed for future work. This work only looked at the classification of neuroblastoma images using low-level feature extraction combined with a range of data sampling techniques. While results achieved show useful improvements to existing methods, the use of high-level feature extraction techniques can provide even more significant improvements. Use of techniques such as deep neural networks have the potential to extract even more robust and useful features and need to be explored thoroughly in the future. Furthermore, the work only focuses on the classification of patched 300×300 stained tissue images into different categories of neuroblastoma. Additionally, the images are carefully curated by expert pathologists and only contain areas that best represent each class. In, a practical scenario, pathologists examine whole tissue images under different magnifications. So, the proposed techniques need to be evaluated with whole slide tissue images under different magnifications.

5 Conclusion

This study began with the objective of improving existing approaches in the histopathological classification of early childhood tumour called neuroblastoma. To improve the results, the study looked at two main approaches. The first approach was to explore alternative low-level feature extraction techniques which could extract more relevant features and improve accuracy. However, this approach did not provide any significant improvements to existing approaches in neuroblastoma. This was because the main problem associated with the dataset was the class imbalance problem. To address the class imbalance in the dataset, both undersampling and oversampling approaches were thoroughly explored. The use of these approaches improved the accuracy of the models to over 90% in the given dataset. More importantly, these methods improved the classification of the minority classes by up to 35%. Such improvements have a high significance in the medical domain because the classification of the neuroblastoma types determine the treatments that patients receive. Pathologist classification using the Shimada system is still considered the gold standard in neuroblastoma and this work only intends to show role computer-based approaches can have to support decision making and act a tool for second reference. Further research is required to classify whole tissue images under different magnifications and predict patient outcomes for practical applications of computer-based approaches in neuroblastoma.

References

1. Park, J.R., Eggert, A., Caron, H.: Neuroblastoma: biology, prognosis, and treatment. Pediatr. Clin. North Am. **55**(1), 97–120 (2008)
2. Maris, J.M.: Recent advances in neuroblastoma. N. Engl. J. Med. **362**(23), 2202–2211 (2010)
3. Shimada, H., et al.: The international neuroblastoma pathology classification (the Shimada system). Cancer Interdisc. Int. J. Am. Cancer Soc. **86**(2), 364–372 (1999)
4. Maris, J.M., Matthay, K.K.: Molecular biology of neuroblastoma. J. Clin. Oncol. **17**(7), 2264 (1999)
5. Gheisari, S., Catchpoole, D.R., Charlton, A., Kennedy, P.J.: Patched completed local binary pattern is an effective method for neuroblastoma histological image classification. In: Boo, Y.L., Stirling, D., Chi, L., Liu, L., Ong, K.-L., Williams, G. (eds.) AusDM 2017. CCIS, vol. 845, pp. 57–71. Springer, Singapore (2018). https://doi.org/10.1007/978-981-13-0292-3_4
6. Gheisari, S., Catchpoole, D.R., Charlton, A., Kennedy, P.J.: Convolutional deep belief network with feature encoding for classification of neuroblastoma histological images. J. Pathol. Inf. **9**, 17 (2018)

7. Gheisari, S., Catchpoole, D., Charlton, A., Melegh, Z., Gradhand, E., Kennedy, P.: Computer aided classification of neuroblastoma histological images using scale invariant feature transform with feature encoding. Diagnostics **8**(3), 56 (2018)
8. Lowe, D.G.: Distinctive image features from scale-invariant keypoints. Int. J. Comput. Vision **60**(2), 91–110 (2004)
9. Yang, J., Jiang, Y.-G., Hauptmann, A.G., Ngo, C.-W.: Evaluating bag-of-visual-words representations in scene classification. In: Proceedings of the International Workshop on Workshop on Multimedia Information Retrieval, pp. 197–206. ACM (2007)
10. Bay, H., Tuytelaars, T., Van Gool, L.: SURF: speeded up robust features. In: Leonardis, A., Bischof, H., Pinz, A. (eds.) ECCV 2006. LNCS, vol. 3951, pp. 404–417. Springer, Heidelberg (2006). https://doi.org/10.1007/11744023_32
11. Panchal, P., Panchal, S., Shah, S.: A comparison of SIFT and SURF. Int. J. Innov. Res. Comput. Commun. Eng. **1**(2), 323–327 (2013)
12. Lenc, L., Král, P.: A combined SIFT/SURF descriptor for automatic face recognition. In: Sixth International Conference on Machine Vision (ICMV 2013), vol. 9067. International Society for Optics and Photonics, p. 90672C (2013)
13. Chawla, N.V., Japkowicz, N., Kotcz, A.: Special issue on learning from imbalanced data sets. ACM SIGKDD Explor. Newsl. **6**(1), 1–6 (2004)
14. Ramyachitra, D., Manikandan, P.: Imbalanced dataset classification and solutions: a review. Int. J. Comput. Bus. Res. (IJCBR) **5**(4) (2014)
15. Kotsiantis, S., Kanellopoulos, D., Pintelas, P.: Handling imbalanced datasets: a review. GESTS Int. Trans. Comput. Sci. Eng. **30**(1), 25–36 (2006)
16. Barlow, H., Mao, S., Khushi, M.: Predicting high-risk prostate cancer using machine learning methods. Data **4**, 129 (2019)
17. Khushi, M., Clarke, C.L., Graham, J.D.: Bioinformatic analysis of cis-regulatory interactions between progesterone and estrogen receptors in breast cancer. PeerJ **2**, e654 (2014). https://doi.org/10.7717/peerj.654

Data Mining ENCODE Data Predicts a Significant Role of SINA3 in Human Liver Cancer

Matloob Khushi[1(✉)], Usman Naseem[1], Jonathan Du[1], Anis Khan[2], and Simon K. Poon[1]

[1] School of Computer Science, The University of Sydney, Sydney, NSW, Australia
matloob.khushi@sydney.edu.au

[2] The Westmead Institute of Medical Research, The University of Sydney, Sydney, Australia

Abstract. Genomic experiments produce large sets of data, many of which are publicly available. Investigating these datasets using bioinformatics data mining techniques may reveal novel biological knowledge. We developed a bioinformatics pipeline to investigate Chip-seq DNA binding proteins datasets for HepG2 liver cancer cell line downloaded from ENCODE project. Of 276 datasets, 175 passed our proposed quantity control testing. A pair-wise DNA co-location analysis tool developed by us revealed a cluster of 19 proteins significantly collocating on DNA binding regions. The results were confirmed by tools from other labs. Narrowing down our bioinformatics analysis showed a strong enrichment of DNA-binding protein SIN3A to activator (H3K79me2) and repressor (H3K27me3) indicating SIN3A plays has an important regulatory role in vital liver functions. Whether increased enrichment varies in liver infection we compared histone modification between HepG2 and HepG2.2.15 cells (HepG2 derived hepatitis B virus (HBV) expressing stable cells) and observed an increase SIN3A enrichment in promoter regions (H3K4me3) confirming a known biological phenotype. The mechanistic role of SIN3A protein in case of liver injury or insult during liver infection warrants further dry and wet lab investigations.

Keywords: Cancer · HepG2 · Transcription factor binding sites · Bioinformatics · ENCODE

1 Introduction

Many cellular functions are mediated by sequence-specific DNA-binding proteins commonly known as transcription factors (TF). Transcription factor binding to DNA, either alone or as part of a complex with other proteins, is influenced by epigenetic modifications and small sequence variations or single nucleotide polymorphisms. Perturbation of transcription factor binding plays a role in many diseases, including but not limited to cancer. Therefore, the characterisation

H. Yang et al. (Eds.): ICONIP 2020, LNCS 12534, pp. 15–25, 2020.
https://doi.org/10.1007/978-3-030-63836-8_2

of transcription factor bind-ing sites (TFBS), or otherwise known as transcription factor binding regions (TFBR) across the genome using next-generation technologies has been a focus of research in recent years [8,11,18,22,23]. Much of these data are now available in public repositories but is under-utilized. In this study, we aim to develop a workflow to effectively collate, query, and mine publicly available next-generation sequencing (ChIP-seq) datasets to discover novel and biologically relevant insight. Liver cancer is one of the leading cause of cancer-related death worldwide [6]. In this study, we aimed to investigate human liver cancer by sourcing data about HepG2 cells (cell line derived from human hepatoblastoma of a 15-year-old male [15]) from the National Human Genome Research Institute's ENCODE project [3].

2 Methods

A total of 276 publicly available ChIP-seq datasets for the HepG2 cell line were downloaded. Datasets in other formats were converted to FASTQ format using the sratoolkit. The quality metrics of all datasets were extracted using FastQC (v0.11.3) [1]. Datasets with less than 10,000,000 reads with a minimum average Phred quality score of 20 were removed. The control of remaining datasets were checked and im-proved using Cutadapt (v1.9.dev4) [16]. Reads were trimmed using a quality thresh-old of 20 and reads which were shorter than 25bp after trimming were removed. The processed sequencing reads were then mapped to the human genome (GRCh38) with Bowtie (v1.1.2) [13]. Subsequently, duplicated reads were removed using PicardTools' MarkDuplicates tool.

Peak calling on all datasets were performed with MACS (2.1.0) using a q-value cutoff of 0.05 and the nomodel parameter to prevent downscaling of the dataset in accordance with its controls [25]. In addition to this, all histone mark datasets used the broad parameter with a cutoff of 0.1. The following MACS2 parameters for his-tone mark datasets "-g 3.0e9 -q 0.05 –nomodel –broad –broad-cutoff 0.1" and "-g 3.0e9 -q 0.05 –nomodel" were used for transcription factor binding site (TFBS).

To merge biological replicates, Irreproducibility Discovery Rate (IDR) analysis was performed and all overlapping peaks were kept with no threshold applied [14]. The number of peak overlaps between pair-wise datasets was determined using bedtools' intersectBed [17]. To minimise artefacts due to sequencing depth, only datasets with more than 15,000,000 aligned reads and more than 5,000 peaks were compared (ref. intersect_filtered_peaks.xlsx). A pair-wise comparison of all transcription factor binding sites was performed using GenometriCorr [4]. The Jaccard measure was used to quantify the correlation between pairwise binding sites. From these results, a heatmap was generated using R's gplots package. In addition the clustered heatmap was generated for all transcription factors using Overlap Correlation Value (OCV) calculated below. A prox-imity ratio pr is calculated using the following Eq. 1:

$$pr = \frac{2l + 2d + r}{L} \tag{1}$$

Where l is the length of the query region (Fig. 1), d is the distance between the query region and the reference region of length r and L is the total chromosome length under observation. In Eq. 1 Length l and distance d are multiplied by 2 to consider the binding locations on both sides of the reference region. In the Fig. 2 query DNA C is closest to the reference DNA B, whereas, C' region is shown as a possible location of B keeping the same distance d apart from the reference. From Eq. 1 we can see the bigger the l and d the closer the value of pr will be to 1. Hence lower the pr, more significant of this proximity. We considered all values lower than 0.05 as significant. The OCV is calculated as:

$$OCV = \frac{n}{q} \tag{2}$$

Where n is the number of regions in query dataset having pr values less than 0.05 and q is the total number of regions in query dataset. Using EpiMINE's Enrich tool [7], a graphical depiction illustrating the presence and absence of preferential en-richment of selected transcription factor binding sites on each histone mark was generated.

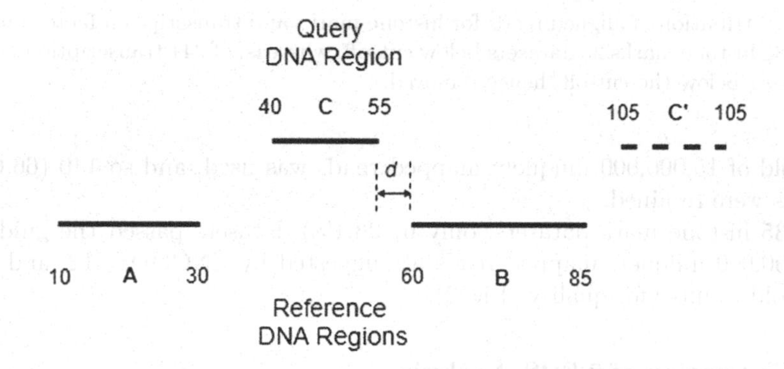

Fig. 1. An example of query region C in a close proximity of a reference region B

ChIP-seq data of HepG2.2. 15 cells (stably expressing HBV) were also obtained for comparison of any differences in the histone modification H3K4me3 between the infected and non-infected cells [20]. To allow for comparison without sequencing depth bias, the HBV ChIP-seq data was downsampled to approximately 10,000,000 reads to match the non-infected H3K4me3 data.

3 Results

The number of aligned reads after quality trimming of the sequencing data was rec-orded and analysed. Out of 241 ChIP-seq datasets targeting transcription factor bind-ing sites, 166 (68.88%) of the datasets met the guideline of 10,000,000 uniquely mapped reads as suggested by ENCODE in 2012 [12]. For this study, a

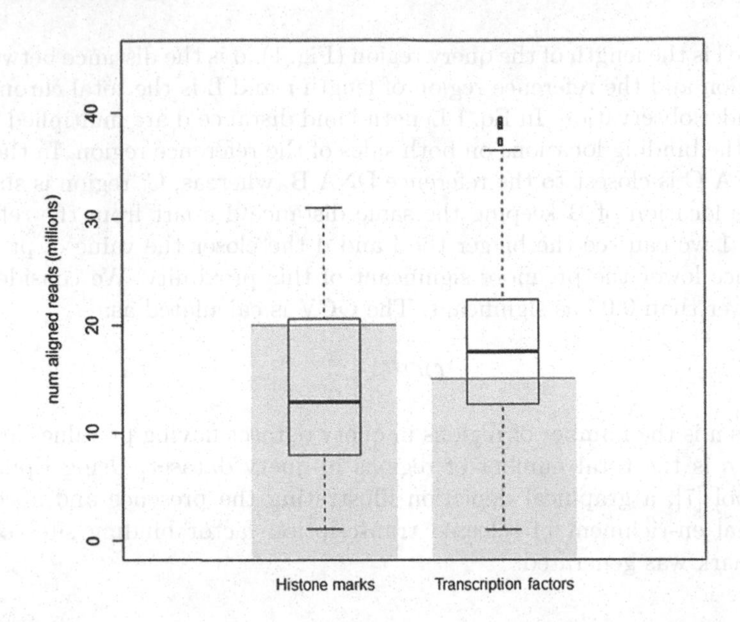

Fig. 2. Distribution of aligned reads for histone marks and transcription factor binding sits. Of 35 histone marks 25 datasets below cut-off, whereas, of 241 transcription factors 80 datasets below the cut-off, hence removed.

threshold of 15,000,000 uniquely mapped reads was used, and so 140 (66.66%) datasets were retained.

Of 35 histone mark datasets, only 6 (23.1%) datasets passed the guideline of 20,000,000 uniquely mapped reads as suggested by ENCODE [12] and used thresh-old to filter for quality (Fig. 2).

3.1 Co-location of TFBS Analysis

The Jaccard measure was used to quantify pairwise correlation of all combinations of transcription factors and identify possible binding site similarities. Figure 3 shows the distribution of Jaccard values for all pair-wise combinations. Since the data is skewed toward the positive direction, the colour key was adjusted to show more contrast around the mean. The resulting matrix was used as input in gplots' heatmapping function.

Figure 4 heat map reveals a cluster at lower-most/left-most dendogram branch of 21 transcription factors (TBP, ELF1, YY1, SIN3A, MXI1, MAZ, CHD2, TAF1, CEBPD, MAX, HCFC1, SIN3B, ZHX2, GABPA, ZNF143, TBL1XR1, CUX1, RCOR1, FOS, RFX5, ZBTB7A) significantly collocating and showing a strong cor-relation. This subset was further visualised as a separate heatmap as shown in the Fig. 5.

A second correlation heatmap was then generated using BiSA OCV (Fig. 6). This heatmap revealed a cluster of 29 factors of which 19 were the same as that of previous Jaccard analysis. Only MAX and ZNF143 were not present in

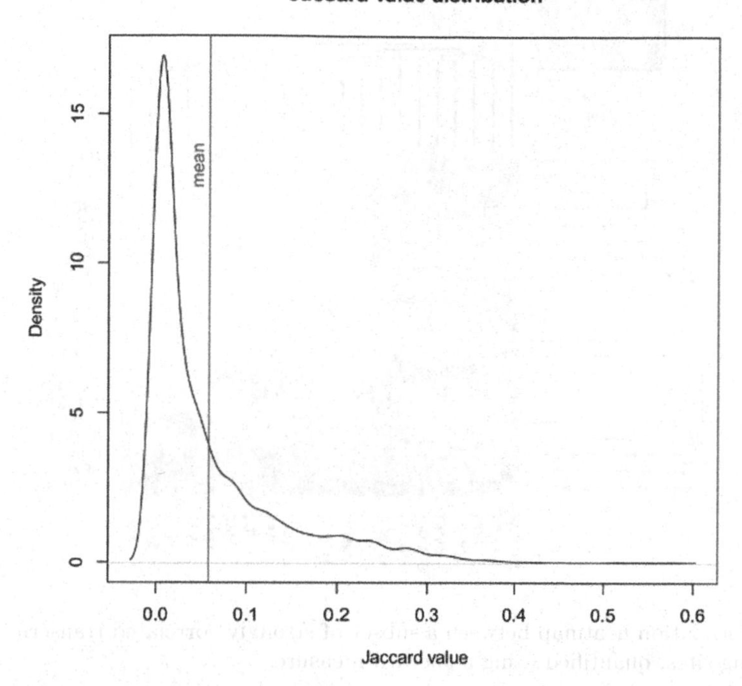

Fig. 3. Density plot showing distribution of Jaccard values for all transcription factor combinations (n = 2211)

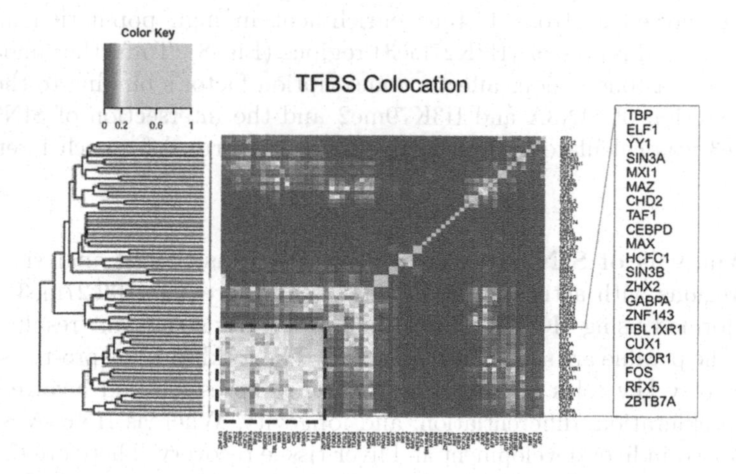

Fig. 4. Correlation heatmap between all transcription factor binding sites, quantified using a Jaccard measure.

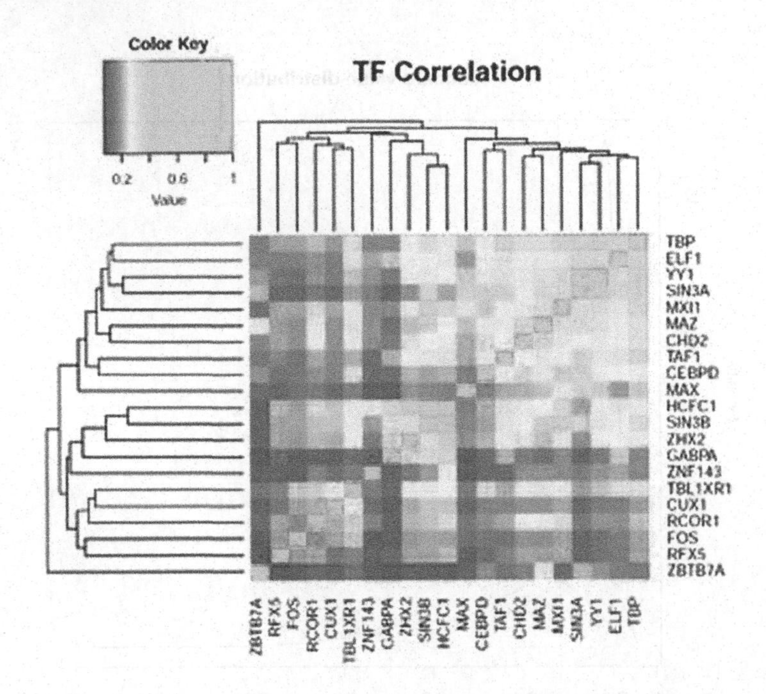

Fig. 5. Correlation heatmap between a subset of strongly correlated transcription factor binding sites, quantified using a Jaccard measure.

this cluster. This confirmed the presence of a strong 19 TFs cluster in HepG2. Enrichment of this subset transcription factors on histone marks was quantified and visualised using EpiMINE's Enrich tool (Fig. 7).

SIN3A showed a strong histone enrichment in high population activator (H3K79me2) and repressor (H3K27me3) regions (Fig. 8). To further understand how differing histone regions affect a transcription factor's binding to the DNA, the inter-section of SIN3A and H3K79me2 and the intersection of SIN3A and H3K27me3 was calculated. Motif analysis was performed for each intersecting set.

Motif Analysis of SIN3A Overlapping Regions. Motif analysis of overlapping regions with activator (H3K79me2) and repressor (H3k27me3) regions were performed using HOMER tool and top 4 most significant results ranked based on the p-value are shown in the Table 1. The forkhead box pro-teins (FOX) known to play key roles in regulating the gene expression and involved in cell growth, proliferation, differentiation, and longevity. Whereas HNF4A is known to be involved in liver development and liver tissue recovery. Therefore the significance presence of FOX and HNF4A motifs in the common regions of activator marker (H3K79me2) identify the possibility of key role of SIN3A in the functioning of liver.

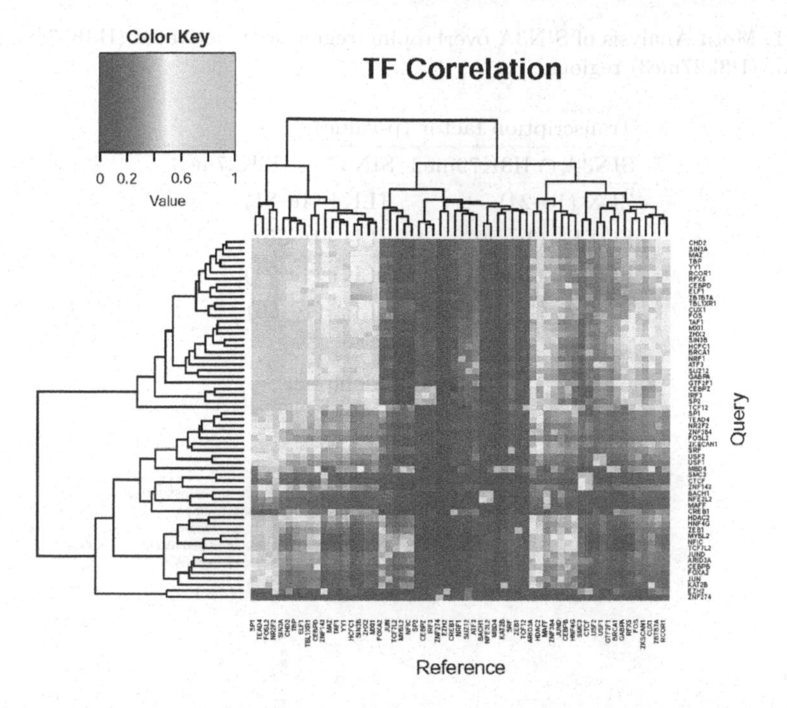

Fig. 6. Correlation heatmap between all transcription factor binding sites, quantified using BiSA OCV.

3.2 HBV-transfection Effect on Histone Modification

In this section we validated our bioinformatics colocation analysis by comparison with known biological knowledge. HBV infection is a known risk factor for liver cancer [6,19,24]. As histone modification H3K4me3 (trimethylation of Histone H3 lysine K4) is known to be associated with gene activation, cell division and usually referred as promoter regions, we analysed HBV infection effects in the expression of H3K4me3 markers. H3K4me3 dataset for stably expressing HBV cells 'HepG2.2. 15' was sourced from Tropberger et al. [20] while two non-infected datasets where sourced from ENCODE project. The dataset from Broad Institute was labelled as H3K4m3 (Broad) and the dataset from University of Washington was labelled as H3K4me3 (UW) as shown in the Fig. 3. There were significantly H3K4me3 en-richment observed across the HBV genome in stably expressing HBV cells 'HepG2.2. 15' when compared to the control HepG2 cells. The H3K4me3 dataset from the HepG2.2. 15 had around 80,000 unique peaks (77.5% and 78.5%) when overlapped with the two HepG2 control datasets (Fig. 9).

Table 1. Motif Analysis of SIN3A overlapping region with activator (H3K79me2) and repressor (H3k27me3) regions.

Transcription factor (p-value)	
SIN3A ∩ H3K79me2	SIN3A ∩ H3K27me3
ZFX (1e–24)	ELK4 (1e–81)
FOXP1 (1e–20)	FLI1 (1e–75)
FOXA2 (1e–18)	ELK1 (1e–71)
HNF4a (1e–16)	ELF1 (1e–56)

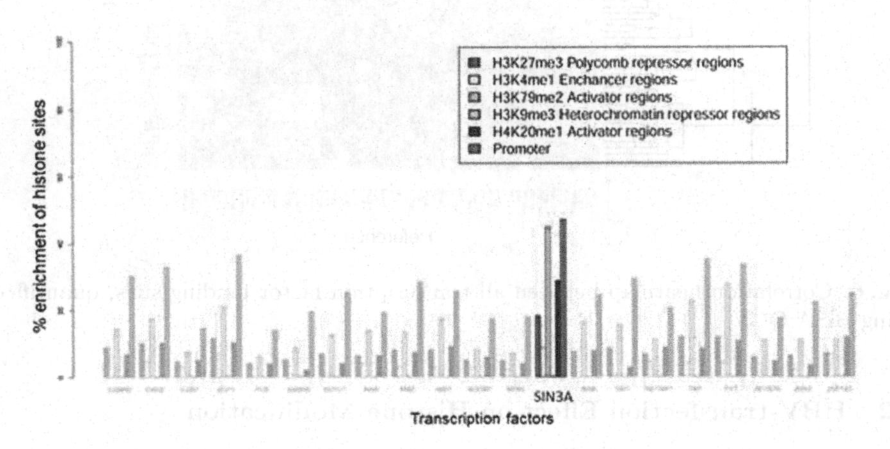

Fig. 7. Preferential enrichment of selected transcription factor binding sites on histone marks.

4 Discussion

Bioinformatics tools and methods has helped predict many genomic functions [2, 26]. Previously we have provided detailed insight on how estrogen receptors (ER) and progestin receptor (PR) binding on DNA converge on certain locations in breast cancer [10]. In this study, we focused on analysis of DNA-binding proteins and his-tone modifications in liver cancer cell line 'HepG2'.

Pair-wise correlations among the datasets were measure using Jaccard coefficient. The data clustered into three disjoint groups having similar correlation values be-tween them. There was a strong positive clustering in lower left corner. The coloca-tion significance was also studied by BiSA and visualized by a heatmap which con-firmed the similar 3-set clustering pattern seen in the earlier heatmap based on the Jaccard index. There were 19 DNA-binding proteins namely, TBP, ELF1, YY1, SIN3A, MXI1, MAZ, CHD2, TAF1, CEBPD,

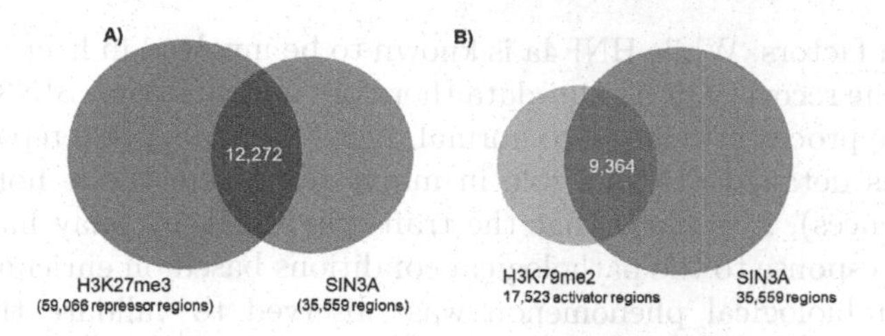

Fig. 8. Overlap of Sin3A with histone marks. A) Nearly 35% Sin3A regions overlap with H3K27me3 repressor regions. Nearly 26% of Sin3A regions overlap with K3K79me2 activa-tor regions..

HCFC1, SIN3B, ZHX2, GABPA, TBL1XR1, CUX1, RCOR1, FOS, RFX5 and ZBTB7A, appeared in both BiSA and Jaccard based clustering.

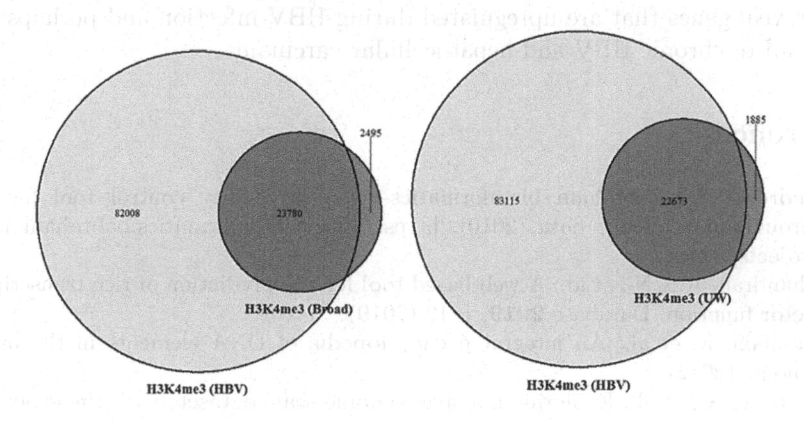

Fig. 9. Venn diagrams of the number of peaks from H3K4me3 in stably expressing HBV cells 'HepG2.2. 15' (blue) and non-infected HepG2 cells (red, Broad Institute and University of Washington datasets). (Color figure online)

The enrichment of above 19 factors in histone modification was studied using ENRICH plot which gives an indication of histone coverage among transcription factors. SIN3A shown a strong enrichment in most histone marks. Overlap of SIN3A was studied against the histone dataset with largest number of peaks for activator regions and repressor regions. SIN3A's 35% (12,272) DNA-binding regions were found overlapping with H3K27me3 repressor regions and 26% (9,363) regions were overlapping with activator H3K79me2.

Motif analysis is the tool to find known and unknown DNA sequences where pro-teins could bind (References). Motif analysis was performed on SIN3A's intersect-ing regions with H3K27me3 and H3K79me2. Similarly, in the activator region over-laps, we found the presence of FOX and HNF4a protein motifs. FOX protein is a member of the forkhead family of transcription factors, known to bind and reconfig-ure condensed chromatin and enable the binding of other transcription factors. While HNF4a is known to be involved in liver development and liver tissue recovery [5,9]. Our data therefore indicates that SIN3A has a role to play in the processes relat-ed to normal liver functioning and repair. Previous literature has detailed SIN3A's role in maintaining skin tissue homeostasis in mice (References), it is likely that the transcription factor may have a similar role in liver response to the pathological conditions based on enrichment results.

A known biological phenomenon was observed to validate this bioinformatics colocation analysis. Analysis of the histone marker H3K4me3 between 'HepG2.2. 15' (stably expressing HBV) cells and the control 'HepG2 cells' showed over 80,000 unique peaks in 'HepG2.2. 15' cells. H3K4me3 is associated with transcriptional activation; significantly greater peaks suggest greater transcriptional activation dur-ing HBV infection. Our results agree with previous literature suggesting that the expression of DNA methyltransferase genes is significantly upregulated in response to HBV which leads to methylation of host CpG islands [21]. Functional and gene enrichment of these unique peaks could possibly reveal genes that are upregulated during HBV-infection and perhaps ones that lead to chronic HBV and hepatocellular carcinoma.

References

1. Andrews, S.: Babraham bioinformatics-fastqc a quality control tool for high throughput sequence data (2010). https://www.bioinformatics.babraham.ac.uk/projects/fastqc/
2. Chandran, A.K.N., et al.: A web-based tool for the prediction of rice transcription factor function. Database **2019**, 1–12 (2019)
3. Dunham, I., et al.: An integrated encyclopedia of DNA elements in the human genome (2012)
4. Favorov, A., et al.: Exploring massive, genome scale datasets with the genometricorr package. PLoS Comput. Biol. **8**(5), e1002529 (2012)
5. Fekry, B., et al.: Incompatibility of the circadian protein bmal1 and hnf4α in hepatocellular carcinoma. Nat. Commun. **9**(1), 1–17 (2018)
6. Flecken, T., et al.: Mapping the heterogeneity of histone modifications on hepatitis B virus DNA using liver needle biopsies obtained from chronically infected patients. J. Virol. **93**(9), e02036 (2019)
7. Jammula, S., Pasini, D.: Epimine, a computational program for mining epigenomic data. Epigenetics Chromatin **9**(1), 42 (2016)
8. Jing, F., Zhang, S., Cao, Z., Zhang, S.: An integrative framework for combining sequence and epigenomic data to predict transcription factor binding sites using deep learning. IEEE/ACM Trans. Comput. Biol. Bioinf. (2019)
9. Khushi, M.: Benchmarking database performance for genomic data. J. Cell. Biochem. **116**(6), 877–883 (2015)
10. Khushi, M., Clarke, C.L., Graham, J.D.: Bioinformatic analysis of cis-regulatory interactions between progesterone and estrogen receptors in breast cancer. PeerJ **2**, e654 (2014)
11. Khushi, M., Liddle, C., Clarke, C.L., Graham, J.D.: Binding sites analyser (bisa): software for genomic binding sites archiving and overlap analysis. PloS one **9**(2), e87301 (2014)

12. Landt, S.G., et al.: Chip-seq guidelines and practices of the ENCODE and mod-ENCODE consortia. Genome Res. **22**(9), 1813–1831 (2012)
13. Langmead, B., Trapnell, C., Pop, M., Salzberg, S.L.: Ultrafast and memory-efficient alignment of short DNA sequences to the human genome. Genome Biol. **10**(3), R25 (2009)
14. Li, Q., Brown, J.B., Huang, H., Bickel, P.J., et al.: Measuring reproducibility of high-throughput experiments. Ann. Appl. Stat. **5**(3), 1752–1779 (2011)
15. López-Terrada, D., Cheung, S.W., Finegold, M.J., Knowles, B.B.: Hep g2 is a hepatoblastoma-derived cell line. Human Pathol. **40**(10), 1512 (2009)
16. Martin, M.: Cutadapt removes adapter sequences from high-throughput sequencing reads. EMBnet. J. **17**(1), 10–12 (2011)
17. Quinlan, A.R., Hall, I.M.: Bedtools: a flexible suite of utilities for comparing genomic features. Bioinformatics **26**(6), 841–842 (2010)
18. Schmidt, F., Kern, F., Ebert, P., Baumgarten, N., Schulz, M.H.: Tepic 2-an extended framework for transcription factor binding prediction and integrative epigenomic analysis. Bioinformatics **35**(9), 1608–1609 (2019)
19. Shibata, T., Aburatani, H.: Exploration of liver cancer genomes. Nat. Rev. Gastroenterol. Hepatol. **11**(6), 340 (2014)
20. Tropberger, P., Mercier, A., Robinson, M., Zhong, W., Ganem, D.E., Holdorf, M.: Mapping of histone modifications in episomal HBV cccDNA uncovers an unusual chromatin organization amenable to epigenetic manipulation. Proc. Natl. Acad. Sci. **112**(42), E5715–E5724 (2015)
21. Vivekanandan, P., Daniel, H.D.J., Kannangai, R., Martinez-Murillo, F., Torbenson, M.: Hepatitis B virus replication induces methylation of both host and viral DNA. J. Virol. **84**(9), 4321–4329 (2010)
22. Wang, W., et al.: Analyzing the surface structure of the binding domain on DNA and RNA binding proteins. IEEE Access **7**, 30042–30049 (2019)
23. Wei, H.: Construction of a hierarchical gene regulatory network centered around a transcription factor. Briefings Bioinfo. **20**(3), 1021–1031 (2019)
24. Wong, M.C., et al.: The changing epidemiology of liver diseases in the Asia-pacific region. Nat. Rev. Gastroenterol. Hepatol. **16**(1), 57–73 (2019)
25. Zhang, Y., et al.: Model-based analysis of chip-seq (macs). Genome Biol. **9**(9), 1–9 (2008)
26. Zheng, X., Zhong, S.: From structure to function, how bioinformatics help to reveal functions of our genomes (2017)

Diabetic Retinopathy Detection Using Multi-layer Neural Networks and Split Attention with Focal Loss

Usman Naseem[1](✉), Matloob Khushi[1], Shah Khalid Khan[2], Nazar Waheed[3], Adnan Mir[4], Atika Qazi[5], Bandar Alshammari[6], and Simon K. Poon[1]

[1] School of Computer Science, University of Sydney, Sydney, Australia
usman.naseem@sydney.edu.au
[2] School of Engineering, RMIT University, Melbourne, Australia
[3] School of Electrical Engineering, University of Technology Sydney, Sydney, Australia
[4] School of Computer Sciences, Western Sydney University, Sydney, Australia
[5] Centre for Lifelong Learning, Universiti Brunei Darussalam, Gadong, Brunei
[6] College of Computer and Information Sciences, Jouf University, Sakakah, Saudi Arabia

Abstract. Diabetic retinopathy (DR) is the most common eye threatening micro-vascular complication of diabetes. It develops and grows without arbitrary symptoms and can ultimately lead to blindness. However, 90% of the DR-attributed blindness is preventable but needs prompt diagnosis and appropriate treatment. Presently, DR detection is time and resource-consuming, i.e., required qualified ophthalmologist technician to examine the retina colour fundus for investigating the existence of vascular anomaly associated lesions. Nevertheless, an automatic DR scanning with specialised deep learning algorithms can overcome this challenge. In this paper, we present an automatic detection of DR using Multi-layer Neural Networks and Split Attention with Focal Loss. Our method outperformed state-of-the-art (SOTA) networks in early-stage detection and achieved 85.9% accuracy in DR classification. Because of high performance, it is believed that the results obtained in this paper are of great importance to the medical and the relevant research community.

Keywords: Diabetic Retinopathy · Deep learning · Computer-aided diagnosis · ResNet · Split attention · Ophthalmoscopy

1 Introduction

The occurrence of vision impairment owing to Diabetic Retinopathy (DR) is on the rise, and such incidences are projected to hit epidemic levels worldwide in the next few decades. DR still had been the primary cause of adult blindness. Globally, there were 425 million diabetes patients in 2017, with its projected rise to around 642 million by the end of 2040 [2]. DR develops and grows without

H. Yang et al. (Eds.): ICONIP 2020, LNCS 12534, pp. 26–37, 2020.
https://doi.org/10.1007/978-3-030-63836-8_3

arbitrary symptoms and can ultimately lead to blindness. Such an unprecedented rise in the number of DR patients' needs prompt identification to cure its lethal consequences as blindness.

DR classification includes the weighting of various characteristics and position of retina colour fundus image. Its identification is typically time and resource-consuming. This is based upon a trained ophthalmologist examination of the patient's retina colour fundus image to ascertain the existence of lesions consistent with the vascular anomaly. However, developments and advancements in AI technologies have created massive opportunities to promote DR identification on time. Deep learning (DL) subset, i.e., the algorithms of the Convolution Neural Network (CNN) have an excellent rationale for using image recognition and representation, which include medical imaging as well such as pathology, dermatology, and radiology [5].

Customizability and accessibility of CNNs algorithm such as rectified linear units (ReLU), drop-out implementation, and high computational capability graphical processing units (GPUs) are powerful tools in swift identification of DR. The consistency and timing of DR diagnosis using this technique is critical for cost-effectiveness and defectiveness; the simpler it is to treat, the sooner it is found. In this paper, we proposed the transfer learning methodology and an automatic method using single human fundus imaging for the DR identification. This approach learns useful features from noisy and limited data sets, which would be used to detect and track various DR phases. Moreover, due to the worldwide proliferation of both associated retinal diseases and DR, the automatic DR detection and tracking should be able to keep pace with rising demand for screening. The findings reported in this article are of considerable significance for the scientific industry and in particular to the biomedical community.

The remainder of this paper is structured as follows: Sect. 2 presents related work. Section 3 explains proposed methodology, the experimental results and discussion are presented in Sect. 4, and Sect. 6 summarizes the conclusion.

2 Related Work

Many research efforts have been devoted to the problem of early DR detection. First, researchers have attempted to use traditional computer vision and machine learning approaches to provide an effective solution to this problem. For instance, Priya et al. [21] Presented an approach focused on computer vision for the identification of DR stages using colour fundus images. They managed to extract features from the raw image by using image processing and passed them to the SVM for binary classification and got a sensitivity of 98%, specificity 96%, and accuracy of 97.6% on a testing set of 250 images. Quellec et al. [22] used a traditional KNN algorithm with optimal filters on two classes to achieve an AUC of 0.927. Also, Sinthanayothin et al. [25] proposed an automated DR detection system on morphological features using the KNN algorithm and obtained sensitivity and specificity of 80.21% and 70.66%, respectively. Larsen et al. [12] demonstrated an automatic diagnosis of DR in fundus photographs with a visibility threshold.

They reported accuracy of 90.1% for true cases detection and 81.3% for the detection of the false case.

DL algorithms have become popular and applied to various areas in the last few years [9,10,15–18]. Pratt et al. [14,20] Built a CNN-architecture And data augmentation capable of identifying the complex features involved in the classification process, such as micro-aneurysms, exudates and retina haemorrhages, and thus delivering a diagnosis automatically and without user input. There also some research work has been carried out based on pre-trained CNN models. Since the pre-trained CNN based models are previously trained with datasets such as ImageNet, they are well suited for this kind of image classification task. Also, some of the models provide variation in architecture which benefits in many cases to solve different image classification problems. Marıa A. Bravo and Pablo A. Arbelaez worked in such a similar way to detect DR [1]. They used both VGG-16 and Inception-V4 pre-trained models and developed classifiers with both of these. Shaohua et al. have developed classifiers with multiple CNN models [30]. Different developments in deep CNNs further improved the prediction accuracy for the tasks of efficient image detection/classification. CNN pretrained models have dramatically strengthened the next annual challenges, including ImageNet Large-scale Visual Recognition Competition (ILSVRC). Many pre-trained models were introduced, such as VGG-16, VGG-19 [13], GoogleNet [27], ResNet [6], Xception [3], Inception-V3 [29] and DenseNet [7] highly reliable and effective to train if they have shorter interactions between input and output layers.

Previous studies had mostly focused on the binary classification of DR, which restricted the scope of DR Detection studies. The purpose of this work was to predict the severity level of DR fundus photography images. The purpose of this work was to predict the severity level of DR fundus photography images among five classes-No DR, Mild DR, Moderate DR, Severe DR, and Proliferative DR. In this study, we have used multi-layer neural network with split attention with the focal loss for DR classification. Our network trained on APTOS2019 Kaggle's dataset outperformed other SOTA networks in early-stage detection and achieved 85.9% accuracy.

3 Methodology

3.1 Proposed Approach

In this paper, we propose a classification architecture by a modified ResNet50 architecture. We choose ResNet50 as the backbone network for deep feature extraction. This architecture is made up of three distinct components; Outer structure in which ResNet50 architecture is used followed by Intermediate structure where ResNeXt architecture is used and finally Inner structure where split Attention architecture is used. ResNeXt and split-attention widened our network architectures by using multi-channel parallel extraction and channel-wide re-weighting. Besides these components, we also enhance the accuracy of DR classification by using focal loss in our model. To achieve this, we implement active re-weighting to solve the problem of imbalanced distribution. The diagram of the framework proposed is given in the Fig. 1.

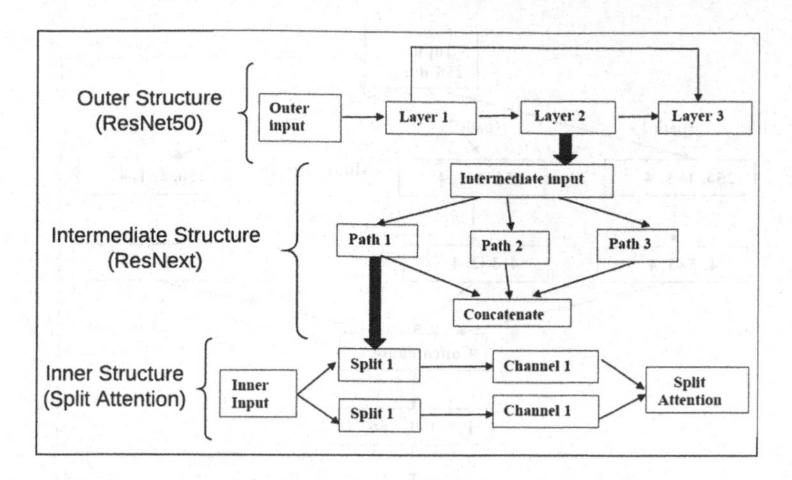

Fig. 1. Illustration of the proposed network architecture

The results of our model is evaluated with the VGG, ResNet50, ResNet50 with self and split attention. We also compared the results of our proposed model with previous studies conducted on DR classification. The proposed framework of the architectures chosen is as follows:

Outer Structure (ResNet50): Deep-residual learning network (ResNet) presents the idea of a residual block. The residual blocks are constructed to provide a relation between the first block input and the second block output. This process of adding lets the residual block learn about the residual function and prevents outburst of parameters. ResNet50 structure is a 50-layer residual block composed of a convolutionary layer, 48 residual blocks, and a 1×1 and 3×3 classification layer with small filters. This model won the classification challenge for ILSVRC 2015 and obtained excellent results on ImageNet and MS-COCO object detection competitions.

Intermediate Structure (ResNext): Xie et al. [32] proposed a variant of ResNet that is codenamed ResNeXt with the following building block:

As shown in Fig. 3 ResNeXt Applies several channels but preserves the cardinality of the system (the size of the set of transformations), reduces the design complexity and improves the scalability. ResNext can extract parallel features from the input during the procedure, while all paths have the same system design and filter scale. It is quite identical to the Inception module of [28], both adopt the concept of split-transform-merge, but in this version, the outputs of separate paths are combined by adding them combined, whereas in [28] they are concatenated with depth. Another distinction is that each path in [28] is unique from one another (1×1, 3×3, and 5×5 convolution), whereas in this design all paths share the identical architecture (Fig. 2).

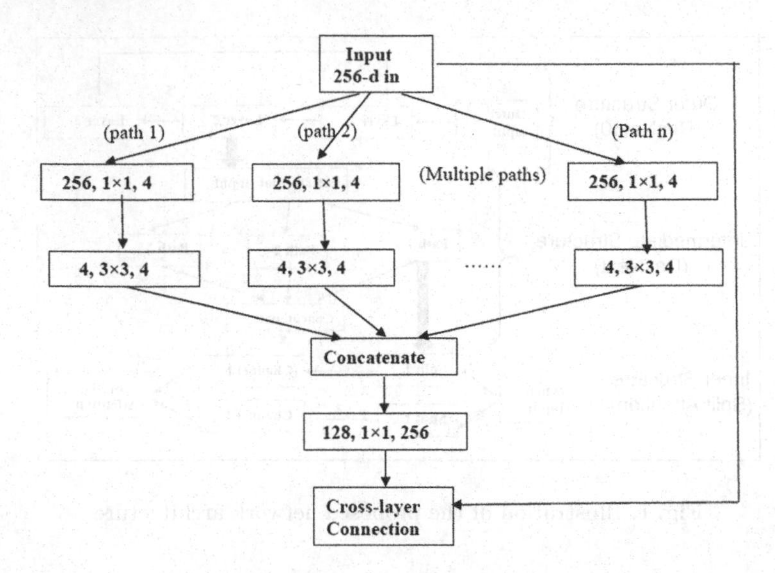

Fig. 2. Architecture of ResNeXt

Inner Structure (Split Attention for Channel-Wide Reweighting). We used split attention which allows channel-wide re-weighing and effectively adjusting the filter in multiple channels allowing the network to accommodate the accepting field size.

When humans looks at objects of various sizes and distances, the scale of the visual cortical neuronal receptive field is changed according the stimulus. Thus, in general, the size of the convolution kernel for a particular model for a specific task is calculated for the CNN-based network, so it is possible to construct a method that enables the network to adjust the size of the acceptance field as per multiple input scales.

Such design is basically divided into three parts of operation: Break, Fuse, and Pick. Split refers to a complete convolution operation with different kernel sizes on the input (including effective grouped/depth-wise convolutions, batch normalisation, ReLU function). Fuse component is a global average pooling process and the Select operation refers to the size, and Select uses two weight matrix sets to conduct a weighting action and then the aggregate to get the final output.

As shown in Fig. 1, our overall deep model is the integrated use of emerging DL techniques including ResNet, ResNeXt and Split Attention. This design can be seen as architecture of a tripartite scale, where:

- Outer architecture: utilizing cross-layer connections to build deep and stable multi-layer networks.
- Intermediate architecture: utilizing multiple parallel paths to better extract features while maintaining the cardinality.

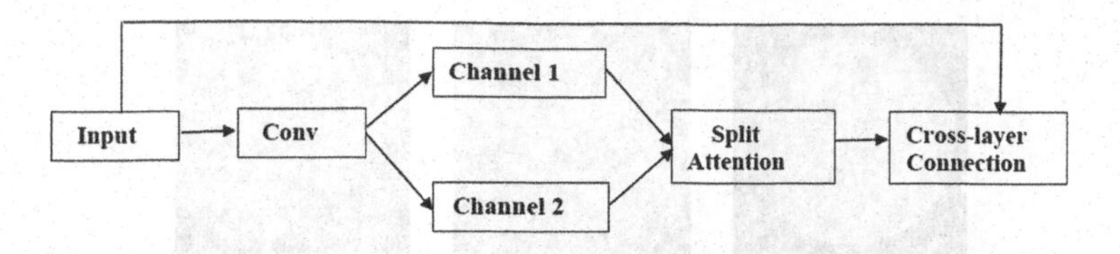

Fig. 3. Utilization of split attention for channel-wide re-weighting

- Inner architecture: utilizing Split-Attention, which enables channel-wide reweighing and actively change the filters in different channels that allow the network to adjust the size of the acceptance field adaptively.
- Finally, utilizing focal loss to solve image class imbalance problem and implementing data pre-processing and augmentation for better training efficiency and performance.

3.2 Dataset

The dataset utilized in this study is provided by the Asia Pacific Tele-Ophthalmology Society's APTOS 2019 competition on DR classification. The purpose of this competition is to build ML techniques to screen fundus images automatically for early classification of DR in rural regions where medical examination is edious and hard to carry out. The dataset contains a total of 3662 retina images obtained from various labs utilizing fundus photography, under a range of imaging conditions. The fundus images given in this dataset are divided into five categories. Class distribution of dataset used is given in Table 1.

Table 1. Distribution of DR datasets

DR Grade	Grade name	# Train set	# Test set	Total
0	No DR	1625	180	1805
1	Mild DR	333	37	370
2	Moderate DR	899	100	999
3	Severe DR	174	19	193
4	Proliferative DR	265	30	295

3.3 Data Pre-processing

The fundus images of the given dataset were gathered with different cameras from various clinics. The input images that have been acquired differ considerably in image intensity. So we conducted various methods of pre-processing to simplify the training phase.

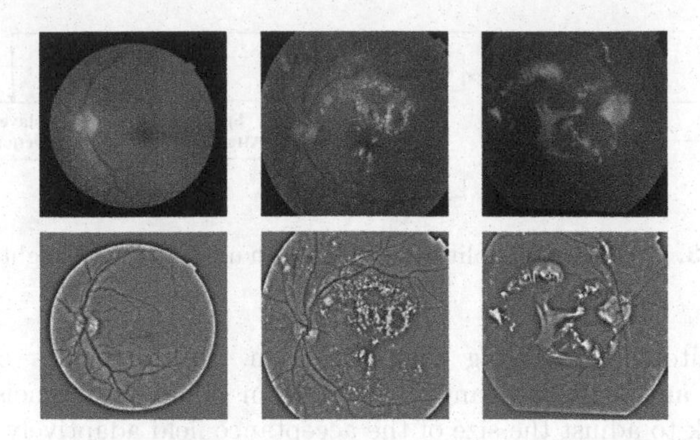

Fig. 4. Examples of actual fundus images (upper row), and associated images processed (bottom row)

- Resizing: With regard to the dimensional variations of the actual images, we resampled images to 819×614 depending on the aspect rate, cropped from the middle to the final resolution of the 600×600 pixels utilising bicubic interpolation to make sure that each retinal circle is located at the centre of the image.
- Min-pooling: To improve the clarity of the blood vessels and regions of lesion, we used Graham's technique. As per this technique, the black pixels are first excluded from the background, and then the process of image normalisation is carried out on the basis of the min-pooling filtering. Figure 4 displays actual image examples and the matching image pre-processed by using min pooling filtering technique.
- Image normalization: The cross-channel strength amount of images were normalised from $[0, 255]$ to $[-1, 1]$. This aids eliminate bias from the features and maintain a consistent distribution throughout the dataset. Further, ImageNet was used to standardise the images using ImageNet as a pre-processing stage means subtraction.

3.4 Metrics for Performance Evaluation

We conduct a series of experiments to analyze the performance of our proposed framework and is assessed and compared based on evaluation metrics, namely, accuracy. These measures are mathematically expressed as follows, provided the number of true positives (TP), false positives (FP), true negatives (TN) and false negatives (FN):

$$Accuracy = \frac{TP + TN}{TP + FP + FN + TN}$$

Table 2. Comparison of proposed model with others

Model	Accuracy
CNN[a]	0.740
CNN [20]	0.750
CNN [4]	0.770
GoogleNet [24]	0.450
AlexNet [31]	0.374
VGG16 [31]	0.500
InceptionNet V3 [31]	0.632
$\text{CNN}_{baseline}$ [11]	0.745
$\text{DenseNet}_{baseline}$[b]	0.764
VGG [19]	0.820
GNN [23]	0.793
Xception [8]	0.795
InceptionV3 [8]	0.787
MobileNet [8]	0.790
ResNet50 [8]	0.746
Modified Xception [8]	0.746
ResNet [26]	0.850
Proposed Model	**0.859**

[a]https://www.kaggle.com/kmader/inceptionv3-for-retinopathy-gpu-hr
[b]https://dragonlong.github.io/publications/automatic-image-grading.pdf

4 Experimental Results and Discussion

The computations were done on Kaggle kernel having 4 CPU cores with 17 GB RAM and 2 CPU cores with 14 GB RAM. Several cutting-edge methodologies and their published results are extensively compared to assess the performance of our model; published results being obtained from each of their original publications. Results are given in Table 2 which confirms that the proposed model beats all peers and attains the highest DR classification accuracy. Referring to Table 2, we note that the model proposed provided high precision (85.90%). Also, loss and accuracy curves were plotted to keep track of the performance of the model concerning the number of epochs and confusion matrix, to describe the performance of a classification model by comparing the true labels with the predicted labels are presented in Fig. 5.

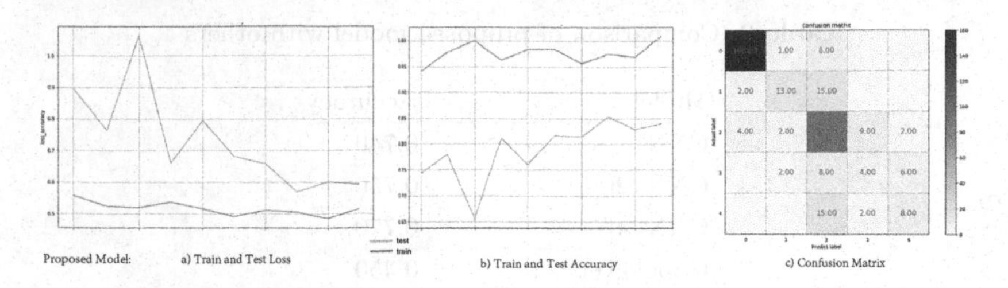

Proposed Model:
a) Train and Test Loss
b) Train and Test Accuracy
c) Confusion Matrix

Fig. 5. Proposed model: (a) Training loss v/s Test loss, (b) Training accuracy v/s Test accuracy and (c) Confusion Matrix

Table 3. Comparison of proposed architecture with others

Model	Accuracy
ResNet50	0.746
ResNet50+ Self Attention (Layer 2)	0.763
ResNet50+ Self Attention (Layer 3)	0.787
ResNet50+ Self Attention (Layer 4)	0.752
ResNet50+ Split Attention	0.818
Proposed Model	**0.859**

5 Analysis

To support the effectiveness of the proposed model, it tests and contrasts the performance of ResNet50, ResNext, and split attention models with our proposed algorithm. As shown in Table 3, the performance of our method increases ResNet50 up to 11.3%, ResNet+ Self attention at layer 2 up to 9.6%, ResNet+ Self attention at layer 3 up to 7.2%, ResNet+ Self attention at layer 4 up to 7.2% and ResNet+ Split attention up to 4.1% in terms of accuracy when compared to our proposed model which is a significant improvement.

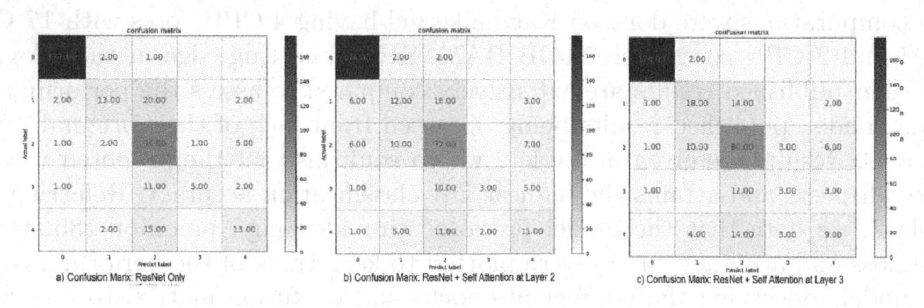

a) Confusion Marix: ResNet Only
b) Confusion Marix: ResNet + Self Attention at Layer 2
c) Confusion Marix: ResNet + Self Attention at Layer 3

Fig. 6. Confusion Matrix: (a) ResNet, (b) ResNet+ Self Attention at Layer 2 and (c) ResNet+ Self Attention at Layer 3

Moreover, the ResNet50 architecture, when used with self-attention at different layers gives a better performance at layer 2 when compared to the original architecture of ResNet50 and other variants tested in our study. Also, accuracy is improved when self-attention is replaced with split attention which highlights the importance of split attention. The classifier with lowest performance is ResNet50 with an accuracy of 74.6%. The results obtained shows that the combination of these models together plays a vital role and increases accuracy. Confusion matrix, train and test loss and accuracy are shown in Fig. 6 and Fig. 7, respectively.

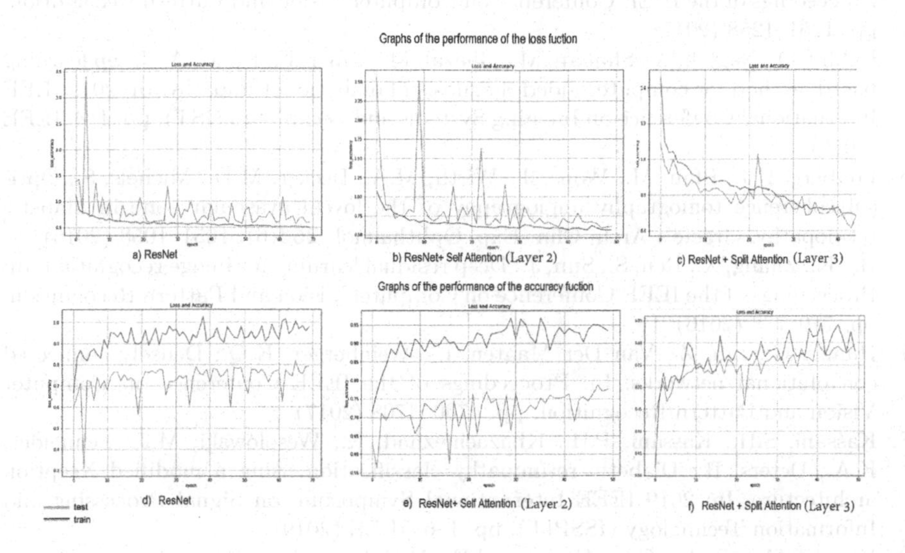

Fig. 7. Comparison of loss and accuracy of tested models

6 Conclusion

In this research, we proposed a new framework based on a modified ResNet50. We presented a three-layer model which is composed of ResNET50, ResNext and split attention for classification of severity DR. The proposed model can significantly fuse feature maps of varying depths and provide an effective and highly inexpensive methodology for classifying DR severity. Pre-processing min-pooling is used to enhance input image colour contrast. In addition, addressing the strongly imbalanced groups in the dataset, we used focal, which further improved the performance of our model for DR classification. The modified ResNet50 deep extractor accomplished substantially better results compared to the original ResNet50 and other variants tested in our study and a few other SOTA algorithms on the APTOS dataset. We demonstrated that the concatenation of ResNet50 components and the ResNext with split attention contribute to impressive results despite the insufficient number of training images.

References

1. Bravo, M.A., Arbeláez, P.A.: Automatic diabetic retinopathy classification. In: 13th International Conference on Medical Information Processing and Analysis, vol. 10572, p. 105721E. International Society for Optics and Photonics (2017)
2. Chaturvedi, S.S., Gupta, K., Ninawe, V., Prasad, P.S.: Automated diabetic retinopathy grading using deep convolutional neural network. arXiv preprint arXiv:2004.06334 (2020)
3. Chollet, F.: Xception: deep learning with depthwise separable convolutions. In: Proceedings of the IEEE Conference on Computer Vision and Pattern Recognition, pp. 1251–1258 (2017)
4. Dekhil, O., Naglah, A., Shaban, M., Ghazal, M., Taher, F., Elbaz, A.: Deep learning based method for computer aided diagnosis of diabetic retinopathy. In: 2019 IEEE International Conference on Imaging Systems and Techniques (IST), pp. 1–4. IEEE (2019)
5. Freiberg, F.J., Pfau, M., Wons, J., Wirth, M.A., Becker, M.D., Michels, S.: Optical coherence tomography angiography of the foveal avascular zone in diabetic retinopathy. Graefe's Arch. Clin. Exp. Ophthalmol. **254**(6), 1051–1058 (2016)
6. He, K., Zhang, X., Ren, S., Sun, J.: Deep residual learning for image recognition. In: Proceedings of the IEEE Conference on Computer Vision and Pattern Recognition, pp. 770–778 (2016)
7. Huang, G., Liu, Z., Van Der Maaten, L., Weinberger, K.Q.: Densely connected convolutional networks. In: Proceedings of the IEEE Conference on Computer Vision and Pattern Recognition, pp. 4700–4708 (2017)
8. Kassani, S.H., Kassani, P.H., Khazaeinezhad, R., Wesolowski, M.J., Schneider, K.A., Deters, R.: Diabetic retinopathy classification using a modified Xception architecture. In: 2019 IEEE International Symposium on Signal Processing and Information Technology (ISSPIT), pp. 1–6. IEEE (2019)
9. Khushi, M., Clarke, C.L., Graham, J.D.: Bioinformatic analysis of cis-regulatory interactions between progesterone and estrogen receptors in breast cancer. PeerJ **2**, e654 (2014)
10. Khushi, M., Liddle, C., Clarke, C.L., Graham, J.D.: Binding sites analyser (BiSA): software for genomic binding sites archiving and overlap analysis. PLoS ONE **9**(2), e87301 (2014)
11. Lam, C., Yi, D., Guo, M., Lindsey, T.: Automated detection of diabetic retinopathy using deep learning. AMIA Summits Transl. Sci. Proc. **2018**, 147 (2018)
12. Larsen, N., Godt, J., Grunkin, M., Lund-Andersen, H., Larsen, M.: Automated detection of diabetic retinopathy in a fundus photographic screening population. Invest. Ophthalmol. Vis. Sci. **44**(2), 767–771 (2003)
13. Liu, S., Deng, W.: Very deep convolutional neural network based image classification using small training sample size. In: 2015 3rd IAPR Asian Conference on Pattern Recognition (ACPR), pp. 730–734. IEEE (2015)
14. Naseem, U., Musial, K., Eklund, P., Prasad, M.: Biomedical named-entity recognition by hierarchically fusing BioBERT representations and deep contextual-level word-embedding. In: 2020 International Joint Conference on Neural Networks (IJCNN), pp. 1–8. IEEE, July 2020
15. Naseem, U., Razzak, I., Eklund, P., Musial, K.: Towards improved deep contextual embedding for the identification of irony and sarcasm. In: 2020 International Joint Conference on Neural Networks (IJCNN), pp. 1–7. IEEE, July 2020

16. Naseem, U., Razzak, I., Hameed, I.A.: Deep context-aware embedding for abusive and hate speech detection on Twitter. Aust. J. Intell. Inf. Process. Syst. **15**(3), 69–76 (2019)
17. Naseem, U., Razzak, I., Musial, K., Imran, M.: Transformer based deep intelligent contextual embedding for Twitter sentiment analysis. Future Gener. Comput. Syst. **113**, 58–69 (2020)
18. Rehman, A., Naz, S., Naseem, U., Razzak, I., Hameed, I.A.: Deep autoencoder-decoder framework for semantic segmentation of brain tumor. Aust. J. Intell. Inf. Process. Syst. **15**(3), 53–60 (2019)
19. Nguyen, Q.H., et al.: Diabetic retinopathy detection using deep learning. In: Proceedings of the 4th International Conference on Machine Learning and Soft Computing, ICMLSC 2020, pp. 103–107. Association for Computing Machinery, New York (2020). https://doi.org/10.1145/3380688.3380709
20. Pratt, H., Coenen, F., Broadbent, D.M., Harding, S.P., Zheng, Y.: Convolutional neural networks for diabetic retinopathy. Procedia Comput. Sci. **90**, 200–205 (2016)
21. Priya, R., Aruna, P.: SVM and neural network based diagnosis of diabetic retinopathy. Int. J. Comput. Appl. **41**(1), 6–12 (2012)
22. Quellec, G., Russell, S.R., Abràmoff, M.D.: Optimal filter framework for automated, instantaneous detection of lesions in retinal images. IEEE Trans. Med. Imaging **30**(2), 523–533 (2010)
23. Sakaguchi, A., Wu, R., Kamata, S.i.: Fundus image classification for diabetic retinopathy using disease severity grading. In: Proceedings of the 2019 9th International Conference on Biomedical Engineering and Technology, pp. 190–196 (2019)
24. Sathiya, G., Gayathri, P.: Automated detection of diabetic retinopathy using GLCM, January 2014
25. Sinthanayothin, C., Kongbunkiat, V., Phoojaruenchanachai, S., Singalavanija, A.: Automated screening system for diabetic retinopathy. In: 3rd International Symposium on Image and Signal Processing and Analysis 2003, Proceedings of the ISPA 2003, vol. 2, pp. 915–920. IEEE (2003)
26. Smailagic, A., et al.: O-MedAL: online active deep learning for medical image analysis. Wiley Interdiscip. Rev. Data Min. Knowl. Discov. **10**(4), e1353 (2020)
27. Szegedy, C., et al.: Going deeper with convolutions. In: Proceedings of the IEEE Conference on Computer Vision and Pattern Recognition, pp. 1–9 (2015)
28. Szegedy, C., et al.: Going deeper with convolutions. CoRR abs/1409.4842 (2014). http://arxiv.org/abs/1409.4842
29. Szegedy, C., Vanhoucke, V., Ioffe, S., Shlens, J., Wojna, Z.: Rethinking the inception architecture for computer vision. In: Proceedings of the IEEE Conference on Computer Vision and Pattern Recognition, pp. 2818–2826 (2016)
30. Wan, S., Liang, Y., Zhang, Y.: Deep convolutional neural networks for diabetic retinopathy detection by image classification. Comput. Electr. Eng. **72**, 274–282 (2018)
31. Wang, Z., Yang, J.: Diabetic retinopathy detection via deep convolutional networks for discriminative localization and visual explanation. In: Workshops at the Thirty-Second AAAI Conference on Artificial Intelligence (2018)
32. Xie, S., Girshick, R., Dollár, P., Tu, Z., He, K.: Aggregated residual transformations for deep neural networks (2016)

Enhancer-DSNet: A Supervisedly Prepared Enriched Sequence Representation for the Identification of Enhancers and Their Strength

Muhammad Nabeel Asim[1,2](✉), Muhammad Ali Ibrahim[1,2], Muhammad Imran Malik[3], Andreas Dengel[1], and Sheraz Ahmed[1]

[1] German Research Center for Artificial Intelligence (DFKI), 67663 Kaiserslautern, Germany
Muhammad_Nabeel.Asim@dfki.de
[2] University of Kaiserslautern (TU Kaiserslautern), 67663 Kaiserslautern, Germany
[3] National Center for Artificial Intelligence (NCAI), National University of Sciences and Technology, Islamabad, Pakistan

Abstract. Identification of enhancers and their strength prediction plays an important role in gene expression regulation and currently an active area of research. However, its identification specifically through experimental approaches is extremely time consuming and labor-intensive task. Several machine learning methodologies have been proposed to accurately discriminate enhancers from regulatory elements and to estimate their strength. Existing approaches utilise different statistical measures for feature encoding which mainly capture residue specific physico-chemical properties upto certain extent but ignore semantic and positional information of residues. This paper presents "Enhancer-DSNet", a two-layer precisely deep neural network which makes use of a novel k-mer based sequence representation scheme prepared by fusing associations between k-mer positions and sequence type. Proposed Enhancer-DSNet methodology is evaluated on a publicly available benchmark dataset and independent test set. Experimental results over benchmark independent test set indicate that proposed Enhancer-DSNet methodology outshines the performance of most recent predictor by the figure of 2%, 1%, 2%, and 5% in terms of accuracy, specificity, sensitivity and matthews correlation coefficient for enhancer identification task and by the figure of 15%, 21%, and 39% in terms of accuracy, specificity, and matthews correlation coefficient for strong/weak enhancer prediction task.

Keywords: Enhancer identification · Strong enhancer · Weak enhancers · Enhancer classification · Deep enhancer predictor · Enhancer strength identification · Enriched k-mers

H. Yang et al. (Eds.): ICONIP 2020, LNCS 12534, pp. 38–48, 2020.
https://doi.org/10.1007/978-3-030-63836-8_4

1 Introduction

Enhancers are functional cis elements which belong to diverse subgroups (e.g. strong enhancer, weak enhancers, poised enhancers, and inactive enhancers), where each type of enhancer is associated with multifarious biological activities [13]. Mainly, in gene expression regulation, enhancers play an indispensable role for the generation of proteins and RNA [19] and ensure very close relationship between biological processes [12]. Enhancers impact cell growth, cell differentiation, cell carcinogenesis, virus activity, and tissue specificity through enhancing genes transcription [12]. Enhancer may be located in separate chromosome or 20 kb far away from genes [15] as compared to promoters which are usually located around start transcriptional sites of genes. Building on these locational differences, identifying enhancers is widely considered far more challenging than promoters. Discriminating enhancers from regulatory elements, estimating their location and overall strength are few most promising tasks which can facilitate deeper comprehension of eukaryotic spatiotemporal gene regulation and evolution of diseases [15].

Initially, enhancers were discovered through typical experimental approaches [3,11]. Former approach used to identify enhancers by utilizing their association with transcriptional factor [22], whereas, latter approach leveraged DNase-I hypersensitivity. While former approach under detected enhancers [6] as all enhancers are not occupied by transcription factors, latter approach over detected as it classified even DNA segments or non-enhancers as enhancers [15,17]. Although subsequent methodologies of genome wide mapping of histone modifications [7,8,20] decently alleviated high false positive and false negative rate of initial experimental techniques for the discovery of promoters and enhancers. However, these approaches are rigorously expensive, time, and resource consuming. Due to these shortcomings and with the influx of high throughput biological data related to enhancers, demand of robust computational methodologies capable to differentiate enhancers from regulatory elements and estimate their strength got significantly rocketed.

Up to this date, several computational methodologies have been proposed to discriminate enhancers from non-enhancers in genome such as CSI-ANN [9], RFECS [20], EnhancerFinder [8], EnhancerDBN [5], and BiRen [24]. Proposed predictors differ in terms of feature encoding and classifier. For example, CS1-ANN [9] utilized data transformation approach for samples formulation and Artificial Neural Network (ANN) for classification. Likewise, EnhancerFinder [8] incorporates evolutionary conservation knowledge into sample formulation and a combination of several kernel learning approaches for classification. EnhancerDBN [5] makes use of deep belief network (DBN), RFEC [20] utilizes random forest classifier [4], whereas BiRen [24] leverages deep learning approaches to accelerate predictive performance. These approaches only capable to discriminate enhancers from regulatory elements in genome. Therefore, robust enhancer determinant and strength prediction approaches are still scarce. iEnhancer-2L [15] is the very first tool developed to discover enhancer along with their strength using solely sequence information and it has been extensively utilized for genome analysis. To further improve the performance at both layers, more computational methodologies have been developed afterward which have improved iEnhancer-2L [15] methodology further by using the combination of statistical measures to better represent physico-chemical properties such as EnhancerPred [12], iEnhancer-PsedeKNC [15] iEnhancer-EL [16], Tan et al. Enhancer [21], and EnhancerPred2.0 [10]. Up to date, only one recently proposed approach namely "iEnhancer-5Step" [14] makes use of SVM classifier and unsupervisedly prepared neural k-mer embeddings to better capture local patterns for the task of enhancer determinant and strength prediction.

Nevertheless, still a lot of improvement in performance is required as these approaches produce confined performance especially in distinguishing strong enhancers from weak enhancers. To develop an optimal machine learning model for enhancer identification and strength prediction task, most crucial step is to encode biomedical sequence into fixed-size low dimensional vectors. In this context, few sequence encoding approaches including Local Descriptor, Conjoint Triad (CT), Auto Covariance (AC), and PSE-KNC [16] have been utilized where residual oriented physico-chemical properties are taken into account. But, the major downfalls for such manually curated feature vectors are, these approaches fail to take semantic information of residues into account (such as residues order) in sequences and also neglect noteworthy information from large number of unlabelled biomedical sequences that can assist the classifier to better identify class boundaries. To overcome these shortcomings upto certain extent, Le et al. [14] have recently employed neural word embeddings prepared in an unsupervised manner. Although unsupervised k-mer embeddings capture semantic information of k-mers, however they still lack to associate inherent k-mer relationships with sequence type keeping within low-dimensional vector space. To fully reap the benefits of neural word embeddings for creating an optimal representation of k-mers present in sequences, we present a novel k-mer based sequence representation scheme which prepares the sequence embeddings in a supervised manner where we fuse the alliance of k-mers with sequence type. To evaluate the effectiveness of presented enriched sequence representation, we present a two-layer classification methodology (Enhancer-DSNet) based on linear classifier and perform experimentation over a publicly available benchmark dataset and independent test set for the task of enhancer determinant and strength prediction task. We have obtained excellent predictive accuracy, outperformed various combinations of machine learning algorithms, commonly-used sequence encoding schemes, and unsupervisedly prepared k-mer embeddings with significant margins.

2 Materials and Methods

This section discusses proposed two-layer classification methodology "Enhancer-DSNet", benchmark dataset and independent test set used for experimentation, and evaluation measures.

3 Proposed Enhancer-DSNet Methodology

With the huge success of pre-trained neural word embeddings over diversified NLP tasks [1], biomedical researchers have extensively utilized distributed representations in different biomedical tasks [23]. These embeddings are usually prepared in an unsupervised manner by training a shallow neural network on gigantic sequence corpora. Pre-trained neural k-mer embeddings are semantically meaningful low dimensional dense representation of k-mers present in the sequences. Although neural k-mer embeddings prepared in an unsupervised manner create proximal representation of highly similar k-mer groups in embedding space and have shown good performance in different biomedical tasks such as sequences structural similarity estimation [2], and transmembrane prediction [18]. However, these embeddings still lack to associate class information with distinct arrangements of nucleotides present in sequences, a phenomena that can significantly raise the classifier performance [23].

Considering relationships between distinct k-mers largely depend on k-mer size and sequence type, we have generated k-mer embeddings in a supervised manner. Unlike trivial neural k-mer embeddings, here, we improve k-mer representation by creating associations between k-mers positions and sequence type (Fig. 1).

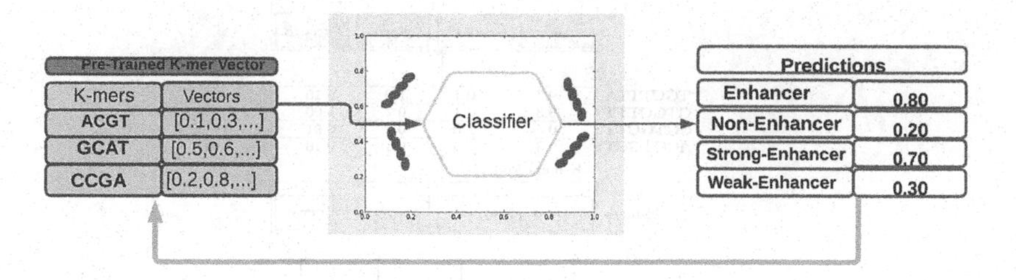

Fig. 1. Supervisedly prepared neural k-mer embeddings

To generate sequence embeddings, k-mer embeddings are concatenated through summation. In this manner, we are accurately capturing semantic information and local patterns present in sequences. Also, we are computing sequences similarity correctly within low dimensional space revealing functional relationship, while making sure that computation relies on set of features pertinent to hand on problem. Architecture of proposed two-layer Enhancer-DSNet approach is illustrated in Fig. 2. Where firstly, overlapped k-mers of each sequence is generated by sliding a window across the sequence with stride size of 1. Afterward, overlapped k-mers of sequences are passed to embedding layer, where 100 dimensional vectors are generated for each overlapped k-mers. All k-mer vectors are then aggregated to generate 100 dimensional vector for the whole sequence. In order to avoid over fitting the model, a dropout layer with dropout rate of 0.5% is utilized. After dropout layer, softmax classifier is used to incorporate label information into the sequence vectors by updating model parameters. In this manner, we ensure that, on independent test set, model is able to extract meaningful patterns through which classifier will better discriminate the sequences at both layers.

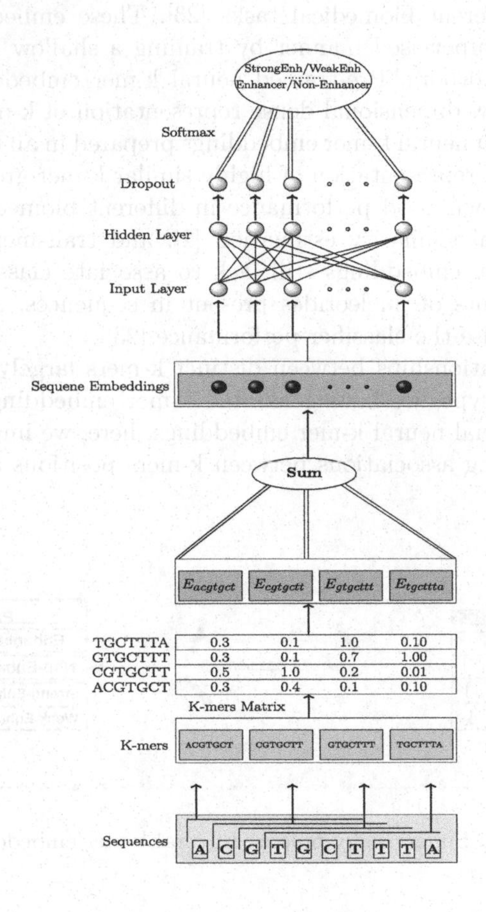

Fig. 2. Architecture of proposed two-layer classification methodology "Enhancer-DSNet"

4 Benchmark Dataset

To evaluate the integrity of proposed Enhancer-DSNet approach, experimentation is performed on a publicly available benchmark dataset and independent test set [15]. These resources have been utilized in previous studies to evaluate enhancer determinant and strength prediction approaches [10,12,15,16,21]. Enhancer and non-Enhancer discrimination benchmark dataset has 2968 samples, out of which 1484 samples are enhancers and 1484 samples are non-enhancers. Out of 1484 enhancer samples, 742 samples are strong enhancers and remaining 742 samples are weak enhancers. While enhancer/non-enhance dataset is used to discriminate enhancers from non-enhancers, strong/weak enhancer subset formulated using enhancer samples is further used to estimate the strength of enhancers. Besides benchmark dataset, an independent test set is also publicly available which contain 400 samples, out of which 200 samples are enhancers and remaining 200 samples are non-enhancers. From 200 enhancer samples, 100 samples are strong enhancers and remaining 100 samples are weak enhancers. Just like benchmark dataset, enhancer/non-enhancer independent test set is used to for enhancer/non-enhancer prediction task, whereas strong/weak enhancer subset formulated using enhancer samples of independent test is used to estimate the strength of enhancers. Detailed formulation of benchmark and independent test set have been clearly elaborated in Liu et al. [16] work, hence there in no need to repeat here.

5 Evaluation Metrics

Following evaluation criteria of previous studies related to the classification of enhancer and other regulatory elements, and estimating the strength of enhancers [10,12,15,16,21], here we have used 4 different evaluation measures (sensitivity, specificity, accuracy, and matthews correlation coefficient) to perform a fair performance comparison of proposed approach with state-of-the-art approaches. As these measures are briefly described in previous studies [12,15] so here we just give a short description. To provide intuitive understanding for readers, evaluation metrics along with mathematical expressions are briefly described below:

$$f(x) = \begin{cases} \text{Accuracy (ACC)} = 1 - (O_{-}^{+} + (O_{+}^{-})/(O^{+} + O^{-}) & 0 \leq \text{Acc} \leq 1 \\ \text{Specificity (SP)} = 1 - (O_{+}^{-}/O^{-}) & 0 \leq \text{SP} \leq 1 \\ \text{Sensitivity (SN)} = 1 - -(O_{-}^{+}/O^{+}) & 0 \leq \text{SN} \leq 1 \\ \text{MCC} = 1 - (O_{-}^{+}/O^{+} + O_{+}^{-}/O^{-})/\sqrt{(1 + O_{+}^{-} - O_{-}^{+}/O^{+})(1 + O_{-}^{+} - O_{+}^{-}/O^{-}} & -1 \leq \text{MCC} \leq 1 \end{cases} \tag{1}$$

Here, O^{+} infers total positive class observations investigated, O^{-} represents total negative class observations investigated. While, number of positive class observations predicted correctly and are negative class observations predicted correctly. Whereas, represent positive class observation incorrectly predicted as negative and are negative class observations mis-classified as positive.

6 Experimental Setup and Results

This section illustrates experimental details and briefly describes Results of proposed Enhancer-DSNet approach.

To generate sequence embeddings of benchmark dataset in a supervised manner, and to perform experimentation over benchmark dataset and independent test set, we have used Pytorch API. To generate supervised sequence vectors, we have trained the newly developed skip-gram model for 30 epochs with 0.008

learning rate and adam optimizer. Experimentation for both enhancer identification and strength prediction tasks is performed using 7-mer enriched sequence vectors.

6.1 Results

Here, we briefly describe and compare the performance of proposed Enhancer-DSNet methodology with state-of-the-art Enhancer determinant and strength prediction approaches using cross validation and benchmark independent test set.

Cross-validation. In order to better evaluate the performance of a classifier by eliminating biasness towards the split of dataset, most widely used re-sampling approach is called cross-validation. In k-fold cross validation, one can split a dataset into k number of groups, for example, 5-fold cross validation will segregate entire dataset into 5 groups where each group will be splitted into train, and test sets to train and test the model. In this manner, each group of limited data samples take part in training and testing processes. Another similar unbiased performance estimator is jackknife test where training is performed over entire dataset except one observation of a dataset which is iteratively used to test the model. In comparison to cross-validation, jackknife test is quite expensive to compute especially for large datasets and it has also high variance as datasets used to estimate classifier performance are quite similar. Hence, k-fold cross validation is widely considered a better estimator of bias and variance as it is a well compromise among computational requirements and impartiality. Existing enhancer and non-enhancer discriminator and enhancer strength predictor approaches (EnhancerPred [12], iEnhancer-PsedeKNC [15] iEnhancer-EL [16], EnhancerPred2.0 [10]) utilized jackknife test to evaluate the performance of their models on a benchmark dataset. However, most recent Tan et al. predictor [21] performance is evaluated using 5 fold cross validation. Following Tan et al. [21] work, in our experimentation, we have also used 5-fold cross validation on a benchmark dataset. So here, using 5-fold cross validation, we perform performance comparison of Enhancer-DSNet with most recent Tan et al. predictor [21].

Figures 3a and b illustrate the performance of Enhancer-DSNet across 5-folds on a benchmark dataset of enhancer/non-enhancer and strong/weak enhancer prediction task. To sum up, performance of Enhancer-DSNet remains consistent across 5-folds when evaluated in terms of 4 distinct evaluation metrics.

Table 1 reports the average of performance figures produced by 5-fold cross validation at layer 1 and 2 in terms of accuracy, specificity, sensitivity and matthews correlation coefficient (mcc). As is indicated by the Table 1, for enhancer/non-enhancer prediction task (layer-1), proposed Enhancer-DSNet outshines Tan et al. Enhancer [21] by the figure of 3% in terms of sensitivity, 2%

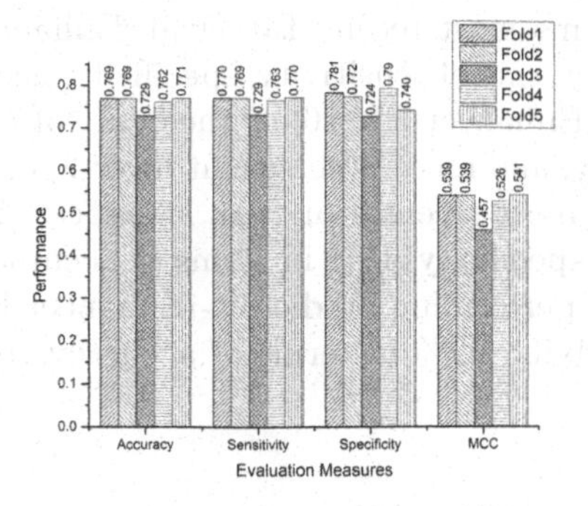

(a) Enhancer/Non-Enhancer Prediction

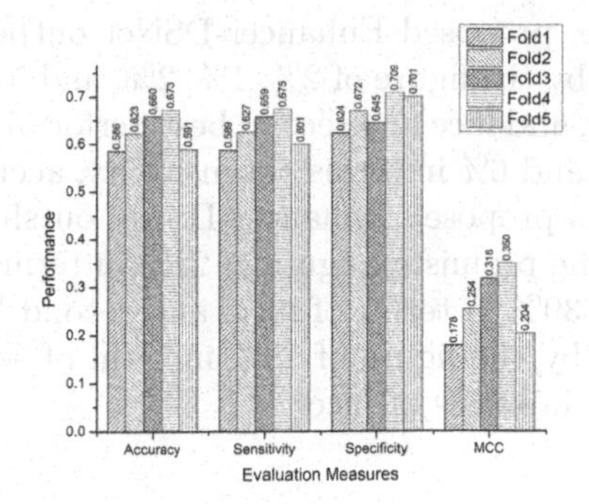

(b) Strong/Weak Enhancer Prediction

Fig. 3. Performance of Enhancer-DSNet produced over 5-folds for layer 1 and 2 in terms of accuracy, specificity, sensitivity, and MCC

Table 1. Performance comparison of Enhancer-DSNet with most recent Tan et al. Enhancer [21] using 5-fold cross validation for enhancer/non-enhancer and strong/weak enhancer prediction task

Classifiers	Sensitivity	Specificity	Accuracy	MCC
1st layer (enhancer/non-enhancer)				
Enhancer-DSNet	**0.76**	**0.76**	**0.76**	**0.52**
Tan et al. Enhancer [21]	0.73	0.76	0.74	0.50
2nd layer (strong enhancer/weak enhancer)				
Enhancer-DSNet	**0.63**	**0.67**	**0.63**	**0.26**
Tan et al. Enhancer [21]	0.80	0.38	0.59	0.20

in terms of accuracy and 2% in terms of matthews correlation coefficient. However, for strong/weak enhancer prediction task (layer-2), proposed Enhancer-DSNet outperforms Tan et al. Enhancer [21] with a huge margin across 4 different evaluation metrics. Enhancer-DSNet significantly superior performance overshadows most recent Tan et al. Enhancer [21] performance by the figure of 17% in terms of sensitivity, 29% in terms of specificity, 4% in terms of accuracy, and 6% in terms of mcc.

Performance over Benchmark Independent Test Set. Table 2 reports the performance of proposed Enhancer-DSNet and existing predictors produced over independent test set for enhancer/non-enhancer and independent subset for strong/weak enhancer prediction tasks in terms of accuracy, specificity, sensitivity, and matthews correlation coefficient. According to the Table 2, at layer-1, among all existing predictors excluding most recent Tan et al. Enhancer [21], and -iEnhancer-EL [16] mark better performance across most evaluation metrics.

Here, proposed Enhancer-DSNet outperforms most recent Tan et al. Enhancer [21] by the figure of 2%, 1%, 2%, and 5% in terms of sensitivity, specificity, accuracy, and mcc and second best performing -iEnhancer-EL [16] by the figure of 7%, 3%, and 6% in terms of sensitivity, accuracy, and mcc. Whereas, at layer-2, once again proposed Enhancer-DSNet outshines most recent Tan et al. Enhancer [21] by the promising figure of 21% in terms of specificity, 15% in terms of accuracy, and 39% in terms of mcc, and second best performing predictor -iEnhancer-EL [16] by the figure of 29% in term of sensitivity, 22% in terms of accuracy, and 48% in terms of mcc.

Table 2. Performance comparison of Enhancer-DSNet with existing enhancer/non-enhancer and strong/weak enhancer predictors over independent test set

Classifiers	Sensitivity	Specificity	Accuracy	MCC
1st layer (enhancer/non-enhancer)				
Enhancer-DSNet	**0.78**	**0.77**	**0.78**	**0.56**
Tan et al. Enhancer [21]	0.76	0.76	0.76	0.51
iEnhancer-EL [16]	0.71	0.79	0.75	0.50
iEnhancer-2L [15]	0.71	0.75	0.73	0.46
EnhancerPred [12]	0.74	0.75	0.74	0.48
2nd layer (strong enhancer/weak enhancer)				
Enhancer-DSNet	**0.83**	**0.67**	**0.83**	**0.70**
Tan et al. Enhancer [21]	0.83	0.46	68.49	0.31
iEnhancer-EL [16]	0.54	0.68	0.61	0.22
iEnhancer-2L [15]	0.47	0.74	0.61	0.22
EnhancerPred [12]	0.45	0.65	0.55	0.10

Results Reproduce Ability Issue. It is important to mention that recent enhancer determinant and strength prediction approach namely "i-Enhancer-5Step" is given by Lee et al. [14]. Authors have utilized unsupervisedly prepared sequence embeddings by treating each nucleotide as word and entire sequence as sentence. Then, these embeddings are passed to SVM classifier. To re-produce reported results [14], we have performed rigorous experimentation using all mentioned parameters [14], but the performance figures we attained are reasonably low than the reported ones [14]. Also authors of most recent predictor namely iEnhancer-EL [16] did not compare their performance figures with Lee et al. i-Enhancer-5Step [14]. Building on this, we consider the results reported in Lee et al. [14] work are fraudulent. Therefore, similar to iEnhancer-EL [16], we also do not compare the performance of proposed Enhancer-DSNet with Lee et al. i-Enhancer-5Step [14].

7 Conclusion

In the marathon of improving the performance of Enhancer identification and their strength prediction, researchers have predominantly employed physico-chemical properties based, bag-of-words based and unsupervisedly prepared k-mer embeddings with different classifiers. Considering these approaches fail to utilize association of inherent sequence relationships with sequence type, we have fused such association by generating sequence embeddings in a supervised fashion which are later fed to a two-layer classification methodology Enhancer-DSNet based on linear classifier. Over a benchmark dataset, proposed Enhancer-DSNet approach outperforms most recent predictor by the figure of 2%, 3%, 2% in terms of accuracy, sensitivity and mcc for enhancer identification task and by the figure of 29%, 4%, 6% in terms of accuracy, specificity, and mcc for strong/weak enhancer prediction task. This studly findings has opened new doors of further research where biomedical researchers can utilize supervisedly prepared sequence embeddings to enhance the performance of multifarious biomedical tasks.

References

1. Almeida, F., Xexéo, G.: Word embeddings: a survey. arXiv preprint arXiv:1901.09069 (2019)
2. Bepler, T., Berger, B.: Learning protein sequence embeddings using information from structure. arXiv preprint arXiv:1902.08661 (2019)
3. Boyle, A.P., et al.: High-resolution genome-wide in vivo footprinting of diverse transcription factors in human cells. Genome Res. **21**(3), 456–464 (2011)
4. Breiman, L.: Random forests. Mach. Learn. **45**(1), 5–32 (2001)
5. Bu, H., Gan, Y., Wang, Y., Zhou, S., Guan, J.: A new method for enhancer prediction based on deep belief network. BMC Bioinformatics **18**(12), 418 (2017)
6. Chen, J., Liu, H., Yang, J., Chou, K.C.: Prediction of linear b-cell epitopes using amino acid pair antigenicity scale. Amino Acids **33**(3), 423–428 (2007)
7. Ernst, J., et al.: Mapping and analysis of chromatin state dynamics in nine human cell types. Nature **473**(7345), 43–49 (2011)
8. Erwin, G.D., et al.: Integrating diverse datasets improves developmental enhancer prediction. PLoS Comput. Biol. **10**(6), e1003677 (2014)
9. Firpi, H.A., Ucar, D., Tan, K.: Discover regulatory DNA elements using chromatin signatures and artificial neural network. Bioinformatics **26**(13), 1579–1586 (2010)
10. He, W., Jia, C.: EnhancerPred2.0: predicting enhancers and their strength based on position-specific trinucleotide propensity and electron-ion interaction potential feature selection. Mol. Biosyst. **13**(4), 767–774 (2017)
11. Heintzman, N.D., Ren, B.: Finding distal regulatory elements in the human genome. Curr. Opin. Genet. Dev. **19**(6), 541–549 (2009)
12. Jia, C., He, W.: EnhancerPred: a predictor for discovering enhancers based on the combination and selection of multiple features. Sci. Rep. **6**, 38741 (2016)
13. de Lara, J.C.F., Arzate-Mejía, R.G., Recillas-Targa, F.: Enhancer RNAs: insights into their biological role. Epigenetics Insights **12**, 2516865719846093 (2019)
14. Le, N.Q.K., Yapp, E.K.Y., Ho, Q.T., Nagasundaram, N., Ou, Y.Y., Yeh, H.Y.: iEnhancer-5Step: identifying enhancers using hidden information of DNA sequences via Chou's 5-step rule and word embedding. Anal. Biochem. **571**, 53–61 (2019)

15. Liu, B., Fang, L., Long, R., Lan, X., Chou, K.C.: iEnhancer-2L: a two-layer predictor for identifying enhancers and their strength by pseudo k-tuple nucleotide composition. Bioinformatics **32**(3), 362–369 (2016)
16. Liu, B., Li, K., Huang, D.S., Chou, K.C.: iEnhancer-EL: identifying enhancers and their strength with ensemble learning approach. Bioinformatics **34**(22), 3835–3842 (2018)
17. Liu, B., Yang, F., Huang, D.S., Chou, K.C.: iPromoter-2L: a two-layer predictor for identifying promoters and their types by multi-window-based PseKNC. Bioinformatics **34**(1), 33–40 (2018)
18. Ng, P.: dna2vec: consistent vector representations of variable-length k-mers. arXiv preprint arXiv:1701.06279 (2017)
19. Omar, N., Wong, Y.S., Li, X., Chong, Y.L., Abdullah, M.T., Lee, N.K.: Enhancer prediction in proboscis monkey genome: a comparative study. J. Telecommun. Electron. Comput. Eng. (JTEC) **9**(2–9), 175–179 (2017)
20. Rajagopal, N., et al.: RFECS: a random-forest based algorithm for enhancer identification from chromatin state. PLoS Comput. Biol. **9**(3), e1002968 (2013)
21. Tan, K.K., Le, N.Q.K., Yeh, H.Y., Chua, M.C.H.: Ensemble of deep recurrent neural networks for identifying enhancers via dinucleotide physicochemical properties. Cells **8**(7), 767 (2019)
22. Visel, A., et al.: ChIP-seq accurately predicts tissue-specific activity of enhancers. Nature **457**(7231), 854–858 (2009)
23. Wang, Y., et al.: A comparison of word embeddings for the biomedical natural language processing. J. Biomed. Inform. **87**, 12–20 (2018)
24. Yang, B., et al.: BiRen: predicting enhancers with a deep-learning-based model using the DNA sequence alone. Bioinformatics **33**(13), 1930–1936 (2017)

Machine Learned Pulse Transit Time (MLPTT) Measurements from Photoplethysmography

Philip Mehrgardt[1(✉)], Matloob Khushi[1], Anusha Withana[1,2], and Simon Poon[1]

[1] School of Computer Science, The University of Sydney, Sydney, NSW 2008, Australia
pmeh3648@uni.sydney.edu.au

[2] Cardiovascular Initiative, The University of Sydney, Sydney, NSW 2008, Australia

Abstract. Pulse transit time (PTT) provides a cuffless method to measure and predict blood pressure, which is essential in long term cardiac activity monitoring. Photoplethysmography (PPG) sensors provide a low-cost and wearable approach to obtain PTT measurements. The current approach to calculating PTT relies on quasi-periodic pulse event extractions based on PPG local signal characteristics. However, due to inherent noise in PPG, especially at uncontrolled settings, this approach leads to significant errors and even missing potential pulse events. In this paper, we propose a novel approach where global features (all samples) of the time-series data are used to develop a machine learning model to extract local pulse events. Specifically, we contribute 1) a new noise resilient machine learning model to extract events from PPG and 2) results from a study showing accuracy over state of the art (e.g. HeartPy) and 3) we show that MLPTT outperforms HeartPy peak detection, especially for noisy photoplethysmography data.

Keywords: Pulse transit time · Pulse Arrival Time · Blood pressure · Medical · Data analysis · Machine learning

1 Introduction

Continuous monitoring of blood pressure over long periods is essential to prevent critical cardiovascular events that can cause irreversible health damages and even death. The classical approach to measuring blood pressure uses pressure cuffs, requiring mechanical apparatuses which are undesired in long-term monitoring. Researchers have proposed alternative methods to measure blood pressure using pulse transit time (PTT), which does not need mechanical cuffs [1–7]. An extensive review of blood pressure monitoring theory and practice using PTT can be found in [8]. Researchers also found a strong correlation among PTT, the measurement of respiratory effort [9] and the detection of microarousals [9]. Other

H. Yang et al. (Eds.): ICONIP 2020, LNCS 12534, pp. 49–62, 2020.
https://doi.org/10.1007/978-3-030-63836-8_5

applications included the indication of cardiovascular changes during obstetric spinal anesthesia [10], myocardial performance [11], respiratory events [12] and hypertension detection [13,14].

PTT is the duration that it takes for a pulse wave (PW) to travel between two arterial points [13]. It can be measured with different methods, such as arterial probes [15], electrocardiogram (ECG) or contactless photoplethysmography (PPG). ECG electrically records a graph of voltage versus time, which is typically acquired from electrodes attached to the chest or limbs whereas PPG optically records blood volume changes vs time.

While ECG may have suffered from measurement artefacts [16] such as electromagnetic interference or loose leads, it was still considered the gold standard for continuous heart rate monitoring [17]. Despite the high signal quality, its usability was limited by the requirement of leads attached to the skin. In comparison, optical PPG sensors were found to be the least invasive and therefore overall desirable. However, they were also sensitive to sensor pressure and artefacts caused by motion [18]. Pulse transit times can also be measured between different physical locations and temporal parts of the cardiac cycle. The cardiac cycle is a repeating series of pressure changes within the heart of living organisms and was often described to have several peaks, particularly the P wave, QRS complex and T wave [19]. No standardised PTT definition had been established due to various measurement possibilities regarding

(i) different sensor types
(ii) sensor or lead locations and
(iii) temporal reference points.

Van Velzen, et al. identified 43 different methods to determine PTT [20] between an ECG R-peak and a PPG signal [21–25]. A PTT for ECG-PPG measurements is called Pulse Arrival Time (PAT). All reviewed methods [20] used anchor points to calculate the PAT (Fig. 1A).

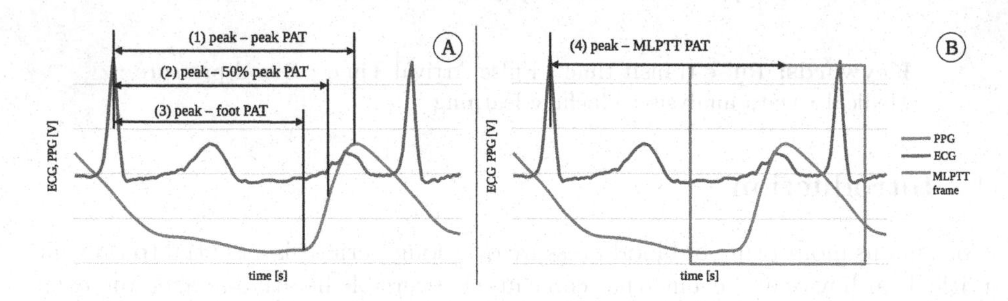

Fig. 1. A: Existing PAT definitions, B: Proposed MLPTT PAT measurement

These anchor points could be the foot (or onset), peak or a percentage (e.g. 50%, 25%, etc.) of the pulse wave. PAT could be calculated by

$$PAT_{peak-peak} = t_{PPG_{peak}}(n) - t_{ECG_{R-peak}}(n), \tag{1}$$

$$PAT_{peak-50\%peak} = t_{PPG_{50\%peak}}(n) - t_{ECG_{R-peak}}(n), \tag{2}$$

$$PAT_{peak-foot} = t_{PPG_{foot}}(n) - t_{ECG_{R-peak}}(n), \tag{3}$$

where n was the individual heartbeat. Many PPG sensor signal quality-related challenges were identified in previous research. Their readings were found to contain artefacts [18,26] or malformed segments induced by sensor motion or attachment pressure variations [20]. To compensate for undesired signal anomalies, at least 19 of the 43 previously reviewed methods [20] used signal filtering for PTT calculated from anchor points. Although PTT is the duration that it takes for a pulse wave to travel between two arterial points, it was so far considered as the time between two selected points on two curves, with $PAT_{peak-foot}$ and $PAT_{peak-peak}$ most widely used. Table 1 provides a summary of previously used anchor points in literature.

Table 1. Anchor point prevalence in the literature [20].

Anchor point	Foot	Onset	Upstroke	5% peak	25% peak	50%	90%	Peak
Count	15	6	7	1	4	6	1	8

Machine learning and particularly neural networks had shown great potential at extracting spatial information from data [27]. We hypothesized that while filtering had refined PTT measurements, accuracy could be improved further by not considering an individual point on the curve but the shift of the shape of the curve using all points. We therefore propose Machine Learned Pulse Transit Time (MLPTT), which was trained on a sliding frame to detect curve properties relative to a virtual anchor point (Fig. 1B). The proposed PAT measurement could be defined with the following equation

$$PAT_{peak-MLPTT} = t_{PPG_{virtual\ anchor}}(n) - t_{ECG_{R-peak}}(n), \tag{4}$$

where n was an individual heartbeat. PAT is a particular type of PTT and was used to validate the proposed method by comparing the predicted PAT with PAT calculated by HeartPy, a toolkit that was designed to handle (noisy) PPG data [28,29]. While the advantages of PAT are that R-peaks of the cardiovascular QRS complex can be detected with numerous algorithms [30] such as Pan-Tompkins [31], it requires conductive electrodes attached to the skin in selected locations and is, therefore, more disruptive than PPG, which measures optically. Gao et al. compared the pulse transit time estimates of PAT and PPG-PTT with invasive I-PTT using arterial probes as a reference [32]. They concluded that PPG-PTT correlated well with all blood pressure levels. To validate MLPTT, the PAT of three BIDMC PPG and Respiration Dataset [33] patients was calculated with HeartPy and MLPTT. Both methods provided independent results from each other, the overall validation workflow is shown in Fig. 2.

Contribution: We propose a machine learning based PTT calculation method and show it is more accurate than the existing signal processing approaches. The specific contributions can be summarized as follows:

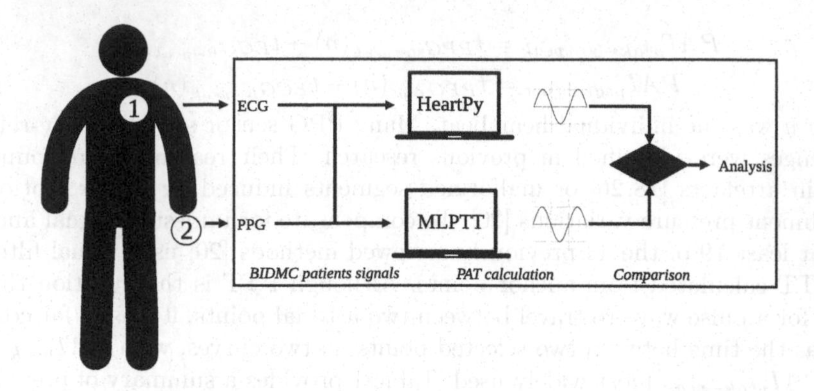

Fig. 2. End-to-end PTT validation workflow (1: ECG, 2: finger PPG).

a) We propose a new noise resilient machine learning model to extract events from PPG and demonstrate it particularly on PAT extraction.
b) Results from a study showing accuracy over state of the art (i.e. HeartPy) with statistical significance.
c) We show that MLPTT outperforms HeartPy peak detection especially for noisy photoplethysmogram data.
d) We discuss and evaluate the results.

2 Design

Pulse transit time was previously measured between two anchor points on ECG, PPT or arterial probe pressure signals [32]. Since all these signals are quasi-periodic, we hypothesized that PTT can be considered signal phase shift per heartbeat period. While phase shift can be calculated between the same anchor point on different signals, we propose to use machine learning to consider all points on the signal to estimate phase shift and therefore PTT.

2.1 Machine Learned Pulse Transit Time

The idea to consider all points on the signal was implemented in MLPTT. MLPTT consisted of a sequence of 4 processes: frame segmentation, waveform binary classifier, frame segmentation, anchor point classifier (Fig. 3: 3, 4, 7, 8).

MLPTT Frame Segmentation for Binary Classification. PPG and ECG data were loaded from a dataset (Fig. 3: 1, 2). To avoid discontinuities in the time series data, 70% training and 30% test ratio were used for all patients instead of k-fold verification. The first two MLPTT processes (Fig. 3: 3 and 4) aimed at finding a periodic pattern in the PPG signal. This was achieved by using the known periods from ECG signals R-peaks, detected with the Pan-Tompkins algorithm (Fig. 3: 5 and [31]), as virtual anchor point labels. Frames of 60 samples

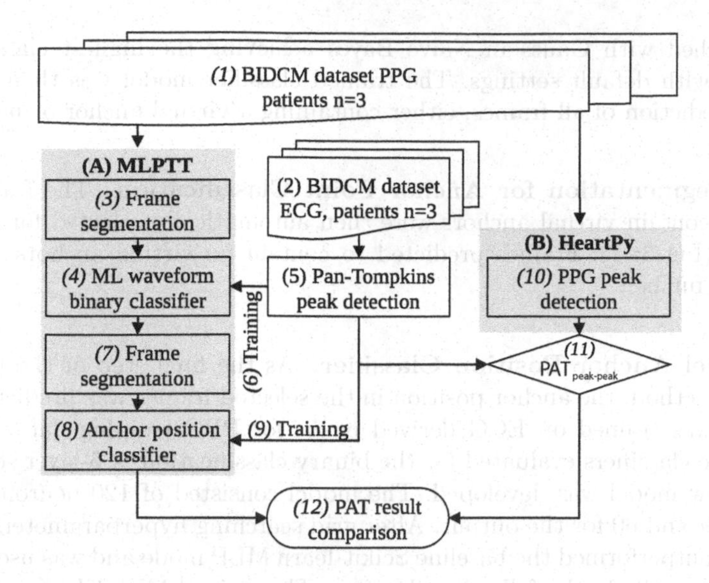

Fig. 3. MLPTT and HeartPy process diagram

each were created by shifting the frame in increments of one sample. In the case of the BIDMC dataset sampled 125 Hz, one sample was 8ms long. Each frame was labelled respective containing a virtual anchor or not.

Waveform Binary Classifier. The segmented and by anchor points labelled frames were subsequently used as training input for the ML waveform binary classifier (Fig. 4A: 4).

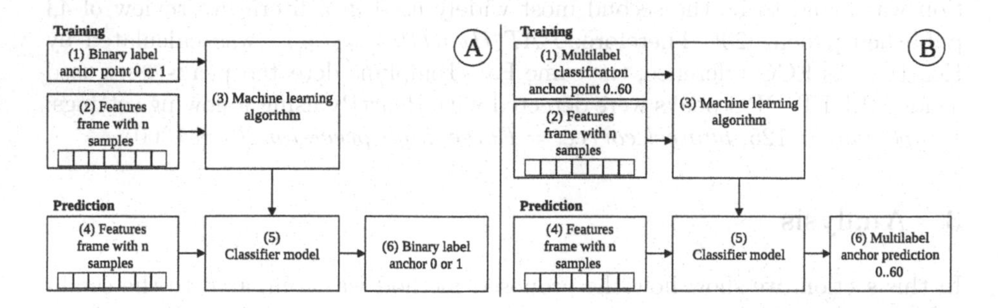

Fig. 4. A Binary waveform classifier. **B** Virtual anchor multilabel anchor position classifier.

The goal of this step was to train a classifier that could detect quasi-periodic waveforms in longer PPG time series which contained one virtual anchor. This could also be considered a heart rate classification based on PPG data, with the detected consecutive virtual anchors manifested as one heartbeat each. Settings for Python Scikit-learn 0.22.1 KNN, SVC, Gaussian process, decision tree, random forest, MLP, AdaBoost and Gaussian process Naïve Bayes classifiers were grid searched with Gaussian Naïve Bayes achieving the highest classification accuracy with default settings. The trained classifier model was then used for binary prediction of all frames, either containing a virtual anchor or not.

Frame Segmentation for Anchor Point Classification. The frames predicted to contain virtual anchors were then automatically selected for a second classifier (Fig. 3: 7). Frames predicted to contain no virtual anchors were not evaluated further.

Multilabel Anchor Position Classifier. As the final step of the proposed MLPTT method, the anchor position in the selected frames was predicted. This classifier was trained on ECG derived peaks for PPG training data. In addition to the classifiers evaluated for the binary classification, a 5-layer sequential TensorFlow model was developed. The model consisted of 120 neurons for the input layer and 60 for the output. After grid searching hyperparameters for this model, it outperformed the baseline Scikit-learn MLP mode and was used for the prediction results in the following chapters. The trained MLPTT was then used to predict $PAT(MLPTT)_{peak-virtual\_anchor}$ by predicting the virtual anchors for test data extracted from the BIDMC dataset, which were then downsampled to one anchor per period.

2.2 HeartPy Pulse Transit Time

HeartPy is a toolkit designed to handle noisy PPG data and was used to detect peaks in PPG signals. For three out of four tested PAT definitions, Rajala et al. reported the smallest relative error for $PAT_{peak-peak}$ [34] and this definition was found to be the second most widely used in a literature review of 43 published papers [20]. Therefore, $PAT(HeartPy)_{peak-peak}$ was calculated by HeartPy. As ECG reference, the same Pan-Tompkins detected peaks were used as for MLPTT. PPG peaks were detected with HeartPy using following settings: $sample\_rate = 125$, $hampel\_correct = False$, $high\_precision\_fs = 1000$.

3 Analysis

In this section, we show how the proposed method was compared to the established toolkit.

3.1 Methods

For PTT and PAT, there is no ground truth readily available. We therefore initially compared MLPTT PAT with HeartPy PAT as a reference, to disprove the null hypothesis that their correlation is not statistically relevant. PAT was used instead of PPG PTT because ECG R-peak detection accuracy was found to be higher than for PPG, ECG R-peak detection has been standard practice and for some populations, comparatively better signal to noise ratios had been observed [35] as well as lower morphological variance [36]. Because our eventual goal was to measure PTT and therefore phase shift between signals, we considered only the AC component of the time series PTT estimations for comparison. Accordingly, the PAT of 196–240 ECG-detected heartbeats was calculated for 5 patients with HeartPy and MLPTT.

We then disproved that there is no statistically significant correlation between both curves. If the following relationship was true

$$\sum_{k=0}^{n} MLPTT_{PAT}(hb_k) \neq \sum_{k=0}^{n} HeartPy_{PAT}(hb_k), \tag{5}$$

where hb_k were the heartbeats of each patient from the beginning of the test data ($k = 0$) to the last heartbeat ($k = n$), MLPTT PAT and HeartPy PAT were not correlated. If they were linearly correlated, plotting $x = MLPTTPAT(hb_k)$ and $y = HeartPyPAT(hb_k)$ would have formed a straight line. The correlation was to be proven with Pearson correlation and its respective p-value <0.05.

4 Results

We measured performance by showing a linear relation between our proposed method PAT results and HeartPy PAT results and plotted resulting PAT curves plotted on top of each other with the respective difference filled (Fig. 5).

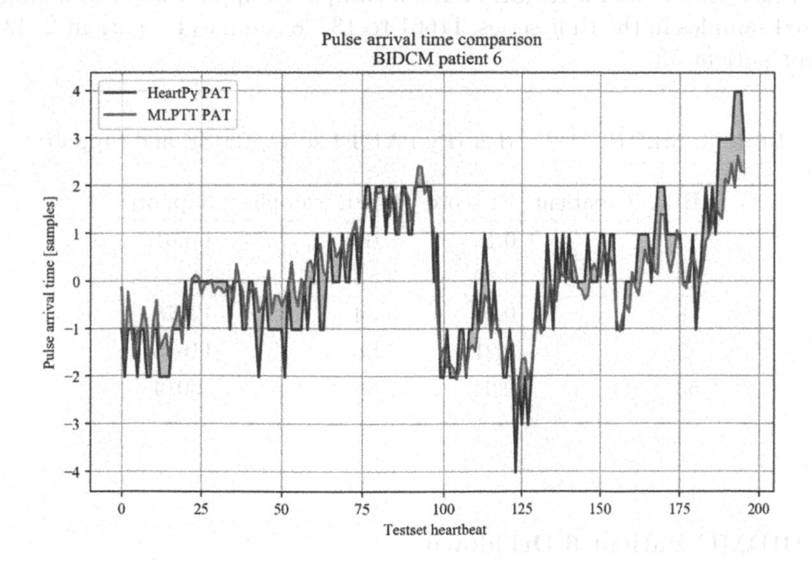

Fig. 5. HeartPy and MLPTT AC components, difference filled in red. (Color figure online)

Out of 5 tested patients (BIDMC patients 2, 6, 8, 42, 53), HeartPy failed to detect exactly one PPG peak per ECG period for patients 2 and 53, in some instances detecting 0 and in some instances 2. These patients were removed and only patients 6, 8 and 42 used for comparison. The Pearson correlation between HeartPy PAT and MLPTT PAT was the following (Table 2).

Table 2. MLPTT PAT - HeartPy PAT Pearson correlation and p-value.

BIDMC patient	Test set heartbeats	Pearson correlation coefficient	Pearson p-value
2	219	−0.0093	0.89
6	196	0.87	4.20E−60
8	240	0.83	1.20E−62
42	201	0.85	2.50E−56
53	222	−0.67	0.32

F1 score and RMSE were calculated between the ECG signals R-peaks and the MLPTT detected virtual anchor from PPG. The predicted R-peak from PPG data was used as a reference to benchmark the stability of the method, but it should not be considered as ground truth, which is indeterminable for PPG with current methods and can only be approximated intravenously [32]. For patients 2–42 the F1 score was in the range of 0.15–0.30 with RMSE extending between 3.83 and 8.01 samples (Table 3). Patient 53 showed the lowest accuracy with an F1 score of 0.04 and a RMSE of 15.66 samples. Support was the number of analysed samples in the time series, 11661 to 13278 samples for patient 2–42 and 9014 for patient 53.

Table 3. MLPTT PAT - HeartPy PAT F1 score, RMSE and support.

BIDMC patient	F1 score	RMSE [samples]	Support
2	0.15	8.01	11661
6	0.30	3.83	11335
8	0.20	5.54	13278
42	0.21	5.88	11645
53	0.04	15.66	9014

4.1 BIDMC Patient 6 Drilldown

These were the MLPTT subprocess results sampled from one out of the 3/5 BIDMC patients of whom HeartPy detected all PPG peaks.

HeartPy and MLPTT PAT Patient 6 Pearson Correlation. A scatterplot with HeartPy calculated PAT on the x-axis and MLPTT PAT on the y-axis was produced (Fig. 6). A histogram on top of each axis showed the respective PAT distribution. The Pearson correlation coefficient p was listed for all tested patients in Table 2.

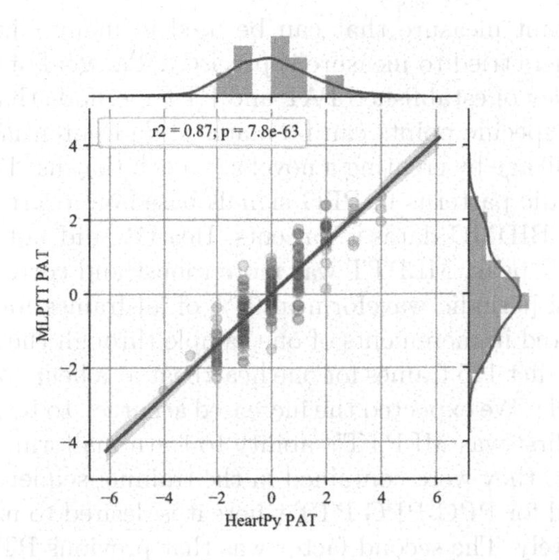

Fig. 6. HeartPy MLPTT Pearson correlation.

Waveform Binary Classifier. The waveform binary classifier achieved at an overall anchor prediction accuracy of 97% for 17977 frames of 60 samples each. Precision, recall, f-1 score and support were listed in Table 4. Precision was the ratio of correctly positively predicted virtual anchors to all positive virtual anchors ($Precision = TP/TP + FP$, where TP was True Positive and FP was False Positive). The calculated precision was 0.92 for frames without anchors and 0.99 for frames with anchors. Recall, the ratio of correctly predicted virtual anchors to all virtual anchors in the class was 0.99 for frames without anchor and 0.95 for frames containing a virtual anchor ($Recall = TP/TP + FN$, where TP was True Positive and FN False Negative). Support was the number of samples in the respective class (Table 4).

Table 4. Waveform binary classifier results for BIDMC patient 6.

	Precision	Recall	F1 score	Support
Anchor = 0	0.92	0.99	0.95	6253
Anchor = 1	0.99	0.95	0.97	11724
Accuracy			0.97	17977

Multilabel Anchor Position Classifier. The multilabel anchor position classifier achieved an overall prediction accuracy of 30% (Table 3), with most predictions scattered closely around a confusion matrix diagonal.

5 Discussion

PTT is an important measure that can be used in many clinical applications and many researchers tried to measure it precisely. The goal of this study was to assess if the accuracy of established PAT and PTT methods that relied on signal filtering and fixed specific points can be matched and potentially exceeded. We addressed this challenge by creating a novel approach that used machine learning to find quasi-periodic patterns in PPG signals based on a series of samples.

For five tested BIDMC dataset patients, HeartPy did not detect one PPG peak for every ECG peak. MLPTT was more robust and correctly classified the presence of a quasi-periodic waveform in 97% of all frames for patient 6. Since the frame was shifted in increments of one sample through the entire signal, the classifier would predict 125 frames for one heartbeat at a heart rate of 60bpm and a sample rate 125 Hz. We expected the increased accuracy to be driven by at least three factors, the first was MLPTT's ability to learn malformed PPG shapes of any form as long as they were contained in the training sequence. This could be particularly helpful for PPG-PPG PTT where it is desired to measure the signal phase shift accurately. The second factor was that previous PTT measurements were dependent on high signal quality at a specific location during each period. For example when measuring the peak-foot PAT, a pronounced foot would have to occur in the PPG signal which could be corrupted by motion artefacts or other noise. If the peak was still in its true position for a period with a corrupted foot, HeartPy would not benefit while MLPTT could learn and still predict the correct virtual anchor point. The third expected reason was that MLPTT was stepping through the entire signal in frames of 1 sample interval. Therefore, the multilabel classifier had the chance to predict every virtual anchor point for the number of intervals in one frame, 60 times in our implementation.

The calculated HeartPy- and MLPTT PAT correlated with statistical significance for in total 637 tested heartbeats of BIDMC patients 6, 8 and 42. Although PAT for patients 2 and 53 could not be compared due to HeartPy's PPG peak detection inconsistencies compared to ECG, the MLPTT PAT prediction did not show any significant variations in the AC component that could have been caused by anchor point misdetection. MLPTT showed a pronounced confusion matrix diagonal for the prediction of virtual anchors. Overall, MLPTT correlated with HeartPy with statistical significance for the tested dataset, required no adjustments for individual patients and showed more robust PAT measurements for patients with noisy measurements.

5.1 Limitations

Despite the strong Pearson correlation with HeartPy PAT measurements, the method and current implementation are not without limitations. The behaviour for medically relevant outliers was not tested and patients' background was not investigated. Some BIDMC signals also showed apparent sawtooth and potentially sine type noise. Lin et al. reported a sawtooth pattern in a different PPG dataset [37], which indicates that this type of noise may be prevalent in PPG recordings of medical equipment. Furthermore, despite our attempts to find the ideal parameters for HeartPy, HeartPy could offer additional built-in signal filters that we did not use. These might have improved HeartPy performance further. No custom filtering was applied for MLPTT for different patients.

Theoretical limitations included that the tested implementation with a frame length of 0...60 samples allowed for a maximum of 125bpm heart rate. Exceeding this heart rate would have led to more than one period per frame, which the implementation was not designed to handle in its first revision. This could be addressed in further revisions. Another theoretical limitation were edge cases in which the virtual anchor point is at either edge of the frame. It can be expected that these frames were more difficult to classify, which resulted in a slight decrease in classification accuracy for the outermost classes. This could be mitigated in further revisions by not considering those frames and only processing frames with a predicted minimum distance from the outer frame limits. Foremost, although both tested methods showed a statistically significant correlation, there were no readily available ground truth measurements for PAT and PTT. Invasive arterial probe measurements were found to produce the smallest errors for blood pressure prediction [32] and can be considered to be more reliable due to measuring the arterial pressure directly. Testing PPG-PTT against I-PTT would be more meaningful, but no dataset was available at the time of writing. We recognize the risks that incorrect PTT measurements may cause if used in clinical applications. All PTT measurements based on the new method should be revalidated against methods such as I-PTT with statistical significance before clinical deployment.

6 Conclusion

The key contribution of this paper is to provide a novel approach where global features (all samples) of the time-series data are used to develop a machine learning model to extract local pulse events. We evaluated the performance of MLPTT for more than 50000 samples of a reference dataset and validated the performance in comparison to a reference method for over 1000 heartbeats of 5 patients. The analyses show that MLPTT copes significantly better with inherently noisy PPG data than the reference method. The proposed technique is suitable for analysis of other medical recordings and for application in many other domains that rely on time series data.

Acknowledgements. This work was supported by the University of Sydney Cardiovascular Initiative funding. Dr Withana is the recipient of an Australian Research Council Discovery Early Career Award (DE200100479) funded by the Australian Government.

References

1. Geddes, L.A., Voelz, M.H., Babbs, C.F., Bourland, J.D., Tacker, W.A.: Pulse transit time as an indicator of arterial blood pressure. Psychophysiology **18**(1), 71–74 (1981). https://doi.org/10.1111/j.1469-8986.1981.tb01545.x. ISSN 0048-5772
2. Poon, C.C.Y., Zhang, Y.T.: Cuff-less and noninvasive measurements of arterial blood pressure by pulse transit time. In: 2005 IEEE Engineering in Medicine and Biology 27th Annual Conference, pp. 5877–5880 (2005). https://doi.org/10.1109/IEMBS.2005.1615827
3. Lane, J.D., Greenstadt, L., Shapiro, D., Rubinstein, E.: Pulse transit time and blood pressure: an intensive analysis. Psychophysiology **20**(1), 45–49 (1983). https://doi.org/10.1111/j.1469-8986.1983.tb00899.x. ISSN 0048-5772
4. Pitson, D.J., Stradling, J.R.: Value of beat-to-beat blood pressure changes, detected by pulse transit time, in the management of the obstructive sleep apnoea/hypopnoea syndrome. Eur. Respir. J. **12**(3), 685–692 (1998)
5. Gesche, H., Grosskurth, D., Küchler, G., Patzak, A.: Continuous blood pressure measurement by using the pulse transit time: comparison to a cuff-based method. Eur. J. Appl. Physiol. **112**, 309–315 (2011). https://doi.org/10.1007/s00421-011-1983-3
6. Ripoll, V.R., Vellido, A.: Blood pressure assessment with differential pulse transit time and deep learning: a proof of concept. Kidney Dis. **5**(1), 23–27 (2019). https://doi.org/10.1159/000493478
7. Sklarsky, A., Garvin, N.M., Pawelczyk, J.A.: Limitations of pulse transit time to estimate blood pressure. FASEB J. **33**(1 supplement), 562.13 (2019). https://doi.org/10.1096/fasebj.2019.33.1_supplement.562.13
8. Mukkamala, R., et al.: Toward ubiquitous blood pressure monitoring via pulse transit time: theory and practice. IEEE Trans. Biomed. Eng. **62**(8), 1879–1901 (2015). https://doi.org/10.1109/tbme.2015.2441951. ISSN 0018-9294
9. Smith, R.P., Argod, J., Pépin, J.-L., Lévy, P.A.: Pulse transit time: an appraisal of potential clinical applications. Thorax **54**(5), 452–457 (1999). https://doi.org/10.1136/thx.54.5.452
10. Sharwood-Smith, G., Bruce, J., Drummond, G.: Assessment of pulse transit time to indicate cardiovascular changes during obstetric spinal anaesthesia. BJA Br. J. Anaesth. **96**(1), 100–105 (2005). https://doi.org/10.1093/bja/aei266. ISSN 0007-0912
11. Obrist, P.A., Light, K.C., McCubbin, J.A., Hutcheson, J.S., Hoffer, J.L.: Pulse transit time: relationship to blood pressure and myocardial performance. Psychophysiology **16**(3), 292–301 (1979). https://doi.org/10.1111/j.1469-8986.1979.tb02993.x. ISSN 0048-5772
12. Pèpin, J.-L., et al.: Pulse transit time improves detection of sleep respiratory events and microarousals in children. CHEST **127**(3), 722–730 (2005). https://doi.org/10.1378/chest.127.3.722. ISSN 0012-3692
13. García, M.T.G., et al.: Can pulse transit time be useful for detecting hypertension in patients in a sleep unit? Archivos de Bronconeumología (English Ed.) **50**(7), 278–284 (2014). https://doi.org/10.1016/j.arbr.2014.05.001. ISSN 15792129
14. Elgendi, M., et al.: The use of photoplethysmography for assessing hypertension. NPJ Digital Med. **2**(1), 60 (2019). https://doi.org/10.1038/s41746-019-0136-7. ISSN 2398-6352
15. Nabeel, P.M., Joseph, J., Sivaprakasam, M.: Arterial compliance probe for local blood pulse wave velocity measurement, vol. 2015 (2015). https://doi.org/10.1109/EMBC.2015.7319689

16. Samaniego, N.C., Morris, F., Brady, W.J.: Electrocardiographic artefact mimicking arrhythmic change on the ECG. Emerg. Med. J. **20**(4), 356–357 (2003). https://doi.org/10.1136/emj.20.4.356
17. Anton, O., Fernandez, R., Rendon-Morales, E., Aviles-EspinosaÂ, R., Jordan, H., Rabe, H.: Heart rate monitoring in newborn babies: a systematic review. Neonatology **116**(3) (2019). https://doi.org/10.1159/000499675
18. Pollreisz, D., TaheriNejad, N.: Detection and removal of motion artifacts in PPG signals. Mob. Netw. Appl (2019). https://doi.org/10.1007/s11036-019-01323-6. ISSN 1572-8153
19. Ponnle, A., Ogundepo, O.: Development of a computer-aided application for analyzing ECG signals and detection of cardiac arrhythmia using back propagation neural network-part I: model development. Int. J. Appl. Inf. Syst. **9** (2015). https://doi.org/10.5120/ijais15-451378
20. van Velzen, M.H.N., Loeve, A.J., Niehof, S.P., Mik, E.G.: Increasing accuracy of pulse transit time measurements by automated elimination of distorted photoplethysmography waves. Med. Biol. Eng. Comput. **55**(11), 1989–2000 (2017). https://doi.org/10.1007/s11517-017-1642-x
21. Kortekaas, M., Niehof, S., Van Velzen, M., Galvin, E., Huygen, F., Stolker, R.: Pulse transit time as a quick predictor of a successful axillary brachial plexus block. Acta Anaesthesiologica Scandinavica **56**, 1228–1233 (2012). https://doi.org/10.1111/j.1399-6576.2012.02746.x
22. Foo, J.Y.A., et al.: Effects of poorly perfused peripheries on derived transit time parameters of the lower and upper limbs. Generic (2008). https://doi.org/10.1515/BMT.2008.023
23. Foo, J.Y.A., Lim, C.S.: Difference in pulse transit time between populations: a comparison between Caucasian and Chinese children in Australia. J. Med. Eng. Technol. **32**(2), 162–166 (2008). https://doi.org/10.1080/03091900600632694. ISSN 0309-1902
24. Foo, J.Y.A., Lim, C.S.: Difference in pulse transit time between populations: a comparison between Caucasian and Chinese children in Australia. Generic (2007). https://doi.org/10.1515/BMT.2007.043
25. Foo, J.Y.A.: Normality of upper and lower peripheral pulse transit time of normotensive and hypertensive children. J. Clin. Monit. Comput. **21**, 243–248 (2007). https://doi.org/10.1007/s10877-007-9080-1
26. Singha Roy, M., Gupta, R., Chandra, J.K., Das Sharma, K., Talukdar, A.: Improving photoplethysmographic measurements under motion artifacts using artificial neural network for personal healthcare. IEEE Trans. Instrum. Meas. **67**(12), 2820–2829 (2018). https://doi.org/10.1109/TIM.2018.2829488. ISSN 1557-9662
27. Mehrgardt, P., Zandavi, S.M., Poon, S.K., Kim, J., Markoulli, M., Khushi, M.: U-Net segmented adjacent angle detection (USAAD) for automatic analysis of corneal nerve structures. Data **5**(2) (2020). https://doi.org/10.3390/data5020037
28. van Gent, P., Farah, H., Nes, N., Arem, B.: Analysing noisy driver physiology real-time using off-the-shelf sensors: heart rate analysis software from the taking the fast lane project (2018). https://doi.org/10.13140/RG.2.2.24895.56485
29. van Gent, P., Farah, H., Nes, N., Arem, B.: Heart rate analysis for human factors: development and validation of an open source toolkit for noisy naturalistic heart rate data (2018)
30. Kohler, B., Hennig, C., Orglmeister, R.: The principles of software QRS detection. IEEE Eng. Med. Biol. Mag. **21**(1), 42–57 (2002). https://doi.org/10.1109/51.993193. ISSN 1937-4186

31. Pan, J., Tompkins, W.J.: A real-time QRS detection algorithm. IEEE Trans. Biomed. Eng. **BME–32**(3), 230–236 (1985). https://doi.org/10.1109/TBME.1985.325532
32. Gao, M., Bari Olivier, N., Mukkamala, R.: Comparison of noninvasive pulse transit time estimates as markers of blood pressure using invasive pulse transit time measurements as a reference. Physiol. Rep. **4**(10), e12768 (2016). https://doi.org/10.14814/phy2.12768. ISSN 2051-817X
33. Pimentel, M.A.F., et al.: Toward a robust estimation of respiratory rate from pulse oximeters. IEEE Trans. Biomed. Eng. **64**(8), 1914–1923 (2017). https://doi.org/10.1109/TBME.2016.2613124. ISSN 1558-2531
34. Rajala, S., Ahmaniemi, T., Lindholm, H., Taipalus, T.: Pulse arrival time (PAT) measurement based on arm ECG and finger PPG signals - comparison of PPG feature detection methods for PAT calculation, vol. 2017 (2017). https://doi.org/10.1109/EMBC.2017.8036809
35. Becker, D.: Fundamentals of electrocardiography interpretation. Anesth. Prog. **53**, 53–63 (2006). https://doi.org/10.2344/0003-3006(2006)53[53:FOEI]2.0.CO;2. quiz 64
36. Orphanidou, C., Bonnici, T., Charlton, P., Clifton, D., Vallance, D., Tarassenko, L.: Signal-quality indices for the electrocardiogram and photoplethysmogram: derivation and applications to wireless monitoring. IEEE J. Biomed. Health Inform. **19**(3), 832–838 (2015). https://doi.org/10.1109/JBHI.2014.2338351. ISSN 2168-2208
37. Lin, Y.-T., Lo, Y.-L., Lin, C.-Y., Frasch, M.G., Wu, H.-T.: Unexpected sawtooth artifact in beat-to-beat pulse transit time measured from patient monitor data. PLOS ONE **14**(9), e0221319 (2019). https://doi.org/10.1371/journal.pone.0221319

Weight Aware Feature Enriched Biomedical Lexical Answer Type Prediction

Keqin Peng[1,2], Wenge Rong[1,3(✉)], Chen Li[1,2], Jiahao Hu[1,2], and Zhang Xiong[1,3]

[1] State Key Laboratory of Software Development Environment, Beihang University, Beijing 100191, China
{keqin.peng,chen.li,hujiahao,xiongz}@buaa.edu.cn
[2] Sino-French Engineer School, Beihang University, Beijing 100191, China
[3] School of Computer Science and Engineering, Beihang University, Beijing, China
w.rong@buaa.edu.cn

Abstract. Lexical Answer Type (LAT) prediction is an essential part of question classification. It aims to assign certain lexical answer type to the questions to narrow down the search space and improve the classifier's performance. LAT prediction is a challenge in the biomedical domain since it is more of a multi-label classification question, which means each question has more than one label. In this paper, we employ the Label Powerset method to transform multi-label classification problems into multi-classification problems. Afterwards we introduced a random forest based mechanism to partition the features into used (important) and unused (unimportant) sets with corresponding weights. Furthermore, by assuming that the unimportant features are not useless, we employ principal components analysis to get the information from the unused feature set. By combing these two types of features, the experimental study on the BioMedLAT dataset has demonstrated our method's potential.

Keywords: Biomedical question classification · Lexical answer type prediction · Random forest · PCA · Feature weight

1 Introduction

With the development of biomedical techniques, biomedical scientific literature has been exploded rapidly. The researchers have met a challenge in seeking information given a certain problem to avoid repeated experiments [9]. To overcome this problem, a lot of approaches have been proposed, and question answering (QA) system has gained widespread attention in the community [12].

A QA system normally consists of three components, i.e., question processing, candidate retrieval, and answer processing. Question classification is a fundamental step of question processing and one of its challenges is to determine the Lexical Answer Type (LAT) of the question [11]. LAT prediction is to mark

H. Yang et al. (Eds.): ICONIP 2020, LNCS 12534, pp. 63–75, 2020.
https://doi.org/10.1007/978-3-030-63836-8_6

expected biomedical entities to a question, thereby narrowing the search space in the answer processing step. For instance, for the question "*How many genes are imprinted in the human genome?*", LAT label *Quantity* will be assigned to this question, it can detect the relevance of questions and answers and then improve the overall performance.

Nowadays, many researchers have done a lot of work on LAT prediction and have achieved great success in the general QA systems, while in the biomedical field it is still a challenging task. This is mainly because unlike in the open domain, biomedical questions generally correspond to more than one biomedical term, hence the LAT prediction problem is more of a multi-label classification problem than a multi-class classification problem [19]. Meanwhile, in the biomedical field, the scale of data is far less than that of the open domain applications, as such it is more difficult to conduct large-scale training as in the other domains.

One of the most widely employed methods for multi-label classification is data transformation [25], which transforms multi-label data to single-label data [20], thereby converting complex multi-label classification into a multi-class problem or multiple binary problems. Among the popular data transformation methods, Label Powerset [4] has been attracted much attention as it does not ignore the relevance between labels since it treats multiple labels existed in one instance as one label by considering the relevance between labels.

Label Powerset based methods have proven the effectiveness in biomedical LAT prediction [21]. In such approaches, there is a critical challenge in selecting proper features. Conventional methods relied on choosing the features that can improve the performance one by one and the features related to the focus word. This kind of mechanism is easy to be implemented, while the influence of different features on the final prediction performance is not fully investigated to some extent. Therefore it is interesting to ask if we can effectively select the most important features with corresponding weights in the LAT prediction. To this end, in this research, we proposed a random forests based feature partition mechanism to solve this problem since it is tolerant of noise and robust to overfitting [8]. All features are firstly partitioned into used (important) and unused (unimportant) categories. Furthermore, it is also interesting to ask if the unselected features are really useless. There might be still implicit information among them. As such we further employ Principal Component Analysis (PCA) [1] to extract information from unselected features as a set of new orthogonal variables.

The contribution of this research is two folders: 1) we use random forests to divide the features into selected and unselected parts with corresponding weights. 2) we developed a PCA based unselected feature information extraction mechanism to enrich inputs of LAT prediction process. The experimental study has shown the potential of this framework.

2 Related Work

Inspired by the challenge BioASQ, the QA systems have been extensively studied in the biomedical domain [14]. As an essential part of the QA system, question classification has been attached much attention in the community and it consists of two challenges, i.e., question type classification and lexical answer type prediction (LAT) [7]. Question type classification aims to categorize the submitted questions into several pre-defined types [18]. For example, Sarrouti et al. categorize questions into three types questions (Yes/No, Factoid and Summary) by using a syntactic and rule based approach [15], and the system achieves high accuracy in 1433 Biomedical Questions from BioASQ challenges [18]. Besides question type classification, recently the lexical answer type prediction has also received much attention. In the general QA applications, Ferrucci et al. [5] analyzed a random sample of 20,000 questions extracting the LATs, and found that the most frequent 200 explicit LATs cover less than 50% of the data while it can improve 20% accuracy. Similarly Alfio et al. [6] used the unsupervised method based on PRISMATIC to produce results for 504 cases out of 813 test questions, and it achieves high accuracy for Coarse-Grained Evaluation.

Inspired by the LAT prediction in general QA systems, the biomedical field has also begun to use LAT prediction to improve the performance of the bioQA system. Weissenborn et al. [22] argued that the phrase containing the LAT was found directly after the question word (what/which), or after the question word followed by a form of "be". They proposed to use the dependency parse to process the question and used the pattern to find the LAT. However, they found that most of LAT will not appear in the questions, thereby not working well in most questions. Later on Yang et al. [23] introduced two additional answer types, i.e., CHOICE and QUANTITY besides the UMLS semantic types to improve the performance. Zhang et al. [13] tried to automatically classify the LATs into six types, i.e., disease, drug, gene/protein, mutation, number and choice. HPI system [16] used machine learning to predict LAT for a limited number of biomedical semantic types, while it only has a LAT term as the expected answer.

Neves and Kraus [11] annotated the headword and assigned UMLS semantic types to 643 factoid/list questions from BioASQ dataset. They extracted the excepted LAT by defining headword and assigned one or more semantic types to the identified headword. Then in the paper of Wasim et al. [21], they argued that most biomedical questions have more than one label, while most data have only one label in the BioMedLAT corpus annotated by Neves et al. [11]. Hence, based on the BioMedLAT corpus, Wasim et al. developed a multi-label biomedical LAT corpus. They reduced the number of features through feature engineering and used the method Label Powerset with logistic regression to train the LAT prediction system, which achieved better performance than OAQA.

3 Methodology

Figure 1 is the proposed LAT prediction process. Firstly we will process the questions to abandon some irrelevant information. Afterwards we will extract the features from the processed questions by using the Random Forests to select the most important features and get the feature weights and assign the information of feature weights to the used (important) and unused (unimportant) features.

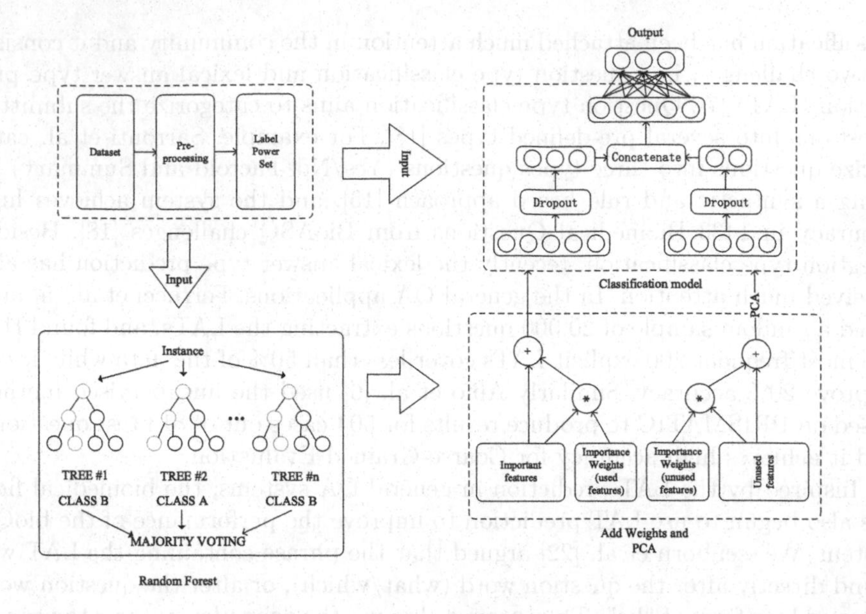

Fig. 1. Pipeline: based on random forest and PCA neural network framework

Next, we will apply PCA on the unused features to further filter some implicit information. Finally, the multi-label classification task is conducted[1].

Here we need to obtain the LAT terms for each input question. The dataset is defined as $D = \{Q, L\}$, Q is a question dataset defined as $Q = \{q_1, q_2, ...q_i, ..., q_m\}$, where the q_i denotes the i-th question and L is the LAT labels set defined as $L = \{ls_1, ls_2, ..., ls_i, ..., ls_m\}$ where ls_i represent the LAT labels set of the i-th question, the form of ls_i is $\{l_1, l_2, ...\}$ where l_i denotes the i-th LAT label. We need to use the question set Q to obtain the corresponding label set L.

3.1 Pre-processing

Following Wasim et al.'s work [21], we input our dataset $Q = \{q_1, q_2, ...q_i, ..., q_m\}$ to ClearNLP[2] to perform lemmatization. In addition, the questions are processed by POS tagger, parser and ClearNLP Bioinformatics model, and then we get the new dataset $Q_1 = \{q_{n1}, q_{n2}, ...q_{ni}, ..., q_{nm}\}$. Afterwards we extract features from the new dataset Q_1 using OAQA system [23]. Now we get the feature set $F = \{f_1, f_2, ...f_i, ..., f_p\}$ where f_i denotes the i-th feature. Meanwhile for the i-th question, we get the values of all the features as the form $v_i = \{v_{i1}, v_{i2}, ...v_{ii}, ..., v_{ip}\}$. After extracting features from questions, we use data transformation to process the data. We use the *Label Powerset* technique to

[1] Our source code is available at https://github.com/Romainpkq/LATPrediction.
[2] https://github.com/clir/clearnlp.

Instance	Attribute	Label Set
1	a1	L1, L2
2	a2	L3, L4
3	a3	L1, L3
4	a4	L1, L2
5	a5	L3, L4

Label Power Set

Instance	Attribute	Label Set
1	a1	L12
2	a2	L34
3	a3	L13
4	a4	L12
5	a5	L34

Fig. 2. The result of label powerset

transform the multi-label classification problem to the multi-class classification problem in order to make use of multiple techniques of multi-class classification. Here is an example to explain the processing, as shown in Fig. 2.

Label Powerset regards the labels per-instance as one label and it considers the relationship between labels. We use that method to process the L and get the new label set L_1 defined as $L_1 = \{l_{n1}, l_{n2}, ...l_{ni}, ...\}$ where l_{ni} represent the i-th new label. We will train and test the classifier using these new labels.

3.2 Feature Selection

After extracting features from the questions and considering not all features are equally important for the LAT prediction, here we propose three techniques to process the features to get the most useful information.

1) **Select important features based on random forest**

In this research, we use the random forest to distinguish between important and unimportant features. Random forests are a combination of tree predictors such that each tree depends on the values of a random vector sampled independently and with the same distribution for all trees in the forest [2]. It can capture the relation between features and labels, and can reduce the correlation between tree models by generating different training sets. Meanwhile, it uses the bagging method to train multiple trees, each tree will produce a classification result and finally, it produces the final result through the majority vote. The random forest can measure the weights of features by Gini index, which is defined as:

$$Gini(p) = \sum_{i=1}^{n} p_i * (1 - p_i) \tag{1}$$

where i represents the i-th class, and p_i denotes represents the sample weight of class i. We use the random forests to get the importance of all the features, and judge the most important ones from them.

2) **Add features weights to feature vector**

After using the random forests, we get all features' importance weights, which have important information that can reflect the impact of different features on the classifier. We use them to highlight the differences between features. On the other hand, we cannot let the feature weights to damage the information of the features. We add the information of weights through the formula:

$$f_{new}(x) = f_{old} * (1 + n * w) \tag{2}$$

where f_{new} represents the new feature vector after adding the weights, f_{old} represents the old feature vector. w denotes the feature weights and n is a multiple that we choose.

Now we get the important feature set $F_{imp} = \{f_{imp_1}, f_{imp_2}, ..., f_{imp_q}\}$, the unused feature set $F_{un} = \{f_{un_{q+1}}, f_{un_{q+2}}, ..., f_{un_p}\}$. Therefore for the i-th question, we get the important features values $v_{i_{imp}} = \{v_{i_{imp_1}}, v_{i_{imp_2}}, ..., v_{i_{imp_q}}\}$, the unused features values $v_{i_{un}} = \{v_{i_{un_{q+1}}}, v_{i_{un_{q+2}}}, ..., v_{i_{un_p}}\}$.

3) **Obtain information from unused features based on PCA**

Though we have chosen the most important features, we still have a lot of feature information not used. In order to use the information in the unused features, we firstly select some unused (unimportant) features according to the weights obtained from random forest, and then use PCA to process them and get the covariance matrix. Afterwards the PCA uses Singular Value Decomposition (SVD) to calculate the eigenvectors and eigenvalues of the covariance matrix, which is defined as follows:

$$A_{m*n} = U_{m*m} * D_{m*n} * V_{n*n}^T \tag{3}$$

where A_{m*n} denotes the data matrix, U_{m*m} and V_{n*n} are orthogonal matrices, D_{m*n} represents a diagonal matrix. After SVD, we will get the matrix D_{m*n}, and then we get the most important eigenvectors and project the data into the space of the eigenvectors. After PCA, for the i-th question, we transform the selected unused features values $v_{i_{un}} = \{v_{i_{un_{q+1}}}, v_{i_{un_{q+2}}}, ..., v_{i_{un_o}}\}$ to the new features values $v_{i_{nun}} = \{v_{i_{nun_1}}, v_{i_{nun_2}}, ..., v_{i_{nun_r}}\}$.

3.3 LAT Prediction

Here we employed a neural network structure in supervised learning to predict the LAT terms. The model has two inputs. For the i-th question, one of them is the feature vector $v_{i_{imp}}$ obtained through the random forest, while the other one is obtained from the unused features $v_{i_{nun}}$ through PCA. Meanwhile, we use the dropout technique to the input layers to avoid overfitting. We get the hidden representation $h_1(x)$, $h_2(x)$ through the forward propagation:

$$h_1(x) = f(W_1 * x_1 + b_1) \tag{4}$$

$$h_2(x) = f(W_2 * x_2 + b_2) \tag{5}$$

where x_1 denotes the feature vector obtained through the random forest, x_2 denotes the other input, $f(x) = max(0, x)$ denotes the Rectified Linear Unit

(ReLU) [10], W_1 and W_2 are weight matrices, b_1 and b_2 are the bias. The ReLU activation function can better avoid the vanishing gradient problem than sigmoid and tanh function [17]. Because the importance of the two inputs is different, we set the dimension of the first hidden layer higher.

We then concatenate the two vectors $h_1(x)$, $h_2(x)$ and get a combined vector $h_3(x)$ whose dimension is the sum of the two vectors:

$$h_3(x) = concatenate(h_1(x), h_2(x)) \tag{6}$$

Finally, the hidden vector $h_3(x)$ are mapped to the output vector $o(t)$ using transformation:

$$o(x) = g(W'h(x) + b') \tag{7}$$

where the activation function $g(x_i) = \frac{e^{x_i}}{\sum_{j=1}^{n} e^{x_j}}$ is the Softmax function, W' is weight matrix, and b' is the bias. The output of the model is in the form of label powerset, and then we restore the output to the form of L. We use the categorical cross entropy loss function as the cost function. This function can well present the similarity between the predicted value and the real value:

$$\text{Cost} = -\frac{1}{n}\sum_{i=1}^{n}[y_i \log \hat{y}_i + (1 - y_i)\log(1 - \hat{y}_i)] \tag{8}$$

where $\hat{y}_i$ is the predicted result and y_i is the label of the instance i, n is the number of the instances. Here we use Adam to optimise the back propagation.

4 Experimental Study

4.1 Dataset and Experiment Configuration

Here we use MLBioLAT corpus[3] as the dataset [21], which has 780 instances. In this research, we use 10-fold cross validation and split the dataset as the CrossValidation class does in Meka tool to test the model, and the number of the labels is 85. The number of questions for each label is shown in Fig. 3 [21].

During the random forest, it is configured with 101 estimators, and its random state is 1. We choose the most important 1500 features as the first input and also use the 900 most important features among the unused features to the PCA model and set the number of PCA components to 50. Because the weights of the used features and unused features are different, we add weights of different multiples to these two feature vectors. For the used feature vector, we set the multiple equal 2. For the unused feature vector the multiple is 1. In the LAT predication process, we use a neural network model with a 1500-dimension input layer and a 50-dimension input layer, and there is a 512-dimension hidden layer for the first input and a 32-dimension hidden layer for the second one. We adopt 10 epochs and 20 instances per batch to train the model.

[3] https://github.com/wasimbhalli/Multi-label-Biomedical-QC-Corpus.

Table illustrating number of questions in each question class after applying 65% threshold.

Class	Questions	Class	Questions	Class	Questions	Class	Questions
umls:aapp	213	**choice**	14	umls:tmco	4	umls:humn	1
umls:gngm	136	umls:mbrt	14	umls:orgm	4	umls:dora	1
umls:dsyn	116	umls:celc	13	umls:inpr	3	tmtool:Species	1
umls:enzy	69	umls:biof	13	umls:comd	3	umls:grup	1
umls:bacs	61	umls:fndg	12	umls:emod	3	umls:phsf	1
umls:phsu	59	umls:bpoc	12	umls:amas	3	umls:antb	1
umls:sosy	42	umls:mobd	10	umls:fngs	3	umls:clna	1
umls:orch	40	umls:resa	10	umls:elii	3	umls:inpo	1
umls:qnco	37	umls:bact	8	umls:patf	3	umls:hlca	1
quantity	34	umls:lbpr	8	umls:tisu	3	umls:food	1
umls:clnd	34	umls:moft	7	umls:lbtr	2	umls:ftcn	1
umls:genf	29	umls:horm	7	umls:orga	2	umls:resd	1
umls:nusq	24	umls:phpr	7	umls:orgf	2	umls:popg	1
umls:rcpt	20	umls:qlco	7	umls:inch	2	umls:carb	1
umls:neop	20	umls:clas	7	umls:acty	2	umls:grpa	1
tmtool:Gene	18	umls:cnce	7	umls:blor	2	umls:medd	1
tmtool:Disease	18	umls:nnon	7	umls:geoa	2	umls:ortf	1
umls:mnob	17	umls:diap	6	umls:euka	2	umls:bdsu	1
umls:celf	16	umls:chem	6	umls:spco	2	tmtool:ProteinMutation	1
umls:topp	16	umls:cell	5	umls:npop	2		
tmtool:Chemical	15	umls:virs	5	umls:bsoj	1		
umls:imft	14	umls:cgab	5	umls:irda	1		

Fig. 3. Dataset: number of questions in each label class [21]

4.2 Evaluation Metrics and Baseline

In the multi-label classification problem, the prediction of a classifier can be more than one label. Following [21], the micro F1 [24] score is used to evaluate the performance of the classifier. For the micro F1 score, there are four categories for each label. For i-th label, tp_i, fp_i, fn_i and tn_i represent True Positives, False Positives, False Negatives and True Negatives [3]. The micro F1 score is defined:

$$precision_{mi} = \frac{\sum_{i=1}^{N} tp_i}{\sum_{i=1}^{N} tp_i + \sum_{i=1}^{N} fp_i} \tag{9}$$

$$recall_{mi} = \frac{\sum_{i=1}^{N} tp_i}{\sum_{i=1}^{N} tp_i + \sum_{i=1}^{N} fn_i} \tag{10}$$

$$F1_score_{mi} = 2 * \frac{recall_{mi} * precision_{mi}}{recall_{mi} + precision_{mi}} \tag{11}$$

In this paper, we compare our methods against the previous methods, especially the STOA methods proposed by Wasim et al. [21], and also three more methods, i.e., Label Powerset with logistic regression (LPLR), Structured SVM (SSVM) and Restricted Boltzmann Machine (RBM). We also compared our methods against the method proposed in OAQA system [23], which is used Copy Transformation based logistic regression (CLR).

4.3 Results and Discussion

The overall comparison of the proposed model against the baseline methods is displayed in Table 1. It is found that the proposed model outperforms the baseline methods. Our model can better capture the relationship between features and labels to achieve better performance. To illustrate the effectiveness of the

Table 1. Micro F1 Score: (RF: Random Forest, FW: Feature Weights, PCA: Principal Component Analysis)

Method	Micro F1 score
RBM (Restricted Boltzmann Machine)	0.28
SSVM (structured support vector machine)	0.42
CLR (Copy Logistic Regression)	0.43
LPLR (Label Powerset Logistic Regression)	0.47
LPLR + FDFS (feature-driven semantic features)	0.50
Our Framework (with RF)	0.497
Our Framework (RF + FW)	0.499
Our Framework (RF + PCA)	0.510
Our Framework (RF + PCA + FW)	**0.515**

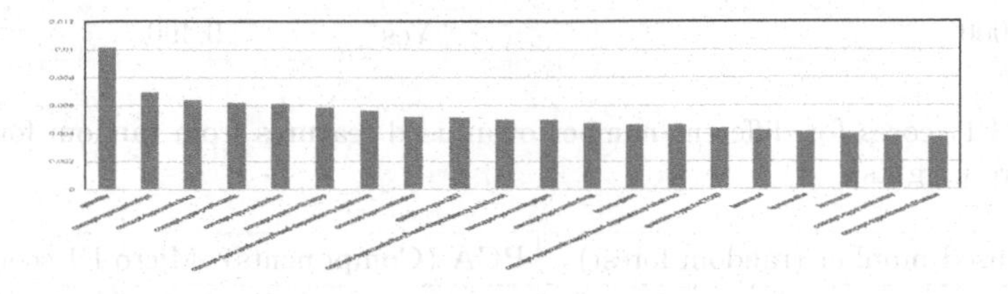

Fig. 4. Weights of the 20 most important features obtained through random forest

weighted feature, we also present the feature weights obtained by Random Forest in Fig. 4, it is observed that different features have different weights.

Besides, to evaluate the effect of weights and also PCA on the final LAT predication task, we further perform some tests, as shown in Tables 2 and 3. In Table 2, we get the best performance when we choose 1500 as the used (important) features through the random forest method, and adding weights features can improve the performance of the model. The experiments in Table 3 follows the test in Table 2 with 1500 selected as used features, while we do not add the weights into feature vectors since we just want to prove the PCA's effectiveness. We further choose several unused features according to their weights and employ PCA to improve the performance of the classifier. We also present the influence of PCA in Fig. 5. It is found that the samples in the same labels have similar feature values, which means that PCA can help predict the LAT terms.

Comparing with other baseline methods, we find the proposed model gained better performance. There are three main reasons. 1) From the Table 2, we can notice that not all the features are useful, and even some features may harmful. We need to filter features to get the most important ones. 2) At the same time, when we filter the features, different features should have different weights to the

Table 2. F1 scores for different number of features selected from random forest

Used features number (random forest)	Add weights	Micro F1 score
4189 (All feature selected)	No	0.488
4189 (All feature selected)	Yes	0.489
3000	No	0.494
3000	Yes	0.495
2500	No	0.489
2500	Yes	0.497
2000	No	0.491
2000	Yes	0.496
1500	No	0.496
1500	Yes	**0.499**
1000	No	0.489
1000	Yes	0.490

Table 3. F1 scores for different number of unused features from random forest (not add feature weights)

Unused number (random forest)	PCA (Components)	Micro F1 score
2689 (All unused feature selected)	No	0.488
2689 (All unused feature selected)	Yes (50)	0.501
1500	No	0.488
1500	Yes (50)	0.506
1000	No	0.489
1000	Yes (50)	0.506
900	No	0.491
900	Yes (50)	**0.510**
500	No	0.487
500	Yes (50)	0.494

classifier as Fig. 4 suggests. Here we add the weights to achieve this distinction; 3) After we select the most important features, there is still a lot of information in the unused features, the PCA method can extract information well and avoid overfitting in the neural network based predication process.

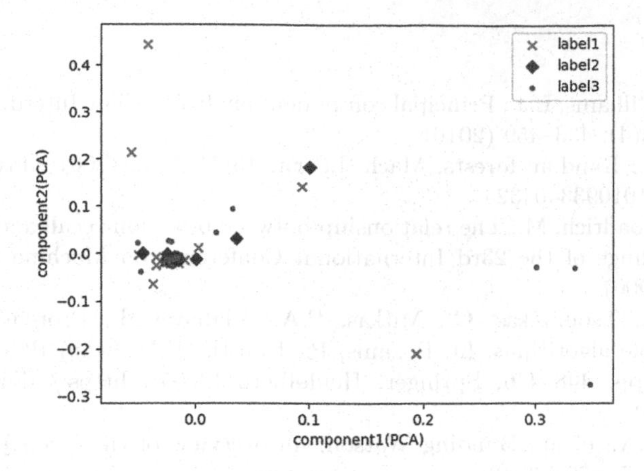

Fig. 5. Visualization of the relation between components of PCA and labels (Component1, Component2: the Most Two Important Components; Label1, Label2, Label3: three Labels After Label Powerset)

5 Conclusion and Future Work

In this paper, we have studied the lexical answer type (LAT) prediction problem in the biomedical domain. As the LAT problem is more of a multi-label classification problem than a multi-class classification problem, we adopt a multi-label classification technique to this problem. The conventional *Copy* data transformation technique used in the OAQA system is an effective method to treat the multi-label data, but it does not grab the relationship between labels. In our research, we use *Label Powerset* to obtain the relationship between labels and we use a neural network to better capture the relationship between features and labels. During this research, we observe that not all the features are useful and the feature weight is helpful. Therefore, we use the random forest to filter features and add the weights information obtained through the random forest. On the other hand, we argue that unused features may also contain some useful information and they can help to determine the Lexical Answer Type. To this end, the PCA method is used to extract implicit information from this feature set. Finally, we develop a neural network model as a classifier to make use of these filtered features. The experimental result has shown the method's potential.

Although the proposed methods have achieved promising performance in the experimental study, there are also some limitations that deserve further investigation. As shown in the experiment, our research does not consider the imbalance of data, while in the dataset some labels have only one or two instances. It will seriously affect the accuracy of the classification. In future work, it is important to solve the problem of data imbalance and to find a more effective method to well represent the questions without information perturbation.

Acknowledgement. This work was partially supported by State Key Laboratory of Software Development Environment of China (No. SKLSDE-2019ZX-16).

References

1. Abdi, H., Williams, L.J.: Principal component analysis. Wiley Interdisc. Rev. Comput. Stat. **2**(4), 433–459 (2010)
2. Breiman, L.: Random forests. Mach. Learn. **45**(1), 5–32 (2001). https://doi.org/10.1023/A:1010933404324
3. Davis, J., Goadrich, M.: The relationship between precision-recall and ROC curves. In: Proceedings of the 23rd International Conference on Machine Learning, pp. 233–240 (2006)
4. Diplaris, S., Tsoumakas, G., Mitkas, P.A., Vlahavas, I.: Protein classification with multiple algorithms. In: Bozanis, P., Houstis, E.N. (eds.) PCI 2005. LNCS, vol. 3746, pp. 448–456. Springer, Heidelberg (2005). https://doi.org/10.1007/11573036_42
5. Ferrucci, D.A., et al.: Building Watson: an overview of the DeepQA project. AI Mag. **31**(3), 59–79 (2010)
6. Gliozzo, A.M., Kalyanpur, A.: Predicting lexical answer types in open domain QA. Int. J. Semant. Web Inf. Syst. **8**(3), 74–88 (2012)
7. Li, Y., Su, L., Chen, J., Yuan, L.: Semi-supervised learning for question classification in CQA. Nat. Comput. **16**(4), 567–577 (2016). https://doi.org/10.1007/s11047-016-9554-5
8. Liaw, A., Wiener, M.: Classification and regression by RandomForest. R News **2**(3), 18–22 (2002)
9. Mollá, D., González, J.L.V.: Question answering in restricted domains: an overview. Comput. Linguist. **33**(1), 41–61 (2007)
10. Nair, V., Hinton, G.E.: Rectified linear units improve restricted Boltzmann machines. In: Proceedings of the 27th International Conference on Machine Learning, pp. 807–814 (2010)
11. Neves, M., Kraus, M.: BioMedLAT corpus: annotation of the lexical answer type for biomedical questions. In: Proceedings of the Open Knowledge Base and Question Answering Workshop, pp. 49–58 (2016)
12. Neves, M., Leser, U.: Question answering for biology. Methods **74**, 36–46 (2015)
13. Peng, S., You, R., Xie, Z., Wang, B., Zhang, Y., Zhu, S.: The Fudan participation in the 2015 BioASQ challenge: large-scale biomedical semantic indexing and question answering. In: Proceedings of Working Notes of CLEF 2015 Conference and Labs of the Evaluation Forum (2015)
14. Sarrouti, M., Alaoui, S.O.E.: SemBioNLQA: a semantic biomedical question answering system for retrieving exact and ideal answers to natural language questions. Artif. Intell. Med. **102**, 101767 (2020)
15. Sarrouti, M., Lachkar, A., Ouatik, S.E.A.: Biomedical question types classification using syntactic and rule based approach. In: Proceedings of 7th International Joint Conference on Knowledge Discovery, Knowledge Engineering and Knowledge Management, vol. 1, pp. 265–272. IEEE (2015)
16. Schulze, F., et al.: HPI question answering system in BioASQ 2016. In: Proceedings of the 4th BioASQ Workshop, pp. 38–44 (2016)
17. Shin, M., Jang, D., Nam, H., Lee, K.H., Lee, D.: Predicting the absorption potential of chemical compounds through a deep learning approach. IEEE/ACM Trans. Comput. Biol. Bioinform. **15**(2), 432–440 (2016)
18. Tsatsaronis, G., et al.: An overview of the BioASQ large-scale biomedical semantic indexing and question answering competition. BMC Bioinform. **16**, 138:1–138:28 (2015)

19. Tsoumakas, G., Katakis, I.: Multi-label classification: an overview. Int. J. Data Warehouse. Min. **3**(3), 1–13 (2007)
20. Tsoumakas, G., Katakis, I., Vlahavas, I.: A review of multi-label classification methods. In: Proceedings of the 2nd ADBIS workshop on Data Mining and Knowledge Discovery, pp. 99–109 (2006)
21. Wasim, M., Asim, M.N., Khan, M.U.G., Mahmood, W.: Multi-label biomedical question classification for lexical answer type prediction. J. Biomed. Inform. **93** (2019)
22. Weissenborn, D., Tsatsaronis, G., Schroeder, M.: Answering factoid questions in the biomedical domain. In: Proceedings of the 1st Workshop on Bio-Medical Semantic Indexing and Question Answering (2013)
23. Yang, Z., Gupta, N., Sun, X., Xu, D., Zhang, C., Nyberg, E.: Learning to answer biomedical factoid & list questions: OAQA at BioASQ 3B. In: Proceedings of Working Notes of CLEF 2015 Conference and Labs of the Evaluation Forum (2015)
24. Yao, Y., Zhou, B.: Micro and macro evaluation of classification rules. In: Proceedings of 7th IEEE International Conference on Cognitive Informatics, pp. 441–448 (2008)
25. Zhang, M.L., Zhou, Z.H.: A review on multi-label learning algorithms. IEEE Trans. Knowl. Data Eng. **26**(8), 1819–1837 (2013)

Neural Data Analysis

Decoding Olfactory Cognition: EEG Functional Modularity Analysis Reveals Differences in Perception of Positively-Valenced Stimuli

Nida Itrat Abbasi[1], Sony Saint-Auret[2], Junji Hamano[3], Anumita Chaudhury[3], Anastasios Bezerianos[1], Nitish V. Thakor[1], and Andrei Dragomir[1](✉)

[1] N.1 Institute for Health, National University of Singapore, Singapore, Singapore
andrei.drag@gmail.com

[2] Faculty of Basic and Biomedical Sciences, Université de Paris, Paris, France

[3] Procter & Gamble, International Operations SA Singapore Branch, Singapore, Singapore

Abstract. Investigating the functional modular organisation of the brain provides a deeper insight into the complex network phenomena that govern cognitive processes like olfactory perception. In recent years, understanding the neural mechanisms associated with this unique sensory modality has been gaining traction, due to increasing applications in various clinical and non-clinical research areas. Anatomically distinct, but functionally interconnected brain regions, organized as communities (or functional modules) enable high-order cognitive processes by providing support for the integration of several localized, highly specialized processing functions. In this work, to understand the elicited neuronal communication pathways in response to fragrance stimuli of varying positive valence, graph theoretical network metrics were calculated to quantify differences in brain's functional networks modular organization estimated from source localised EEG signals. We found that inter-modular connectivity differences in neural responses to olfactory stimuli of different pleasantness levels may be linked to inhibitory processes in the frontal and central-occipital regions. Moreover, our results indicate that significant intra-modular connectivity changes may be linked to emotional processing of fragrance stimuli of varying pleasantness.

Keywords: Olfaction · Graph theory · Brain networks · Functional modularity

1 Introduction

The brain is the most complex organ in the human body, operating through interconnected neuronal ensembles (or functional modules) that collaborate to

N. I. Abbasi and S. Saint-Auret—Equal Contribution.

H. Yang et al. (Eds.): ICONIP 2020, LNCS 12534, pp. 79–89, 2020.
https://doi.org/10.1007/978-3-030-63836-8_7

achieve high-order cognitive function [1]. It is for this reason that the functional network paradigm and network modularity have become popular tools in understanding the functioning of the brain, especially in the area of sensory perception [2,3]. The investigation of sensory encoding has numerous applications ranging from clinical to consumer research. Out of the five sensory modalities, vision, audition and somatosensation have been studied in great detail and previous studies have investigated the brain networks responsible for their processing [4–6]. However, the neuronal pathways related to olfactory processing, and particularly their interplay with other cognitive pathways, still remain to be understood. This is partly owing to the fact that the olfactory pathway is the only sensory pathway which bypasses the thalamus to reach the neorcortex [7] and partly because of its complex intrinsic interconnections with emotion, memory and reward processing pathways [8].

As seen from previous fMRI literature, several regions in the cortex as well as sub-cortical entities participate in olfactory decoding. The olfactory bulb, upon receipt of an odor stimulus, initiates a signal pathway which cascades downstream to cortical entities like the piriform cortex, amygdala, and entorhinal cortex. These regions have been previously associated with emotion and memory processing, alluding to the fact that tertiary encoding of olfactory stimuli predominantly involves high-order cognitive processes [8,9]. Literature evidence points that olfaction can modulate mood and emotion and that the strength of olfaction-dependent brain evoked responses is linked to recollection memory processing [10].

Despite relatively well documented findings in primary olfactory processing, olfactory encoding research has mostly focused on characterizing and discriminating between the response to positively and negatively-valenced (pleasant and unpleasant) odour stimuli [11–13]. However, how does the human brain differentiate between fragrances that induce varying responses on the finer-graded positively-valenced scalc ("loving" vs "liking") remains obscure. For this purpose, the main objective of this study is to investigate the changes in the functional modular organisation of the brain while perceiving fragrances of different levels of pleasantness, which enables us to account for the interplay with other functional processes, such as the emotion-related component of olfaction.

Traditionally, power spectrum analysis has been the most widely used EEG decoding methodology [14,15]. However, the use of functional connectivity network paradigm, which enables investigating the functional interplay between brain regions in order to achieve high-order cognitive function, is becoming an essential approach in neuroscience research [1]. In this context, understanding the modular organisation of the brain functional connectivity networks have revealed that the formation of densely-connected cohesive modules also known as "communities" [1,16]. Thus, the underlying functional segregation can be well-understood due to topological modularity that reflects the dynamic and adaptive brain networks [16]. The unique mediation between segregation (localized function) and integration (distributed function) of the functionally connected brain regions give a deeper insight into the brain function, particularly when investigating cognitive processing [1,16,17]. In this study, our hypothesis was that the functional modular organization resulting in response to positively-valenced fragrance stimuli of various pleasantness levels, reveals differences both in terms of localized (intra-modular) and distributed (inter-modular) functional connectivity.

This work is a continuation of our previous work studying olfactory perception using power spectrum density analysis methods [18]. However, in this study, we aim to identify the neuronal ensembles (functional modules) and how they interact with each other in response to fragrance stimuli. To this goal, we have used weighted-Phase Lag Index (wPLI) to analyse the phase-synchronisation and information flow between brain regions in both alpha and gamma frequency bands, which have been previously implicated in olfactory, emotional and memory-related processing [19,20].

2 Methodology

2.1 Recruitment of Participants

For this study, thirty-two female subjects between the ages of 21 to 45 were recruited. Exclusion criteria encompassed any respiratory dysfunctions, neurological disorders or presence of any metallic implants that could affect the data acquisition procedure. The research has been approved by the Institution Review Board (IRB) of the National University of Singapore. The participants were provided with monetary remuneration at the end of their participation.

2.2 Experimental Protocol

Four pleasant fragrances (pleasantness was monitored through behavioural questionnaires) have been used as olfactory stimuli in our study. The subjects were blindfolded during the experiment to avoid influence of any visual stimuli while the experiment was conducted in an isolated laboratory room to avoid noise and other confounding stimuli to affect the data acquisition. During the experiment, the subjects underwent a sequence of trials (10 for each fragrance), where each trial lasted for about 8 s, following which the subjects were asked to instantly rate on a scale of 0–10 the pleasantness and intensity of the fragrance. Odour masking was avoided by briefly presenting coffee beans to the subjects after the trial and the inter-trial interval was kept as 2 min. During the beginning (1st trial), middle (5th trial) and end (10th trial) of each fragrance, the subjects were asked two additional questions regarding the self-evaluated emotion level and the fragrance characteristics. In this work, we have considered for further analysis the trials corresponding to the most loved (sample with highest pleasantness ratings) and the least-liked fragrances (sample with lowest pleasantness rating), as determined from the self-reported behavioural ratings.

2.3 EEG Data Acquisition and Preprocessing

EEG recording was performed using a standard 64-channel ANT- Neuro cap at a sampling rate of 512 Hz. Ag/AgCl electrodes were used and a conductive electrolyte gel was applied between the scalp and the electrodes to lower the impedance (<=5 k ohms). The raw and continuous data was pre-processed on EEGLAB [21]. At first, the data was downsampled 256 Hz and then band-pass filtered from 0.3 Hz to 40 Hz. Automatic artifact rejection (AAR) was performed to remove artifacts from blinks and other eye movements as well as muscle activity that could affect the EEG recording [22]. The data was epoched in accordance with the trials (8 s from start of stimuli) and Independent Component Analysis (ICA) was performed to remove any remaining elements that might not be arising from the cortex.

Source localisation was performed using standardised-Low Resolution Electromagnetic Tomography (s-LORETA) to estimate the current source density at the cortical level in the brain [23]. The s-LORETA software was used to extract voxel-based activation information from the brain and this data was further parcellated into established anatomical regions (116 ROIs) based on the Automatic Anatomical Labeling (AAL) atlas. Twenty six cerebellar and ten sub-cortical regions were removed from further analysis.

Weighted Phase lag index (wPLI) was employed to investigate the functional connectivity using phase-synchronisation of the current source density values, yielding adjacency matrices [24]. wPLI was followed up with thresholding using the Orthogonal Minimum Spanning Tree (OMST) method to obtain orthogonal trees [25]. This approach ensures filtering the connectivity matrix by prioritizing topological criteria, while accounting for optimization between the global efficiency of the network and the cost of preserving its connections.

2.4 EEG Data Analysis - Community Detection

The Louvain algorithm [26] was performed on the adjacency matrices corresponding to the functional connectivity networks to estimate community (functional modularity) partitions at each epoch. This algorithm is based on the maximization of a modularity index

$$Q = 1/2m \sum_{i,j} [W_{i,j} - \gamma(k_i k_j)/2m]\delta(c_i, c_j) \quad (1)$$

where $W_{i,j}$ is the edge weight between node i and node j, k_i is the sum of edge weights associated with node i and γ is the resolution parameter, that was modulated between 0.5 and 1.6 with a stepsize of 0.1 [26]. For each resolution parameter, the modularity index Q was measured, and the resolution parameter associated to the highest Q was kept. For each subject's epoch, the Louvain algorithm was run a hundred times and consensus partitioning was performed across them to obtain consistent communities for each epoch, as this algorithm is based on a heuristic principle [16,17]. Then, consensus clustering was obtained, this time across all subjects and epochs, to establish the representative community structure. For consensus clustering, the threshold τ was set at a level of 0.2 [16]. For visualisation of the communities on a brain anatomical sketch, Brain Net Viewer software was used [27].

Finally, to characterize changes in the modularity of the functional connectivity networks between different conditions, several community-based metrics were estimated (Table 1): Intra-module metrics (Density D_{intra}, Efficiency E_{intra} and Clustering coefficient C_{intra}) and inter-module metrics (Density D_{inter}, Efficiency E_{inter}). In the mathematical representations given below, m denotes a specific community and a_{ij} represents the edge for nodes i and j. Such metrics can provide effective summarizing statistics on the differences in modular structure of functional networks [1,16,17]. The inter-module efficiency provides an estimate of how efficiently information is being exchanged between different functional modules, while the inter-module density quantifies the number of actual functional connections between pairs of modules over the number of total possible connections. Similarly, the intra-module efficiency and density metrics describe how efficiently information is being exchanged within specific modules, and how many possible connections are being formed within modules, respectively. Intra-modular clustering coefficient describes how tight connections form around nodes within a specific module, providing information on localized, highly specific functional processing [1].

Table 1. Intra- and inter-module metrics used in this study

Inter-module metrics	Mathematical representation
Density (D_{inter})	$\frac{1}{N_m N_n} \sum_{i \in m} \sum_{j \in n} a_{ij}$
Efficiency (E_{inter})	$\frac{1}{N_m N_n} \sum_{i \in m} \sum_{j \in n} \frac{1}{l_{ij}}$
Intra-module metrics	Mathematical representation
Clustering coefficient (C_{intra})	$\frac{1}{N_m} \sum_{i \in m} \frac{sum_{j,k \in m} (w_{ij} w_{jk} w_{kl})^{1/3}}{k_i(k_i - 1)}$
Density (D_{intra})	$\frac{1}{N_m(N_m - 1)} \sum_{i,j \in m, i \neq j} a_{ij}$
Efficiency (E_{intra})	$\frac{1}{N_m(N_m - 1)} \sum_{i,j \in m, j \neq i} \frac{1}{l_{ij}}$

These were computed for two conditions, namely the high pleasantness and the low pleasantness, as determined by the behavioural ratings from the instantaneous behavioural responses collected after every trial. For each metric, statistical analysis was performed using Student's t-test.

3 Results

3.1 Partitioning the Functional Connectivity Networks

Consensus community detection was performed to identify the highly-interconnected modular clusters in the functional connectivity network. This resulted in four communities (functional modules) being formed for both the alpha ([8–13] Hz) and the gamma ([30–40] Hz) bands (see Appendix tables for the list of brain regions in each detected community). Figure 1(a) shows that the topological organisation of these communities in the alpha band. As seen in the figure, Community$_\alpha$ 1 consists of cortical entities in the central-occipital region, Community$_\alpha$ 2 consists of cortical entities in the frontal region, Community$_\alpha$ 3 consists of cortical entities in the left temporal region while Community$_\alpha$ 4 consists of cortical entities in the right temporal region.

However, in the Gamma band, the organisation of communities is slightly different as compared with the alpha band. As seen in Fig. 2(a), these communities consists of neuronal ensembles from the central-occipital regions (Community$_\gamma$ 1), frontal and right temporal regions (Community$_\gamma$ 2), left temporal region (Community$_\gamma$ 3) and right temporal region (Community$_\gamma$ 4).

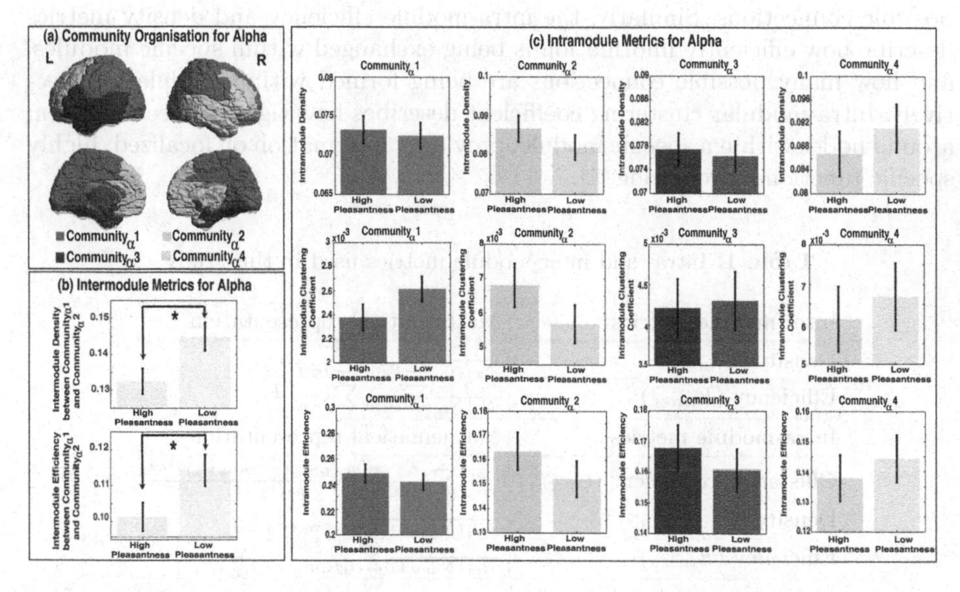

Fig. 1. (a) Four distinct communities were identified comprising of cortical entities from central-occipital region (Community$_\alpha$ 1), frontal region (Community$_\alpha$ 2), left temporal region (Community$_\alpha$ 3) and right temporal region (Community$_\alpha$ 4). (b) Inter-module metrics (between communities) like D_{inter} and E_{inter}. (c) Intra-modular (within communities) metrics like C_{intra}, D_{intra} and E_{intra}. $^*p < 0.05$.

3.2 Inter-module Metrics

In order to study the impact of olfactory perception on the interactions between communities, inter-module density (D_{inter}) and efficiency (E_{inter}) were computed across the high pleasantness and low pleasantness conditions for both alpha and gamma bands across all community pairs. Figure 1(b) shows that in the alpha band, both D_{inter} and E_{inter} are statistically significantly higher for the low pleasantness fragrance than for the high pleasantness fragrance between the Community 1 and 2 (D_{inter}:t(31)= −2, 435, p = 0.015; E_{inter}:t(31)= −2, 252, p = 0.025). Thus, brain entities in the frontal and central-occipital regions have statistically significant increase in the number of information channels while perceiving the low pleasantness odour stimuli as compared with the high-pleasantness odour stimuli. Moreover, there is also a statistically significant increase in the efficiency of the information exchange for the low pleasantness fragrance in comparison with the high pleasantness fragrance between community 1 and community 2 in the alpha band. These suggest that low pleasantness odours elicit higher efficiency in the flow of information and wider connectivity between the frontal, central and occipital regions in the alpha band. The D_{inter} and E_{inter} for other community pairs in the alpha band were not found to be statistically significant. Moreover, we also did not find any statistical significance in the intermodule metrics between all community pairs in the gamma band.

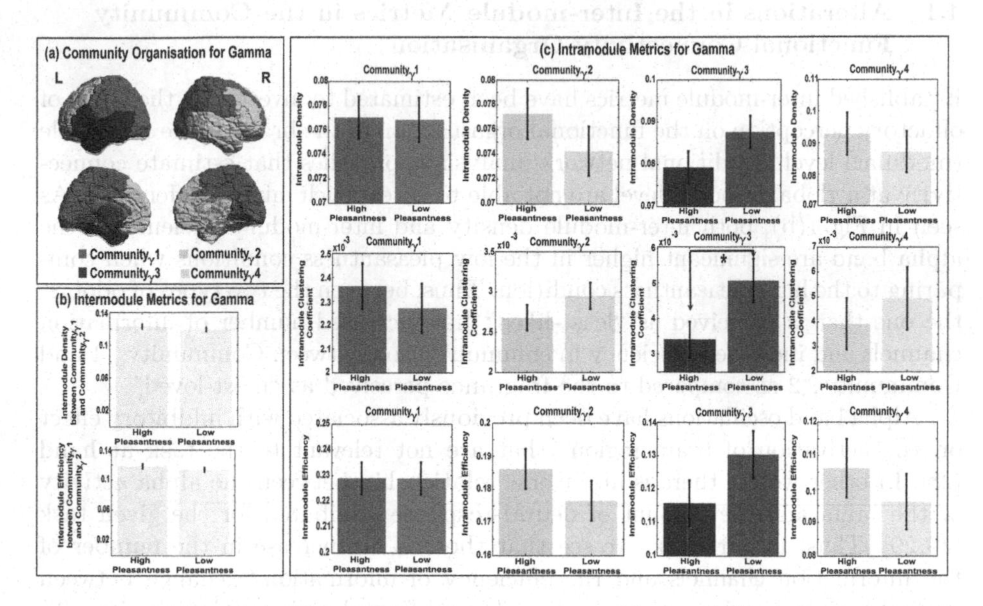

Fig. 2. (a) Four distinct communities were identified comprising of cortical entities from central-occipital region (Community$_\gamma$ 1), frontal-right temporal (Community$_\gamma$ 2), left temporal region (Community$_\gamma$ 3) and right temporal region (Community$_\gamma$ 4). (b) Inter-module metrics (between communities) like D_{inter} and E_{inter}. (c) Intra-modular (within communities) metrics like C_{intra}, D_{intra} and E_{intra}. $^*p < 0.05$.

3.3 Intra-module Metrics

For both pleasantness conditions, three intra-module metrics (C_{intra}, D_{intra} and E_{intra}) were computed to understand the information transfer within the modular community organisation. As shown in Fig. 1(c), the intra-module metrics in the alpha band do not show any statistically significant differences in perception of the two types of odour stimuli. However, Fig. 2(c), in the gamma band, C_{intra} in Community$_\gamma$ 3 (left temporal region) was found to have undergone statistically significant increase (t(31) = −2, 132, p = 0.034) between the high pleasantness and the low-pleasantness conditions. Thus, the two different pleasantness conditions seem to be characterized by significant differences only the intra-module metrics in the gamma band and not the alpha band. Moreover, the effect is predominant only in the localized areas through formation of tight functional clusters (as indicated by the clustering coefficient metric) rather than in the number of information channels or in the efficiency of the information exchange.

4 Discussion

4.1 Alterations in the Inter-module Metrics in the Community Functional Connectivity Organisation

Established inter-module metrics have been estimated to investigate the effect of olfactory perception on the functional organisation of the brain at the mesoscale (modular) level. Traditional network analysis approaches that estimate connectivity at a global or nodal level are not able to detect such minute differences. As seen in Fig. 1(b), both inter-module density and inter-module efficiency in the alpha band are significant higher in the low pleasantness conditions when comparing to the high pleasantness condition. Thus, between the two types of odours, the one that is perceived as "least-liked" has increased number of information channels and increased efficiency in communication between Community$_\alpha$ 1 and Community$_\alpha$ 2 as compared to the fragrance perceived as "most-loved".

Alpha band oscillations have been previously associated with inhibitory effect on the activation of brain regions that are not relevant to the task at hand [28]. In other words, there is an inverse relationship between the alpha activity of the brain and the amount of neural response attributed for the given task [28,29]. Thus, in our work, we see that there is an increase in the number of the information channels and the efficiency of information exchange between central-occipital regions (Community$_\alpha$ 1) and frontal regions (Community$_\alpha$ 2). The frontal region of the brain contains centers that function as modulators of emotion related process [8] while the central-occipital region contains entities that specialise in working memory, conscious awareness and selective attention [30]. These are reinforced by our findings that Community$_\alpha$ 1 includes regions such as amygdala and anterior cingulate cortex, with well established emotion processing roles [31], as well as the olfactory cortex (please see Appendix for description of brain regions within each community for the two frequency bands). Due to the inverse relationship between alpha oscillation and cortical activity, we can speculate based on our results that the low-pleasantness condition induces increased inter-modular connectivity, leading to inhibitory activity in emotion related cortical machinery.

Furthermore, Gamma band cortical activity has been associated with positive emotion regulation [19]. However, we can see from Fig. 2(b) that the inter-module metrics do not show any statistically significant changes between the two pleasantness conditions. Thus, collectively, these results suggest that, at the inter-modular level, evoked cortical response to odours of different positive valence levels occurs primarily due to alteration in the inhibitory processes rather than excitatory processes in the brain.

4.2 Variations in the Intra-module Metrics in the Community Functional Connectivity Organisation

Intra-module metrics have been computed to determine the differences in functional modularity between the two conditions in terms of localized (within module) processing. As seen in Fig. 2(c), statistically significant differences have been observed in the intra-module clustering coefficient for Community 3 in the gamma band. Thus, for the low pleasantness condition, there is increased formation of segregated and specialised clusters in the left temporal lobe. The left temporal lobe contains neuronal sites that specialise in semantic detailing of emotion processing [32]. A study on dementia patients revealed that patients with left temporal lobe atrophy had impaired responses towards the semantic detailing of the presented stimulus while provided normal responses in emotional tasks [33]. Thus, we speculate that the increased clustering coefficient in the self-reported low pleasantness condition could be due to the increase in the semantic processing of the emotions for that condition. In other words, while the subject is behaviourally reporting low pleasantness scores for the fragrance, the increased formation of localized functional connections around nodes in the left temporal lobe suggests increase in functionally segregated clusters for semantic processing of the emotions.

The intra-module metrics in the alpha band (Fig. 1(c)) do not show any statistically significant changes, alluding to the fact that, at the intra-module scale, the processing of the olfactory stimuli of positively-valenced stimuli of different pleasantness levels does not differ in terms of inhibitory brain responses. In other words, the intra-modular brain activations differ at the semantic processing of the emotions (left temporal lobe in the gamma band) and not in inhibitory response (alpha band).

5 Conclusions

In this work, we have investigated the changes in functional connectivity in the mesoscale architecture (communities) of the brain in response to fragrances of varying pleasantness. It was observed that at the inter-modular level, the difference in brain responses towards fragrances of varying hedonic value occurs because of changes in the inhibitory neuronal processes rather than excitatory processes. These inhibitory responses occur in regions responsible for emotion processing. However, at the intra-modular level, changes in the brain responses, corresponding to olfactory perception of fragrances with varying pleasantness, occur due to excitatory processes responsible for semantic processing of positive emotions. Further concurrent EEG-fMRI studies need to be conducted to visualize the changes in the functional connectivity of the modules at a finer spatio-temporal scale.

A Appendix: Anatomical Segregation of ROIs in Communities

Segregation of Anatomical Regions Into Communities- Alpha Band	
Community No	Regions of Interest (AAL)
Communityα 1	Left Precentral gyrus; Right Precentral gyrus; Right Superior Frontal gyrus, dorsolateral; Left Supplementary Motor Area; Right Supplementary Motor Area; Left Median cingulate and paracingulate gyri; Right Median cingulate and paracingulate gyri; Left Posterior cingulate and paracingulate gyri; Right Posterior cingulate and paracingulate gyri; Left Calcarine fissure and surrounding cortex; Right Calcarine fissure and surrounding cortex; Left Cuneus; Right Cuneus; Left Lingual gyrus; Right Lingual gyrus; Left Superior Occipital gyrus; Right Superior Occipital gyrus; Left Middle Occipital gyrus; Right Middle Occipital gyrus; Left Inferior Occipital gyrus; Right Inferior Occipital gyrus; Left Postcentral gyrus; Right Postcentral gyrus; Left Superior Parietal gyrus; Right Superior Parietal gyrus; Left Inferior Parietal supramarginal and angular gyri; Right Inferior Parietal supramarginal and angular gyri; Left Supramarginal gyrus; Right Supramarginal gyrus; Left Angular gyrus; Right Angular gyrus; Left Precuneus; Right Precuneus; Left Paracentral lobule; Right Paracentral lobule.
Communityα 2	Left Superior Frontal gyrus, dorsolateral; Left Superior Frontal gyrus, orbital part; Right Superior Frontal gyrus, orbital part; Left Middle Frontal gyrus; Right Middle Frontal gyrus; Left Middle Frontal gyrus,orbital part; Right Middle Frontal gyrus,orbital part; Left Superior Frontal gyrus, medial; Right Superior Frontal gyrus, medial; Left Superior Frontal gyrus, medial orbital; Right Superior Frontal gyrus, medial orbital; Left gyrus rectus;Right gyrus rectus; Left Anterior cingulate and paracingulate gyri; Right Anterior cingulate and paracingulate gyri.
Communityα 3	Left Inferior Frontal gyrus, opercular part; Left Inferior Frontal gyrus, triangular part;Left Inferior Frontal gyrus, orbital part; Left Rolandic Operculum; Left Olfactory Cortex;Right Olfactory Cortex; Left Insula; Left Parahippocampal gyrus; Right Parahippocampal gyrus; Left Amygdala; Right Amygdala; Left Fusiform gyrus; Right Fusiform gyrus; Left Heschl gyrus; Left Superior temporal gyrus; Left Temporal pole: superior temporal gyrus; Left Middle temporal gyrus; Left Temporal pole: middle temporal gyrus; Left Inferior temporal gyrus.
Communityα 4	Right Inferior Frontal gyrus, opercular part; Right Inferior Frontal gyrus, triangular part; Right Inferior Frontal gyrus, orbital part; Right Rolandic Operculum; Right Insula; Right Heschl gyrus; Right Superior temporal gyrus; Right Temporal pole: superior temporal gyrus; Right Middle temporal gyrus; Right Temporal pole: middle temporal gyrus; Right Inferior temporal gyrus.

Segregation of Anatomical Regions Into Communities- Gamma Band	
Community No	Regions of Interest (AAL)
Communityγ 1	Left Precentral gyrus; Right Precentral gyrus; Right Superior Frontal gyrus, dorsolateral; Left Supplementary Motor Area; Right Supplementary Motor Area; Left Median cingulate and paracingulate gyri; Right Median cingulate and paracingulate gyri; Left Posterior cingulate and paracingulate gyri; Right Posterior cingulate and paracingulate gyri; Left Calcarine fissure and surrounding cortex; Right Calcarine fissure and surrounding cortex; Left Cuneus;Right Cuneus; Left Lingual gyrus;Right Lingual gyrus; Left Superior Occipital gyrus; Right Superior Occipital gyrus; Left Middle Occipital gyrus; Right Middle Occipital gyrus; Left Inferior Occipital gyrus; Right Inferior Occipital gyrus; Left Postcentral gyrus;Right Postcentral gyrus; Left Superior Parietal gyrus; Right Superior Parietal gyrus; Left Inferior Parietal, but supramarginal and angular gyri; Right Inferior Parietal, but supramarginal and angular gyri; Left Supramarginal gyrus; Right Supramarginal gyrus; Left Angular gyrus; Right Angular gyrus; Left Precuneus; Right Precuneus; Left Paracentral lobule; Right Paracentral lobule
Communityγ 2	Left Superior Frontal gyrus, dorsolateral; Left Superior Frontal gyrus, orbital part; Right Superior Frontal gyrus, orbital part; Left Middle Frontal gyrus; Right Middle Frontal gyrus; Left Middle Frontal gyrus, orbital part; Right Middle Frontal gyrus,orbital part; Right Inferior Frontal gyrus, opercular part; Right Inferior Frontal gyrus, triangular part; Right Inferior Frontal gyrus, orbital part; Left Olfactory Cortex; Right Olfactory Cortex; Left Superior Frontal gyrus, medial; Right Superior Frontal gyrus, medial; Left Superior Frontal gyrus, medial orbital; Right Superior Frontal gyrus, medial orbital; Left gyrus rectus; Right gyrus rectus; Right Insula; Left Anterior cingulate and paracingulate gyri; Right Anterior cingulate and paracingulate gyri; Left Parahippocampal gyrus; Right Parahippocampal gyrus; Left Amygdala; Right Amygdala; Right Temporal pole: superior temporal gyrus; Right Temporal pole: middle temporal gyrus
Communityγ 3	Left Inferior Frontal gyrus, opercular part; Left Inferior Frontal gyrus, triangular part; Left Inferior Frontal gyrus, orbital part; Left Rolandic Operculum; Left Insula; Left Fusiform gyrus; Left Heschl gyrus; Left Superior temporal gyrus; Left Temporal pole: superior temporal gyrus; Left Middle temporal gyrus; Left Temporal pole: middle temporal gyrus; Left Inferior temporal gyrus.
Communityγ 4	Right Rolandic Operculum; Right Fusiform gyrus; Right Heschl gyrus; Right Superior temporal gyrus; Right Middle temporal gyrus; Right Inferior temporal gyrus.

References

1. Garcia, J.O., Ashourvan, A., Muldoon, S., Vettel, J.M., Bassett, D.S.: Proc. IEEE **106**(5), 846 (2018)
2. Abbasi, N.I., Harvy, J., Bezerianos, A., Thakor, N.V., Dragomir, A.: 2019 9th International IEEE/EMBS Conference on Neural Engineering (NER), pp. 635–638. IEEE (2019)

3. Ding, K., et al.: J. Neural Eng. (2020)
4. Dijkstra, N., Zeidman, P., Ondobaka, S., van Gerven, M.A., Friston, K.: Sci. Rep. **7**(1), 1 (2017)
5. Gurtubay-Antolin, A., León-Cabrera, P., Rodríguez-Fornells, A.: Eneuro **5**, 6 (2018)
6. Hutton, J.S., Dudley, J., Horowitz-Kraus, T., DeWitt, T., Holland, S.K.: Brain Connect. **9**(7), 580 (2019)
7. Shepherd, G.M.: Neuron **46**(2), 166 (2005)
8. Courtiol, E., Wilson, D.A.: Perception **46**(3–4), 320 (2017)
9. Howard, J.D., Plailly, J., Grueschow, M., Haynes, J.D., Gottfried, J.A.: Nat. Neurosci. **12**(7), 932 (2009)
10. Sullivan, R.M., Wilson, D.A., Ravel, N., Mouly, A.M.: Front. Behav. Neurosci. **9**, 36 (2015)
11. Grabenhorst, F., Rolls, E.T., Margot, C., da Silva, M.A., Velazco, M.I.: J. Neurosci. **27**(49), 13532 (2007)
12. Rolls, E.T., Kringelbach, M.L., De Araujo, I.E.: Eur. J. Neurosci. **18**(3), 695 (2003)
13. Yazdani, A., Kroupi, E., Vesin, J.M., Ebrahimi, T.: 2012 Fourth International Workshop on Quality of Multimedia Experience, pp. 272–277. IEEE (2012)
14. Anokhin, A.P., Van Baal, G., Van Beijsterveldt, C., De Geus, E., Grant, J., Boomsma, D.: Behav. Genet. **31**(6), 545 (2001)
15. Kumar, N., Kumar, J.: Procedia Comput. Sci. **84**, 70 (2016)
16. Taya, F., et al.: Hum. Brain Mapp. **39**(9), 3528 (2018)
17. Abbasi, N.I., et al.: 2019 9th International IEEE/EMBS Conference on Neural Engineering (NER), pp. 631–634. IEEE (2019)
18. Abbasi, N.I., Bose, R., Bezerianos, A., Thakor, N.V., Dragomir, A.: 2019 41st Annual International Conference of the IEEE Engineering in Medicine and Biology Society (EMBC), pp. 5160–5163. IEEE (2019)
19. Aydemir, O.: Neural Comput. **29**(6), 1667 (2017)
20. Flores-Gutiérrez, E.O., et al.: Int. J. Psychophysiol. **71**(1), 43 (2009)
21. Delorme, A., Makeig, S.: J. Neurosci. Methods **134**(1), 9 (2004)
22. Gómez-Herrero, G., et al.: Proceedings of the 7th Nordic Signal Processing Symposium – NORSIG 2006, pp. 130–133. IEEE (2006)
23. Pascual-Marqui, R.D., et al.: Methods Find. Exp. Clin. Pharmacol. **24**(Suppl D), 5 (2002)
24. Vinck, M., Oostenveld, R., Van Wingerden, M., Battaglia, F., Pennartz, C.M.: Neuroimage **55**(4), 1548 (2011)
25. Dimitriadis, S.I., Salis, C., Tarnanas, I., Linden, D.E.: Front. Neuroinform. **11**, 28 (2017)
26. Blondel, V.D., Guillaume, J.L., Lambiotte, R., Lefebvre, E.: J. Stat. Mech. Theory Exp. **2008**(10), P10008 (2008)
27. Xia, M., Wang, J., He, Y.: PLoS ONE **8**(7), e68910 (2013)
28. Uusberg, A., Uibo, H., Kreegipuu, K., Allik, J.: Int. J. Psychophysiol. **89**(1), 26 (2013)
29. Dai, Z., et al.: Front. Hum. Neurosci. **11**, 237 (2017)
30. Bor, D., Seth, A.K.: Front. Psychol. **3**, 63 (2012)
31. Soudry, Y., Lemogne, C., Malinvaud, D., Consoli, S.M., Bonfils, P.: Eur. Ann. Otorhinolaryngol. Head Neck Dis. **128**(1), 18 (2011)
32. Kiehl, K.A., Smith, A.M., Mendrek, A., Forster, B.B., Hare, R.D., Liddle, P.F.: Psychiatry Res. Neuroimaging **130**(1), 27 (2004)
33. Perry, R.J., et al.: Neurocase **7**(2), 145 (2001)

Identifying Motor Imagery-Related Electroencephalogram Features During Motor Execution

Yuki Kokai(✉), Isao Nambu(✉), and Yasuhiro Wada(✉)

Graduate School of Engineering, Nagaoka University of Technology, 1603-1 Kamitomioka, Nagaoka, Niigata 940-2188, Japan
yuuki.s.hormspipo@gmail.com, inambu@vos.nagaokaut.ac.jp, ywada@nagaokaut.ac.jp

Abstract. Brain–computer interface technology facilitates communication and control of computers with brain signals. This technique uses motor imagery to enable a robotic arm to function as a third arm for the subject. During the process, the robotic arm must move in synchrony with the two human arms, and consequently motor imagery and motor execution must be performed simultaneously. In this study, we examined whether information related to motor imagery could be detected with an electroencephalogram during simultaneous measurement of motor imagery and motor execution. Our experiment included five participants who performed motor execution, and motor execution with motor imagery. To identify motor imagery-related features, we initially extracted event-related spectrum perturbation (ERSP) data and performed a t-test to examine significant differences using averaged-trial ERSP data. Subsequently, the data were classified with Fisher's linear discriminant as the single-trial classification. Results revealed significant differences between the two movement conditions and the motor imagery-related features for each subject. The single-trial classification analysis demonstrated slightly higher accuracy than the chance level classification, but the difference was not significant. These results suggest that information related to motor imagery could possibly be decoded during motor execution, however performance improvement at the single-trial level will be necessary in future studies.

Keywords: EEG · BCI · Motor imagery

1 Introduction

Many studies on the relationship between human brain function and human mobility have been conducted. Brain–machine interfaces and brain–computer interfaces control external devices (e.g., computers or robots) using brain activity as an input signal. This technology has the ability to use a robot arm as our third arm. However, we need to perform motor imagery of the moving third

H. Yang et al. (Eds.): ICONIP 2020, LNCS 12534, pp. 90–97, 2020.
https://doi.org/10.1007/978-3-030-63836-8_8

arm. Further, when we use it in daily life, we need to move the third arm along with our original two arms. However, electroencephalogram (EEG) of motor execution and motor imagery appear in similar places [1]. Thus, the EEG of motor imagery hides behind the EEG of motor execution if both motor tasks are performed simultaneously. However, this does not mean that the features of motor imagery cannot be found. In some previous studies, features of motor imagery were found with one condition, i.e., no motor execution while measuring motor imagery. For example, Herman P et al. [1] performed feature extraction of motor imagery of a moving hand, and Takahashi M et al. [2] detected event-related desynchronization (ERD) online. Therefore, we formulated a hypothesis that motor imagery-related EEG features can be found while simultaneously performing motor imagery and motor execution. In this study, we performed an experiment to measure two conditions: "motor execution" and "motor execution with motor imagery." To prove this hypothesis, we examined the difference of trial-averaged event-related spectral between two movements, by performing a t-test using trial-averaged event-related spectrum perturbation (ERSP). Subsequently, we performed the classification analysis using Fisher's linear discriminant as the single-trial classification to check if we can distinguish between two conditions in a single trial.

2 Method

2.1 Experiment and Preprocessing Procedure

Participant. Five healthy right-handed men (aged 22–23 years) participated in this experiment, which was conducted after obtaining approval from the Ethics Committee of the Nagaoka University of Technology and in accordance with the Declaration of Helsinki. In addition, the details of this experiment were clearly explained to the participants; all participants provided informed consent. (None of the subjects had performed motor imagery before this experiment.)

Experimental Procedure. Participants sat on a chair facing a monitor, which displayed an arrow in the right or left direction. If the monitor displayed a right arrow, we asked the subjects to move their right hand. In contrast, if the monitor displayed a left arrow, we asked the subjects to move their left hand. The movement conditions were as follows: (1) motor execution, (2) motor imagery, and (3) motor execution plus motor imagery. In this paper, we compared data from two conditions, motor execution and motor execution plus motor imagery. In this experiment, motor execution refers to finger tapping, and motor imagery refers to stretching arms. This experiment was conducted for three consecutive days, and the subjects completed three sessions per movement condition. One session consisted of 20 trials. One trial was composed of a task and rest. The task was performed for 6 s, and the rest was for 2 s (Fig. 1).

Recording. An EEG measurement system (Biosemi ActiveTwo; Biosemi, Amsterdam, The Netherlands) was used to measure the brain and muscle activities. Sixty-four electrodes were arranged based on the international 10/10 method; one electrode was placed under the right eye to perform electrooculography. In addition, electromyography (EMG) of the biceps and extensor digitorum brevis was performed to check for muscle activity during movements. The sampling frequency of the EEG, EMG, and EOG was 1024 Hz (Fig. 1).

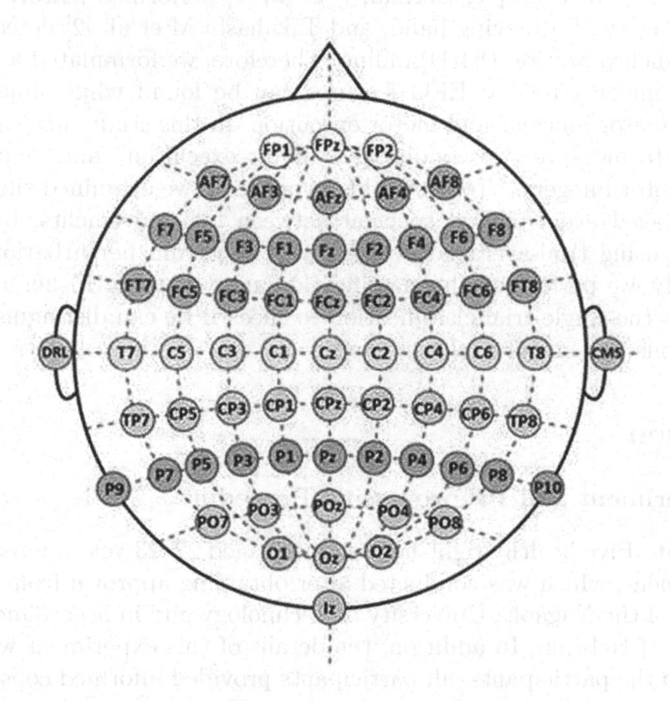

Fig. 1. 64-channel EEG arrangement.

2.2 Data Analysis

Data Preprocessing and Feature Extraction. The preprocessing and feature extraction of EEG and EMG respectively explained.

- EEG
  First, a high-pass filter (cut-off frequency: 2 Hz) was applied. Subsequently, Baseline corrections were performed on the extracted data to ensure that the average of the baseline (300–0 ms before task) became 0 using the average of the baseline. Furthermore, a low-pass filter (cut-off frequency: 40 Hz) was applied. Following this, blinking artifacts were removed from the brain signals using electrooculography measured from under the right eye using the independent component analysis (ICA) of EEGLAB [3]. ERDS was verified in the frequency domain of the brain activity before actual motor function or

during motor imagery [4] as follows:
Event-related desynchronization (ERDS) was verified in the frequency domain of the brain activity before actual motor function or during motor imagery [4] as follows:

$$ERS(f,t) = |F(f,t)|^2 \tag{1}$$

where $F(f,t)$ is the power spectrum calculated by the short-time Fourier transform at frequency f and time t. Thus, it may be possible to visually observe changes in brain activity by calculating event-related spectrum perturbation (ERSP), which was calculated using the time-frequency analysis in EEGLAB after ICA. The frequency power change (from the baseline) was chosen as the feature value for the following classification analysis. The frequency power during the baseline is μ_β:

$$\mu_\beta = \frac{1}{m} \sum_{t \in Base} |F(f,t)|^2 \tag{2}$$

where base is the time interval of the baseline and m is the number of time samples at the baseline. ERSP is defined as:

$$ERSP(f,t) = 10 \log_{10} \frac{ERS(f,t) - \mu_\beta}{\mu_\beta} \tag{3}$$

- EMG
First, baseline corrections were performed on the extracted data to ensure that the average of the baseline (300–0 ms before task time) was 0 using the average of the baseline. Subsequently, a high-pass filter (2 Hz) was applied. Furthermore, full-wave rectification was applied. Since the positive and negative EMG signals are determined by the stretching direction of the muscle, the positive and negative values are meaningless considering the magnitude of ENG. Following this, normalization was applied. To see the movement of muscles, we set the maximum value to 1.
For feature extraction, we calculated the average value of EMG during task time for each trial.

Checking EMG Difference. First, we performed the t-test on the EMG data for two movement tasks. If there was no significant difference in EMG between "motor execution" and "motor execution with motor imagery," movements performed by the subject were the same in the two movement conditions. In other words, movement conditions did not affect the results of the t-test on EEG.

ERSP Difference Between of Motor Execution and Motor Imagery. In order to confirm if we could find new features, we performed the *t*-test as a significant difference verification using averaged-trial ERSP. Since the EEG was performed in the frequency/time direction of a certain channel, this method could judge which place showed a significant difference using the t-value.

Classification of Motor Imagery and Motor Execution a Single Trial. In order to confirm if we could perform single-trial classification, we used Fisher's linear discriminant. Cross-validation was performed under the following conditions to calculate the accuracy. A 10-fold cross-validation was performed by randomly selecting 90% of the entire trial for training data and 10% for test data.

3 Result

3.1 Checking Muscle Activity During Motor Imagery EMG

In order to confirm if there was no significant difference among EMG, motor execution, and motor execution with motor imagery, we performed the t-test on EMG. The results of t-tests, EMG of motor execution, and EMG of motor execution + motor imagery are as follows: t-values of subjects 1 and 2: 0.068 and 0.77, respectively, in the right hand, and 0.77 and 0.053, respectively, in the left hand. There were no significant differences in EMG. In addition, the average t-values for the other subjects were 0.56 and 0.79, respectively, in the right hand and 0.63 and 0.083, respectively, in the left hand. There were no significant differences in EMG. This result indicates that if we find a significant difference in the t-test of EEG, it appears on adding motor imagery.

3.2 Difference Between Motor Imagery and Execution EEG

To find new features, we performed the t-test for significant difference verification using averaged-trial ERSP data, EEG of motor execution, and EEG of motor execution with motor imagery. Figure 2 shows the results and the color map of the t-value. The places where a significant difference occurred depending on the subject were as follows: Subject 1, approximately 5 Hz in the front of the head; Subject 2, approximately 20 Hz in the motor area; Subject 3, approximately 10 Hz in the motor area; Subject 4, approximately 10 Hz at the back of the head; and Subject 5, approximately 0–10 Hz at the back of the head.

3.3 Classification Accuracy a Single Trial

As a single-trial classification, results of the Fisher liner discriminant are presented in Table 1. Within Table 1, the left column indicates test data accuracy, while the right column indicates training data accuracy. The chance level was 0.5. Therefore, the accuracy of training was high. The accuracy of the test data was slightly higher than that of the chance level. We believe that there was over-fitting; therefore, learning with test data was not successful. Each channel number indicates the next position. (1) P + number = back head. (2) C + number = motor area. (3) FC, T + number = upper motor area.

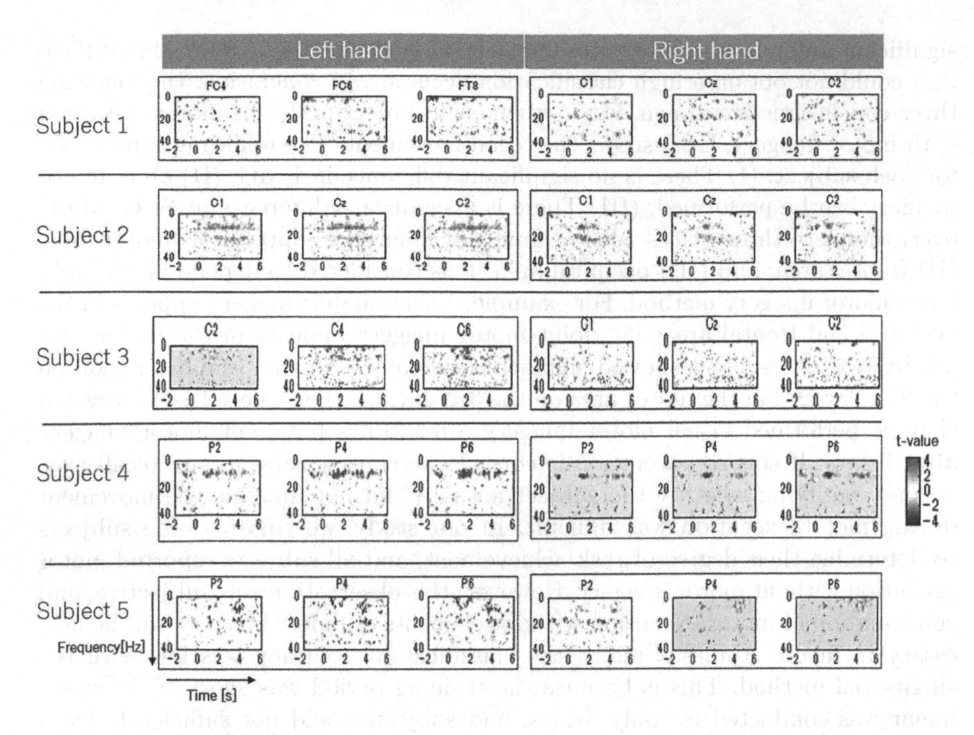

Fig. 2. Result of t-value by t-test to ERSP of two movement conditions, "motor execution" and "motor execution with motor imagery". The color shows a significant difference. The darker the color, the larger the t-value; the lighter the color, the smaller the t-value. (Color figure online)

Table 1. Result of classification accuracies using Fisher's linear discriminant. Chance level is 0.5.

	Test Accuracy	
	Left hand	Right hand
Subject 1	0.56	0.52
Subject 2	0.60	0.59
Subject 3	0.56	0.54
Subject 4	0.58	0.56
Subject 5	0.58	0.57

	Training Accuracy	
	Left hand	Right hand
Subject 1	0.98	0.74
Subject 2	0.98	0.93
Subject 3	0.98	0.98
Subject 4	0.99	0.84
Subject 5	0.97	0.96

4 Conclusion

In this study, we examined whether new features could be found during the simultaneous measurement of motor imagery and motor execution using EEG. We conducted an experiment in which five participants performed two movements: "motor execution" and "motor execution with motor imagery." The results show significant differences and features for each subject, but the single-trial classification could not obtain a high classification accuracy. In conclusion, the following three conclusions are drawn. First, when each subject performs motor execution with motor imagery, EEG shows the recall movement. The following can be said for each subject: (I) There is no significant difference in EMG. (II) Only motor imagery can be performed. (III) There is a significant difference in EEG. Moreover, there are three places where significant differences appear: (I) motor area, (II) frontal area, and (III) occipital area. It is considered to depend on the subject's motor imagery method. For example, Visual motor imagery appears in the occipital and frontal areas [5]. Joint motor imagery appears in the motor area [2]. In fact, one subject showed significant differences in the occipital region on the 2nd day and in the motor area on the 3rd day. Further, one subject answered that he performed visual motor imagery after 2 days but joint motor imagery after 3 days. If this consideration is correct, we can measure in any recall way. Second, verifying whether the subject actually did not imagine the movement during motor execution was difficult. In our study, we surveyed the subjects to determine their degree of task achievement, and all subjects reported motor execution without motor imagery. However, this observation was subjective, and confirmation of motor execution without elements of motor imagery will be necessary in future studies. Finally, the classification accuracy was low with the single-trial method. This is because the training period was short. This experiment was conducted for only 3 days, and subjects could not sufficiently learn movement and motor execution with motor imagery. Some subjects answered that it was more difficult than performing motor imagery alone. However, some subjects said that movement became easier with each passing day. Even though the training period was short, we found significant differences using averaged-trial ERSP. These results show that information related to features might be decoded from the simultaneous measurement of motor execution with motor imagery, although it is necessary to improve the performance at the single-trial level in future works.

Acknowledgments. This work was partly supported by JSPS KAKENHI Grant Numbers 18K19807 and 18H04109, KDDI foundation, and Nagaoka University of Technology Presidential Research Grant. We would like to thank Editage (www.editage.com) for English language editing.

References

1. Herman, P., Prasad, G., McGinnity, T.M., Coyle, D.: Comparative analysis of spectral approaches to feature extraction for EEG-based motor imagery classification. IEEE Trans. Neural Syst. Rehabil. Eng. **16**(4), 317–326 (2008)
2. Takahashi, M., Gouko, M., Ito, K.: Online motor imagery training effect for the appearance of event related desynchronization (ERD). Trans. Inst. Syst. Control Inf. Eng. **22**(5), 199–205 (2009)
3. Delorme, A., Makeig, A.: EEGLAB: an open source toolbox for analysis of single-trial EEG dynamics including independent component analysis. J. Neurosci. Methods **134**(1), 9–21 (2004)

4. Chen, X., Bin, G., Daly, I., Gao, X.: Event-related desynchronization (ERD) in the alpha band during a hand mental rotation task. Neurosci. Lett. **541**, 238–242 (2013)
5. Graimann, B., Huggins, J.E., Levine, S.P., Pfurtscheller, G.: Visualization of significant ERD/ERS patterns in multichannel EEG and ECoG data. Clin. Neurophysiol. **113**(1), 43–47 (2002)

Inter and Intra Individual Variations of Cortical Functional Boundaries Depending on Brain States

Zhen Zhang[1], Junhai Xu[1], Luqi Cheng[2], Cheng Chen[1], and Lingzhong Fan[2,3](✉)

[1] Tianjin Key Laboratory of Cognitive Computing and Application, College of Intelligence and Computing, Tianjin University, Tianjin 300350, China
[2] Brainnetome Center, National Laboratory of Pattern Recognition, Institute of Automation, Chinese Academy of Sciences, Beijing 100190, China
lzfan@nlpr.ia.ac.cn
[3] School of Artificial Intelligence, University of Chinese Academy of Sciences, Beijing 100190, China

Abstract. The topography of human functional brain network not only differs between individuals but also reconfigures according to the specific brain state. However, it remains unknown how the fine-grained functional boundaries change in different individuals and states. Instead of using within-parcel features, we proposed an avenue to directly investigate the individual boundary differences and state-specific reconfigurations of functional topography using the boundary mapping method. By quantitatively calculating the inter-subject and intra-subject boundary variation rankings of different states and networks, we observed that the individual variation of functional boundaries is higher in the resting state compared to other task states, and is particularly variable in high-order association networks. In addition, we also proved that the parcel boundaries within individuals are significantly more similar than those between individuals from the view of boundary variation. Our results reveal the spatio-temporal distribution of the inter and intra individual functional topography variations and emphasize the importance of considering the individualized functional parcellation for understanding the dynamic of brain organization.

Keywords: Functional parcellation · Boundary variation · Task states · Networks

1 Introduction

The topography of human functional brain network not only differs between individuals but also reconfigures according to the specific brain state [1,2]. Most

Funded by the Beijing Advanced Discipline Fund, the Strategic Priority Research Program of Chinese Academy of Sciences (XDB32030200) and the National Natural Science Foundation of China (61703302).

H. Yang et al. (Eds.): ICONIP 2020, LNCS 12534, pp. 98–109, 2020.
https://doi.org/10.1007/978-3-030-63836-8_9

of the previous studies placed too much emphasis on the individual variations and dynamic reconfigurations from the perspective of functional connectivity and modular architecture of the large-scale brain networks [2,3]. However, the dynamic boundaries of the functional brain regions are often not considered, and fixed brain parcellations or atlases regardless of brain state are often used in such studies. Recent studies found that the human brain functional boundaries reconfigure across different cognitive states [2], and comparing with the intrinsic networks at resting state, the reconfigurations of task-related networks are modest [1]. Therefore, it remains unknown how the fine-grained functional boundaries change in different individuals and cognitive states. In other words, the spatio-temporal distribution of the functional cortical boundary reconfiguration and the corresponding interindividual variability are poorly understood.

Brain parcellation is important for understanding the organizational principles of the human brain. Among different parcellation methods, the data-driven functional parcellation using functional features with BOLD (Blood Oxygen Level-Dependent) MRI is very popular. There are two major methods in this field, clustering and boundary mapping [4,5]. In the clustering method, spatial elements (voxels or vertices) are grouped based on their functional similarity, while boundary mapping selects sharply changing positions as parcel borders. In recent research, Salehi et al. found that functional parcellation will reconfigure substantially and reproducibly depending upon brain task state when transiting from resting state [2]. Although the parcels of the brain are not defined statically, the number of parcels is fixed in the clustering mapping. Complementary to this, the border generated from boundary mapping does not have fixed parcels and intuitive within-parcel features that could be obtained in clustering easily, which make it difficult researching by boundary mapping. Some researches studied reconfiguration of networks consisted of parcels generated by boundary mapping method between resting and task states [3], but not at the level of parcels. What we did was to research the differences in functional parcellation results based on the boundary mapping method among resting state and all task states.

In the current study, we used a functional parcellation method based on boundary mapping proposed by Gordon et al. [6], and extended this method to individuals and task states, generating functional parcellation results for each subject's each state, including resting and task. 108 subjects in the Human Connectome Project (HCP) data S500 release was used. A new feature characterizing parcellation result generated by Gordon's boundary mapping method was selected and a new index quantifying boundary variation was proposed. We observed whether there is a significant difference between the results from resting and task states or not in the view of boundary mapping. Furthermore, we measured the boundary variation of different task states by leveraging the new index, looking for the variation regularity and relationship with the existing network index [7]. We leveraged 7 networks model proposed by Yeo et al. [8] to measure the changes of boundary variation in different cortical areas and its relation with brain function in certain regions. Besides, we evaluated which has higher boundary variance between inter- and intra-subject from the new perspective.

2 Method

2.1 Material and Preprocessing

We chose 108 subjects in the open dataset HCP S500. All subjects for analysis completed 7 conventional tasks provided by HCP, which are emotion (EMOT.), gambling (GAMB.), language (LANG.), motor (MOTO.), social (SOCI.), relational (RELA.), and working memory (WM). The HCP data were already preprocessed by the minimal processing pipeline [9], including artifact removal, motion correction, and registration to MNI standard space. Furthermore, the band-pass filter ($0.008\,\text{Hz} < \text{f} < 0.1\,\text{Hz}$) was applied to reduce low-frequency drift and high-frequency noise.

2.2 Feature of Boundary and the Index Measuring Boundary Variation

Instead of focusing on parcels or within-parcel characteristics, we researched on the cortical spatial distribution of functional boundary and its variation. In Gordon's method, the final parcellation result is created from an edge-density map with a given threshold, which is a representation of transition in functional connectivity [6]. We could observe that shapes of high values in edge-density map and of parcel boundary are the same in the same subject and states (Fig. 1a, 1b), because the density map contains complete boundary information, and expresses the latter abundantly and accurately. As a result, the chosen feature characterizing final parcellating boundary is the edge-density map, which has more advantages on quantifying boundary difference in large or small scopes than spatially discrete and discontinuous final parcellation boundary.

We extended Gordon's pipeline to individuals and task states. To research the boundary variation on specific cortical area and avoid the averaging effect of the whole cortex, we leveraged 7 functional networks proposed by Yeo to represent certain brain functional areas: visual network (VIN), somatomotor network (SMN), dorsal attention network (DAN), ventral attention network (VAN), limbic network (Limbic), frontoparietal network (FPN) and default mode network (DMN) [8]. The functional boundary distribution of a specific region in a certain state can be represented by the spatial-corresponding edge-density map (Fig. 1c). For convenience, situations in resting state and all task states are collectively called task-conditions, and those in specific network and task-condition are generally called task-network conditions. All work in this paper was dealing with the left and right brain separately and got consistent results.

We used root mean square error (RMSE) to measure the difference between 2 edge-density maps:

$$RMSE(X,Y) = \sqrt{\frac{1}{n}\sum_{i=1}^{n}(X_i - Y_i)^2} \tag{1}$$

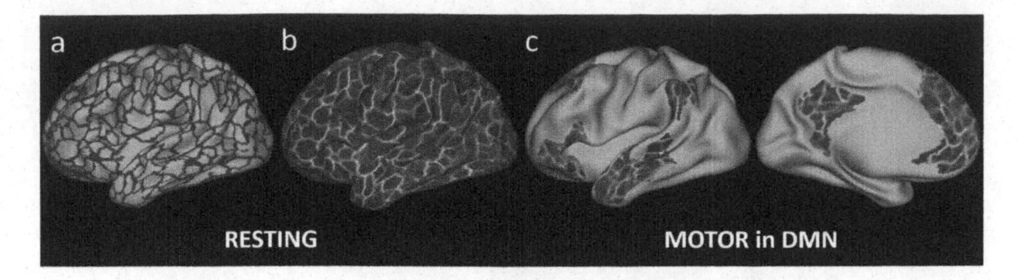

Fig. 1. Functional parcellation boundary and corresponding edge-density map. a) Parcellation result in resting state of No. 102816 subject. b) Edge-density map of the boundary in a). The lighter area has a greater frequency to be marked as boundary. c) Edge-density map under the motor task state in DMN of the same subject.

where X, Y are spatial-corresponding edge-density maps, n means the number of cortical vertices inside and i means one of them. Lower RMSE score means a lower difference between 2 maps. Index RMSE has a great property that it is sensitive to outliers. If peak values from 2 edge-density maps, which is a light line in Fig. 1c and has the probability to become the final boundary, overlap, the RMSE score would decrease obviously. This indicator well assesses the degree of spatial overlap between peak values of edge-density maps as well as final boundaries.

In summary, the edge-density map was used to characterize the functional parcellation boundary. The difference of edge-density maps within subjects or between subjects in certain cortical areas was utilized to represent boundary variation, which is quantified by the RMSE index. Furthermore, we could explore which task-condition or which cortical area has lower boundary variation by comparing RMSE data on statistics.

2.3 Statistical Differences of Boundary Variation Distribution

In the following sections, we compared parcellation boundaries by pairs across subjects or across task-conditions, which produce a batch of RMSE scores for a certain condition. We called them an RMSE data group for this certain condition, which represents boundary variation and is convenient for our succeeding discussion. Matched by subject-pairs or by task-pairs, different RMSE data groups were compared for statistically significant differences to research if there are differences in boundary variation between corresponding conditions. Since measurement of boundary variation takes 0 as the lower limit, the distribution of RMSE data groups is not normal and is more similar to a chi-square distribution with the degree of freedom bigger than 1 (Fig. 2a). To our surprise, differences between paired RMSE data groups could not pass all normal tests in MATLAB, even it looks very similar to the normal curve (Fig. 2b). This result prevented us from using the paired t-test to determine the significant difference between RMSE data groups. Therefore, we used the sign test instead, a method not as rigid as paired t-test but not require a normal population.

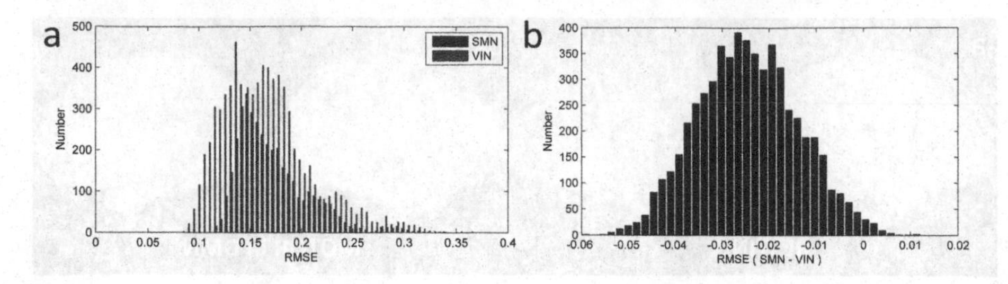

Fig. 2. Distribution of RMSE data groups and their difference. a) Distribution of RMSE data groups from SMN and VIN under the emotion task state. b) Distribution of difference of 2 groups in a).

The method of sign test is to test the sign of the difference between paired data groups. Specifically, if two samples have the same distribution, the number of positive and negative differences should account for about half for each sign. On the contrary, if one of the '+' and '−' signs increases significantly, it proves that there is a significant difference between distributions and we can infer which RMSE data group has a smaller value as a whole. For example, we propose the null hypothesis that scores distribution of RMSE data group A is NOT less than that of group B, which means '−' sign probability of difference A-B is less than 0.5. The events of '−' sign obey binomial distribution and probability are given according to the number of events in the test. When the number of '−' sign is large enough, the probability is low enough that we can negate the null hypothesis and conclude that the distribution of A is less than B with a statistically significant difference.

2.4 Inter-subject Boundary Variation

For the purpose of quantifying boundary variation between subjects in different task-network conditions, regional edge-density maps were chosen from each subject in specific task-condition and network (Fig. 1c) and were compared across subjects by pairs. An RMSE data group was got and represented boundary variation in this task-network condition, with the length equal to paired-comparison number. To briefly show the tendency of inter-subject boundary variation in different conditions, averaged RMSE values across subjects in each task-network condition were calculated, and shown in matrices and left cortical map (Fig. 3). Then, leveraging the statistical method of comparing boundary variation in different conditions, statistical distribution differences of boundary variation between 7 networks or 8 task-condition were analyzed respectively.

First, we researched the ranking of inter-subject boundary variation of 7 networks under the same task-condition. We compared these 7 RMSE data groups from each network and got all size relations by pairing each other. The ranking of boundary variation from 7 networks under this task-condition was generated by collecting all these relations. We have made many comparisons, so FDR correction was applied before inferring the ranking. Similar comparisons were carried under other task-conditions and rankings of networks in each task-condition were obtained. Second, using the same method, we also could generate rankings of 8 task-conditions in specific networks. Fixing the network, there are 8 RMSE data groups of 8 task-conditions. Compare these groups for pairs and get statistical size relations. The ranking of task-conditions in this network could be generated and so were rankings in other networks.

2.5 Intra-subject Boundary Variation

Regional edge-density maps were chosen as Sect. 2.4 and each map was compared with another map that belongs to the same subject and same network but in different task-conditions by the RMSE index. RMSE data group generating by this way reflect boundary variation of task-condition pair from the same subjects in the specific cortical region. So, there is an 8*8 task-condition variation matrix for each subject and network. Each term in the matrix represents the boundary variation between two task-conditions represented by RMSE. We could see the tendency of the intra-subject variation between task-conditions by averaging all 8*8 task-condition variation matrices across subjects and networks. The result is shown in Fig. 4.

After having a task-condition variation matrix, we averaged it into a task-condition variation vector, in which each term represents average boundary variation of specific task-condition compared with other task-conditions and is what we called intra-subject boundary variation. There was a task-condition variation vector for each subject and network, so there was an RMSE data group for each task-network condition with the length equal to the subject number. Fixing a functional network, we could test if there is a statistical difference between intra-subject variation from arbitrary 2 task-conditions and get intra-subject variation rankings of task-conditions in each network. The generation of rankings of networks used a similar method.

2.6 Comparison of Boundary Variation Between Inter- and Intra-subject

We proposed a method based on RMSE measurement to compare relative size between variation of inter- and intra-subject. A hypothesis was proposed that boundary variation with the same subject and different task-conditions is less than that in different subjects. All comparisons, both within subject and between subjects, consist of comparison units. A comparison unit is made of two pairs of edge-density maps, belong to combinations of 2 subjects (i and j) and 2 task-conditions (A and B). We marked their edge-density maps as Ai, Bi, Aj, Bj respectively. Map-paired comparisons were processing between these 4 maps when map-pair had the same subject or task-condition, and 4 RMSE scores were generated. 2 of them were calculated across (Ai, Bi) and (Aj, Bj), representing variation between the same subject and different task-conditions, which we called t1, t2, meaning values across tasks of the same subject. Others were calculated across (Ai, Aj) and (Bi, Bj), representing variation between different subjects and the same task-condition. We called them s1, s2, meaning values across subjects. If the average value of t1 and t2 is smaller than that of s1 and s2, the conclusion of this comparison unit is following our hypothesis. In this situation, if the minimum variation is a value across subjects, we thought the persuasion was insufficient and marked this result as an ambiguous result, otherwise a positive result. If the average value of t1 and t2 is bigger, marked as a negative result. Method on the representative comparison unit was extended to comparisons among all subjects and task-conditions. We observed the tendency of all results to get the conclusion. Comparisons were conducted in each network and hemisphere respectively.

3 Result

3.1 Tendency of Inter- and Intra-subject Boundary Variation

Inter-subject boundary variation represents the difference of function boundary feature of different subjects, while intra-subject boundary variation represents the boundary difference between a task-condition with others from the same subject. The thinking to get inter- and intra-subject boundary variation has been introduced in the Method section. Here, we use the mean value to show the general tendencies of these 2 variations.

Average inter-subject variation values were calculated across subjects and were shown in the matrices in Fig. 3a, 3b. Looking at these two matrices, we could observe variation patterns in different networks with the same task-condition on a row of the matrix, and also could observe patterns in the same network with different task-conditions on a column. Average boundary variation values of the left hemisphere were also demonstrated on the left cerebral cortex map for a more intuitive presentation.

Averaged intra-subject variation values across networks and subjects were shown at matrices of both hemispheres (Fig. 4a, 4b), and averaged intra-subject variation across subjects was demonstrated on the cortical map of the left hemisphere (Fig. 4c).

Inter- and intra-subject boundary variations were visually consistent on the cortical map. From Fig. 3 and 4, we concluded that the resting state has greater boundary variation than any task state. In all pictures in Fig. 3 and 4, the corresponding parts of the resting state have the color that represents the highest variation. In different task states, the emotion task has the smallest boundary variation and the gambling task has the 2nd smallest. From 2 maps on the left cortex, SMN is the network that has the most minimum variation values across all task-conditions. Further analysis based on statistics were in the next section.

3.2 Rankings of Inter- and Intra-subject Boundary Variation on Statistics

Ranking of Networks. We compared all inter-subject boundary variation of all networks in pairs for the statistically significant difference ($p(FDR) < 0.01$) by

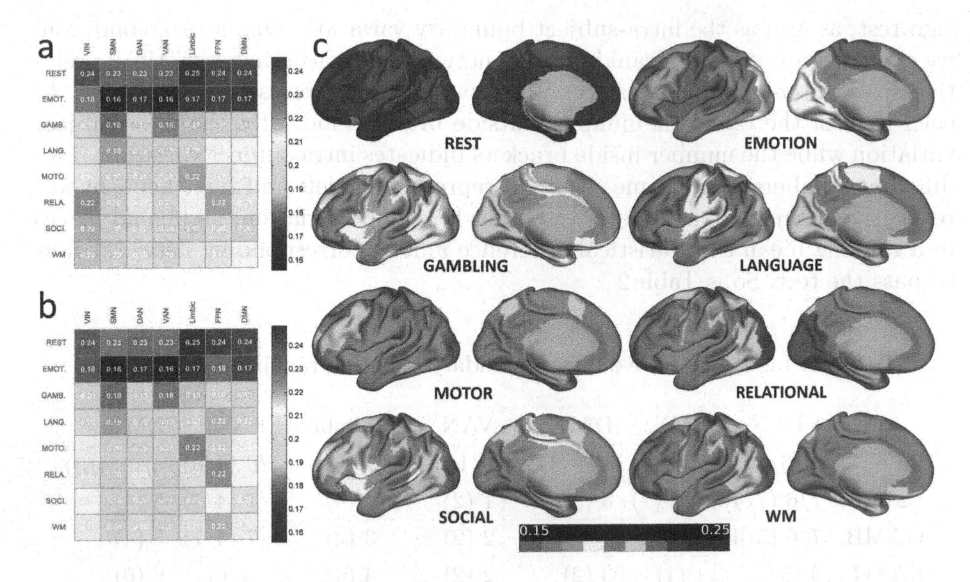

Fig. 3. Tendency of inter-subject boundary variation. a) Matrix of inter-subject boundary variation averaged across subjects in every task-network condition in left hemisphere. b) Matrix in right. c) Average boundary variation values across subjects in every task-condition shown on left cerebral cortex.

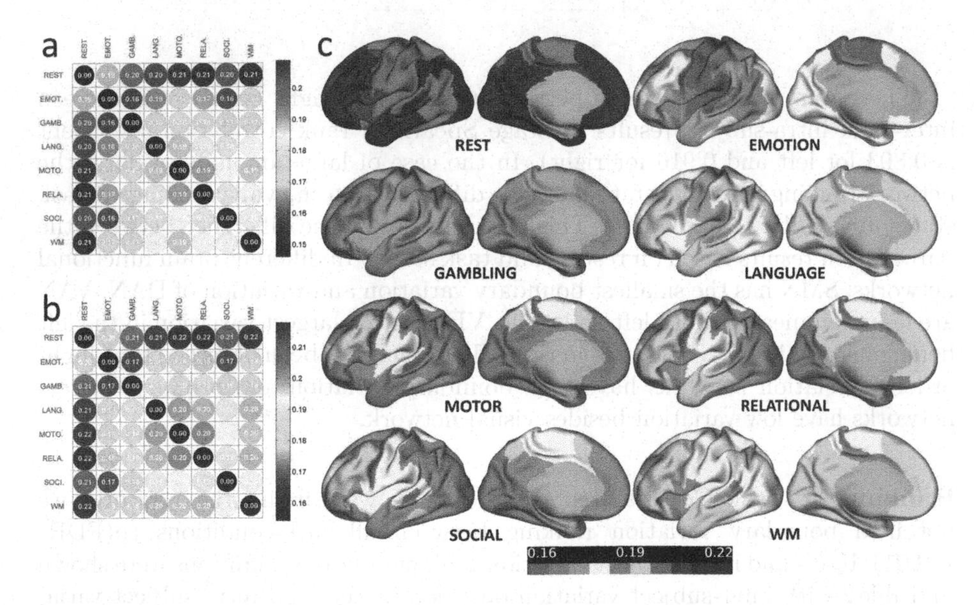

Fig. 4. Tendency of intra-subject boundary variation. a) Matrix of intra-subject boundary variation averaged across subjects and networks in every task-condition pair in left hemisphere. b) Matrix in right hemisphere. c) Intra-subject boundary variation values averaged across subjects in every task-condition shown on left cerebral cortex.

sign test, as well as the intra-subject boundary variation. Fixing task-condition, we could obtain variation ranking in all networks by gathering statistically relative sizes. Inter- and intra-subject variation rankings were shown in Table 1. In each item of the table, the number outside brackets indicates the inter-subject variation while the number inside brackets indicates intra-subject variation. Two different numbers in the same situation represent the left and right hemispheres respectively. Smaller ranking means lower boundary variation, and equal places in a ranking mean the statistical difference among corresponding networks fails to pass the test. So is Table 2.

Table 1. Inter- and intra-subject boundary variation rankings of networks.

	VIN	SMN	DAN	VAN	Limbic	FPN	DMN
REST	4 (6/4)	2/1 (1)	1/3 (1/2)	3/1 (3/2)	7 (6)	4/6 (4/6)	4/5 (4/5)
EMOT.	7/6 (7/4)	1/2 (1)	3 (3)	1 (2)	4 (4)	6/7 (4/7)	5 (4)
GAMB.	7/6 (5/4)	1 (1)	3/4 (3)	2 (2)	3 (4)	6/7 (4/7)	5 (4)
LANG.	5 (5/4)	1 (1)	3 (2)	2 (2)	4 (4)	7 (5/6)	6 (6)
MOTO.	6/4 (6/4)	1 (1)	2/3 (2)	3/1 (2)	7 (6/7)	4/6 (4/6)	5/4 (5/4)
RELA.	7/6 (4)	1 (1)	3 (2/3)	2 (2)	4/3 (4)	6/7 (4/7)	5 (4)
SOCI.	7/6 (7/4)	2 (1)	5/4 (3)	1 (2)	3 (4)	6 (4/7)	4/5 (4)
WM	7/5 (5/4)	1 (1)	3 (2)	2 (2)	4 (5/6)	6/7 (4/6)	5 (5)

There were similar patterns between the left and right brains and between inter- and intra-subject results (average Spearman rank correlation coefficient is 0.893 for left and 0.916 for right). In the case of large amounts of data, the network ranking in resting state showed a different pattern, while rankings in task states are similar to each other. This is a reflection of the difference between the parcellation results based on resting and task states. In different brain functional networks, SMN has the smallest boundary variation and variation of DAN, VAN are less than mean in both left and right. VIN has the largest variation in the left hemisphere while FPN is largest in right. The trend has been observed that high-order association networks have high boundary variation and primary cortical networks have low variation besides visual network.

Ranking of Task-Conditions. In the same way, fixing the network, we obtained boundary variation ranking between all task-conditions (p(FDR) < 0.01). Inter- and intra-subject variation rankings of task-conditions were shown in Table 2 with inter-subject variation outside brackets and intra-subject variation inside.

Statistically, we could get the same conclusion as previous. There is high accordance between results of inter- and intra-subject variation (average Spearman rank correlation coefficient is 0.897 for left and 0.921 for right). Left and

Table 2. Inter- and intra-subject boundary variation rankings of task-conditions.

	REST	EMOT.	GAMB.	LANG.	MOTO.	RELA.	SOCI.	WM
VIN	8 (8)	1 (1)	2 (2)	3/4 (3/4)	5/4 (5/4)	5/4 (5/4)	5/4 (5/4)	3(3)
SMN	8 (8)	1 (1)	2 (2)	3 (3)	7/4 (4)	4 (4)	4 (4)	6/4 (4)
DAN	8 (8)	1 (1)	2 (2)	3 (3)	3/4 (5)	7 (5)	5 (5)	5 (3)
VAN	8 (8)	1 (1)	2 (2)	4/3 (3/4)	6 (3/4)	5 (3/4)	3 (3)	6 (3/4)
Limbic	8 (8)	1 (1)	2 (2)	4/3 (3/4)	7 (7/6)	6/5 (3/4)	3 (3)	6 (3/6)
FPN	8 (8)	1 (1)	2 (2)	5 (3/4)	4 (3/4)	7 (3/4)	3 (3/4)	5 (3)
DMN	8 (8)	1 (1)	2 (2)	4/7 (6/4)	4/5 (6/4)	4/5 (3/4)	3 (3)	4 (3/4)

right hemispheres have the same variation pattern and all task-condition rankings are almost the same in each network. There is the biggest boundary variation in resting state, the smallest variation in the emotion task, and the second smallest variation in the gambling task both in inter- and intra-subject situations.

The resting state has the highest intra-subject boundary variation, which is significantly greater than the variation of any task state in terms of ranking and cortical map. It indicates that there is a large difference between functional parcel boundaries of resting state and task state, which is consistent with the conclusion found by the clustering method that functional parcellation will reconfigure while transiting from resting to task state [2]. For further research on the accordance between results of inter- and intra-subject variation between task-conditions, we did another experiment to explain this phenomenon. We superimposed the final parcellation boundary of all subjects according to the task-condition on the left cerebral cortex. Then frequency that each cortical vertex was identified as the boundary was calculated. We observed the distribution of the frequency of all left cortical vertices and calculated the mean and variance, shown in Table 3.

Table 3. Mean value and standard variation of boundary frequency.

	REST	EMOT.	GAMB.	LANG.	MOTO.	RELA.	SOCI.	WM
Mean value	0.257	0.262	0.261	0.259	0.259	0.258	0.259	0.259
Standard variation	0.139	0.167	0.154	0.147	0.147	0.146	0.148	0.147

On the premise that the mean value is basically identical, we found that task-condition with lower boundary variance has a greater variance, which means some cortical vertices are identified as border more often and some not. In other words, the functional parcellation boundary of task-condition with lower variance is preferred to converging on some vertices. It can be inferred that the functional boundary of task-condition with lower variance is closer to the intrinsic and putative border of the human cerebral cortex than those with higher variation, which causes appearance that this task-condition has lower boundary variation no matter within-subject or between-subject.

Relationship with Network Efficiency. The boundary variation of task states is related to the global network efficiency after functional network reconfiguration caused by activation. Network efficiency is the quantification of communication capability between vertices and global efficiency is calculated as the averaged value of all vertices' efficiencies. Zuo et al. found that accompanied with the reconstruction of the functional network when entering a task state, global efficiency improves greatly [7]. The global network efficiency for each task state shows a significant negative correlation with boundary variation, meaning task state with lower boundary variation has higher global network efficiency. This phenomenon maybe because they have a common internal motivation factor. One of the possible explanations is the degree of task engagement, and relevant research is underway.

3.3 Comparison of Boundary Variation Between Inter- and Intra-subject

Results were similar in the left and right hemispheres. In different cortical regions, the common percentage of negative cases was about 2%, while the maximum percentage did not exceed 9%. Even if we considered ambiguous cases, the maximum percentage of cases that did not strongly support our hypothesis was less than 26%. Limbic network, default mode network, and visual network were cortical areas that have the largest, middle, and the smallest percentage of negative cases in both hemispheres. From all results revealed, we verified our hypothesis and concluded that intra-subject parcel boundaries are significantly more similar than inter-subject parcel boundaries. Shown in Table 4.

Table 4. Results of comparison of boundary variation between inter- and intra-subject.

	VIN	SMN	DAN	VAN	Limbic	FPN	DMN
Positive cases (L)	91.74%	84.23%	90.69%	79.54%	73.84%	87.94%	84.51%
Negative cases (L)	0.60%	2.53%	0.74%	5.19%	9.33%	2.04%	2.01%
Positive cases (R)	91.75%	85.32%	90.42%	80.91%	74.19%	90.66%	86.19%
Negative cases (R)	0.60%	1.94%	1.06%	5.33%	9.44%	0.66%	1.92%

4 Conclusion

In this paper, inspired by the conclusion based on the clustering method that functional parcellation will reconfigure substantially and reproducibly when entering task state, we gave a new perspective of boundary variation based on the boundary mapping method. Rather than analyzing parcellation difference by features within the parcels, we preferred to focus on the change of edge-density maps, which characterize functional boundary, in different individuals, cortical networks and task-conditions, and proposed RMSE index to quantify the variation. We found results similar to cluster-based method when researching on the intra-subject boundary variation, proving the significant parcellation difference between resting state and task states. Furthermore, the boundary variations of task-network conditions were quantified in inter- and intra-subject methods respectively, and patterns are demonstrated by cortical maps and rankings. Consistency between results of inter- and intra-subject indicated that the boundary of task state with lower variation is closer to the intrinsic boundary of the human brain. The boundary variation of task states is consistent with the efficiency of the network after reconfiguration, indicating they have the same principle. There are higher boundary variations in high-order association networks. In addition, we also proved that the parcel boundaries within individuals are significantly more similar than those between individuals from the view of boundary variation. To conclude, our results revealed the spatio-temporal distribution of the inter and intra individual functional topography variations, and emphasize the importance of considering the individualized functional parcellation for understanding the dynamic of brain organization. In the future, the next stage of building the functional brain atlas will be dynamic instead of static by including information on spatio-temporal brain state changes during normal development or aging as well as disease-related effects.

References

1. Cole, M.W., Bassett, D.S., Power, J.D., Braver, T.S., Petersen, S.E.: Intrinsic and task-evoked network architectures of the human brain. Neuron **83**, 238–251 (2014)
2. Salehi, M., Greene, A.S., Karbasi, A., Shen, X., Scheinost, D., Constable, R.T.: There is no single functional atlas even for a single individual: functional parcel definitions change with task. Neuroimage **208**, 116366 (2020)
3. Gratton, C., et al.: Functional brain networks are dominated by stable group and individual factors, not cognitive or daily variation. Neuron **98**, 439–452 (2018). e435
4. Eickhoff, S.B., Yeo, B.T.T., Genon, S.: Imaging-based parcellations of the human brain. Nat. Rev. Neurosci. **19**, 672–686 (2018)
5. Fan, L., et al.: The human brainnetome atlas: a new brain atlas based on connectional architecture. Cereb. Cortex **26**, 3508–3526 (2016)
6. Gordon, E.M., Laumann, T.O., Adeyemo, B., Huckins, J.F., Kelley, W.M., Petersen, S.E.: Generation and evaluation of a cortical area parcellation from resting-state correlations. Cereb. Cortex **26**, 288–303 (2016)
7. Zuo, N., Yang, Z., Liu, Y., Li, J., Jiang, T.: Both activated and less-activated regions identified by functional MRI reconfigure to support task executions. Brain Behav. **8**, e00893 (2018)
8. Yeo, B.T., et al.: The organization of the human cerebral cortex estimated by intrinsic functional connectivity. J. Neurophysiol. **106**, 1125–1165 (2011)
9. Glasser, M.F., et al.: The minimal preprocessing pipelines for the Human Connectome Project. Neuroimage **80**, 105–124 (2013)

Phase Synchronization Indices for Classification of Action Intention Understanding Based on EEG Signals

Xingliang Xiong[1], Xuesong Lu[2], Lingyun Gu[1], Hongfang Han[1], Zhongxian Hong[2], and Haixian Wang[1(✉)]

[1] Key Laboratory of Child Development and Learning Science of Ministry of Education, School of Biological Science and Medical Engineering, Southeast University, Nanjing 210096, Jiangsu, People's Republic of China
hxwang@seu.edu.cn

[2] Department of Rehabilitation, Zhongda Hospital, Southeast University, Nanjing 210009, Jiangsu, People's Republic of China

Abstract. The classification of action intention understanding based on EEG signals is very important for human-robot and social interaction studies. In order to classify the action intention understanding brain signals efficiently, we first use three kinds of phase synchronization indices, phase locking value (PLV), phase lag index (PLI) and weight phase lag index (WPLI), to construct functional connectivity matrices in multiple micro time windows, and then extract the sum of significant edge values of each time window matrix as the classification feature, finally apply support vector machine (SVM) classifier to implement action intention understanding data classification task. Classification result shows that new method performs well on three datasets (alpha, beta and fusion frequency bands), and brain network statistical analysis demonstrates that many significant edges appear on the alpha frequency band. We conclude that the phase synchronization indices are extremely useful for the classification task, the sum of significant edge values is an effective classification feature, and the action intention understanding closely correlates with the alpha frequency band.

Keywords: Phase synchronization · Classification · Action intention understanding

1 Introduction

In recent years, the study of action intention understanding has attracted extensive attention [1–6]. The classification of action intention understanding based on EEG signals is one of the most important branches of the study, which is viewed as a key factor for human-robot interaction [1,7–9]. Many researchers carried out a lot of experiments on the action intention understanding classification by different methods [1,5,9]. However, the classification accuracy is often

H. Yang et al. (Eds.): ICONIP 2020, LNCS 12534, pp. 110–121, 2020.
https://doi.org/10.1007/978-3-030-63836-8_10

unsatisfactory. This is due to two important reasons: (1) it is difficult to extract the most useful classification features, (2) it is hard to collect a large number of training samples. As for the former question, some researchers use different brain signal collection techniques to solve it [1,9], and some other researchers mainly focus on extracting further feature [5]. Due to the popularity of deep learning, many people begin to consider using neural network (e.g., convolution neural network (CNN) and recurrent neural network (RNN)) technique to extract features. As for the latter question, people usually spend much time and money to recruit subjects and collect a certain amount of neural information data.

Brain network is an efficient tool to study neuroscience, which has some comprehensive merits (e.g., feature extraction, brain region position) [1,5]. There are a lot of algorithms to construct the brain network [10]. The synchronous oscillation of a neural network is the main potential mechanism of brain information integration and processing, and the synchronization of multiband signals is the key feature of information exchange between different brain regions [11]. In the EEG synchronization analysis algorithms, the phase synchronization analysis method can directly separate the signal phase information in a given frequency range from the amplitude of nonstationary information. Then, the signal phase information can be used for the synchronization analysis of EEG narrowband signals, such as mu rhythm [11–13]. It is noteworthy that some previous studies point out the action intention understanding correlates with the alpha and beta rhythms [14,15]. The advantage of phase synchronization is that it has nothing to do with the amplitudes of two neural oscillatory activities, but only with the phases.

Considering that, we first use three phase synchronization indices, phase locking value (PLV) [11], phase lag index (PLI) [12] and weighted phase lag index (WPLI) [13], to construct brain networks in this study, then apply these brain networks to solve the former two problems (feature extraction and sample collection). As for the first problem (feature extraction), we adopt t-test and FDR correction to select the edges of a functional connectivity matrix and sum for these edges as a feature. And as for the second problem (sample collection), inspired by the literature [16], we use an idea that one divides into three to model more samples. Each subject has three kinds of brain networks that are calculated by the PLI, WPLI and PLV under a certain stimulus condition. We view the different brain networks as different samples. Because our final goal is to classify action intention understanding, the sample model method is feasible. More details about how to solve these two problems are presented out in next section.

The main contribution of this study is that we design a novel method which can effectively solve the problem in feature selection for classifying action intention understanding EEG signals. Many other neural information data classification studies (e.g., epilepsy, emotion, and mathematical genius classification tasks) also can draw lessons from the novel feature extraction method. The training sample collection idea based on the phase synchronization indices shows its advantage under the condition of limited manual sampling. With the analyses of signals classification and brain network statistical test, we found that the alpha frequency band easily obtains significant achievements in experiments, which further supports the conclusion that action intention understanding closely correlates with the alpha frequency band in some previous studies [5,14,15].

2 Materials and Method

2.1 Subjects

After deleting 5 subjects of which EEG data were seriously noised (e.g., someone has abnormal channels that there were no signals in them), we totally retained 25 healthy subjects (17 males, 8 females; aged 19–25 years, mean $\pm$ SD: 22.96 ± 1.54; right-handed). This research was approved by the Academic Committee of the School of Biological Sciences and Medical Engineering, Southeast University, China.

2.2 Experimental Paradigm

In the progress of EEG data acquisition, all subjects were asked to see three kinds of hand-cup interaction pictures that were performed by an actor. They only needed to judge the actor's intention, but not to implement any concrete operations. The three action intentions were drinking water (Ug), moving the cup (Tg) and simply contact the cup (Sc). Figure 1 shows the experimental stimuli and procedure. This design comes from Ortigue et al.'s experimental paradigm [17].

2.3 Data Collection and Preprocessing

In this study, we use 64 AgCl electrodes that were arranged with the international 10–20 system to record the EEG signals. The sampling rate was set to 500 Hz, and the reference electrode was set as M1 that was placed on the left mastoid. All the data collection tasks were completed with the Neuroscan 4.3.

In order to obtain useful data, we clean the raw EEG signals by Neuroscan 4.3 and EEGLAB 14.0 [18]. In the light of some previous data preprocessing experiences, it can't efficiently clear noises of the raw EEG data by independent component analysis (ICA). We adopted ocular processing in the Neuroscan to replace the ICA in the EEGLAB. Mastoid reference is efficient in somatosensory evoked potentials, which correlates with action behaviors. Thereupon, we transformed the unilateral mastoid reference (M1) into bilateral mastoid (M1, M2) re-reference. Both the ocular processing and re-reference are completed with the Neuroscan.

When completed the ocular processing and re-reference, we used the EEGLAB to select 60 scalp electrodes that cover frontal, partial, central, occipital and temporal areas. Then, we adopted the Basic FIR filter in the EEGLAB to extract the 1–30 Hz data. And then, we segmented the data with event types

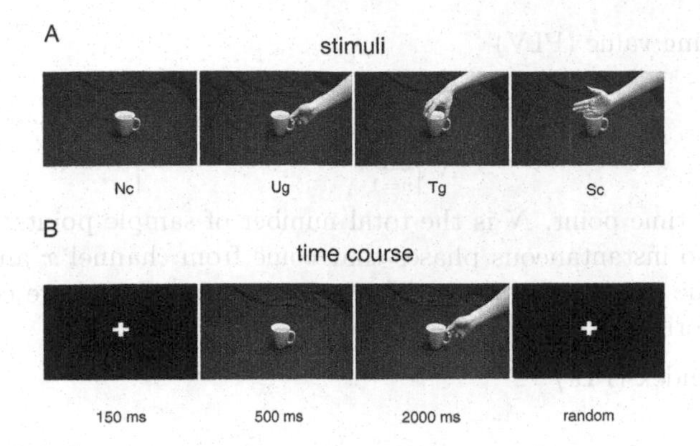

Fig. 1. Experimental stimuli. (A) An example of stimulus materials. Ug (use grip), Tg (transport grip), and Sc (simple contact) denote that the actor grasps the cup for drinking water, moving it, and touching it without any clear intention, respectively. (B) An example of the experimental procedure in a trial. In the stimuli, a symbol '+' was first presented on the screen with 150 ms. Then, a cup was shown for 500 ms to keep consistent sense. The formal action intention stimulus that was denoted by a hand-cup interaction picture starts after 650 ms, which was sustained for 2000 ms. When the hand-cup interaction stimulus appears, the subjects need to guess what the actor want to do immediately. Before the next trial, the '+' was presented again with a random time that varied from 1000 to 2000 ms.

in a time window (−0.65 s to 2.5 s) and removed the baseline by setting the baseline at −0.65 s. In the end, we deleted artifacts with a threshold range that varied from −75 to 75 μV. A total of 679 trials were removed and an average of 267 trials were kept for each subject. It is noteworthy that the alpha and beta frequency sub-bands used in this study were extracted by low resolution electromagnetic tomography (LORETA) in the source space.

2.4 Phase Synchronization Indices

The advantage of phase synchronization is that it has nothing to do with the amplitudes of any two neural oscillatory activities, but only with the phases. Considering that, we use three phase synchronization indices to construct the functional connectivity matrices. Before giving out the concrete formulas of the three indices, it needs to explain two important concepts in this study. One is that each brain network is represented by a functional connectivity matrix. Another is that the node in the brain network is defined by the region of interest (RIO) in the source space, i.e., each node corresponds to a RIO. Our experiments mainly based on the whole brain, which has 84 RIOs that are defined in the LORETA. The preprocessed 60 channels EEG data were converted into 84 RIO time series with the LORETA. Hence, the size of a functional connectivity matrix (brain network) is 84×84. The mathematical expressions of the three algorithms that are used to construct the functional connectivity matrices are defined as follows:

- Phase locking value (PLV)

$$PLV_{xy} = \frac{1}{N}\left|\sum_{n=1}^{N} e^{j\{\Phi_{n,x}(t)-\Phi_{n,y}(t)\}}\right| \tag{1}$$

where t is the time point, N is the total number of sample points, $\Phi_{n,x}(t)$ and $\Phi_{n,y}(t)$ are two instantaneous phases that come from channel x and channel y at the nth time point respectively. The instantaneous phases are computed by Hilbert transform.

- Phase lag index (PLI)

$$PLI_{xy} = \frac{1}{N}\left|\sum_{n=1}^{N} sign(sin(\triangle\Theta(t_n)))\right| \tag{2}$$

where $\triangle\Theta$ is also the instantaneous phase difference that is between the time series $x(t)$ and $y(t)$ at the nth sample time point as demonstrated in the Eq. (1).

- Weight phase lag index (WPLI)

$$WPLI_{xy} = \frac{\left|\langle\tilde{S}(w)\rangle\right|}{\left\langle|\tilde{S}(w)|\right\rangle} = \frac{\left|\langle|\tilde{S}(w)|sign(\tilde{S}(w))\rangle\right|}{\left\langle|\tilde{S}(w)|\right\rangle} \tag{3}$$

where $\tilde{S}(w)$ denotes the imaginary component of cross-spectrum between time series $x(t)$ and $y(t)$. The symbols $\langle\cdot\rangle$, $|\cdot|$ and function $sign$ denote mean, absolute, signum function, respectively.

After obtaining the functional connectivity matrices that are calculated by PLV, PLI and WPLI, we first use paired t-test and false discovery rate (FDR) correction ($p < 0.05$) to find the positions of which edges are significantly different between two kinds of action intention understandings (Ug-vs-Tg, Ug-vs-Sc and Tg-vs-Sc). The final positions are determined by the edges that are all significantly different on the three paired comparisons. Then, we use the final positions to select the weighted edges in each time window. Finally, we use the sum of the selected edges in each time window as the classification feature, i.e., each time window matrix is corresponding to one feature. Because there are 63 dynamical time windows in this study (We divide the full-length time series into 63 sub time series, each has a length of 50 ms.), hence, it totally has 63 features for a single frequency band dataset. Figure 2 shows out the flow chart of our novel method. It is important to note that the feature selection is completed on the condition of 5-fold cross validation.

As for the problem of increasing data samples that is mentioned in the introduction, we let the feature vectors which are from the three functional connectivity matrices (PLV, PLI and WPLI) of one subject be three samples. Therefore, we can make 75 data samples for the original 25 subjects on one kind of action intention stimulus. A total of 225 data samples (Ug, Tg, and Sc all have 75 samples) are collected in this experiment.

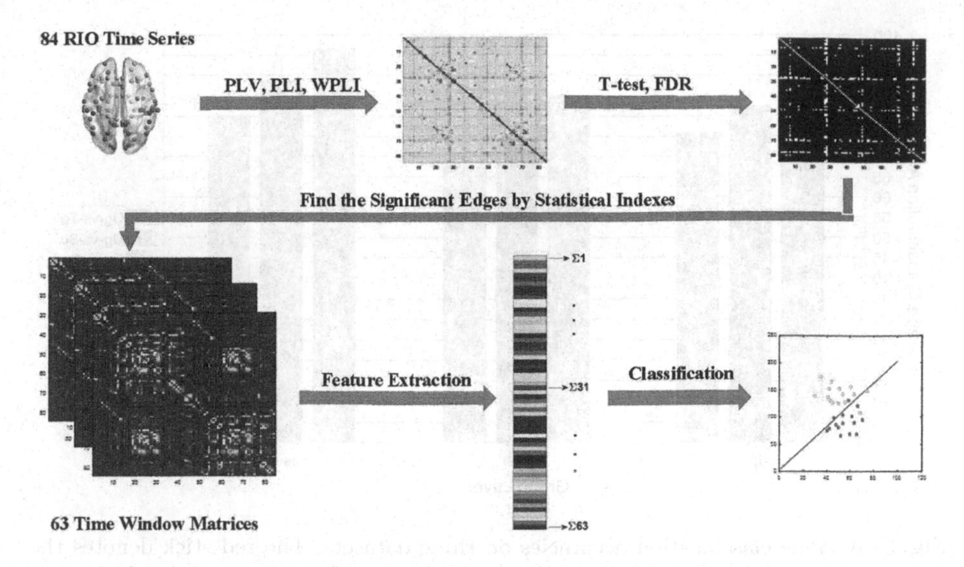

Fig. 2. The flow chart of new method.

3 Results

The experimental results mainly include two parts: action intention understanding classification and brain network statistics. In the following, we give out the demonstrations of the results.

3.1 Action Intention Understanding Classification

The classification pattern is one-versus-one, i.e., Ug-vs-Tg, Ug-vs-Sc and Tg-vs-Sc. The classifier is the classical support vector machine (SVM), of which kernel function and order number parameters are set as polynomial, 1, respectively. We use three datasets, alpha, beta and fusion, to carry out the action intention understanding classification. That is to say, the action intention understanding based on EEG signals come from the alpha, beta frequency sub-bands. Fusion dataset is constructed by merging the features of alpha and beta datasets into a big dataset. The classification tasks based on the group level, not a single subject. In order to avoid random factors, we implemented 1000 times 5-fold cross validation and calculated the mean of these 1000 times experiments as the final classification accuracy.

Figure 3 shows the average classification accuracies on the alpha, beta and fusion datasets. We can see that the lowest accuracy is over 65%, while the highest accuracy is even over 95%. The classification accuracies on the Ug-vs-Tg are all more than 85%, which performs the best compared with the other two one-versus-one patterns. Among the three datasets, beta performs the worst and

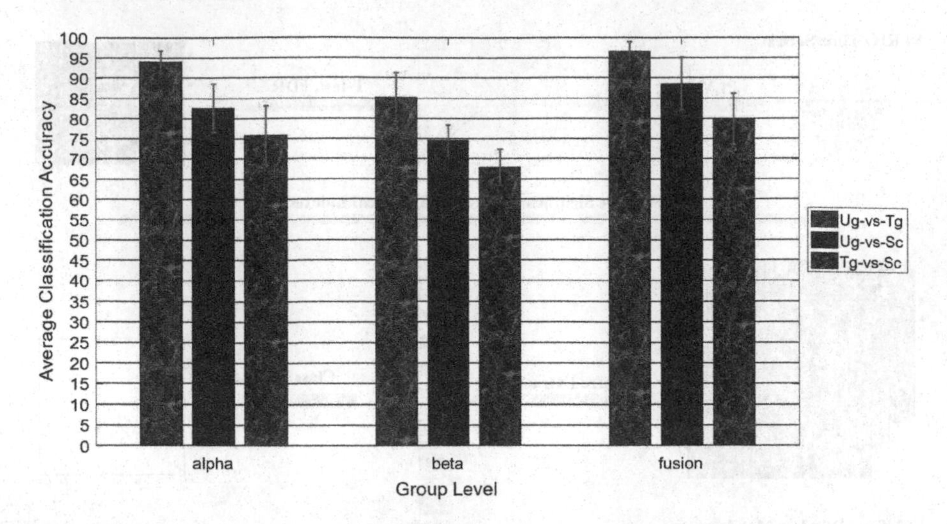

Fig. 3. Average classification accuracies on three datasets. The red stick denotes the standard deviation. The alpha, beta and fusion are three different datasets, which come from the alpha, beta and fusion frequency bands, respectively. (Color figure online)

fusion performs the best. Especially, the average classification accuracies on the fusion dataset are all greater than or equal to 80%. Table 1 displays the details of four classification estimation metrics under different conditions. We can see that most of the numerical values of mean (average classification accuracy), sensitivity and specificity are very high, while the numerical values of standard deviation are very low.

3.2 Brain Network Statistics

In this part, we give out the edge value differences of the average brain network that is based on 63 dynamical time windows. Each functional connectivity edge is tested by serious t-test and FDR correction. The statistics are carried out on the three one-versus-one patterns.

Figure 4 shows the statistical results on the alpha and beta frequency subbands. We can see that there are many significant connectivity edges in the alpha band, while the beta band is very sparse, especial in both Ug-vs-Sc and Tg-vs-Sc. Additionally, we also can see that many important nodes appear in the temporal, frontal and occipital lobes. The positions of the important nodes are uncommon on different frequency bands and one-versus-one patterns.

Table 1. Classification estimation metrics. The mean denotes the average classification accuracy.

Condition		Estimation metrics			
Frequency band	Pattern	Mean	Standard deviation	Sensitivity	Specificity
Alpha	Ug-vs-Tg	94.00	2.79	93.33	94.67
	Ug-vs-Sc	82.67	5.96	76.00	89.33
	Tg-vs-Sc	76.00	7.23	61.33	90.67
Beta	Ug-vs-Tg	85.33	6.06	89.33	81.33
	Ug-vs-Sc	74.67	3.8	66.67	82.67
	Tg-vs-Sc	68.00	4.47	44.00	92.00
Fusion	Ug-vs-Tg	96.67	2.36	98.67	94.67
	Ug-vs-Sc	88.67	6.50	81.33	96.00
	Tg-vs-Sc	80.00	6.24	64.00	96.00

4 Discussion

From the experimental results of action intention understanding classification and brain network statistics, we have some important findings. In this section, we present out the elaborations of these findings.

The results of classification accuracies display the novel method performs well on the alpha, beta and fusion datasets (see Fig. 3 and Table 1), which suggest that the feature extraction and sample collection that are dealt with by the phase synchronization indices are effective and satisfactory. Some previous studies [1,5] which use a single index of phase synchronization (PLV, PLI or WPLI) to decode action intention understanding point out that the phase synchronization is a useful tool. The experimental results based on multiple indices of phase synchronization (PLV, PLI and WPLI) further prove the previous viewpoint. Both the alpha and beta frequency sub-bands show extremely high classification accuracies, which indicate that the action intention understanding brain activities easily occur at these two bands. These are consistent with some previous studies [5,15]. Figure 3 demonstrates that the fusion dataset obtains the best classification accuracy. Because the fusion dataset contains twice the classification features compared with a single alpha or beta dataset (both alpha and beta have 63 features, fusion has 126 features), it more easily obtains a better result. Another feasible explanation is that the fusion dataset not only captures the alpha but also contains the beta frequency band information, therefore, the fusion dataset can take the advantages of both frequency bands at the same time. Briefly speaking, feature fusion is an effective method for classification. It is noteworthy that when compared with the previous studies about action intention understanding classification [1,5,9], our novel method obtains more higher classification accuracies.

In the Fig. 4, many functional connective edges are still retained after serious FDR correction ($p < 0.05$). Because the difference brain networks are

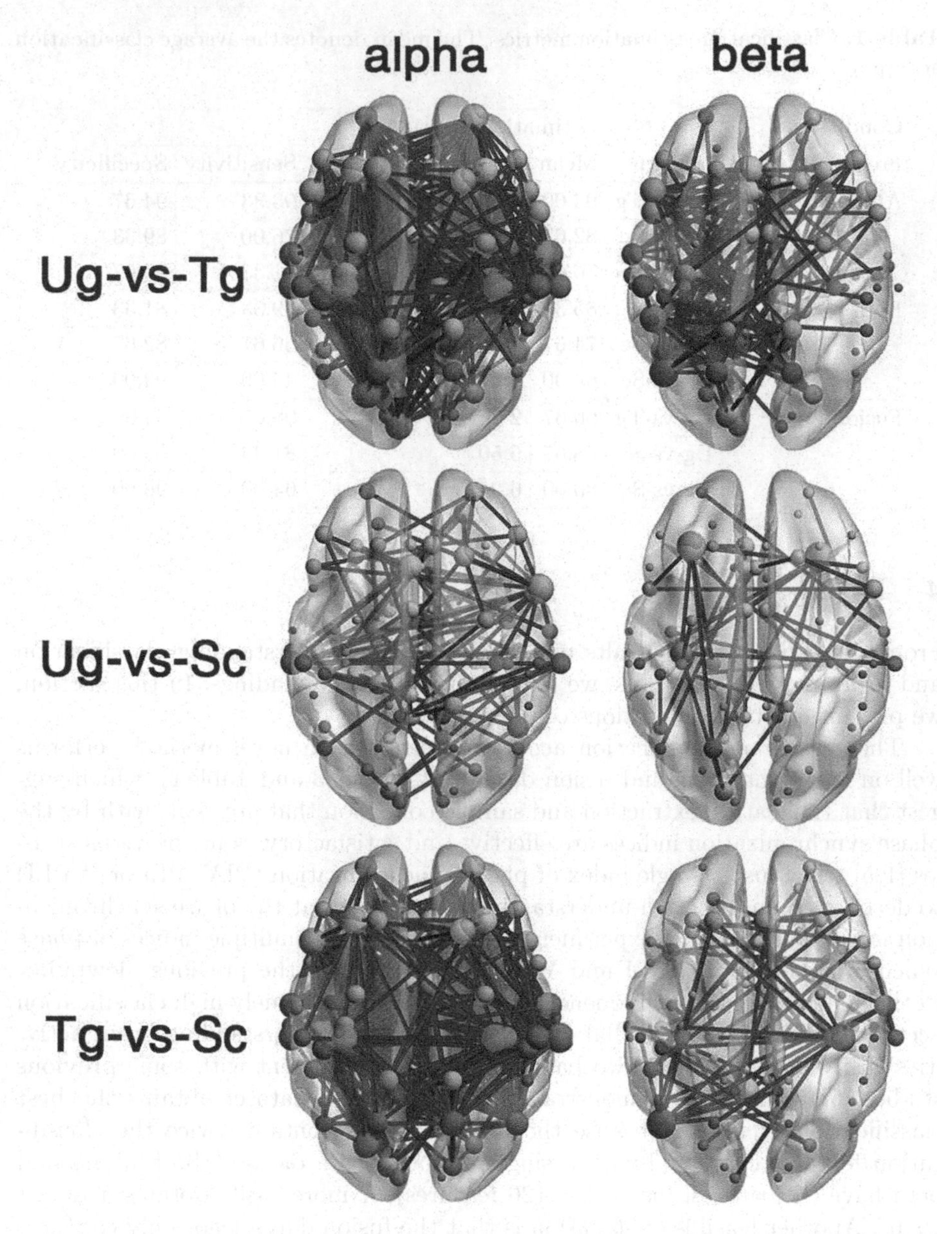

Fig. 4. Difference brain network. Each edge is determined by T-test and FDR correction ($p < 0.05$). The red, yellow, green, cyan, blue and purple-red nodes are from the temporal lobe, limbic lobe, frontal lobe, occipital lobe, sub-lobar and parietal lobe, respectively. The size of the node denotes the degree, the larger the size is, the higher the degree is. (Color figure online)

constructed by the mean of the 63 sub brain networks, hence, we conclude that the dynamical network effectively catches the difference information of action intention understanding brain activities in each microstate (50 ms). The alpha frequency band displays more significant edges than the beta frequency band, which suggests that the action intention understanding brain activities more easily occur at the alpha band [5,16]. The classification accuracies in the alpha dataset are better than the results in the beta dataset (see Fig. 3) also support the viewpoint that action intention understanding brain activities closely correlate with the alpha frequency band [5]. Some big size nodes appear at the temporal, partial, frontal and occipital lobes, which indicate that different action intention stimuli easily lead to different extents brain activities in these areas. These are consistent with some previous studies [6,16,17,19]. According to distribution of the big size nodes in both alpha and beta frequency bands, we conclude that human beings need the cooperation of multiple brain regions to complete the correct understanding of action intention. If one big size node were destroyed, the structure of brain network were destroyed, then the corresponding function will be changed. This is why some people with brain injuries can not understand other people's intention perfectively, e.g., can not understand others' language and gesture.

Although the novel method obtains very high classification accuracies, there are still some limitations in this study. For instance, the classifications are implemented on the group level (we view each subject from one stimulus condition, Ug, Tg or Sc, as a classification sample), while the number of raw subjects is only 25. Hence, it still needs to expand the sample size in the future. Additionally, the classification accuracy based on integral level, i.e., the classification accuracy is equal to the number of correctly classified samples divided by the total number of samples in the test set. For the prediction of a real sample with unknown stimulus condition, the integral level calculate strategy can not complete effectively. The reason is that the classification task is carried out under the 5-fold cross validation condition, which easily leads the sample labels to be confused. Of course, we have a satisfactory way to solve this problem. With the idea of Leave-One-Out (LOO) cross validation, we can use the real sample to be predicted as the test set and all other samples as the training set. Because the real sample is divided into three (they come from PLV, PLI and WPLI, respectively), hence, it needs to predict three sample labels. When the labels of the three indices are consistent, then we can get the predict class of the real sample.

5 Conclusion

In summary, the novel method improves the classification accuracies of action intention understanding based on EEG signals effectively. It has some merits of generality, e.g., it can be applied in other state-of-the-art neuro informatics data recording technologies, such as MRI, fMRI, MEG, NIRS and so on, meanwhile, it also can be introduced into other neuroscience study fields, such as emotion, motor imagery, Alzheimer's disease, epilepsy, autism, alcohol addiction, etc. When adopt the LOO cross validation strategy, the novel method will have some potential practical application values for the human-robot interaction. In the future, we will try our best to weaken the deficiencies in the novel method so as to obtain more satisfactory classification results and decode more complex neural mechanisms about the action intention understanding brain signals.

Acknowledgement. This work was supported in part by the National Nature Science Foundation of China under Grant 61773114, the Foundation of Hygiene and Health of Jiangsu Province under Grant H2018042, and the Key Research and Development Plan (Industry Foresight and Common Key Technology) of Jiangsu Province under Grant BE2017007-3.

References

1. Zhang, Z., Yang, Q., Leng, Y., Yang, Y., Ge, S.: Classification of intention understanding using EEG-NIRS bimodal system. In: 12th International Computer Conference on Wavelet Active Media Technology & Information Processing, Chengdu, pp. 67–70. IEEE (2015)
2. Brune, C., Woodward, A.L.: Social cognition and social responsiveness in 10-month-old infants. J. Cogn. Dev. **8**(2), 133–158 (2007)
3. Catmur, C.: Understanding intentions from actions: direct perception, inference, and the roles of mirror and mentalizing systems. Conscious. Cogn. **36**, 426–433 (2015)
4. Avanzini, P., Fabbridestro, M., Volta, R.D., et al.: The dynamics of sensorimotor cortical oscillations during the observation of hand movements: an EEG study. PLoS ONE **7**, e37534 (2012)
5. Zhang, L., Gan, J.Q., Zheng, W., Wang, H.: Spatiotemporal phase synchronization in adaptive reconfiguration from action observation network to mentalizing network for understanding other's action intention. Brain Topogr. **31**(3), 447–467 (2017). https://doi.org/10.1007/s10548-017-0614-7
6. Ge, S., Ding, M., Zhang, Z., et al.: Temporal-spatial features of intention understanding based on EEG-fNIRS bimodal measurement. IEEE Access **5**, 14245–14258 (2017)
7. Dindo, H., Lopresti, L., Lacascia, M., Chella, A., Dedić, R.: Hankelet-based action classification for motor intention recognition. Rob. Auton. Syst. **94**, 120–133 (2017)
8. Bundy, D.T., Pahwa, M., Szrama, N., et al.: Decoding three-dimensional reaching movements using electrocorticographic signals in humans. J. Neural Eng. **13**(2), 026021.1–026021.18 (2016)
9. Liu, H., Zheng, W., Sun, G., et al.: Action understanding based on a combination of one-versus-rest and one-versus-one multi-classification methods. In: 10th International Congress on Image & Signal Processing, Shanghai, pp. 1–5. IEEE (2017)
10. Niso, G., et al.: HERMES: towards an integrated toolbox to characterize functional and effective brain connectivity. Neuroinformatics **11**(4), 405–434 (2013). https://doi.org/10.1007/s12021-013-9186-1
11. Lachaux, J.P., Rodriguez, E., Martinerie, J., et al.: Measuring phase synchrony in brain signals. Hum. Brain Mapp. **8**, 194–208 (1999)
12. Stam, C.J., Nolte, G., Daffertshofer, A.: Phase lag index: assessment of functional connectivity from multichannel EEG and MEG with diminished bias from common sources. Hum. Brain Mapp. **28**(11), 1178–1193 (2007)

13. Vinck, M., Oostenveld, R., Wingerden, M.V., Battaglia, F., Pennartz, C.M.A.: An improved index of phase-synchronization for electrophysiological data in the presence of volume-conduction, noise and sample-size bias. Neuroimage **55**(4), 1548–1565 (2011)
14. Hari, R.: Action-perception connection and the cortical mu rhythm. Prog. Brain Res. **159**(1), 253–260 (2006)
15. Avanzini, P., Fabbri-Destro, M., Volta, R.D., et al.: The dynamics of sensorimotor cortical oscillations during the observation of hand movements: an EEG Study. PLoS ONE **7**(5), e37534 (2012)
16. Xiong, X., Yu, Z., Ma, T., et al.: Weighted brain network metrics for decoding action intention understanding based on EEG. Front. Hum. Neurosci. **14**, 232 (2020)
17. Ortigue, S., Sinigaglia, C., Rizzolatti, G., Grafton, S.T.: Understanding actions of others: the electrodynamics of the left and right hemispheres. A high-density EEG neuroimaging study. PLoS ONE **5**(8), e12160 (2010)
18. Arnaud, D., Scott, M.: EEGLAB: an open source toolbox for analysis of single-trial EEG dynamics including independent component analysis. J. Neurosci. Methods **134**(1), 9–21 (2004)
19. Fogassi, L., Ferrari, P.F., Gesierich, B., et al.: Parietal lobe: from action organization to intention understanding. Science **308**(5722), 662–667 (2005)

The Evaluation of Brain Age Prediction by Different Functional Brain Network Construction Methods

Hongfang Han[1], Xingliang Xiong[1], Jianfeng Yan[1], Haixian Wang[1](✉), and Mengting Wei[2](✉)

[1] Key Laboratory of Child Development and Learning Science of Ministry of Education, School of Biological Science and Medical Engineering, Southeast University, Nanjing 210096, Jiangsu, People's Republic of China
hxwang@seu.edu.cn

[2] Institute of Psychology, Chinese Academy of Sciences, Beijing 100101, People's Republic of China
nosanny@vip.163.com

Abstract. Brain functional network (BFN) analysis based on functional magnetic resonance imaging (fMRI) has proven to be a value method for revealing organization architectures in normal aging brains. However, a comprehensive comparison of different BFN methods for predicting brain age remains lacking. In this paper, we introduce a novel method to establish the BFN by using the Schatten-0 (S_0) and ℓ_0-regularized low rank sparse representation (S_0/ℓ_0 LSR) method. Moreover, the performance of different BFN methods in the brain age prediction with different feature extraction methods is evaluated. A support vector regression (SVR) is applied to the BFN data to predict brain age. Experimental results for resting state fMRI data sets show that compared with the Pearson correlation (PC), sparse representation (SR), low rank representation (LR), and low rank sparse representation (LSR) methods, the LSR method can achieve better modularity and predict brain age more accurately. The novel approach can enhance our understanding of the functional network of the aging brain.

Keywords: Brain functional network (BFN) · Brain age prediction · Schatten-0 (S_0) and ℓ_0-regularized low rank sparse representation (S_0/ℓ_0 LSR) · fMRI

1 Introduction

Brain functional network (BFN) analysis is a valuable approach for understanding the complex functional system of the brain and for studying brain development, cognitive ability, disease, and other phenomena [1]. The study of brain age prediction can provide theoretical analysis of changes in behavior or cognitive function caused by brain aging. Recent studies have revealed age-related changes

H. Yang et al. (Eds.): ICONIP 2020, LNCS 12534, pp. 122–134, 2020.
https://doi.org/10.1007/978-3-030-63836-8_11

in connectivity at the whole-brain level across the entire lifespan [2–4]. However, methods for predicting changes in the BFN of the normal aging brain suffer from the following limitations: 1) they ignore the accuracy of the constructed BFN and 2) they ignore the effects of different BFN methods on age prediction.

Accurately constructing the BFN is an important prerequisite for brain function prediction. According to graph theory, the BFN consists of nodes and edges, where nodes represent the brain regions and edges represent the associations between nodes in fMRI study. Pearson correlation (PC) [5] is commonly used for constructing the BFN. However, the PC method only focus on the pairwise associations between nodes and ignores the influences of other nodes. Some studies have found that both brain activity and brain functional connectivity have sparsity [6]. ℓ_1-*norm* is commonly employed [7] as a sparse solution for the sparse representation (SR) of fMRI data. However, the BFN commonly has more type of structures than sparsity [8], and low rank representation (LR) [9] can capture the modular structure of the data. Considering the sparsity and structures of the brain network, Qiao et al. [8] propose the sparse and low rank representation (LSR) method, which considers nuclear norm and as the measures of low rank and sparsity, respectively. However, the optimal solution of the nuclear and $\ell_1 - norm$ regularized objective just achieves an approximate solution [10], and does not provide exact measures of rank and sparsity. Brbic et al. [11] use the direct solution Schatten-0 (S_0) and ℓ_0-regularized low rank sparse representation (S_0/ℓ_0 LSR) method to solve the over-penalized problem of LSR, which has demonstrated good performance in speech recognition and handwriting recognition. Motivated by the highly correlated fMRI data of spatially adjacent brain regions, we introduce the S_0/ℓ_0 LSR to construct the BFN in this study.

There are many studies of brain age prediction [3,12,13]. However, few have explored the effects of different the BFN methods on age prediction. In this paper, we evaluate the performance of PC, SR, LR, LSR and S_0/ℓ_0 LSR on age prediction. The five methods are used to construct brain functional networks after processing the fMRI data. Then, the functional connection (FC) values and network metrics are extracted as features. Finally, a SVR is employed to evaluate the prediction performance of the five different BFN methods. Note that actual age is used in this study to optimize the parameters of the method. Thus, we make the implicit assumptions that 1) age-related changes in connectivity are similar and unimodal in a population; 2) all subjects in the cohort age at the same pace. The overall procedure of our study is shown in Fig. 1.

The main contributions of this paper are two-fold. First, we use a novel method to construct a sparse and modular structure and statistically robust brain functional network with the aim of accurately predicting brain age. Second, we evaluate the effects of five different BFN construction methods on age prediction in the case of two different feature extraction methods.

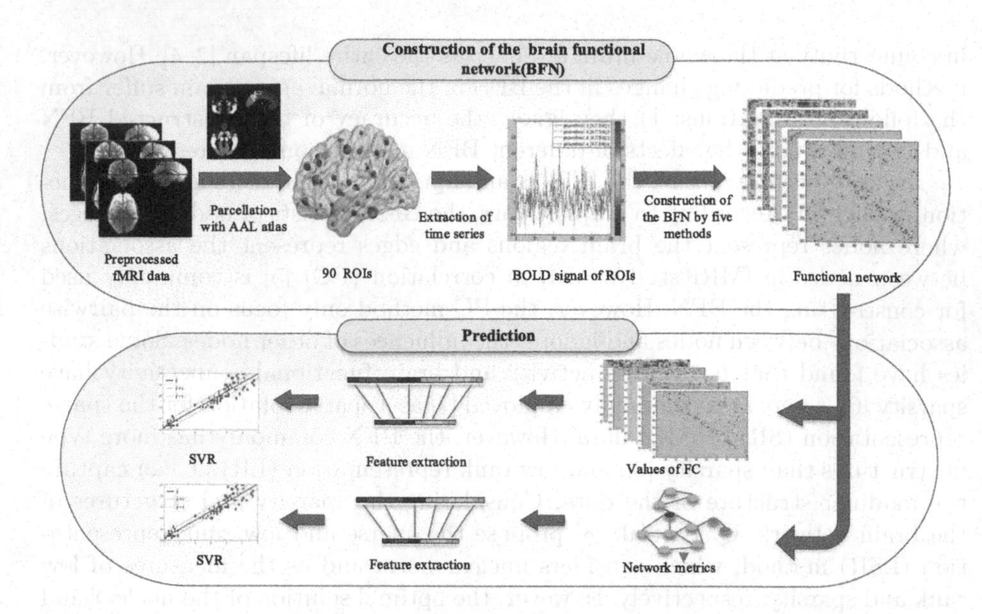

Fig. 1. A schematic of our method for evaluating age prediction.

2 Materials and Methods

2.1 Participants and Data Preprocessing

Data from ninety-two participants (age range: 13–85; 49 males and 43 females) are obtained from the Nathan Kline Institute and can be downloaded from http://www.fcon.com/_1000.projects.nitrc.org/indi/pro/nki.html [14]. The subjects underwent a scan session using a 3T Siemens Trio scanner. Resting state fMRI scans are collected using an echo-planar imaging (EPI) sequence with the following parameters: $time\,repetition(TR)/time\,echo(TE) = 2500/30\,\text{ms}$, $voxel\,size = 3.0*3.0*3.0\,\text{mm}^3$, 38 interleaved slices. Each scan session comprises 260 functional volumes. Inside the scanner, subjects are instructed to keep their eyes closed and not move.

The fMRI data ware preprocessed by the Data Processing Assistant for Resting-State fMRI (DPARSF) [15] in MATLAB R2013b. The first ten images of each subject ware discarded. The remaining 250 images ware processed as follows: 1) slice timing and realignment ware conducted; 2) The six motion parameters, whole brain, cerebrospinal fluid (CSF) and white matter (WM) signals as covariates ware regressed out by using the Friston-24 parameter model; 3) spatial normalization to standard MNI space was conducted by DARTEL procedure; and 4) images ware smoothed with a full width half maximum (FWHM = 4 mm) Gaussian kernel and temporal bandpass filtering(0.01–0.08 Hz). Then, we used the Automated Anatomical Labeling (AAL) template of 90 regions of interest (ROIs) as nodes of the brain network, and the time series of 90 nodes of each subject ware obtained.

2.2 Construction of the BFN

- Pearson correlation (PC)

$$W_{ij}^{(pc)} = \frac{\sum_{t=1}^{n}(i_t - \bar{i})^T(j_t - \bar{j})}{\sqrt{\sum_{t=1}^{n}(i_t - \bar{i})^2 \sum_{t=1}^{n}(j_t - \bar{j})^2}}, \quad (1)$$

where $W_{ij}^{(pc)}$ represents the correlation coefficient of the time series between node i and j, and i_t and j_t denote the fMRI signal value between node i and j at time point t, respectively. n is the total number of time points, and $\bar{i}$ and $\bar{j}$ are the average time series of the nodes.

- Low rank sparse representation (LSR)
Let $x = (x_1, ..., x_n) \in \mathbb{R}^{d \times n}$ be the normalized fMRI time matrix of a subject, which has n nodes with d time points. For the time series of a node, the time series of all the other nodes $X = (x_1, ..., x_{i-1}, x_{i+1}, ..., x_n) \in \mathbb{R}^{d \times (n-1)}$ are used as a dictionary for representation, which contains all samples except x_i itself, with a coding coefficient w_i. LSR solves the following problem:

$$\min_{w} \lambda_1 \|W\|_* + \lambda_2 \|W\|_1, s.t. X - XW = 0, \quad (2)$$

where λ_1 and λ_2 are the rank and sparsity regularization parameters, $W = (w_1, w_2, ..., w_n) \in \mathbb{R}^{n \times n}$ is the coding coefficient matrix, nuclear norm $\|W\|_*$ denotes the low rank constraint on matrix, ℓ_1-*norm* $\|W\|_1$ in W is for sparsity constraint. The combination of the two constraints $\|W\|_1$ the BFN can produce module structure. For data with noise, problem (2) can be extended to the objective function as follows:

$$\min_{w} \frac{1}{2} \|X - XW\|_F^2 + \lambda_1 \|W\|_* + \lambda_2 \|W\|_1, \quad (3)$$

where $\|X - XW\|_F^2$ indicates the data-fitting term. Note that when $\lambda_1 = 0$, Eq. (3) is the problem of low rank representation (LR); when $\lambda_2 = 0$, Eq. (3) reduces to the sparse representation (SR). The optimization problem of the objective function can be solved by a proximal method [16]. Once the W representation matrix is obtained, we could replace W with $W^{LSR} = (W + W^T)/2$ to produce the affinity matrix. The elements of the matrix W^{LSR} represents the FC values between node i and node j.

- Schatten-0 (S_0) and ℓ_0-regularized low rank sparse representation (S_0/ℓ_0 LSR)
In this study, we introduce Schatten-0 (S_0) and ℓ_0-regularized low rank sparse representation (S_0/ℓ_0 LSR) [11] to construct the BFN, in which S_0 and ℓ_0 quasi-norm regularizations are used for the low rank and sparse constraint. The objective function of the S_0/ℓ_0 LSR method can be formulated as follows:

$$\min_{w} \frac{1}{2} \|X - XW\|_F^2 + \lambda_1 \|W\|_{S_0} + \lambda_2 \|W\|_0, \quad (4)$$

where $\|W\|_{S_0}$ is the Schatten-0 quasi-norm of W defined as $\|W\|_{S_0} = \|diag(\Sigma)\|_0$, and $W = U\Sigma V^T$ is the singular value decomposition of matrix

W for the low rank constraint. diag(Σ) represents the vector of diagonal elements of matrix Σ. $\|W\|_0$ is the ℓ_0 quasi-norm of W, which represents the number of nonzero elements in W for the sparsity constraint.

The objective function (4) can be optimized via the ADMM method [17]. For the proximal maps of low rank and sparsity regularizations, the proximal average method [18] is used to average the solutions. And the hard thresholding function is used as a proximity operator of the penalties. The iteration of the hard thresholding operator [19,20] on the singular values and coefficients solve the problems of the low rank and sparsity constraints. The definition entry-wise of the proximity operator $H : \mathbb{R} \rightarrow \mathbb{R}$ of $\|x\|_0$ is shown as follows [19]:

$$H(y; \lambda) = \arg\min_{x \in \mathbb{R}} \left\{ \frac{1}{2}(y - x)^2 + \lambda \|x\|_0 \right\}. \tag{5}$$

The hard thresholding function [18] $H : \mathbb{R} \rightarrow \mathbb{R}$ of $\|x\|_0$ defined in (6) is used to solve the closed-form problem of (5) at $y \in \mathbb{R}$.

$$H(x; \lambda) = \begin{cases} x, & \text{if } |x| > \sqrt{2\lambda} \\ \{0, x\}, & \text{if } |x| = \sqrt{2\lambda} \\ 0, & \text{if } |x| = \sqrt{2\lambda} \end{cases} \tag{6}$$

The proximity operator of $\|W\|_{S_0}$ is obtained by the hard thresholding function applied entry-wise to Σ [20]. Finally, the S_0/ℓ_0 LSR is constructed by symmetric association matrix with $W^{s_0/l_0 LSR} = (W + W^T)/2$.

2.3 Evaluation of Brain Age Prediction

The basic procedure of age prediction is illustrated Fig. 1, and includes three main steps. The leave-one-out-cross-validation (LOOCV) method is used to estimate the prediction accuracy in the three steps. First, the features are extracted by two common methods. In the first method, the FC values of the constructed BFN are directly used as the features. In this study, there are 90 nodes in the network, thus the feature dimensionality is 4005. In the second method, the effective network metrics are extracted from the weighted matrices as features. For second method, we obtain eight network metrics from five aspects: functional integration(FI), functional segregation(FS), modularity index(MI), nodal centrality(NC), and network resilience(NR), which can be implemented by the brain connectivity toolbox [21]. The formulas and definitions are presented in Table 1. The feature vectors based on network metrics features will have a size of 453 for each subject. Second, the feature selection aims to remove irrelevant or redundant features and retain discriminative features, which can lead to better prediction performance. Here, the Pearson correlation is employed to select useful features. The correlation between the feature and the ages is computed. The larger Pearson correlation score indicates the more useful discriminative feature. Finally, the SVR [22] is employed to evaluate the prediction performance of the

Table 1. Features based on network metrics.

Network metrics		Formula	Definitions
FI	Global efficiency	$E_{\text{gol}}(G) = \frac{1}{N(N-1)} \sum_{i \neq j \in N} \frac{1}{d_{ij}}$	Where d_{ij} is the shortest path length from node i to j in network G with G nodes
FS	Local efficiency	$E_{loc}(G) = \frac{1}{N} \sum_{i \in G} E_{glob}(G_i)$	Where $E_{loc}(G)$ is the local efficiency of Where $E_{\text{gol}}(G_i)$ is the subgraph of the neighbors of the node i
	Clustering coefficient	$C_i = \frac{2e_i}{k_i(k_i-1)}$	Where e_i is the number of triangles around the node i, and k_i is the degree of node i
MI	Modularity	$M = \frac{1}{L} \sum_{ij} [A_{ij} - \frac{k_i k_j}{L}] \delta(\delta_i, \delta_j)$	Where A_{ij} is the elements of the matrix, L is the total number of edges, δ_i and δ_j are the labels of the group
	Participation coefficient	$P_i = 1 - \sum_{m \in M} (\frac{k_{im}}{k_i})^2$	Where M is the set of modules, and k_{im} is the number of edges of node i and the other nodes in module m
NC	Betweenness	$B_i = \sum_{s \neq t \neq i \in N} \frac{\delta_{st}(i)}{\delta_{st}}$	Where $\delta_{st}(i)$ is the number of shortest paths between node s and t that pass through node i, and denotes the number of the shortest paths between node s and t
	Degree centrality	$k_i = \sum_{j \in N} l_{ij}$	Where l_{ij} is the connection status between the node i and j
NR	Average neighbor degree	$k_{nn,i} = \frac{\sum_{j \in N} l_{ij} k_j}{k_i}$	Average neighbor degree reflects the network vulnerability to insult

five BFN methods. The kernel function and tolerance of SVR are set as linear kernel and 0.1, respectively. The SVR is implemented using the LIBSVM toolbox [23]. The whole process is iterated 1000 times with the permutation test method to avoid random prediction.

2.4 Evaluation Metrics

The evaluation metrics focus on two aspects. The Newman's spectral algorithm [24] is employed to calculate the modularity scores of the BFN. The formula and definitions are listed in the 'modularity' row of Tables 1. The higher the modularity, the higher the structural quality of the network. Two metrics are used to evaluate the effectiveness of the prediction: 1) the accuracy R is the correlation between the true age and predicted age and 2) the mean square error (MSE) can be calculated as follows:

$$MSE = (|A_1 - P_1|^2 + |A_2 - P_2|^2 + ... + |A_n - P_n|^2)/n \tag{7}$$

where A_n and P_n are the true and predicted values, respectively, for each fold in the LOOCV. A high R values and low MSE indicate good prediction performance.

3 Results and Discussion

The experimental results and discussion are presented in three sections. The five BFNs and their modularity are presented in Sect. 3.1. The performance in age prediction of the five BFN methods is described in Sect. 3.2. The consensus features for SVR of the S_0/ℓ_0 LSR methods by FC features are presented in Sect. 3.3.

3.1 The Brain Functional Network (BFN)

We use the PC, LR, SR, LSR and S_0/ℓ_0 LSR to construct the BFN, and the regularized parameters are selected for the five methods. For the SR, LR and LSR, the regularized parameter λ is selected in the range $[2^{-5}, 2^{-4}, ..., 2^0, ..., 2^4, 2^5]$. For the S_0/ℓ_0 LSR method, λ is optimized in the range $[0.1, 0.9]$ with step 0.1. The regularization parameter λ is empirically set to 2^3 for LR, 2^0 for SR, $\lambda_1 = 2^{-1}$ and $\lambda_2 = 2^{-3}$ for LSR, and $\lambda_1 = 0.5$ and $\lambda_1 = 0.5$ for LSR. For fair comparison, PC method is set a series of thresholding parameters [100%, 10%] with step 10%, we also empirically preserve 30% of the strong edge weights.

Figure 2 shows the FC matrices of a randomly selected subject constructed by the PC, LR, SR, LSR and S_0/ℓ_0 LSR methods. It is apparent from the inferred FC matrices that SR, LSR and S_0/ℓ_0 LSR can automatically remove some weak connections and achieve better sparsity than PC and LR. In other words, SR, LSR and S_0/ℓ_0 LSR lead to a more sparsely BFN in contrast to the PC and LR. Compared with the PC and SR, the LR, LSR and S_0/ℓ_0 LSR methods can obtain a modular structure in the BFN. Both the LSR and S_0/ℓ_0 LSR methods lead to sparse and modular structure network, and the S_0/ℓ_0 LSR method can more clearly capture modular structures.

The modularity scores of the five BFNs with different thresholds are calculated. First, the networks are processed to remove weak connections based on different thresholds, which vary from 0 to 0.9 with a step size of 0.1. The larger the threshold values are, the more edges are preserved. Second, the absolute FC values are obtained since negative FC values are invalid for Newman's algorithm. Finally, the average modularity scores for all subjects for the five methods are obtained. The average modularity scores with different thresholds are shown in Fig. 3.

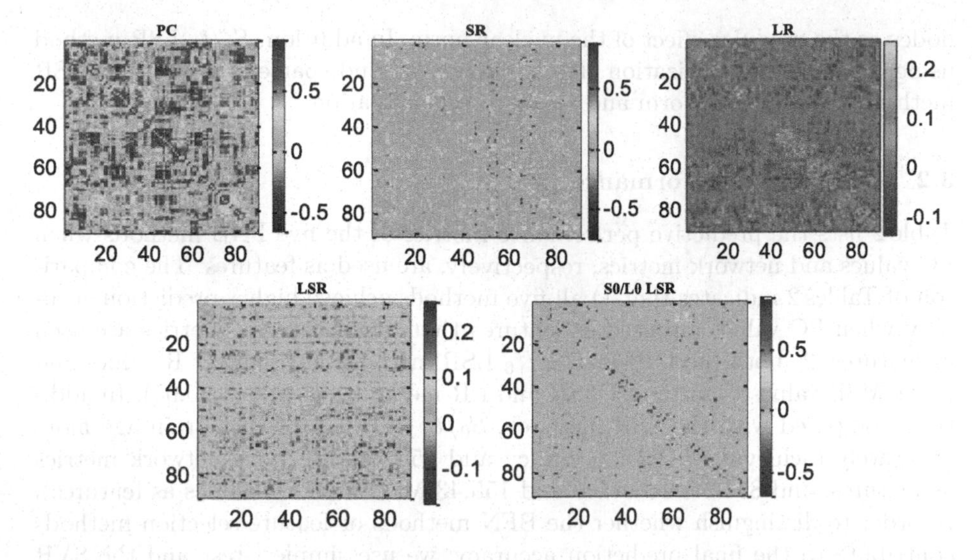

Fig. 2. The FC matrices of the same subject estimated by five different methods.

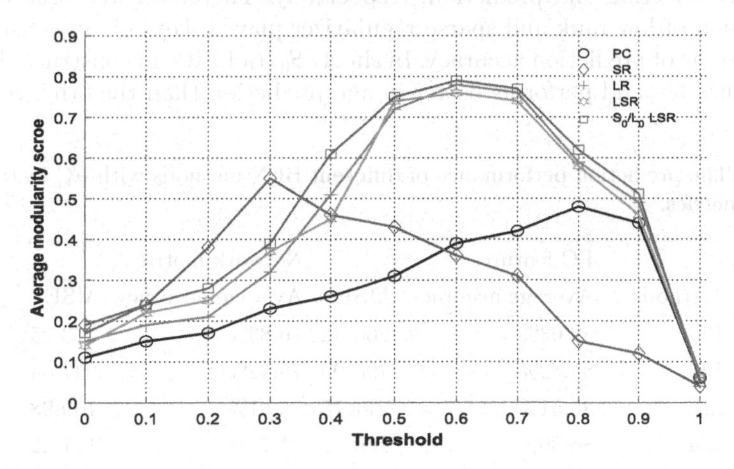

Fig. 3. Average modularity scores of networks constructed by five BFN methods.

From Fig. 3, it can be observed that the LR, LSR and S_0/ℓ_0 LSR methods obtain higher average modularity scores and larger areas under the curve than PC and SR methods. In particular, S_0/ℓ_0 LSR has a peak value 0.803 at a threshold value of 0.6 and the largest area under the curve, compared with the other methods. The PC and SR methods yield lower average modularity scores than the other methods. The high modularity score can be partly due to the modular effect of LR, LSR and S_0/ℓ_0 LSR methods, which can select functionally correlated nodes altogether by using the nuclear norm. In theory, the low rank and sparse regularizer help LSR stand out from the PC, LR and SR, owing to it obtain a sparse solution as the and select the highly correlated nodes as the modular effect of the nuclear norm. In addition, S_0/ℓ_0 LSR method using S_0 and ℓ_0 regularization closer to low rank and sparse problem than LSR method using nuclear norm and ℓ_1-*norm* regularization.

3.2 Prediction Performance

Table 2 lists the predictive performance metrics of the five FNB methods when FC values and network metrics, respectively, are used as features. The comparison of Tables 2 indicates that 1) all five methods achieve higher prediction accuracy when FC values are used as features than when network metrics are used as features; 2) both the LSR and S_0/ℓ_0 LSR methods have higher R values and lower MSE values than the PC, SR and LR methods (P-value < 0.05). In addition, compared with the LSR methods, S_0/ℓ_0 LSR can predict brain age more accurately (achieving 88.62% accuracy and 155.43 MSE with network metrics as features, and 85.32% accuracy and 155.43 MSE with FC values as features). In order to distinguish whether the BFN methods or feature selection methods contribute to the final prediction accuracy, we use simple t-test and the SVR for feature selection and prediction, respectively. Therefore, we argue that the combination of low rank and sparse regularizer plays a key role in constructing BFN in terms of prediction accuracy. In short, S_0/ℓ_0 LSR can construct the BFN more accurately and performs better in age prediction than the other methods.

Table 2. The prediction performance of different BFN methods with FC features and network metrics.

	FC features		Network metrics	
Methods	Average accuracy	MSE	Average accuracy	MSE
PC	82.63%	260.42	76.83%	273.25
LR	85.82%	198.93	78.42%	247.54
SR	85.51%	211.78	79.81%	199.98
LSR	86.30%	160.42	81.78%	171.22
S_0/ℓ_0 LSR	88.62%	135.15	85.32%	155.43

3.3 Consensus Features and Discriminative Brain Regions

When the FC values are adopted as features in the SVR, those that survive in each fold throughout the cross-validation process are called the consensus features and the corresponding nodes are called the discriminative brain regions. The consensus features retained in each fold by the five BFN construction methods are obtained. Since the S_0/ℓ_0 LSR method achieves the best prediction performance, we show the consensus features and discriminative brain regions of only the S_0/ℓ_0 LSR method in Fig. 4 and Fig. 5.

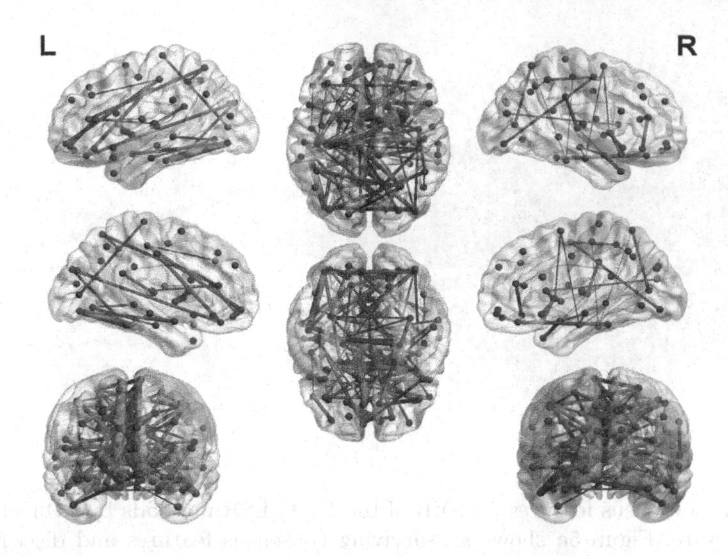

Fig. 4. The representation of all surviving consensus features for the SVR of the S_0/ℓ_0 LSR method by FC features using BrainNet Viewer software.

Figure 4 and Fig. 5a reveal the consensus features and discriminative brain regions, including 108 negative edges and 110 positive edges. In Fig. 4, the weights of the consensus features are proportional to the functional connection thickness. In Fig. 5, the positive relationships between nodes are represented by the red lines, the negative relationship between nodes are represented by the blue lines. We then select the top 20 consensus features, representing 10 negative edges, 10 positive edges and 35 brain regions, as shown in Fig. 5b shows the top brain regions which contribute to age prediction include the left inferior temporal gyrus, left superior temporal gyrus, right superior parietal gyrus, occipital superior gyrus, left superior frontal right superior frontal medial postcentral, right postcentral, right superior orbital frontal, and right middle frontal. Some discriminative brain regions are located in the default mode network (left posterior cingulate gyrus, right posterior cingulate gyrus, precuneus); and salience network (insula, anterior cingulate and paracingulate gyri). The results are consistent with some previous studies [3]. In addition, other brain regions with high discrimination such as the left superior temporal gyrus and right superior frontal medial, are detected in our study, which were also reported in [25]. These consensus features and discriminative brain regions may provide insight into the mechanisms of brain aging.

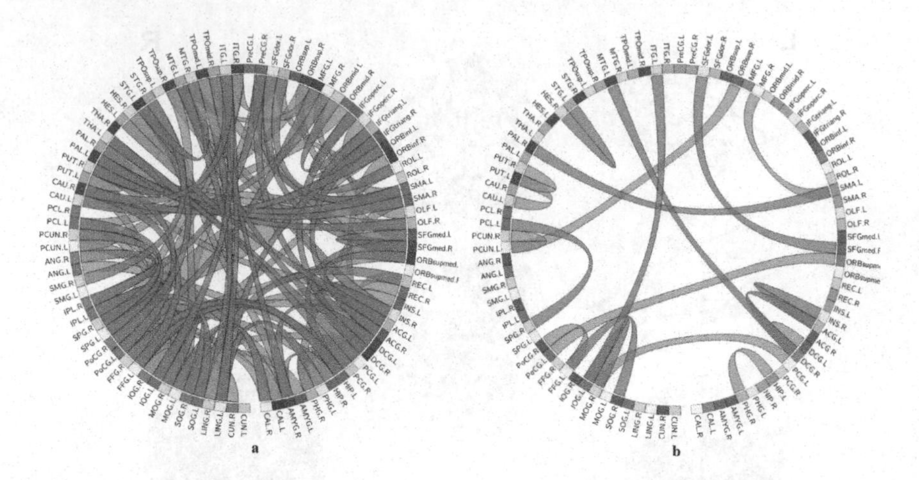

Fig. 5. The consensus features for SVR of the S_0/ℓ_0 LSR methods by FC features via Circos software. Figure 5a shows all surviving consensus features and discriminative brain regions, Fig. 5b shows the top 20 consensus features and discriminative brain regions. (Color figure online)

4 Conclusion and Limitation

In this study, we introduce a novel method to construct the BFN by using the S_0/ℓ_0 LSR, and evaluate the performance in age prediction by five different BFN construction methods. Compared with PC, SR, LR, and LSR, S_0/ℓ_0 LSR achieves better modularity and performs the best in age prediction with fMRI data sets, when FC values and network metrics are extracted as features. The issue of how to select the parameters automatically for the BFN methods deserves further exploration. Our introduced S_0/ℓ_0 LSR method is a promising method for constructing the brain functional network and performs well in age prediction.

Acknowledgments. This work was supported in part by the National Nature Science Foundation of China under Grant 61773114 and the Key Research and Development Plan (Industry Foresight and Common Key Technology) of Jiangsu Province under Grant BE2017007-3.

References

1. Sporns, O., Chialvo, D.R., Kaiser, M., et al.: Organization, development and function of complex brain networks. Trends Cogn. Sci. **8**(9), 418–425 (2004)
2. Muetzel, R.L., Blanken, L.M.E., Thijssen, S.F., et al.: Resting-state networks in 6-to-10 year old children. Hum. Brain Mapp. **37**(12), 4286–4300 (2016)
3. Vij, S.G., Nomi, J.S., Dajani, D.R., et al.: Evolution of spatial and temporal features of functional brain networks across the lifespan. Neuroimage **173**(2018), 498–508 (2018)

4. Sole-Padulles, C., Castro-Fornieles, J., de la Serna, E., et al.: Intrinsic connectivity networks from childhood to late adolescence: effects of age and sex. Cogn. Neurosci. **17**, 35–44 (2016)
5. Li, K., Guo, L., Li, G., et al.: Cortical surface based identification of brain networks using high spatial resolution resting state fMRI data. In: 2010 IEEE International Symposium on Biomedical Imaging: From Nano to Macro, pp. 656–659. IEEE (2010)
6. Lee, K., Tak, S., Ye, J.C.: A data-driven sparse GLM for fMRI analysis using sparse dictionary learning with MDL criterion. IEEE Trans. Med. Imaging **30**(5), 1076–1089 (2011)
7. Li, X., Hu, Z., Wang, H.: Overlapping community structure detection of brain functional network using non-negative matrix factorization. In: Hirose, A., Ozawa, S., Doya, K., Ikeda, K., Lee, M., Liu, D. (eds.) ICONIP 2016. LNCS, vol. 9949, pp. 140–147. Springer, Cham (2016). https://doi.org/10.1007/978-3-319-46675-0_16
8. Qiao, L., Zhang, H., Kim, M., et al.: Estimating functional brain networks by incorporating a modularity prior. Neuroimage **141**, 399–407 (2016)
9. Liu, G., Lin, Z., Yan, S., et al.: Robust recovery of subspace structures by low rank representation. Pattern Anal. Mach. Intell. **35**(1), 171–184 (2013)
10. Donoho, D.L., Elad, M.: Optimally sparse representation in general (nonorthogonal) dictionaries via ℓ_0 minimization. Proc. Nat. Acad. Sci. **100**(5), 2197–2202 (2003)
11. Brbic, M., Kopriva, I.: ℓ_0-motivated low rank sparse subspace clustering. IEEE Trans. Cybern. **50**(4), 1711–1725 (2020)
12. Mwangi, B., Hasan, K.M., Soares, J.C.: Prediction of individual subject's age across the human lifespan using diffusion tensor imaging: a machine learning approach. Neuroimage **75**(2013), 58–67 (2013)
13. Zhai, J., Li, K.: Predicting brain age based on spatial and temporal features of human brain functional networks. Front. Hum. Neurosci. **13**(2019), 62 (2019)
14. Nooner, K.B., Colcombe, S., Tobe, R., et al.: The NKI-Rockland sample: a model for accelerating the pace of discovery science in psychiatry. Front. Neurosci. **6**, 152 (2012)
15. Yan, C., Zang, Y.: DPARSF: a MATLAB toolbox for "pipeline" data analysis of resting-state fMRI. Front. Neurosci. **4**, 13 (2010)
16. Combettes, P.L., Pesquet, J.C.: Proximal splitting methods in signal processing. In: Bauschke, H., Burachik, R., Combettes, P., Elser, V., Luke, D., Wolkowicz, H. (eds.) Fixed-Point Algorithms for Inverse Problems in Science and Engineering. SOIA, vol. 49, pp. 185–212. Springer, New York (2011). https://doi.org/10.1007/978-1-4419-9569-8_10
17. Boyd, S., Parikh, N., Chu, E., et al.: Distributed optimization and statistical learning via the alternating direction method of multipliers. Trends Mach. Learn. **3**(1), 1–122 (2011)
18. Yu, Y.L.: Better approximation and faster algorithm using the proximal average. In: Advances in Neural Information Processing Systems, pp. 458–466 (2013)
19. Blumensath, T., Davies, M.E.: Iterative thresholding for sparse approximations. J. Fourier Anal. Appl. **14**(5–6), 629–654 (2008). https://doi.org/10.1007/s00041-008-9035-z
20. Liang, J., Fadili, J., Peyré, G.: A multi-step inertial forward-backward splitting method for non-convex optimization. In: Advances in Neural Information Processing Systems, vol. 2, no. 5, pp. 99–110 (2016)
21. Rubinov, M., Sporns, O.: Complex network measures of brain connectivity: uses and interpretations. Neuroimage **52**(3), 1059–1069 (2010)

22. Drucker, H., Burges, C.J.C., Kaufman, L., et al.: Support vector regression machines. In: Advances in Neural Information Processing Systems, vol. 9, pp. 155–161 (1997)
23. Chang, C.C., Lin, C.J.: LIBSVM: a library for support vector machines. ACM Trans. Intell. Syst. Technol. **2**(3), 1–27 (2011)
24. Newman, M.E.J.: Modularity and community structure in networks. Proc. Natl. Acad. Sci. **103**(23), 8577–8582 (2006)
25. Vergun, S., Deshpande, A.S., et al.: Characterizing functional connectivity differences in aging adults using machine learning on resting state fMRI data. Front. Comput. Neurosci. **7**, 38 (2013)

Transfer Dataset in Image Segmentation Use Case

Anna Wróblewska[1,2](✉), Sylwia Sysko-Romańczuk[1,2], and Karol Prusinowski[1]

[1] Warsaw University of Technology, Warsaw, Poland
a.wroblewska@mini.pw.edu.pl,
{anna.wroblewska1,sylwia.sysko.romanczuk}@pw.edu.pl
[2] Synerise, Warsaw, Poland

Abstract. The most labour-intensive stage of machine learning (ML) modelling is the appropriate preparation of correct dataset. This paper aims to show transfer dataset approach in image segmentation use case to lower labour intensity. Moreover, we test the effectiveness of this approach by training deep learning models on our prepared dataset. The models achieved high-performance metrics, even on very hard test data.

Keywords: Datasets · Transfer dataset · Data preprocessing · Data augmentation · Image segmentation · Convolutional neural networks · Transfer learning.

1 Introduction

In an online marketplace, product images play a crucial role in capturing user attention and motivating them to make a purchase [1–3]. User needs fulfilment and their satisfaction by buying are e-commerce's main goals. Exploring and comparing the full range of products impact user satisfaction. Hence, precise and exact product presentation is critical in communication with potential buyers.

What is an online marketplace business concept? It is about connecting supply and demand and bringing economic advantage, at least for one side of the market. Different sellers offer a wide range of products. Their images are usually of a wide range of quality and with various background, from professional studios to home-made photos with a mobile phone. To place clear and meaningful product photos in search listings and offer descriptions, one must remove background with noisy additional information [1]. This approach requires the introduction of segmentation techniques.

In this segmentation task, the most challenging issues are additional logotypes, texts, natural background, and overall lousy quality, i.e. image resolution.

The work was supported by the EU co-funded Smart Growth Operational Programme 2014–2020 (project no. POIR.01.01.01-00-0695/19) and the dataset was provided by Allegro, Warsaw, Poland.

H. Yang et al. (Eds.): ICONIP 2020, LNCS 12534, pp. 135–146, 2020.
https://doi.org/10.1007/978-3-030-63836-8_12

They often appear in product images because sellers frequently want to create their recognition and capture user attention in the marketplace by making their images full of branding texts and coloured frames following their brand design [3].

High-quality segmentation techniques are useful for image retrieval tasks and preparing good recommendations based on clear images or even sheer product segments. Segmentation is crucial especially in visual product categories like fashion or jewellery, in which it is challenging to describe patterns and styles in text attributes. On the other hand, imprecise segmentation from the background can mislead product visual representation for finding similarities between offers.

This paper aims to show transfer dataset approach in image segmentation use case. We had access to a massive dataset of noisy offer images gathered from Allegro, the most prominent online marketplace in Eastern Europe. The primary annotation task for the dataset was to classify three product images' defects, i.e. noisy background consisting of promotional texts, logotypes, and additional coloured frames. These classification annotations were much less laborious to gather than outlining products in each image for training segmentation ML models.

We will show an approach on how to transfer the dataset from image classification tasks to the segmentation one. After preparing a suitable dataset, we benchmarked the most popular neural network architectures. The results approved the effectiveness of our transfer data approach, having high performance even on a very hard test dataset. Finally, we achieved very accurate models – in terms of segmentation metrics – thus lowering labour intensity of preparing dataset.

We define a new term and approach as the effect of this study: transfer dataset (TD), adequate and complementary to transfer learning (TL). Hence, transfer dataset is a research approach in machine learning that focuses on dataset enrichment for one problem to a different but related task (see a definition of transfer learning in [12] and Wikipedia). Transfer dataset means a data enrichment that comprises data augmentation and task change; in our case from classification to segmentation.

TD aims to lower the labour intensity of ML dataset annotation and preprocessing. Another more obvious, yet crucial goal of TD is to improve models' accuracy. In this case, the goals are the same as the goals of TL; however, the approach is complementary, because we change data, not model parameters. Noteworthy, currently many studies focus on dataset influence on models, e.g. data drift problem. The current research questions are: how to catch the data drift and observation outliers?; when gathered dataset is sufficient to the task (see [11])?. The primary and general goal of our research is to minimise the workload and make the process more automatic. The transfer dataset approach shares these objectives.
Our main contributions in this work are as follows:

- We define transfer dataset term and research approach.
- We demonstrate how to utilise this approach on an image segmentation task.

- We experiment with several techniques of data augmentation to transfer dataset.
- We show how to change a task from classification to segmentation in the marketplace product segmentation task (segmentation is an arduous and time-consuming task for humans, thus our approach can significantly lower human workload).
- We test a few state-of-the-art neural network architectures to segment products and achieved high performance on test datasets (even on the very hard one).
- We experimentally demonstrate the neural models' abilities to surpass results of the previously popular non-neural GrabCut algorithm in terms of performance metrics and response time.

In the following section, we will describe our transfer dataset use case (Sect. 2). Then we experiment with a region of interest segmentation models (Sect. 3). In the last sections, we discuss and conclude the paper and show directions for future research.

2 Transfer Dataset: Our Approach

Our approach to transfer a dataset from an image classification tasks to the segmentation one is a process with the following steps (see also Fig. 1):

1. Acquiring a preliminary classification dataset with defined classes of product image noise and the following classes/labels indicating annotated defects: 'dark background', 'frames', 'additional marketing text'.
2. Filtering clear images without defected labels from the preliminary dataset.
3. Extracting foreground of proper product objects with a standard technique GrabCut to have ground truth product masks.
4. Generating larger sets (train and test sets) using the product masks and different backgrounds with data augmentation techniques (e.g. translation, scaling, blurring the background, adding artificial frames and texts) to build a segmentation dataset.
5. Collecting hard cases (with all the noisy labels in classification) and preparing product segment masks manually to have a final hard test ready and audit our approach.

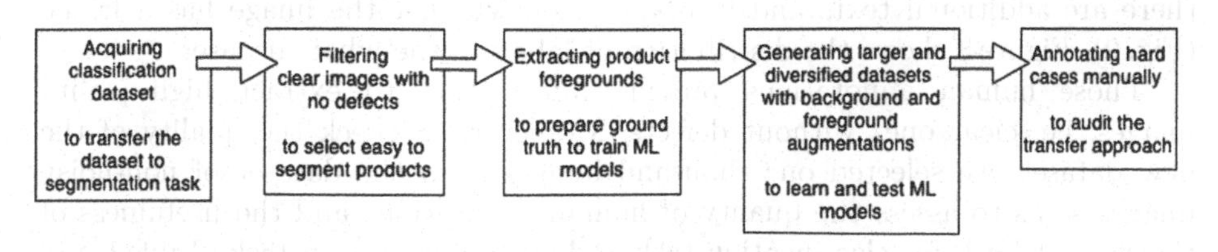

Fig. 1. Our process for transferring dataset from classification to image segmentation task

We need to prepare a broad set of images to learn neural networks and test the product segmentation reliably. This section describes the initial dataset and the work done to create regions of interest (ROIs). The ROIs are depicted as binary product masks on an image matrix.

We describe data augmentation methods we used to generate various representative images to enlarge our collection with product images and product segments (ROIs). These segments are product masks that designate pixels containing the offered product at each image in a set.

2.1 Our Preliminary Dataset

We collected our preliminary dataset from offer images used for product visualisation at the online marketplace Allegro from a vast range of product categories, e.g. mobile phones and other electronic accessories, fashion, products for home and garden, antics. This set comprises thousands of images of very different quality and with additional objects besides the product, e.g. natural background or coloured logotypes and other peripheral accessories (see Fig. 2). Sizes of images vary to a large extent: from tiny images of 64 × 48 pixels up to very large photos of 2560 × 2560 pixels. Shapes also differ and come in square or rectangular image matrices. This is a set of complex and diverse data, much more complicated than open sets available on the Internet, e.g. ImageNet, CIFAR-10, and other sets listed in [3].

Fig. 2. Examples of e-commerce offer images from the preliminary dataset

We supported each image in the dataset with labels that designate previously defined defects, i.e. whether an image has a dark/noisy background, whether there are additional texts and logotypes, and whether the image has a frame (Fig. 2). Figure 3 shows the distribution of labels in the whole dataset.

Those human annotations provide information to extract high-quality images, i.e. clear ones without defects. Moreover, to check the quality of the new dataset, we selected one thousand images from the dataset of non-noisy images so as to assess the quality of human annotations and the usefulness of the dataset both for classification task and our segmentation task. Table 1 and Fig. 4 reveal a few problems and their occurrences in the final annotations. We can see that only about 65% of images are without defects.

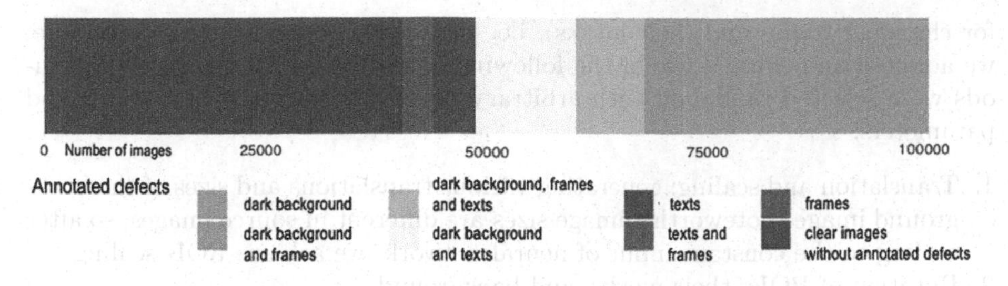

Fig. 3. Distribution of labels in our preliminary dataset

Table 1. Error distribution in labelling estimated on one thousand clear images assessed as having no defects (without background, text, and frames).

	Good images	Images with shadows	Images with subtitles	Images with background
The number of images	647	174	131	48

Fig. 4. Wrong human annotations: images annotated as without any defects but having shadows in the background, marketing texts, etc.

2.2 The Change of Task and Data Augmentation

At first, we filtered clear images from the preliminary dataset based on information about annotated defects. Then we used the GrabCut technique to segment proper product objects [3,4]. The parameters of the algorithms are several iterations: default is 5, bounding box adjustment way. We adaptively chose the bounding box (BB) for the product in each image after binary image thresholding and morphological closing. Then, we set BB rectangle as parallel to the image frame and enlarged by 2% in each direction.

This way, we achieved binary product masks, as in masks one states for pixels within the ROI and zero for background. Finally, we prepared one thousand images with product masks and an additional test set with 120 images with ROIs, not used during the model training phase.

In the next step, we generated more massive sets with images with product masks using data augmentation techniques for noisy background generation and for changing ROIs (and their masks), i.e. their sizes, rotations, etc. Each time, we adapted and chose a few of the following techniques, while particular methods were selected randomly with arbitrary probabilities that were our method parameters:

1. Translation and scaling: generating various translations and sizes of the background image (noteworthy, image sizes are different in source images, so after scaling to the constant input of neural network, we achieve ROIs scaling).
2. Rotation of ROIs, their masks, and background.
3. Blurring the product in the final image, e.g. Gaussian filtering, which affects the same ROI mask but with different contrast with the background.
4. Adding artificial background: blending monochromatic colours, gradient colours, real images used as backgrounds, e.g. carpet images, bricks, grass (see Fig. 5).
5. Adding promotional texts, e.g. 'gratis', 'special offer' in random places around the primary product mask (and also with random letter sizes and fonts).
6. Adding frames using the techniques and patterns meant for the background.

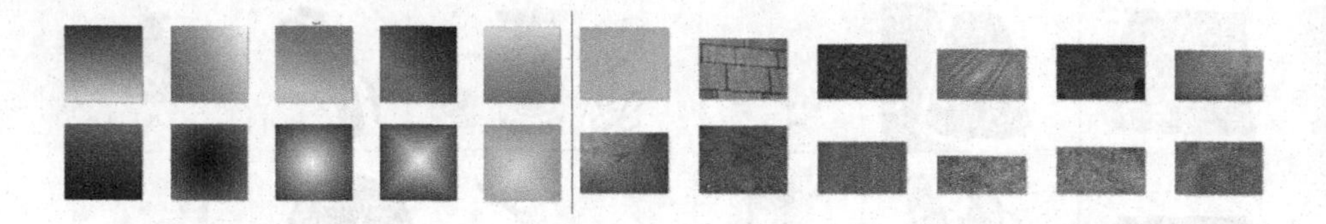

Fig. 5. Sample backgrounds: stable and gradient colour and real images imitating natural scenes

We used the techniques to add a random number of augmented clear and defected images (with the mask from original images). Figure 6 presents an example of the data augmentation steps in our approach.

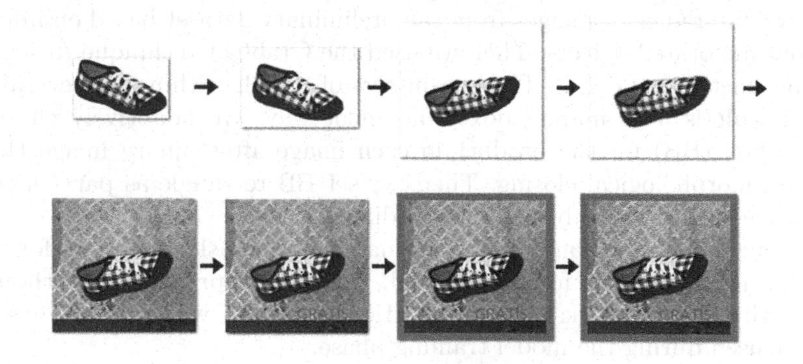

Fig. 6. An example of data augmentation in our settings

The defected images are images with noisy background, marketing texts and frames. Table 2 listed our final train and test sets. This approach allows for transmitting a variety of examples to the neural networks to train them on how to segment products in noisy and clear images.

Table 2. Train and test sets for our segmentation task.

Dataset name	The number of images	Characteristics
Train-1	1959	A smaller train dataset made from 206 source original images (with ROI masks done with manually adjusted GrabCut to each image) adding from four to 14 new augmented images to each original image (original meaning taken from the preliminary dataset with no augmentation)
Test-1	86	A smaller validation set (Image are separated between Train-1 and Test-1 sets.)
Train-2	41794	A bigger train set made from 1000 source original images (with ROIs masks done with GrabCut with constant parameters)
Test-2	1505	A bigger validation set (see Fig. 7)
Test-AD	184	A distinct test set of average difficulty
Hard-Test	15	A distinct test set with very hard cases (Fig. 8). Each ROI was segmented from the background manually because GrabCut technique failed on these images

Fig. 7. Sample images from Test-2 dataset

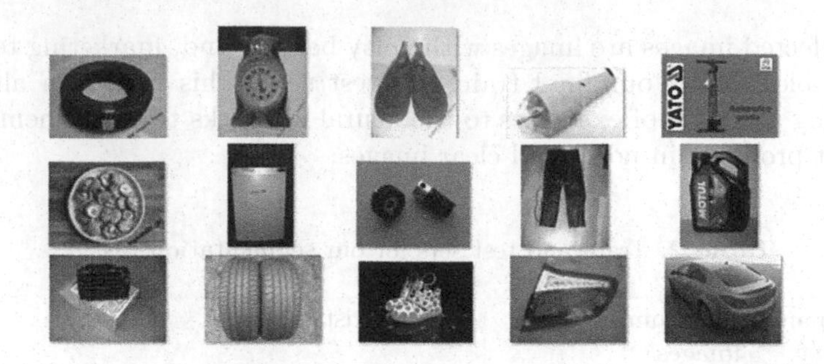

Fig. 8. Sample images from a collection of difficult images (Hard-Test)

3 Experiments: Settings and Results

3.1 Initial Test of Many Networks

Our goal was to create the best possible network taken from established research for our segmentation task. We tested several networks (see Table 3) based on VGG-16 and VGG-19 architectures [5] (as backbones), having 16 and 19 weight layers, respectively. The final architectures' input and output consisted of 448 × 448 colour images. We scaled our images to this size. Table 3 describes architectures for our segmentation task in this experiment. Each network ended with a pixel classification map as an output layer (binary mask of a product object).

We chose intersection over union (IoU) as the primary metric. IoU measures the overlap between two regions of model output and ground truth [9,10]. IoU is built using pixel set operations in proportion of an intersection to a union on segments of output binary masks (P) and ground truth gathered from prepared image masks (G), expressed in the following relationship:

$$IoU = area(G \cap P)/area(G \cup P) \tag{1}$$

One model prediction (an output region) is considered to be True Positive (TP) if $IoU > 0.5$ and False Positive (FP) if $IoU < 0.5$. We use accuracy based on the above assumptions on TPs and FPs. Moreover, we measure IoU to better visualize intermediate results while tuning networks.

The results of a few given architectures (listed in Table 3) are gathered in Table 4. Neural networks are much better than GrabCut. Results of VGG SegNet are slightly better than FCN-8, while SegNet was trained significantly faster, so in the following test, we concentrate only on adjusting VGG SegNet.

Table 3. Architectures of tested neural networks.

Network name	Description & Parameters
UNet	Four parts for encoding and decoding, each comprising three conv layers and then max-pooling or up-sampling, respectively for encoding and decoding, each encoder's part additionally copies their activations to respective parts of decoder (the network architectures resembles U-shape) as in [7]
VGG UNet	Four first parts of encoding and decoding fragments are from VGG, each part contains at least two (to three) convolutional layers of VGG with max-pooling of 2 × 2 as in [5]; the other UNet parts are the same as in [7]
FCN	FCN architecture as in [8], with added five conv layers of VGG as a first part of the network
SegNet-Basic	Four decoding and four encoding layers; first two parts with two conv layers each and final two big layers with three conv layers each; each encoding part separated with max-pooling (mask 2 × 2); each decoding part separated with up-sampling; furthermore, each convolutional layer is used with batch normalisation and ReLU activation such as in [6]
VGG SegNet	Encoding with VGG and decoding with SegNet-Basic
VGG FullSegNet	Encoding with the first four layers of VGG and full decoding part of SegNet as in [6]

Table 4. Preliminary results of networks based on VGG-16 architecture on Train-1 and Test-1 datasets.

Model	Train Epochs	Train-1 Accuracy	Test-1 Accuracy	Test-1 IoU
VGG SegNet	**9**	**98.39**	**97.85**	**83.83**
FCN-8	8	97.73	96.85	80.06
FCN-32	8	96.27	95.52	70.66
VGG UNet	5	93.58	94.33	69.15
GrabCut	–	–	–	25.95

3.2 More Precise Tests of the Best Architecture

Further tests for VGG SegNet – which performed the best in the first tests – were conducted on a larger and more diverse collection of images, i.e. Train-2 and Test-2. Figure 9 shows an example of how segmentation looks for subsequent epochs during training. We gathered the results of parameter tuning in Table 5. Final results and benchmark comparison with GrabCut technique was done on datasets Test-AD and Test-Hard, available in Table 6 and Table 7, respectively.

Table 5. VGG SegNet tuning on Train-2 and Test-2.

Model	Epochs	Test-2 IoU
VGG-16 SegNet (optimizer – adadelta: lr = 1.0, rho = 0.95, decay = 0.0; batch = 2)	418	92.21
VGG-19 SegNet (optimizer – adadelta: lr = 1.0, rho = 0.95, decay = 0.0; batch = 2)	441	92.56
VGG-19 SegNet (optimizer – adamax: lr = 0.002, beta_1 = 0.9, beta_2 = 0.999, decay=0.0; batch = 2)	73	93.63
VGG-19 SegNet (optimizer – adamax: lr = 0.002, beta_1 = 0.9, beta_2 = 0.999, decay = 0.0; batch = 20 + diversification of images within one batch)	100	93.02
VGG FullSegnet (optimizer – adamax: lr = 0.002, beta_1 = 0.9, beta_2 = 0.999, decay = 0.0; batch = 2)	300	*96.01*

Table 6. Final results on Test-AD.

Method	IoU	min IoU	Accuracy	Response time (sek)
Trivial method (all image is ROI)	18.78	1.06	18.78	0.000
GrabCut (high threshold = 240)	64.09	0.00	76.57	2.8714
GrabCut (low threshold = 25)	62.15	0.00	88.70	1.5083
VGG-19 FullSegNet	95.70	63.57	99.35	0.0487
Smoothed SegNet output	95.72	59.10	99.36	0.0675

Table 7. Final results on Test-Hard.

Method	IoU	min IoU	Accuracy	Response time (sek)
Trivial method (all image is ROI)	19.12	1.06	19.12	0.000
GrabCut (threshold high = 240)	63.73	0.00	76.30	2.8823
GrabCut (threshold lot = 25)	61.93	0.00	88.39	1.5342
VGG-19 FullSegNet	95.39	34.64	99.22	0.0487
Smoothed SegNet results	95.41	34.64	99.23	0.0676

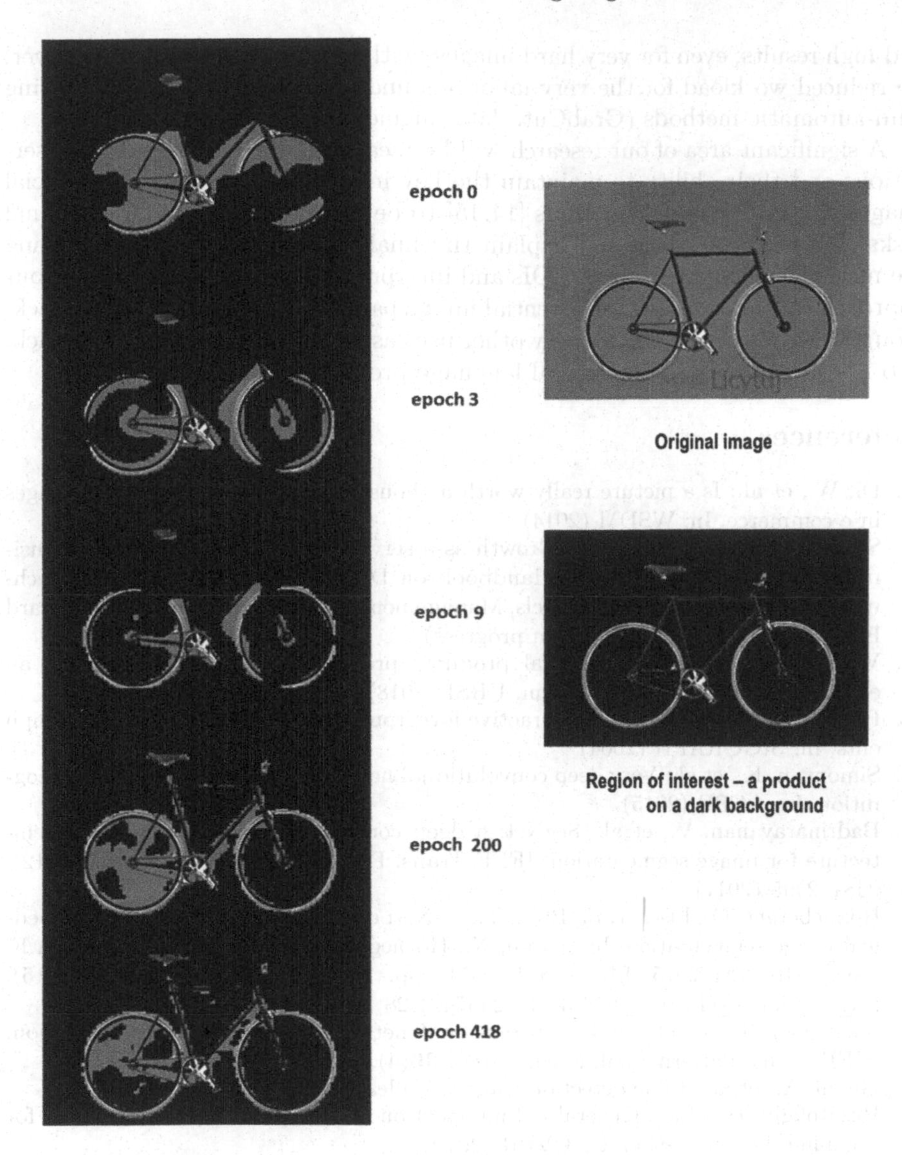

Fig. 9. Sample image, ground truth: ROI and network output during training

4 Conclusions

We defined and presented our transfer dataset approach which is complementary to transfer learning. Transfer dataset in machine learning modelling focuses on a dataset enrichment for one problem to a different but related task. This enrichment comprises data augmentation and task change.

We showed that the approach achieves its goals. In our use case – transferring dataset from image classification to image segmentation task – we achieved stable and high results, even for very hard images, with a low response time. Moreover, we reduced workload for the very laborious image segmentation labelling using semi-automatic methods (GrabCut, data augmentation).

A significant area of our research will be measuring the importance of observations and their ability to maintain the key information to choose the crucial images [13] and their salient ROIs [14,15] to be transferred from source to final tasks. We will also check and explain the final task results and models using the methods for finding salient ROIs and interpretability techniques to see if our approach concentrated on the essential image parts, not on the augmented backgrounds. Moreover, we will also try other use cases for transfer dataset approach, also in other domains, e.g. natural language processing and time series.

References

1. Di, W., et al.: Is a picture really worth a thousand words? On the role of images in e-commerce. In: WSDM (2014)
2. Sysko-Romańczuk, S., et al.: Growth-as-a-service in marketplace powered businesses. In: Baumann, S. (ed.) Handbook on Digital Business Ecosystems: Technologies, Markets, Business Models, Management, and Societal Challenges. Edward Elgar Publishing, UK (2021). (in progress)
3. Wróblewska, A., et al.: Optimal products presentation in offer images for an e-commerce marketplace platform. URSI (2018)
4. Rother, C., et al.: GrabCut: interactive foreground extraction using iterated graph cuts. In: SIGGRAPH (2004)
5. Simonyan, K., et al.: Very deep convolutional networks for large-scale image recognition. In: ACPR (2015)
6. Badrinarayanan, V., et al.: SegNet: a deep convolutional encoder-decoder architecture for image segmentation. IEEE Trans. Pattern Anal. Mach. Intell. **39**(12), 2481–2495 (2017)
7. Ronneberger, O., Fischer, P., Brox, T.: U-Net: convolutional networks for biomedical image segmentation. In: Navab, N., Hornegger, J., Wells, W.M., Frangi, A.F. (eds.) MICCAI 2015. LNCS, vol. 9351, pp. 234–241. Springer, Cham (2015). https://doi.org/10.1007/978-3-319-24574-4_28
8. Shelhamer, E., et al.: Fully convolutional networks for semantic segmentation. IEEE Trans. Pattern Anal. Mach. Intell. **39**(4), 640–651 (2016)
9. Gilani, A., et al.: Table detection using deep learning. In: ICDAR (2017)
10. Rezatofighi, H., et al.: Generalized intersection over union: a metric and a loss for bounding box regression. In: CVPR (2019)
11. Hohman, F., et al.: Understanding and visualizing data iteration in machine learning. In: CHI (2020)
12. Yang, Q., et al.: Transfer Learning. Cambridge University Press (2020)
13. Zabalza, J., et al.: Novel folded-PCA for improved feature extraction and data reduction with hyperspectral imaging and SAR in remote sensing. ISPRS J. Photogrammetry Remote Sens. **93**, 112–122 (2014)
14. Zabalza, J., et al.: Novel segmented stacked autoencoder for effective dimensionality reduction and feature extraction in hyperspectral imaging. Neurocomputing **185**, 1–10 (2016)
15. Yan, Y., et al.: Unsupervised image saliency detection with Gestalt-laws guided optimization and visual attention based refinement. Pattern Recogn. **79**, 65–78 (2018)

Neural Network Models

A Gaussian Process-Based Incremental Neural Network for Online Regression

Xiaoyu Wang[1(✉)], Lucian Gheorghe[2], and Jun-ichi Imura[1]

[1] Graduate School of Engineering, Tokyo Institute of Technology, Tokyo, Japan
xiaoyuwanganddl@gmail.com, imura@sc.e.titech.ac.jp

[2] Nissan Research Center, Nissan Motor Co., Ltd., Kanagawa, Japan
lucian@mail.nissan.co.jp

Abstract. This paper proposes a Gaussian process-based incremental neural network algorithm to handle the online regression problem. It can extract prototypes by an incremental neural network, where 1) Gaussian process approximations are adopted to update the threshold regions and the posterior distribution of the dependent variable at the weight vectors of nodes and 2) the optimal bandwidth matrix is derived for adapting to network structure. Besides, we discuss some properties of the proposed approach, and the experimental results show that our approach achieves remarkable accuracy improvement in extracting prototypes for online regression on noisy data.

Keywords: Incremental neural network · Gaussian process · Online regression

1 Introduction

Classical parametric regression has the problem that if the assumed form is misspecified, the accuracy decreases severely. Thus, it is undesirable to learn data without giving much prior knowledge. In contrast, nonparametric regression that attempts to directly generate the regression model has become increasingly important, among which regressogram and kernel regression are the two most popular approaches [1]. Regressogram assumes that the data points falling into the same block of a partition of the input space have the same weight and takes the average value as the prediction. As a modification, kernel regression gives higher weights to the nearby data points of a query point x_s as $\hat{f}(x_s) = \sum_{i=1}^{n} \frac{K_h(x_i - x_s) f_{x_i}}{\sum_{i=1}^{n} K_h(x_i - x_s)}$, where f_{x_i} represents the value of regression function f at x_i, k is a kernel function and h is the kernel bandwidth parameter. However, directly taking training sample points as the kernel components only works in batch mode. Hence, it is significantly necessary to propose an online learning approach that can extract prototypes automatically.

As an inherently online learning and topology-preserving approach, self-organizing map (SOM) generally adopts a 1-D or 2-D lattice to represent the

H. Yang et al. (Eds.): ICONIP 2020, LNCS 12534, pp. 149–161, 2020.
https://doi.org/10.1007/978-3-030-63836-8_13

input space of samples. However, the fixed number of nodes and the discontinuity associated with the boundaries in SOM [2] limit its performance on data sets of unknown complicated distributions. Recently, based on growing neural gas theory, several self-organizing incremental neural network (SOINN) algorithms [3–5] have been introduced. These algorithms calculate a threshold region for each node, the network structure can be adaptive to input data and due to the local concept drift property [6], the intermediate fit model can be incorporated well into the incremental neural network learning process. Various experimental results in [4,5] show that SOINN achieves performance comparable to the current state-of-the-art algorithms.

However, the threshold region in ESOINN [3], a sphere with the maximum Euclidean distance of a node to its neighbors as the radius, does not consider carefully the information of the underlying local distributions and ignores the knowledge that data belonging to the same subclass lie on a much lower nonlinear dimensional manifold [7,8]; the Mahalanobis distance employed in KDESOINN [5] generates threshold regions of extremely high fractional anisotropy [9,10] and results in many inappropriate edges.

On the other hand, Gaussian process (GP) [11,12], for its modeling flexibility and robustness to overfitting, has become a standard approach to solve many machine learning tasks. As further extensions, several Gaussian process-based incremental algorithms have been proposed for online density estimation [9,10]. However, as far as we know, an incremental regression model that combines the advantages of incremental neural network and GP has yet to be explored.

This paper proposes a novel Gaussian process-based incremental neural network for online regression that can incrementally update the weight vectors in input space and the network structure in map space. It employs sparse GP regression approximations [13,14] to calculate threshold regions and the posterior distribution of the dependent variable at nodes (prototypes).

The remainder of this paper is organized as follows. We show the general learning framework of SOINN in Sect. 2.1, and then derive the optimal bandwidth matrix, GP posterior approximation and threshold region determination in Sect. 2.2, 2.3, and 2.4, respectively. Section 2.5 discusses further its property and Sect. 2.6 shows the algorithm. At last, Sect. 3 presents some experimental results to demonstrate the performance of our approach.

2 Proposed Method

In this part, we explain comprehensively a Gaussian process-based incremental neural network for online regression and analyze some of its properties.

2.1 Overview of SOINN

Algorithm 1 describes the general learning process of SOINN, and Table 1 gives the definitions of the involved variables, parameters, and symbols.

When a new sample s_i is observed, first, based on the employed distance measure D (Euclidean distance in ESOINN [3]; Mahalanobis distance in KDESOINN [5]), line 2 searches the nearest 2 nodes $w_{1,2}$ (called the 1st and 2nd winning nodes of s_i, respectively), and line 3–7 calculate their similarity threshold $r_{w_{1(2)}}$. Then, depending on whether $D(s_i, w_1) > r_{w_1} \parallel D(s_i, w_2) > r_{w_2}$, line 8–16 update the network. It is noteworthy that since the initially formed edges may become inappropriate as the input of samples, one effective method is to delete the initially formed edges when their ages reach an upper bound age_{max} at line 15. At last, when the number of input samples comes to an integer multiple of λ, outlier nodes defined with degree 0 are deleted at line 18.

Algorithm 1: SOINN

Initialization: $\mathcal{V} \leftarrow \{s_1, s_2\}, \mathcal{E} \leftarrow \emptyset$

1 **while** $\sim isempty(s_i)$ **do**
2 $\quad w_1 \leftarrow \arg\min_{n_i \in \mathcal{V}} D(s_i, n_i), w_2 \leftarrow \arg\min_{n_i \in \mathcal{V} \setminus w_1} D(s_i, n_i);$
3 $\quad$ **if** $\sim isempty(\mathcal{N}_{w_{1(2)}})$ **then**
4 $\quad\quad r_{w_{1(2)}} = \max_{n_j \in \mathcal{N}_{w_{1(2)}}} D(n_j, w_{1(2)});$
5 $\quad$ **else**
6 $\quad\quad r_{w_{1(2)}} = \min_{n_j \in \mathcal{V} \setminus \mathcal{N}_{w_{1(2)}}} D(n_j, w_{1(2)});$
7 $\quad$ **end**
8 $\quad$ **if** $D(s_i, w_1) > r_{w_1} \parallel D(s_i, w_2) > r_{w_2}$ **then**
9 $\quad\quad \mathcal{V} \leftarrow \mathcal{V} \cup \{s_i\};$
10 $\quad$ **else**
11 $\quad\quad \mathcal{E} \leftarrow \mathcal{E} \cup \{(w_1, w_2)\};$
12 $\quad\quad age(w_1, w_2) = 0, age(w_1, n_j) = age(w_1, n_j) + 1 \quad \% \, n_j \in \mathcal{N}_{w_1};$
13 $\quad\quad W_{w_1} \leftarrow W_{w_1} + 1, w_1 = w_1 + \frac{1}{W_{w_1}+1}(s_i - w_1);$
14 $\quad\quad n_j = n_j + g(W_{w_1})(s_i - n_j);$
15 $\quad\quad \mathcal{E} \leftarrow \mathcal{E} \setminus \{e \mid age(e) > age_{max}\}, \mathcal{V} \leftarrow \mathcal{V} \setminus \{n_i \mid |\mathcal{N}_{n_i}| = 0\};$
16 $\quad$ **end**
17 $\quad$ **if** $mod(i, \lambda) = 0$ **then**
18 $\quad\quad \mathcal{V} \leftarrow \mathcal{V} \setminus \{s \mid |\mathcal{N}_s| = 0\};$
19 $\quad$ **end**
20 **end**

2.2 Gaussian Process and Bandwidth Matrix Optimization

Gaussian process (GP) [11,12] can represent the underlying function rigorously in a nonparametric form, where the values of the dependent variable $\mathbf{f} = \{f_{x_1}, f_{x_2}, ..., f_{x_n}\}$ at n values of independent variable $\mathbf{x} = \{x_1, x_2, ..., x_n\} (x_i \in \mathbb{R}^d)$ generally follow a zero-mean $n-$ dimensional Gaussian distribution with a

Table 1. Definitions of Variables, Parameters, and Symbols

$\mathcal{V}(\mathcal{E})$	The set of nodes (edges)
$\mathcal{N}_{n_i}$	The set of the nodes connected with node n_i $\mathcal{N}_{n_i} := \{n_{1,i}, ..., n_{\lvert\mathcal{N}_{n_i}\rvert,i}\}$, $\lvert\mathcal{N}_{n_i}\rvert$ is its cardinality
r_{n_i}	The similarity threshold of node n_i
$D(x_1, x_2)$	The distance between x_1 and x_2
T_{n_i}	The threshold region of node n_i, $T_{n_i} := \{x \mid D(x, n_i) \leq r_{n_i}\}$
W_{n_i}	The number of times n_i is the 1st winner in competitive learning, that is the winning times of n_i
W	The set of the winning times of all nodes $\{W_{n_i}\}$
$age(n_i, n_j)$	The age of the edge linking n_i and n_j
age_{max}	A predetermined upper bound of age for deleting initially formed edges
$g(W_{n_j})$	The coefficient for updating the weight vectors of the neighbors of n_j, generally $g(W_{n_j}) := \frac{1}{100W_{n_j}}$
λ	The number of inputs in one learning period

kernel matrix as its covariance. So, given $\mathbf{x}$ and $\mathbf{f}$, the prediction at a query point x_0 (denoted as $\hat{f}_{x_0}$) is subject to a Gaussian distribution

$$\hat{f}_{x_0}|\mathbf{x}, \mathbf{f} \sim \mathcal{N}(K_{x_0,\mathbf{x}} K_{\mathbf{x}}^{-1}\mathbf{f}, K_* - K_{x_0,\mathbf{x}} K_{\mathbf{x}}^{-1} K_{x_0,\mathbf{x}}^T), \tag{1}$$

where k is a kernel function: $\mathbb{R}^d \times \mathbb{R}^d \rightarrow \mathbb{R}$, $K_* = k(x_0, x_0)$, $[K_{x_0,\mathbf{x}}]_p = k(x_0, x_p)$ and $[K_{\mathbf{x}}]_{p,q} = k(x_p, x_q)$.

Since using a fixed bandwidth parameter leads to a sub-optimal rate of convergence [15], we adopt an adaptive kernel matrix M_{n_i}, that is, $\forall n_j, n_k \in \mathcal{N}_{n_i}$,

$$k_i(n_j, n_k) = \sigma_1^2 \exp\{-\frac{1}{2}(n_j - n_k)^T M_{n_i}^{-1}(n_j - n_k)\} + \sigma_2^2 \delta(j, k) \tag{2}$$

where $\delta(j, k)$ is the Kronecker delta function. And different from [9], we consider the distribution of dependent values in calculating the optimal M_{n_i}, and for the reason that directly minimizing $\sum_{n_j \in \mathcal{N}} W_{n_j}(f_{n_j} - \hat{f}_{n_j})^2$ can cause overfitting, the leave-one-out method is employed here to achieve the optimal M_{n_i} by minimizing the least squares cross-validation error as follows

$$M_{n_p} = \arg\max_{M_{n_p}} \sum_{n_j \in \mathcal{V}} W_{n_j}(f_{n_j} - \mathbb{E}(\hat{f}_{n_j}|\mathcal{V} \setminus n_j, M \setminus M_{n_j}))^2 \tag{3}$$

$\Rightarrow$

$$\frac{\partial(\sum_{n_j\in\mathcal{V}} W_{n_j}(f_{n_j}-\mathbb{E}(\hat{f}_{n_j}|\mathcal{V}\setminus n_j, M\setminus M_{n_j}))^2}{\partial M_{n_p}^{-1}}$$

$$=\frac{1}{2}\sum_{j\neq p} W_{n_p}W_{n_j}\frac{M_{n_p}k_p(n_j,n_p)|M_{n_p}|^{-\frac{1}{2}}}{\sum_{t\neq j} M_{n_t}k_t(n_j,n_t)|M_{n_t}|^{-\frac{1}{2}}}\left[f_{n_j}-\frac{\sum_{t\neq j} f_{n_t}W_{n_t}k_t(n_j,n_t)}{\sum_{t\neq j} W_{n_t}k_t(n_j,n_t)}\right] \tag{4}$$

$$\left[f_{n_p}-\frac{\sum_{t\neq j} f_{n_t}W_{n_t}k_t(n_j,n_t)}{\sum_{t\neq j} W_{n_t}k_t(n_j,n_t)}\right](M_{n_p}-(n_j-n_p)(n_j-n_p)^T).$$

Consider the underlying lower dimensional manifold and the idea of the manifold Parzen density estimator [7], the optimal kernel component at n_p is approximately aligned with the plane locally tangent to this underlying manifold and the information about this tangent plane can be gathered from the neighbors of n_p. So, from Eq. (4) and Bayes theorem, we have

$$M_{n_p}\approx\frac{\sum_{n_j\in\mathcal{N}_{n_p}} W_{n_j}M_{p,j}\left(\frac{f_{n_p}+f_{n_j}}{2}-\frac{\sum_{n_t\in\mathcal{N}_{n_j}} W_{n_t}f_{n_t}}{\sum_{n_t\in\mathcal{N}_{n_j}} W_{n_t}}\right)^2}{\sum_{n_j\in\mathcal{N}_{n_p}} W_{n_j}\left(\frac{f_{n_p}+f_{n_j}}{2}-\frac{\sum_{n_t\in\mathcal{N}_{n_j}} W_{n_t}f_{n_t}}{\sum_{n_t\in\mathcal{N}_{n_j}} W_{n_t}}\right)^2}+\rho\mathbf{I}_d \tag{5}$$

where $M_{p,j}=(n_p-n_j)(n_p-n_j)^T$ and a small isotropic (spherical) Gaussian noise of variance in all directions, $\rho\mathbf{I}_d$, is added to guarantee that M_{n_p} is positive definite.

2.3 Posterior Approximation

GP is employed to achieve the closed form of the posterior approximation, for a new sample point s_m, assume that the previous posterior at $\mathcal{N}_{w_i}^0$ (the neighbors of the winning node w_i $(i=1,2)$ before updating the network) is subject to a Gaussian distribution $q_0(f_{\mathcal{N}_{w_i}^0})=\mathcal{N}(f_{\mathcal{N}_{w_i}^0}|m_{w_{i0}},v_{w_{i0}})$, the posterior approximation at $\mathcal{N}_{w_i}^1$ (the neighbors of w_i after updating the network), $q_1(f_{\mathcal{N}_{w_i}^1})$, can be derived from Bayesian variational inference [13,14] as the following steps:

1). Calculate the Kullback-Leibler divergence from $P(f|\{s_{1\leq i\leq m}\})$ to $q_1(f)$ (denoted as $\mathrm{KL}[q_1(f)\|P(f|\{s_{1\leq i\leq m}\})]$) and its derivative w.r.t $q_1(f_{\mathcal{N}_{w_i}^1})$, we have

$$\frac{\partial\mathrm{KL}[q_1(f)\|P(f|\{s_{1\leq i\leq m}\})]}{\partial q_1(f_{\mathcal{N}_{w_i}^1})}=\frac{\partial(\int q_1(f)\log\frac{P(f_{\mathcal{N}_{w_i}^0})q_1(f_{\mathcal{N}_{w_i}^1})}{P(f_{\mathcal{N}_{w_i}^1})q_0(f_{\mathcal{N}_{w_i}^0})\mathcal{N}(s_m|f,\sigma_2^2\mathbf{I})}df)}{\partial q_1(f_{\mathcal{N}_{w_i}^1})} \tag{6}$$

2). Set the above derivative to 0, one can obtain

$$q_1(f_{\mathcal{N}^1_{w_i}}) = \mathcal{N}\Bigg(f_{\mathcal{N}^1_{w_i}} \Bigg| \begin{bmatrix} K_{s_m \mathcal{N}^1_{w_i}} \\ K_{\mathcal{N}^0_{w_i} \mathcal{N}^1_{w_i}} \end{bmatrix}^T \Bigg(\begin{bmatrix} K_{s_m \mathcal{N}^1_{w_i}} \\ K_{\mathcal{N}^0_{w_i} \mathcal{N}^1_{w_i}} \end{bmatrix} K^{-1}_{\mathcal{N}^1_{w_i}} \begin{bmatrix} K_{s_m \mathcal{N}^1_{w_i}} \\ K_{\mathcal{N}^0_{w_i} \mathcal{N}^1_{w_i}} \end{bmatrix}^T + \begin{bmatrix} \sigma_2^2 \mathbf{I} & 0 \\ 0 & (v^{-1}_{w_{i0}} - K^{-1}_{\mathcal{N}^0_{w_i}})^{-1} \end{bmatrix} \Bigg)^{-1} \begin{bmatrix} f_{s_m} \\ (v^{-1}_{w_{i0}} - K^{-1}_{\mathcal{N}^0_{w_i}})^{-1} v^{-1}_{w_{i0}} m_{w_{i0}} \end{bmatrix}, v_{w_{i1}} \Bigg), \tag{7}$$

where

$$v_{w_{i1}} = K_{\mathcal{N}^1_{w_i}} - \begin{bmatrix} K_{s_m \mathcal{N}^1_{w_i}} \\ K_{\mathcal{N}^0_{w_i} \mathcal{N}^1_{w_i}} \end{bmatrix}^T \Bigg(\begin{bmatrix} K_{s_m \mathcal{N}^1_{w_i}} \\ K_{\mathcal{N}^0_{w_i} \mathcal{N}^1_{w_i}} \end{bmatrix} K^{-1}_{\mathcal{N}^1_{w_i}} \begin{bmatrix} K_{s_m \mathcal{N}^1_{w_i}} \\ K_{\mathcal{N}^0_{w_i} \mathcal{N}^1_{w_i}} \end{bmatrix}^T + \begin{bmatrix} \sigma_2^2 \mathbf{I} & 0 \\ 0 & (v^{-1}_{w_{i0}} - K^{-1}_{\mathcal{N}^0_{w_i}})^{-1} \end{bmatrix} \Bigg)^{-1} \begin{bmatrix} K_{s_m \mathcal{N}^1_{w_i}} \\ K_{\mathcal{N}^0_{w_i} \mathcal{N}^1_{w_i}} \end{bmatrix}. \tag{8}$$

2.4 Threshold Region Determination

The threshold region of node n_i is the set of points that have high similarity with $\mathcal{N}_{n_i}$, that is

$$T_{n_i} = \{(x_s, f_s) | v^i_s \le v^i_{n_i}, |f_s - m^i_s| \le \alpha \sqrt{v^i_s}\} \tag{9}$$

where α is a positive number (3 in the paper) and

$$\begin{cases} v^i_s &= K_* - K_{s\mathcal{N}^1_{n_i}} K^{-1}_{\mathcal{N}^1_{n_i}} K_{\mathcal{N}^1_{n_i} s} + K_{s\mathcal{N}^1_{n_i}} K^{-1}_{\mathcal{N}^1_{n_i}} v_{n_i 1} K^{-1}_{\mathcal{N}^1_{n_i}} K_{\mathcal{N}^1_{n_i} s} \\ m^i_s &= K_{s\mathcal{N}^1_{n_i}} K^{-1}_{\mathcal{N}^1_{n_i}} \mu_i \\ \mu_i &= v_{n_i 1} K^{-1}_{\mathcal{N}^1_{n_i}} \begin{bmatrix} K_{s_m \mathcal{N}^1_{n_i}} \\ K_{\mathcal{N}^0_{n_i} \mathcal{N}^1_{n_i}} \end{bmatrix}^T \begin{bmatrix} \sigma_2^2 \mathbf{I} & 0 \\ 0 & (v^{-1}_{n_i 0} - K^{-1}_{\mathcal{N}^0_{n_i}})^{-1} \end{bmatrix}^{-1} \\ & \begin{bmatrix} f_{s_m} \\ (v^{-1}_{n_i 0} - K^{-1}_{\mathcal{N}^0_{n_i}})^{-1} v^{-1}_{n_i 0} m_{n_i 0} \end{bmatrix}. \end{cases} \tag{10}$$

One can find that $v^i_s \le v^i_{n_i}$ is an ellipse in an implicit feature space, and so is $|f_s - m^i_s| \le \alpha \sqrt{v^i_s}$ from the following theorem

Theorem 1.

$$\begin{aligned} & |f_s - m^i_s| \le \alpha \sqrt{v^i_s} \\ \Leftrightarrow \quad & \|Q(K^{-1}_{\mathcal{N}^1_{n_i}} K_{\mathcal{N}^1_{n_i} s} - f_s (Q^T Q)^{-1} \mu_i)\|_2^2 \le \alpha^2 (K_* - \frac{f_s^2}{\alpha^2 + \mu_i^T \beta_i^{-1} \mu_i}) \end{aligned} \tag{11}$$

where $Q^T Q = \alpha^2 \beta_i + \mu_i \mu_i^T$ *and* $\beta_i = K_{\mathcal{N}^1_{n_i}} - v_{n_i 1}$.

Proof.

$$\begin{aligned}
&|f_s - m_s^i| \leq \alpha\sqrt{v_s^i} \\
\Leftrightarrow\ & K_{s\mathcal{N}_{n_i}^1} K_{\mathcal{N}_{n_i}^1}^{-1} (\alpha^2 K_{\mathcal{N}_{n_i}^1} - \alpha^2 v_{n_i 1} + \mu_i \mu_i^T) K_{\mathcal{N}_{n_i}^1}^{-1} K_{\mathcal{N}_{n_i}^1 s} \\
&\leq 2 f_s \mu_i^T K_{\mathcal{N}_{n_i}^1}^{-1} K_{\mathcal{N}_{n_i}^1 s} + \alpha^2 K_{ss} - f_s^2 \\
\Leftrightarrow\ & \|Q(K_{\mathcal{N}_{n_i}^1}^{-1} K_{\mathcal{N}_{n_i}^1 s} - f_s (Q^T Q)^{-1} \mu_i)\|_2^2 \leq \alpha^2 K_* + f_s^2 \mu_i^T (Q^T Q)^{-1} \mu_i - f_s^2 \\
&= \alpha^2 (K_* - \frac{f_s^2}{\alpha^2 + \mu_i^T \beta_i^{-1} \mu_i}).
\end{aligned} \tag{12}$$

It shows that, as an extension of the threshold regions in ESOINN and KDESOINN that are simply a sphere or an ellipse in the input space, our model gives threshold regions in a reproducing kernel Hilbert space.

2.5 Property Discussion

In this section, we explore further the property of the proposed model.

Theorem 2.

$$\lim_{\substack{\|v_{n_i 0}\|_{\max} \to 0 \\ \sigma_2 \to 0}} \|v_{n_i 1}\|_{\max} = 0. \tag{13}$$

Proof. Regarding the relationship between $|\mathcal{N}_{n_i}^1|$ and $|\mathcal{N}_{n_i}^0|$, there are 2 cases as follows.

1). When $|\mathcal{N}_{n_i}^1| = |\mathcal{N}_{n_i}^0| + 1$, we have $\lim_{\|v_{n_i 0}\|_{\max} \to 0} (v_{n_i 0}^{-1} - K_{\mathcal{N}_{n_i}^0}^{-1})^{-1} = \lim_{\|v_{n_i 0}\|_{\max} \to 0} v_{n_i 0} - v_{n_i 0} (K_{\mathcal{N}_{n_i}^0} - v_{n_i 0})^{-1} v_{n_i 0} = 0$, then from Eq. (8), (13) is proven.

2). When $|\mathcal{N}_{n_i}^1| = |\mathcal{N}_{n_i}^0|$,

$$\begin{aligned}
v_{n_i 1}^{-1} &= K_{\mathcal{N}_{n_i}^1}^{-1} + \frac{1}{\sigma_2^2} K_{\mathcal{N}_{n_i}^1}^{-1} K_{\mathcal{N}_{n_i}^1 s_m} K_{s_m \mathcal{N}_{n_i}^1} K_{\mathcal{N}_{n_i}^1}^{-1} \\
&+ K_{\mathcal{N}_{n_i}^1}^{-1} K_{\mathcal{N}_{n_i}^1 \mathcal{N}_{n_i}^0} (v_{n_i 0}^{-1} - K_{\mathcal{N}_{n_i}^0}^{-1}) K_{\mathcal{N}_{n_i}^0 \mathcal{N}_{n_i}^1} K_{\mathcal{N}_{n_i}^1}^{-1}.
\end{aligned} \tag{14}$$

Suppose that the singular value decomposition of $K_{\mathcal{N}_{n_i}^0 \mathcal{N}_{n_i}^1} K_{\mathcal{N}_{n_i}^1}^{-1} := U \Sigma V$, then for any eigenvalue λ_i of $K_{\mathcal{N}_{n_i}^1}^{-1} K_{\mathcal{N}_{n_i}^1 \mathcal{N}_{n_i}^0} (v_{n_i 0}^{-1} - K_{\mathcal{N}_{n_i}^0}^{-1}) K_{\mathcal{N}_{n_i}^0 \mathcal{N}_{n_i}^1} K_{\mathcal{N}_{n_i}^1}^{-1}$,

$$\lim_{\|v_{n_i 0}\|_{\max} \to 0} \frac{1}{\lambda_i} \leq \lim_{\|v_{n_i 0}\|_{\max} \to 0} \|(v_{n_i 0}^{-1} - K_{\mathcal{N}_{n_i}^0}^{-1})^{-1}\| \|U \Sigma^2 U^T\| = 0. \tag{15}$$

From the property that for two Hermitian matrices A and B, $\lambda_{\min}(C := A + B) \geq \lambda_{\min}(A) + \lambda_{\min}(B)$, it follows that $\lim_{\|v_{n_i 0}\|_{\max} \to 0} \lambda_{\min}(v_{n_i 1}^{-1}) = \infty$, so $\lim_{\|v_{n_i 0}\|_{\max} \to 0} \|v_{n_i 1}\|_{\max} = 0$, (13) is proven.

When the dependent variable is constant, $|f_s - m_s^i| \leq \alpha\sqrt{v_s^i}$ always holds and the threshold region determination is consistent with the results in [16] that infers cluster set based on the variance function of Gaussian process.

2.6 Algorithm: A Gaussian Process-Based Incremental Neural Network for Online Regression

Based on the above analysis of optimal bandwidth matrix, posterior approximation, and threshold region determination, we present a novel algorithm to extract prototypes for online regression using Gaussian process-based incremental neural network (GPINN) in Algorithm 2.

First, we check whether a new sample $s_i \in T_{w_1} \cap T_{w_2}$ using Eq. (9) at line 3 after calculating the winner candidate set at line 2. Second, update the network in line 3–13 (e.g. $\mathcal{V}$, weight vectors, edge, kernel bandwidth matrices (Eq. (5)) and the posterior approximation (Eq. (7)) of $w_{1(2)}$. Third, similar to [9,10], at the end of each learning period, some edges are added between the nodes that have duplex edges in a k-NN graph on $\mathcal{V}$ (denote the set of the added edges as $\mathcal{A}$) at line 16 to achieve a more robust network structure, and then the kernel bandwidth matrices and posterior approximations are updated in line 17–20.

Algorithm 2: GPINN for Online Regression

Initialization: $\mathcal{V} \leftarrow \{s_1, s_2\}, \mathcal{E} \leftarrow \emptyset$

1 **while** $\sim isempty(s_i)$ **do**

2 $\quad \{w_1, w_2\} \leftarrow arg \min_{\{n_{\phi_1}, n_{\phi_2}\} \subset \mathcal{V}} \sum_{j=1}^{2} \|s_i, n_{\phi_j}\|_2$;

3 $\quad$ **if** $s_i \notin T_{w_1} \cap T_{w_2} \leftarrow Eq.(9)$ **then**

4 $\quad\quad \mathcal{V} \leftarrow \mathcal{V} \cup \{s_i\}$;

5 $\quad$ **else**

6 $\quad\quad \mathcal{E} \leftarrow \mathcal{E} \cup \{(w_1, w_2)\}$;

7 $\quad\quad age(w_1, w_2) = 0, age(w_1, n_j) = age(w_1, n_j) + 1 \quad \% \, n_j \in \mathcal{N}_{w_1}$;

8 $\quad\quad W_{w_1} \leftarrow W_{w_1} + 1, w_1 = w_1 + \frac{1}{W_{w_1}+1}(s_i - w_1)$;

9 $\quad\quad n_j = n_j + g(W_{w_1})(s_i - n_j) \quad \% \, n_j \in \mathcal{N}_{w_1}$;

10 $\quad\quad \mathcal{E} \leftarrow \mathcal{E} \setminus \{e \mid age(e) > age_{max}\}, \mathcal{V} \leftarrow \mathcal{V} \setminus \{n_i \mid |\mathcal{N}_{n_i}| = 0\}$;

11 $\quad\quad M_{n_k \in w_1 \cup \mathcal{N}_{w_1}} \leftarrow$ Eq.(5);

12 $\quad\quad q_1(f_{\mathcal{N}_{w_1}}), q_1(f_{\mathcal{N}_{w_2}}) \leftarrow$ Eq.(7);

13 $\quad$ **end**

14 $\quad$ **if** $mod(i, \lambda) = 0$ **then**

15 $\quad\quad \mathcal{V} \leftarrow \mathcal{V} \setminus \{s \mid |\mathcal{N}_s| = 0\}$;

16 $\quad\quad \mathcal{E} \leftarrow \mathcal{E} \cup \mathcal{A}$;

17 $\quad\quad$ **for** $\forall n_j \in \{n_i | \exists n_k, (n_i, n_k) \in \mathcal{A}\}$ **do**

18 $\quad\quad\quad M_{n_j} \leftarrow$ Eq.(5);

19 $\quad\quad\quad q_1(f_{\mathcal{N}_j}) \leftarrow$ Eq.(7);

20 $\quad\quad$ **end**

21 $\quad$ **end**

22 **end**

Complexity: 1). In calculating the winning nodes and their corresponding threshold regions, the time complexity of line 2 is $O(|\mathcal{V}|)$ and the time complexity of calculating $\{K_{w_{1(2)}}^{-1}\}$ in determining whether $s_i \in T_{w_1} \cap T_{w_2}$ at line 3 is

$O(age^3_{max})$. 2). In the network adjustment process, the complexity of updating $\{M_{n_j}\}$ and $q(f_{\mathcal{N}_j})$ at line 18 and line 19 are both $O(|\mathcal{V}|)$. The complexity of k-NN graph is $O(|\mathcal{V}|^2)$. The complexity of other lines is $O(1)$.

3 Experimental Results

In order to demonstrate the performance of our method, we have conducted experiments on both synthetic data and real-word data.

3.1 Synthetic Data

We first extracted prototypes (nodes) using three models: (1) ESOINN, (2) KDESOINN, and (3) our model (GPINN), and then adopted the prevalent Gaussian kernel regression with parameter suggested by Bowman and Azzalini [17] on the nodes given by each model. Here, the normalized root-mean-square error (NRMSE) is employed to compare their regression accuracies. Besides, the classical batch learning kernel smoothing regression (denoted as Kernel in Fig. 1(a) and 1(b)) is used as the benchmark.

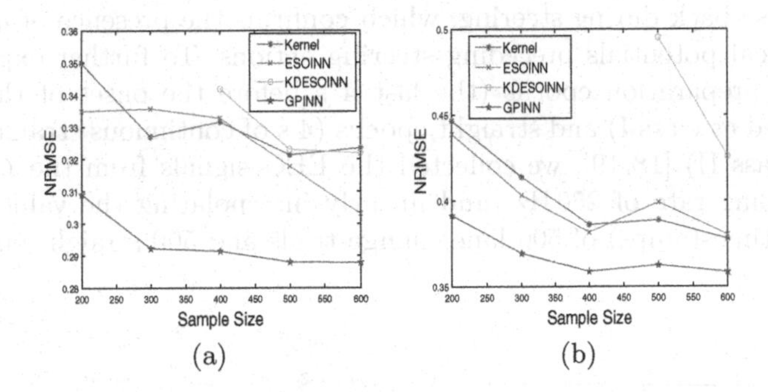

Fig. 1. Changes in NRMSE with sample size on (a) data set I and (b) data set II

The true regression functions consist of 5 sine functions with dependent variable f and independent variable x (of different amplitudes in data set I and discontinuous in data set II). In addition, we add as noise a contamination distribution f_{con} (the Gaussian distribution with mean 0 and variance 100), that is,

Data set I

$$\begin{cases} f &= (2.5 + 0.5 * \lceil x \rceil) \sin(2\pi * (x - \lfloor x \rfloor)) \cdot \mathbf{1}_{\alpha<0.95} + f_{con} \cdot \mathbf{1}_{0.95\leq\alpha}, \\ 0 &< x < 5, \end{cases}$$

Data set II

$$\begin{cases} f &= 5\sin(2\pi * (t - \lfloor t \rfloor)) \cdot \mathbf{1}_{\alpha<0.95} + f_{con} \cdot \mathbf{1}_{0.95\leq\alpha}, \\ x &= t + \lceil t \rceil \\ 0 &< t < 5, \end{cases}$$

where $\alpha \sim \text{unif}(0, 1)$. The training samples of the two data sets are randomly sampled from the domains of the functions.

In order to compare these four approaches comprehensively, we tested on different training sample sizes (for the three incremental neural network models, $age_{max} = 3, \lambda = 150$ and $\frac{\sigma_2^2}{\sigma_1^2} = 0.1$), and for each size, we tested 20 times and recorded the average NRMSEs depicted in Fig. 1(a) and 1(b). It is noteworthy that since there are no nodes left in KDESOINN when the training sample size is less than 400 on data set I and less than 500 on data set II, the corresponding average NRMSEs of KDESOINN are not shown in the figure.

3.2 Real-World Data

Recent research [18–20] showed that there is a decrease in the amplitude of electroencephalography (EEG) signal before the start of the steering action followed by an increase back during steering, which confirms the presence of movement-related cortical potentials preceding steering actions. To further explore brain activities in preparation epochs (the last 4 s before the onset of the steering action, named as class I) and straight epochs (4 s of continuous straight driving, named as class II) [18,19], we collected the EEG signals from the *Cz* channel at the sampling rate of 256 Hz (and linearly interpolating the values between consecutive timestamps) of 500 lane change trials and 500 straight epochs.

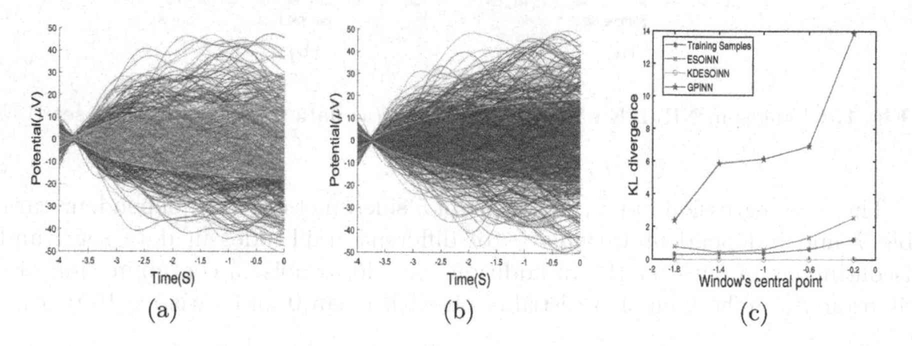

Fig. 2. The EEG signals and the one standard deviation intervals of (a) Preparation Epochs and (b) Straight Epochs; (c) Changes in the average KL divergence with the time window position

Figure 2(a) and 2(b) shows the signals and the one standard deviation intervals. Some epochs with amplitudes close to 50 μV are regarded as contamination epochs (noise data) and can lead to large intersections between the standard deviation intervals of these two classes [18,19]. So, it is desired that the extracted prototypes should be class-specific, that is, for a time window w_t (t is the timestamp of the window's central point) before the onset of the movement, the distributions of the dependent values of the prototypes falling into w_t (denoted as P_{pre,w_t} and P_{str,w_t} for the preparation epoch and the straight epoch, respectively) should have a large divergence value, $\mathrm{KL}(P_{pre,w_t} \| P_{str,w_t})$.

Table 2. Paired-sample t-test

Window's central point	t-test(GPINN, Training samples)	t-test(GPINN, ESOINN)
−1.8 s	0.0145	0.0477
−1.4 s	0.0056	0.0071
−1.0 s	0.0023	0.0042
−0.6 s	<0.001	<0.001
−0.2 s	<0.001	0.0013

In this experiment, for each class, 2000 training samples were drawn randomly from the last 2 s of the epochs. For ESOINN, KDESOINN and GPINN (test times = 15, $age_{max} = 3, \lambda = 250$ and $\frac{\sigma_2^2}{\sigma_1^2} = 0.1$), Fig. 2(c) shows the changes in the average $\mathrm{KL}(P_{pre,w_t} \| P_{str,w_t})$ as the time window of length 400 ms slides in steps of 400 ms. Besides, the divergences calculated directly from the training samples are given as the benchmark. Table 2 gives the paired-sample t-test results and shows that our approach achieves the statistically significantly largest divergences at the 5% significance level.

The results on both synthetic data and real-world data show that, due to the high fractional anisotropy of threshold region, there are very few nodes in the network generated by KDESOINN; the threshold regions in ESOINN do not carefully consider the distribution of neighbors that can lead to many inappropriate edges and significantly decrease its accuracy. In contrast, by mapping input space to an implicit feature space, our approach adopts Gaussian process to calculate the probability of generating an edge between winning nodes and can update the threshold region and approximate the posterior with the advantage of robust to inappropriate edges.

4 Conclusion

This paper has introduced a novel Gaussian process-based incremental neural network for online regression. It addresses the shortcomings of the previous Euclidean distance-based and Mahalanobis distance-based self-organizing incremental neural network algorithms; in detail, it allows online updates of threshold regions and the posterior approximation of the dependent variable at nodes as the adjustment of the network in both input space and map space. Moreover, we derived the optimal bandwidth matrix adaptive to the learned network structure and discussed some of its properties. The experimental results show that the proposed approach achieves improvement in extracting prototypes for online regression on noisy data.

References

1. Wolfgang, H.: Applied Nonparametric Regression (Econometric Society Monographs). Cambridge University Pressk, Cambridge (1990)
2. Mount, N.J., Weaver, D.: Self-organizing maps and boundary effects: quantifying the benefits of torus wrapping for mapping SOM trajectories. Pattern Anal. Appl. **14**, 139–148 (2011)
3. Shen, F., Hasegawa, O.: Self-organizing incremental neural network and its application. In: International Conference on Artificial Neural Networks, pp. 535–540 (2010)
4. Shen, F., Hasegawa, O.: A fast nearest neighbor classifier based on self-organizing incremental neural network. Neurocomputing **21**, 1537–1547 (2008)
5. Nakamura, Y., Hasegawa, O.: Nonparametric density estimation based on self-organizing incremental neural network for large noisy data. IEEE Trans. Neural Netwo. Learn. Syst. **28**, 25–32 (2016)
6. Alexey, T.: The problem of concept drift: definitions and related work. Tech. rep., Computer Science Department, Trinity College Dublin (2004)
7. Vincent, P., Bengio, Y.: Manifold parzen windows. In: Advances in Neural Information Processing Systems, pp. 825–832 (2002)
8. Bengio, Y., Delalleau, O., Le Roux, N.: The curse of dimensionality for local kernel machines. Tech rep. 1258, Université de Montréal (2005)
9. Wang, X., Casiraghi, G., Zhang, Y., Imura, J.: A Gaussian process-based self-organizing incremental neural network. In: International Joint Conference on Neural Networks, pp. 1–8 (2019)
10. Wang, X., Hasegawa, O.: Adaptive density estimation based on self-organizing incremental neural network using Gaussian process. In: International Joint Conference on Neural Networks, pp. 4309–4315 (2017)
11. Rasmussen, C.E., Williams, C.K.L.: Gaussian Processes for Machine Learning. MIT Press, Cambridge (2006)
12. Williams, C.K.L., Barber, D.: Bayesian classification with Gaussian processes. IEEE Trans. Pattern Anal. Mach. Intell. **20**, 1342–1351 (1998)
13. Broderick, T., Boyd, N., Wibisono, A., Wilson, A.C., Jordan, M.I.: Streaming variational bayes. In: Advances in Neural Information Processing Systems, pp. 1727–1735 (2013)
14. Bui, T.D., Nguyen, C.V., Turner, R.E.: Streaming sparse Gaussian process approximations. In: Advances in Neural Information Processing Systems, pp. 3299–3307 (2017)
15. Terrell, G.R., Scott, D.W.: Estimation, variable kernel density. Ann. Stat. **20**, 1236–1265 (1992)
16. Kim, H.C., Lee, J.: Clustering based on Gaussian processes. Neural Comput. **19**, 3088–3107 (2007)
17. Bowman, A.W., Azzalini, A.: Applied Smoothing Techniques for Data Analysis. Clarendon Press, Oxford (1997)

18. Gheorghe, L., Chavarriaga, R., Millán, J.D.R.: Steering timing prediction in a driving simulator task. In: 35th Annual International Conference of the IEEE Engineering in Medicine and Biology Society, pp. 6913–6916 (2013)
19. Gheorghe, L.: Detecting EEG correlates during preparation of complex driving maneuvers. Doctoral thesis, École Polytechnique Federale de Lausanne (2016)
20. Chavarriaga, R., et al.: Decoding neural correlates of cognitive states to enhance driving experience. IEEE Trans. Emerg. Top. Comput. Intell. **2**, 288–297 (2018)

Analysis on the Boltzmann Machine with Random Input Drifts in Activation Function

Wenhao Lu[1], Chi-Sing Leung[1](✉), and John Sum[2]

[1] Department of Electronic Engineering, City University of Hong Kong, Kowloon Tong, Hong Kong
wenhaolu3-c@my.cityu.edu.hk, eeleungc@cityu.edu.hk

[2] Institute of Technology Management, National Chung Hsing University, Taichung, Taiwan
pfsum@nchu.edu.tw

Abstract. The Boltzmann machine (BM) model is able to learn the probability distribution of input patterns. However, in analog realization, there are thermal noise and random offset voltages of amplifiers. Those realization issues affect the behaviour of the neurons' activation function and they can be modelled as random input drifts. This paper analyzes the activation function and state distribution of BMs under the input random drift model. Since the state of a neuron is also determined by its activation function, the random input drifts may cause a BM to change the behaviour. We show that the effect of random input drifts is equivalent to raising temperature factor. Hence, from the Kullback–Leibler (KL) divergence perspective, we propose a compensation scheme to reduce the effect of random input drifts. In our derive of compensation scheme, we assume that the input drift follows the Gaussian distribution. Surprisedly, from our simulations, the proposed compensation scheme also works very well for other distributions.

Keywords: Activation function · State distribution · Boltzmann machine · Noise

1 Introduction

The Boltzmann machine (BM) model is a kind of stochastic Hopfield network models [1,2]. With its energy-based model, it can learn the internal representations of inputs. The BM model can be used many applications, such as image and speech recognitions [3–5].

When implementing a BM by analog hardware, there some imperfect issues. For instance, the random offset voltage and thermal noise in electronic components may lead to noisy behaviors like random input drifts in activation function [6–8]. For traditional non-stochastic neural network models, many noise aware learning algorithms, like the algorithms developed in [9–11], have been

H. Yang et al. (Eds.): ICONIP 2020, LNCS 12534, pp. 162–171, 2020.
https://doi.org/10.1007/978-3-030-63836-8_14

developed for feedfoward neural networks. However, very few results have been presented for the BM model [12–14]. Recently, Sum et al. [14] have investigated how the additive weight and bias noise affects the conditional probability and state distribution during training and operation.

In this paper, we consider a problem regarding the implementation of a Boltzmann machine. Suppose the Boltzmann machine fitting the distribution of a set of binary patterns have been attained by numerical simulation. The model parameters are thus used as reference for the design of the analog hardware BM. However, it is known that the analog components have non-ideal properties, like random input drifts to the threshold logic unit. Hence, there is a loss in Kullback–Leibler (KL) divergence between the drifted BM (i.e. the implemented model) and the drift-free BM (i.e. the ideal model). When a trained BM is running on hardware, their behaviours are also affected by the thermal noise and random offset drifts of operational amplifiers. Here, we investigate the effect of random input drifts in activation function on a BM.

If the random drift follows zero-mean Gaussian distribution, a compensation method is derived for setting the temperature factor for the Boltzmann learning. The BM attained is then best-fit for hardware implementation. Experimental results show that the compensation method still works well for some zero-mean symmetric distributions other than Gaussian distribution. The main contributions can be summarized as follows

- We present the activation function and state distribution for BM under the random input drift condition. We theoretically prove that the effect of random input drifts is approximately the same as increasing the temperature factor.
- We propose the compensated method. With the proposed method, in term of KL divergence metric, the property of the drifted network is approximately the same as that of drift-free BM.

This brief is organized as follows. Section 2 reviews the background of BM and presents the drifted BM model. Section 3 presents our analytical results and proposes the compensation method, respectively. Section 4 shows the experimental results. Section 5 gives the concluding remark.

2 Background

A BM is a stochastic recurrent neural network with symmetrical connections. Assume there are p neurons in the BM. Each neuron is fully connected with the other neurons. Let s_k, where $k = 1, 2, \cdots, p$, be the state of the kth neuron. Let $\mathbf{S_{-k}}$ be the state vector like $[s_1, \cdots, s_{k-1}, s_{k+1}, \cdots, s_p]^T$., i.e., the collection of the states of neurons except the kth neuron.

The state of the kth neuron can be on (node output is 1) or off (node output is 0), which is controlled by a stochastic activation function. The probability that s_k becomes 1 is given by

$$P(s_k = 1|\mathbf{S}_{-k}) = \frac{1}{1 + e^{-u_k/T}}, \tag{1}$$

where u_k is the input for this activation function, and it is defined as

$$u_k = \sum_{l=1}^{p} w_{kl} s_l + d_k. \tag{2}$$

In (2), w_{kl} is the interconnection weight between the kth and the lth neuron, and d_k is the input bias. Note that a BM is symmetrical connected, i.e., $w_{kl} = w_{lk}$ and not self-connected, $w_{kk} = 0$. In (1), T is called temperature factor. In the BM concept, the states of neurons are updated in an asynchronous way.

We use $\mathbf{v}$ to denote visible neurons and $\mathbf{h}$ to denote hidden neurons. For a noise-free trained BM, its state distribution is given by

$$\tilde{P}(\mathbf{v}, \mathbf{h}) = \frac{e^{-E(\mathbf{v},\mathbf{h})/T}}{\sum_{\mathbf{v},\mathbf{h}} e^{-E(\mathbf{v},\mathbf{h})/T}}, \tag{3}$$

where $E(\mathbf{v}, \mathbf{h})$ is the energy of state $(\mathbf{v}, \mathbf{h})$:

$$E(\mathbf{v}, \mathbf{h}) = -\sum_{k<l} w_{kl} s_k s_l - \sum_{k} s_k b_k. \tag{4}$$

For the probability distribution over the visible neurons which is reconstructed by the drift-free BM, it is defined as

$$\begin{aligned} \tilde{P}(\mathbf{v}) &= \sum_{\mathbf{h}} \tilde{P}(\mathbf{v}, \mathbf{h}) \\ &= \frac{\sum_{\mathbf{h}} e^{-E(\mathbf{v},\mathbf{h})/T}}{\sum_{\mathbf{v},\mathbf{h}} e^{-E(\mathbf{v},\mathbf{h})/T}}. \end{aligned} \tag{5}$$

When a trained BM is running on hardware, their behaviours are also affected by the thermal noise and random offset drifts of operational amplifiers. Those external imperfect issues can be modelled as random input drifts in activation function, stated in (1).

We use the notation Δu to represent drifts. Under the noisy condition, the activation function in (1) and (2) are modelled as

$$P_n(s_k = 1 | \mathbf{S}_{-k}, \Delta u_k) = \frac{1}{1 + e^{-\widetilde{u}_k/T}} \tag{6}$$

$$\widetilde{u}_k = u_k + \Delta u_k. \tag{7}$$

Here, the value of Δu_k is time-varying but its statistics are time-invariant. It follows zero-mean Gaussian distribution with variance σ^2. From (6)–(7), it can be seen that the random input drift Δu_k affects the stochastic decision.

Figure 1 depicts that when σ increases, the BM with random drifts in activation function has degradation on the KL divergence metric.

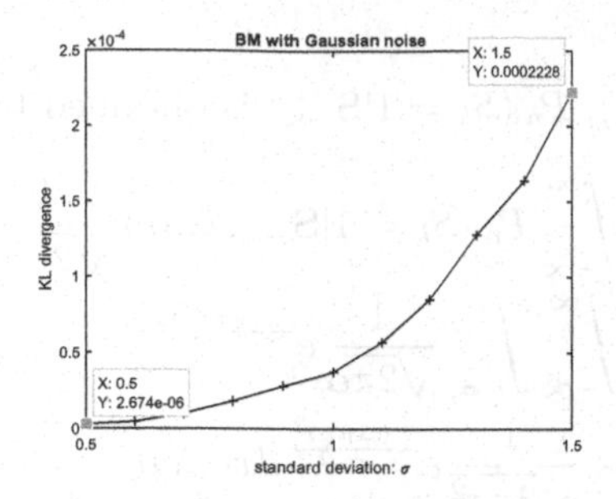

Fig. 1. The loss in KL divergence for the Gaussian random drift case. The BM is a 4 − 3 − 4 encoder. It is trained with 16 states from 0000 to 1111. We measure the KL divergence between the drifted BM and drift-free BM.

3 Main Result

In this section, we first theoretically analyze how random input drifts affect the activation function and state distribution. Then, we propose a compensation method to suppress the effect of input drifts based on the KL divergence perspective.

In our analysis, we use the following important lemma.

Lemma 1: *A logit model can be approximate to the normal integral [15], given by*

$$\frac{1}{1+e^{-\beta z}} \approx \int_{-\infty}^{z} \frac{1}{\sqrt{2\pi}} e^{-\frac{x^2}{2}} dx, \tag{8}$$

where $\beta = 1.702$.

From (6) and (7), input drifts lead to the situation that the activation function works improperly. Theorem 1 presents how the input drifts affect the conditional probability.

Theorem 1. *For a drifted BM, the effective activation function is given by*

$$P_n(S_k = 1|\mathbf{S}_{-k}) \approx \frac{1}{1+e^{-u_k/(\eta T)}} \tag{9}$$

where $\eta = \sqrt{1+\frac{\sigma^2}{\beta^2 T^2}}$.

Proof: Based on the Lemma 1, we can obtain that

$$\begin{aligned} P_n(S_k = 1|\mathbf{S_{-k}}, \Delta u_k) &= \frac{1}{1+e^{-\widetilde{u}_k/T}} \\ &\approx \int_{-\infty}^{\widetilde{u}_k} \frac{1}{\sqrt{2\pi\alpha}} e^{-\frac{x^2}{2\alpha}} dx \\ &= \int_{-\infty}^{u_k} \frac{1}{\sqrt{2\pi\alpha}} e^{-\frac{(x+\Delta u_k)^2}{2\alpha}} dx \end{aligned}$$

where $\alpha = \beta^2 T^2$.

Since $\Delta u_k \sim N(0, \sigma^2)$, $P_n(S_k = 1|\mathbf{S_{-k}})$ is obtained by

$$\begin{aligned}
P_n(S_k = 1|\mathbf{S}_{-k}) &= \int_{-\infty}^{\infty} P_n(S_k = 1|\mathbf{S}_{-k}, \Delta u_k) \frac{1}{\sqrt{2\pi\sigma^2}} e^{-\frac{(\Delta u_k)^2}{2\sigma^2}} dx d\Delta u_k \\
&= \int_{-\infty}^{\infty} \int_{-\infty}^{u_k} \frac{1}{\sqrt{2\pi\alpha}} e^{-\frac{(x+\Delta u_k)^2}{2\alpha}} \\
&\times \frac{1}{\sqrt{2\pi\sigma^2}} e^{-\frac{(\Delta u_k)^2}{2\sigma^2}} dx d\Delta u_k \\
&= \frac{1}{\sqrt{2\pi\alpha}} \frac{1}{\sqrt{2\pi\sigma^2}} \int_{-\infty}^{u_k} \int_{-\infty}^{\infty} e^{-\frac{x^2}{2(\alpha+\sigma^2)}} \\
&\times e^{-\frac{(\Delta u_k + \frac{\sigma^2}{\sigma^2+\alpha} x)^2}{2\sigma^2\alpha/(\sigma^2+\alpha)}} d\Delta u_k dx \\
&= \frac{1}{\sqrt{2\pi(\sigma^2+\alpha)}} \int_{-\infty}^{u_k} e^{-\frac{x^2}{2(\sigma^2+\alpha)}} dx \\
&\approx \frac{1}{1+e^{-u_k/(\eta T)}}
\end{aligned}$$

where $\eta = \sqrt{1 + \frac{\sigma^2}{\beta^2 T^2}}$. The proof is completed. ■

The interpretation of Theorem 1 is that when a trained BM with operational temperature T is affected by input drifts, it is equivalent to the situation that the operational temperature is increased from T to $\sqrt{1 + \frac{\sigma^2}{\beta^2 T^2}}$, where σ^2 is the variance of the random input drifts.

With the new activation function on the neurons, we can obtain Theorem 2 that shows the new state distribution.

Theorem 2. *For a BM with random input drifts, the state distribution is given by*

$$\tilde{P}_n(\mathbf{v}, \mathbf{h}) = \frac{e^{-E(\mathbf{v},\mathbf{h})/(\eta T)}}{\sum_{\mathbf{v},\mathbf{h}} e^{-E(\mathbf{v},\mathbf{h})/(\eta T)}}. \tag{10}$$

Proof: Compared (1) with (9), it is observed that the effect of random input drifts for activation function is the same as increasing temperature from T to ηT ($\eta > 1$). Hence, for the state distribution in (3), under the random drift condition, it is modified as

$$\tilde{P}_n(\mathbf{v}, \mathbf{h}) = \frac{e^{-E(\mathbf{v},\mathbf{h})/(\eta T)}}{\sum_{\mathbf{v},\mathbf{h}} e^{-E(\mathbf{v},\mathbf{h})/(\eta T)}}. \tag{11}$$

The proof is completed. ■

Based on Theorem 2, the probability distribution over the visible neurons which is reconstructed by the drifted BM is given by

$$\begin{aligned}\tilde{P}_n(\mathbf{v}) &= \sum_{\mathbf{h}} \tilde{P}_n(\mathbf{v},\mathbf{h}) \\ &= \frac{\sum_{\mathbf{h}} e^{-E(\mathbf{v},\mathbf{h})/(\eta T)}}{\sum_{\mathbf{v},\mathbf{h}} e^{-E(\mathbf{v},\mathbf{h})/(\eta T)}}.\end{aligned} \tag{12}$$

Furthermore, the KL divergence between the drift-free probability distribution $\tilde{P}(\mathbf{v})$ and the drifted probability distribution $\tilde{P}_n(\mathbf{v})$ is

$$D_1 = \sum_{\mathbf{v}} \tilde{P}(\mathbf{v})\, ln \frac{\tilde{P}(\mathbf{v})}{\tilde{P}_n(\mathbf{v})}. \tag{13}$$

Since the temperature factor in $\tilde{P}(\mathbf{v})$ and $\tilde{P}_n(\mathbf{v})$ are different, it is clear that the random input drifts introduce a loss in KL divergence.

The following theorem gives us a way to cancel out the loss in the KL divergence caused by random input drifts in the activation function.

Theorem 3. *Based on the KL divergence perspective, in realization, to make the probability distribution for the drifted BM to be the same as that for drift-free BM, we should set the temperature factor of the drifted BM to*

$$T' = T\sqrt{1 - \frac{\sigma^2}{\beta^2 T^2}} \tag{14}$$

such that the updated activation function is given by

$$\check{P}_n(S_k = 1|\mathbf{S}_{-k}) = \frac{1}{1 + e^{-\tilde{u}_k/T'}}. \tag{15}$$

Proof: Based on Theorem 1,

$$P_n(S_k = 1|\mathbf{S}_{-k}) \approx \frac{1}{1 + e^{-u_k/\sqrt{T^2 + \frac{\sigma^2}{\beta^2}}}}. \tag{16}$$

We modify temperature factor from T to T'. As a result,

$$\begin{aligned}\check{P}_n(S_k = 1|\mathbf{S}_{-k}) &= \frac{1}{1 + e^{-u_k/\sqrt{T'^2 + \frac{\sigma^2}{\beta^2}}}} \\ &\approx \frac{1}{1 + e^{-u_k/\sqrt{T^2 - \frac{\sigma^2}{\beta^2} + \frac{\sigma^2}{\beta^2}}}} \\ &= \frac{1}{1 + e^{-u_k/T}} \\ &= P(S_k = 1|\mathbf{S}_{-k}).\end{aligned}$$

Thus,

$$\check{P}_n(\mathbf{v}, \mathbf{h}) \approx \tilde{P}(\mathbf{v}, \mathbf{h}). \tag{17}$$

The visible unit distribution is

$$\check{P}_n(\mathbf{v}) = \sum_{\mathbf{h}} \check{P}_n(\mathbf{v}, \mathbf{h}) \tag{18}$$

$$\approx \sum_{\mathbf{h}} \tilde{P}_n(\mathbf{v}, \mathbf{h}). \tag{19}$$

The KL divergence between the drifted BM with (15) and the drift-free BM is

$$D_2 = \sum_{\mathbf{v}} \tilde{P}(\mathbf{v}) \ln \frac{\tilde{P}(\mathbf{v})}{\check{P}_n(\mathbf{v})}. \tag{20}$$

Since $\tilde{P}(\mathbf{v}) \approx \check{P}_n(\mathbf{v})$, $D_2 \approx 0$. The proof is completed. ■

Theorem 3 shows that by reducing the temperature factor from T to T', the effect of random input drifts can be canceled out. However, this method has the limitation. From (14), we can obtain that

$$1 - \frac{\sigma^2}{\beta^2 T^2} \geq 0$$

$$\sigma \leq \beta T. \tag{21}$$

In other words, to apply Theorem 3, the standard deviation of the random input drifts must be smaller than $1.702T$.

4 Simulations

In this section, we evaluate the behaviour of drifted BMs after using the compensated activation function. A drift-free BM is trained first. Its configuration is like 4 visual neurons as inputs, 3 hidden neurons and 4 visual neurons as outputs. Without loss of generality, the temperature factor T is set as 1. The BM is trained to do two tasks. Task 1 is to be a $4-3-4$ auto-encoder. Task 2 is to achieve bitwise negation. For example, the input is 1001 and the corresponding output is 0110. For both tasks, the input sets include 16 states from 0000 to 1111. After training the BM can achieve those two tasks.

During testing, three types of noise as random input drifts are added into the activation function: (1) zero-mean Gaussian noise with different standard deviation; (2) zero-mean uniform noise with different maximum noise level; (3) for beta distribution: $\text{Beta}_{a,b}(x) = \frac{\Gamma(a+b)}{\Gamma(a)\Gamma(b)} x^{a-1} (1-x)^{b-1}$, where $\Gamma(.)$ is the Gamma function and $a = b = 2$, we shift it to zero mean and stretch it to different maximum noise level. Although our compensation method is designed for the Gaussian case, we also evaluate its ability for other zero-mean symmetric distributions. In the simulation, for each drift level and each input pattern, we run the BM long enough. Afterwards, we measurement the state distribution by using 1×10^6 samples.

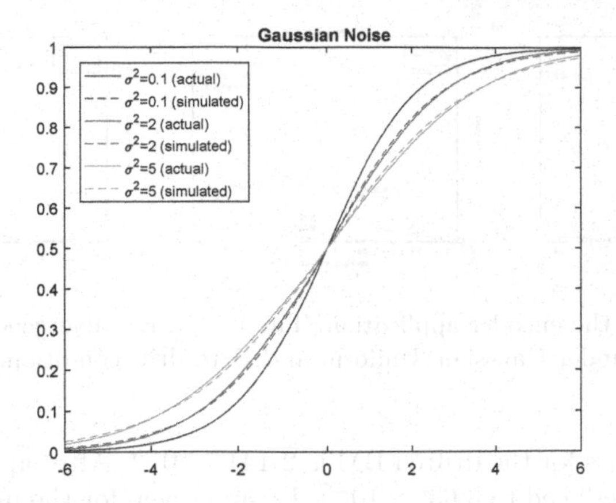

Fig. 2. The actual and simulated activation function for zero-mean Gaussian noise with different variance.

4.1 Use of Theorem 1

Since the compensated activation function is derived from the approximation in Theorem 1, we investigate how well the approximation is. Figure 2 shows the approximated and actual probability under the Gaussian drifts. The input for activation function, U_G, is in $\{-6, -5.99, \cdots, 5.99, 6\}$. For each variance setting $(\sigma^2 = 0.1,\ 2,\ 5)$, 100,000 zero-mean Gaussian variance ΔU_G are generated. The actual probability for each given U_G is obtained by averaging $\left(1 + e^{-(U_G + \Delta U_R)/T}\right)^{-1}$. The approximated probability is based on (9). It can be seen that the approximated probability well approximates the actual one.

4.2 Use of Theorem 3

The advantage of using Theorem 3 is that the loss in KL divergence due to random input drifts can be cancelled out by only reducing the temperature factor in activation function. The KL divergence between the drifted BM and dirft-free BM in (13) is taken as the baseline, denoted as D_{base}. The KL divergence between the revised and noise-free BMs in (20) is denoted as D_{revise}. Figures 3 and 4 illustrate the behaviour of our compensation method under different noise conditions based on the KL divergence metric.

For the Gaussian case, in the auto-encoder application (the first column of Fig. 3), when $\sigma = 1.5$, the loss for drifted BM is 2.268×10^{-4}. After applying Theorem 3, the loss is reduced to 9.809×10^{-6}. Similarly, in the bitwise negation application (the first column of Fig. 4), using the revised activation function can effectively decreases the loss in KL divergence from 8.219×10^{-4} to 2.781×10^{-5}.

One concern of our results is that the theory is derived based on the Gaussian assumption. As shown in the second and third columns of Figs. 3, 4. Our compensation method still works very well. For instance, for the encoder application with random drifts being uniformly distributed, when the drift level is

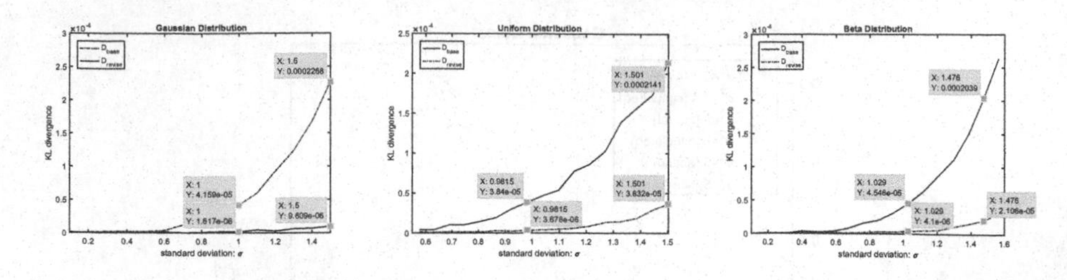

Fig. 3. KL Loss in the encoder application. The loss in KL divergence for drifted and compensated BM under Gaussian, Uniform and Beta drift condition.

$\sigma = 1.5$, the KL loss for the drifted BM is 2.141×10^{-4}. After applying Theorem 3, the KL loss is reduced to 3.632×10^{-5}. In sum, even for the uniform and beta noise cases, D_{revise} is smaller than D_{base}. That means although our method is designed for canceling out the effect of Gaussian drifts, it still works for other drift distributions.

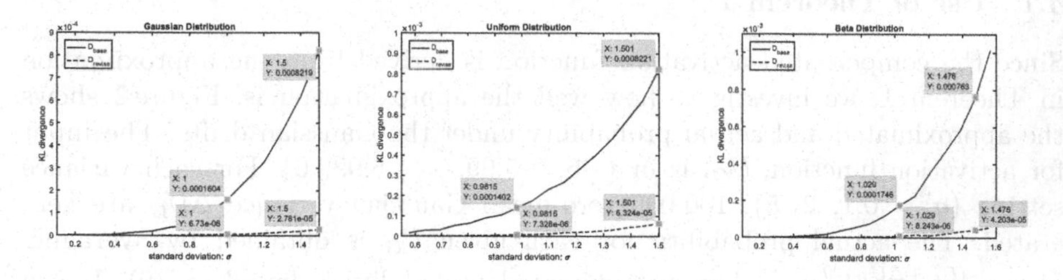

Fig. 4. KL loss in the bitwise negation application. The loss in KL divergence for drifted and compensated BM under Gaussian, Uniform and Beta drift conditions.

5 Conclusion

This paper investigates the behaviors of BM with the Gaussian noise as random input drifts in the activation function. We show that the effect of random input drifts is to increase the temperature factor in activation function (Theorem 1) and state distribution (Theorem 2). Since the random input drifts cause the loss in the KL divergence for drifted BMs, we propose a compensation method (Theorem 3). We prove that after applying Theorem 3, the behaviours of the drift-free BM and the drifted BM are approximately same based on the KL divergence metric. Our simulation shows that our compensation method works not only for the Gaussian drift but also for other distributions.

As a final note, the results presented in this paper can be equally applied to the model with other noise conditions.

(1) $s_k \in \{0,1\}$ and $\widetilde{d}_k = d_k + \Delta d_k$ for $k = 1, \cdots, p$.
(2) $s_k \in \{-1,1\}$ and $\widetilde{u}_k = u_k + \Delta u_k$ for $k = 1, \cdots, p$.

(3) $s_k \in \{-1, 1\}$ and $\widetilde{d}_k = d_k + \Delta d_k$ for $k = 1, \cdots, p$.

However, the results presented in this paper are not applicable to the model with $s_k \in \{-1, 1\}$ and additive weight noise, i.e. $\widetilde{w}_{kl} = w_{kl} + \Delta w_{kl}$ for $k, l = 1, \cdots, p$.

Acknowledgement. The work presented in this paper is supported by a research grant from the Taiwan MOST No. 108-2221-E-005-036 and a research grant from City University of Hong Kong (9610431).

References

1. Ackley, D.H., Hinton, G.E., Sejnowski, T.J.: A learning algorithm for Boltzmann machines. Cogn. Sci. **9**(1), 147–169 (1985)
2. Hinton, G.E., Sejnowski, T.J., Ackley, D.H.: Boltzmann machines: constraint satisfaction networks that learn. Carnegie-Mellon University, Pittsburgh (1984)
3. Prager, R.W., Harrison, T.D., Fallside, F.: Boltzmann machines for speech recognition. Comput. Speech Lang. **1**(1), 3–27 (1986)
4. Ma, H.: Pattern recognition using Boltzmann machine. In: Proceedings IEEE Southeastcon 1995. Visualize the Future, pp. 23–29 (1995)
5. Tang, Y., Salakhutdinov, R., Hinton, G.: Robust Boltzmann machines for recognition and denoising. In: IEEE Conference on Computer Vision and Pattern Recognition, pp. 2264–2271 (2012)
6. Lee, P.: Low noise amplifier selection guide for optimal noise performance. In: Analog Devices Application Note AN-940 (2009)
7. He, J., Zhan, S., Chen, D., Geiger, R.L.: Analysis of static and dynamic random offset voltages in dynamic comparators. IEEE Trans. Circuits Syst. I Regul. Pap. **56**(5), 911–919 (2009)
8. Redoute, J.-M., Steyaert, M.: Measurement of emi induced input offset voltage of an operational amplifier. Electronics Letters. **43**(20), 1088–1090 (2007)
9. Wang, H., Feng, R., Han, Z.F., Leung, C.S.: ADMM-based algorithm for training fault tolerant RBF networks and selecting centers. IEEE Trans. Neural Netw. Learn. Syst. **29**(8), 3870–3878 (2018)
10. Leung, C.S., Wan, W.Y., Feng, R.: A regularizer approach for RBF networks under the concurrent weight failure situation. IEEE Trans. Neural Netw. Learn. Syst. **28**(6), 1360–1372 (2017)
11. Xiao, Y., Feng, R., Leung, C.S., Sum, J.: Objective function and learning algorithm for the general node fault situation. IEEE Trans. Neural Netw. Learn. Syst. **27**(4), 863–874 (2016)
12. Chen, H., Murray, A.F.: Continuous restricted Boltzmann machine with an implementable training algorithm. IEE Proc. Vis. Image Signal Process. **150**(3), 153–158 (2003)
13. Chen, H., Fleury, P.C.D., Murray, A.F.: Continuous-valued probabilistic behavior in a VLSI generative model. IEEE Trans. Neural Netw. Learn. Syst. **17**(3), 755–770 (2006)
14. Sum, J., Leung, C.S.: Learning algorithm for Boltzmann machines with additive weight and bias noise. IEEE Trans. Neural Netw. Learn. Syst. **30**(10), 3200–3204 (2019)
15. Haley, D.C.: Estimation of the dosage mortality relationship when the dose is subject to error. Tech. rep. Applied Mathematics and Statistics Labs. Standford University (1952)

Are Deep Neural Architectures Losing Information? Invertibility is Indispensable

Yang Liu[1,2(✉)], Zhenyue Qin[1], Saeed Anwar[1,2], Sabrina Caldwell[1], and Tom Gedeon[1]

[1] Australian National University, Canberra, Australia
{yang.liu3,zhenyue.qin,saeed.anwar,sabrina.caldwell}@anu.edu.au, tom@cs.anu.edu.au

[2] Data61-CSIRO, Canberra, Australia

Abstract. Ever since the advent of AlexNet, designing novel deep neural architectures for different tasks has consistently been a productive research direction. Despite the exceptional performance of various architectures in practice, we study a theoretical question: what is the condition for deep neural architectures to preserve all the information of the input data? Identifying the information lossless condition for deep neural architectures is important, because tasks such as image restoration require keep the detailed information of the input data as much as possible. Using the definition of mutual information, we show that: a deep neural architecture can preserve maximum details about the given data if and only if the architecture is invertible. We verify the advantages of our Invertible Restoring Autoencoder (IRAE) network by comparing it with competitive models on three perturbed image restoration tasks: image denoising, JPEG image decompression and image inpainting. Experimental results show that IRAE consistently outperforms non-invertible ones. Our model even contains far fewer parameters. Thus, it may be worthwhile to try replacing standard components of deep neural architectures with their invertible counterparts. We believe our work provides a unique perspective and direction for future deep learning research.

1 Introduction

Ever since AlexNet won the ImageNet challenge in 2012 [13], deep learning has been revolutionizing research in many industries. One key factor to account for the success of deep learning is the transferability of deep learning architectures [2]. That is, a neural architecture that exhibits excellent performance for one task can also excel in a variety of other related tasks. However, the majority of deep learning architectures were initially proposed to address high-level computer vision tasks. Recently, researchers also explored applying these deep architectures used for high-level tasks to tackle low-level vision tasks.

Empirical results suggest the plausibility of transferring the neural architectures for high-level vision tasks to addressing low-level image-processing tasks.

Y. Liu and Z. Qin—Equal contribution.
Code: https://github.com/Lillian1082/IRAE_pytorch.

H. Yang et al. (Eds.): ICONIP 2020, LNCS 12534, pp. 172–184, 2020.
https://doi.org/10.1007/978-3-030-63836-8_15

Nonetheless, there is a division between the requirements for deep models to solve high- and low-level vision tasks. To specify, it may be acceptable to miss image details for high-level tasks, as long as it captures the most salient features. However, missing details of images can be unsupportable when dealing with low-level vision tasks. Instead of primarily concentrating on conceptual vision features, models for low-level tasks require specific minutiae such as colors and textures to be able to restore original images [16].

Inspired by the division between the requirements for high- and low-level vision tasks, we study whether it is proper to apply deep architectures of high-level vision tasks to tackle low-level tasks. From the perspective of mutual information, we show that: in order to let a neural architecture to preserve all the information of the given input, the neural architecture needs to be invertible. In this paper, we evaluate the performance of invertible neural architectures on image restoration tasks. Invertible neural architectures exhibit excellence experimental results. Thus, we believe it is a promising avenue to replace non-invertible neural components with their invertible counterparts. In summary, our contributions are three-fold:

1. Deriving from the definition of mutual information, we show non-invertible deep neural architectures lead to loss of information concerning the input.
2. Inspired by the need for invertibility, we develop an Invertible Restoring Autoencoder (IRAE) network via invertible flow-based generative algorithms.
3. We test IRAE with a series of experiments, finding that we achieve superior performance on both image denoising and inpainting tasks. Moreover, our model has fewer parameters than the baseline information-lossy models.

2 Preliminaries

2.1 Residual Blocks

Deep neural networks suffer from the problem of vanishing gradient [8] when the depth of the network increases. To address this problem, Residual Networks and Highway Networks [24] use additional pathways to connect the input with the output of a layer directly. Such residual paths facilitate back-propagation, bypassing the multiplication with the layer weight to alleviate the vanishing gradient. These residual blocks are common features in image restoration models. For example, REDNet [18] uses symmetric residual blocks. Zhang *et al.* [32] employ a large number of residual blocks to preserve detailed information.

2.2 Flow-Based Generative Models

The arguably most important cornerstone of generative models is maximum likelihood estimation (MLE). Generative models aim to maximize the probabilities of producing results that look similar to the given data. Unlike variational autoencoders (VAE) [10] and generative adversarial networks (GANs) [7] that bypass accurately estimating densities, flow-based generative models directly maximize log probabilities of the given data. Therefore, flow-based generative models require all the model components to be invertible. Pioneering flow-based generative models include NICE [3] and RealNVP [4]. However, they suffer from poor generation quality. Recently, Glow was proposed by Kingma *et al.* [11]. Glow can generate realistic-looking images, achieving similar and sometimes better performance than other generative algorithms like VAE and GAN.

2.3 Mutual Information

Mutual Information (MI) is a quantity measuring the dependencies between two variables. Intuitively, it estimates the amount of information that one can obtain about one variable when observing the other [1]. MI is more powerful than correlations. Correlations can only measure dependencies between two linearly dependent variables. MI can tackle non-linearity among variables [20]. Thus, MI is employed to investigate how learning is achieved in a deep neural network with many non-linear layers. Examples include pc-softmax and the information-bottleneck theory [20,22]. The formula for MI $\mathbb{I}(\mathbf{x};\mathbf{y})$ between variables $\mathbf{x}$ and $\mathbf{y}$ is:

$$\mathbb{I}(\mathbf{x};\mathbf{y}) = \mathbb{E}_{(\mathbf{x},\mathbf{y})}\left[\log\left(\frac{P(\mathbf{x},\mathbf{y})}{P(\mathbf{x})P(\mathbf{y})}\right)\right].$$

3 Conditions When Deep Architectures Lose Information

In this section, we show interesting circumstances under which a deep neural architecture loses information about the given input data. To this end, we first present the definition of an invertible deep neural architecture:

Definition 1. *A deep neural architecture is invertible if and only if:*

1. *It satisfies the function property of being deterministic.*
2. *It meets the definition of a one-to-one function.*

Each input $\mathbf{x}$ corresponds to a unique resultant variable $\mathbf{z}$, We consider the cases of the variables being discrete and continuous separately. If the variables are discrete, one immediate implication of Definition 1 is $P(\mathbf{x}|\mathbf{z}) = 1$, *i.e.*, the conditional probability of the input data $\mathbf{x}$ given the output $\mathbf{z}$ is one, where the output $\mathbf{z}$ can either be the intermediate features or the final output. In contrast, for discrete variables, a non-invertible deep neural architecture has $P(\mathbf{x}|\mathbf{z}) < 1$, since multiple different input $\mathbf{x}$ can lead to the same output $\mathbf{z}$. Furthermore, if the variables are discrete, we consider $P(\mathbf{x}|\mathbf{z})$ as a probability, thus it cannot exceed 1. As to the cases of continuous variables, we use a conclusion from [12], stating that MI is invariant under the smooth invertible transformations of the variables, to show invertible can preserve information about the input data. We then have the following proposition:

Proposition 1. *If a deep neural architecture is not invertible, then it will lose information from the input during the feed-forward process.*

Proof. Following the above definition, we use $\mathbf{x}$ and $\mathbf{z}$ to respectively denote the input data and the middle or final features of $\mathbf{x}$ processed by a neural network. We employ MI, denoted as $\mathbb{I}(\mathbf{x};\mathbf{z})$, to represent the information that $\mathbf{z}$ carries about $\mathbf{x}$. From the definition of MI, we have:

$$\mathbb{I}(\mathbf{x};\mathbf{z}) = \mathbb{E}_{(\mathbf{x},\mathbf{z})}\left[\log\left(\frac{P(\mathbf{x},\mathbf{z})}{P(\mathbf{x})P(\mathbf{z})}\right)\right] = \mathbb{E}_{(\mathbf{x},\mathbf{z})}\left[\log\left(\frac{P(\mathbf{x}|\mathbf{z})}{P(\mathbf{x})}\right)\right].$$

From the definition of invertible deep neural architectures, if variables $\mathbf{x}$ and $\mathbf{z}$ are discrete, we have $p(\mathbf{x}|\mathbf{z}) = 1$ if and only if the architecture is invertible. In contrast, when not invertible, we have: $p(\mathbf{x}|\mathbf{z}) < 1$. That is, the MI between $\mathbf{x}$ and $\mathbf{z}$ is larger when the network is invertible as compared to not being invertible.

When variables $\mathbf{x}$ and $\mathbf{z}$ are continuous, we utilize the fact that MI is invariant under the smooth invertible transformations of the variables [12]. That is, given two continuous variables $\mathbf{m}$ and $\mathbf{n}$, we have:

$$\mathbb{I}(\mathbf{m};\mathbf{n}) = \mathbb{I}(f(\mathbf{m});g(\mathbf{n})), \tag{1}$$

where both functions f and g are smooth and invertible functions. Therefore, if function $f(\mathbf{x}) = \mathbf{x}$, and function $g(\mathbf{x})$ is an invertible neural architecture, outputting $\mathbf{z}$, *i.e.*, $g(\mathbf{x}) = \mathbf{z}$. Then, we have:

$$\mathbb{I}(\mathbf{x};\mathbf{x}) = \mathbb{I}(f(\mathbf{x});g(\mathbf{x})) = \mathbb{I}(\mathbf{x};\mathbf{z}). \tag{2}$$

That is, the MI between input $\mathbf{x}$ and output $\mathbf{z}$ is the same as the MI between input $\mathbf{x}$ and itself, implying output $\mathbf{z}$ preserves information of input $\mathbf{x}$.

From Proposition 1, we note that we require invertible deep neural networks to maintain all the detailed information about the input data.

4 Flow-Based Image Restoration Models

In Subsect. 2.2, we have described the requirement of flow-based generative models. Flow-based generative models require an invertible mapping between the input and the latent tensors, directly conducting maximum likelihood estimation for the given data. Due to the invertibility between inputs and outputs, as Proposition 1 suggests, flow-based models are information-lossless.

We also require a image-restoration model to be information-lossless. Therefore, we investigate the empirical performance of applying flow-based invertible deep architectures to address image restoration tasks.

4.1 Architecture Overview

Figure 1 presents an overview of our deep architecture for image restoration. Our primary requirement is to make the architecture completely invertible to preserve all information about the given data, as we have described in Sect. 3. To fulfill the invertible requirement, we aim for an encoding-decoding symmetric image-restoration deep architecture. In the architecture, every component is invertible. In the subsequent sections, we describe each component of our architecture in detail.

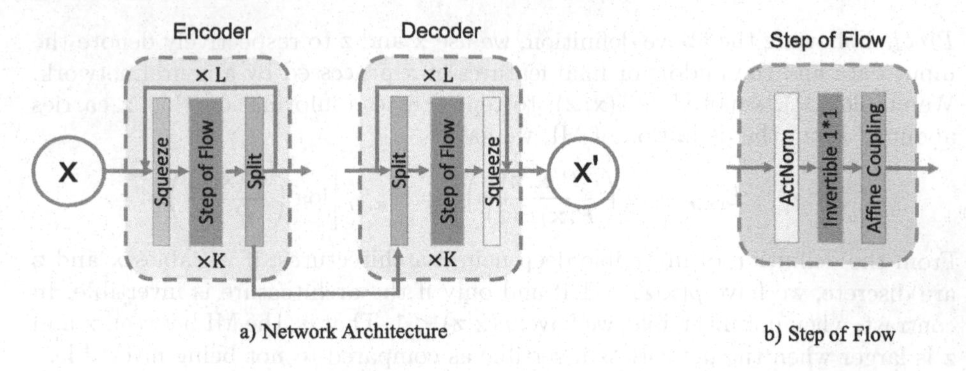

Fig. 1. Architecture of our Invertible Restoring Autoencoder (IRAE) network. "ActNorm" means "activation normalization".

4.2 Encoder and Decoder

Our auto-encoding architecture is inspired by Glow [11], a flow-based generative model. Two motivations inform this choice: 1) We require an invertible deep architecture to preserve all the information about the input data, and 2) among flow-based models, Glow can generate the highest quality images. However, our architecture is considerably different from the Glow architecture. While Glow contains only an encoder and relies on the inverse function of the encoder to reconstruct the images, we utilize an additional decoder, and hence our architecture is symmetric, as Fig. 1 depicts. We argue that the decoder is mandatory because, unlike reconstructing the given images, our model aims to convert the given perturbed data to the original forms. Nonetheless, the entire architecture is still invertible since encoders and decoders are invertible.

4.3 Invertible Local Spatial Feature Extraction

The great success of CNNs can be attributed to their ability to leverage local spatial features. Specifically, CNN filters can exploit the 2D-spatial structure of images via employing spatial convolutions to extract the local information around each pixel. Unfortunately, this operation is non-invertible due to dimension reduction, which leads to further information loss. To address this dimension reduction, we borrow ideas from a flow-based model called Real-NVP and utilize the spatial checkerboard pattern within Real-NVP to apply the local spatial feature aggregation. Particularly, we squeeze a $1 \times 4 \times 4$ tensor, where the order is "channel $\times$ width $\times$ height", into a $4 \times 2 \times 2$ tensor. Consequently, each resultant channel corresponds to a 4×4 region of the original image. We then perform a 1×1 convolution to aggregate channel information together. In this squeezing and 1×1 convolving fashion, we facilitate extracting local spatial features invertibly.

4.4 Steps of Flow

We follow the same set of flow steps as the Glow model, consisting of three invertible sub-steps:

1. **Activation Normalization** abbreviated as ActNorm, performs an affine transformation with a scale and a bias parameter per channel. The intention is to initialize the first minibatch to have mean zero and standard deviation of one after ActNorm to address covariate shift, similar to batch normalization.
2. 1×1 **Convolution** can be viewed as a linear transformation without shrinking dimensions. Thus, it is invertible.
3. **Affine Coupling** aims to mix information of different dimensions. It consists solely of invertible functions, so the composite function is still invertible.

4.5 Loss Function

Assume the training pairs we use are $\{\boldsymbol{x}_i, \boldsymbol{y}_i\}_{i=1}^N$, where $\boldsymbol{x}_i$ is the ground truth image and $\boldsymbol{y}_i$ is the corresponding corrupted image,

$$\boldsymbol{y}_i = \boldsymbol{A} \otimes \boldsymbol{x}_i + \boldsymbol{n}_i, \tag{3}$$

where $\otimes$ is the element-wise multiplication. $\boldsymbol{A}$ is the degradation matrix (which is an identity matrix for image denoising tasks). We follow the standard loss function for image restoration tasks and use the ℓ_1 loss as the objective function,

$$L_p = \frac{1}{N}\sum_{i=1}^{N} ||IRAE(\boldsymbol{y}_i) - \boldsymbol{x}_i||_1, \tag{4}$$

where L_p is the pixel loss and Invertible Restoring Autoencoder (*IRAE*) is our network.

5 Experiments

To evaluate the image restoration performance of our invertible deep neural architecture, we conduct experiments on three tasks: 1) image denoising, 2) JPEG compression, and 3) image inpainting. All of these experiments show that our model can consistently restore images to their original forms, and better than other competitive methods. The quantitative results of noise removal even show a large-margin improvement. More pleasantly, despite superior performance, our invertible architecture contains fewer parameters than other competitive models. Specifically, our model has 1.33×10^6 parameters, whilst DnCNN [30] and U-Net [21] have 1.48×10^6 and 7.70×10^6 parameters respectively.

Table 1. Quantitative denoising performance of our model compared against others. The column σ stands for different noise levels. "BLD" represents blind denoising. Values are average PSNR(dB). The best results are highlighted in bold.

Dataset	σ	DnCNN [30]	FFDNet [31]	U-Net [21]	**Ours**
CelebA	15	31.4597	31.8551	31.7719	**33.0812**
	25	29.4340	28.9438	30.5915	**31.0795**
	50	26.7126	25.5980	26.4775	**28.0887**
	BLD	30.3496	30.8492	30.8807	**31.6158**
Bird	15	30.0825	31.1383	31.9578	**33.0805**
	25	28.4882	28.3442	30.3652	**31.0111**
	50	26.4529	26.4082	27.7812	**28.0976**
	BLD	28.7662	28.4686	31.4173	**31.4448**
Flower	15	29.8116	30.8773	32.0690	**32.6267**
	25	28.8444	28.2578	30.2644	**30.5044**
	50	26.1073	26.1131	27.1729	**27.5092**
	BLD	28.4645	28.7165	31.3378	**31.6123**

5.1 Experimental Settings

For our architecture *IRAE* (see Fig. 1) (Invertible Restoring Autoencoder), we apply $K = 16$ and $L = 2$ for the encoding and decoding layers. We adopt Adam [9] as the optimizer and the learning rate as 10^{-3} initially, which decays to 2×10^{-4} after the first 50 epochs. Afterward, if the peak signal-to-noise ratio (PSNR) does not improve for 10 epochs, the learning rate decays to a fifth of the original rate. The training terminates when the learning rate decreases below 10^{-6}.

The evaluation metrics we use for comparison are the average peak signal-to-noise ratio (PSNR) and the structural similarity index (SSIM). Higher values indicate better performance for both metrics.

5.2 Comparison on Image Restoration Tasks

To show the competitive performance of our *IRAE* model, we compare its performance on the following three image restoration tasks.

- **Image Noise Removal.** We first evaluate the denoising capability of *IRAE*on three types of images: CelebA (human faces) [17], Flower (natural plants) [19] and Bird [27]. The synthetic noise added to these datasets is standard additive white Gaussian noise (AWGN) [26], with three distinct standard deviations: $\sigma = 15, 25, 50$. We also conduct experiments for blind denoising with random noise levels between 0 to 55. Table 1 quantitatively demonstrate that our model performs denoising better than other approaches by a large margin. Figure 2 exhibits qualitative results of the denoising performance of our model compared against others.

Fig. 2. Qualitative visualization of image denoising of our model compared with other methods. The noise level $\sigma = 50$.

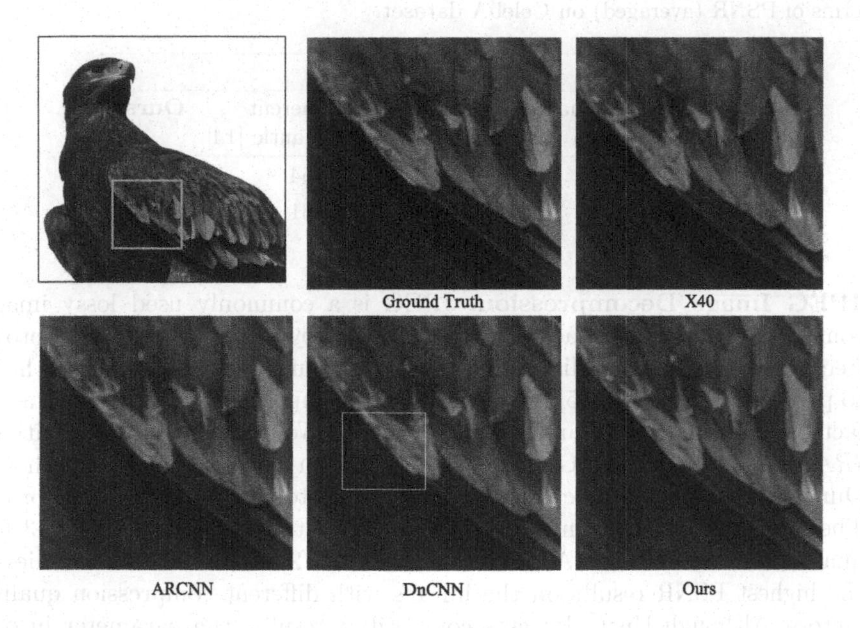

Fig. 3. Qualitative visualization of JPEG decompression of our model compared with other methods. Our model preserves details more clearly and does not have artifacts as in the yellow box in DnCNN while remaining invertible and requiring few parameters. (Color figure online)

Table 2. Quantitative JPEG decompression performance of our model compared against others. 'QF' means "quality factor". Higher QFs represent less compression loss. PSNR values are average (in dBs). The best results are highlighted in bold.

QF	Mod			
	AR-CNN [5]	DnCNN [30]	U-Net [21]	**Ours**
10	27.8221	27.5406	28.7041	**28.7322**
20	29.9814	28.8870	30.9274	**30.9984**
30	31.1935	28.6428	32.0025	**32.1832**
40	31.9283	31.0741	32.8755	**33.0835**

(a) Quantitative JPEG decompression results in PSNR (dB).

QF	Mod			
	AR-CNN [5]	DnCNN [30]	U-Net [21]	**Ours**
10	0.9368	0.9326	**0.9479**	0.9476
20	0.9568	0.9469	0.9647	**0.9652**
30	0.9658	0.9350	0.9712	**0.9715**
40	0.9702	0.9648	0.9757	**0.9765**

(b) Quantitative JPEG decompression results in SSIM.

Table 3. Quantitative results of our model compared with others on image inpainting, in terms of PSNR (averaged) on CelebA dataset.

Metrics	Models			
	Contextual attention [29]	Shift-Net [28]	Coherent semantic [14]	**Ours**
PSNR	23.93	26.38	26.54	**27.14**
SSIM	0.882	0.926	0.931	**0.975**

– **JPEG Image Decompression.** JPEG is a commonly used lossy image compression method. It achieves compression by converting images into a frequency domain, then discarding the high-frequency regions that are hard to perceive for humans [15]. However, JPEG compression often leads to artefacts, such as blockiness and 'mosquito noise'. We evaluate the capability of *IRAE*to decompress JPEG images in comparison with competitive methods. Our model can reconstruct JPEG images back to their near-original forms. The effectiveness is shown in Table 2 for quantitative results and Fig. 3 for qualitative visualization. As illustrated in Table 2, our architecture achieves the highest PSNR results on the images with different compression quality factors. Although Unet also gets competitive results, the parameter in our network is 1.33×10^6, which is far less than the parameter used in Unet (7.70×10^6). For factor 10, we achieved almost the same SSIM result but higher PSNR value than Unet, which also demonstrates our method's superiority in denoising.

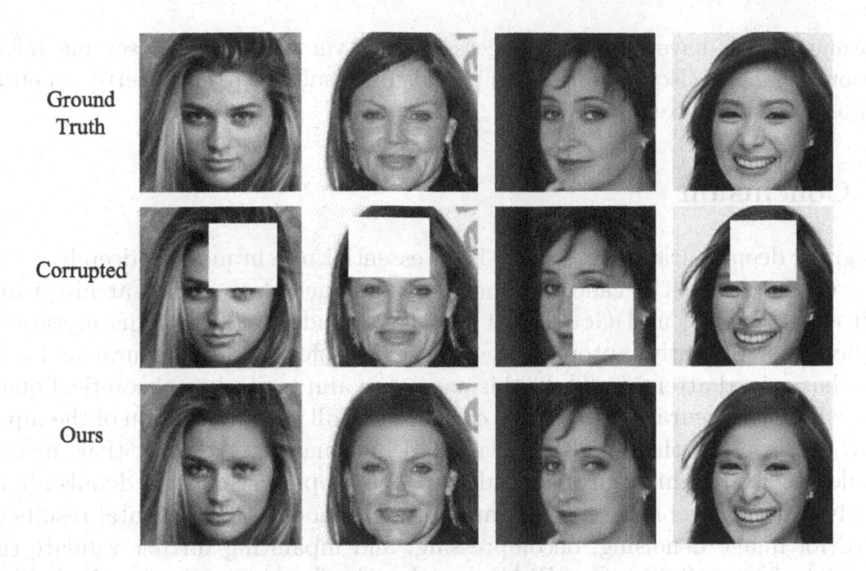

Fig. 4. Image inpainting. Our model is general and has not been customized for image inpainting like the state-of-the-art methods.

- **Image Inpainting.** Lastly, we show that our invertible deep neural architecture also performs better for image inpainting. We employ the CelebA dataset with a size of 256 × 256, followed by randomly generating masks of size 128 × 128, overlapped on each image. We ignore masking parts that are outside the central region of images. We are pleased to find that our model outperforms, even by a large margin, other methods that are specifically designed for image inpainting, such as the ones based on adversarial training, as Table 3 indicates. Figure 4 presents a qualitative visualization of our model's results on image inpainting.

6 Discussion and Future Work

One may concern that: preserving all the information of the given images is not logically plausible. To specify, the noises on an image is also a part of the information. Then, how can we conduct denoising if we also preserve the information of noises? However, preserving the noise information is not equivalent to keeping noises visible. That is, if a model maps noises to values that are extremely close to zero, then noises on images become invisible. Consequently, the model performs denoising well. Also, from the information-theoretic perspective, the model preserves all the information of the given image. Nevertheless, if the model loses the information of the visually salient regions, then even the model can remove all the noises, it is still not acceptable.

This paper can be inception of investigating whether performance improves if neural network components become invertible. With the definition of mutual information, we have shown the necessity of invertibility for preserving information. We have also demonstrated promising results of using invertible neural networks for image restoration.

7 Conclusion

Designing deep neural architectures is an essential role in modern deep learning research. In the past decade, the manually designed deep neural architectures such as VGGs [23] and GoogLeNet [25] have made breakthroughs in various applications. Recently, automatic searching for effective deep neural architectures has gained attention [6]. In this paper, we aim to study a theoretical question: what deep neural architectures can preserve all the information of the input data? We leverage the definition of mutual information. We show that: invertible deep neural architectures are indispensable to preserve all the details about the given data. We propose IRAE, an invertible model. Experimental results of IRAE for image denoising, decompressing, and inpainting further validate the necessity of invertibility. Our IRAE even has far fewer parameters. We believe our theoretical results and practical demonstration in this paper imply that: making deep neural architectures invertible can be a promising future direction for deep learning research.

References

1. Belghazi, M.I., et al.: Mutual information neural estimation. In: International Conference on Machine Learning, pp. 531–540 (2018)
2. Bengio, Y.: Deep learning of representations for unsupervised and transfer learning. In: Proceedings of ICML Workshop on Unsupervised and Transfer Learning, pp. 17–36 (2012)
3. Dinh, L., Krueger, D., Bengio, Y.: Nice: non-linear independent components estimation. In: ICLR (2015)
4. Dinh, L., Sohl-Dickstein, J., Bengio, S.: Density estimation using real NVP. In: International Conference on Machine Learning (2016)
5. Dong, C., Deng, Y., Change Loy, C., Tang, X.: Compression artifacts reduction by a deep convolutional network. In: Proceedings of the IEEE International Conference on Computer Vision, pp. 576–584 (2015)
6. Elsken, T., Metzen, J.H., Hutter, F.: Neural architecture search: a survey. J. Mach. Learn. Res. **20**, 1–21 (2019)
7. Goodfellow, I., et al.: Generative adversarial nets. In: Advances in Neural Information Processing Systems, pp. 2672–2680 (2014)
8. Hochreiter, S.: The vanishing gradient problem during learning recurrent neural nets and problem solutions. Int. J. Uncertainty Fuzziness Knowl. Based Syst. **6**(02), 107–116 (1998)
9. Kingma, D.P., Ba, J.: Adam: A method for stochastic optimization. arXiv preprint arXiv:1412.6980 (2014)
10. Kingma, D.P., Welling, M.: Auto-encoding variational bayes. In: International Conference of Machine Learning (2013)

11. Kingma, D.P., Dhariwal, P.: Glow: generative flow with invertible 1×1 convolutions. In: Advances in Neural Information Processing Systems, pp. 10215–10224 (2018)
12. Kraskov, A., Stögbauer, H., Grassberger, P.: Estimating mutual information. Phys. Rev. E **69**(6), 066138 (2004)
13. Krizhevsky, A., Sutskever, I., Hinton, G.E.: Imagenet classification with deep convolutional neural networks. In: Advances in Neural Information Processing Systems, pp. 1097–1105 (2012)
14. Liu, H., Jiang, B., Xiao, Y., Yang, C.: Coherent semantic attention for image inpainting. In: Proceedings of the IEEE International Conference on Computer Vision, pp. 4170–4179 (2019)
15. Liu, X., Lu, W., Zhang, Q., Huang, J., Shi, Y.Q.: Downscaling factor estimation on pre-jpeg compressed images. IEEE Trans. Circ. Syst. Video Technol. **30**(3), 618–631 (2019)
16. Liu, Y., Anwar, S., Zheng, L., Tian, Q.: Gradnet image denoising. In: Proceedings of the IEEE/CVF Conference on Computer Vision and Pattern Recognition Workshops, pp. 508–509 (2020)
17. Liu, Z., Luo, P., Wang, X., Tang, X.: Large-scale celebfaces attributes (celeba) dataset (2018). Accessed 15 Aug 2018
18. Mao, X., Shen, C., Yang, Y.B.: Image restoration using very deep convolutional encoder-decoder networks with symmetric skip connections. In: Advances in Neural Information Processing Systems, pp. 2802–2810 (2016)
19. Nilsback, M.E., Zisserman, A.: Automated flower classification over a large number of classes. In: 2008 Sixth Indian Conference on Computer Vision, Graphics & Image Processing, pp. 722–729. IEEE (2008)
20. Qin, Z., Kim, D.: Rethinking softmax with cross-entropy: Neural network classifier as mutual information estimator. arXiv preprint arXiv:1911.10688 (2019)
21. Ronneberger, O., Fischer, P., Brox, T.: U-Net: convolutional networks for biomedical image segmentation. In: Navab, N., Hornegger, J., Wells, W.M., Frangi, A.F. (eds.) MICCAI 2015. LNCS, vol. 9351, pp. 234–241. Springer, Cham (2015). https://doi.org/10.1007/978-3-319-24574-4_28
22. Saxe, A.M., et al.: On the information bottleneck theory of deep learning. J. Stat. Mech. Theory Exp. **2019**(12), 124020 (2019)
23. Simonyan, K., Zisserman, A.: Very deep convolutional networks for large-scale image recognition. arXiv preprint arXiv:1409.1556 (2014)
24. Srivastava, R.K., Greff, K., Schmidhuber, J.: Highway networks. arXiv preprint arXiv:1505.00387 (2015)
25. Szegedy, C., et al.: Going deeper with convolutions. In: Proceedings of the IEEE Conference on Computer Vision and Pattern Recognition, pp. 1–9 (2015)
26. Thangaraj, A., Kramer, G., Böcherer, G.: Capacity bounds for discrete-time, amplitude-constrained, additive white gaussian noise channels. IEEE Trans. Inf. Theory **63**(7), 4172–4182 (2017)
27. Wah, C., Branson, S., Welinder, P., Perona, P., Belongie, S.: The caltech-ucsd birds-200-2011 dataset (2011)
28. Yan, Z., Li, X., Li, M., Zuo, W., Shan, S.: Shift-net: image inpainting via deep feature rearrangement. In: Proceedings of the European Conference on Computer Vision (ECCV), pp. 1–17 (2018)
29. Yu, J., Lin, Z., Yang, J., Shen, X., Lu, X., Huang, T.S.: Generative image inpainting with contextual attention. In: Proceedings of the IEEE Conference on Computer Vision and Pattern Recognition, pp. 5505–5514 (2018)

30. Zhang, K., Zuo, W., Chen, Y., Meng, D., Zhang, L.: Beyond a gaussian denoiser: residual learning of deep CNN for image denoising. IEEE Trans. Image Process. **26**(7), 3142–3155 (2017)
31. Zhang, K., Zuo, W., Zhang, L.: Ffdnet: toward a fast and flexible solution for CNN-based image denoising. IEEE Trans. Image Process. **27**(9), 4608–4622 (2018)
32. Zhang, Y., Tian, Y., Kong, Y., Zhong, B., Fu, Y.: Residual dense network for image restoration. IEEE Trans. Pattern Anal. Mach. Intell. (2020)

Automatic Dropout for Deep Neural Networks

Veena Dodballapur[1], Rajanish Calisa[2], Yang Song[3](✉), and Weidong Cai[1]

[1] School of Computer Science, University of Sydney, Sydney, Australia
[2] School of Electrical and Data Engineering, University of Technology Sydney, Ultimo, Australia
[3] School of Computer Science and Engineering, University of New South Wales, Kensington, Australia
yang.song1@unsw.edu.au

Abstract. A greater demand for accuracy and performance in neural networks has led to deeper networks with a large number of parameters. Overfitting is a major problem for such deeper networks. Dropout is a popular regularization strategy used in deep neural networks to mitigate overfitting. However, dropout requires a hyperparameter to be chosen for every dropout layer. This process becomes tedious when the network has several dropout layers. In this paper, we introduce a method of sampling a dropout rate from an automatically determined distribution. We further build on this automatic selection of dropout rate by clustering the activations and adaptively applying different rates to each cluster. We have evaluated both our approaches using the CIFAR-10, CIFAR-100, and Fashion-MNIST datasets, using two state-of-the-art Wide ResNet variants as well as a simpler network. We show that our methods outperform standard dropout across all datasets and neural networks.

Keywords: Dropout regularization · Convolution neural networks

1 Introduction

Deeper and wider convolution neural networks (CNNs) have been gaining popularity over the last few years in the quest for better feature representation. This has resulted in neural networks with a greater number of network parameters. It is well known that such networks are typically prone to over-fitting. Hence, in practice, regularization methods have been proposed to improve generalization and ultimately network performance. A number of regularization methods have been proposed, such as L2 regularization, early stopping, data augmentation, and dropout, to achieve better model generalization and prevent model over-fitting.

Of all the regularization techniques, the standard dropout [3] and its proposed improvements [9] have been prolific in the advancement of neural network performance since the application of standard dropout in 2012 to win the

H. Yang et al. (Eds.): ICONIP 2020, LNCS 12534, pp. 185–196, 2020.
https://doi.org/10.1007/978-3-030-63836-8_16

Large Scale Visual Recognition Challenge [8]. The popularity of dropout has also been helped by advances in the theoretical understanding of why dropout works. According to the seminal paper [3], dropout is one of the regularization methods which work by reducing co-adaptation of the neurons.

A limitation of the dropout method is that it typically requires a dropout rate hyper-parameter, which defines the probability of a neuron being dropped or retained during training. This parameter, if set too high, results in many neurons being dropped, affecting the convergence of training and, if set too low, does not generalize the model well. In addition, most of the deep network implementations have a fixed dropout rate for all layers of the network. In the original dropout implementation [3], the rate was set to 0.5. Usually the optimal dropout rate for a network is determined through cross-validation.

There have been several approaches proposed in the literature which improve on standard dropout by focussing on improving regularization [5,13,18] and rate of convergence during training [15,20]. The proposed methods are broadly categorized into adaptive and stochastic dropout approaches [9]. Whilst some adaptive dropout techniques are based on structural changes to the network [5], other adaptive techniques have proposed to use additional networks [24] during training. The Gaussian dropout method [14] is a stochastic technique which achieves the dropout effect by adding a predetermined Gaussian noise to the neurons. With this method, no neuron is actually dropped. A further extension of this method automatically learns the parameters of a Gaussian noise distribution as part of the network training process by learning a variational objective [6]. This in essence, leads to learning a dropout rate for each layer of the network. A more recent stochastic approach, named jumpout [18], proposes a method sampling the dropout rate from a Gaussian distribution.

Fast dropout [20] proposes a method for faster training convergence by showing that activations during training with dropout can be approximated as a Gaussian distribution. Further extending fast dropout, the variational dropout [23] introduces a Bayesian feature noising model to infer dropout noise for neurons (or features). Targeted dropout [2] proposes a method of dropping a certain proportion of low-valued weights to reduce degradation of performance due to dropout towards the end of training. Curriculum dropout [11] proposes a method of varying dropout to make learning more difficult as the network training progresses.

[4] proposes stochastic depth where a subset of layers are dropped during training, while retaining the entire depth during testing. Swapout [13] is another method which generalizes on both dropout and stochastic depth. In this method, the activations are randomly assigned the value zero, input value, output value, or sum of input and output values. There is also a family of adaptive dropout techniques that rely on additional neural networks or neural network elements. Standout is an adaptive method, where a deep belief network and deep neural network (DNN) with shared parameters are trained jointly to obtain the dropout rate adaptively across layers [1]. In fraternal dropout [24], the network

is trained to be invariant to dropout by training two identical networks with shared parameters but with different dropout masks.

In this paper, we propose two methods of improving the performance of the neural networks using dropout as follows:

- In our first method, we automatically determine the dropout rate to be used for a layer by using the distribution of neuron activations from the previous layer.
- In our second method, we cluster the activations using k-means and apply different dropout rates to activations in each cluster.

Our methods are widely applicable to a variety of neural networks such as auto-encoders, deep neural networks for classification and segmentation where dropout is used for model regularization. They do not rely on making structural changes to the network or additional network elements during training. Unlike the standard dropout, we eliminate the need to search the hyper-parameter space for an optimal dropout rate. Compared to the most recent development, jumpout [18], our method differs by estimating the parameters of the Gaussian automatically from the activations of each layer.

Our method differs from all the dropout methods cited above. Standard dropout [3] requires a hyper-parameter for every layer with dropout function enabled but our method removes the need for this hyper-parameter. Unlike [1] and [24], our method does not introduce any structural changes to the network or need an additional network during training. Our method is different from drop connect [16] in that it does not apply sparsity to weights and biases of neurons. Further, unlike Bayesian methods such as the fast dropout [20] and variational dropout [6], our method does not involve modeling of noise for each neuron as a substitute for dropout. Finally, all of the above methods need a hyper-parameter which is either in the form of a dropout rate or parameters of a Gaussian for their respective implementation.

Our automatic dropout rate method is inspired by jumpout [18], in that we too sample a dropout rate from a Gaussian distribution but unlike jumpout, which requires cross validation to establish the parameters of the Gaussian distribution, our proposed automatic dropout rate selection method determines it from the activation of the neurons at each individual layer.

Our second method that adapts dropout rates to clusters of activations is inspired by a recent method called guided dropout [5]. Our method, instead of learning the strength of a neuron as they do in [5], applies clustering to the activations at training time and applies different dropout rates to each cluster. In our implementation, we modify the dropout rate determined automatically in accordance with the cluster centre of each cluster.

2 Methods

We propose two methods of choosing a dropout rate automatically, eliminating the need to choose a hyper-parameter for a dropout for every given layer

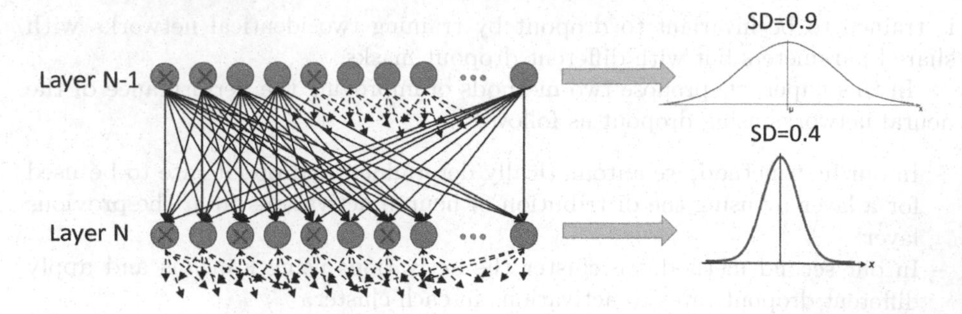

Fig. 1. Illustration of automatic dropout rate calculation at two layers of a neural network and its effect on how dropout rate is sampled and applied to activations in individual layers. Note the shape of the Gaussian affecting the probability of a dropout rate selected.

of the network. In our first method, we choose a dropout rate for a layer by analysing the distribution of neuron activations from the previous layer. We further enhance this method by clustering activations and applying an adaptive dropout for the activations based on their membership to a cluster. Both these methods are described in more detail in following sections.

2.1 Automatic Dropout

As observed by existing studies [12,17,19], a DNN which uses ReLU activations can be thought of as piecewise linear weight function which divides the input plane into regions. For non-zero bias terms which are usually the case with neural networks, the regions form a convex polyhedra. The shape of the polyhedra is based on the activation patterns, each represented by a set of linear constraints.

Our automatic dropout rate method relies on finding smooth approximations to such convex polyhedra. Since ReLU is a cheap and efficient activation function used in state-of-the-art deep networks, our research is focused on improving the performance of such deep networks. Dropping out some activations changes the polyhedral accordingly. Thus, the dropout mechanism trains a family of DNNs, each represented by a slightly different polyhedron. A constant dropout rate can be considered as the average distance between the polyhedra created by dropout. So, the weights of the final model can be thought of as an average of these individual polyhedral regions created as a result of dropout. To obtain good generalization performance, the resulting averaging should result in locally smooth regions. However, with fixed dropout, it can only be achieved when the dropout rate is low.

As in jumpout [18], we sample the dropout rate from a Gaussian distribution which ensures that the probability of dropping out more activations reduces as the number of units dropped out increases. We further extend the work of jumpout [18] by automatically choosing a Gaussian distribution based on the spread of the activated neurons. This ensures that the probability of obtaining a smooth polyhedra not only increases but it is also locally optimized by sampling from a Gaussian distribution of an appropriate shape.

In our method we automatically determine the dropout rate for a dropout layer based on the standard deviation of the distribution of activations of the neurons from the earlier layer. We then sample a dropout rate from this distribution and determine a dropout mask based on the dropout rate. To ensure that during the training phase neither too many activations are dropped out nor too few activations are dropped out, we further truncate $|p|$ so that $|p| \in [s_{min}, s_{max}]$, where $0 \leq s_{min} < s_{max} \leq 1$. We note that the standard deviation σ will have a lower value if the activations are all similar to one another when compared to the case where there is a large variation in the magnitude of the activations.

The details of the automatic dropout rate method are illustrated in Algorithm 1. In our implementation, this algorithm is implemented in a separate custom dropout layer into which the activations a_j from the previous j^{th} layer are input. Hyper-parameters s_{min},s_{max} are set such that the sampled dropout rate is clipped to be within acceptable bounds. The details of automatic dropout rate algorithm are also illustrated in Fig. 1. In this figure, we show a part of a neural network with 2 layers – $N-1$ and N. For each layer, during training, we determine the standard deviation of the activations. In this example, for layer $N-1$ the computed standard deviation is 0.9, for layer N it is 0.4. The computed standard deviation is used to determine the parameters of the Gaussian distribution to sample a dropout rate from. In all our experimental results, we refer to this method by name as M_1.

Algorithm 1. Automatic Dropout (M_1)

Input a_j,s_{min},s_{max} where a_j are activations of the jth layer
Compute the mean of the activations
$a_{ave} = (\sum a_j)/n$
Compute standard deviation of activations
$a_{std} = \sqrt{(\sum (a_j - a_{ave})^2)/(n-1)}$
Compute dropout probability
$N(0, a_{std}), p_{trunc} = min(s_{min} + p, s_{max})$
Apply dropout to activations with a rate p_{trunc}

2.2 Adaptive Dropout Based on Clustering

Inspired by [5], we hypothesize that varying dropout rates based on the activation itself would allow the network to generalize better. Thus, we propose an enhancement over standard dropout to adaptively dropout neurons based on the strength of the activation. Our adaptive dropout strategy is based around the clustering of activations and applying varying levels of dropout rates for each cluster. The details of the adaptive dropout based on clustering are illustrated in Algorithm 2. In our implementation, this algorithm is implemented in a separate custom dropout layer along with the automatic dropout method to which the activations a_j from the previous j^{th} layer are input. Hyper-parameters s_{min},s_{max} are set to values so that the sampled dropout rate is clipped to be within acceptable bounds.

Algorithm 2. Adaptive dropout using clustering (M_2)

Input a_j,s_{min},s_{max} where a_j are activations of the jth layer
Compute the mean of the activations
$c_k = kmeans(a_j)$
for $k = 1$ to 3 **do**
  Determine the cluster with the highest mean
  $c_{max} = max(c_k)$
end for
Compute p_{trunc} according to either algorithm M_1 or use a fixed dropout rate
for $k = 1$ to 3 **do**
  Compute dropout probability for each cluster k
  $p_k = p_{trunc} * (c_k/c_{max})$
end for
Apply dropout p_k to activations belonging to cluster k

In [5], a separately learned parameter called strength of a node, a scalar value indicating the importance of the node/neuron learned during training. In our method, we simplify this formulation by using the magnitude of the activation as a proxy for the strength. Further, we perform k-means clustering of the activations at training time and group the activations into N clusters. We then order the clusters in descending order of the cluster mean and apply a dropout rate which is inversely proportional to the cluster mean (details later in this section). The intuition underlying this dropout strategy is as follows. We want to progressively decrease the dropout rate as the strength of activation decreases. Similar to [5], we argue that by dropping higher strength activations at a higher dropout rate, the network generalizes better by relying on activations of a lower strength. Note that this method can use a fixed dropout rate (dropout as a hyper-parameter) as well as the automatically determined dropout rate proposed earlier and progressively adapt to different clusters of activations.

We rely on [10] to provide a theoretical understanding of why the adaptive dropout using clustering works. The process of dropout as a regularization method can be understood by considering a neural network with input $x \in \mathbb{R}^{d_2}$. Let $y \in \mathbb{R}^{d_1}$ be the output feature vector obtained by the neural network represented by the matrix M. Then, $y = Mx$ for some $M \in \mathbb{R}^{d_1 \times d_2}$. Now if we assume that the neural network has two hidden layers with respective weight matrices $U \in \mathbb{R}^{d_1 \times r}$ and $V \in \mathbb{R}^{d_2 \times r}$, the goal of learning is to determine the weight matrices corresponding to the hidden layers U, V such that the following objective in Eq. (1) is minimized:

$$f(U, V) = \mathbb{E}[\| (y - UV^T x) \|^2]. \quad (1)$$

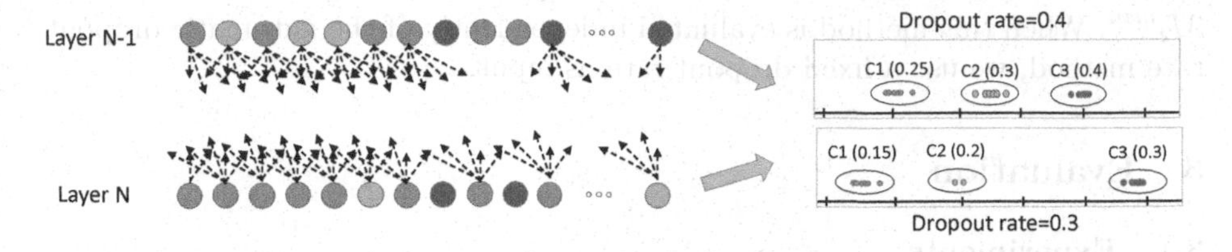

Fig. 2. Illustration of how adaptive dropout based on clustering works in combination with automatic dropout rate selection. The clustering of activation allows different dropout rates to be applied selectively to activations within a cluster based on the strength of the activation.

In the presence of dropout, Eq. (1) can be rewritten as follows:

$$f(U,V) = \mathbb{E}[\| (y - \frac{1}{\theta}Udiag(b)V^T x) \|^2], \tag{2}$$

where $b \in Bernoulli(\theta)$, which means the indicator function b is sampled from Bernoulli trials. In [10], Lemma A.1 is used to show that Eq. (2) can be rewritten as follows:

$$f(U,V) = l(U,V) + \lambda \sum \| u_i \|^2 \| v_i \|^2, \tag{3}$$

where u_i, v_i are the columns of the matrices U, V respectively and λ is a regularization constant equal to $\frac{1-\theta}{\theta}$. The term $\lambda \sum \| u_i \|^2 \| v_i \|^2$ increases in value, hence increasing the overall loss $f(U,V)$ when higher activations are dropped. This allows the network to rely on activations of a lower magnitude to improve their contribution to the overall network generalization. A similar principal is used in [5] with the strength parameter.

The details of the adaptive dropout based on the clustering algorithm are illustrated in Fig. 2. A key feature of this method is to use the magnitude of a neuron activation as a guide to regularizing the network. We argue that just as using large weights leads to network overfitting, large activations also contribute to overfitting. The adaptive dropout through clustering addresses this issue.

As illustrated in Fig. 2, we cluster the activations for each layer independently. In our implementation, we use three clusters which map to low, medium, and high strength activations. We adaptively apply the dropout rate to each cluster as follows. For the cluster with the largest centroid we apply the dropout rate determined according to either Algorithm 1 or using fixed dropout rate. For the next highest cluster, we reduce the dropout rate by an amount c_k/c_{max} where c_k is the cluster mean and c_{max} is the maximum cluster mean. This process is repeated for all clusters. This allows the network to not overly rely on higher strength activations during training and leads to better generalization. In all our experimental results, we refer to this method by name as M_2. For ablation study, we also design a variation of this method where we adaptively apply the dropout rate with the highest dropout rate assigned to the cluster with the lowest centroid. In all our experimental results, we refer to this method as

$M_2{}^{min}$. When this method is evaluated independently of the automatic dropout rate method, we use a fixed dropout rate as input.

3 Evaluation

3.1 Experiments

All our experiments were performed using two NVIDIA GeForce GTX 1080 Ti GPUs and the Keras/TensorFlow programming platform. We used Wide residual network (Wide ResNet) [22] and v3 network[1] as the baseline. We chose Wide ResNet [22] because it has demonstrated state-of-the-art results over other popular networks. The variants of Wide ResNet, namely $width = 8$, $depth = 16$ and $width = 10$, $depth = 28$ of Wide ResNet, have fewer parameters than ResNet but have performed better on CIFAR [7]. In our experiments involving Wide ResNet and CIFAR datasets, we used stochastic gradient descent (SGD) with Nesterov momentum and cross-entropy loss. The initial learning rate was set to 0.1, weight decay to 0.0005, dampening to 0, and momentum to 0.9. We trained the network for 200 epochs, where the learning rate was dropped by 0.2 at 60, 120, and 160 epochs. In our experiments involving the Fashion-MNIST dataset and Wide ResNet, the initial learning rate was set to 0.001, weight decay to 0.0005, dampening to 0, and momentum to 0.9. We used a time-based learning decay of 1e−5. We trained the network for 50 epochs. For both CIFAR and Fashion-MNIST experiments, mini-batch size was set to 64 and horizontal flip was the only data augmentation used. The training protocol was very similar to [22].

We also performed experiments on a simpler network similar to VGG called the v3 network. This was to apply our methods to a very different architecture when compared to Wide ResNet. For our experiments with the v3 network, we used SGD with an initial learning rate of 0.01 and momentum of 0.9. We trained the network for 50 epochs, where we dropped the learning rate by 0.5 and used momentum decay of 0.05 every 10 epochs. This training protocol is similar to the implementation of v3 network.

For standard dropout, we empirically determined the best performing dropout (=0.3) for Wide ResNet across all datasets and used this value for all the layers of Wide ResNet across all our experiments. For the proposed automatic dropout rate, we limited the automatically determined dropout rate such that it always lay in the interval [0.01, 0.6]. We compared our implementation with the standard dropout [3] as well as jumpout [18]. For the v3 network, we have compared our method with the standard dropout only since the jumpout method did not converge for the datasets we experimented with.

To get a better understanding of our proposed methods, we performed several ablation experiments by combining our two proposed methods and introducing variations on our proposed methods. For these ablation experiments we used the three variants of Wide ResNet namely $width = 8$, $depth = 16$, $width = 10$,

[1] The "v3" network from https://github.com/jseppanen/cifa_lasagne.

Table 1. Results for Wide ResNet

Method	Wide ResNet:16_8			Wide ResNet:28_10		
	CIFAR10	CIFAR100	FASHION	CIFAR10	CIFAR100	FASHION
Dropout	0.945	0.762	0.911	0.949	0.776	0.910
Jumpout	0.939	0.752	0.912	0.926	0.735	0.921
M_1	0.949	0.769	0.925	0.953	0.780	0.921
$M_1 + M_2$	0.951	0.771	0.931	0.956	0.785	0.932

Table 2. Results for v3 network

Method	CIFAR10	CIFAR100	FASHION
Dropout	0.845	0.481	0.920
M_1	0.874	0.495	0.923
$M_1 + M_2$	0.884	0.565	0.932

$depth = 28$, and $width = 4$, $depth = 40$. We chose these variants because they explore the network architecture both in terms of width and depth. We considered a simple training protocol with no data augmentation, an initial learning rate of 0.01, SGD with Nesterov momentum, cross-entropy loss, and time based learning rate decay of 1e−5. We trained for 50 epochs. We chose to use no data augmentation to measure the regularization effect of the methods alone.

3.2 Results

For our experiments we chose the CIFAR-10 [7], CIFAR-100 [7], and Fashion-MNIST [21] datasets which are mainly image classification datasets. To measure the improvement provided by using dropout and its variants, we kept data augmentation to a minimum. We ran the evaluation of each method for each dataset and network 3 times and reported the median results.

Table 1 shows the results of our method using the CIFAR-10, CIFAR-100, and Fashion-MNIST datasets for the two variants of the Wide ResNet we evaluated on. We used the same augmentation method and training protocol [22] for all methods and hence can compare the relative performance of our proposed methods against both standard dropout and jumpout. We note that across both variants of Wide ResNet networks tested, our methods outperform the standard dropout and jumpout for the CIFAR-10, CIFAR-100, and Fashion-MNIST using the automatic dropout rate method. The combination of the first and second methods gave a small improvement over the first method alone.

Table 2 shows the results on the v3 network. For the v3 network, we ran experiments for dropout and our proposed methods only. We note that, as in the case of Wide ResNet, our automatic dropout rate method, as well as the combination of automatic dropout rate and adaptive dropout with clustering, outperform the standard dropout.

Table 3. Ablation experiments

Method	Result for Wide ResNet:16_8			Result for Wide ResNet:28_10		
	FASHION	CIFAR10	CIFAR100	FASHION	CIFAR10	CIFAR100
Dropout	0.910	0.856	0.549	0.910	0.810	0.612
Jumpout	0.910	0.834	0.548	0.920	0.820	0.623
M_1	0.925	0.874	0.632	0.920	0.850	0.636
$M_1 + M_2$	0.932	0.878	0.641	0.932	0.889	0.644
M_2	0.934	0.870	0.600	0.929	0.886	0.649
$M_1 + {M_2}^{min}$	0.920	0.870	0.600	0.929	0.870	0.607
${M_2}^{min}$	0.915	0.858	0.589	0.929	0.862	0.575

Table 4. Results for Wide ResNet:40_4

Method	FASHION	CIFAR10	CIFAR100
Dropout	0.910	0.780	0.591
Jumpout	0.920	0.830	0.592
M_1	0.920	0.862	0.612
$M_1 + M_2$	0.923	0.870	0.618
M_2	0.921	0.868	0.611
$M_1 + {M_2}^{min}$	0.920	0.844	0.547
${M_2}^{min}$	0.918	0.833	0.546

Table 3 and Table 4 show the results of our ablation experiments. Firstly, we note that the results across different datasets are slightly lower than the state-of-the-art because data augmentation and highly tuned learning rates were not used. Here we show that across all networks and datasets, our first method (M_1) outperforms the standard dropout as well as jumpout [18]. Further combining the first and second methods ($M_1 + M_2$) performs better than the first method alone. However, we note that introducing M_2 slows down training performance by a factor of 5 when compared to M_1 alone because the k-means algorithm is computationally expensive. We hope to address this in a future work. We also note that in some instances our second method (M_2) with a carefully chosen dropout rate performs as well as the first and second methods combined ($M_1 + M_2$). This confirms the utility of both methods in their own right. Finally, the variation on the second method as described earlier (${M_2}^{min}$) on its own and in combination with first method (M_1) performs worse but is still comparable to jumpout for fashion and CIFAR-10 datasets, confirming our original hypothesis about dropping high strength activations more often for better generalization.

4 Conclusion

In this paper, we have proposed two methods for improving dropout. Our first method removes the need for choosing a dropout rate hyper-parameter. Our second method applies dropout rate adaptively to activations based on clustering. We have demonstrated that both our methods outperform standard dropout when applied individually as well as when combined. We have demonstrated the versatility and applicability of our methods of our results by evaluating on well-known datasets such as CIFAR and Fashion-MNIST using three variants of Wide ResNet and a substantially different network such as the v3 network.

References

1. Ba, J., Frey, B.: Adaptive dropout for training deep neural networks. In: Advances in Neural Information Processing Systems, pp. 3084–3092 (2013)
2. Gomez, A.N., Zhang, I., Swersky, K., Gal, Y., Hinton, G.E.: Targeted dropout (2018)
3. Hinton, G.E., Srivastava, N., Krizhevsky, A., Sutskever, I., Salakhutdinov, R.R.: Improving neural networks by preventing co-adaptation of feature detectors. arXiv preprint arXiv:1207.0580 (2012)
4. Huang, G., Sun, Yu., Liu, Z., Sedra, D., Weinberger, K.Q.: Deep networks with stochastic depth. In: Leibe, B., Matas, J., Sebe, N., Welling, M. (eds.) ECCV 2016. LNCS, vol. 9908, pp. 646–661. Springer, Cham (2016). https://doi.org/10.1007/978-3-319-46493-0_39
5. Keshari, R., Singh, R., Vatsa, M.: Guided dropout. In: Proceedings of the AAAI Conference on Artificial Intelligence, vol. 33, pp. 4065–4072 (2019)
6. Kingma, D.P., Salimans, T., Welling, M.: Variational dropout and the local reparameterization trick. In: Advances in Neural Information Processing Systems, pp. 2575–2583 (2015)
7. Krizhevsky, A., Nair, V., Hinton, G.: CIFAR-10 and CIFAR-100 datasets. https://www.cs.toronto.edu/kriz/cifar.html (2009)
8. Krizhevsky, A., Sutskever, I., Hinton, G.E.: ImageNet classification with deep convolutional neural networks. In: Advances in Neural Information Processing Systems, pp. 1097–1105 (2012)
9. Labach, A., Salehinejad, H., Valaee, S.: Survey of dropout methods for deep neural networks. arXiv preprint arXiv:1904.13310 (2019)
10. Mianjy, P., Arora, R., Vidal, R.: On the implicit bias of dropout. arXiv preprint arXiv:1806.09777 (2018)
11. Morerio, P., Cavazza, J., Volpi, R., Vidal, R., Murino, V.: Curriculum dropout. In: Proceedings of the IEEE International Conference on Computer Vision, pp. 3544–3552 (2017)
12. Raghu, M., Poole, B., Kleinberg, J., Ganguli, S., Dickstein, J.S.: On the expressive power of deep neural networks. In: Proceedings of the 34th International Conference on Machine Learning, vol. 70, pp. 2847–2854. JMLR. org (2017)
13. Singh, S., Hoiem, D., Forsyth, D.: Swapout: learning an ensemble of deep architectures. In: Advances in Neural Information Processing Systems, pp. 28–36 (2016)
14. Srivastava, N., Hinton, G., Krizhevsky, A., Sutskever, I., Salakhutdinov, R.: Dropout: a simple way to prevent neural networks from overfitting. J. Mach. Learn. Res. **15**(1), 1929–1958 (2014)
15. Tompson, J., Goroshin, R., Jain, A., LeCun, Y., Bregler, C.: Efficient object localization using convolutional networks. In: Proceedings of the IEEE Conference on Computer Vision and Pattern Recognition, pp. 648–656 (2015)

16. Wan, L., Zeiler, M., Zhang, S., Le Cun, Y., Fergus, R.: Regularization of neural networks using DropConnect. In: International Conference on Machine Learning, pp. 1058–1066 (2013)
17. Wang, S., et al.: Analysis of deep neural networks with extended data Jacobian matrix. In: International Conference on Machine Learning, pp. 718–726 (2016)
18. Wang, S., Zhou, T., Bilmes, J.: JumpOut: improved dropout for deep neural networks with rectified linear units (2018)
19. Wang, S., Zhou, T., Bilmes, J.: Bias also matters: bias attribution for deep neural network explanation. In: International Conference on Machine Learning, pp. 6659–6667 (2019)
20. Wang, S., Manning, C.: Fast dropout training. In: International Conference on Machine Learning, pp. 118–126 (2013)
21. Xiao, H., Rasul, K., Vollgraf, R.: Fashion-MNIST: a novel image dataset for benchmarking machine learning algorithms. arXiv preprint arXiv:1708.07747 (2017)
22. Zagoruyko, S., Komodakis, N.: Wide residual networks. arXiv preprint arXiv:1605.07146 (2016)
23. Zhuo, J., Zhu, J., Zhang, B.: Adaptive dropout rates for learning with corrupted features. In: Twenty-Fourth International Joint Conference on Artificial Intelligence (2015)
24. Zolna, K., Arpit, D., Suhubdy, D., Bengio, Y.: Fraternal dropout. arXiv preprint arXiv:1711.00066 (2017)

Bayesian Randomly Wired Neural Network with Variational Inference for Image Recognition

Pegah Tabarisaadi(✉), Abbas Khosravi, and Saeid Nahavandi

Institute for Intelligent Systems Research and Innovation (IISRI), Deakin University, Geelong, VIC, Australia
pegah.tabarisaadi@gmail.com,
{abbas.khosravi,saeid.nahavandi}@deakin.edu.au

Abstract. The architecture of a neural network (NN) plays a significant role in its performance. Recently, automating the process of choosing the architecture has received huge attention. While neural architecture search (NAS) methods mainly concern on searching, the searching space is constrained and designed manually. Stochastic network generators will loosen the constraint as well as automating the process. In this work an uncertainty quantification (UQ) for randomly wired neural networks (RWNN) is investigated. The classical Watts-Strogatz (WS) random graph is utilized as the random generator and a bayes by backprop algorithm is introduced to measure the epistemic and aleatoric uncertainties of RWNN for image recognition tasks. The RandAlexNet architecture is proposed and the algorithm is applied on. We test the algorithm on FashionMNIST, CIFAR10 and MNIST datasets and the good results achieved illustrate the effectiveness of our proposed method.

Keywords: Uncertainty quantification (UQ) · Bayes by backprop · Randomly wired neural network · Epistemic uncertainty · Aleatoric uncertainty

1 Introduction

The machine learning research was revolutionised in 2012 by the introduction of deep learning (DL) algorithms. Compared to conventional machine learning techniques, DL models achieve best-in-class performance, scale effectively with data, are fully transferable, and automatically extract useful information for decision making [1]. DL models implement the hierarchical feature extraction in an end-to-end fashion which does not require manual design. The performance of DL models is always bounded by their engineered architecture. Accordingly, there has been a desire to automate the process of architecture engineering without human intervention. This has made the research in the field of neural architecture search (NAS) in the spotlight.

H. Yang et al. (Eds.): ICONIP 2020, LNCS 12534, pp. 197–207, 2020.
https://doi.org/10.1007/978-3-030-63836-8_17

NAS as a new area of research, is an optimization problem, trying to automate the process of choosing the architecture with the lowest validation error [2–13]. Literally NAS eventually shifts the paradigm from architecture engineering to architecture learning. Reinforcement learning and evolutionary algorithms have been both applied to NAS in recent years [2,12,14–18]. In both categories of algorithms, a set of sequential actions generate a network architecture whose performance metrics are applied as the reward.

The human expert first makes several restricting assumptions on the network type and its components such as the bounds for the number of layers, neurons, filters, learning rate, and so on. These values are often set based on prior knowledge about the performance of the state-of-the-art models. For instance, the filter size for convolutional neural networks is set to values such as 3, 5, and 7 which have been used in models like ResNet [19] or DenseNets [20]. Similar restrictions apply to how the network neurons and layers are connected. Thus, the search space in the controlled NAS is always biased.

The random network generators aim to make the NAS conditions more relaxed as well as producing NNs with reasonable accuracies for tasks such as image recognition [21]. Network generator consist of all possible wiring order that the network is sampled from. Considering stochastic network generators leads to advent of randomly wired neural networks [21]. Traditionally, the network generators have been hand-designed and their searching space was limited. Three random graph generators are used in [21] to minimize the humans involvement in NAS. These are Erdos-Renyi (ER) [22], Barabasi-Albert (BA) [23] and Watts-Strogatz (WS) [24]. According to result reported in [21], WS produces networks obtaining higher or competitive accuracy in comparison to their peers. Authors in [21] also mention that the networks produced by a single generator often have a low accuracy variance.

Estimating uncertainties associated with NN predictions is fundamental in real world applications such as healthcare or autonomous driving [25–28]. In particular, the epistemic uncertainty represents how much the end user can trust network predictions on new samples must be properly quantified [29]. The epistemic uncertainty is mainly related to the data insufficiency and can be reduced through collection of more quality data. The aleatoric uncertainty that represents the inherent uncertainty in data is also considered in UQ. It is mainly caused by the measurement noise or image quality. It is important to mention that the focus of researches on RWNNs have been mainly on the network performance metrics such as forecasting accuracy. UQ and its proper evaluation for RWNNs (in general for NAS problems) has been largely overlooked. The research community has been trying to develop RWNNs with a competitive performance without paying attention to the reliability of predictions made by those networks. In recent years, the so-called Bayesian Deep Learning (BDL) was introduced as an effective UQ method to apply bayesian frameworks to deep NNs [30]. Bayes by backprop introduced by [31] as an effective method to estimate the true posterior probability. This algorithm automatically regularize the network and prevents overfitting as well as making the NN more reliable.

The bayes by backprop algorithm was first applied to feedforward NNs [31]. Later this technique was extended to recurrent [32] and convolutional NNs [33] as well.

In this paper we propose a bayes by backprop algorithm for quantifying the epistemic and aleatoric uncertainties of RWNNs for image recognition tasks. Our method for quantifying uncertainties of RWNN uses WS random graph as the random generator for constructing the NN. However our method generalizes to ER and BA or other random graphs as well. A random architecture should also be considered. We propose RandAlexNet here which is inspired by AlexNet [34]. It should be mentioned that our proposed algorithm can also be applied on other random architectures. The aleatoric and epistemic uncertainties are estimated separately to determine who we should put the blame on (the model or dara) and specify whether we can reduce it or not. Finally we test the proposed algorithm on CIFAR10, FashionMNIST and MNIST datasets and the uncertainties are estimated in all cases.

The rest of this paper is organised as follows. Section 2 provides some background information. The proposed method for uncertainty quantification of RWNN predictions is discussed in Sect. 3. Sections 4 and 5 present the experiments and conclusion respectively.

2 Background

2.1 Variational Inference

The common probabilistic algorithms such as variable elimination [35] are mainly complicated and computationally demanding and slow. Many researches have developed algorithms to approximate the solution of the inference problem. Two main approximation families are sampling and variational methods [36].

While sampling inferences methods collect data samples to estimate the marginal probability, variational inference methods focus on presenting the inference as an optimization problem [37–39]. In this work we use variational inference method as it is of a good reason effective and less computationally demanding than classic sampling methods [40].

A variational distribution q is used to approximate p. The Kullback-Leibler (KL) divergence is applied for this approximation by finding q as close as possible to p. KL divergence is defined as follows:

$$KL(q||p) = \sum_{x} q(x) log \frac{q(x)}{p(x)} \tag{1}$$

Considering the x^* and y^* as the new input and predicted output respectively. Equation 2 predicts the new output.

$$p(y^*|x^*, X, Y) = \int p(y^*|f^*)p(f^*|x^*, X, Y)df^* \tag{2}$$

As the integral Eq. 2 is intractable, a random variable w is chosen to approximate 2 as follows:

$$p(y^*|x^*, X, Y) = \int p(y^*|f^*)p(f^*|x^*, w)p(w|X, Y)df^*dw \tag{3}$$

where $p(w|X,Y)$ is intractable. It will be then approximated by $q(w)$ and KL divergence will be applied:

$$q(y^*|x^*) = \int p(y^*|f^*)p(f^*|x^*, w)q(w)df^*dw \tag{4}$$

We finally get the KL as this:

$$KL_{VI} = \int q(w)p(F|X, w)log\, p(Y|F)dFdw - KL(q(w)||p(w)) \tag{5}$$

This is known as the variational inference technique [33].

2.2 Bayes by Backprop

Variational inference methods are used to approximate the posterior distribution. Bayes by backprop [31,41] as a variational inference technique minimizes the KL divergence to make q_θ as similar as possible to $p(w|D)$. So one has the following optimization problem:

$$\theta^{opt} = arg_\theta min\, KL\, [q_\theta(w|D)||p(w|D)] \tag{6}$$

$$\theta^{opt} = arg_\theta minKL[q_\theta(w|D)||p(w)] - E_{q(w|D)}[log\, p(D|w)] + log\, p(D) \tag{7}$$

Considering Eq. 1, we need to deal with another intractability. First, the true posterior p is approximated with q where θ is a learnable parameter. Then using sampling methods such as Monte Carlo, we sample from the variational posterior q. So the cost function is defined as follows:

$$F(D, \theta) \approx \sum_{i=1}^{n} log\, q_\theta(w^{(i)}|D) - log\, p(w^i) - log\, p(D|w^{(i)}) \tag{8}$$

where n is the number of draws and $w^{(i)}$ is sampled from $q_\theta(w|D)$. The cost function 8 is aimed to be minimized with respect to θ during training. The bayes by backprop has been used for feedforward [31,42], recurrent [43] and convolutional NN [33].

2.3 Randomly Wired Neural Networks

Consider the network generator g as a transformation from the parameter space Θ to the NN architecture space N, $g : \Theta \Rightarrow N$. In system analysis, a network generator can be considered as a function that receives some parameters (Θ) and returns a network architecture [21]. A random parameter s is used in stochastic network generators. $g(\Theta, s)$ for a fixed Θ and different values of s represent a uniform probability distribution over all possible s values [21].

The NN architecture will be achieved as follows:

- A graph is considered (a set of nodes and edges). There is no restriction and any graph from the graph theory can be considered (this step is done by network generator).
- The graph is mapped to a NN. The mapping is arbitrary in general.

In this paper we use WS random graph model for the network generator. The WS random graph [24] is a graph for defining small-world networks. Small-world networks have low average distance between nodes and high clustering. The steps of generating a WS random graph are defined as follows [24]:

- All the nodes (N) are placed in a circle.
- Each node is connected to m nearest neighbor on either side of it. This network is known as a regular network.
- Randomly chose $M \times \beta$ links then rewire one end of the chosen links to another randomly chosen node. β is the probability of rewiring (Fig. 1).

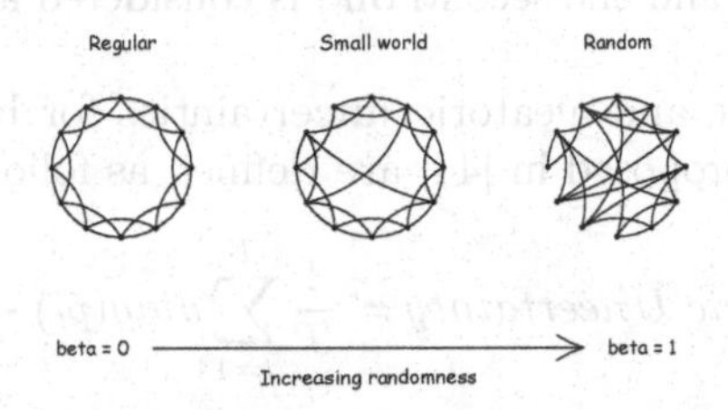

Fig. 1. Regular, small world and random graphs. The regular graph has the high average distance and high clustering. The random graph has small average distance and low clustering. The small world network has small average distance and high clustering. β (probability of rewiring) for regular graph is 0 and for random graph is 1.

3 Methods

3.1 Validation Accuracy and Uncertainty Estimation

Consider x^* is an unseen data and y^* is the predicted output (class), the predictive distribution is defined as follows:

$$p_{\mathfrak{D}}(y^*|x^*) = \int p_w(y^*|x^*)p_{\mathfrak{D}}(w)dw \tag{9}$$

where $p_{\mathfrak{D}}(y^*|x^*)$ In Bayes by Backprop q is considered to have a Gaussian distribution $q_\theta(w|\mathfrak{D}) \backsim \mathcal{N}(w|\mu, \sigma^2)$ and $\theta = \mu, \sigma$ is learned with datasets $\mathfrak{D}$. Assuming categorical distribution for predictive distribution one has:

$$p_{\mathfrak{D}}(y^*|x^*) = \int Cat(y^*|f_w(x^*))\mathcal{N}(w|\mu,\sigma^2) = \int \prod_{c=1}^{C} f(x_c^*|w)^{y_c^*} \frac{1}{\sqrt{2\pi\sigma^2}} e^{-\frac{(w-\mu)^2}{2\sigma^2}} dw \tag{10}$$

In Eq. 10, C is the number of all classes and $\sum_c f(x_c^*|w) = 1$. Due to intractability, sampling method is used to estimate the predictive distribution:

$$E_q[p_D(y^*|x^*)] = \int q_\theta(w|D)p_w(y|x)dw \approx \frac{1}{T}\sum_{t=1}^{T} p_{w_t}(y^*|x^*) \tag{11}$$

The predictive variance is defined as follows:

$$\begin{aligned} Var_q(p(y^*|x^*)) &= E_q[yy^T] - E_q[y]E_q[y]^T \\ &= \frac{1}{T}\sum_{t=1}^{T} diag(\hat{p}_t) - \hat{p}_t\hat{p}_t^T + \frac{1}{T}\sum_{t=1}^{T}(\hat{p}_t - \bar{p})(\hat{p}_t - \bar{p})^T \end{aligned} \tag{12}$$

where $\bar{p} = \frac{1}{T}\sum_{t=1}^{T}\hat{p}_t$ and $\hat{p}_t = Softmax(f_{w_t}(x^*))$ In Eq. 12, the first term is considered as aleatoric and the second one is considered as epistemic uncertainties.

The mean epistemic and aleatoric uncertainties for bayesian RandAlexNet based on the method proposed in [44] are defined as follows:

$$Aleatoric\ Uncertainty = \frac{1}{T}\sum_{t=1}^{T} diag(\hat{p}_t) - \hat{p}_t\hat{p}_t^T \tag{13}$$

$$Epistemic\ Uncertainty = \frac{1}{T}\sum_{t=1}^{T}(\hat{p}_t - \bar{p})(\hat{p}_t - \bar{p})^T \tag{14}$$

The aleatoric or statistical uncertainty represents the inherent probabilistic variability. The aleatoric uncertainty is irreducible and represents how noisy the data are. On the other hand epistemic or systematic uncertainty refers to the uncertainty of the model. The epistemic uncertainty can be reduced by preparing more data. Calculating the epistemic and aleatoric uncertainties can be helpful to distinguish if the quality of the data is not good enough (high aleatoric uncertainty) or the model itself should be blamed (high epistemic uncertainty).

3.2 Activation Function

Also the most popular activation function in NN architecture is ReLU [33], in this paper we use *Softplus* activation function. The main advantage of *Softplus* function defined in Eq. 15 is that it never becomes zero for $x \to -\infty$, but the ReLU function becomes zero when $x \to -\infty$. The *Softplus* function is defined as follows:

$$Softplus(x) = \frac{1}{\beta}.log(1 + exp(\beta.x)) \tag{15}$$

The most common value for β is 1.

3.3 Network Architecture

We propose RandAlexNet architecture to implement the uncertainty quantification of RWNN on. The specification of RandAlexNet architecture that is inspired from AlexNet [34] is reported in Table 1.

4 Experiments

In this section the details of our proposed algorithm are illustrated and the results and figures are reported. Based on the WS random graph model, the network generator is automated. A bayes by backprop algorithm is utilized on the randomly wired neural network. In all the experiments we conduct our algorithm on RandAlexNet, that is inspired form AlexNets [34]. The exact architecture is represented in Table 1. The proposed algorithm is examined on CIFAR10, FashionMNIST and MNIST datasets. The number of epochs in all cases is 100. The initial learning rate is set as 0.1 and the weight decay and momentum are chosen as 5e−4 and 0.9 respectively. The epistemic and aleatoric uncertainties for CIFAR10, FashionMNIST and MNIST datasets are reported in Table 2. The aleatoric uncertainty of CIFAR10 is larger (almost twice) of the aleatoric uncertainties of FashionMNIST and MNIST. The lower validation accuracy of CIFAR10 in comparison to FashionMNIST and MNIST datasets also corroborate it. The small number of training examples of CIFAR10 can be considered as a contributing factor. The validation accuracies for these datasets are also reported in Table 3. Figures 2, 3 and 4 illustrate the epistemic and aleatoric and total uncertainties per epochs for CIFAR10, MNIST and FashionMNIST dataset. It should be noticed that randomly wired neural networks, proposed in [21], do not suppose to reach noticeably higher accuracies in comparison to other architectures. It is claimed that randomly wired neural networks in most cases reach competitive accuracies in comparison to other image classification architectures. In [21] the best accuracy for image classification on CIFAR10 dataset is reported as 74.7%. However using the information reported in [21], the best accuracies we reached in our simulations was 66.8%. It is also worth mentioning that these results achieved in different situations in comparison to ours. For instance the number of epoch, optimizer, learning rate, momentum and etc are totally different. In one word, there are lots of contributing factors for reaching better accuracies but our work do not focus on improving the accuracies. Our main concern is presenting a bayesian framework for randomly wired neural network and estimating the uncertainties of the predictions. However our accuracy results are competitive and acceptable in comparison to others. The idea in this paper is applied on RandAlexNet (Table 1) architecture, that is inspired by AlexNet [34]. This architecture can also be improved and better results can be achieved. The optimization of the parameters can be achieved by trial and error.

Table 1. The RandAlexNet architecture.

Layer Type	Width	Stride	Padding	Nonlinearity
B-Convolution (3 × 3)	10	1	1	SoftPlus
Max-Pooling (2 × 2)		2	0	
B-Convolution (3 × 3)	10	1	1	SoftPlus
Max-Pooling (2 × 2)		2	0	
Randwire				
Randwire				
B-Convolution (1 × 1)	1280	1	1	SoftPlus
Max-Pooling (2 × 2)		2	0	
Flatten-Layer				
B-Fully Connected				

Table 2. Epistemic and aleatoric uncertainty for bayesian RandAlexNet calculated for MNIST, FashionMNIST and CIFAR10.

	Epistemic uncertainty	Aleatoric uncertainty
MNIST	0.0413	0.1123
FashionMNIST	0.0402	0.1073
CIFAR10	0.0367	0.2113

Table 3. Validation accuracies (in percentage) for RandAlexNet for MNIST, Fashion-MNIST and CIFAR10.

Dataset	Validation accuracy (%)
MNIST	87.5
FashionMNIST	87.5
CIFAR10	66.88

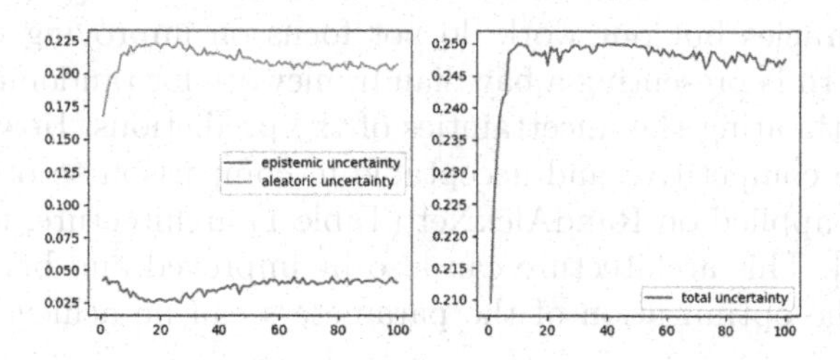

Fig. 2. The epistemic, aleatoric and total uncertainties per epochs for CIFAR10 dataset.

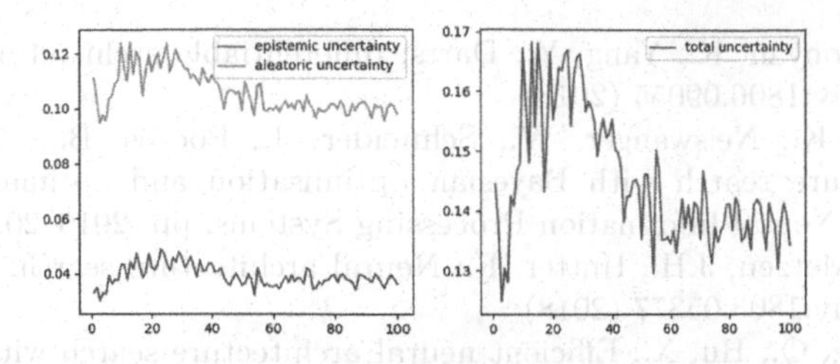

Fig. 3. The epistemic, aleatoric and total uncertainties per epochs for MNIST dataset.

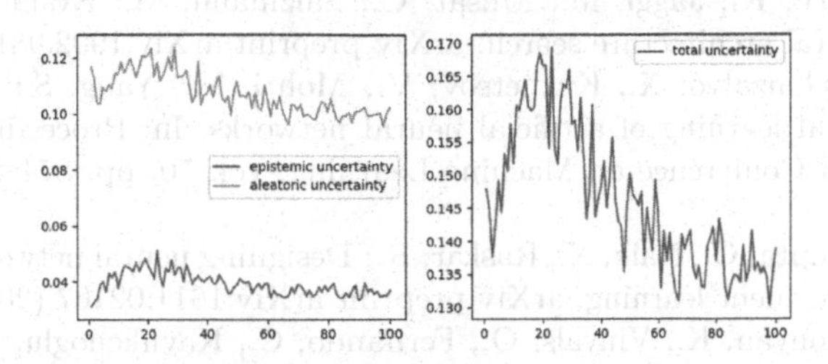

Fig. 4. The epistemic, aleatoric and total uncertainties per epochs for FashionMNIST dataset.

5 Conclusion

In this paper, the uncertainty quantification problem for randomly wired neural networks for the task of image classification is investigated. The WS random graph model is applied as the random network architecture generator. The bayes by back prop algorithm is proposed and applied for tuning parameters of RandAlexNet to compute predictive uncertainty estimates using randomly wired neural network. The epistemic and aleatoric uncertainties are estimated separately. The simulation results on MNIST, FashionMNIST, and CIFAR10 datasets demonstrate the competency of proposed framework for capturing epistemic and aleatoric.

References

1. Zeiler, M.D., Fergus, R.: Visualizing and understanding convolutional networks. In: Fleet, D., Pajdla, T., Schiele, B., Tuytelaars, T. (eds.) ECCV 2014. LNCS, vol. 8689, pp. 818–833. Springer, Cham (2014). https://doi.org/10.1007/978-3-319-10590-1_53
2. Zoph, B., Le, Q.V.: Neural architecture search with reinforcement learning. arXiv preprint arXiv:1611.01578 (2016)
3. Liu, C., et al.: Progressive neural architecture search. In: Ferrari, V., Hebert, M., Sminchisescu, C., Weiss, Y. (eds.) ECCV 2018. LNCS, vol. 11205, pp. 19–35. Springer, Cham (2018). https://doi.org/10.1007/978-3-030-01246-5_2
4. Pham, H., Guan, M.Y., Zoph, B., Le, Q.V., Dean, J.: Efficient neural architecture search via parameter sharing. arXiv preprint arXiv:1802.03268 (2018)

5. Liu, H., Simonyan, K., Yang, Y.: Darts: differentiable architecture search. arXiv preprint arXiv:1806.09055 (2018)
6. Kandasamy, K., Neiswanger, W., Schneider, J., Poczos, B., Xing, E.P.: Neural architecture search with Bayesian optimisation and optimal transport. In: Advances in Neural Information Processing Systems, pp. 2016–2025 (2018)
7. Elsken, T., Metzen, J.H., Hutter, F.: Neural architecture search: a survey. arXiv preprint arXiv:1808.05377 (2018)
8. Jin, H., Song, Q., Hu, X.: Efficient neural architecture search with network morphism. arXiv preprint arXiv:1806.10282 (2018)
9. Sciuto, C., Yu, K., Jaggi, M., Musat, C., Salzmann, M.: Evaluating the search phase of neural architecture search. arXiv preprint arXiv:1902.08142 (2019)
10. Cortes, C., Gonzalvo, X., Kuznetsov, V., Mohri, M., Yang, S.:: AdaNet: adaptive structural learning of artificial neural networks. In: Proceedings of the 34th International Conference on Machine Learning, vol. 70, pp. 874–883. JMLR. org (2017)
11. Baker, B., Gupta, O., Naik, N., Raskar, R.: Designing neural network architectures using reinforcement learning. arXiv preprint arXiv:1611.02167 (2016)
12. Liu, H., Simonyan, K., Vinyals, O., Fernando, C., Kavukcuoglu, K.: Hierarchical representations for efficient architecture search. arXiv preprint arXiv:1711.00436 (2017)
13. White, C., Neiswanger, W., Savani, Y.: BANANAS: Bayesian optimization with neural architectures for neural architecture search. arXiv preprint arXiv:1910.11858 (2019)
14. Zoph, B., Vasudevan, V., Shlens, J., Le, Q.V.: Learning transferable architectures for scalable image recognition. In: Proceedings of the IEEE Conference on Computer Vision and Pattern Recognition, pp. 8697–8710 (2018)
15. Cai, H., Chen, T., Zhang, W., Yu, Y., Wang, J.: Reinforcement learning for architecture search by network transformation. arXiv preprint arXiv:1707.04873 (2017)
16. Xie, L., Yuille, A.: Genetic CNN. In: Proceedings of the IEEE International Conference on Computer Vision, pp. 1379–1388 (2017)
17. Real, E., et al.: Large-scale evolution of image classifiers. In: Proceedings of the 34th International Conference on Machine Learning, vol. 70, pp. 2902–2911. JMLR. org (2017)
18. Miikkulainen, R., et al.: Evolving deep neural networks. In: Artificial Intelligence in the Age of Neural Networks and Brain Computing, pp. 293–312. Elsevier (2019)
19. He, K., Zhang, X., Ren, S., Sun, J.: Deep residual learning for image recognition. In: Proceedings of the IEEE Conference on Computer Vision and Pattern Recognition, pp. 770–778 (2016)
20. Huang, G., Liu, Z., Van Der Maaten, L., Weinberger, K.Q.: Densely connected convolutional networks. In: Proceedings of the IEEE Conference on Computer Vision and Pattern Recognition, pp. 4700–4708 (2017)
21. Xie, S., Kirillov, A., Girshick, R., He, K.: Exploring randomly wired neural networks for image recognition. In: Proceedings of the IEEE International Conference on Computer Vision, pp. 1284–1293 (2019)
22. Erdős, P., Rényi, A.: On the evolution of random graphs. Publ. Math. Inst. Hung. Acad. Sci **5**(1), 17–60 (1960)
23. Albert, R., Barabási, A.-L.: Statistical mechanics of complex networks. Rev. Mod. Phys. **74**(1), 47 (2002)
24. Watts, D.J., Strogatz, S.H.: Collective dynamics of 'small-world' networks. Nature **393**(6684), 440 (1998)

25. Shoeibi, A., et al.: Automated detection and forecasting of covid-19 using deep learning techniques: a review. arXiv preprint arXiv:2007.10785 (2020)
26. Jokandan, A.S., et al.: An uncertainty-aware transfer learning-based framework for covid-19 diagnosis. arXiv preprint arXiv:2007.14846 (2020)
27. Alizadehsani, R., et al.: Hybrid genetic-discretized algorithm to handle data uncertainty in diagnosing stenosis of coronary arteries. Expert Syst. (2020)
28. Alizadehsani, R., et al.: Model uncertainty quantification for diagnosis of each main coronary artery stenosis. Soft Comput. **24**, 1–12 (2019)
29. Postels, J., Ferroni, F., Coskun, H., Navab, N., Tombari, F.: Sampling-free epistemic uncertainty estimation using approximated variance propagation. In: Proceedings of the IEEE International Conference on Computer Vision, pp. 2931–2940 (2019)
30. Phan, B.T.: Bayesian deep learning and uncertainty in computer vision. Master's thesis, University of Waterloo (2019)
31. Blundell, C., Cornebise, J., Kavukcuoglu, K., Wierstra, D.: Weight uncertainty in neural networks. arXiv preprint arXiv:1505.05424 (2015)
32. Fortunato, M., Blundell, C., Vinyals, O.: Bayesian recurrent neural networks. arXiv preprint arXiv:1704.02798 (2017)
33. Shridhar, K., Laumann, F., Maurin, A.L., Olsen, M., Liwicki, M.: Bayesian convolutional neural networks with variational inference. arXiv preprint arXiv:1806.05978 (2018)
34. Krizhevsky, A., Sutskever, I., Hinton,G.E.: ImageNet classification with deep convolutional neural networks. In: Advances in Neural Information Processing Systems, pp. 1097–1105 (2012)
35. Cozman, F.G., et al.: Generalizing variable elimination in Bayesian networks. In: Workshop on Probabilistic Reasoning in Artificial Intelligence, pp. 27–32. Citeseer (2000)
36. Roche, A.: Approximate inference via variational sampling. arXiv preprint arXiv:1105.1508 (2011)
37. Locatello, F., Khanna, R., Ghosh, J., Rätsch, G.: Boosting variational inference: an optimization perspective. arXiv preprint arXiv:1708.01733 (2017)
38. Regier, J., Jordan, M.I., McAuliffe, J.: Fast black-box variational inference through stochastic trust-region optimization. In: Advances in Neural Information Processing Systems, pp. 2402–2411 (2017)
39. Saddiki, H., Trapp, A.C., Flaherty, P.: A deterministic global optimization method for variational inference. arXiv preprint arXiv:1703.07169 (2017)
40. Jankowiak, M., Pleiss, G., Gardner, J.R.: Sparse Gaussian process regression beyond variational inference. arXiv preprint arXiv:1910.07123 (2019)
41. Graves, A.: Practical variational inference for neural networks. In: Advances in Neural Information Processing Systems, pp. 2348–2356 (2011)
42. Houthooft, R., Chen, X., Duan, Y., Schulman, J., De Turck, F., Abbeel, P.: Curiosity-driven exploration in deep reinforcement learning via Bayesian neural networks. arXiv preprint arXiv:1605.09674 (2016)
43. Rumelhart, D.E., Hinton, G.E., Williams, R.J.: Learning representations by back-propagating errors. Nature **323**(6088), 533–536 (1986)
44. Kwon, Y., Won, J.-H., Kim, B.J., Paik, M.C.: Uncertainty quantification using Bayesian neural networks in classification: application to ischemic stroke lesion segmentation (2018)

Brain-Inspired Framework for Image Classification with a New Unsupervised Matching Pursuit Encoding

Shiming Song, Chenxiang Ma, and Qiang Yu(✉)

Tianjin Key Laboratory of Cognitive Computing and Application,
College of Intelligence and Computing, Tianjin University, Tianjin, China
{songshiming,machenxiang,yuqiang}@tju.edu.cn

Abstract. The remarkable object recognition ability of biological systems allows individuals to have prompt and reliable responses to different stimuli. Despite many implementations, an efficient and effective one is still under exploring. Spiking neural networks (SNNs), following brain-like processing, provide a potential solution for efficient object recognition. The existing SNNs can benefit an efficient feature extraction from a temporal code, but they are vulnerable to noise, less adaptive and vitally poor in recognition accuracy. How could one make full use of the biological plausibility to improve their performance? In this paper, we propose a new temporal-based encoding method with unsupervised matching pursuit. Additionally, a unified SNN framework for image recognition is designed by integrating our encoding with recently advanced synaptic learning. We evaluate our approach on MNIST, with systematic insights into encoding capabilities, robustness to noise, learning efficiency and classification performance. The results highlight the effectiveness and efficiency of our spike-based approach. To date and the best of our knowledge, our approach achieves the best temporal-based accuracy performance. Moreover, our approach requires and consumes fewer number of neurons and spikes, making it significantly advantageous to fast and efficient computation. Our work also contributes to motivating new brain-inspired developments on image classification.

Keywords: Temporal encoding · Spiking neural networks · Image classification · Multi-spike

1 Introduction

Object recognition is an important cognitive ability that enables human to quickly respond to various visual stimuli and then make proper decisions accordingly. Inspired by this, it has been successfully applied to various visual tasks such as medical diagnosis, face recognition and autonomous driving [1,4]. However, as compared to human brain, most of the current approaches for object recognition are less biologically plausible and inefficient. Recently, spiking neural networks (SNNs) attract increasing attention as they are more biologically

H. Yang et al. (Eds.): ICONIP 2020, LNCS 12534, pp. 208–219, 2020.
https://doi.org/10.1007/978-3-030-63836-8_18

plausible and computationally powerful than previous models due to the use of time dimension [8,17]. Therefore, SNNs provide a potential solution for efficient and biologically plausible object recognition.

A unified SNN framework model for object recognition mainly includes two parts: spike encoding and learning. SNNs have an aptitude for dealing with spatiotemporal spike patterns owing to the temporally-dynamic characteristics, but poor at feature extraction [9]. Therefore, an appropriate encoding frontend that converts visual stimuli into spikes is an essential step required by SNNs for object recognition. Biological experiments show that the visual systems use a hierarchical structure for information processing [16], which motivates the development of hierarchical models resembling information processing in the mammalian brain [19]. In a typical hierarchical model, simple cells (SCs) will integrate information from their receptive fields in response to an external stimulus. Then, complex cells (CCs) extract information with nonlinear refinement on the outputs of SCs. Several temporal-based hierarchical encoding methods have been proposed according to this, such as HMAX [20] and S1C1 [26]. However, predefined filters are selected for SCs in these methods, thus limiting their capabilities to freely adapt to various tasks for a better extraction of features.

In recent years, convolutional neural networks (CNNs) gain a series of successes in machine vision and have demonstrated remarkable capabilities for feature extraction[11]. This motivates a CNN-based temporal encoding frontend [23] being proposed to improve the performance of SNNs by resorting to the powerful feature extraction ability of CNNs. Compared with previous hierarchical models, SCs' weights can be learned through the backpropagation method in this approach. However, its recognition accuracy is still poor as compared to other state-of-the-art SNN-based approaches, and thus leaving more room for improvement. Moreover, this approach depends on massive labeled data and powerful computing platforms [21], being inferior to the efficiency of biological systems.

How could one benefit the efficiency from the temporal-based neural code while making full use of the biological plausibility to improve the performance of SNNs? In this paper, we propose a new temporal-based encoding method with unsupervised matching pursuit (UMP). Our approach follows a similar routine to the retina, where neurons with stronger activations will fire earlier, while weaker ones spike later or not at all [14]. In our encoding scheme, SCs integrate information from their receptive fields in a hierarchical model, followed by a competing strategy to select the neuron with the strongest activation to elicit a spike. Afterward, a lateral inhibition is sent from the winner neuron to those silent ones. Iterating these steps, input information is thus encoded into a series of spikes. Importantly, our UMP provides an unsupervised learning scheme to adjust the selectivity of SCs, which can lead to a more sparse and effective representation of images and thus make our method competent for a broad range of tasks.

After spike encoding, learning rules are employed to train SNNs to have desired responses to different inputs. Recently, a new multi-spike learning rule called TDP has been proposed and demonstrates various learning advantages such as simplicity, efficiency, robustness, and more importantly, the remarkable applicability to various tasks [24]. In this work, we therefore develop an alternative framework for image classification by integrating our encoding method with TDP.

Our main contributions are proposing a new unsupervised matching pursuit temporal encoding method and designing a unified SNN framework for the challenging task of image classification. Several experiments are conducted to evaluate the performance of our method. Our approach obtains 98.56% classification accuracy on the MNIST task. To the best of our knowledge, our approach is the best among temporal-based approaches, and even achieves comparable performance to the rate-based ones. Notably, our temporal-based approach requires less network resource and computation, which is significantly beneficial to low-power and high-speed implementations. Thus, our work also contributes to paving a way towards the new paradigm of brain-like computation and processing.

2 Methodology

In this section, we will introduce the methods used in our framework, including UMP temporal encoding and TDP multi-spike learning. Details are presented in the following.

2.1 UMP Encoding

When external stimuli are presented, it is widely believed that the retina plays a critical role for feature extraction in the nervous systems [8]. The ganglion cells (GCs) in the retina will integrate information from their receptive fields in response to visual stimulus [18]. Inspired by this, the hierarchical model [16] is thus proposed to emulate the information processing in visual systems. In a typical hierarchical model, simple cells (SCs) are introduced to collect information from a local position, and the process is given as

$$r = \sum_{i} \sum_{j} I(x+i, y+j) \cdot w(i,j) \tag{1}$$

where (x, y) and (i, j) indicates the position and the range of SCs' receptive field in the input image I, respectively. w represents the weights of SCs, and r denotes neuron's activation value. Afterward, complex cells (CCs) are introduced to extract information with nonlinear pooling operation on the outputs of SCs. This hierarchical scheme has inspired several temporal-based encoding methods for images, such as HMAX [20], S1C1 [26] and Focal [15]. However, these encoding frontends are vulnerable to noise, less adaptive and relatively poor in performance of accuracy.

In this paper, we propose a new temporal encoding method with unsupervised matching pursuit (UMP). Following a similar process to Eq. (1), the activation value A_i of SCs can be calculated as

$$A_i = \sum_{\boldsymbol{l} \in R_i} I(\boldsymbol{l}) \cdot \phi_i(\boldsymbol{l}) \tag{2}$$

where $I(\boldsymbol{l})$ is the pixel value of an input image I in position $\boldsymbol{l}$. ϕ_i and R_i represent the weight and receptive field of neuron i, respectively.

Afterward, a competing strategy is introduced to select the neuron with the strongest activation to elicit a spike. Subsequently, a lateral inhibition is sent from the winner neuron to others. The selection and inhibition processes are iterated until none one of neurons is strong enough to fire. We implement this by successively removing the best matching unit from the input image. More specifically, the initial image I^0 and activation values A_i^0 at step $t = 0$ are set to I and A_i, respectively. Then, the neuron to fire is chosen as the one with the highest activation value.

$$i^0 = ArgMax_i(|A_i^0|) \tag{3}$$

where i^0 denotes the index of the fired neuron, and its corresponding activation value is defined as $A_{i^0}^0$. Afterward, inhibition is applied in a way to remove features of the selected neuron from previous input I^0, yielding the follow

$$I^1 = I^0 - \frac{< I^0, \phi_{i^0} >}{\|\phi_{i^0}\|^2} \cdot \phi_{i_0} = I^0 - \frac{A_{i^0}^0}{N_{i^0}^2} \cdot \phi_{i^0} \tag{4}$$

where I^1 is the remaining information at step $t = 1$, and $N_{i^0}^2$ is the squared norm of ϕ_{i^0}. $< I^0, \phi_{i^0} >$ indicates the integration process described by Eq. (1). Merging Eq. (2)–(4), we can get a clear inhibition effect of the fired neuron on the others.

$$A_i^1 =< I^1, \phi_i >= A_i^0 - \frac{A_{i^0}^0}{N_{i^0}^2} \cdot < \phi_{i^0}, \phi_i > \tag{5}$$

According to Eq. (3) and (5), the iteration process in later time $t > 0$ can thus be given as

$$\begin{cases} i^t = ArgMax_i(|A_i^t|) \\ A_i^{t+1} = A_i^t - m^t \cdot < \phi_{i^t}, \phi_i > \end{cases} \tag{6}$$

with $m^t = A_{i^t}^t / N_{i^t}^2$ denoting UMP coefficient.

Iterating these steps, the input image is thus converted into a series of spikes. In our UMP scheme, we bin these spikes with a temporal precision of 1 ms before feeding them to downstream spiking neurons.

2.2 Unsupervised Kernel Learning

How could one set proper weights for SCs such that they could be adaptive to various tasks with a good selectivity of features? In this part, we propose an unsupervised learning scheme to adjust their weights according to their responding activation and the input. Our learning rule is given as

$$\Delta\phi_{i^t} = \lambda m^t \cdot I_{i^t}^t \tag{7}$$

where λ is the learning factor. $I_{i^t}^t$ indicates the residual error of weight ϕ_{i^t} in step t. Equation (7) thus provides a way to estimate the gradient $\Delta\phi_i$. In order to have better discriminative weights for the learning, we add a regularization term for selecting the winner neuron during learning. The modified selection scheme for learning at time t is given as

$$i^t = ArgMax_i(|A_i^t| + \gamma\|\phi_i - \phi_{i^{t-1}}\|) \tag{8}$$

where γ is a regularization constant and $\|\phi_i - \phi_{i^{t-1}}\|$ denotes the distance between two neurons.

With the above learning rule to set proper weights, the input image can thus be converted into a spare spatiotemporal spike pattern that can be further used for multi-spike learning and classification.

2.3 Neuron Model

In this paper, we use the leaky integrate-and-fire (LIF) neuron model [3] due to its simplicity and computational efficiency. When an input spike pattern is presented, each afferent will result in a post-synaptic potential (PSP). The neuron continuously integrates PSP into its membrane potential $V(t)$, and it will elicit a spike whenever $V(t)$ crosses the threshold. According to the model, neuron's membrane potential is calculated by integrating synaptic currents from its N afferents as

$$V(t) = \sum_{i}^{N} \omega_i \sum_{t_i^j < t} K(t - t_i^j) - \vartheta \sum_{t_s^j < t} \exp\left(-\frac{t - t_s^j}{\tau_m}\right) \tag{9}$$

where ω_i is the synaptic efficacy. t_i^j denotes the time of the j-th spike from the i-th afferent, and t_s^j is the time of the j-th output spike. ϑ represents the neuron's threshold. The last term in Eq. (9) is a reset dynamic that allows neuron to continuously integrate following input spikes and elicit output ones after firing. $K(t - t_i^j)$ is a normalized kernel function, which is defined as

$$K(t - t_i^j) = V_0 \left[\exp\left(-\frac{t - t_i^j}{\tau_m}\right) - \exp\left(-\frac{t - t_i^j}{\tau_s}\right)\right] \tag{10}$$

where τ_m and τ_s are the decay time constants of the membrane integration and synaptic currents, respectively. V_0 normalizes $K(t - t_i)$ such that the maximum value of the kernel function is unity.

2.4 Multi-spike Learning

Learning rules are employed to train neurons to make proper responses to certain spatiotemporal spike patterns. In recent years, various learning methods have been introduced. We adopt the TDP [24] rule in this paper due to its computational advantages on efficiency, feature selectivity and classification.

TDP is developed based on spike-threshold-surface (STS) function [6] which describes the relationship between neuron's threshold and the number of output spikes. With other conditions being fixed, the responses of a neuron can be determined by its thresholds. STS defines a series of critical threshold values ϑ^*_{k+1} where the number of output spike n_o jumps from k to $k+1$. Based on this, the critical threshold ϑ^* is employed to tune neuron's weight ω_i. According to [24], its gradient with respect to ω_i is given as

$$\vartheta_i^{*'} = \frac{\partial V(t^*)}{\partial \omega_i} - \sum_{j=1}^{m} \frac{\partial V(t^*)}{\partial t_s^j} \frac{1}{\dot{V}(t_s^j)} \frac{\partial V(t_s^j)}{\partial \omega_i} \tag{11}$$

where m denotes the total number of output spikes before the critical time t^*.

According to the relation between the actual n_o and the target n_d number of spikes, the learning rule can be given as

$$\Delta\omega = \begin{cases} -\lambda \frac{d\vartheta^*_{n_o}}{d\omega} & if\ n_o > n_d \\ \lambda \frac{d\vartheta^*_{n_o+1}}{d\omega} & if\ n_o < n_d \\ 0 & otherwise. \end{cases} \tag{12}$$

where $d\vartheta^*_{n_k}/d\omega$ is the derivative of critical threshold with respect to synaptic weights, which is evaluated by Eq. (11).

In our image classification task, the neurons are trained to elicit at least 6 spikes in response to their target categories and to keep silent otherwise.

3 Experiments

In this section, several experiments are conducted and comparisons between ours and other baseline models are provided. The accuracies are averaged over ten independent runs.

3.1 Dataset

The MNIST is a large handwritten digit dataset that contains 60,000 training and 10,000 test images of digits 0–9, with a pixel size of 28×28 each. It is widely used in spike-based research [17]. We thus adopt MNIST to provide a clear comparison between our work and other baselines. The MNIST database is available from http://yann.lecun.com/exdb/mnist.

3.2 Experimental Settings

In both our UMP and convolution-based methods [14,23], encoding neurons sharing the same filter are organized in one feature map but with a specific location focus each. Therefore, the number of encoding neurons will linearly increase with the number of feature maps (or filters), resulting in an inefficient representation. To control the number of encoding neurons as well as the computational efficiency, we set the number of filters to eight. Following a similar procedure as previous hierarchical models, we employ a 4×4 pooling operation on the outputs of SCs. Our method thus results in 400 encoding neurons. We find this approach can lead to a sparse and effective representation of images.

In the synaptic learning, we set $\tau_m = 40$ ms, $\tau_s = 10$ ms and $\lambda = 10^{-4}$. We use a single neuron to learn each one target category. In our readout, the final decision is made by the neuron with the maximal number of output spikes.

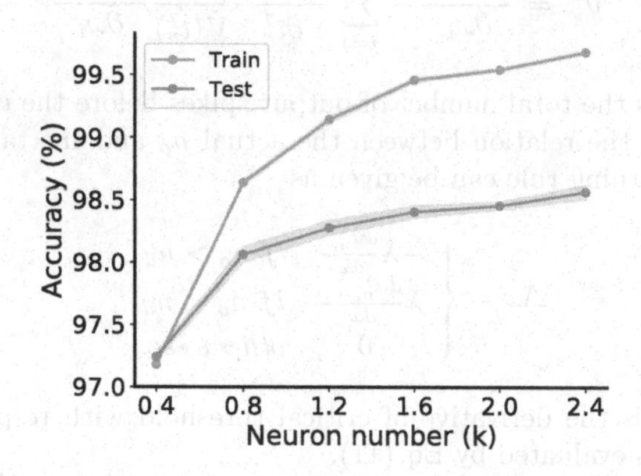

Fig. 1. The train and test accuracies versus neuron numbers used for encoding.

3.3 Effect of the Number of Encoding Neurons

In this section, we conduct experiments with different numbers of encoding neurons to clearly show their effects on recognition performance. Encoding neurons in UMP are mutually inhibited, and thus could affect the learning performance if insufficient information is extracted. To overcome this, we use a grouping scheme where neurons in the same group mutually inhibit each other, while the ones in a different group are not bound. In this way, we can balance exploration and exploitation. In our experiments, we set 400 neurons for each group while increase the number of groups one by one.

The train and test accuracies with respect to the neuron number are shown in Fig. 1. These experimental results are based on the TDP learning rule. We observe that both the train and test accuracies increase with the number of neuron groups. Our method achieves 97.24% and 98.06% test accuracy when the neuron number is 400 and 800, respectively. When the encoding number is increased to 1200, the performance is growing slowly, indicating that the features from input images have been extracted well that facilitate the learning. This experiment highlights that our UMP can effectively learn and extract useful features from the input data. A larger number of encoding neurons leads to a better accuracy, but could also increase the computational cost.

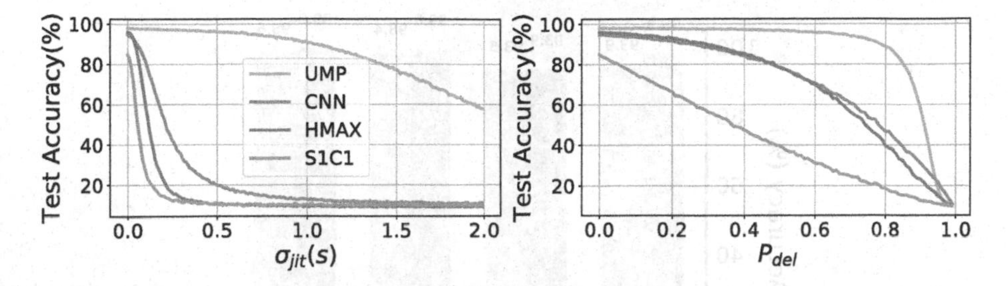

Fig. 2. Test accuracies of different temporal-based encoding methods against spike jitter noise σ_{jit} (the top panel) and spike deletion noise P_{del} (the bottom).

3.4 Comparison with Different Temporal-Based Encoding

To show the feature extraction ability, we compare our UMP with S1C1 [26], HMAX [19] and CNN-based [23] temporal encoding methods.

S1C1 employs two scales difference of Gaussian (DoG) filters ($\sigma = 1$ for 5×5 pixels as scale 1, and $\sigma = 2$ for 7×7 pixels as scale 2) to encode input images. The number of encoding neurons used in S1C1 is 200. HMAX adopts four orientations ($\pi/8$, $\pi/4 + \pi/8$, $\pi/2 + \pi/8$, and $3\pi/4 + \pi/8$) 7×7 Gabor filters to convert image into spikes. CNN-based encoding method uses the convolutional and pooling layers of a trained CNN as the encoding frontend. The number of neurons used in HMAX and CNN is 800. In order to make a fair comparison between the four temporal encoding methods with respect to their feature extraction capabilities, we use 800 neurons, i.e. two neuron groups, to encode the input image with our UMP. Additionally, we examine the robustness of different encoding methods by adding two types of different random noises to the spatiotemporal spike pattern: spike jitter σ_{jit} and spike deletion P_{del} noises.

As can be seen from Fig. 2, UMP outperforms the other three encoding methods under multiple test conditions, highlighting the advanced feature extraction capabilities and good robustness of our UMP, which can even tolerate severe noises up to 0.5 s of jitter and 60% deletion with a subtle loss of accuracy. The improved performance of our UMP is attributed to the competition mechanism and unsupervised learning from data samples. The effectiveness and robustness of our approach may be beneficial for implementing hardware systems that are prone to perturbations.

Notably, in the above experiments, both CNN and HMAX generate 800 spikes with each encoding neuron firing one, while only 400 spikes are elicited with our UMP thanks to our competition scheme as it depresses both redundancy and weak activations. This suggests that under an event-driven scheme [24], our UMP saves at least half of the computation. Therefore, as is compared to CNN and HMAX, our method is more efficient, accurate and robust.

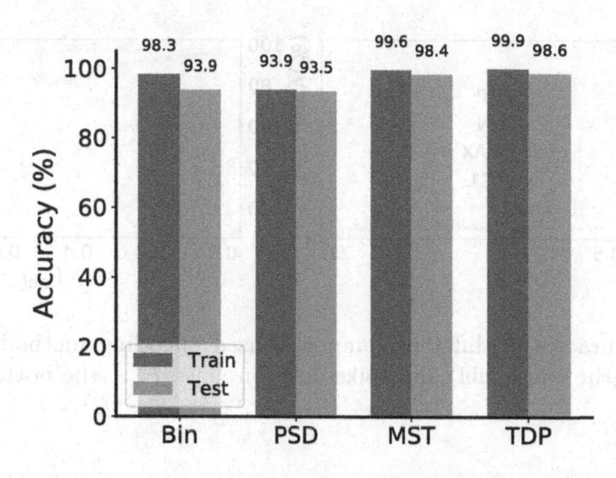

Fig. 3. Performance comparison of different synaptic learning rules: Bin, PSD, MST and TDP. The same UMP encoding is used for all rules.

3.5 Comparison Between Different Learning Methods

In this part, we focus on the performance comparison between TDP and other learning rules.

We select the tempotron ('Bin') [7], Precise-Spike-Driven Synaptic Plasticity ('PSD') [25] and MST [6] learning rules for comparison in this experiment. The Bin trains neuron to elicit a spike when the input spike pattern belongs to the target class and to keep silent otherwise. During readout, the neuron's status of firing or not is used for making decision. The PSD trains neuron to emit spikes at the specified times for corresponding patterns. For simplicity, we set the desired spike train as 20, 40, 60, 80 and 100 ms in this work. In decision, we choose the neuron with the maximum number of desired spikes as the classification result. For MST, we use the same experimental setup as TDP.

As can be seen from Fig. 3, the two multi-spike learning rules, i.e. MST and TDP, achieve better performance than others, highlighting the advanced ability of multi-spike rule for processing spatiotemporal spike patterns. MST and TDP achieve similar performance on both the training and test datasets. However, as shown in [24], TDP performs better than MST in terms of learning speed due to its simplicity and efficiency. The reason for the relatively poor performance of PSD may because that restricting neurons to fire exactly at specified times will

Table 1. Test accuracies (%) of SNNs on MNIST

Methods	Neurons (Structure)	Coding scheme	Accuracy
S1C1-SNN [26]	200 + 10	Temporal	78.00
CSNN [23]	800 + 10	Temporal	87.00
Multi-Net [2]	784 + 3136 + 150	Temporal	91.26
Our work	**800 + 10**	**Temporal**	**98.06**
Dendritic neurons [10]	5000 + 10	Rate	90.26
Spiking RBM [13]	6470 + 1010	Rate	94.09
Unsupervised STDP [5]	784 + 6400	Rate	95.00
Our work	**2400 + 10**	**Temporal**	**98.56**
Spiking NN [12]	784 + 800 + 10	Rate	98.64
Spiking CNN [22]	784 + 15C5 + P2 + 40C5 + P2 + 300 + 10	Rate	99.42

result in insufficient learning as these times could hardly cover useful features. The above experiments demonstrate the powerful learning capability of TDP rule. We thus propose a unified framework for image classification by integrating our encoding method with TDP and compare it with other SNN models in the following.

3.6 Performance Comparison with Other SNNs

In this part, we compare our framework with other baseline SNNs. The performance comparisons on MNIST dataset are shown in Table 1. We first compare our system with three state-of-the-art temporal-based frameworks, i.e. S1C1-SNN [26], CSNN [23] and Multi-Net [2]. As shown in Table 1, the recognition accuracies of previous temporal-based systems are relatively poor as compared to ours. Moreover, our work only depends on a light network structure. Our work thus significantly improves the performance of a temporal-based spiking framework for a practical task.

Besides, we also compare ours with the rate-based models where firing rates of spikes are used to represent information. In terms of recognition accuracy, our framework outperforms many of them, while achieves a comparable performance with the deep rate-based models such as Spiking NN [12] and Spiking CNN [22]. Despite their high accuracy, the use of high-density spikes and multi-layer structures reduces their biological plausibility and computational efficiency. Oppositely, our method uses the efficient temporal code and has a very light network structure. Compared with these frameworks, our approach uses fewer number of spikes, spiking neurons and computation resource, which greatly improves the computational efficiency. The efficient and effective performance of our framework would provide an alternative approach for image classification, being potentially beneficial to low-power and high-speed developments.

4 Conclusion

In this work, we proposed a new unsupervised matching pursuit (UMP) temporal encoding method and designed a unified SNN framework for the challenging task of image classification. Firstly, we introduced UMP method to emulate the efficient and effective encoding procedure in the retina. Then, TDP multi-spike learning rule was adopted to adjust neuron's weights such that it will respond appropriately to different inputs. Finally, we developed an alternative framework for image classification by integrating UMP with TDP. Several experiments were conducted to benchmark our system, and the performance comparisons between ours and several baseline ones were provided. Our approach achieves significantly better accuracy than previous temporal-based SNN implementations, and was even comparable to rate-based SNNs, while only using lighter network structure and fewer computation resource. Our framework highlights the advantageous potential of more brain-like temporal-based SNNs on practical developments. The outstanding performance of our approach would be a step forward towards closing the gap between artificial neural network and the brain. It would also pave the way for more research efforts to be made to the new paradigm of brain-like computation.

Acknowledgments. This work was supported in part by the National Natural Science Foundation of China under Grant 61806139, and in part by the Natural Science Foundation of Tianjin under Grant 18JCYBJC41700.

References

1. Amato, F., López, A., Peña-Méndez, E.M., Vaňhara, P., Hampl, A., Havel, J.: Artificial neural networks in medical diagnosis. J. Appl. Biomed. **11**(2), 47–58 (2013). https://doi.org/10.2478/v10136-012-0031-x. ISSN 1214-021X
2. Beyeler, M., Dutt, N.D., Krichmar, J.L.: Categorization and decision-making in a neurobiologically plausible spiking network using a STDP-like learning rule. Neural Netw. **48**, 109–124 (2013)
3. Burkitt, A.N.: A review of the integrate-and-fire neuron model: I. Homogeneous synaptic input. Biol. Cybern. **95**(1), 1–19 (2006)
4. Chen, C., Seff, A., Kornhauser, A., Xiao, J.: DeepDriving: learning affordance for direct perception in autonomous driving. In: 2015 IEEE International Conference on Computer Vision, Chile, pp. 2722–2730. IEEE (2015)
5. Diehl, P.U., Cook, M.: Unsupervised learning of digit recognition using spike-timing-dependent plasticity. Front. Comput. Neurosci. **9**, 99 (2015)
6. Gütig, R.: Spiking neurons can discover predictive features by aggregate-label learning. Science **351**(6277), aab4113 (2016)
7. Gütig, R., Sompolinsky, H.: The tempotron: a neuron that learns spike timing-based decisions. Nat. Neurosci. **9**(3), 420–428 (2006)
8. Hopfield, J.J.: Pattern recognition computation using action potential timing for stimulus representation. Nature **376**(6535), 33–36 (1995)
9. Hu, J., Tang, H., Tan, K.C., Li, H.: How the brain formulates memory: a spatio-temporal model research frontier. IEEE Comput. Intell. Mag. **11**(2), 56–68 (2016)

10. Hussain, S., Liu, S.C., Basu, A.: Improved margin multi-class classification using dendritic neurons with morphological learning. In: 20th IEEE International Symposium on Circuits and Systems (ISCAS), Australia, pp. 2640–2643. IEEE (2014)
11. LeCun, Y., Bengio, Y., Hinton, G.: Deep learning. Nature **521**(7553), 436–444 (2015)
12. Lee, J.H., Delbruck, T., Pfeiffer, M.: Training deep spiking neural networks using backpropagation. Front. Neurosci. **10**, 508 (2016)
13. Merolla, P., Arthur, J., Akopyan, F., Imam, N., Manohar, R., Modha, D.S.: A digital neurosynaptic core using embedded crossbar memory with 45pJ per spike in 45nm. In: 2011 IEEE Custom Integrated Circuits Conference (CICC), USA, pp. 1–4. IEEE (2011)
14. Perrinet, L., Samuelides, M.: Sparse image coding using an asynchronous spiking neural network. In: 10th European Symposium on Artificial Neural Networks, Computational Intelligence and Machine Learning (ESANN), Belgium, pp. 313–318 (2002)
15. Perrinet, L., Samuelides, M., Thorpe, S.: Coding static natural images using spiking event times: do neurons cooperate? IEEE Trans. Neural Networks **15**(5), 1164–1175 (2004)
16. Riesenhuber, M., Poggio, T.: Hierarchical models of object recognition in cortex. Nat. Neurosci. **2**(11), 1019–1025 (1999)
17. Roy, K., Jaiswal, A., Panda, P.: Towards spike-based machine intelligence with neuromorphic computing. Nature **575**(7784), 607–617 (2019)
18. Rullen, R.V., Thorpe, S.J.: Rate coding versus temporal order coding: what the retinal ganglion cells tell the visual cortex. Neural Comput. **13**(6), 1255–1283 (2001)
19. Serre, T., Oliva, A., Poggio, T.: A feedforward architecture accounts for rapid categorization. Proc. Natl. Acad. Sci. **104**(15), 6424–6429 (2007)
20. Serre, T., Wolf, L., Bileschi, S., Riesenhuber, M., Poggio, T.: Robust object recognition with cortex-like mechanisms. IEEE Trans. Pattern Anal. Mach. Intell. **29**(3), 411–426 (2007)
21. Simonyan, K., Zisserman, A.: Very deep convolutional networks for large-scale image recognition. arXiv preprint arXiv:1409.1556 (2014)
22. Wu, Y., Deng, L., Li, G., Zhu, J., Shi, L.: Spatio-temporal backpropagation for training high-performance spiking neural networks. Front. Neurosci. **12**, 331 (2018)
23. Xu, Q., Qi, Y., Yu, H., Shen, J., Tang, H., Pan, G.: CSNN: an augmented spiking based framework with perceptron-inception. In: 27th International Joint Conferences on Artificial Intelligence (IJCAI), Sweden, pp. 1646–1652 (2018)
24. Yu, Q., Li, H., Tan, K.C.: Spike timing or rate? Neurons learn to make decisions for both through threshold-driven plasticity. IEEE Trans. Cybern. **49**(6), 2178–2189 (2018)
25. Yu, Q., Tang, H., Tan, K.C., Li, H.: Precise-spike-driven synaptic plasticity: learning hetero-association of spatiotemporal spike patterns. PLoS ONE **8**(11), e78318 (2013)
26. Yu, Q., Tang, H., Tan, K.C., Li, H.: Rapid feedforward computation by temporal encoding and learning with spiking neurons. IEEE Trans. Neural Netw. Learn. Syst. **24**(10), 1539–1552 (2013)

Estimating Conditional Density of Missing Values Using Deep Gaussian Mixture Model

Marcin Przewięźlikowski, Marek Śmieja(✉), and Łukasz Struski

Faculty of Mathematics and Computer Science, Jagiellonian University, Kraków, Poland
m.przewie@gmail.com, {marek.smieja,lukasz.struski}@uj.edu.pl

Abstract. We consider the problem of estimating the conditional probability distribution of missing values given the observed ones. We propose an approach, which combines the flexibility of deep neural networks with the simplicity of Gaussian mixture models (GMMs). Given an incomplete data point, our neural network returns the parameters of Gaussian distribution (in the form of Factor Analyzers model) representing the corresponding conditional density. We experimentally verify that our model provides better log-likelihood than conditional GMM trained in a typical way. Moreover, imputation obtained by replacing missing values using the mean vector of our model looks visually plausible.

Keywords: Missing data · Density estimation · Imputation · Gaussian mixture model · Neural networks.

1 Introduction

Estimating missing values from incomplete observations is one of the basic problems in machine learning and data analysis [7]. A typical approach relies on replacing missing values with a single vector based on available information contained in observed inputs [9,32]. While imputation techniques are frequently used by practitioners, they only give point estimate instead of a probability distribution. Quantifying the probability distribution of missing values plays an important role in generative models [12], uncertainty prediction [5] as well as is useful in applying classification models to incomplete data [3,28,33].

While deep generative models such as VAE, GAN or WAE [6,10,30] are capable of modeling the distribution of complex high dimensional data, such as images, it may be difficult to use them to estimate the uncertainty contained in missing data due to the nonlinear form of decoder (generator) [18]. The authors of [15,23] define a sampling procedure based on pseudo-Gibbs sampling and

---

A preliminary version of this paper appeared as an extended abstract [21] at the ICML Workshop on The Art of Learning with Missing Values.

H. Yang et al. (Eds.): ICONIP 2020, LNCS 12534, pp. 220–231, 2020.
https://doi.org/10.1007/978-3-030-63836-8_19

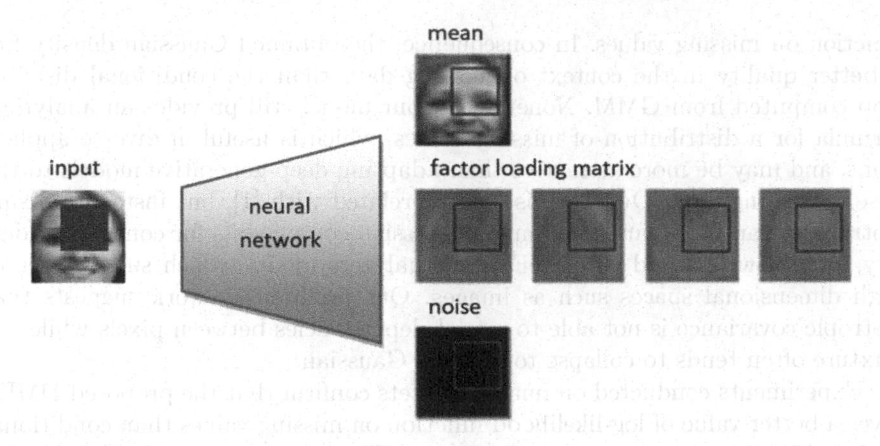

Fig. 1. The idea of the proposed DMFA. Given a missing data point, our model returns the parameters of conditional Gaussian density: mean, factor loading matrix (we use 4 factors) and noise matrix for the model of Factor Analyzers, which describes a distribution of missing data (the area inside the blue square). (Color figure online)

Metropolis-within-Gibbs algorithm for filling missing values by iterative auto-encoding of incomplete data. Śmieja et al. [25,26] propose iterative algorithm for maximizing conditional density based on the dynamics of auto-encoder models. Mattei and Frellsen use importance sampling for training VAE on incomplete data as well as for replacing missing values by single or multiple imputation [16]. Flow models can also be trained to represent a conditional density as a neural network transformation [13,31]. Nevertheless, the constructed density cannot be maximized analytically. One can only produce samples or attempt to maximize the corresponding density numerically.

In the case of shallow density models, such Gaussian mixture models (GMMs), we can easily calculate a conditional density function related to missing values in a closed-form [2,5] as well as to maximize it analytically. Moreover, simple Gaussian form of the conditional density function allows us to combine conditional GMM with other machine learning techniques that can process missing data without using any imputations at preprocessing stage [27,28]. Another related line of work has explored autoregressive models for conditional data generation or density estimation [19].

In this paper, we propose DMFA (Deep Mixture of Factor Analyzers) for estimating the probability density function of missing values, which combines the features of deep learning models and GMMs. We construct a neural network, which takes an incomplete data point and returns the parameters of Gaussian density (represented as Factor Analyzers model) modeling the distribution of missing values, see Fig. 1. Since the proposed network returns an individual Gaussian density for every missing data point, its expressiveness is higher than using GMM with a fixed number of components. In contrast to classical GMM, which estimates a density of the whole data, DMFA directly maximizes the likelihood function on missing values. In consequence, the obtained Gaussian density has a better quality in the context of missing data than the conditional distribution computed from GMM. Nonetheless, our model still provides an analytical formula for a distribution of missing values, which is useful in diverse applications, and may be more attractive than adapting deep generative models to the case of missing data. Our work is strictly related with [1], but instead of using isotropic covariance matrix and many Gaussian components for conditional density, we follow [24] and employ Factor Analyzers model, which suits better to high dimensional spaces such as images. Our preliminary work suggests that isotropic covariance is not able to model dependencies between pixels while the mixture often tends to collapse to a single Gaussian.

Experiments conducted on image datasets confirm that the proposed DMFA gives a better value of log-likelihood function on missing values than conditional GMMs [24]. Moreover, imputations obtained by replacing missing values with the mean vector of returned Gaussian density look visually plausible. The paper also contains a visualization of produced density function, which gives an insight into the proposed DMFA.

2 Density Model for Missing Data

In this section, we introduce DMFA model. First, we recall basic facts concerning GMM and MFA in high dimensional data. Next, we show how to compute conditional density from GMM. Finally, we present the proposed DMFA – a deep learning model for estimating conditional Gaussian density on missing values.

2.1 Gaussian Mixture Model for High Dimensional Data

GMM is one of the most popular probabilistic models for describing a density of data [17]. A density function of GMM is given by:

$$p(x) = \sum_{i=1}^{k} p_i N(\mu_i, \Sigma_i)(x),$$

where $p_i > 0$ is the weight of i-th Gaussian component with mean vector μ_i and covariance matrix Σ_i (we have $\sum_{i=1}^{k} p_i = 1$). Given a dataset $X \subset \mathbb{R}^n$, GMM is estimated by minimizing the negative log-likelihood:

$$l(x) = -\sum_{x \in X} \log p(x).$$

While theoretically GMM can be estimated using EM or SGD, this procedure may fail in the case of high dimensional data, such as images. Observe that for color images of size 32×32, the covariance matrix of a single component has $4.7 \cdot 10^6$ free parameters. In training phase, we need to store and invert these covariance matrices, which is computationally inefficient and may cause many numerical problems [24].

It is widely believed that high-dimensional data, such as images, are embedded in a lower-dimensional manifold and using full covariance matrix may not be necessary. For this reason, it is recommended to use the Mixture of Factor Analyzers (MFA) [4] or Probabilistic PCA (PPCA) [29], in which every Gaussian density is spanned on a lower-dimensional subspace. In contrast to the typical GMM, the covariance matrix in MFA is given by $\Sigma = AA^T + D$, where $A = A_{n \times l}$ is a factor loading matrix, which is composed of l vectors $a^1, \ldots, a^l \in \mathbb{R}^n$ such that $l \ll n$, and $D = D_{n \times n} = \mathrm{diag}(d)$ is a diagonal matrix representing noise[1] defined by $d \in \mathbb{R}^n$. The set of vectors a^i defines a linear subspace, which spans a Gaussian density $N(\mu, \Sigma)$, while adding a noise matrix guarantees that Σ is invertible. The use of MFA drastically reduces the number of parameters in a covariance matrix as well as avoids problems with inverting large matrices. Recent studies show that MFA can be effectively estimated from image data and is able to describe a higher spectrum of data density than GAN models, see [24] for details.

2.2 Conditional Gaussian Density

It is imporant to note that GMM can not only describe a density of data, but is also useful for quantifying the uncertainty of missing data. A missing data point is denoted by $x = (x_o, x_m)$, where x_o represents known values, while x_m describes absent attributes. Given a missing data point x, a natural question is: *what is the distribution of missing values given the observed ones?* In the case of density models, the answer is given by a conditional density [5]:

$$p(x_m|x_o) = \frac{p(x_o, x_m)}{p(x_o)} = \frac{p(x)}{p(x_o)}.$$

In contrast to many deep generative models, e.g. GANs or VAE, the formula for conditional density can be found analytically for GMM. For a single Gaussian density $N(\mu, \Sigma)$ with $\mu = \begin{pmatrix} \mu_o \\ \mu_m \end{pmatrix}$ and $\Sigma = \begin{pmatrix} \Sigma_{oo} & \Sigma_{om} \\ \Sigma_{om}^T & \Sigma_{mm} \end{pmatrix}$, the conditional Gaussian density is given by:

$$p(x_m|x_o) = N(\hat{\mu}_m, \hat{\Sigma}_m),$$

where

$$\begin{aligned} \hat{\mu}_m &= \mu_m + \Sigma_{om}\Sigma_{oo}^{-1}(x_o - \mu_o), \\ \hat{\Sigma}_m &= \Sigma_{mm} - \Sigma_{om}\Sigma_{oo}^{-1}\Sigma_{om}^T. \end{aligned} \tag{1}$$

Note that $N(\hat{\mu}_m, \hat{\Sigma}_m)$ is a Gaussian density in a lower dimensional space, where the dimension equals the number of missing values.

To extend these formulas to the mixture of Gaussian densities, we need to find a conditional density of every Gaussian component and recalculate the weights p_i.

[1] PPCA uses spherical matrix D.

Since the tails of Gaussian densities decrease exponentially, the resulting conditional GMM in high dimensional spaces typically reduces to a single Gaussian. Other components become irrelevant, because their weights (in the conditional density) are close to zero.

2.3 Deep Conditional Gaussian Density for Missing Data

An important advantage of GMM is that the conditional densities can be calculated and maximized analytically, which may be appealing in the context of missing data. However, GMM is not trained to estimate a density of missing data – its objective is the log-likelihood computed on all data points. In consequence, there are no guarantees that the resulting conditional density gives optimal log-likelihood for missing values.

In this paper, we are motivated by typical deep learning models used for image inpainting [8,35]. Let us recall that context-encoder [20] first generates missing values by selecting random masks for images in every mini-batch. Next, missing values are replaced by zeros and such images together with corresponding masks are processed by the auto-encoder neural network. The model is trained to minimize the mean-square error on missing values. Since the loss covers only the missing part, the context-encoder should find a better replacement for missing values than the model, which is trained to reconstruct the whole image.

Following the above motivation, the proposed DMFA creates a Gaussian density, which minimizes the negative log-likelihood on missing values. More precisely, given a data point $x \in \mathbb{R}^n$, we first generate a random binary mask M to simulate missing attributes. The pair (x, M) induces a missing data point (x_o, x_m). DMFA defines a neural network f, which takes (x_o, x_m) together with M and returns the parameters of conditional Gaussian density $p(x_m|x_o)$. Following MFA model, we represent covariance matrix using factor loading matrix $A = A_{n \times l} = (a^1, \dots, a^l)$, and the noise matrix $D = D_{n \times n} = \text{diag}(d)$. In the case of images, f simply returns the mean image μ and the covariance matrix $\Sigma = AA^T + D$ represented by l images spanning a Gaussian density supplied with the noise image ($l+2$ images in total) .

Given the output μ and Σ of the neural network f, we define a conditional Gaussian density as

$$p(x_m|x_o) = N(\mu_m, \Sigma_{mm}),$$

where μ_m and Σ_{mm} denote the restrictions of μ and Σ to missing coordinates, see Fig. 1 for illustration. Since the number of missing values can be different for subsequent data points, f has to output the parameters of n-dimensional Gaussian density $N(\mu, \Sigma)$. However, $N(\mu, \Sigma)$ does not need to estimate a density of the whole data. In consequence, we do not have to use the formulas for conditional density (1), but we can simply restrict μ and Σ to missing attributes in order to define a conditional density $p(x_m|x_o)$. In our case, $\Sigma_{mm} = A_{m\cdot}A_{m\cdot}^T + D_{mm}$, where $A_{m\cdot}$ denotes the restriction of matrix $A = A_{n \times l}$ to the rows indexed by m.

DMFA is trained to minimize the negative log-likelihood of conditional density $p(x_m|x_o)$ which is given by:

$$l(x_o, x_m) = -\log p(x_m|x_o) = -\log N(\mu_m, \Sigma_{mm})(x_m).$$

Observe that the above objective is calculated only on the parameters of μ, Σ corresponding to missing values (other entries are not used by the model). This means that f can theoretically return irrelevant values on coordinates related to the observed values. The most important thing is that DMFA directly minimizes the log-likelihood of $p(x_m|x_o)$ and thus should provide a better estimate of missing values than using conditional density obtained by a typical GMM.

Let us highlight that we do not need to specify the number of mixture components as in the classical GMM. Once the neural network is fed with a missing data point, it generates an individual density for this data point. In the case of the classical mixture model, conditional density is formed by restricting the most probable Gaussian components (from the set of mixture components) to missing values. In consequence, our conditional density should be more expressive than the one obtained from the classical GMM, where the number of components is fixed.

3 Experiments

In this section, we compare the quality of a density produced by DMFA with a conditional density obtained from GMM. For this purpose, we use three typical image datasets: MNIST [11], Fashion-MNIST [34] and CelebA [14].

3.1 Gray-Scale Images

First, we consider two datasets of gray-scale images: MNIST and Fashion-MNIST. For each test image of the size 28×28, we drop a patch of size 14×14, at a (uniformly) random location. DMFA is instantiated using 4 convolutional layers. This is followed by a dense layer, which produces the final output vectors (the number of latent dimensions l determining the covariance matrix equals 4). Our model is trained with a learning rate $4 \cdot 10^{-5}$ for 50 epochs. As a baseline, we use the implementation of MFA [24] trained in a classical way[2]. The number of components $k = 50$ and the number of latent dimensions $l = 6$ in every Gaussian following the authors' code.

We examine the imputation constructed by replacing missing values with the mean vector of corresponding conditional density. Sample results presented in Fig. 2 for MNIST show that MFA produces sharper imputations than DMFA. However, the results returned by MFA do not always agree with ground-truth (7th and 9th rows). This is confirmed by verifying the mean-square error (MSE) of imputations, Table 1. Since DMFA usually gives images more similar to the ground-truth, it obtains lower MSE values than MFA. It is also evident from

[2] The code was taken from https://github.com/eitanrich/torch-mfa.

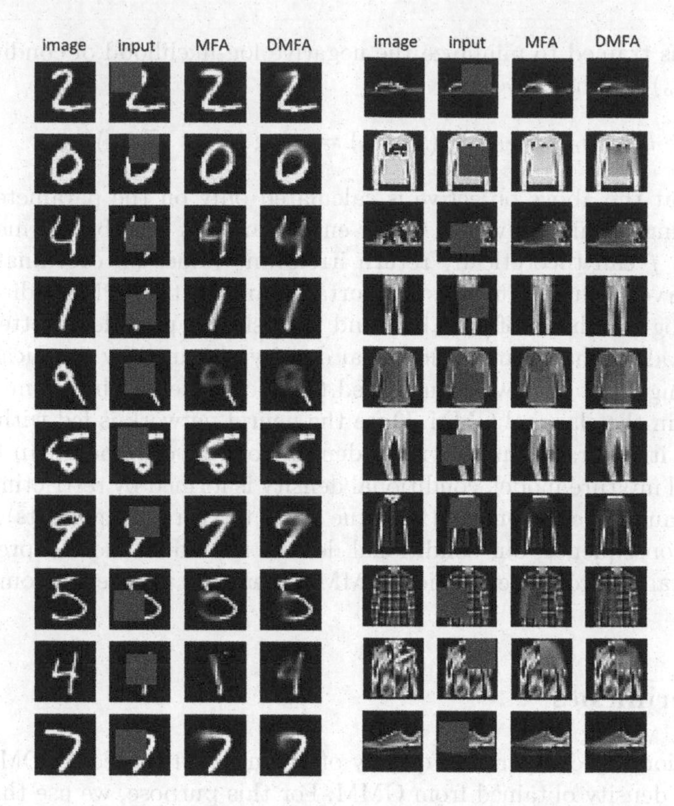

Fig. 2. Sample imputation results produced by DMFA and MFA on MNIST (left) and Fashion-MNIST (right) datasets.

Table 1 that a density returned by DMFA has significantly higher log-likelihood, which means that DMFA finds a better solution to the underlying problem.

It is worth noting that the imputation generated by MFA is in fact similar to nearest neighbor imputation. Indeed, we first select a Gaussian density which has the highest conditional probability and next project its mean vector onto the linear subspace corresponding to the missing data point (with respect to the covariance matrix). Replacing missing values using nearest neighbor usually gives sharp results, but may completely disagree with true values on the missing region. On the other hand, more blurry image corresponding to the mean vector of conditional density produced by DMFA may suggest that DMFA focuses on estimating a high-quality density function rather than finding a single value for imputation. This hypothesis is supported by high values of log-likelihood function in Table 1.

Imputations generated for Fashion-MNIST show that MFA does not pay too much attention to details. While it reflects the shape of cloth items reasonably well, it is not able to predict a texture at all (compare 2nd, 5th 8th and 9th rows). On the other hand, while DMFA sometimes gives blurry results, it is more effective at discovering more detailed description of the texture. It may be caused by the fact that DMFA does not fix the number of components and returns an individual conditional density for every input image using a neural network. In consequence, its expressiveness is significantly better than MFA. While MSE values of both models are similar for Fashion-MNIST, a disproportion between log-likelihoods is again huge.

Table 1. Negative log-likelihood (NLL) and mean-square error (MSE) of the most probable imputation obtained by DMFA and MFA (lower is better).

Dataset	Measure	MFA	DMFA	DMFA full-conv
MNIST	NLL	58.10	–244.81	–
	MSE	18.59	12.96	–
Fashion-MNIST	NLL	–85.15	–252.49	–
	MSE	6.12	6.03	–
CelebA	NLL	–882.54	–1222.85	–1325.13
	MSE	9.82	7.73	4.14

3.2 CelebA Dataset

We also use the CelebA dataset (aligned, cropped and resized to 32×32), which is composed of color face images, with missing regions of size 16×16. Processing of CelebA images is more resource demanding than working with MNIST and Fashion-MNIST datasets. Therefore, in addition to the convolutional neural network with a dense layer from the previous example, we also examine a fully-convolutional neural network based on DCGAN [22], which does not contain any dense layer and, in consequence, suits better to large data. Our preliminary experiments suggested that it is difficult for the fully-convolutional model to find a good candidate for the mean vector of returned density from scratch. To cope with this problem, we supply the negative log-likelihood with MSE loss[3] for the first 10 epochs, which is later turned off. Again, we put $l = 4$ and train DMFA with a learning rate $4 \cdot 10^{-5}$ for 50 epochs. The baseline MFA model uses $k = 300$ components and $l = 10$ latent dimensions.

The Fig. 3 shows that the fully convolutional version of DMFA leads to the best looking imputations (last column). The second version of DMFA also coincides with ground-truth, but its quality is worse. The results obtained by MFA are not satisfactory. Quantitative assessment, Table 1, confirms that DMFA implemented using fully convolutional network outperforms standard DMFA both in terms of MSE and log-likelihood.

[3] In fact, minimizing MSE leads to fitting a Gaussian density with isotropic covariance, so this form of loss function still optimizes a log-likelihood.

3.3 Parameters of Conditional Density

Finally, we analyze a density estimated by DMFA. Figure 4 shows images corresponding to the mean vector, the factors determining the covariance matrix and the noise vector.

Note that DMFA returns the parameters of n-dimensional Gaussian density, but the conditional density is obtained by restricting them to missing attributes. Interestingly that the model with an additional dense layer (1st-9th rows) gives a reasonable estimate on the whole image. Note however that the mean vector outside the mask is not exactly the same as the input data – it is especially evident for CelebA datasets. Introducing a dense layer allows the neural network to easily fuse the information from the whole image, which may help the neural network to fit better to the changing area of missing data. On the other hand, it is evident that fully convolutional architecture focuses only on predicting values at missing coordinates (and small area that surrounds it). It is generally difficult (or even impossible) to fully convolutional networks to mix the information from distant areas of the image and thus it concentrates only on estimating a density on the required missing region.

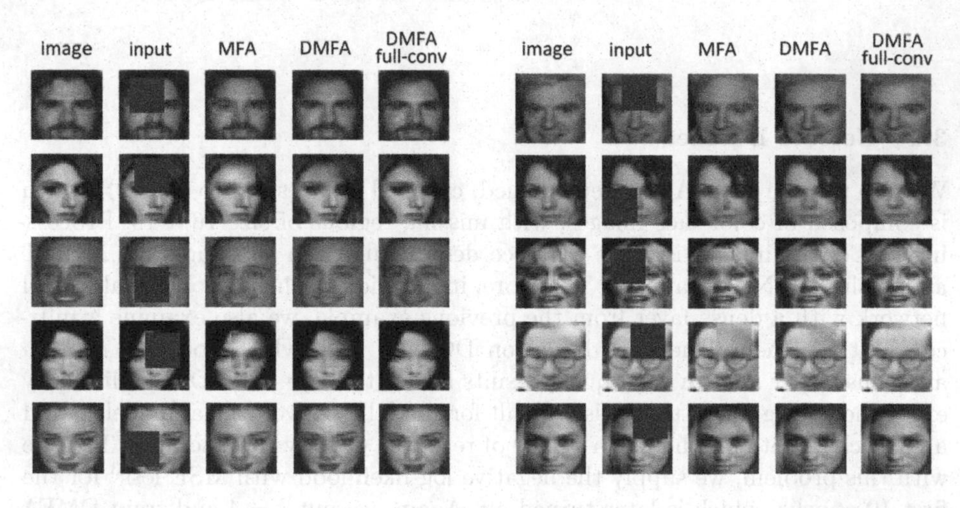

Fig. 3. Sample imputation results produced by MFA and two versions of DMFA model on CelebA.

It is evident that the factors determining the covariance matrix contain diverse shapes, which allows the obtained density to cover a wide spectrum of possible values. For example, the first three factors in the first row correspond to digit “7" while the last one is more similar to the digit “9". Factors in the third row determine different writing styles of digit “4". In the case of Fashion-MNIST, factors are mainly responsible for adding brightness intensity to a given shape. Observe that the factors for fully convolutional architecture have lower variance than using additional dense layer. It is partially confirmed by lower MSE and

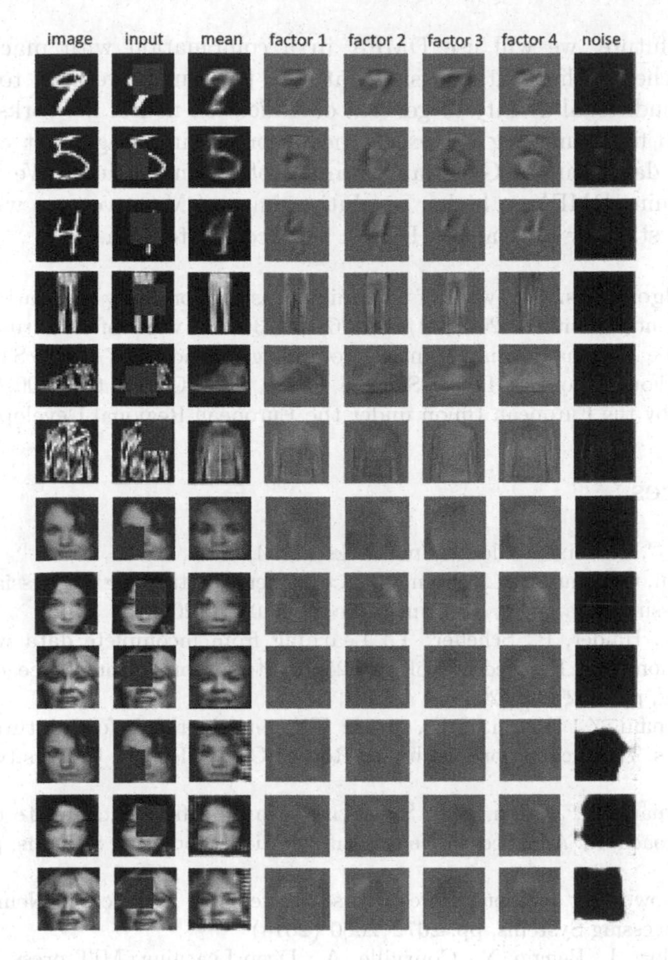

Fig. 4. Parameters of Gaussian distribution returned by DMFA (the last three rows correspond to the fully convolutional architecture).

negative log-likelihood values. The magnitude of the noise (last column) is very low (except for MNIST), which is a positive effect, because the noise is added only to guarantee the invertibility of covariance matrix.

4 Conclusion and Future Work

We proposed a deep learning approach for estimating the conditional Gaussian density of missing values given the observed ones. Experiments showed that the obtained density has significantly lower value of negative log-likelihood function than conditional GMM trained in a classical way. Moreover, imputations produced by replacing missing values with the mean vector of resulting Gaussian look visually plausible.

In the future, we will use DMFA in a combination with machine learning approaches dealing with missing data. In particular, we plan to apply the obtained conditional density to general classification neural networks, which do not need to fill in missing values at the preprocessing stage, but can process incomplete data using a Gaussian estimate of missing values. We would also like to examine DMFA on higher resolution images. Moreover, we will focus on designing a strategy for training DMFA on incomplete data.

Acknowledgements. The work of M. Śmieja was supported by the National Science Centre (Poland) grant no. 2018/31/B/ST6/00993. The work of Ł. Struski was supported by the National Science Centre (Poland) grant no. 2017/25/B/ST6/01271 as well as the Foundation for Polish Science Grant No. POIR.04.04.00-00-14DE/18-00 co-financed by the European Union under the European Regional Development Fund.

References

1. Bishop, C.M.: Mixture density networks (1994)
2. Delalleau, O., Courville, A., Bengio, Y.: Efficient em training of gaussian mixtures with missing data. arXiv preprint arXiv:1209.0521 (2012)
3. Dick, U., Haider, P., Scheffer, T.: Learning from incomplete data with infinite imputations. In: Proceedings of the 25th International Conference on Machine Learning, pp. 232–239 (2008)
4. Ghahramani, Z., Hinton, G.E., et al.: The em algorithm for mixtures of factor analyzers. Technical report, Technical Report CRG-TR-96-1, University of Toronto (1996)
5. Ghahramani, Z., Jordan, M.I.: Supervised learning from incomplete data via an em approach. In: Advances in Neural Information Processing Systems, pp. 120–127 (1994)
6. Goodfellow, I., et al.: Generative adversarial nets. In: Advances in Neural Information Processing Systems, pp. 2672–2680 (2014)
7. Goodfellow, I., Bengio, Y., Courville, A.: Deep Learning. MIT press, Cambridge (2016)
8. Iizuka, S., Simo-Serra, E., Ishikawa, H.: Globally and locally consistent image completion. ACM Trans. Graph. (ToG) **36**(4), 1–14 (2017)
9. Jerez, J.M., et al.: Missing data imputation using statistical and machine learning methods in a real breast cancer problem. Artif. Intell. Med. **50**(2), 105–115 (2010)
10. Kingma, D., Welling, M.: Auto-encoding variational Bayes. In: International Conference on Learning Representations (2014)
11. LeCun, Y., Bottou, L., Bengio, Y., Haffner, P.: Gradient-based learning applied to document recognition. Proc. IEEE **86**, 2278–2324 (1998)
12. Li, S.C.X., Jiang, B., Marlin, B.: Misgan: learning from incomplete data with generative adversarial networks. arXiv preprint arXiv:1902.09599 (2019)
13. Li, Y., Akbar, S., Oliva, J.B.: Flow models for arbitrary conditional likelihoods. arXiv preprint arXiv:1909.06319 (2019)
14. Liu, Z., Luo, P., Wang, X., Tang, X.: Deep learning face attributes in the wild. In: International Conference on Computer Vision (2015)
15. Mattei, P.A., Frellsen, J.: Leveraging the exact likelihood of deep latent variable models. In: Advances in Neural Information Processing Systems. pp. 3855–3866 (2018)

16. Mattei, P.A., Frellsen, J.: Miwae: deep generative modelling and imputation of incomplete data sets. In: International Conference on Machine Learning, pp. 4413–4423 (2019)
17. McLachlan, G.J., Peel, D.: Finite Mixture Models. John Wiley & Sons, Hoboken (2004)
18. Nazabal, A., Olmos, P.M., Ghahramani, Z., Valera, I.: Handling incomplete heterogeneous data using vaes. Pattern Recogn., 107501 (2020)
19. Van den Oord, A., Kalchbrenner, N., Espeholt, L., Vinyals, O., Graves, A., et al.: Conditional image generation with pixelcnn decoders. In: Advances in Neural Information Processing Systems, pp. 4790–4798 (2016)
20. Pathak, D., Krahenbuhl, P., Donahue, J., Darrell, T., Efros, A.: Context encoders: feature learning by inpainting. In: IEEE Conference on Computer Vision and Pattern Recognition, pp. 2536–2544 (2016)
21. Przewięźlikowski, M., Śmieja, M., Struski, Ł.: Estimating conditional density of missing values using deep gaussian mixture model. In: ICML Workshop on the Art of Learning with Missing Values (Artemiss), p. 7 (2020)
22. Radford, A., Metz, L., Chintala, S.: Unsupervised representation learning with deep convolutional generative adversarial networks. arXiv preprint arXiv:1511.06434 (2015)
23. Rezende, D.J., Mohamed, S., Wierstra, D.: Stochastic backpropagation and approximate inference in deep generative models. arXiv preprint arXiv:1401.4082 (2014)
24. Richardson, E., Weiss, Y.: On GANs and GMMs. In: Advances in Neural Information Processing Systems, pp. 5847–5858 (2018)
25. Śmieja, M., Kołomycki, M., Struski, L., Juda, M., Figueiredo, M.A.T.: Can autoencoders help with filling missing data? In: ICLR 2020 Workshop on Integration of Deep Neural Models and Differential Equations, p. 6 (2020)
26. Śmieja, M., Kołomycki, M., Struski, L., Juda, M., Figueiredo, M.A.T.: Iterative imputation of missing data using auto-encoder dynamics. In: International Conference on Neural Information Processing, p. 12. Springer, Cham (2020)
27. Śmieja, M., Struski, Ł., Tabor, J., Marzec, M.: Generalized RBF kernel for incomplete data. Knowl.-Based Syst. **173**, 150–162 (2019)
28. Śmieja, M., Struski, Ł., Tabor, J., Zieliński, B., Spurek, P.: Processing of missing data by neural networks. In: Advances in Neural Information Processing Systems, pp. 2719–2729 (2018)
29. Tipping, M.E., Bishop, C.M.: Probabilistic principal component analysis. J. Roy. Stat. Soc. Ser. B (Stat. Methodol.) **61**(3), 611–622 (1999)
30. Tolstikhin, I., Bousquet, O., Gelly, S., Schölkopf, B.: Wasserstein auto-encoders (2017). arXiv:1711.01558
31. Trippe, B.L., Turner, R.E.: Conditional density estimation with bayesian normalising flows. arXiv preprint arXiv:1802.04908 (2018)
32. Van Buuren, S.: Flexible Imputation of Missing Data. CRC Press, Boca Raton (2018)
33. Williams, D., Carin, L.: Analytical kernel matrix completion with incomplete multi-view data. In: Proceedings of the International Conference on Machine Learning (ICML) Workshop on Learning with Multiple Views, pp. 80–86 (2005)
34. Xiao, H., Rasul, K., Vollgraf, R.: Fashion-mnist: a novel image dataset for benchmarking machine learning algorithms. arXiv preprint arXiv:1708.07747 (2017)
35. Yu, J., Lin, Z., Yang, J., Shen, X., Lu, X., Huang, T.S.: Generative image inpainting with contextual attention. In: IEEE Conference on Computer Vision and Pattern Recognition, pp. 5505–5514 (2018)

Environmentally-Friendly Metrics for Evaluating the Performance of Deep Learning Models and Systems

Sorin Liviu Jurj(✉), Flavius Opritoiu, and Mircea Vladutiu

Advanced Computing Systems and Architectures (ACSA) Laboratory,
Computers and Information Technology Department,
"Politehnica" University of Timisoara, Timisoara, Romania
jurjsorinliviu@yahoo.de,
{flavius.opritoiu,mircea.vladutiu}@cs.upt.ro

Abstract. Climate change is considered to be one of the most important issues we are facing right now as a specie and existent metrics and benchmarks used to evaluate the performance of different Deep Learning (DL) models and systems are currently focused mainly on their accuracy and speed, without also considering their energy consumption and cost. In this paper, we introduce four novel DL metrics, two regarding inference called Accuracy Per Consumption (APC) and Accuracy Per Energy Cost (APEC) and two regarding training called Time To Closest APC (TTCAPC) and Time To Closest APEC (TTCAPEC), which take into account not only a DL model's accuracy but also its energy consumption, energy cost and the time it takes to train it up to that point. Experimental results prove that all four DL metrics are promising, encouraging future DL researchers to make use of models and platforms that require low power consumption as well as of green energy when powering their DL-based systems.

Keywords: Deep Learning · Metrics · Energy consumption · Energy cost · Green energy

1 Introduction

With unprecedented growth in the number of platforms, e.g. CPUs, GPUs and FPGAs as well as in the number of DL algorithms, architectures, and frameworks, the need for a fair comparison between DL-based systems when performing training or inference by using appropriate metrics is crucial.

Until recently, it was difficult to fairly compare DL models due to the inexistent standard evaluation criteria. In the last years, efforts to deliver efficient tools for benchmarking DL implementations were made by various researchers from both academia and industry, an example in this direction being the MLPerf Benchmark [1] introduced initially (in 2018) only for training but recently (in November 2019) also regarding inference [2] and being supported by a group of 40 organizations like e.g. Google and Microsoft. Regarding training, when measuring the performance of DL implementations, there were many types of metrics used in prior DL benchmarks, i.e.

H. Yang et al. (Eds.): ICONIP 2020, LNCS 12534, pp. 232–244, 2020.
https://doi.org/10.1007/978-3-030-63836-8_20

throughput (samples per second), but recently Time-To-Accuracy (TTA), an end-to-end training time to a specified validation accuracy level, is the accepted metric in the DL community, being also the main metric used in MLPerf. A consequence of this race towards occupying the first place in a Benchmark with the TTA as a metric for training is that the state-of-the-art DL models consume an enormous amount of energy, affecting the climate change and limiting the Artificial Intelligence (AI) innovation, with a report from Allen Institute for AI [3] arguing that energy efficiency should be considered a more common evaluation criterion for AI papers, at least as important as accuracy and that the focus on a single metric is detrimental to our society, economy, and environment, with recent work in [4] even concluding that there is a very significant carbon footprint to DL. Despite there being many available DL benchmarks [1, 2, 5, 6] that consider various metrics like time, cost, utilization, memory footprint, throughput, timing breakdown, strong scaling and communication as well as latency and load balancing, only MLPerf Benchmark is considered [5] to have energy as a metric for training (power measurement spec for inference is expected only in a future update).

Considering these aspects, we strongly believe in the necessity of incorporating in the next generation DL benchmarks the ability to take into account the energy consumption that a DL system has when training or running inference. Furthermore, we think that it should be taken into account also the autonomy of such a system, i.e. its ability to work independently of a traditional power grid source and instead is able to use 100% green energy such as solar energy [7]. For a more scalable and sustainable future, especially considering the emerging focus of Green AI [3], we propose two DL metrics for inference called APC and APEC and two metrics for training called TTCAPC and TTCAPEC.

The paper is organized as follows. In Sect. 2 we present the related work. Section 3 describes the proposed APC, APEC, TTCAPC and TTCAPEC metrics. Section 4 presents the experimental setup and results. Finally, Sect. 5 concludes this paper.

2 Related Work

The question of energy consumption to be used as a metric when evaluating the performance of DL models or DL-based systems is of high importance for many papers in the literature.

An example is a work in [3] where the authors advocate for a simple and compute-efficient metric, suggesting the use of energy efficiency as a metric when evaluating a DL model instead of "Red AI" which refers to the kind of AI research that uses extreme computational power and costs to achieve state-of-the-art results regarding accuracy. A comprehensive analysis of important metrics such as accuracy, inference time, and power consumption is made in [8] where the authors show the importance of these metrics when designing 14 efficient DNNs. Similarly, the authors in [9] expand the analysis to over 40 DNN architectures, highlighting the importance of metrics when evaluating the performance of a neural network. Also, the authors in [10] contribute to the challenge of estimating the energy consumption in machine learning by providing useful guidelines and a large selection of the latest software tools for a machine learning expert who wants to design and estimate energy for future DL systems.

Some arguments against using only TTA as a metric when evaluating DL systems on the MLPerf Benchmark are presented by the work in [11] where the authors propose the Time-To-Multiple-Thresholds (TTMT) curves and the Average-Time-To-Multiple-Thresholds (ATTMT) metric. By comparison, their metric targets the training part, without taking into consideration the energy efficiency whereas our metrics target both the training and the inference parts and take into consideration the energy consumption as well as the energy cost of a DL-based system. Additionally, the TTA and ATTMT metrics are able to compare only different systems, whereas our metrics are able to compare also different models trained and executed in different systems, e.g. to identify on which hardware is better to train a DL model and on which hardware is better to run an inference with the same DL model.

3 The Proposed Deep Learning Metrics

In order to solve the problem of lacking in accountability in energy consumption and costs, in this section we will describe the 4 proposed metrics.

We want the APC and APEC metrics to comply with the following important properties: Output range from 0 to 1; 100% accuracy and 0 energy consumption/cost imply the value of the metric is 1; 0% accuracy implies the metric is 0 regardless of energy consumption/cost; The value of the metric increases with accuracy and decreases with energy consumption/cost; Consumption/cost from inaccurate inferences are weighted more heavily. We consider these to be the most important requisites for a combination of two measures into one metric. Since it is a metric, it is desirable that it ranges from 0 to 1, so that it can be expressed in terms of percentage and give some sense of how close or distant the value of the metric is from the ideal (i.e. 1) result. When combining two measures into a single metric it is important to consider how we want each measure to influence the metric. Since lower consumption is desirable, consumption should lower the final metric, and since higher accuracy is desirable, accuracy should increase the final metric. We also want it to convey some common-sense properties: if the DL-based system running inference has 0% accuracy, it doesn't matter how much or little it costs because we won't use it, and an inaccurate inference is a complete waste of energy by itself, so it makes sense to penalize its cost more heavily.

With the previous properties in mind, we define a common function presented in Eq. (1) for both metrics. It is a prerequisite in order to be able to create the final APC and APEC metrics.

$$WC_{\alpha}(c, acc) = 2c((1-\alpha)(1-acc) + \alpha.acc) \tag{1}$$

This is a function $WC_{\alpha}(c, acc)$ to **weight energy consumption/cost** differently between accurate inferences and inaccurate ones, where c is the energy consumption/cost of a system, which could be measured per inference or per unit of time, acc is the accuracy of the model and α is a parameter (ranges from 0 to 0.5) that controls how much weight is assigned to accurate inferences (i.e. if $\alpha = 0$ the weight assigned to accurate inferences is 0; if $\alpha = 0.5$, the weight assigned is the same in all cases/for accurate as well as inaccurate inferences). The function WC_α has the following properties: If system *a* has a higher energy consumption/cost than system *b* and both have the same accuracy, the weighted consumption/cost of *b* is lower or the same; If system *a* has better accuracy than system *b* and both consume/cost the same the weighted consumption/cost of *a* is lower or the same; If energy consumption/cost of a system is 0 the weighted consumption/cost is 0; Consumption/cost from inaccurate inferences are weighted more heavily.

The APC metric is a function that takes into account not only the accuracy of a system (*acc*) but also the energy consumption of the system (*c*), as can be seen in Eq. (2):

$$APC_{\alpha,\beta}(c, acc) = \frac{acc}{\beta.WC_\alpha(c, acc) + 1} \tag{2}$$

where *c* stands for the energy consumption of the system and it's measured in watt-hour (Wh) and *acc* stands for accuracy; α is the parameter for the WC_α function, the default value is 0.1; β is a parameter (ranges from 0 to infinity) that controls the influence of the cost in the final result: higher values will lower more heavily the value of the metric regarding the cost. The default value is 1. It is important to mention here that, as a rule of thumb, our recommendation is to use a value for β in the ballpark of $1/avg$ where *avg* is the average cost of the systems to evaluate. This average cost is among different systems that perform the same task, not each individual cost average from a system to measure. For example, if the commonly used methods to solve a task have an average cost of B, then, when measuring the APC for these systems, in order to compare them to our own, we would use as β the value 1/B.

In the APC metric "*c*" means **consumption** and is proposed to be a measure of the energy consumption of a single inference in a system, having its value always greater than 0. The APC metric's properties are the following: Ranges from 0 to 1; 100% accuracy and 0 energy consumption imply the APC is 1; 0% accuracy implies an APC of 0 regardless of energy consumption; APC increases with accuracy and decreases with energy consumption; Consumption from inaccurate inferences are weighted more heavily. In order to see how accuracy and consumption affect the APC value, we plot APC over the consumption for different values of accuracy. In Fig. 1 we can see most of the properties demonstrated in this section. Where α is 0, the consumption is not measured for correct inferences which imply that a model with 100% accuracy will not be penalized by its consumption (e.g. the constant pink colored line that is seen on the top-left side of Fig. 1) as compared to where α is 0.25 and 0.5. Similarly, in order to show how different values of β affect the APC metric, some variations in β are presented on the right side of Fig. 1 where β is 10, 100, and 1000. We can see how the higher the β, the heavier the impact of the energy consumption is in the value of the APC metric.

The APEC metric is a function presented in Eq. (3) and is in appearance the same as the APC function.

$$APEC_{\alpha,\beta}(c, acc) = \frac{acc}{\beta.WC_{\alpha}(c, acc) + 1} \quad (3)$$

However, in practice, the two metrics are fundamentally different. Here, the main difference lays in the meaning of the input "c".

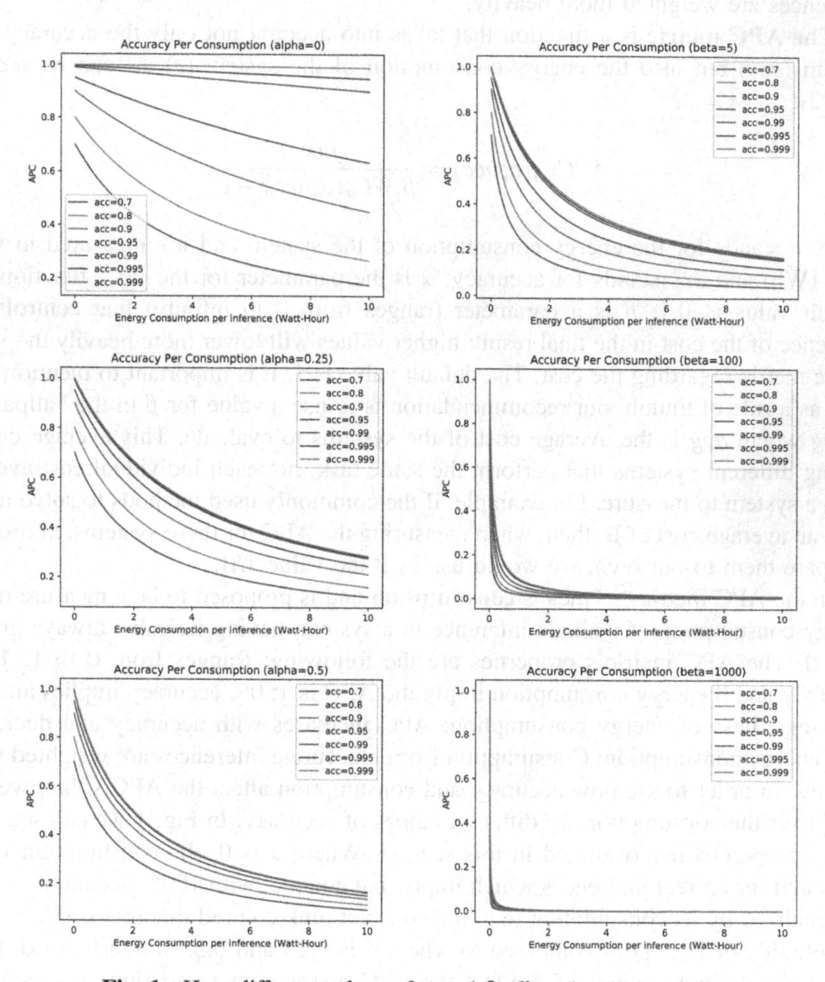

Fig. 1. How different values of α and β affect the APC metric.

In APEC, "c" means **cost** and is proposed to be a measure of the energy cost of a single inference in a system, therefore, it is measured in different units and in different ranges. In Germany, for example, 1 kWh of energy costs 30.5 cents EUR [12]. Therefore, if our system pays 100Wh of energy for each inference, the cost "c" of our system will be 3.05 (cents EUR). However, it is possible to set up a system in which one doesn't pay for the energy, for example, if the energy it consumes is a renewable type of energy such as e.g. the green energy (solar energy) that comes from the sun with the help of a solar tracker. In these kinds of systems, the cost of electricity would be 0, and the APEC of these systems would be the same as the accuracy. Only in these cases would it be theoretically possible to obtain 100% APEC. The APEC metric's properties are all the same as the APC metric's properties presented earlier. The only difference is the meaning of c, which here means cost, thus the impact that different values for α and β have on the APEC metric is similar to the APC metric, as seen earlier in Fig. 1.

Following, we will define a metric called **Time To Closest Accuracy Per Measured Energy (TTCAPME)** that takes into account the energy consumption/cost of a model, its accuracy, and the time it takes to train it up to that point. We also want to be able to compare with this metric for the same problem both different models and different systems. For this, we define a delta in accuracy and another one in energy consumption/cost for each problem, such that variations within that delta are considered negligible. For example, if the accuracy delta (δ_a) is 0.01 and the energy delta (δ_e) is 0.1, then a model with 0.924 accuracy and 1.12 energy consumption/cost and a model with 0.921 accuracy and 1.18 energy consumption/cost would be considered equally good. Having defined both deltas, the grid is formed by the intervals of accuracy and energy consumption/cost, and the value in each element of the grid is the Accuracy Per Measured Energy (APME) of the lowest value in that element of the grid, e.g. the element on the accuracy interval (0.25, 0.26) and energy interval (1.5, 1.6) would be APME (1.5, 0.25). APME is a function that increases with accuracy and decreases with energy consumption/cost. An example of this type of grid can be seen in Fig. 2 where redder colors represent higher values of APC. The grid maps accuracies and energies to the "closest" APC values. This metric compares training times of models within the same grid interval, considering better the model that takes less time to fall into that interval. For models on different APME values, we consider better the one with higher APME value. Then, the metric effectively maps the ternary of values

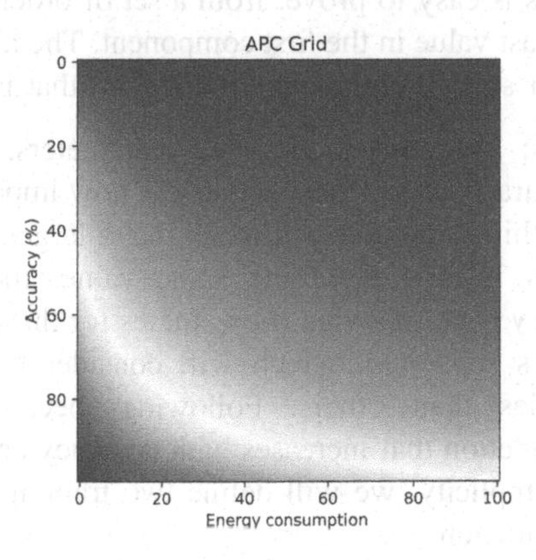

Fig. 2. APC Grid with energy delta (δ_e) = 1 and accuracy delta (δ_a) = 0.01. Redder colors represent higher values of APC.

(accuracy, energy consumption/cost, training time) to the ordered pair of values ("closest" APC, training time), and offers us a way to compare between these outputs.

Ordinality: In order to be able to compare between values of our metric's outputs, we need the mathematical tools to define <, = ,>.

Definition 1.1: Let a_1, a_2 be real numbers between 0 and 1 and b_1, b_2 real positive numbers. Then we define the relationships between the ordered pairs $(a_1, b_1), (a_2, b_2)$ as follows:

If $a_1 = a_2$ then $(a_1, b_1) < (a_2, b_2)$ if and only if $b_1 > b_2$, $(a_1, b_1) = (a_2, b_2)$ if and only if $b_1 = b_2$ and $(a_1, b_1) > (a_2, b_2)$ if and only if $b_1 < b_2$.
if $a_1 < a_2$ then $(a_1, b_1) < (a_2, b_2)$, and if $a_1 > a_2$ then $(a_1, b_1) > (a_2, b_2)$ regardless of b_1, b_2.

We will prove that the set of ordered pairs with the previously defined ordinality is **well-ordered.**

Trichotomy: Since we defined the relations case by case, for two pairs only one and exactly one of the relations is true.

Transitivity: We want to prove that if v, w, and z are ordered pairs with the previously defined ordinality and v< w and w < z then v < z:

$v = (v_a, v_b), w = (w_a, w_b), z = (z_a, z_b)$
$v < w$, then either $v_a < w_a$ or $v_a = w_a$ and $v_b > w_b$
$w < z$, then either $w_a < z_a$ or $w_a = z_a$ and $w_b > z_b$
If $v_a < w_a$ then $v_a < z_a$ therefore $v < z$
If $v_a = w_a$ *and* $v_b > w_b$ and $w_a < z_a$ then $v_a < z_a$ therefore $v < z$
If $v_a = w_a$ and $v_b > w_b$ and $w_a = z_a$ and $w_b > z_b$ then $v_a = z_a$ and $v_b > w_b > z_b$, therefore $v < z \diamond$

Well-Foundedness: We want to prove that every nonempty set of ordered pairs has a least element, that is, it has an element x such that there is no other element y in the subset where x > y. This is easy to prove: from a set of ordered pairs we can find the elements that have the least value in the first component. Then, from these elements, we find the one with greater second component value, and that is the least element.

Parameters Properties: This metric has two parameters, energy delta (δ_e) and accuracy delta (δ_a). Accuracy delta reflects inversely how important accuracy is for the model. High values for this parameter will mean that a larger range of accuracies will be grouped together as if they were the same value, therefore making smaller improvements in accuracy is not relevant. Low values for these parameters will tend to keep different accuracies separated, which will consider better models those with slightly better accuracies than others. Following, because the training metric TTCAPME requires a function that increases with accuracy and decreases with energy consumption/cost, for simplicity, we will define two training metrics by using either APC or APEC as this function.

Regarding **the TTCAPC metric**, the objective of this metric is to combine training time and the APC inference metric in an intuitive way, as can be seen in formula (4). The formula for TTCAPC is:

$$\begin{aligned} & TTCAPC_{\beta,\delta_e,\delta_a}(trainingTime, energy, acc) \\ & = \left(trainingTime, APC_{\alpha,\beta}(rounded_{\delta_e}(energy), rounded_{\delta_a}(acc))\right) \end{aligned} \tag{4}$$

where *trainingTime* is the training time in seconds for the model, *energy* is the **energy consumption** in Wh of inferencing with the model, and *acc* is the accuracy of the model. This will mean that higher accuracies will be celebrated and higher net energy consumptions and higher training times will be penalized.

Regarding **the TTCAPEC metric**, the objective of this metric is to combine training time and the APEC inference metric, as can be seen in formula (5).

$$\begin{aligned} & TTCAPEC_{\beta,\delta_e,\delta_a}(trainingTime, energy, acc) \\ & = \left(trainingTime, APEC_{\alpha,\beta}(rounded_{\delta_e}(energy), rounded_{\delta_a}(acc))\right) \end{aligned} \tag{5}$$

The formula for TTCAPEC is the same as the one presented earlier for the TTCAPC, but here the meaning of *energy* is different, meaning the **energy cost** in Euro cents of inferencing with the model. Similar to TTCAPC, this will mean that higher accuracies will be celebrated, but higher energy costs and training times will be penalized.

4 Experimental Setup and Results

In order to realize the experiments with the above-defined metrics, we needed to measure and extract two types of data: the accuracy of the DL models and the energy consumption of the system they run training and inference on.

For this tasks, regarding the inference, we make use of one of our previously trained DL models from the work in [13], namely the MobileNetV2, as well as of the systems (i.e. Nvidia Jetson TX2 and a laptop containing an Nvidia GTX 1060 GPU) on which this DL model was running inference in real-time [7]. Regarding the training, in this case, we make use of all four DL models from [13]. It is important to mention that regarding the training time (seconds) for the Nvidia Jetson TX2, the values are simulated. We naturally want to measure the APC for different values of accuracy and power consumption.

For this, first, we run experiments for 2 h on both the laptop containing the Nvidia GTX 1060 GPU as well as on the Nvidia Jetson TX2 platform and feed their power consumption values into the APC equation presented in (2), where "*c*" in this case stands for the power consumption of the system running the MobileNetV2 DL model in real-time using motion detection [7]. Because both platforms run Linux Ubuntu, these power consumption values are taken 12 times (one power consumption value every 10 min) with the help of "sudo powerstat" for the laptop containing the Nvidia GTX 1060 GPU and with the help of a power measurement script [14] as well as "sudo./tegrastats" for the Nvidia Jetson TX2 platform.

Secondly, we noted the accuracy values also every 10 min for a total of 12 times (2 h), but in this case, instead of measuring the inference accuracy for both platforms, we presented them only once, since presenting them for both doesn't influence our experimental results at all. Because of the weather, lighting, and image quality conditions, to name only a few, it resulted in many different accuracy values, as seen in Table 1. This situation was very helpful in our experiment because it can be easily seen how well our metrics perform beside only with big differences in power consumption values. We used alpha = 0.1 as the default and beta = 0.1 since the average consumption is close to 10 and the inverse of this number is 0.1. In Table 1 we can see these results.

As we can see, the APC metric succeeds in unifying the two metrics of accuracy and energy consumption into one, and therefore it is a better metric in the cases where both accuracy and energy consumption are required to be taken into account in the final result.

Table 1. APC with alpha = 0.1 and beta = 0.1 for our MobileNetV2 DL model [7, 13] running inference in real-time for 2 h, with 12 samples taken every 10 min.

Power consumption [W]		Inference accuracy [%]	APC [%]	
Laptop	Nvidia Jetson TX2		Laptop	Nvidia Jetson TX2
50.07	8.85	99.7	65.84	91.39
50.51	9.01	92.11	49.42	79.81
47.16	9.01	91.32	49.63	78.69
49.11	9.07	94.54	54.57	83.27
49.6	6.94	50.25	13.51	36.4
49.12	8.96	25.57	5.34	15.13
49.15	9.19	80.69	34.39	64.46
48.51	9.11	47.31	12.49	31.06
47.9	9.23	60.14	18.8	42.25
47.03	9.05	85.86	41.5	71.21
48.15	9.15	99.42	65.98	90.68
46.01	9.3	98.31	64.25	88.79

We also want to measure the APEC of our DL models in order to see how they stand against each other and more importantly to see the difference between the two types of energy: green energy (solar power) and energy grid. For simplicity and because it is out of the scope of this paper to experiment with data regarding electricity costs for all the countries in the world, we will just take Germany as an example. According to "Strom Report" (based on Eurostat data) [12], German retail consumers paid 0.00305 Euro cents for a Wh of electricity in 2017. We will use that value to calculate the cost of energy by plugging it in the equation presented in (3)", where "c" in this case stands for the energy cost. In Table 2 we can see these results. We used alpha = 0.1 as the default and beta = 50 since the average cost is close to 0.03 and the inverse of this number is rounded up to 50.

Table 2. APEC with alpha = 0.1 and beta = 50 for our MobileNetV2 DL model [7, 13] running inference in real-time for 2 h with regular (paid) energy as well as with solar (free) energy.

Power cost [Cents EUR]		Inference accuracy [%]	APEC [%]		APEC (Green energy) [%]
Laptop	Nvidia Jetson TX2		Laptop	Nvidia Jetson TX2	
0.1527	0.0269	99.7	55.87	87.56	99.7
0.1540	0.0274	92.11	39.74	74.58	92.11
0.1438	0.0274	91.32	40.03	73.36	91.32
0.1497	0.0276	94.54	44.65	78.37	94.54
0.1512	0.0211	50.25	9.77	31.80	50.25
0.1498	0.0273	25.57	3.77	12.46	25.57
0.1499	0.0280	80.69	26.43	58.31	80.69
0.1479	0.0277	47.31	9.01	26.31	47.31
0.1460	0.0281	60.14	13.82	36.54	60.14
0.1434	0.0276	85.86	32.64	65.36	85.86
0.1468	0.0279	99.42	56.08	86.69	99.42
0.1403	0.0283	98.31	54.36	84.50	98.31

As we see in Table 2, in the cases where we use green (solar) energy to power our DL-based systems, the APEC is in every case around 20% higher for the Nvidia Jetson TX2 platform and around 50% higher for the laptop containing the Nvidia GTX 1060 GPU in terms of absolute values. As can be observed, the APEC metric is superior in cases where not only the accuracy but also the energy cost matter in the final result.

Regarding the results presented in Table 3, for the models trained on different systems, we can see that if we choose an accuracy delta of 0.1 and energy delta of 1, the APC is different for each of them, therefore, this means that the best system is the one with the higher APC and training time is not considered. However, with the same models but with larger deltas we see that two models result in the same APC, and therefore the deciding factor is the training time.

Regarding the experiments for the TTCAPEC metric, we use the same country and price for electricity as mentioned earlier regarding the experiments with the APEC metric. Similarly to the results regarding TTCAPC presented in Table 3, on Table 4 we can see that for the models we trained on different systems, if we choose an accuracy delta of 0.1 and energy delta of 0.001, the APEC is different for each of them, therefore the best system is the one with the best APEC and training time is not considered. However, with the same models but with larger deltas, all models result in the same APEC, and therefore the deciding factor is the training time.

It is important to mention that despite using the term accuracy in our APC and APEC metrics, both metrics can work well also by using another metric in place of accuracy (as long as it ranges from 0 to 1, meaning that 0 represents a negative score and 1 represents a positive one), such as the ones used by MLPerf Benchmark [1, 2].

Table 3. TTCAPC values for four different DL models (V = VGG-19, I = InceptionV3, R = ResNet-50, M = MobileNetV2) in two different hardware platforms. EC = Energy Consumption; RA = Rounded Accuracy; REC = Rounded Energy Consumption; TT = Training Time.

	Laptop				Nvidia Jetson TX2			
	V	I	R	M	V	I	R	M
Accuracy	90.56	93.41	93.49	94.54	90.56	93.41	93.49	94.54
EC (Wh)	49.95	53.05	50.26	48.45	11.61	10.33	9.97	8.90
Accuracy delta = 0.1, Energy delta = 1, beta = 0.1, alpha = 0.1								
RA	90.55	93.45	93.45	94.55	90.55	93.45	93.45	94.55
REC	48.5	53.5	50.5	48.5	8.5	10.5	9.5	8.5
Closest APC	47.72	50.50	51.83	54.87	78.24	80.08	81.19	**83.91**
TT (seconds)	20.27	38.85	21.39	38.84	20.27	38.85	21.39	38.84
Accuracy delta = 5, Energy delta = 10, beta = 0.1, alpha = 0.1								
RA	92.5							
REC	45	55		45	15		5	
Closest APC	52.74	48.14		52.74	73.92		85.35	
TT (seconds)	20.27	38.85	21.39	38.84	20.27	38.85	**21.39**	38.84

Also, the metrics proposed in this paper can work for any DL-based system; all that is needed is to have the training time, the consumption, the cost, and the accuracy measured.

Table 4. TTCAPEC values for four different DL models (V = VGG-19, I = InceptionV3, R = ResNet-50, M = MobileNetV2) in two different hardware platforms. EC = Energy Cost; REC = Rounded Energy Cost; RA = Rounded Accuracy; TT = Training Time.

	Laptop				Nvidia Jetson TX2			
	V	I	R	M	V	I	R	M
Accuracy	90.56	93.41	93.49	94.54	90.56	93.41	93.49	94.54
EC (cents)	0.1524	0.1618	0.1533	0.1478	0.0354	0.0315	0.0304	0.0272
Accuracy delta = 0.1, Energy delta = 0.001, beta = 50, alpha = 0.1								
REC	0.1525	0.1615	0.1535	0.14775	0.0355	0.0315	0.0305	0.0275
RA	90.55	93.45		94.55	90.55	93.45		94.55
Closest APEC	37.557	40.924	42.096	45.000	68.161	74.739	75.217	78.468
Closest APEC (Green Energy)	90.55	93.45		94.55	90.55	93.45		**94.55**
TT (seconds)	20.273	38.853	21.396	38.847	20.273	38.853	21.396	38.847
Accuracy delta = 5, Energy delta = 0.1, beta = 50, alpha = 0.1								
REC	0.15				0.05			
RA	92.5							
Closest APEC	40.997				65.198			
Closest APEC (Green Energy)	92.5							
TT (seconds)	20.273	38.853	21.396	38.847	**20.273**	38.853	21.396	38.847

5 Conclusions and Future Work

Currently, the performance evaluation of a DL model is mainly based on its accuracy, but not also on its energy consumption or its energy cost.

In this paper, we introduce four metrics, two for inference called APC and APEC and two for training called TTCAPC and TTCAPEC for evaluating the performance of DL models and systems not only regarding their accuracy but also their energy consumption and cost. In our experimental results, we succeeded to prove that all four metrics are efficient, showing, to the best of our knowledge, for the first time in literature, that by using high accuracy together with low power consumption, especially green energy (e.g. solar energy) during training and inference, a DL model or system is evaluated as being much more performant than one that, despite having the same accuracy, consumes more energy and uses a traditional power grid (paid electricity).

We believe that these metrics will encourage future researchers to develop and use greener energy-based systems and that they will evaluate their performance only based on how "green" they are and how less negative impact they have on our planet.

References

1. Mattson, P., et al.: MLPerf Training Benchmark. arXiv:191001500 (2019)
2. Reddi, V.J., et al.: MLPerf Inference Benchmark. arXiv:191102549 (2019)
3. Schwartz, R., Dodge, J., Smith, N.A., Etzioni, O.: Green AI. arXiv:1907.10597v3 (2019)
4. Strubell, E., Ganesh, A., McCallum, A.: Energy and policy considerations for deep learning in NLP. arXiv:1906.02243 (2019)
5. Ben-Nun, T., et al.: A modular benchmarking infrastructure for high-performance and reproducible deep learning. arXiv:190110183 (2019)
6. Zhu, H., et al.: TBD: benchmarking and analyzing deep neural network training. arXiv: 180306905 (2018)
7. Jurj, S.L., Rotar, R., Opritoiu, F., Vladutiu, M.: Efficient implementation of a self-sufficient solar-powered real-time deep learning-based system. In: Iliadis, L., Angelov, P.P., Jayne, C., Pimenidis, E. (eds.) EANN 2020. PINNS, vol. 2, pp. 99–118. Springer, Cham (2020). https://doi.org/10.1007/978-3-030-48791-1_7
8. Canziani, A., Paszke, A., Culurciello, E.: An analysis of deep neural network models for practical applications. arXiv:160507678 (2017)
9. Bianco, S., Cadene, R., Celona, L., Napoletano, P.: Benchmark analysis of representative deep neural network architectures. IEEE Access **6**, 64270–64277 (2018). https://doi.org/10.1109/ACCESS.2018.2877890
10. García-Martín, E., et al.: Estimation of energy consumption in machine learning. J. Parallel Distrib. Comput. **134**, 75–88 (2019). https://doi.org/10.1016/j.jpdc.2019.07.007
11. Verma, S., et al.: Metrics for machine learning workload benchmarking. In: International Workshop on Performance Analysis of Machine Learning Systems (FastPath) in Conjunction with ISPASS, Texas, The University of Texas (2019)
12. Strom-Report. https://1-stromvergleich.com/electricity-prices-europe/. Accessed 25 Apr 2020

13. Jurj, S.L., Opritoiu, F., Vladutiu, M.: Real-time identification of animals found in domestic areas of Europe. In: Proceedings of SPIE 11433, Twelfth International Conference on Machine Vision (ICMV 2019), p. 1143313 (2020). https://doi.org/10.1117/12.2556376
14. Convenient Power Measurement Script on the Jetson TX2/Tegra X2. https://embeddeddl.wordpress.com/2018/04/25/convenient-power-measurements-on-the-jetson-tx2-tegra-x2-board/. Accessed 03 May 2020

Hybrid Deep Shallow Network for Assessment of Depression Using Electroencephalogram Signals

Abdul Qayyum[1], Imran Razzak[2(✉)], and Wajid Mumtaz[3]

[1] University of Burgundy, Dijon, France
abdul.qayyum@u-bourgogne.fr
[2] School of Information Technology, Deakin University, Geelong, Australia
imran.razzak@deakin.edu.au
[3] National University of Sciences and Technology, Islamabad, Pakistan
wajidmumtaz@gmail.com

Abstract. Depression is a mental health disorder characterised by persistently depressed mood or loss of interest in activities resulting impairment in daily life significantly. Electroencephalography (EEG) can assist with the accurate diagnosis of depression. In this paper, we present two different hybrid deep learning models for classification and assessment of patient suffering with depression. We have combined convolutional neural network with Gated recurrent units (RGUs), thus the proposed network is shallow and much smaller in size in comparison to its counter LSTM network. In addition to this, proposed approach is less sensitive to parameter settings. Extensive experiments on EEG dataset shows that the proposed hybrid model achieve highest accuracy, f1 score 99.66%, 99.93% and 98.87%, 99.12% for eye open and eye close dataset respectively in comparison to state of the art methods. Based on high performance, the proposed hybrid approach can be used for assessment of depression for clinical applications and can deployed remotely in hospital or private clinics for clinical evaluation.

Keywords: EEG · Depression · Anxiety · Electroencephalographic · Mental disorder

1 Introduction

Mental health conditions such as depression and anxiety can cause distress, impact our daily life functioning and relationships, and are associated with poor physical health and premature death from suicide. Depression is a major cause of morbidity worldwide. It is estimated that above 3 million in Australians and 16.2 million Americans are living with anxiety or depression. Recent study showed that COVID-19 worsened the mental health condition. Approximately, 14% of the global burden of disease is attributed to mental health disorders. Among these, unipolar depression is the second leading cause.

H. Yang et al. (Eds.): ICONIP 2020, LNCS 12534, pp. 245–257, 2020.
https://doi.org/10.1007/978-3-030-63836-8_21

Depression can lead to serious health complications if left untreated, thus, continuous monitoring is required since the patient may experience frequent depressive episodes [7,8,11]. Fortunately, effective treatments such as therapy, medication, diet, and exercise can be used to effetely treat patient suffering from depression, however, it requires efficient early diagnosis [9,10]. Recently, deep learning has escalated automatic diagnosis of unipolar depression to a next level [12].

Complex and nonlinear nature of EEG signals require the development efficient methods to achieve high diagnostic performance. Recently, several deep learning based methods have been applied for EEG based depression diagnosis. Acharaya et al. employed convolutional neural network (CNN) consisting of 13 layers of abstraction for the classification of depression and reported 93.5% and 96% classification performance on 30 (15 depressed and 15 healthy controls) for left and right hemispheres, respectively [1]. Similarly, Ay et al. combined convolutional neural network and long short term memory (LSTM) and achieved 99.12% and 97.66% diagnostic accuracy for right and left hemisphere, respectively [2]. CNN learns the temporal properties of signal followed by LSTM for sequence learning. Yildirim et al. presented CNN for abnormality detection in EEG signals [13] and achieved detection error rate of accuracy of 79.34%, on TUH EEG abnormal dataset. Li et al. analyzed different aspects of EEG (spectral, spatial, and temporal information) for the diagnosis of mild depression [4]. Results showed that spectral information of EEG signals play major role whereas temporal information showed significant improvement in diagnostic performance. The utilized the pre-trained ConvNet architectures on EEG-based mental load classification task and achieved accuracy of 85.62% for recognition of mild depression and normal controls. Liao et al. applied kernel eigen-filter-bank common spatial pattern to extract features and achieved 80% accuracy with sever depression [5]. Zhang et al. extracted both linear and nonlinear features from EEG signals from 25 subject with closed eye under resting state and achieved 92.9% and 94.2% with KNN and back-propagation neural network respectively [14]. Mahato and Paul compared the performance of multi layered perceptron neural network, radial basis function network, linear discriminant analysis and quadratic discriminant analysis for detection of major depressive disorder [6]. Comparative analysis of nonlinear and linear features on different methods showed that best performance 93.33% is achieved when linear (band power, inter hemispheric asymmetry) and non-linear feature (relative wavelet energy and wavelet entropy) are combined i.e., combination alpha power and relative wavelet energy using multi layered perceptron neural network and radial basis function network. Not only EEG but other video data is also used for depression diagnosis. Huang et al. presented attention-based CNN and LSTM to differentiate between major depressive disorder (MDD) and bipolar disorders (BD). Audio and facial expression in video sequences are used to identify the mood disorder on the basis of response [3].

The main motivation of this work is to develop a completely automated diagnostic system for depression through raw EEG signals. In order to improve the diagnostic performance, we proposed an electroencephalographic (EEG)-based hybrid deep learning framework that automatically discriminated depressed and healthy controls and provided the diagnosis. We have used CNN for automated feature extraction followed by classification through Gated recurrent units/long short term memory. In comparison to LSTM, GRU may be a better choice for EEG signals due to short sequence whereas LSTM is better for larger sequence. In addition to this, it use less training parameters and less memory thus, faster than its counter LSTM network. In this paper, we have used both LSTM and GRU for depression diagnosis and compared their performance. We describe the theoretical and empirical **key contributions** of this work as follows:

- an end-to-end application to detect the depression automatically using raw EEG signals.
- present hybrid deep shallow model for classification and assessment of depression.
- CNN is used for feature extraction and LSTM/GRU for sequence learning.
- The assessment of depression has been measured using various performance metrics for eye open and eye close datasets that significant improvement in diagnostic performance.

2 Hybrid Depression Diagnostic Framework

In this section, we present hybrid deep network for efficient diagnosis of depression using EEG signals. We have performed two experiment. In our first experiment, we used one-dimensional convolutional neural network (1DCNN) combined with gated recurrent units and IDCNN with LSTM in our second experiment. The 1DCNN-GRU has a capability to extract temporal features in an efficient. Figure 1 shows an overview of the proposed deep hybrid framework. There are three main component IDCNN, GRU/LSTM and classifier. The convolutional layers are the same in both experiments. There are three number of convolutional, two Max-pooling, two dropout layers followed by LSTM/GRU layer and fully connected and sigmoid classification layer. We have performed several experiment to select best feature map combination. In order to improve the performance, the features are extracted from last the dense layer in the trained proposed models before the classification layer and passed these features to machine learning models for classification (KNN, RF and SVM are used to differentiate the control and MDD depression classification patient). The convolutional layer is represented by sliding one dimensional kernel filter over the EEG signal with the specified stride. The convolution between EEG input signal and one-dimensional kernel filter is shown in Eq. 1. The output of the convolutional layer is called the feature map. Figure 1 shows an overview of the proposed ML framework including the computation of the classifier performance metrics. According to the proposed framework, the EEG data were segmented with a window length of one second (256 samples). As the sampling rate was 256 samples per second, each EEG segment contains 19 channels and 256 data points (window length). The selection of a one second window size was based on the empirical evaluation of the proposed models. It was observed that a window size of one second provided best results. From the classifier point of view, the input data dimension was 256 × 19 for each instant of class for two EEG datasets (EO and EC). Moreover, the input data were divided into training and testing set with 80 and 20% ratio.

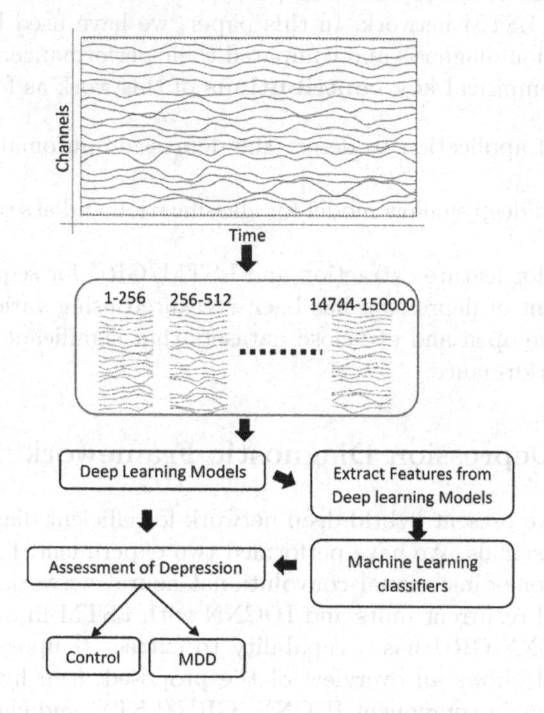

Fig. 1. Proposed hybrid framework for automatic diagnosis of depression

$$O_m = \sum_{t=0}^{N-1} x_t K_{m-t} \text{ such that } t = 0, ...N-1 \quad (1)$$

where x_t is the EEG one-dimensional signal and N is the number of elements in x_t, and t is the data points in EEG signal and O_m is the output of convolutional function. K is the filter or one-dimensional kernel. The subscript t indicates the nth element of the filter vector while m corresponds to the mth output element that is being calculated.

2.1 CNN-Gated Recurrent Units

Gated Recurrent Units (GRU) couples forget gates as well as input gates and training parameters are much smaller than LSTM, thus, it use less memory and faster. The Gated recurrent units have the capability to capture significant features from time series input EEG signal. Figure 2 shows the proposed architecture of IDCNN-GRU. The GRU especially designed for time series data and could bet better choice for EEG classification. The Eq. 2 shows the output activation of GRU gated network and it is computed by linear activation between activation from the previous state and candidate activation function $\hat{h}_t^k$ is represented in Eq. 3. The updated gate is given in Eq. 4 and σ represented the sigmoid activation function, the value of sigmoid function is between 0 and 1. The candidate activation function is controlled by reset gate as shown in Eq. 4. W_z and U_z are trainable weight matrices for the update gate. W_r and U_r are the weight matrices for reset gate. x_t is the input EEG vector and h_{t-1} is the previous state activation. The dimension of x_t is 244×1 vector coming from 1DCNN model pooling layer and the dimension of h_{t-1} previous state hidden function is 128×1. The number of nodes for first hidden layer of GRU is chosen 128 units. The dimensions of W_z, W, W_r are 244×128 and U_z, U_r has dimension equal to square of number of hidden units (128×128). In this way the candidate activation function for the next GRU has the dimension equal to 128×1 number of neurons.

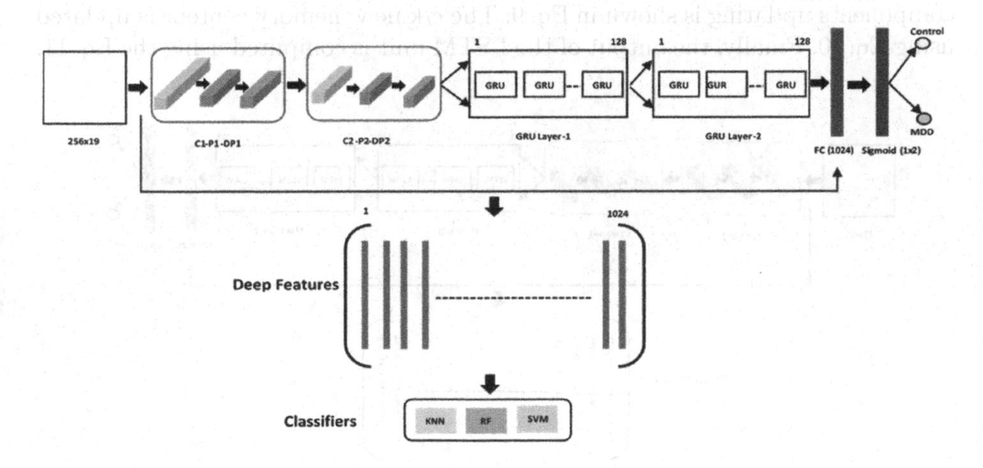

Fig. 2. Hybrid deep network for depression diagnosis using gated recurrent units

$$h_t^k = (1 - z_t^k)h_{t-1}k + z_t^k h_t^k \quad (2)$$

$$z_t^k = \sigma(W_z x_t + U_z h_{t-1}) \quad (3)$$

$$\hat{h}_t^k = tanh(W x_t + U(r_t \odot h_{t-1})) \quad (4)$$

$$r_t^k = \sigma(W_r x_t + U_r h_{t-1}) \quad (5)$$

2.2 1DCNN-Long Short Term Memory

LSTM is one of the widely used deep learning model especially in natural language processing, time series prediction and sequence generation. It has capability to learn features from long time series data and would be better choice in EEG classification and prediction. Figure 3 shows the proposed LSTM based classification of depressed patient. The two cascaded LSTM layers with 128 number of hidden neurons has been proposed with one-dimensional CNN model for classification and assessment of depression. With a greater number of LSTM layers, the parameters have been increased and model could be overfit on test samples. The LSTM model performed better in the time series data due to it only stores the valuable and useful information from the time series data and forget the information that is not useful based on gating system performed in LSTM cells. The mathematically equation used in LSTM layer is given below. The LSTM has three number of gates and these gates are called input, forget and output gates. The Eq. 6 shows the input gate i_t^k and o_t^k is represented output gate and f_t^k represented forget gate is shown in Eq. 7 and 8 respectively. The V_i, V_o and V_f are the trainable internal diagonal matrices and These keeps the memory components within each LSTM unit. The c_t^k represented memory components updating is shown in Eq. 9. The $c_t k$ new memory content is updated using Eq. 10. Finally, the output of the LSTM unit is computed using the Eq. 11.

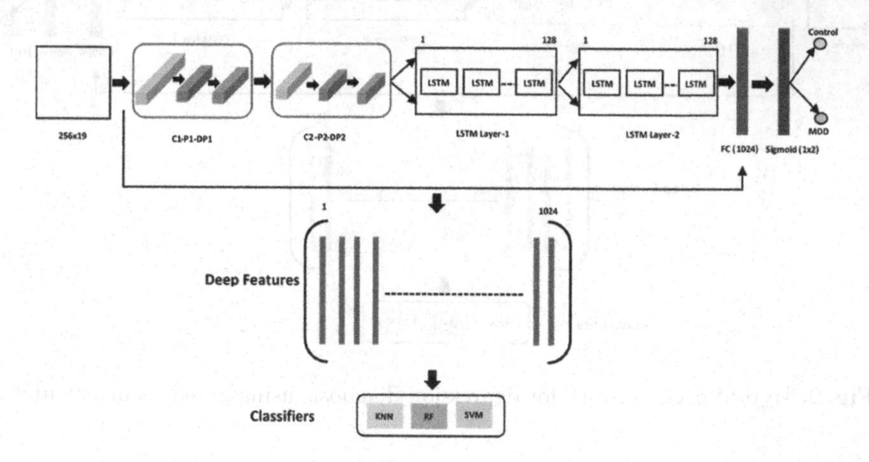

Fig. 3. Hybrid deep network for depression diagnosis using long short term memory

$$i_t^k = \sigma(W_i x_t + U_i h_{t-1} + V_i c_{t-1}) \tag{6}$$

$$o_t^k = \sigma(W_o x_t + U_o h_{t-1} + V_o c_t) \tag{7}$$

$$f_t^k = \sigma(W_f x_t + U_f h_{t-1} + V_f c_{t-1}) \tag{8}$$

$$c_t^k = f_t^k c_{t-1}^k + i_t^k \hat{c}_t^k \tag{9}$$

$$\hat{c}_t^k = tanh(W_c x_t + U_c h_{t-1}) \quad (10)$$

$$h_t^k = o_t^k tanh(c_t^k) \quad (11)$$

Where W_i, W_o, W_f, U_i, U_o, and U_f are the weights matrices and activation functions computed based on the number of hidden units used in LSTM unit. The different number of feature maps in convolutional layer has been used based on experimental evaluation and chooses the best feature maps combination. The number of hidden layers in LSTM has been chosen based on experimental evaluation.

3 Experiment

In this section, we have analyzed and compare the performance of proposed hybrid network with state of the art network. In order to achieved best performance, the features are extracted from last the dense layer in the trained proposed models before the classification layer (sigmoid function for binary classification) and passed these features to machine learning models for classification. The KNN, RF and SVM machine learning models has been chosen in this study for classification of control and MDD depression classification.

3.1 Parameters

There are several training hyperparameters were involved such as parameters learning rate, optimizers, loss-functions. The proposed model based on CNN-LSTM and CNN-GRU were trained using Adam optimizer with learning rate 0.0004. The 50 number of epochs was used to train both proposed models and 100 batch size has been used in our experiment. The binary cross-entropy was used as a loss function and sigmoid was used as activation function for classification layer. The detail of each layer configuration is described in Table 1. The Keras with backend Tensorflow library has been used to trained deep learning models and machine learning algorithms are implemented in Scikit-learn python library. The code is make publicly available[1].

3.2 Results and Discussion

The proposed models were tested using 20% of the available EEG dataset. For further analysis, 10-fold cross validation has been used and EEG input dataset has been divided into ten equal portions. Out of ten portions, nine portions are used for training and one-tenth portion of EEG data are used for testing to check the performance of the system. The results section shows the best scores as for both models. The testing of the proposed models is repeated 10 times. An overall performance was computed by choosing the mean average results from the 10 iterations. The accuracy precision, recall, f1 score performance metrics

[1] Code: https://github.com/RespectKnowledge.

Table 1. The layer configuration based on hybrid GRUs and LSTM with 1DCNN model

Layers	Output size	Number of filter	Feature map
Input	256 × 19	-	-
	GRUs with 1DCNN		
1D convolutional (C1)	252 × 64	5 × 1	64
1D MaxPooling (P1)	250 × 64	3 × 1	64
Dropout (DP1)	250 × 64	-	64
1D convolutional (C2)	246 × 128	5 × 1	128
1D MaxPooling (P2)	244 × 128	3 × 1	128
Dropout (DP2)	244 × 128	-	128
GRU1	244 × 128	-	128
GRU2	244 × 128	-	128
Fully connected (FC)	-	-	1024
Classifier (Sigmoid activation)	2	-	1024 × 2
	LSTM with 1DCNN		
1D convolutional (C1)	252 × 64	5 × 1	64
1D MaxPooling (P1)	250 × 64	3 × 1	64
Dropout (DP1)	250 × 64	-	64
1D convolutional (C2)	246 × 128	5 × 1	128
1D MaxPooling (P2)	244 × 128	3 × 1	128
Dropout (DP2)	244 × 128	-	128
LSTM1	244 × 128	-	128
LSTM2	244 × 128	-	128
Fully connected (FC)	-	-	1024
Classifier (Sigmoid activation)	2	-	1024 × 2

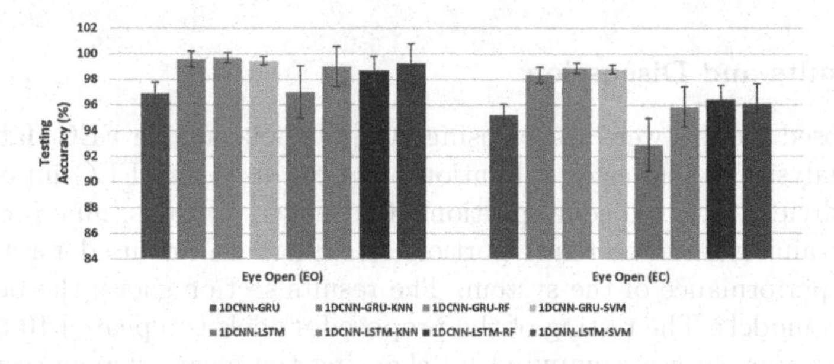

Fig. 4. Comparison of 1DCNN-GRU and 1DCNN-LSTM performance on eye open and eye close dataset.

has been used to evaluate of our proposed models. The performance metrics based on proposed model 1DCNN-GRU is shown given in Table 2 and Fig. 4 and 1DCNN-LSTM is shown in Table 3 and Fig. 4 on both eye open and eye close datasets. In comparison to LSTM based approach, Gated recurrent units showed considerably better performance almost in all parameters for all classifier. The RF classifier with 1DCNN-GRU features produced excellent performance and achieved highest precision, recall and f1 score for both eye open and eye close dataset.

Table 2. Performance metrics based on proposed model 1DCNN-GRU with classification algorithms on eye open and eye close datasets

Algorithms	Classes	Precision	Recall	F1-score
Eye open				
1DCNN-GRU-sigm	Control	96.00	98.00	97.00
	MDD	98.00	95.00	97.00
1DCNN-GRU-KNN	Control	99.00	100.00	99.00
	MDD	100.00	99.00	99.00
1DCNN-GRU-SVM	Control	99.00	99.00	99.00
	MDD	97.00	99.00	98.00
1DCNN-GRU-RF	Control	99.84	100.00	99.92
	MDD	1.00	99.88	99.94
Eye close				
1DCNN-GRU-sigm	Control	94.10	96.10	95.05
	MDD	96.30	94.20	95.25
1DCNN-GRU-KNN	Control	97.10	99.40	99.00
	MDD	99.10	97.60	98.35
1DCNN-GRU-SVM	Control	97.60	99.20	98.40
	MDD	99.10	97.80	98.45
1DCNN-GRU-sigm	Control	99.10	99.40	99.25
	MDD	98.40	99.60	99.00

The ROC curves for proposed models is shown in Fig. 6. The Fig. 6(a) clearly shows that the proposed mode produced robust performance curves for RF classifiers as compared to other models. The proposed 1DCNN-LSTM-RF based on eye open dataset degrade the ROC curves for both classes. The Kolmogorov-Smirnov (KS) test is a nonparametric test and used to discriminate two random variable samples with each other. It is also used for empirical distribution functions of two samples. The KS test determined the value how two samples are far or close to each other. KS Statics based on proposed models is shown in Fig. 5 that validates the robustness of proposed 1DCNN-GRU with RF with KS statistic 0.996 value at threshold 0.50 in comparison to KS statistic 0.983

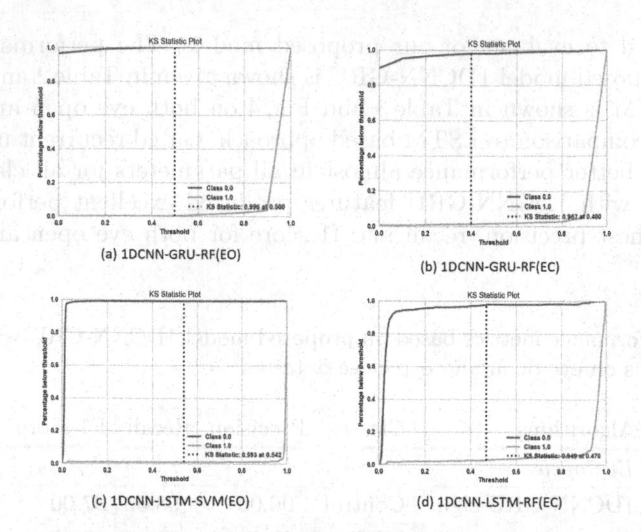

Fig. 5. The KS statistics based on best proposed models. (a) 1DCNN-GRU-RF based on eye open dataset, (b) 1DCNN-GRU-RF based on eye close dataset, (c) 1DCNN-LSTM-SVM based on eye open dataset, (d) 1DCNN-LSTM-RF based on eye close dataset

Table 3. Performance metrics based on proposed model 1DCNN-LSTM with classification algorithms on eye open and eye close dataset

Algorithms	Classes	Precision	Recall	F1-score
Eye open				
DCNN-GRU-sigm	Control	100.00	96.80	98.40
	MDD	96.10	100.00	98.05
1DCNN-GRU-KNN	Control	98.30	98.40	98.35
	MDD	100.00	96.60	98.30
1DCNN-GRU-SVM	Control	100.00	99.60	99.30
	MDD	99.10	100.00	99.05
1DCNN-GRU-RF	Control	98.10	98.30	98.20
	MDD	98.40	98.60	98.50
Eye close				
1DCNN-GRU-sigm	Control	95.00	92.10	93.05
	MDD	91.60	94.00	93.30
1DCNN-GRU-KNN	Control	96.00	96.80	96.40
	MDD	98.00	96.40	97.20
1DCNN-GRU-SVM	Control	98.10	95.60	96.35
	MDD	95.30	97.50	96.40
1DCNN-GRU-sigm	Control	97.40	97.30	97.35
	MDD	98.80	96.10	97.45

value at threshold value .054 using 1DCNN-LSTM. Similarly, 1DCNN-GRU also showed better performance with KS statistic (0.949 at 0.470) value for eye close dataset.

The precision, recall, f1 score and queue rate visualization based on our best proposed models is shown in Figure. The discrimination threshold represented the probability score of the binary classifier that can be chosen between positive and negative class. Usually this is set 50% however, the threshold could be adjusted to decrease or increase the sensitivity to the false positive for some sensitive applications. The F1 score(the harmonic mean of precision and recall) could be tuned for adjusting the threshold of the classifier to best possible fit for the specific applications. By adjusting the discrimination threshold will set the sensitivity to false positives that is the inverse relationship of precision and recall with respect to the threshold. The variability of model can be seen on the visualization curve by running multiple trials using different train and test splits of the data. The band curve shows the variability of each trail based on the median and the band range is from 10th to 90th percentile. The classification threshold could help us the users to determine an appropriate threshold for decision making, particularly in biomedical applications that are using more sensitive data. The visualization curve for threshold classifier based 1DCNN-GRU-RF(EO) produced optimal curve at decision threshold 0.50%. The precision, recall and f1 score are almost completely produced above 99%. Similarly, the threshold curve produced by proposed 1DCNN-GRU-RF based on eye open dataset produced better curve with recall, f1 score deviation. The probability curve for

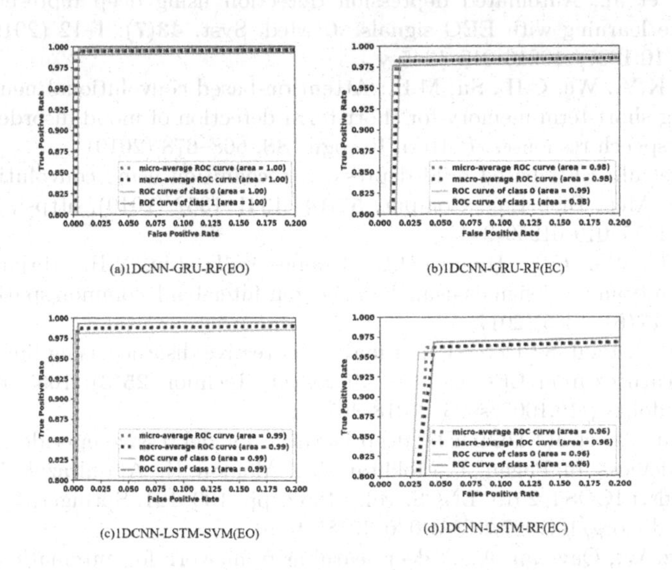

Fig. 6. ROC plot for best proposed models. (a) Eye open using proposed 1DCNN-GRU-RF model, (b) Eye close using proposed 1DCNN-GRU-RF model, (c) Eye open using proposed 1DCNN-LSTM-SVM model, (d) Eye close using proposed 1DCNN-LSTM-RF model. Class 0 represented as Control and class1 denoted as MDD.

1DCNN-LSTM-SVM based on eye open dataset. In this curve, the recall f1 score also has very small deviation from the ideal. Similarly, the 1DCNN-LSTM-RF using eye close dataset produced more deviation in recall, f1 and precision.

4 Conclusion

In this paper, we have presented an end-to-end system based on hybrid deep CNN and Gated recurrent units/LSTM architecture for diagnosis of patient suffering from depression. Comparative evaluation showed that GRU may be a better choice for EEG signals due to short sequence. Extensive experiments on EEG eye open and eye close datasets shows that the proposed hybrid model achieved significantly higher performance in terms of all parameters accuracy, f1 score 99.66%, 99.93% and 98.87%, 99.12%. Based on high performance, the proposed hybrid approach can be used for assessment of depression for clinical applications and can deployed remotely in hospital or private clinics for clinical evaluation.

References

1. Acharya, U.R., Oh, S.L., Hagiwara, Y., Tan, J.H., Adeli, H., Subha, D.P.: Automated EEG-based screening of depression using deep convolutional neural network. Comput. Methods Programs Biomed. **161**, 103–113 (2018)
2. Ay, B., et al.: Automated depression detection using deep representation and sequence learning with EEG signals. J. Med. Syst. **43**(7), 1–12 (2019). https://doi.org/10.1007/s10916-019-1345-y
3. Huang, K.Y., Wu, C.H., Su, M.H.: Attention-based convolutional neural network and long short-term memory for short-term detection of mood disorders based on elicited speech responses. Pattern Recogn. **88**, 668–678 (2019)
4. Li, X., et al.: EEG-based mild depression recognition using convolutional neural network. Med. Biol. Eng. Comput. **57**(6), 1341–1352 (2019). https://doi.org/10.1007/s11517-019-01959-2
5. Liao, S.C., Wu, C.T., Huang, H.C., Cheng, W.T., Liu, Y.H.: Major depression detection from EEG signals using kernel eigen-filter-bank common spatial patterns. Sensors **17**(6), 1385 (2017)
6. Mahato, S., Paul, S.: Detection of major depressive disorder using linear and nonlinear features from EEG signals. Microsyst. Technol. **25**(3), 1065–1076 (2018). https://doi.org/10.1007/s00542-018-4075-z
7. Mdhaffar, A., et al.: DL4DED: deep learning for depressive episode detection on mobile devices. In: Pagán, J., Mokhtari, M., Aloulou, H., Abdulrazak, B., Cabrera, M.F. (eds.) ICOST 2019. LNCS, vol. 11862, pp. 109–121. Springer, Cham (2019). https://doi.org/10.1007/978-3-030-32785-9_10
8. Mumtaz, W., Qayyum, A.: A deep learning framework for automatic diagnosis of unipolar depression. Int. J. Med. Informatics **132**, 103983 (2019)
9. Razzak, I., Blumenstein, M., Xu, G.: Multiclass support matrix machines by maximizing the inter-class margin for single trial EEG classification. IEEE Trans. Neural Syst. Rehabil. Eng. **27**(6), 1117–1127 (2019)

10. Razzak, I., Hameed, I.A., Xu, G.: Robust sparse representation and multiclass support matrix machines for the classification of motor imagery EEG signals. IEEE J. Transl. Eng. Health Med. **7**, 1–8 (2019)
11. Razzak, M.I., Imran, M., Xu, G.: Big data analytics for preventive medicine. Neural Comput. Appl. **32**(9), 4417–4451 (2019). https://doi.org/10.1007/s00521-019-04095-y
12. Razzak, M.I., Naz, S., Zaib, A.: Deep learning for medical image processing: overview, challenges and the future. In: Dey, N., Ashour, A.S., Borra, S. (eds.) Classification in BioApps. LNCVB, vol. 26, pp. 323–350. Springer, Cham (2018). https://doi.org/10.1007/978-3-319-65981-7_12
13. Yıldırım, Ö., Baloglu, U.B., Acharya, U.R.: A deep convolutional neural network model for automated identification of abnormal EEG signals. Neural Comput. Appl. **32**(20), 15857–15868 (2018). https://doi.org/10.1007/s00521-018-3889-z
14. Zhang, X., Hu, B., Zhou, L., Moore, P., Chen, J.: An EEG based pervasive depression detection for females. In: Zu, Q., Hu, B., Elçi, A. (eds.) ICPCA/SWS 2012. LNCS, vol. 7719, pp. 848–861. Springer, Heidelberg (2013). https://doi.org/10.1007/978-3-642-37015-1_74

Iterative Imputation of Missing Data Using Auto-Encoder Dynamics

Marek Śmieja[1(✉)], Maciej Kołomycki[2], Łukasz Struski[1], Mateusz Juda[1], and Mário A. T. Figueiredo[3]

[1] Faculty of Mathematics and Computer Science, Jagiellonian University, Kraków, Poland
{marek.smieja,lukasz.struski,mateusz.juda}@uj.edu.pl

[2] Institute of Applied Informatics, Faculty of Mechanical Engineering, Cracow University of Technology, Kraków, Poland
maciej.kolomycki@mech.pk.edu.pl

[3] Instituto de Telecomunicações, Instituto Superior Técnico, Universidade de Lisboa, Lisbon, Portugal
mario.figueiredo@tecnico.ulisboa.pt

Abstract. This paper introduces an approach to missing data imputation based on deep auto-encoder models, adequate to high-dimensional data exhibiting complex dependencies, such as images. The method exploits the properties of the vector field associated to an auto-encoder, which allows to approximate the gradient of the log-density from its reconstruction error, based on which we propose a projected gradient ascent algorithm to obtain the conditionally most probable estimate of the missing values. Our approach does not require any specialized training procedure and can be used together with any auto-encoder model trained on complete data in a classical way. Experiments performed on benchmark datasets show that imputations produced by our model are sharp and realistic.

Keywords: Missing data imputation · Image inpainting · Auto-encoder · Dynamical system · Auto-encoder's vector field

1 Introduction

Missing data imputation is an important problem in machine learning and data analysis, especially when dealing with real-world applications [5,10,17]. The typical approach is to directly design a specialized model and train it to fill in absent values. By constructing sophisticated architectures, trained under carefully designed loss functions, state-of-the-art models obtain impressive performance, *e.g.*, in image inpainting [11,32]. However, a natural question arises: *can*

The is the extended version of an extended abstract [25] presented at the ICLR Workshop on the Integration of Deep Neural Models and Differential Equations.

H. Yang et al. (Eds.): ICONIP 2020, LNCS 12534, pp. 258–269, 2020.
https://doi.org/10.1007/978-3-030-63836-8_22

we complete missing data at test time using models that were not aware of the imputation task during the training stage?

Our work is motivated by the use of classical parametric (or semi-parametric) density models, such as *Gaussian mixture models* (GMMs) [27], for missing data imputation. In that work, a density is estimated from complete data[1] in a strictly unsupervised way. To apply the model for missing data imputation, the missing values are replaced either by samples or by maximizers of the estimated conditional density of the missing data, given the observed data. Although the use of a shallow density model, such as a GMM, may allow obtaining the conditional density analytically, such a model may be unable to efficiently capture complex dependencies in high-dimensional data, such as images [24].

While deep generative models, *e.g.*, *generative adversarial networks* (GAN) [8], *variational auto-encoders* (VAE) [12], or *Wasserstein auto-encoders* (WAE) [28], are sufficiently expressive do describe complex dependencies in data, it may be impossible to explicitly obtain or maximize the corresponding conditional density of the missing values due to the nonlinear form of decoder (generator) [20]. The authors of [22] define a pseudo-Gibbs sampling procedure for filling missing values by iterative auto-encoding of incomplete data (see also [9, Section 20.11] for more general formulation). In the case of VAE, this procedure can be modified by adding an option to reject the proposal posterior distribution, which results in Metropolis-within-Gibbs algorithm [18]. Mattei and Frellsen use importance sampling for training VAE on incomplete data as well as for replacing missing values by single or multiple imputation [19]. Similarly, it is challenging to obtain a closed-form expression for such a conditional distribution in GAN, but one can design a procedure to sample from it [15,31].

We tackle this problem by exploiting the dynamics of auto-encoders' reconstruction function. Based on theoretical results presented in [1], the reconstruction error of a *denoising auto-encoder* (DAE) [29] yields an approximation of the gradient of the log-probability density function, which is (implicitly) estimated from data. We exploit that fact to maximize the conditional density of the missing values, given the observed ones. The conditionally most probable values are found as the attractors of the iterated reconstruction function. We experimentally demonstrate that, in a place of DAE, any auto-encoder model (*e.g.*, AE, VAE, WAE) can be used in the process of replacing missing data at test time without any additional effort at training stage.

Alternatively, our procedure can be interpreted as a type of pseudo-Gibbs sampling. While the pseudo-Gibbs sampling procedure proposed by [22] directly replaces the input by its reconstruction, which may lead to falling out of the true data manifold, our algorithm adds the reconstruction error to the input with a small weight. Similarly to [18], it improves convergence of the algorithm when the posterior approximation is imperfect. Our procedure is also similar to the algorithm proposed in [6, Section 5.2] for NICE flow model, where the gradient

[1] A GMM can also be learned from incomplete data, but the imputation process does not change.

is given explicitly. Our procedure works for every possible auto-encoder model, even if the gradient is difficult to compute.

We experimentally assess the proposed approach on image datasets, showing that it obtains results comparable to typical deep learning models (with analogous neural network architecture) trained explicitly for missing data imputation. Moreover, by using different initializations in the iterative procedure, we can reach different attractors and, consequently, a diverse set of imputation candidates for the same incomplete input. This makes our model similar to generative models.

The paper is organized as follows. In Sect. 2, we recall known facts concerning AE's vector field and dynamics. Our approach is introduced in Sect. 3 and, next, experimentally assessed in Sect. 4.

2 Auto-Encoder Dynamics

Because they underlie our approach to missing data imputation, this section reviews relevant facts regarding the vector field associated with and auto-encoder reconstruction function and the associated dynamics.

Auto-encoders (AE) have a long history in the field of artificial neural networks, going back at least to the 1980s [7,13]. An AE may be viewed as composition of two maps, an *encoder* $f : \mathbb{R}^d \to Z$ and a *decoder* $g : Z \to \mathbb{R}^d$, such that $Z \subset \mathbb{R}^l$ is the so-called *latent space*. An AE is trained from data with the goal of making the *reconstruction* function $r := g \circ f$ close to identity, *i.e.*, $r(x) \approx x$, for the training data, by capturing the essential features of that data.

Since an AE does not (and should not) achieve perfect reconstructions (specially for input data far from the training data), we can define an *AE vector field* $v : \mathbb{R}^d \to \mathbb{R}^d$ associated to the reconstruction function as $v(x) := r(x) - x$. Analogously, we also define an *AE latent vector field* $u : Z \to Z$, given by $u(z) := f(g(z)) - z$. A natural question arises: what is the structure of the dynamics generated by the vectors fields v and u?

The properties of the vector fields v for a DAE were studied and discussed in [1], where it was shown that the reconstruction error at some point $x \in \mathbb{R}^d$ is approximately equal to the gradient of the log-pdf (*logarithm of the probability density function*) computed at that point, in the low-noise limit, *i.e.*, as $\sigma^2 \to 0$,

$$\nabla_x \log p_X(x) \approx \frac{r_{\sigma^2}(x) - x}{\sigma^2} = \frac{v_{\sigma^2}(x)}{\sigma^2}, \tag{1}$$

where r_{σ^2} is the reconstruction function of the DAE at denoising level σ^2 and v_{σ^2} is the corresponding vector field. Consequently, the point with the highest log-pdf can be found via gradient ascent, *i.e.*, gradient flow, in the limit of infinitesimal steps, by exploiting this equality. In discrete time, with a step-size of the order of σ^2, we thus have: $x_{t+1} = r_{\sigma^2}(x_t)$. Notice that, from Eq. 1, it is clear that fixed points of this iteration correspond to stationary points of the log-pdf, *i.e.*, zeros of its gradient.

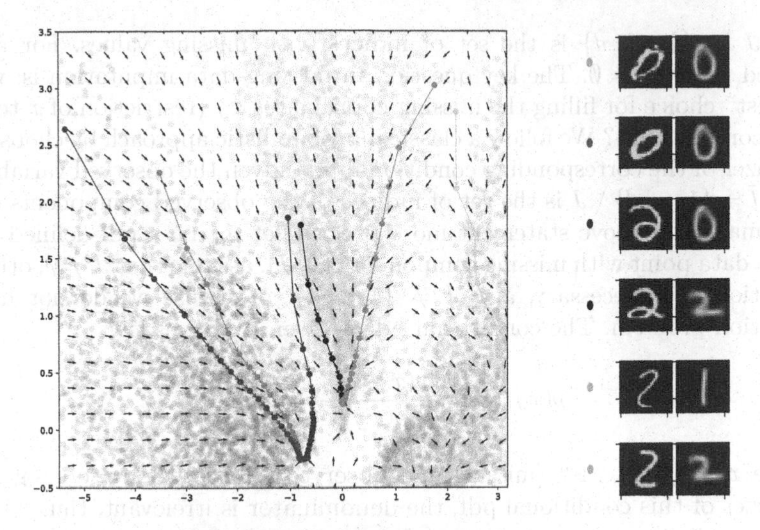

Fig. 1. Example of latent space trajectories (left) for an AE trained on the MNIST dataset (classes 0, 1, 2). Dots represent latent representation of the training examples: 0-red ,1 -green, 2-yellow. On the right hand side, for each trajectory. we present its starting point and the attractor reached after 100 iterations. (Color figure online)

Analyzing the dynamics resulting from the vector field associated with the reconstruction error may be useful in verifying the quality of an AE. The intuition is that this dynamics (and its counterpart in the latent space) should have stationary points, some of which are locally stable, thus are attractors. The basin of attraction of each attractor should consist of a subset of the input space with points with similar features.

As an example, consider an AE trained on digits 0, 1, and 2 of the MNIST dataset, using latent space dimension $l = 2$. Starting from some latent point $z_0 \in Z$, we draw the latent trajectory generated by the iteration $z_{t+1} := f(g(z_t))$, which is the discrete-time counterpart of the gradient flow explained above. In most cases, we observe the behavior shown in Fig. 1: each trajectory travels through the latent space and converges to a fixed point (attractor). For some starting points, a small perturbation may cause the trajectory to converge to different attractor. In Fig. 1, we observe such behaviour for the cyan and pink trajectories; their starting points lie close to the boundary between classes 1 and 2. It is also a low density region, so in some sense the AE is not trained enough there. However, in the case of the blue and black trajectories, we see such behaviour also for starting points in the relatively denser area of class 2.

3 Imputation Method

In this section, we show how to use the discrete-time dynamics above described in the context of missing data to obtain imputations with the highest local probability. A point $x \in \mathbb{R}^d$ with missing components is denoted by a pair (x, J), where $J \subset \{1, \ldots, d\}$ is the set of indices with missing values. For a fully-observed point, $J = \emptyset$. The key question in missing-data imputation is: what is the "best" choice for filling the missing coordinates x_J (restriction of x to unobserved components)? We follow a classical probabilistic approach by choosing the maximizer of the corresponding conditional pdf, given the observed variables $x_{\bar{J}}$, where $\bar{J} = \{1, ..., d\} \setminus J$ is the set of indices of the observed components of x.

To make the above statement more precise, let p_X be a pdf defined on $\mathbb{R}^d$. Given a data point with missing components (x, J), assume that $J \neq \emptyset$, otherwise imputation is unnecessary, and $J \neq \{1, ..., d\}$, otherwise we do not have an imputation problem. The conditional pdf is given by Bayes law,

$$p(x_J | x_{\bar{J}}) = \frac{p(x_J, x_{\bar{J}})}{p(x_{\bar{J}})} = \frac{p_X(x)}{p(x_{\bar{J}})}, \tag{2}$$

because $x_{J \cup \bar{J}} = x \in \mathbb{R}^d$ (missing and observed). Since we are looking for the maximizer of this conditional pdf, the denominator is irrelevant, thus

$$\hat{x}_J = \arg \max_{x_J \in \mathbb{R}^{|J|}} p(x_J | x_{\bar{J}}) = \arg \max_{y \in \mathbb{R}^d : y_{\bar{J}} = x_{\bar{J}}} \log p_X(y). \tag{3}$$

To seek the maximizer of the conditional density defined in Eq. 3, we propose the following procedure (which we show below corresponds to a projected gradient ascent scheme), based on an AE trained on a dataset with characteristics similar to the data on which imputation will be performed:

1. pick an initial filling $\hat{x}_J^0$ of the missing part x_J;
2. iteratively update $\hat{x}_J$ using $\hat{x}_J^{t+1} = \hat{x}_J^t + h\,[r_\sigma(\hat{x}^t) - \hat{x}^t]_J$.

where h is a step size and $\hat{x}^t = (\hat{x}_J^t, x_{J'}) \in \mathbb{R}^d$ denotes a complete point where the observed components are fixed at the observed values and the missing ones are replaced by the current estimate. This procedure corresponds to moving on an (axes-aligned) affine subspace of dimension $\mathbb{R}^{|J|}$ of the data space $\mathbb{R}^d$ in a direction determined by the gradient of the log-density function (see Eq. (1)). Because of the axes-aligned nature of the affine subspace, this coincides with a projected gradient ascent algorithm.

As shown in the Fig. 2, the proposed method depends on the initialization $\hat{x}_J^0$. We observe that, for the smallest missing window, all initializations lead to the same attractor, thus the same imputations. For mid-sized missing regions, our algorithm with random initialization gives different effect from the one obtained using mean and k-NN filling. For the largest hole, we loose the image features and land in the area of a different class regardless on the initialization.

In practice, we can control the final result by careful selecting the starting point. To make the final imputation the most similar to ground truth, we should pick an initialization using simple imputations, *e.g.*, mean or k-NN. To provide more diverse results, we can use add random noise or samples from some prior distribution for the initialization. Consequently, our method has a generative nature and is capable of creating a wide range of imputations depending on the seed (see next section for more results).

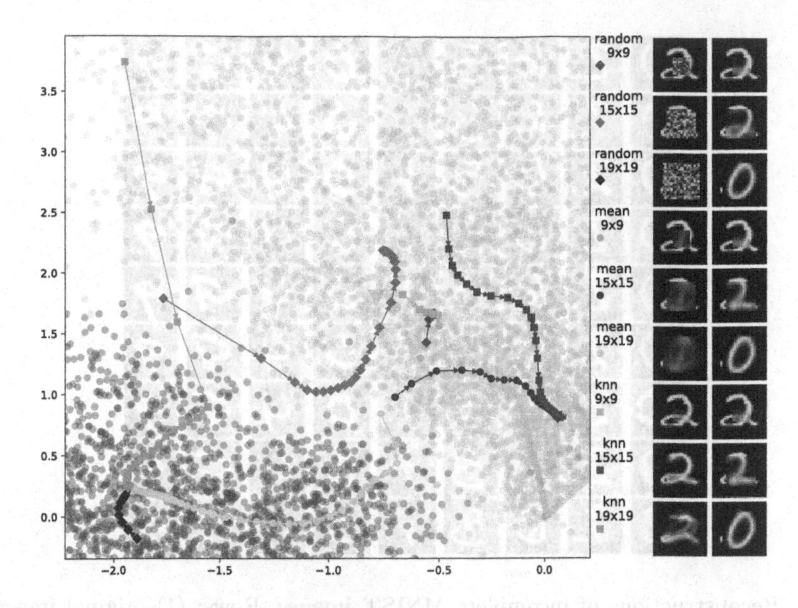

Fig. 2. Illustration of our algorithm (using the same AE as in Fig. 1), for missing regions of three different sizes (9×9, 15×15, and 19×19) on the same data point from class "2" (green point on the left hand side). Three initialization strategies were used: random noise, mean value over the training set, mean value over 5-NN. We show trajectories in latent space (left) and final imputations (rightmost) for different initializations (second column from the right). (Color figure online)

4 Experiments

In this section, we experimentally validate the proposed model. For this purpose, we fit a typical AE on a train set and use it for filling in missing data at test time[2]. To examine the dependence on the initialization, we consider two variants of our model: starting with random noise as initial imputation; initial filling generated using k-NN imputation. We adapted the architecture and the training procedure from [28] (using $\lambda = 0$ to obtain a classical AE).

As a baseline, we apply a pseudo Gibbs sampling (p-Gibbs) [22], where decoded data is directly used as an input to the next iteration. Similarly to our method, we use two variants of initialization: random replacement and k-NN imputation. We also consider a *context encoder* (CE) [21], which is a type of deep AE trained explicitly for filling in missing data. Roughly, a CE takes an incomplete image (with a mask) and focuses on making the output as similar to the original image as possible by minimizing the MSE on the missing area. To make both approaches (p-Gibbs and CE) fully comparable with our method, we use exactly the same architecture for all models. Additionally, we use two typical imputation methods: (a) k-NN [3], which fills missing values with the

[2] For a comparison between different auto-encoder models in the proposed procedure the reader is referred to our workshop paper [25].

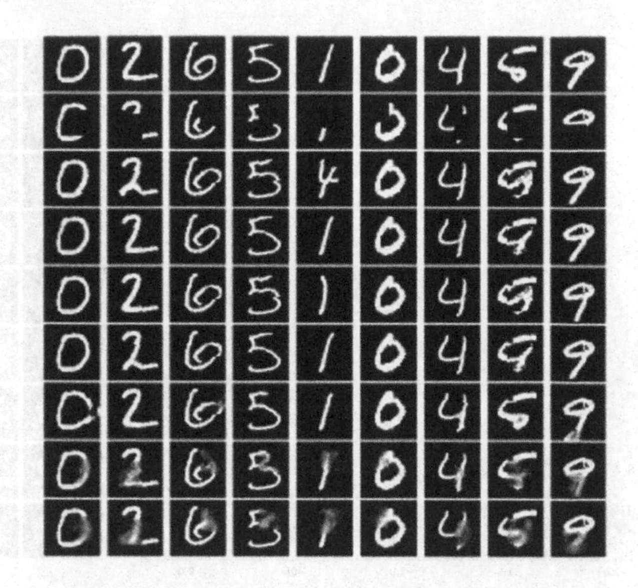

Fig. 3. Reconstructions of incomplete MNIST images. Rows: (1) original image, (2) incomplete image, and imputations using (3) our methods with random initialization, (4) our method with k-NN initialization, (5) p-Gibbs with random initialization, (6) p-Gibbs with k-NN (7) CE (8) k-NN, (9) MICE.

corresponding mean values computed from the k nearest training samples (we used k = 5); (b) MICE [2,4], where several imputations are drawn from the conditional pdf using *Markov chain Monte Carlo* sampling. Rather than presenting state-of-the-art performance, which requires more advanced neural network architecture (and modification of training procedure in our case), we focus on showing that our test procedure obtains similar performance to the typical models trained explicitly for the imputation task.

We consider two datasets of gray-scale images: MNIST [14] and Fashion [30]. For each test image of the size 28×28, we drop a patch of size 13×13, at a (uniformly) random location. We also use the CelebA dataset [16], which is composed of color face images of size 64×64, with missing regions of size 25×25. Analogous missing regions are used for training the CE and MICE.

Figures 3 and 4 present sample results for MNIST and Fashion, respectively. One can observe that the results produced by our method and p-Gibbs with k-NN initialization are visually the most plausible and usually coincide with ground truth. The results obtained by our method with random initialization are also realistic, but differ in some cases from the ground-truth and from p-Gibbs with the analogical initialization (2nd and 5th column for MNIST presents positive effect while 3rd column for Fashion illustrates negative results).

Since a CE is trained to fill in missing data by minimizing the MSE, its results usually coincide with the ground-truth average, but many details are missing (see 7th column for Fashion, where this method failed to complete the

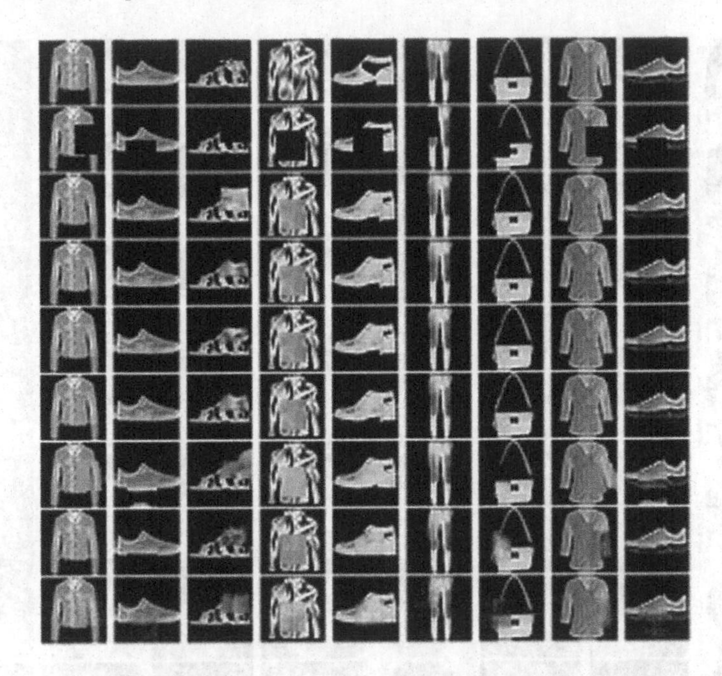

Fig. 4. Reconstructions of incomplete images from Fashion. Each row presents subsequent methods: (1) original image, (2) incomplete image, and imputations using (3) our methods with random initialization, (4) our method with k-NN initialization, (5) p-Gibbs with random initialization, (6) p-Gibbs with k-NN (7) CE (8) k-NN, (9) MICE.

handbag strap). In contrast, our method aims at finding the most probable replacement, usually yielding sharp images, although maybe different from the ground truth. In the case of the more diverse Fashion dataset, the CE produces a lot of artifacts.

While k-NN presents poor performance on MNIST, its results on Fashion are appealing. Since Fashion contains many similar images, k-NN is able to fill in missing regions with analogous shapes. It is evident that MICE fails to complete missing data with reasonable content.

The results for CelebA dataset are presented in Fig. 5. It is clear that using k-NN initialization in our procedure leads to more plausible results than random initialization. As seen in the 1st column, random seed directed a filling trajectory out of true data distribution. On the other hand, the same initialization in the 6th column created forehead bangs, which may be seen as a positive effect. The results produced by p-Gibbs occasionally differ from the ones returned by our method (see 9th column for random initialization as well as 4th and 7th column for k-NN). At first glance, the imputations produced by the CE are visually plausible. However, more detailed inspection reveals that the obtained images are often blurry (5th and 8th columns) and sometimes contain artifacts (2nd column). The use of k-NN imputation alone gives bad results. We were unable to run MICE imputation due to high-dimensionality of data.

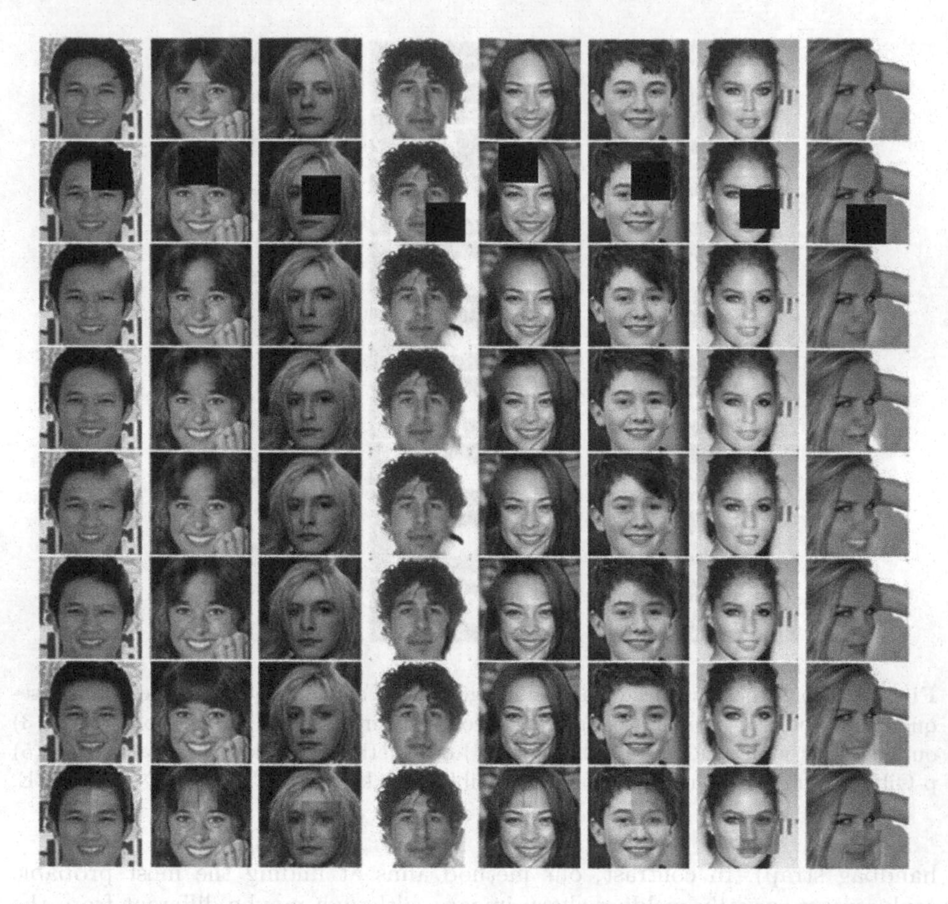

Fig. 5. Reconstructions of partially incomplete images from CelebA. Each row presents subsequent methods: (1) original image, (2) incomplete image, and imputations using (3) our methods with random initialization, (4) our method with k-NN, (5) p-Gibbs with random initialization, (6) p-Gibbs with k-NN (7) CE (8) k-NN.

An interesting aspect is that our method is more "creative" than the others. Varying the initialization, our method can create different styles of the same objects (2nd column for MNIST), examples from different classes (5th column for MNIST) and other data manipulations (longer hairs - 6th column, closed mouths - 8th column for CelebA). This property can be very appealing from a generative perspective, which cannot be easily obtained using typical approaches.

To provide a quantitative assessment of the methods, we measure their *structural similarity* (SSIM) with ground truth. Unlike PSNR or MSE [23], which measure pixel-wise absolute errors, SSIM is based on visible structures in the image. SSIM is calculated for various windows of input images and, for the pixel p, it is defined by [26]:

Table 1. SSIM of imputations.

Method	MNIST	Fashion	CelebA
Ours (random init.)	0.830	0.903	0.901
Ours (k-NN init.)	**0.875**	**0.919**	0.914
p-Gibbs (random init.)	0.828	0.904	0.899
p-Gibbs (k-NN init.)	0.868	0.918	0.907
CE	0.871	0.879	**0.930**
k-NN	0.829	0.841	0.857
MICE	0.797	0.809	-

$$SSIM(p) = \frac{2\mu_x\mu_y + C_1}{\mu_x^2 + \mu_y^2 + C_1} \cdot \frac{2\sigma_{xy} + C_2}{\sigma_x^2 + \sigma_y^2 + C_2},$$

where μ_x, μ_y are the mean values of the patches x and y, respectively, centered at p; σ_x^2, σ_y^2 denote variances; σ_{xy}^2 denotes covariance; C_1 and C_2 are variables intended to stabilize the division with small denominator; its maximal value 1 is attained for identical images.

The results presented in Table 1 show that our method with k-NN initialization gives the highest resemblance with ground-truth on MNIST and Fashion. While the performance of CE on MNIST is only slightly worse, the disproportion between these methods on Fashion is evident. In the case of CelebA, CE is more accurate than our method, but the difference is not high. It can be observed that p-Gibbs performs slightly worse than our method. The disproportion between shallow (k-NN, MICE) and deep methods (AE, CE) is enormous.

5 Conclusion and Future Work

In this paper, we proposed a strategy for filling in missing values based on auto-encoder vector field. Our method does not require a training procedure designed for imputation task, but can be used together with any AE trained in a typical way. The idea is to traverse the AE vector field towards an attractor, which is a local maximum of the probability density of function learned by the AE. Experiments showed that this procedure gives comparable results to typical deep models trained explicitly for imputation tasks.

To increase the performance of the proposed procedure, we plan to modify the training procedure of the AE. One option is to simulate the iterative procedure in the training phase, by reconstructing original images from partial imputations. This should stabilize the test stage and prevent from falling out of the true data distribution when initialized from random noise.

Acknowledgements. The work of M. Śmieja was supported by the National Science Centre (Poland) grant no. 2018/31/B/ST6/00993. The work of Ł. Struski was supported by the National Science Centre (Poland) grant no. 2017/25/B/ST6/01271 as well as the Foundation for Polish Science Grant No. POIR.04.04.00-00-14DE/18-00 co-financed by the European Union under the European Regional Development Fund. The work of M. Juda was supported by the National Science Centre (Poland) grant no. 2014/14/A/ST1/00453 and 2015/19/D/ST6/01215.

References

1. Alain, G., Bengio, Y.: What regularized auto-encoders learn from the data-generating distribution. J. Mach. Learn. Res. **15**, 3563–3593 (2014)
2. Azur, M., Stuart, E., Frangakis, C., Leaf, P.: Multiple imputation by chained equations: what is it and how does it work? Int. J. Methods Psychiatr. Res. **20**, 40–49 (2011)
3. Batista, G., Monard, M.: A study of k-nearest neighbour as an imputation method. Front. Artif. Intell. Appl. **97**, 251–260 (2002)
4. Buuren, S., Groothuis-Oudshoorn, K.: Mice: multivariate imputation by chained equations in R. J. Stat. Softw. **45**(3), 1–68 (2010)
5. Camino, R., Hammerschmidt, C., State, R.: Improving missing data imputation with deep generative models. arXiv preprint arXiv:1902.10666 (2019)
6. Dinh, L., Krueger, D., Bengio, Y.: Nice: non-linear independent components estimation. arXiv preprint arXiv:1410.8516 (2014)
7. Gallinari, P., LeCun, Y., Thiria, S., Fogelman-Soulie, F.: Memoires associatives distribuees. In: COGNITIVA 87, Paris (1987)
8. Goodfellow, I., et al.: Generative adversarial nets. In: Advances in Neural Information Processing Systems, pp. 2672–2680 (2014)
9. Goodfellow, I., Bengio, Y., Courville, A.: Deep Learning. MIT Press, Cambridge (2016)
10. Hwang, U., Jung, D., Yoon, S.: Hexagan: generative adversarial nets for real world classification. arXiv preprint arXiv:1902.09913 (2019)
11. Iizuka, S., Simo-Serra, E., Ishikawa, H.: Globally and locally consistent image completion. ACM Trans. Graph. (ToG) **36**(4), 1–14 (2017)
12. Kingma, D., Welling, M.: Auto-encoding variational Bayes. In: International Conference on Learning Representations (2014)
13. LeCun, Y.: Modeles connexionistes de l'apprentissage. Ph.D. thesis, Ph.D. thesis, Université de Paris VI (1987)
14. LeCun, Y., Bottou, L., Bengio, Y., Haffner, P.: Gradient-based learning applied to document recognition. Proc. IEEE **86**, 2278–2324 (1998)
15. Li, S., Jiang, B., Marlin, B.: MisGAN: learning from incomplete data with generative adversarial networks. arXiv preprint arXiv:1902.09599 (2019)
16. Liu, Z., Luo, P., Wang, X., Tang, X.: Deep learning face attributes in the wild. In: International Conference on Computer Vision (2015)
17. Luo, Y., Cai, X., Zhang, Y., Xu, J., Xiaojie, Y.: Multivariate time series imputation with generative adversarial networks. In: Advances in Neural Information Processing Systems, pp. 1596–1607 (2018)
18. Mattei, P.A., Frellsen, J.: Leveraging the exact likelihood of deep latent variable models. In: Advances in Neural Information Processing Systems, pp. 3855–3866 (2018)
19. Mattei, P.A., Frellsen, J.: Miwae: Deep generative modelling and imputation of incomplete data sets. In: International Conference on Machine Learning, pp. 4413–4423 (2019)

20. Nazabal, A., Olmos, P.M., Ghahramani, Z., Valera, I.: Handling incomplete heterogeneous data using vaes. Pattern Recogn. **107**, 107501 (2020)
21. Pathak, D., Krahenbuhl, P., Donahue, J., Darrell, T., Efros, A.: Context encoders: feature learning by inpainting. In: IEEE Conference on Computer Vision and Pattern Recognition, pp. 2536–2544 (2016)
22. Rezende, D.J., Mohamed, S., Wierstra, D.: Stochastic backpropagation and approximate inference in deep generative models. arXiv preprint arXiv:1401.4082 (2014)
23. Sai Hareesh, A., Chandrasekaran, V.: A novel color image inpainting guided by structural similarity index measure and improved color angular radial transform. In: International Conference on Image Processing, Computer Vision, & Pattern Recognition, pp. 544–550 (2010)
24. Śmieja, M., Struski, Ł., Tabor, J., Zieliński, B., Spurek, P.: Processing of missing data by neural networks. In: Advances in Neural Information Processing Systems, pp. 2719–2729 (2018)
25. Śmieja, M., Kołomycki, M., Struski, L., Juda, M., Figueiredo, M.A.T.: Can autoencoders help with filling missing data? In: ICLR Workshop on Integration of Deep Neural Models and Differential Equations (DeepDiffEq), p. 6 (2020)
26. Stagakis, N., Zacharaki, E.I., Moustakas, K.: Hierarchical image inpainting by a deep context encoder exploiting structural similarity and saliency criteria. In: Tzovaras, D., Giakoumis, D., Vincze, M., Argyros, A. (eds.) ICVS 2019. LNCS, vol. 11754, pp. 470–479. Springer, Cham (2019). https://doi.org/10.1007/978-3-030-34995-0_42
27. Titterington, D., Sedransk, J.: Imputation of missing values using density estimation. Stat. Probab. Lett. **9**(5), 411–418 (1989)
28. Tolstikhin, I., Bousquet, O., Gelly, S., Schölkopf, B.: Wasserstein auto-encoders (2017). arXiv:1711.01558
29. Vincent, P.: A connection between score matching and denoising autoencoders. Neural Comput. **23**(7), 1661–1674 (2011)
30. Xiao, H., Rasul, K., Vollgraf, R.: Fashion-mnist: a novel image dataset for benchmarking machine learning algorithms. arXiv preprint arXiv:1708.07747 (2017)
31. Yoon, J., Jordon, J., Van Der Schaar, M.: Gain: missing data imputation using generative adversarial nets. arXiv preprint arXiv:1806.02920 (2018)
32. Yu, J., Lin, Z., Yang, J., Shen, X., Lu, X., Huang, T.S.: Generative image inpainting with contextual attention. In: IEEE Conference on Computer Vision and Pattern Recognition, pp. 5505–5514 (2018)

Multi-objective Evolution for Deep Neural Network Architecture Search

Petra Vidnerová(✉) and Roman Neruda

The Czech Academy of Sciences, Institute of Computer Science,
Pod Vodárenskou Věží 2, 182 07 Prague 8, Czech Republic
{petra,roman}@cs.cas.cz

Abstract. In this paper, we propose a multi-objective evolutionary algorithm for automatic deep neural architecture search. The algorithm optimizes the performance of the model together with the number of network parameters. This allows exploring architectures that are both successful and compact. We test the proposed solution on several image classification data sets including MNIST, fashionMNIST and CIFAR-10, and we consider deep architectures including convolutional and fully connected networks. The effects of using two different versions of multi-objective selections are also examined in the paper. Our approach outperforms both the considered baseline architectures and the standard genetic algorithm used in our previous work.

1 Introduction

In the last decade, we witness the boom of deep learning. Neural networks are successfully applied in a wide variety of domains, including image recognition, natural language processing, and others [8,14]. There is a variety of efficient learning algorithms to use, however, the performance of the model depends always also on the choice of the right architecture. The choice of architecture is still done manually, requires expert knowledge and a time-demanding trial and error approach.

In recent years, the need for automation of neural architecture search (NAS) is getting more and more apparent. As the accessibility of efficient hardware resources improved significantly, many automatic approaches for the setup of hyper-parameters of learning models appeared. Many machine learning software tools offer automatic search for various hyper-parameters, typically based on grid search techniques. These simple techniques are however applicable on simple hyper-parameters – as real numbers (such as learning rates, dropout rates, etc.) or categorical values from some choice set (such as the type of activation function). The whole architecture, however, is a structured information and has to be searched for as an entire entity. It is hardly possible to use a grid search or similar exhaustive technique for NAS.

This work was partially supported by the Czech Science Foundation project no. 18-23827S and institutional support of the Institute of Computer Science RVO 67985807.

H. Yang et al. (Eds.): ICONIP 2020, LNCS 12534, pp. 270–281, 2020.
https://doi.org/10.1007/978-3-030-63836-8_23

In this paper, we evaluate the possibility of application of multi-objective evolutionary algorithms on NAS. We restrict the problem to feed-forward dense and convolutional neural networks. Evolutionary algorithms have been applied to solve the NAS problem quite often, one of the most successful examples is a work [18] that we mention in the next section. In our previous work [21,22,24] we have applied evolutionary algorithms to search for architectures of simple feed-forward neural networks. The straightforward approach suffers from huge computational requirements, as it is necessary to train and evaluate many candidate networks during the search, and also from the fact that the candidate solutions tend to grow uncontrollably. The resulting networks typically have a good performance but a needlessly huge number of parameters. Therefore, we decided to employ the multi-objective optimization approach and to optimize not only the performance of the network but also its size (number of learnable parameters). In many applications the need for reasonably small models is inherent. This paper shows studies on how the use of multi-objective optimization may help to tackle this problem.

The paper is organized as follows. The next section revises the related work, including available tools for hyper-parameter search and works directly focused on NAS. Section 3 briefly define deep neural networks. Section 4 explains the proposed algorithm. Section 5 describes the results of our experiments. And finally, Sect. 6 contains conclusion.

2 Related Work

Recently, several tools for automatic neural model selection have appeared. The first of them, AutoKeras [11], is a software library for automated machine learning. It provides functions for automatic architecture and hyper-parameters search. The optimization process is based on Bayesian optimization. From our experience with this software, it works well but often produces quite complicated architectures.

The second tool, that deserves to mention, is Talos [20]. It provides a semi-automatic approach to hyper-parameters search for Keras models (Keras [2] is a generally known Python library for neural network implementation, recently it became part of Tensorflow [10]). It enables a user to automatically search for listed hyper-parameters in user-provided ranges, but does not include architecture search. It is based on a grid search.

The above tools appeared quite recently, however, several hyper-optimization frameworks exist already for a longer time. The Python hyperopt [1] library that enables distributed hyper-parameter optimization has been designed in 2013. Although it is limited to hyper-parameters only (not architectures), due to the possibility of conditional hyper-parameters, it enables a user to tune also some architecture properties (however, only in a limited way, such as to tune the number of layers). Last but not least, there is the hyperas [12], a wrapper around the hyperopt library, designed directly for Keras.

The works focused directly on NAS are summarized in a recent survey [6]. The authors classify the NAS approaches based on search space, search strategy, and performance estimation strategy. The search space means what types of architectures are allowed. Often human bias is introduced. In our case, the search space is limited to convolutional neural networks. The search strategy represents the particular optimization algorithm, works vary from random search to evolutionary techniques, Bayesian optimization, or reinforcement learning. The performance estimation strategy is the way how to set up the objective of the optimization. The simplest way is to learn on training data and use the performance on validation data.

Probably the best known evolutionary approaches to NAS for deep neural networks come from Miikkulainen [18]. His approaches are built on the well established NEAT algorithm [19]. While the use of multi-objective optimization for NAS is quite natural, it is not generally used. One of the exceptions is the paper [5] that uses multi-objective Lamarckian evolution for NAS.

Several works concerned with NAS approaches taking some measure of network size or training efficiency into account have appeared recently. Authors of [7] present a solution combining NAS with quantization procedure in order to find the optimal precision that balances the network performance and energy consumption during training. Authors of [17] proposed a multiplexing procedure for channels in convolutional network used for image classification. The positive effect of this modification on three criteria – the classification accuracy, the compactness of resulting networks, and the efficiency of their training – was observed. Authors of the paper [9] proposed the so called Neural Architecture Transformer designed by a Markov decision process that optimizes neural architectures by removing unnecessary computationally intensive operations. Finally, the recent survey paper [25] presents an overview of approaches that include the compactness criterion to NAS algorithms.

The main points of our approach with respect to the above mentioned related work are:

- the search space – we were inspired by the implementation of feed-forward neural networks in Keras and designed the algorithm directly for Keras. So, as in the Keras Sequential model, the network is a list of layers, we consider only networks that can be defined as a list of layers (each layer always fully interconnected with the following layer). In this paper, convolutional and fully connected deep architectures are considered.
- the search strategy – we use multi-objective evolution that optimizes concurrently both the performance of the network and the network size. The state-of-the-art NSGA-2 algorithm [4] and NSGA-3 [3] were chosen for this purpose.
- the performance estimation strategy – since the split for training and validation data always introduces a bias, we use cross-validation to evaluate network performance.

3 Deep and Convolutional Neural Networks

By a deep neural network (DNN) we understand a feed-forward layered neural network architecture with more (typically many) hidden layers. Fully connected DNNs can be seen as deep variants of traditional multi-layer perceptrons contain several dense fully connected layers only. On the other hand, convolutional neural networks (CNN) represent an important subset of DNN containing one or more convolutional layers.

The typical architecture of a CNN is depicted in Fig. 1. The front part of the network is responsible for feature extraction and besides convolutional layers it contains pooling layers (typically max-pooling) that perform down-sampling. The top layers of the network perform the classification itself and are often fully connected dense layers. In this paper, we work with such a network architecture. Further details about the DNN and CNN concepts can be found, e.g. in the book [8].

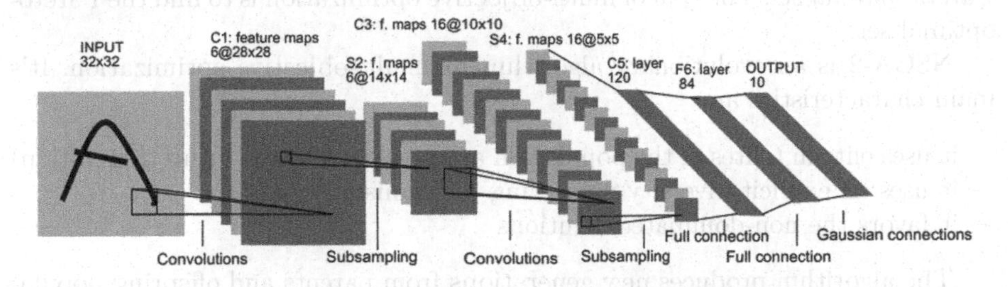

Fig. 1. Convolutional neural network [15].

4 Multi-objective Evolution for Deep Neural Networks

We consider the problem of NAS that aims to find the architecture with good that aims to find the architecture with good performance and a reasonable size at the same time. Therefore, we use the multi-objective optimization approach and optimize both the network performance and its size described by the number of network parameters. To this end, wee chose the NSGA-2 [4] (or NSGA-3 [3], respectively) algorithms that are considered to be the state of the art of the field, to perform the multi-objective evolutionary optimization.

4.1 NSGA-2 and NSGA-3 Algorithms

The abbreviation NSGA stands for non-dominated sorting genetic algorithm, the number 2 (3) stands for the second (third) improved variant of the algorithm.

Multi-objective optimization problems consider optimization problems with more objectives, where the objectives may be conflicting, in the sense that if one objective increases the other decreases. Such problems have no unique solution but a set of solutions.

Our case of NAS can be formalized as follows:

$$max f(x) \text{ and } min\ g(x),$$

where x stands for a particular network architecture, $f(x)$ is the network x accuracy (in case of classification) and $g(x)$ is the network x size expressed by the number of parameters (weights) of the network.

A solution x is said to dominate the other solution x' if

1. x is no worse than x' for all objectives
2. x is strictly better than x' in at least one objective

Among the set of feasible solutions P, the *non-dominated set of solutions* P is such a set that contains all solutions that are not dominated by any other member of P. The non-dominated set of the entire search space is called the *Pareto-optimal set.* The goal of multi-objective optimization is to find the Pareto-optimal set.

NSGA-2 is an evolutionary algorithm for multi-objective optimization. It's main characteristics are

- it uses elitism (elites of the population are preserved to the future generation)
- it uses an explicit diversity preserving mechanism
- it favors the non-dominated solutions

The algorithm produces new generations from parents and offspring population by

1. a non-dominated sorting of all individuals and labeling them by fronts (they are sorted by an ascending level of non-domination)
2. filling the new generation according to front-ranking
   - if the whole front does not fit into the new population, it performs Crowding-sort (crowding distance is related to the density of solutions around each solution), less dense solutions are preferred
3. creating new offspring by crossover and mutation operators, and by crowded tournament selection (selection based on front ranking and crowding distance if ranking equals)

The detailed description of the algorithm is out of the scope of this paper and can be found in the paper [4].

Authors of [3], extended the NSGA-2 to deal with many-objective optimization problem, using a reference point approach with non-dominated sorting mechanism. The NSGA-3 uses a set of reference points to maintain the diversity of the Pareto points during the search. This should improve an even distribution of Pareto points across the objective space.

4.2 Individual Encoding

To apply any evolutionary algorithm to architecture optimization we first have to be able to encode the architecture to an individual.

Our proposal of encoding closely follows the architecture description and implementation in the Keras [2] model *Sequential*. The model implemented as *Sequential* is built layer by layer, similarly, the individual consists of blocks representing individual layers.

Dense Neural Networks. For fully connected feed-forward neural networks (i.e. networks with dense layers only), the individual looks like:

$$I = ([size_1, drop_1, act_1]_1, \ldots, [size_H, drop_H, act_H]_H),$$

where H is the number of hidden layers, $size_i$ is the number of neurons in corresponding layer that is dense (fully connected) layer, $drop_i$ is the dropout rate (zero value represents no dropout), act_i stands for activation function ($act_i \in$ {relu, tanh, sigmoid, hardsigmoid, linear}).

Convolutional Networks. For convolutional network, the individual is slightly modified:

$$\begin{aligned} I &= (I_1, I_2), \\ I_1 &= ([type, params]_1, \ldots, [type, params]_{H1}) \\ I_2 &= ([size, dropout, act]_1, \ldots, [size, dropout, act]_{H2}) \end{aligned}$$

where I_1 and I_2 are the convolutional and dense part, respectively, $H1$, $H2$ is the number of layers in convolutional and dense part, respectively. The blocks in the convolutional part encode $type \in \{\texttt{convolutional}, \texttt{max-pooling}\}$ type of layer and *params* other parameters of the layer (for convolutional layer it is number of filters, size of the filter, and activation function; for max-pooling layer it is only the size of the pool). The blocks in the dense part code dense layers, so they consist of *size* the number of neurons, *drop* the dropout rate (zero value represents no dropout), *act* activation function.

4.3 Genetic Operators

To produce new individuals, we need operators *mutation*, *crossover* and *selection*. We also need to be able to evaluate the quality of individuals by their *fitness*.

Mutation. The *mutation* operator brings random changes to an individual. Each time an individual is mutated, one of the following mutation operators is randomly chosen (each of mutation operators has its own probability):

- mutateLayer - introduces random changes to one randomly selected layer.
- addLayer - one randomly generated block is inserted at a random position. If it is inserted into the first part of the individual, it is either a convolutional layer or max-pooling layer; otherwise, it is a dense layer.

– delLayer - one randomly selected block is deleted.

When *mutateLayer* is performed, again one of the available operators is chosen.

For dense layers they are:

– changeLayerSize - the number of neurons is changed. The Gaussian mutation is used, the final number is rounded (since size has to be an integer).
– changeDropOut - the dropout rate is changed using the Gaussian.
– changeActivation - the activation function is changed, randomly chosen from the list of available activations.

For max-pooling layers:

– changePoolSize - the size of the pooling is changed.

For convolutional layers:

– changeNumberOfFilters - the number of filters is changed. The Gaussian mutation is used, the final number is rounded.
– changeFilterSize - the size of the filter is changed.
– changeActivation - the activation function is changed, randomly chosen from the list of available activations.

Crossover. The operator *crossover* combines two parent individuals and produces two offspring individuals. It is implemented as one-point crossover, where the crossing point is determined at random, but on the border of a block only. Thus, only the whole layers are interchanged between individuals. In the case of CNN, the two parts of the individual are crossed over separately, so if parents are $I = (I_1, I_2)$ and $J = (J_1, J_2)$ we run $crossover(I_1, J_1)$ and $crossover(I_2, J_2)$.

Fitness Function. The fitness function should reflect the quality of the network represented by an individual. To assess the generalization ability of the network represented by the individual we use a cross-validation error. The lower the cross-validation error, the higher the fitness of the individual.

Classical k-fold cross-validation is used, i.e. the training set is split into k-folds and each time one fold is used for testing and the rest for training. The mean loss function on the testing set over k run is evaluated. For the classification tasks, the categorical cross-entropy is used as the loss function, for regression tasks, the mean squared error is used instead. Since the evaluation of the fitness function is the most time demanding step of the algorithm, small values of k are typically used.

5 Experiments

For experimental evaluation we have selected three traditional benchmark classification tasks – MNIST [16], fashionMNIST [26] and Cifar10 [13] datasets. Both MNIST and fashionMNIST datasets contain 70 000 images 28×28 pixels each. 60 000 are used for training, 10 000 for testing. MNIST contains images of handwritten digits, fashionMNIST contains greyscale images of fashion objects. Cifar10 consists of 60000 32×32 color images divided into 10 classification classes. There are 50000 training images and 10000 test images. The datasets are quite small, but this was necessary for performing a proper experimental evaluation. Since all algorithms used (i.e., the evolutionary search, the network gradient training) include random factors, the evaluations had to be repeated several times.

Our algorithm is implemented in Python, and it is publicly available as the nsga-keras library [23].

We run the algorithm on all datasets to search for optimal CNNs and on MNIST dataset to search for optimal dense DNN. In all cases, it was run five times on each task both with NSGA-2 and NSGA-3 variants. The architecture from Pareto front with the best cross-validation accuracy was chosen as the result (i.e. it is typically the largest from the solutions). The corresponding network was trained on the whole training set and evaluated on the test set. This final training and evaluation were done ten times and average values were obtained. Twenty epochs were used for training (the results can be further improved using more training epochs).

In addition, the classical genetic algorithm – with the same individual encoding, crossover and mutation operators, but only a single value fitness function, was run five times on both tasks. The resulting architecture was evaluated in the same way as for the multi-objective algorithm.

Also, for each task, we have chosen one baseline solution, that is the fixed architecture taken from Keras examples.

The Table 1 contains the overview of accuracies of resulting networks for a baseline solution, a solution produced by classical genetic algorithm (GA-CNN), and a solution found by multi-objective NSGA-2 and NSGA-3 algorithms (NSGA2-CNN, NSGA3-CNN). The average values of five runs are recorded.

Table 1. Accuracies of networks found by classical genetic algorithm (GA-CNN), NSGA-2 and NSGA-3 algorithm (NSGA2-CNN and NSGA3-CNN), and a baseline solution. In the case of genetic algorithms, average values, standard deviations and minimal a maximal values over five runs are listed.

Task	Baseline	GA-CNN				NSGA2-CNN				NSGA3-CNN			
		avg	std	min	max	avg	std	min	max	avg	std	min	max
MNIST	98.97	99.33	0.07	99.21	99.43	99.36	0.02	99.33	99.39	99.31	0.06	99.25	99.41
Fashion-MNIST	91.64	93.16	0.16	93.04	93.46	92.67	0.42	91.95	93.07	92.74	0.27	92.42	93.09
Cifar10	74.75	77.40	0.58	76.72	78.47	75.50	0.69	74.13	75.98	75.33	0.56	74.59	75.98

The Table 2 lists the sizes of the resulting networks. Letter K stands for one thousand, i.e. 77K stands for 77 000 learnable parameters.

Table 2. Sizes of networks (number of parameters in thousands) found by classical genetic algorithm (GA-CNN), NSGA-2 and NSGA-3 algorithm (NSGA2-CNN, NSGA3-CNN), and a baseline solution. In case of genetic algorithms, average values, standard deviations and minimal a maximal values over five runs are listed.

Task	Baseline	GA-CNN				NSGA2-CNN				NSGA3-CNN			
		avg	std	min	max	avg	std	min	max	avg	std	min	max
MNIST	600K	1547K	1796K	468K	5123K	77K	48K	28K	168K	242K	270K	67K	778K
Fashion-MNIST	356K	898K	291K	543K	1203K	418K	311K	64K	876K	452K	154K	176K	634K
Cifar10	1250K	321K	87K	198K	426K	207K	87K	97K	363K	301K	176K	129K	638K

The Table 3 contains the overview of accuracies and sizes of resulting dense networks for the MNIST dataset. Again, the average values of five runs are presented.

Table 3. Accuracies and sizes of networks found by classical genetic algorithm (GA-DNN), NSGA-2 and NSGA-3 algorithm (NSGA2-DNN, NSGA3-DNN), and a baseline solution for the MNIST dataset. In the case of genetic algorithms, average values, standard deviations and minimal a maximal values over five runs are listed.

	Baseline	GA-DNN				NSGA2-DNN				NSGA3-DNN			
		avg	std	min	max	avg	std	min	max	avg	std	min	max
Accuracy	98.32	98.30	0.05	98.23	98.35	98.35	0.05	98.29	98.42	98.29	0.06	98.20	98.38
Size	669K	299K	20K	282K	324K	263K	22K	222K	281K	224K	39K	174K	284K

The Fig. 3 records the values of the fitness functions during the evolution (for the NSGA-2 algorithm). We can see that the number of generations needed to converge is not too high.

Regarding the setup of the evolutionary algorithm, the population with 30 individuals was used, networks were trained for 20 epochs (during cross-validation in fitness evaluation), categorical cross-entropy was used as a loss function, and the algorithm was run for 100 generations. In our preliminary tests, we have also tried to use a lower number of epochs to save time, but the results were inferior (Fig. 2).

The experiments were run on GeForce GTX 1080 Ti GPUs. To save the time, the whole population was evaluated at once (on a single card), and also the cross-validation parts were evaluated in parallel. This optimization was not possible for classical genetic algorithm, where larger architectures evolved and therefore it was no more possible to squeeze the whole population to the memory of one card, thus, the fitness evaluation was done in two batches. The time required for the evaluation of one generation was on average 52.24 min for NSGA-2 and NSGA-3 and 178.61 min for the classical GA on MNIST.

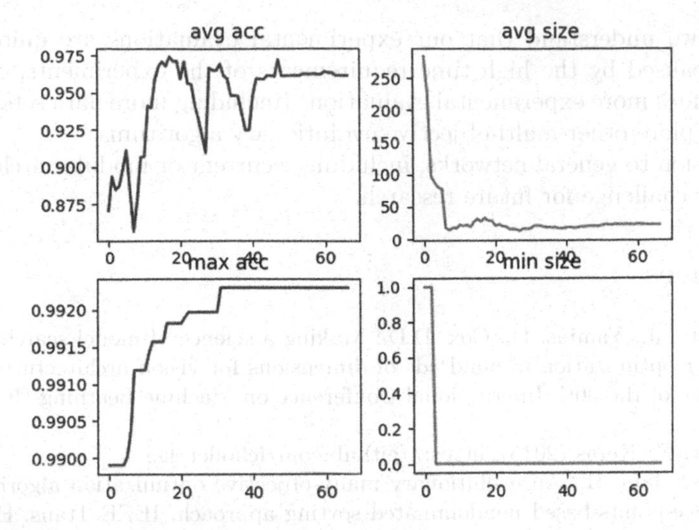

Fig. 2. Example of one run of nsga-keras (NSGA-2 version). On the left, average and maximal accuracy. On the right, average and minimal network size.

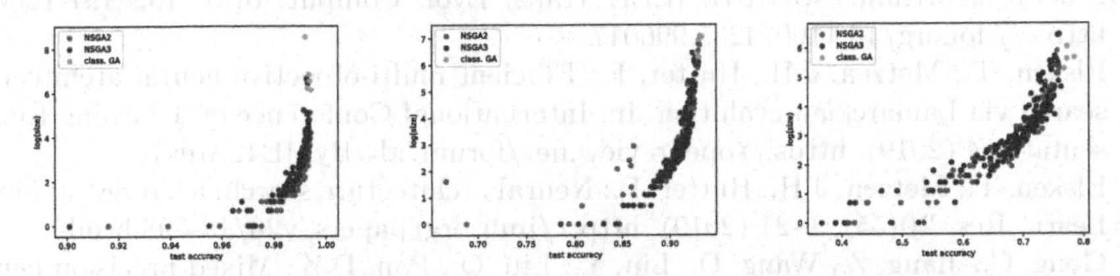

Fig. 3. Resulting pareto fronts for the MNIST, fashionMNIST and Cifar10 datasets.

From the tables, we can see the evolutionary approaches outperform the baseline solution and that our proposed multi-objective approach produces a smaller solution with competitive performance.

6 Conclusion

In this paper, we proposed a novel approach to neural architecture search for deep neural networks. The algorithm is based on multi-objective evolution utilizing the well known NSGA-2 and NSGA-3 algorithms. The performance of the network and the network size are optimized simultaneously, so the algorithm produces competitive networks of reasonable size. The benefits from size optimization are two-fold, first, it prevents the candidate solutions to bloat and slow down the evolution, and second, the smaller solutions learn faster and are suitable for devices with limited memory.

We confirmed the applicability of both evolutionary and multi-objective evolutionary approaches to NAS. The proposed approach is implemented and is publicly available, it is possible to run it both on GPUs and CPUs.

Since we understand that our experimental evaluations are quite limited, which is caused by the high time requirements of the experiments, our future work includes more experimental evaluations (including more data sets). We also plan to explore other multi-objective evolutionary algorithms.

Extension to general networks, including recurrent or modular architectures, is another challenge for future research.

References

1. Bergstra, J., Yamins, D., Cox, D.D.: Making a science of model search: hyperparameter optimization in hundreds of dimensions for vision architectures. In: Proceedings of the 30th International Conference on Machine Learning (ICML 2013) (2013)
2. Chollet, F.: Keras (2015). https://github.com/fchollet/keras
3. Deb, K., Jain, H.: An evolutionary many-objective optimization algorithm using reference-points-based nondominated sorting approach. IEEE Trans. Evol. Comput. **18**(4), 577–601 (2014)
4. Deb, K., Pratap, A., Agarwal, S., Meyarivan, T.: A fast and elitist multiobjective genetic algorithm: NSGA-II. IEEE Trans. Evol. Comput. **6**(2), 182–197 (2002). https://doi.org/10.1109/4235.996017
5. Elsken, T., Metzen, J.H., Hutter, F.: Efficient multi-objective neural architecture search via Lamarckian evolution. In: International Conference on Learning Representations (2019). https://openreview.net/forum?id=ByME42AqK7
6. Elsken, T., Metzen, J.H., Hutter, F.: Neural architecture search: a survey. J. Mach. Learn. Res. **20**(55), 1–21 (2019). http://jmlr.org/papers/v20/18-598.html
7. Gong, C., Jiang, Z., Wang, D., Lin, Y., Liu, Q., Pan, D.Z.: Mixed precision neural architecture search for energy efficient deep learning. In: 2019 IEEE/ACM International Conference on Computer-Aided Design (ICCAD), pp. 1–7 (2019)
8. Goodfellow, I., Bengio, Y., Courville, A.: Deep Learning. MIT Press (2016). http://www.deeplearningbook.org
9. Guo, Y., et al.: NAT: neural architecture transformer for accurate and compact architectures. In: NeurIPS (2019)
10. Goodfellow, I., et al.: TensorFlow: large-scale machine learning on heterogeneous systems (2015). Software available from tensorflow.org. https://www.tensorflow.org/
11. Jin, H., Song, Q., Hu, X.: Auto-Keras: an efficient neural architecture search system. In: Proceedings of the 25th ACM SIGKDD International Conference on Knowledge Discovery & Data Mining, pp. 1946–1956. ACM (2019)
12. Keras + Hyperopt: A very simple wrapper for convenient hyperparameter optimization. http://maxpumperla.com/hyperas/
13. Krizhevsky, A., Nair, V., Hinton, G.: CIFAR-10 (Canadian Institute for Advanced Research). http://www.cs.toronto.edu/kriz/cifar.html
14. Lecun, Y., Bengio, Y., Hinton, G.: Deep learning. Nature **521**(7553), 436–444 (2015). https://doi.org/10.1038/nature14539
15. Lecun, Y., Bottou, L., Bengio, Y., Haffner, P.: Gradient-based learning applied to document recognition. Proc. IEEE **86**(11), 2278–2324 (1998). https://doi.org/10.1109/5.726791
16. LeCun, Y., Cortes, C.: The MNIST database of handwritten digits (2012). http://research.microsoft.com/apps/pubs/default.aspx?id=204699

17. Lu, Z., Deb, K., Boddeti, V.: MUXConv: information multiplexing in convolutional neural networks, pp. 12041–12050, June 2020. https://doi.org/10.1109/CVPR42600.2020.01206
18. Miikkulainen, R., et al.: Evolving deep neural networks. CoRR abs/1703.00548 (2017). http://arxiv.org/abs/1703.00548
19. Stanley, K.O., Miikkulainen, R.: Evolving neural networks through augmenting topologies. Evol. Comput. **10**(2), 99–127 (2002). http://nn.cs.utexas.edu/?stanley:ec02
20. Autonomio Talos [computer software] (2019). http://github.com/autonomio/talos
21. Vidnerová, P., Neruda, R.: Evolution strategies for deep neural network models design. In: Hlaváčová, J. (ed.) Proceedings ITAT 2017: Information Technologies - Applications and Theory. CEUR Workshop Proceedings, vol. 1885, pp. 159–166. Technical University & CreateSpace Independent Publishing Platform, Aachen & Charleston (2017)
22. Vidnerová, P., Neruda, R.: Evolving Keras architectures for sensor data analysis. In: 2017 Federated Conference on Computer Science and Information Systems (FedCSIS), pp. 109–112, September 2017. https://doi.org/10.15439/2017F241
23. Vidnerová, P., Procházka, Š.: NSGA-Keras: Neural architecture search for Keras sequential models (2020). https://github.com/PetraVidnerova/nsga-keras
24. Vidnerová, P., Neruda, R.: Asynchronous evolution of convolutional networks. In: Krajci, S. (ed.) Proceedings of the 18th Conference Information Technologies - Applications and Theory (ITAT 2018), Hotel Plejsy, Slovakia, 21–25 September 2018. CEUR Workshop Proceedings, vol. 2203, pp. 80–85. CEUR-WS.org (2018). http://ceur-ws.org/Vol-2203/80.pdf
25. Xia, W., Yin, H., Jha, N.: Efficient synthesis of compact deep neural networks. CoRR abs/2004.08704 (2020). http://arxiv.org/abs/2004.08704
26. Xiao, H., Rasul, K., Vollgraf, R.: Fashion-MNIST: a novel image dataset for benchmarking machine learning algorithms (2017)

Neural Architecture Search for Extreme Multi-label Text Classification

Loïc Pauletto[1,2](✉), Massih-Reza Amini[2], Rohit Babbar[3], and Nicolas Winckler[1]

[1] Atos, Grenoble, France
{loic.pauletto,nicolas.winckler}@atos.net
[2] University Grenoble Alpes, Grenoble, France
Massih-Reza.Amini@univ-grenoble-alpes.fr
[3] Aalto University, Helsinki, Finland
rohit.babbar@aalto.fi

Abstract. Extreme classification and Neural Architecture Search (NAS) are research topics which have recently gained a lot of interest. While the former has been mainly motivated and applied in e-commerce and Natural Language Processing (NLP) applications, the NAS approach has been applied to a small variety of tasks, mainly in image processing. In this study, we extend the scope of NAS to the task of extreme multilabel classification (XMC). We propose a neuro-evolution approach, which was found to be the most suitable for a variety of tasks. Our NAS method automatically finds architectures that give competitive results with respect to the state of the art (and superior to other methods) with faster convergence. In addition, we perform analysis of the weights of the architecture blocks to provide insight into the importance of different operations that have been selected by the method.

Keywords: Neural architecture search · Machine learning · Extreme multi-label text classification · Evolutionary algorithms

1 Introduction

Neural networks (NNs) have shown impressive performance in many natural language tasks, such as classification, generation, translation and many others. One of the applications that has attracted growing interest in recent years with the availability of large-scale textual data is the extreme multi-label text classification (XMC). The goal in XMC is to classify data to a small subset of relevant labels from a large set of all possible labels [13,16]. A major problem in applying NNs to this task is to design an architecture that can effectively capture the semantics of text. Diverse methods have been employed in the NLP field, such as convolutional neural networks [25], recurrent neural networks [14] as well as a combination of both [26]. However, this design phase is complex and often requires human prior, with a good knowledge of the field and data. Over the

H. Yang et al. (Eds.): ICONIP 2020, LNCS 12534, pp. 282–293, 2020.
https://doi.org/10.1007/978-3-030-63836-8_24

last few years, NAS research has paved the way for the creation of dedicated neural architectures for a given task or even dataset. Most of the NAS studies, have focused on search algorithms for a small number of tasks (eg. image classification) and none of these studies have been applied to XMC before. In this paper, we propose `XMC-NAS` a NAS-based method for automatically designing an architecture for the extreme multi-label text classification task, using a minimum of prior knowledge. In addition, we define a search space with operations (e.g. RNN, Convolution,..) specific to the NLP domain. To evaluate our solution, we have used 3 large scale XMC datasets with an increasing number of labels. Like popular NAS methods we have uses a proxy dataset to train and evaluate architectures during the search phase. The discovered architecture gives competitive results with respect to the state of the art on the proxy dataset with faster convergence. Then we transfer the best performing architecture to other datasets and evaluate it. Our evaluation shows, the discovered architecture also achieves results close to the state of the art. Furthermore, this paper presents a study on the importance of operation types and the network depth with respect to the obtained results.

In the following section, we briefly review some related state-of-the-art. In Sect. 3, we present our solution to extreme multi-label classification with neural architecture search. Experimental results are presented in Sect. 4 and the conclusion and an outcome of this work are presented in Sect. 5.

2 Related Work

In this section, we will present related work on neural architecture search and extreme text classification.

Neural Architecture Search. Studies on the subject of NAS date back to the 1990s [9] and they have gained significant interest in the last few years. In the literature, different approaches have been studied, one of the first approaches was based on Reinforcement Learning (RL) [27]. In these approaches, architectures are first sampled from a controller, typically a RNN, and are further trained and evaluated. The controller is updated from the evaluation results in a feedback loop, improving the sampled architectures over time. Some other approaches use Bayesian Optimization (BO) like [6] in order to predict the accuracy of a new and unseen network and thus select only the best operation or as in [11] which uses Sequential Model-Based Optimization (SMBO) to predict accuracy of a network based on a network with fewer operations. Transfer Learning [22] has also been used in NAS methods [18] allowing more efficient search by weight sharing of overlapping operations, instead of training each new network from scratch. Other NAS methods have used gradient descent based approaches such as in [12] where they use a relaxation, which allows to learn the architecture and the weight of operation simultaneously, using the gradient descent. More recent studies have shown great performances using the well-known evolution algorithms as in [15,20], which consists in starting with a base population and successively apply mutation to the best performing architecture.

Extreme Text Classification. Different methods have been proposed to address the stakes posed by the extreme multi-label text classification [1,2,7]. The most recent of those methods are deep learning based such as XML-CNN [13], a structure of convolutional neural network (CNN) and pooling in order to get a precise text representation. However, it is hard for CNN to capture the most relevant part of a text and the long term dependency. Other methods, more similar, to Seq2seq methods have been applied as discussed in MLC2Seq [16], SGM [23] and AttentionXML [24]. Those methods used recurrent neural network (RNN) to classify the text. Moreover, a significant interest has been given on attention mechanism the last few years [10]. Attention mechanism has demonstrated great performance in sequence modeling, in particular in NLP domain and has therefore been also applied in the context of XMC [23,24].

3 Framework and Model

We propose `XMC-NAS`, a tool to automatically design architecture for the extreme multi-label classification task. Our approach is based on three main components: i) the text embedding, ii) the search of the architecture, and iii) output classification. These three components form a pipeline in which components i) and iii) are fixed and excluded from the search task. Thus, the architecture search task is performed only on the component ii). The first component of our methods consists in transforming the text into word embedding, i.e. Numerical vectors. This embedding step should allow the model to use these representations to produce a more accurate prediction. The second step is the search phase for the most performing architecture, using an *evolutionary algorithm* (c.f. Figure 1). Finally, the last component classifies the output, in several categories. This last component is based on attention mechanism and fully connected layers.

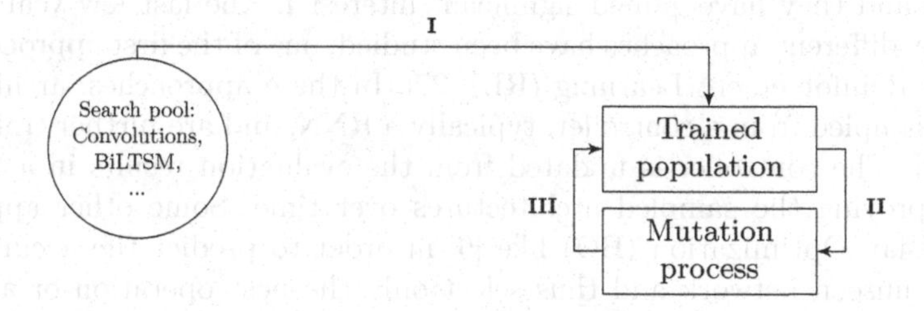

Fig. 1. **I**. Architectures are constructed from randomly sampled operations and then trained and evaluated, **II**. Randomly sample 10 architectures, and rank them by Precision@5 obtained on test set. The most performing one is selected for mutation, **III**. The newly mutated architecture, is trained and evaluated. Then placed in the trained population. The oldest architecture is removed from the population.

In the following section, we describe our approach in detail. First, we present the search space, the search algorithm used and their specificities; and finally, we describe the different parts of the discovered network.

3.1 Architecture Search Phase

The search phase can be broken down as follows. The architecture is searched in a *search space* that defines the possible structure of the final architecture. In this search space we have *candidate operations* that can be used in the architecture. Finally, to research the architecture, we use a *search algorithm* that searches for best architectures in the search space.

Search Space. A neural network architecture is represented in the form of a Direct Acyclic Graph (DAG). Each node in this graph represents a layer. A layer is a single operation, which is chosen among the set of *candidate operations*. The edges of the graph represent the data flow, and each node can have only one input. The graph is constructed as follows: first, the nodes are sampled sequentially (i.e. An operation is selected) to create a graph of N nodes. Then, the input of a node j is selected in the set of previous nodes (i.e. Nodes from 1 to $j-1$). This set is initialized with a node that represents the embedding layer. Finally, when the node j has an operation and an input, it is added to the set of the previous nodes. To have a trade-off between performance and search efficiency, we have set a limit to the maximum number of previous nodes that a layer can take as input. We empirically determined this limit to 5 to achieve reasonable search time on our hardware. Figure 2 illustrates a simple architecture with $N = 6$ nodes (i.e layers).

Candidate Operations. To build our set of *candidate operations*, we have selected the most common operations in NLP field, which consist of a mixture of convolutional, pooling and recurrent layers. We have defined four variants of 1D convolutional layers, with a kernel of different size: 1, 3, 5 and 7 respectively. All convolution layers use a stride of 1 and use padding if necessary to keep a consistent shape. We used the two types of pooling layers that calculate either the average or the maximum on the filter size, this size is set to 3 for both. Similarly to the convolution layers, the pooling layers use a stride of 1 and use padding if necessary. Finally, we used the two popular types of recurrent layers, namely the Gated Recurrent Unit (GRU) [3] and the Long-Short Term Memory (LSTM) [4], which are able to capture long-term dependencies. Specifically, we use bi-directional LSTM and GRU.

Search Algorithm. As NAS algorithm we use the *regularized evolution* as described in [20]. We chose this approach because it allows us to have a fine vision of the impact of each operation on the final result. Regarding the mutation aspect, we use the same configuration as described in [20]:

- Randomly select an input from a node on the network and modify it with a new input.
- Randomly select an operation from the network and change it with a new sample.

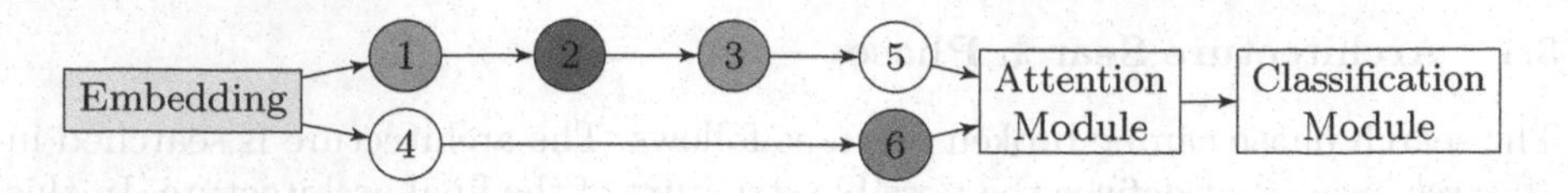

Fig. 2. Illustration of a simple architecture, with 6 layers. The numbers represent the sampling order of the layers. The limit of maximum number of previous layers that can be used as input is set to 5 (e.g. The layer 6 could hence take as input only nodes from 1 to 5). Here, different operations are illustrated with different colors.

Figure 1 illustrates the search algorithm of the *regularized evolution*. In order to see the impact of the number of layers on the final results, a third mutation, corresponding to the addition of a new layer, has been introduced. The choice among these mutations is random. We also seek to better understand how operations perform together, i.e. To evaluate the importance of the various operations with respect to the final results. To do this, we use a linear combination of outputs from each layer where weights are learned during the training process.

3.2 Text Embedding, Attention and Classification Modules

Our network has certain parts fixed, namely the embedding, attention and classification modules. This section will introduce them.

Text Embedding. The embedding layer produces a fixed length representation. This layer is an embedding map, which means each word is mapped to a vector. As initialization, we used the GloVe [17] embedding 840B-300d[1] version, which allows us to skip the step of learning a new embedding from scratch.

Attention Module. We use a self-attention mechanism based on the one demonstrated in [10], similarly to [24]. The attention process helps to grasp the important parts of the text. This mechanism uses a vector $\mathbf{c_t}$ that represents the relevant *context* for the label t, where t is in $1, \ldots, T$. For an input sequence of size N, the context vector is given in Eq. 1.

$$\mathbf{c_t} = \sum_{i=1}^{N} \alpha_{t,i}\mathbf{h_i}, \qquad (1)$$

$$\alpha_{t,i} = \frac{e^{\mathbf{W_t}^T \cdot \mathbf{h_i}}}{\sum_{k=1}^{N} e^{\mathbf{W_t}^T \cdot \mathbf{h_k}}} \qquad (2)$$

Where, $\mathbf{h_i}$ denotes the hidden representation,i.e. The output of RNN encoder states or of the convolution. In the case of BiRNN, layer $\mathbf{h_i}$ is the concatenation of vectors from the forward $\overrightarrow{h_i}$ and backward $\overleftarrow{h_i}$ passes. The term $\alpha_{t,i}$ is called attention factor (Eq. (2)). The set of attention factors $\{\alpha_{t,i}\}$ represents how much of each inputs should be considered for each output. In Eq. (2), $\mathbf{W_t}$ is the attention weight (i.e a learnable weight matrix) for the t-th label.

[1] http://nlp.stanford.edu/data/wordvecs/glove.840B.300d.zip.

Classification Module. The final part of the network is composed of 2 or 3 fully connected layers, which reduces the output of the attention module. The result is then fed into an output layer that classifies it into different labels.

3.3 Analysis of Operations Importance

This section presents an analysis of the weights of the linear combination, particularly the impact of each operation on the final results, and whether different operations combine efficiently. We address this analysis in two steps. In the first step, we focus on how operations combine with each other. In a second step, we analyze the results and the impact of operations as the networks deepen.

First Step: In this step, the base population is randomly initialized, meaning that the input and operation of each node is chosen at random. We try to determine which operation is the most important in the first layers by scaling their outputs with trainable weights of the linear combination. The Fig. 3 shows three examples of the first layers for different combinations of operations as well as the corresponding learned weights assigned to each operations. The block "Rest of the network" represents the attention and classification modules. The blocks in the hatched areas of the Fig. 3 were not part of the mutation process and were "constrained". For each architecture example, the displayed weights are the averages obtained over several runs of the NAS. The different grey scales indicate different experiments. We observe in Fig. 3 that pairs of operations of same type, i.e. BiLSTM, tend to have almost equal weights. However, some trends could be observed in the case of the combination of two convolutions; those with a larger kernel size have higher weights. This effect is particularly pronounced in the case of the kernel size of 1, reflecting the need for sequence modeling blocks at this level. In the case of mixed operations, it turned out that BiLSTM operations systematically have higher weights. An example of a run with mixed operation is presented in the right-hand side of Fig. 3. More generally, our results show that architectures which contain BiLSTM at the first layer, perform better. This first step shows that the result is based mainly on the long-term dependencies captured by BiLSTM rather than on the combinations of local features generated by the convolution.

Second Step: This second stage of analysis aims to quantify the impact of the number of layers on the final results as well as the weights assigned to each operation. According to the results obtained in the first step, which show that the network with BiLSTM layers works better, in this second step, a part of the population has the constraint to start with at least one BiLSTM layer, which takes as input the embedding. For the analysis of the impact of the number of operations, we calculated the average $P@5$ based on the number of operations. The number of operations ranged from 2 to 6. We observed that the average precision is almost constant, regardless of the number of operations in the network, with a range of results close to what we have previously obtained. This result is corroborated by the analysis of the combination of operation weights.

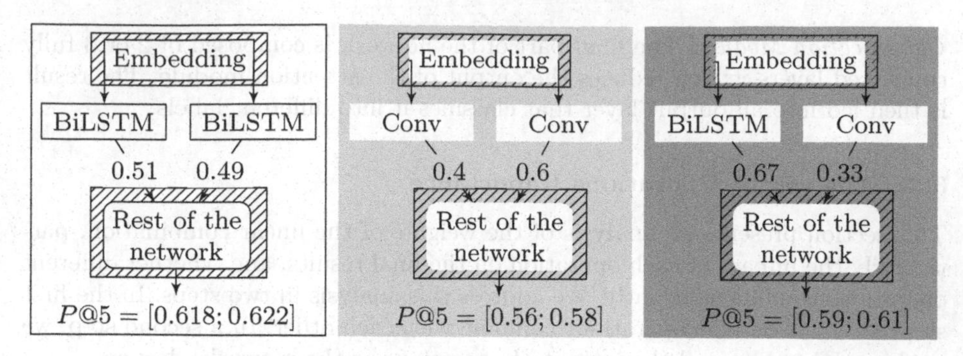

Fig. 3. Visualization of the network architecture with the applied weight on each operation. The weights have been averaged over multiple runs, the range of $P@5$ are obtained on the proxy dataset. For the central case, we also averaged over the kernel size.

The Fig. 4 shows examples of architecture for different combinations of operations with associated weights. This time the operations are assigned sequentially (i.e. One after the other) forming a deeper network. The blocks in hatched areas in Fig. 4 are partially or totally part of a constraint as in the previous subsection. Here, the blocks "Rest of the network" represent the following layers in the network, not shown for the sake of clarity, and still followed by the attention and classification modules. As previously, the weights displayed for each type of architecture are the averages obtained from multiple runs of the NAS. We note on the Fig. 4 that the weights on additional layers are small compared to those that bypass it. This trend has been observed in all experiments and suggests that, given our operations pool, additional layers do not provide much more information. Thus, the information important for the result is extracted by layers that take the embedding as input.

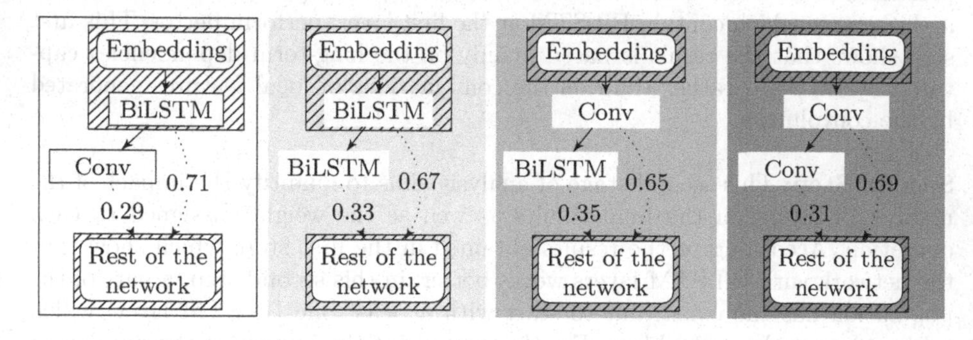

Fig. 4. Visualization of the network architecture, when the network is deeper. The dotted line represents the linear combination of all the layers outputs, with the weight applied on each outputs. The weight indicates that layers which take the embedding layer as input, have a predominant importance on the final result.

4 Experimental Results

We have conducted a number of experiments to evaluate how the proposed XMC-NAS method can help design an efficient neural network model for multi-label text classification.

4.1 Datasets and Evaluation Metrics

We conducted our study on three of the most popular XMC benchmark datasets downloaded from the XMC repository[2]. These datasets are considered large scale, with the number of class labels ranging from 4,000 to 30,000, which are listed from smallest to largest (in term of number of labels) by EURLex-4K, AmazonCat-13K, and Wiki10–31K summarized in Table 1. We followed the same pre-processing pipeline as the one used in [24]. To create the validation set we perform a split with the same initialization seed for all experiments. As evaluation metrics we used the Precision at k denoted by $P@k$, and the normalized discounted cumulative gain at k denoted by $nDCG@k$ [5]. Both metrics are standard and widely used in the state of the art references.

Table 1. Statistics of XMC datasets used in our experiments. L: # of classes.

Dataset	# of Training examples	# of Test examples	L	Avg. of class labels per example	Avg. size of classes
EURLex-4K	15,539	3,809	3,993	25.73	5.31
Wiki10–31k	14,146	6,616	30,938	8.52	18.64
AmazonCat-13K	1,186,239	306,782	13,330	448.57	5.04

4.2 Architecture Search Evaluation

This section presents the data and the hyperparameters that we have used during the search phase of our method. Finally, we present the most performing architecture that has been discovered on the proxy dataset.

Parameters and Data. We performed the search phase on the relatively small EURLex-4K dataset for scalability considerations, we call it the proxy dataset. In each experiment we create a base population of 20 networks. We then apply 50 rounds of mutations. For our experiments we have used the same hyperparameters as in [24], for the training of each sampled network. Namely, the optimizer was Adam [8] with a learning rate set to 0.001, and the maximum number of epochs were set to 30 epochs with early stopping. To be consistent with [24] we have used the same number of hidden states for the LSTM, which are specified in

[2] http://manikvarma.org/downloads/XC/XMLRepository.html.

Table 2. Hyperparameters used for the training of the discovered model.

Dataset	Valid size	BiLSTM Hidden size	Fully connected
EURLex-4K	200	256	[512,256]
Wiki10–31k	200	256	[512,256]
AmazonCat-13K	4000	512	[1024,512,256]

Table 2. The training stops if the performance of the network does not increase during 50 consecutive steps. We have used the cross-entropy loss function as proposed in [13] for training the models.

The Discovered Architecture. The architecture found by XMC-NAS, consists of two BiLSTMs that take the same input and holds their own representation. The outputs of the two BiLSTMs is then concatenated along the hidden dimension, and given as the input of the self-attention block. Finally, we use a chain of fully connected networks to classify the sequence. For the training detail we use the same as presented in the previous paragraph (see also Table 2).

The architecture of the network discovered by XMC-NAS is presented in Fig. 5.

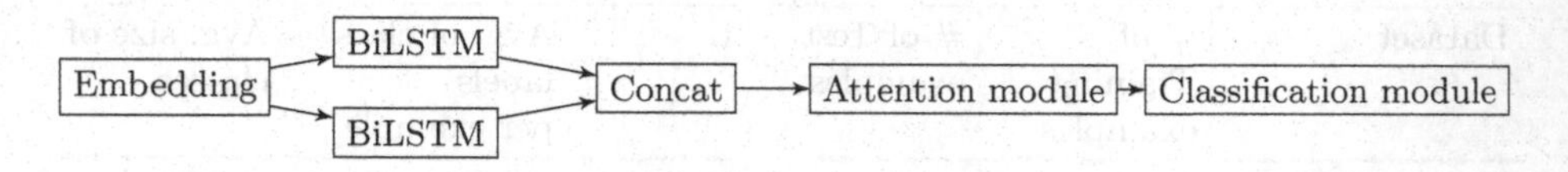

Fig. 5. The discovered network by XMC-NAS, is composed of two BiLSTM whose outputs are concatenated, and then passed in an attention module followed by fully connected layers.

4.3 Performance Evaluation

In this section we will present the results obtained by the XMC-NAS discovered architecture (Fig. 5) on various XMC datasets (Table 1). First we present the results obtained on the proxy dataset (EURLex-4K) used for the search phase. Finally, we evaluate the performance of this discovered architecture, transferred on the other datasets. To train our network, we use 2 Nvidia GV100, with data parallelism training. The search time on the proxy dataset, depending on the configuration, ranges from 6 hours to 5 days. We compare the results of our method to the most representative methods on XMC (with the results provided by the authors in corresponding papers). Some of these techniques are deep learning based like MLC2Seq [16], XML-CNN [13], Attention-XML [24]. The others techniques are, AnnexML [21], DiSMEC [1], PfastreXML [5] and Parabel [19].

EURLex-4K Results. As presented in the left-hand side of the Table 3 we have obtained an improvement on P@1, P@3 and P@5 with respect to the state

Table 3. Comparison performance table on three datasets. Our methods surpass the state of the art in 4 cases and get competitive results that really close to the state of the art otherwise.

	EURLex-4K			Wiki10–31K			AmazonCat-13K		
Methods	P@1	P@3	P@5	P@1	P@3	P@5	P@1	P@3	P@5
AnnexML	0.796	0.649	0.535	0.864	0.742	0.642	0.935	0.783	0.633
DiSMEC	0.832	0.703	0.587	0.841	0.747	0.659	0.938	0.791	0.640
PfastreXML	0.731	0.601	0.505	0.835	0.686	0.591	0.917	0.779	0.636
Parabel	0.821	0.689	0.579	0.841	0.724	0.633	0.930	0.791	0.645
XML-CNN	0.753	0.601	0.492	0.814	0.662	0.561	0.932	0.770	0.614
AttentionXML-1	0.854	0.730	0.611	**0.870**	**0.777**	0.678	**0.956**	**0.819**	**0.669**
XMC-NAS	**0.858**	**0.738**	**0.620**	0.849	0.772	**0.681**	0.951	0.813	0.664

of the art. The shown precisions are the average over 3 different initializations. A significant improvement is obtained on the precision at 3 and 5, where we obtain 0.738 and 0.620 respectively compared to 0.730 and 0.611 before. The Fig. 6a presents the evolution of P@5 and the nDCG@5 over the validation set with respect to the number of epochs. We can observe that our network has a faster convergence. The results is obtained around 15 epochs and after this point, the improvement is relatively small, which indicates that the network might overfit. It is not impossible that the improvement obtained is due to a larger network. However, we have systematically observed faster convergence in all the cases we have experienced. Furthermore, the contribution of the embedding or attention module on the results is not yet clear, as we have not yet studied the impact of these modules.

Architecture Transfer Results. We train and evaluate the discovered architecture following the same training procedure as defined in Sect. 4.2 and using

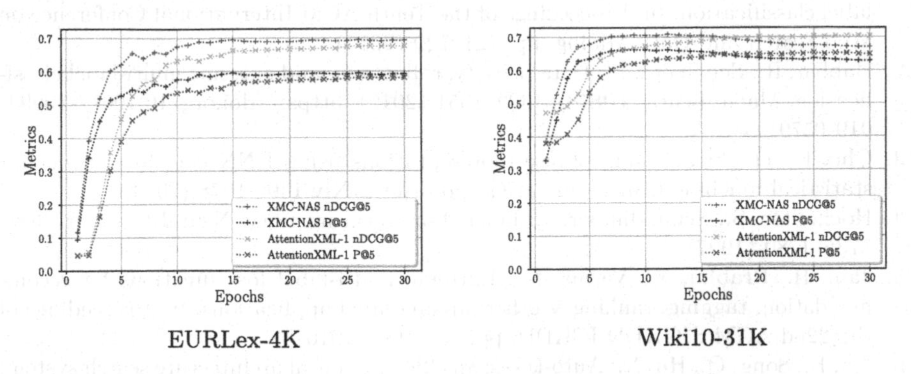

Fig. 6. Plot of the nDCG@5 and P@5 on the validation set, on two different datasets. We notice, the discovered architecture have a faster convergence compared to the current state of the art. In the 6a our method get better final results, in 6b our final results (around epoch 15) are close.

the hyperparameters presented in Table 2 on the two other datasets. The middle and right side of Table 3 show the comparison of the architecture discovered by XMC-NAS on EURLex-4K with others methods. We notice that the best discovered architecture transferred to larger datasets obtains results close to the current state of the art. In some cases we slightly exceed the results as in the case of P@5 on the Wiki10–31K. Moreover we notice in Fig. 6b that our methods still have a faster convergence, the same trend as observed on proxy dataset (cf. Fig. 6a). Moreover, our results on Wiki10–31K and AmazonCat-13K are obtained in half of the epochs required by AttentionXML-1.

5 Summary and Outlook

We have presented in this work an automated method to discover architecture for the specific task of extreme multi-label classification, based on the regularized evolution [20] and with a domain oriented pool of operations. This method has found architectures that have provided competitive results with the existing state of the art methods [24], and in some cases overpassed them. Moreover, our method showed faster convergence rates on all datasets, which are more likely due to a higher complexity of the model. In addition, trainable weights were introduced on each operation of the pool in order to provide more understanding on the impact of each architecture blocks. Many directions are possible as future steps. This includes the tuning of the various hyperparameters, the study of the impact of attention and embedding modules, the development of a method that can handle the scale of datasets, and speed up the search process (e.g. Using transfer learning).

References

1. Babbar, R., Schölkopf, B.: DiSMEC: distributed sparse machines for extreme multi-label classification. In: Proceedings of the Tenth ACM International Conference on Web Search and Data Mining, pp. 721–729 (2017)
2. Babbar, R., Schölkopf, B.: Data scarcity, robustness and extreme multi-label classification. Mach. Learn. **108**(8), 1329–1351 (2019). https://doi.org/10.1007/s10994-019-05791-5
3. Cho, K., et al.: Learning phrase representations using RNN encoder-decoder for statistical machine translation. arXiv preprint arXiv:1406.1078 (2014)
4. Hochreiter, S., Schmidhuber, J.: Long short-term memory. Neural Comput. **9**(8), 1735–1780 (1997)
5. Jain, H., Prabhu, Y., Varma, M.: Extreme multi-label loss functions for recommendation, tagging, ranking & other missing label applications. In: Proceedings of the 22nd ACM SIGKDD ICKDD, pp. 935–944 (2016)
6. Jin, H., Song, Q., Hu, X.: Auto-keras: an efficient neural architecture search system. In: Proceedings of the 25th ACM SIGKDD ICKDD, pp. 1946–1956 (2019)
7. Khandagale, S., Xiao, H., Babbar, R.: Bonsai: diverse and shallow trees for extreme multi-label classification. Mach. Learn. pp. 1–21 (2020)
8. Kingma, D.P., Ba, J.: Adam: a method for stochastic optimization. arXiv preprint arXiv:1412.6980 (2014)

9. Kitano, H.: Designing neural networks using genetic algorithms with graph generation system. Complex Syst. **4**, 461–476 (1990)
10. Lin, Z., et al.: A structured self-attentive sentence embedding. arXiv preprint arXiv:1703.03130 (2017)
11. Liu, C., et al.: Progressive neural architecture search. In: Proceedings of the European Conference on Computer Vision (ECCV), pp. 19–34 (2018)
12. Liu, H., Simonyan, K., Yang, Y.: Darts: differentiable architecture search. arXiv preprint arXiv:1806.09055 (2018)
13. Liu, J., Chang, W.C., Wu, Y., Yang, Y.: Deep learning for extreme multi-label text classification. In: Proceedings of the 40th International ACM SIGIR, pp. 115–124 (2017)
14. Liu, X., Gao, J., He, X., Deng, L., Duh, K., Wang, Y.Y.: Representation learning using multi-task deep neural networks for semantic classification and information retrieval (2015)
15. Maziarz, K., et al.: Evolutionary-neural hybrid agents for architecture search. arXiv preprint arXiv:1811.09828 (2018)
16. Nam, J., Mencía, E.L., Kim, H.J., Fürnkranz, J.: Maximizing subset accuracy with recurrent neural networks in multi-label classification. In: Advances in Neural Information Processing Systems, pp. 5413–5423 (2017)
17. Pennington, J., Socher, R., Manning, C.D.: GloVe: global vectors for word representation. In: Proceedings of the 2014 Conference on Empirical Methods in Natural Language Processing (EMNLP), pp. 1532–1543 (2014)
18. Pham, H., Guan, M.Y., Zoph, B., Le, Q.V., Dean, J.: Efficient neural architecture search via parameter sharing. arXiv preprint arXiv:1802.03268 (2018)
19. Prabhu, Y., Kag, A., Harsola, S., Agrawal, R., Varma, M.: Parabel: partitioned label trees for extreme classification with application to dynamic search advertising. In: Proceedings of the 2018 World Wide Web Conference, pp. 993–1002 (2018)
20. Real, E., Aggarwal, A., Huang, Y., Le, Q.V.: Regularized evolution for image classifier architecture search. In: Proceedings of the AAAI Conference on Artificial Intelligence, vol. 33, pp. 4780–4789 (2019)
21. Tagami, Y.: AnnexML: approximate nearest neighbor search for extreme multi-label classification. In: Proceedings of the 23rd ACM SIGKDD International Conference on Knowledge Discovery and Data Mining, pp. 455–464 (2017)
22. Tan, C., Sun, F., Kong, T., Zhang, W., Yang, C., Liu, C.: A survey on deep transfer learning. In: Kůrková, V., Manolopoulos, Y., Hammer, B., Iliadis, L., Maglogiannis, I. (eds.) ICANN 2018. LNCS, vol. 11141, pp. 270–279. Springer, Cham (2018). https://doi.org/10.1007/978-3-030-01424-7_27
23. Yang, P., Sun, X., Li, W., Ma, S., Wu, W., Wang, H.: SGM: sequence generation model for multi-label classification. arXiv preprint arXiv:1806.04822 (2018)
24. You, R., Dai, S., Zhang, Z., Mamitsuka, H., Zhu, S.: AttentionXML: extreme multi-label text classification with multi-label attention based recurrent neural networks. arXiv preprint arXiv:1811.01727 (2018)
25. Zhang, X., Zhao, J., LeCun, Y.: Character-level convolutional networks for text classification. In: Advances in Neural Information Processing Systems, pp. 649–657 (2015)
26. Zhou, C., Sun, C., Liu, Z., Lau, F.: A C-LSTM neural network for text classification. arXiv preprint arXiv:1511.08630 (2015)
27. Zoph, B., Vasudevan, V., Shlens, J., Le, Q.V.: Learning transferable architectures for scalable image recognition. In: Proceedings of the IEEE Conference on Computer Vision and Pattern Recognition, pp. 8697–8710 (2018)

Non-linear ICA Based on Cramer-Wold Metric

Przemysław Spurek[1], Aleksandra Nowak[1], Jacek Tabor[1], Łukasz Maziarka[1](✉), and Stanisław Jastrzębski[2]

[1] Faculty of Mathematics and Computer Science, Jagiellonian University, Krakow, Poland
przemyslaw.spurek@uj.edu.pl, lukasz.maziarka@ii.uj.edu.pl

[2] Center of Data Science/Department of Radiology, New York University, New York, USA

Abstract. Non-linear source separation is a challenging open problem with many applications. We extend a recently proposed Adversarial Non-linear ICA (ANICA) model and introduce Cramer-Wold ICA (CW-ICA). In contrast to ANICA, we use a simple, closed–form optimization target instead of a discriminator–based independence measure. Our results show that CW-ICA achieves comparable results to ANICA while foregoing the need for adversarial training.

Keywords: Non-linear independent component analysis · Cramer-Wold metric · Disentanglement · Unsupervised learning · Information retrieval

1 Introduction

Linear Independent Components Analysis (ICA) has become an important data analysis technique. For example, it is routinely used for blind source separation in a wide range of signals. The objective of ICA is to identify a *linear* transformation such that after the projection, the components of the dataset are independent. More formally, the aim is to find an *unmixing matrix* W that transforms the observed data $X = (\mathrm{x}_1, \ldots, \mathrm{x}_n)^T$ into maximally independent components $S = WX = (s_1, \ldots, s_n)$ with respect to some measure of independence. Commonly the independence is approximated using a measure of nongaussianity (e.g. kurtosis [3,12], skewness [22]).

An obvious drawback of ICA is the restriction to linear transformations. Unfortunately, in many practical applications, this linearity assumption does not hold, which motivates research into Nonlinear ICA (NICA) [11,13].

One of the key challenges in developing a nonlinear variant of ICA is devising an efficient measure of independence. The currently most popular approach is to constrain the transformation so that independence can be efficiently estimated [1,2,8,26,28]. Another approach is to *learn* the independence measure. This can be achieved using Generative Adversarial Networks (GANs) [9]. In [4] (ANICA

H. Yang et al. (Eds.): ICONIP 2020, LNCS 12534, pp. 294–305, 2020.
https://doi.org/10.1007/978-3-030-63836-8_25

- Adversarial Non-linear ICA) authors demonstrate the efficacy of using GAN for learning an independence measure. They show that GAN based independence measures combined with an autoencoder architecture can be used to solve nonlinear blind source separation problems.

Unfortunately, the use of adversarial training in ANICA comes at the cost of added instability, as also noted by the authors. Our main contribution is *developing an effective independence measure that does not require adversarial training, and matches ANICA performance.* In other words, we found that the adversarial training is not the key contributor to the efficacy of ANICA, and based on this insight we developed a simpler, closed-form independence measure. We demonstrate its efficacy on standard blind source separation problems.

This paper is structured as follows. We start by discussing related work in Sect. 2. In Sect. 3 we describe the key contribution: the independence measure based on *Cramer-Wold metric*. ICA based on the introduced independence measure is described in Sect. 4. Finally, we report experimental results in Sect. 5.

2 Related Work

The fundamental problem in solving NICA is that the solution is in principle non-identifiable. Without any constraints on the space of the mixing functions, there exists an infinite number of solutions [14]. To illustrate, consider that there is an infinite number of possible nonlinear decompositions of a random vector into independent components, and those decompositions are not related to each other in any trivial way. A related problem is that measuring true independence between distributions is often intractable. While ICA can be efficiently solved using approximated independence measures, such as kurtosis, these approaches do not transfer to the nonlinear scenario.

Perhaps the most common approach to solve NICA, which addresses both of the problems, is to pose a constraint on the nonlinear transformation [1,2,8,20,26]. One of the first attempts was to generalize ICA by introducing nonlinear mixing models in which case the solution is still possible to identify [20]. In [19] authors propose Reconstruction ICA (RICA) which requires that mixing matrix W is as close as possible to orthonormal one $WW^T = I$. Thanks to such constraints, one can directly apply independent measures from the classical ICA method.

The aforementioned approaches are arguably limited in their expressive power. In a more recent attempt [8] the authors propose a neural model for modeling densities called Nonlinear Independent Component Estimation (NICE). The authors parameterize the neural network so that it is fully invertible and the output distribution is fully factorized (independent). However, the model incorporates learning the unmixing function using maximum likelihood, which requires specifying a prior density family.

Our work is most closely related to the recently introduced Adversarial Nonlinear ICA model (ANICA) [4]. In contrast to the previous methods, ANICA does not make any strong explicit assumptions on the transformation function.

Instead, a clever adversarial-based measure for estimating and optimizing independence efficiently is proposed. In this work, we will take a closer look at this measure, and argue that the basic premise permits the construction of an effective non-parametrized independence measure.

Finally, let us note that a large process has been made in learning factorized representations using deep neural networks [5,7]. What separates ANICA and our method from the previous work is the direct encouragement of independence in the latent space. A similar path was also taken by [17] where the VAE loss function is augmented with a cost term directly encouraging disentanglement.

3 Independence Measure by Cramer-Wold Distance

In this chapter we develop an efficient independence measure, which contrary to the ANICA model, does not require adversarial training. Our approach can be effectively used to solve nonlinear ICA, in contrast to many other metrics used solely in the context of linear ICA.

In the following, we discuss three independence metrics. Firstly, we consider distance correlation, and adversarial–based metric used in ANICA. In the last part, we introduce our Cramer-Wold based independence metric.

Distance Correlation. One of the most well-known measures of independence of random vectors $\mathbf{X}$ and $\mathbf{Y}$ is the *distance correlation* (dCor) [24], which is applied in [21] to solve the linear ICA problem. Importantly, $dCor(\mathbf{X}, \mathbf{Y})$ equals zero if the random vectors $\mathbf{X}$ and $\mathbf{Y}$ are independent. Moreover, $dCor$ has a closed-form estimator.

However, to ensure the independence of components of a given random vector $\mathbf{X}$ in $\mathbb{R}^D$, one has to compute $dCor(\mathbf{X}_J, \mathbf{X}_{J'})$ for every subset[1] of indexes $J \subset \{1, \ldots, D\}$, where J' denotes the complement set of J and $\mathbf{X}_J$ is the restriction of $\mathbf{X}$ to the set of coordinates given by J. As this procedure has exponential complexity with respect to the number of dimensions, we decided to use a simplified version of $dCor$ which enforces only pairwise independence of the components:

$$dCor_{pairwise}(\mathbf{X}) = \sum_{i<j} dCor(\mathbf{X}_i, \mathbf{X}_j),$$

where $\mathbf{X} = (\mathbf{X}_1, \ldots, \mathbf{X}_D)$.

Adversarial–Based Independence Metric. Now let us describe the adversarial approach used in ANICA. The basic idea is to leverage that a random permutation of features in a sample produces samples that come from a distribution with independent components. More precisely, let $\mathbf{X}$ be a random vector which comes from pdf $f(x_1, \ldots, x_D)$, and let $X = (x_i)_{i=1..n} \subset \mathbb{R}^D$ be a sample from $\mathbf{X}$, where $x_i = (x_i^1, \ldots, x_i^D)$. We will describe how to draw a sample from the density

$$F(x_1, \ldots, x_D) = f_1(x_1) \cdot \ldots \cdot f_D(x_D),$$

[1] Except for the trivial cases when either J or J' is emptyset.

where f_i are the marginal densities of f. To do this, simply randomly choose maps σ_i from $\{1,\ldots,n\}$ into itself, and consider

$$X_{shift} := (y_i)_{i=1..n}, \text{ where } y_i = (x^1_{\sigma_1(i)},\ldots,x^D_{\sigma_D(i)}).$$

Then X_{shift} comes from the pdf F, which has independent components. Consequently, if X and X_{shift} are close, then the same holds for f and F, and consequently f has independent components. In ANICA adversarial training is used to reduce the distance between X and X_σ.

Cramer-Wold Independence Metric. The application of adversarial training in ANICA can lead to instability, as discussed by the authors, and slower training. In this paper, we propose an alternative independence measure. Our main idea is to compute the distance between X and X_σ without resorting to adversarial training.

In order to achieve this, one can choose commonly used metrics, such as the Kullback-Leibler divergence [18] or Wasserstein distance [27]. Instead, due to its simplicity, we have decided to use the recently introduced Cramer-Wold distance d_{cw} [25], which also possesses the advantage of having the closed-form for the distance of two samples[2] $X = (x_i)_{i=1..n}, Y = (y_i)_{i=1..n} \subset \mathbb{R}^D$:

$$d^2_{\mathrm{cw}}(X,Y) = \frac{1}{2n^2\sqrt{\pi\gamma}}\Big(\sum_{ii'}\phi_D(\frac{\|x_i - x_{i'}\|^2}{4\gamma}) + \sum_{jj'}\phi_D(\frac{\|y_j - y_{j'}\|^2}{4\gamma}) - 2\sum_{ij}\phi_D(\frac{\|x_i - y_j\|^2}{4\gamma})\Big).$$

where the bandwidth γ is a hyperparameter, which may be set accordingly to the one-dimensional Silverman's rule of thumb to $\gamma = (\frac{4}{3n})^{1/5}$. The function ϕ_D is computed with the asymptotic formula: $\phi_D(s) \approx (1 + \frac{2s}{D})^{-1/2}$.

As a final step, we normalize each component of X_{shift} to ensure that the Silverman's rule of thumb is optimal, and define our independence metric as:

$$ii_D(X) := d^2_{\mathrm{cw}}(X, cn(X_{shift})), \tag{1}$$

where $cn(Y)$ is the componentwise normalization of Y.

4 Algorithm

We are now ready to define CW-ICA, a nonlinear ICA model based on the Cramer-World independence metric. Following ANICA, we use an Auto-Encoder (AE) architecture.

Let $X \subset \mathbb{R}^N$ denote the input data. An Auto-Encoder is a model consisting of an encoder function $\mathcal{E} : \mathbb{R}^N \to \mathcal{Z}$ and a complementary decoder function $\mathcal{D} : \mathcal{Z} \to \mathbb{R}^N$, aiming to enforce coding of the input variables that minimize the reconstruction error:

$$\text{rec\_error}(X; \mathcal{E}, \mathcal{D}) = \sum_{i=1}^{n} \|x_i - \mathcal{D}(\mathcal{E}x_i)\|^2. \tag{2}$$

[2] In the computation we apply the equality $\phi_D(0) = 0$.

Algorithm 1. (CwICA train loop)

input
  data $\mathbf{X} \in \mathbb{R}^{\mathbf{d}}$, with each sample in a separate row
  encoder $\mathcal{E}$, decoder $\mathcal{D}$
repeat
  sample a batch X of size n from $\mathbf{X}$
  apply encoder $Z = \mathcal{E}X$
  resample to obtain Z_{shift}:
  for $i \in \{1, \dots, n\}$ **do**
    for $j \in \{1, \dots, d\}$ **do**
      $k \sim Uniform(\{1, \dots, d\})$ // sample col. index
      $Z_{shift_{i,j}} = Z_{i,k}$
    end for
  end for
  normalize Z_{shift} by element-wise rescaling

$$\hat{Z}_{shift_{\cdot,j}} = \frac{Z_{shift_{\cdot,j}} - \text{mean}(Z_{shift_{\cdot,j}})}{\text{std}(Z_{shift_{\cdot,j}})} \quad \text{for } j = 1, \dots d$$

  $J = d^2_{\text{cw}}(\hat{Z}_{shift}, Z) \cdot \text{rec\_error}(X; \mathcal{E}, \mathcal{D})$
  Update E and D to minimize J
until converged

The goal of our method is to train an encoder network $\mathcal{E}X$ which maps data to informative, statistically independent features Z. In order to achieve this, we introduce an independence measure on the latent space, by taking advantage of the independence index $\text{ii}_D(\mathcal{E}X)$ defined in (1). We denote this model as the CW-ICA(Cramer-Wold Independent Component Analysis).

To obtain a procedure independent of a possible rescaling of the data, we have decided to use a multiplicative model instead of an additive:

$$\text{cost}(X; \mathcal{E}, D) = \text{ii}_D(\mathcal{E}X) \cdot \text{rec\_error}(X; \mathcal{E}, \mathcal{D}). \tag{3}$$

In contrary to ANICA we do not use an adversarial objective, proposing instead a closed-form solution based on the independence index. However, enforcing independence by itself does not guarantee that the mapping from the observed signals X to the predicted sources Z is informative about the input. Therefore, the decoder constrains the encoder, as proposed in ANICA.

As explained earlier, in the case of the Cramer Wold index it is important to normalize the resampled (permuted) latent variables, which additionally prevents the encoder's output from vanishing or exploding in magnitude.

In addition we implement another AE-based, nonlinear model, which follows the same architecture as CwICA, but substitutes $ii_D(\mathcal{E}X)$ by $dCor(\hat{Z})$. From this point onwards and in all figures and tables, for simplicity, we shall also use the $dCor$ notation instead of the $dCor_{pairwise}$. The $\hat{Z}$ stands for the component-wise normalized features of the encodings of X. We refer to this method as $dCorICA$.

5 Experiments

We evaluate our method on mixed images and synthetic dataset. For comparison we use the nonlinear method ANICA [4] and the PNLMISEP [29], an extension to the MISEP method [1,2]. It should be noted that the PNLMISEP is designed especially for post-nonlinearity, not for the more general nonlinear mixing functions used in presented experiments. We also report the results obtained on the same datasets by four selected linear models. We choose the popular FastICA algorithm [12], the Information-Maximization (Infomax) approach [3], the Joint Approximate Diagonalization of Eigenmatrices (JADE) [6] and the Pearson [23] system PearsonICA [16]. We use the implementations of the linear models in R packages `ica` [10] and `PearsonICA` [15].

5.1 Comparison with ANICA

The CwICA and dCorICA models follow a similar architecture as ANICA, but use a closed-form independence measure on the latent variables, as opposed to the adversarial approach. We compare our algorithms with the ANICA model using the synthetic signals dataset defined in [4].

The dataset in the nonlinear setting consists of $n = 4000$ observations $X \in \mathbb{R}^{n\times 24}$ which are obtained by applying mixing function $X = tanh(tanh(YA)B)$ to the independent sources $Y \in R^{n\times 6}$, where A and B are sampled uniformly from $[-2, 2]$ and $tanh$ is the hyperbolic tangent function. We select the first 500 samples as the test dataset, and train on the remaining 3500 samples. We fit ANICA using the best hyper-parameters setting for this dataset reported by [4]. For CwICA we perform a grid search on the learning rate and bandwidth, using batches of size 256 and choose the model with the smallest total loss on the validation dataset. The validation dataset has a size of 500 and is drawn from the same distribution as the train and test sets. All other model hyper-parameters are set as in ANICA. We also ran a similar grid search on the learning rate and batch size for dCorICA. We do not execute the PNLMISEP, as the implementation of this method is not suitable for input data of this dimensionality.

We also report the performance of the nonlinear methods on linear data. The linear dataset is obtained from the same independent sources Y by a transformation defined by the matrix A. We train the models using the same configuration as in the nonlinear experiment.

We evaluate the methods on test data using the mean $dCor$ distance between all possible pairs of the unraveled latent independent factors Z. In addition, we compute the mean maximum correlation (denoted as $max\ corr$) between the sources Y and the results Z. As ICA extracts the source signals only up to a permutation, we consider all possible pairings of the predicted signals with the source signals and report only the highest *max corr* value. Before computing the *dCor*, the latent variables Z are normalized. The results are presented in Tables 1 and 2. The original sources and the recovered by CwICA signals are presented in Fig. 3.

CwICA behaves very well on the nonlinear dataset, achieving a similar *max corr* value to ANICA, at the same time outperforming it in *MSE* and *dCor* criteria. This makes the method the best choice if a balanced solution is desired.

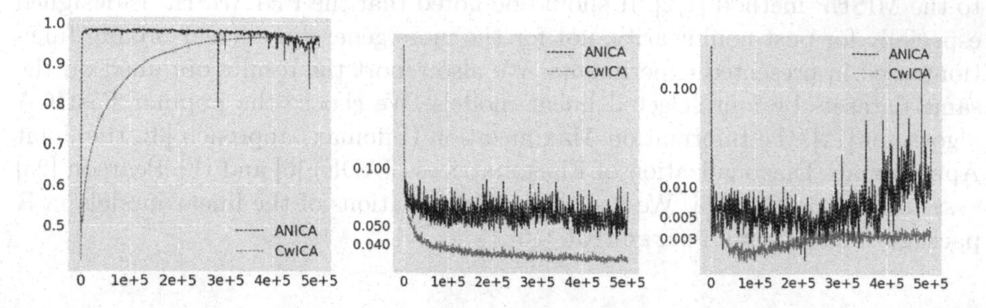

Fig. 1. The number of iterations versus *max corr* (left), *MSE* (middle) and *dCor*(right) for ANICA (black) and CwICA (red) Please note that the *MSE* and *dCor* results are plotted in logarithmic scale on the y-axis. This experiment is separate to the one presented in Table 1, therefore the results may slightly differ. (Color figure online)

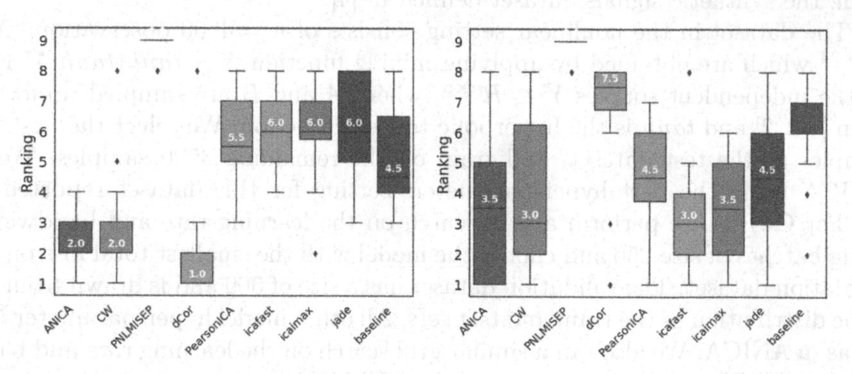

Fig. 2. The ranking (lower is better) of algorithms based on mean maximum correlation between the latent variables and sources (left-hand side) and dCor (right-hand side) in dimension $d = 10$.

Table 1. Results on nonlinear synthetic data

	ANICA	CwICA	PNLMISEP	dCorICA	PearsonICA	icafast exp	icaimax ext	jade
dCor	0.0027	0.0017	—	**0.0000**	0.0017	0.0017	0.0017	0.0017
max corr	**0.9835**	0.9697	—	0.3033	0.8969	0.8926	0.8940	0.9414
MSE	0.0516	**0.0332**	—	0.1475	—	—	—	—

Table 2. Results on linear synthetic data

	ANICA	CwICA	PNLMISEP	dCorICA	PearsonICA	icafast exp	icaimax ext	jade
dCor	0.0027	0.0175	0.0080	**0.0000**	0.0038	0.0038	0.0038	0.0038
max corr	0.8913	0.7805	0.9012	0.2514	0.9997	0.9997	**0.9998**	0.9984
MSE	0.0333	**0.0094**	—	0.1746	—	—	—	—

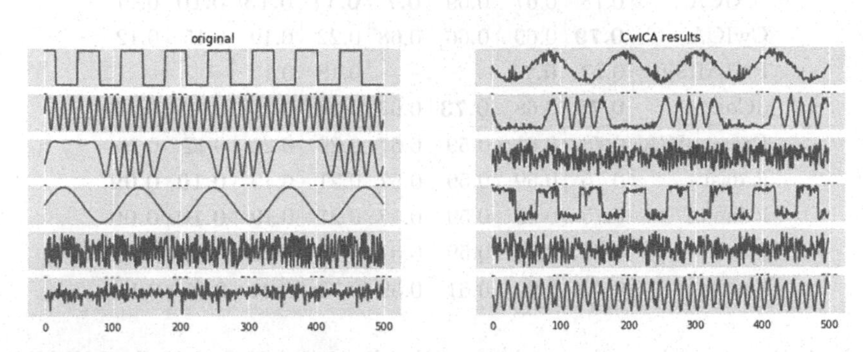

Fig. 3. The original sources (left) and the independent components predicted by CwICA (right) obtained from nonlinear mixtures.

In addition, we run the ANICA and CwICA models 5 times with different seeds. We pick the best model in terms of *dCor* and summarize the reported metrics on the validation dataset during training in Fig. 1. In this experiment, both models were trained using a batch size of 256.

In the linear synthetic data experiments, all non-linear models perform worse than the classical ICA algorithms. This sustains the claim that if the linear characteristic of the mixing function is assumed beforehand, the most efficient is the use of dedicated methods.

The dCorICA algorithm, as expected, achieves the lowest *dCor* cost in a both linear and nonlinear setting; however, fails to recover the original sources. This may suggest that the model focuses on the minimization of the independence loss, disregarding the information in the input.

5.2 Comparison on Image Dataset

One of the most popular applications of ICA is the separation of images. We conduct experiments on a dataset composed of images from the USC-SIPI Image Database, scaled to 100×67 pixels and mixed using $X = f(tanh(YA)B)$, where Y are the original sources, X are the observations, $dim(Y) = dim(X)$, $dim(X) \in \{2, 5, 10, 20\}$, $f(x) = x^2 + x^3$ is applied element-wise, and A and B are sampled uniformly from interval $[-2, 2]$. In addition, we prepare a linear dataset, where the mixing function is defined by the transformation imposed by a random matrix C sampled uniformly from $[-2, 2]$. The components of Y are separate, flattened, gray-scale images, chosen at random from a dataset of size 100. The observations X are normalized before passing to the algorithms. The

Table 3. Results on nonlinear image dataset. For dimension 10 and 20 the PNLMISEP did not converge.

	max corr				*dCor*			
	2	5	10	20	2	5	10	20
ANICA	0.78	0.67	0.69	**0.7**	**0.17**	**0.13**	**0.10**	0.14
CwICA	**0.79**	0.69	0.66	0.68	0.22	0.19	0.15	0.12
PNLMISEP	0.77	**0.71**	–	–	0.18	0.15	–	–
dCorICA	**0.79**	0.68	**0.73**	0.67	0.24	0.20	0.20	0.23
PearsonICA	0.73	0.61	0.59	0.57	0.29	0.21	0.12	0.10
icafast	0.75	0.59	0.59	0.57	0.21	0.19	**0.10**	**0.09**
icaimax	0.75	0.60	0.59	0.57	0.21	0.19	**0.10**	**0.09**
jade	0.74	0.59	0.59	0.57	0.25	0.20	0.11	0.10
baseline	0.70	0.60	0.61	0.59	0.36	0.24	0.15	0.11

Table 4. Results on linear image dataset. For dimension 10 and 20 the PNLMISEP did not converge.

	max corr				*dCor*			
	2	5	10	20	2	5	10	20
ANICA	0.90	0.74	0.73	0.7	0.16	0.11	**0.09**	0.14
CwICA	0.85	0.73	0.74	0.68	0.23	0.15	0.11	0.10
PNLMISEP	0.87	0.74	–	–	**0.14**	**0.09**	–	
dCorICA	0.89	0.74	0.76	0.57	0.30	0.15	0.11	0.28
PearsonICA	0.91	0.82	0.8	0.67	0.25	0.14	0.11	0.18
icafast	0.92	0.83	**0.82**	0.75	0.22	0.15	0.10	**0.08**
icaimax	0.91	**0.84**	**0.82**	**0.77**	0.24	0.14	0.10	0.10
jade	**0.93**	**0.84**	0.79	0.68	0.23	0.14	0.10	0.09
baseline	0.85	0.72	0.71	0.65	0.33	0.16	0.11	0.09

numbers of distinct observation examples for each dimension are 50, 50, 20, 10, respectively.

For each $dim(X)$ we test the ANICA, CW, dCor, PNLMISEP, FastICA, Infomax, Jade and PearsonICA algorithms. All the nonlinear models are trained using the same configurations as in the previous subsection. We report the mean *max corr* and *dCor* distance for each method in Table 3 (nonlinearly mixed data) and in Table 4 (linearly mixed data). We also report the MSE loss for auto-encoders (ANICA, CW, dCor) in Table 5.

CW-ICA achieves high *max corr* on the nonlinearly mixed data, comparable to the other non-linear ICA algorithms (in fact CwICA gets the best results among all ICA algorithms for $dim(X) = 2$) and strongly outperforms ANICA and dCorICA separations on reconstruction loss.

Table 5. Reconstruction loss (MSE) for auto-encoders on the nonlinear image dataset.

dim	ANICA	CwICA	dCorICA
2	0.5839	**0.0097**	0.6041
5	0.5811	**0.0181**	0.5491
10	0.5146	**0.0389**	0.4616
20	0.5299	**0.2748**	0.5079

Additionally, dCorICA gives satisfactory results on the nonlinear setting only for low dimensional data ($dim(X) \in \{2, 5\}$). For $dim(X) \geq 10$ dCorICA still manages to compete with other models in *maxcorr*, but evidently obtains the worst results in *dCor*, although it minimizes this measure directly. This disproportion can be especially observed in Fig. 2, which presents the mean rank of the methods based on the two metrics.

For higher dimensions, the nonlinear methods perform better in *maxcorr*; however, fail to surpass the classical algorithms in terms of *dCor*. An opposite trend in the linear data experiments may be observed for the lower dimensions (up to 10). In general, the linear methods achieve much better max_{corr}, and worse (higher) *dCor*. For $dim(X) = 20$ in both nonlinear and linear setting, the results obtained by auto-encoders are even worse than the baseline scores.

6 Conclusions

In this paper, we have proposed a closed-form independence measure and applied it to the problem of nonlinear ICA. The resulting model, CwICA, achieves comparable results to ANICA, while by using a closed-form formula avoids the pitfalls of adversarial training. Future work could focus on scaling up these approaches to higher-dimensional datasets and applying the developed independence metric in other contexts. Finally, we found that nonlinear methods generally underperform on linearly mixed signals, which could also be addressed in future work.

Acknowledgements. The work of P. Spurek was supported by the National Centre of Science (Poland) Grant No. 2019/33/B/ST6/00894. The work of A. Nowak was supported by the Foundation for Polish Science Grant No. POIR.04.04.00-00-14DE/18-00 co-financed by the European Union under the European Regional Development Fund. The work of J. Tabor was supported by the National Centre of Science (Poland) Grant No. 2017/25/B/ST6/01271. The work of Ł. Maziarka was supported by the National Science Centre (Poland) grant no. 2018/31/B/ST6/00993.

References

1. Almeida, L.B.: MISEP-linear and nonlinear ICA based on mutual information. J. Mach. Learn. Res. **4**(Dec), 1297–1318 (2003)

2. Almeida, L.B.: Linear and nonlinear ICA based on mutual information - the MISEP method. Signal Process. **84**(2), 231–245 (2004)
3. Bell, A.J., Sejnowski, T.J.: An information-maximization approach to blind separation and blind deconvolution. Neural Comput. **7**(6), 1129–1159 (1995)
4. Brakel, P., Bengio, Y.: Learning independent features with adversarial nets for non-linear ica. arXiv preprint arXiv:1710.05050 (2017)
5. Burgess, C.P., et al.: Understanding disentangling in β-vae. CoRR abs/1804.03599 (2018)
6. Cardoso, J.F., Souloumiac, A.: Blind beamforming for non-gaussian signals. In: Radar and Signal Processing, IEE Proceedings F, vol. 140, pp. 362–370. IET (1993)
7. Chen, X., Duan, Y., Houthooft, R., Schulman, J., Sutskever, I., Abbeel, P.: Infogan: interpretable representation learning by information maximizing generative adversarial nets. In: Lee, D.D., Sugiyama, M., Luxburg, U.V., Guyon, I., Garnett, R. (eds.) Advances in Neural Information Processing Systems, vol. 29, pp. 2172–2180. Curran Associates, Inc. (2016)
8. Dinh, L., Krueger, D., Bengio, Y.: Nice: Non-linear independent components estimation. arXiv preprint arXiv:1410.8516 (2014)
9. Goodfellow, I., et al.: Generative adversarial nets. In: Advances in Neural Information Processing Systems, pp. 2672–2680 (2014)
10. Helwig, N.E.: ICA: Independent Component Analysis (2015). http://CRAN.R-project.org/package=ica, r package version 1.0-1
11. Hirayama, J., Hyvärinen, A., Kawanabe, M.: SPLICE: fully tractable hierarchical extension of ICA with pooling. In: Precup, D., Teh, Y.W. (eds.) Proceedings of the 34th International Conference on Machine Learning. Proceedings of Machine Learning Research, vol. 70, pp. 1491–1500. PMLR, International Convention Centre, Sydney (2017)
12. Hyvärinen, A.: Fast and robust fixed-point algorithms for independent component analysis. IEEE Trans. Neural Netw. **10**(3), 626–634 (1999)
13. Hyvarinen, A., Morioka, H.: Unsupervised feature extraction by time-contrastive learning and nonlinear ica. In: Lee, D.D., Sugiyama, M., Luxburg, U.V., Guyon, I., Garnett, R. (eds.) Advances in Neural Information Processing Systems, vol. 29, pp. 3765–3773. Curran Associates, Inc. (2016)
14. Hyvärinen, A., Pajunen, P.: Nonlinear independent component analysis: existence and uniqueness results. Neural Netw. **12**(3), 429–439 (1999)
15. Karvanen, J.: PearsonICA (2008). https://CRAN.R-project.org/package=PearsonICA, r package version 1.2-3
16. Karvanen, J., Eriksson, J., Koivunen, V.: Pearson system based method for blind separation. In: Proceedings of Second International Workshop on Independent Component Analysis and Blind Signal Separation (ICA2000), Helsinki, Finland, pp. 585–590 (2000)
17. Kim, H., Mnih, A.: Disentangling by Factorising. ArXiv e-prints (2018)
18. Kingma, D., Welling, M.: Auto-encoding variational bayes. arXiv:1312.6114 (2014)
19. Le, Q.V., Karpenko, A., Ngiam, J., Ng, A.Y.: ICA with reconstruction cost for efficient overcomplete feature learning. In: Advances in Neural Information Processing Systems, pp. 1017–1025 (2011)
20. Lee, T.W., Koehler, B.U., Orglmeister, R.: Blind source separation of nonlinear mixing models. In: Neural Networks for Signal Processing [1997] VII. Proceedings of the 1997 IEEE Workshop, pp. 406–415. IEEE (1997)
21. Matteson, D.S., Tsay, R.S.: Independent component analysis via distance covariance. J. Am. Stat. Assoc. **112**, 1–16 (2017)

22. Spurek, P., Tabor, J., Rola, P., Ociepka, M.: ICA based on asymmetry. Pattern Recogn. **67**, 230–244 (2017)
23. Stuart, A., Kendall, M.G., et al.: The advanced theory of statistics. Charles Griffin (1968)
24. Székely, G.J., Rizzo, M.L., Bakirov, N.K., et al.: Measuring and testing dependence by correlation of distances. Ann. stat. **35**(6), 2769–2794 (2007)
25. Tabor, J., Knop, S., Spurek, P., Podolak, I., Mazur, M., Jastrzebski, S.: Cramer-wold autoencoder. arXiv preprint arXiv:1805.09235 (2018)
26. Tan, Y., Wang, J., Zurada, J.M.: Nonlinear blind source separation using a radial basis function network. IEEE Trans. Neural Netw. **12**(1), 124–134 (2001)
27. Tolstikhin, I., Bousquet, O., Gelly, S., Schoelkopf, B.: Wasserstein auto-encoders. arXiv:1711.01558 (2017)
28. Zhang, K., Chan, L.: Minimal nonlinear distortion principle for nonlinear independent component analysis. J. Mach. Learn. Res. **9**(Nov), 2455–2487 (2008)
29. Zheng, C.H., Huang, D.S., Li, K., Irwin, G., Sun, Z.L.: MISEP method for post-nonlinear blind source separation. Neural Comput. **19**, 2557–2578 (2007)

Oblique Random Forests on Residual Network Features

Wen Xin Cheng(✉), P. N. Suganthan, and Rakesh Katuwal

School of Electrical and Electronic Engineering, Nanyang Technological University, 50 Nanyang Avenue, Singapore 639798, Singapore
{wenxin001,epnsugan,rakeshku001}@ntu.edu.sg

Abstract. Time series are usually complicated in nature and contains many complex patterns. As such, many researchers work on different ways to pick up such patterns. In this paper, we explore using Residual Networks (a Convolutional Neural Network) as a feature extractor for Oblique Random Forest. Here, we extract features using Residual Networks, and pass the extracted feature set to Oblique Random Forest for classification of time series. Based on the experiments on 85 UCR datasets, we found that using features extracted from Residual Network significantly improves the performance of Oblique Random Forest. In addition, using including intermediate features from Residual Networks significantly improves the performance of Oblique Random Forests.

Keywords: Time series classification · Oblique Random Forests · Feature extraction

1 Introduction

Time series are frequently found in many real-world applications, which includes electronic health records [23], human activity recognition [22,27], acoustic scene classification and cyber-security [25]. Time series can be considered as a series of data where the order which the series is presented is important [7]. This means that useful information will be lost when the values in the series are shuffled.

However, the complexity of time series data and the presence of noise make time series classification difficult. In a survey article, data mining experts had identified Time Series Classification as one of the most challenging problems [29]. Many researchers had made many different attempts to classify time series, with hundreds of different algorithms being proposed to solve such problems since 2015 [1].

A group of classifiers performs classification based on the similarity between the given series and one of the known series. In this case, a similarity measure is used to measure how similar between 2 series and a 1-Nearest Neighbour classifier is usually employed to label the unknown series with the label of the closest known series. Works on such classifiers often proposed different distance measures which compensate for small alignment difference. Such distance measures

H. Yang et al. (Eds.): ICONIP 2020, LNCS 12534, pp. 306–317, 2020.
https://doi.org/10.1007/978-3-030-63836-8_26

includes Dynamic Time Warping (DTW), Derivative DTW [8] and Derivative Transform Distance [10].

Other methods involves the use of feature extractors to classify time series. Many different strategies have been employed. In algorithms such as Time Series Forest [6] and Time Series Bag of Features [2], a shorter segment of the time series that contains the most useful information can be extracted before classification. Another group of classifiers such as Shapelet Transform [3,12] and Learned Shapelets [9] picks up the presence of useful short patterns in the input series before classification. Other methods such as Bag of SFA Symbols [24] picks up the number of occurrence of useful short patterns in the input series before classification.

Ensembles are also popular among researchers as they combine different techniques which can produce superior performance in many datasets. Authors of [18] proposed combining 11 different distance measurement methods to create an Elastic Ensemble. The Collection of Transformation Ensemble (COTE) combines Elastic Ensemble with 3 sets of 8 classifiers, each set of classifiers trained on extracted features from Shapelet Transform, autocorrelation function and power spectrum. However, COTE have high time and memory complexities, which makes such ensembles not practical to be deployed in real-time settings.

Recently, deep learning techniques are getting popular as solutions for time series classification. Convolutional Neural Networks such as Fully Connected Networks and Residual Networks, are used as both a feature extractor and a classifier [28]. Such networks pick up local patterns found in the series and perform classification on the extracted features. In [14], Fully Convolutional Networks features are combined with Long Short Term Memory features to improve the classification performance. Echo Memory Networks employs convolutional layers to extract features from Echo State Networks (a randomised recurrent neural network).

Random Forests [4] are getting popular in other classification problems. Such models are easy to interpret yet can achieve good classification performance. Some researchers have proposed improvements to the original random forests. These include oblique random forest which uses oblique splits to separate the classes [15,30]. In [30], oblique random forest outperforms the original random forest.

In this work, we explore the effect of extracting features from Residual Networks and use them to train Oblique Random Forests. Here, we use Residual Networks (ResNet) as a feature extractor, which will supply features to Oblique Random Forests for classification. We proposed 2 different approaches to extract features from ResNet: First method extracts features only on penultimate layer, and another extracts features from all blocks.

The organisation of the paper is as follows: Sect. 2 gives a brief overview on 2 Convolutional Neural Networks, followed by Random Forests. Section 3 focuses on our proposed method. Section 4 details the experiments setup and results. Finally, Sect. 5 concludes the paper.

2 Convolutional Neural Networks and Random Forests

In this section, we give a brief review on Fully Convolutional Networks, Residual Networks, Random Forests and Oblique Random Forests.

2.1 Fully Convolutional Networks

Fully Convolutional Networks (FCN) has proposed as an effective solution to semantic segmentation on images [19]. In the original problem, the classifiers perform classification on every pixel in the image, which can be used to separate objects from the background.

For time series classification, the time series can be treated as an image with a width of 1. FCN is used as a feature extraction technique in Time Series Classification. In [28], FCN is constructed with 3 convolutional blocks. Each convolutional block comprises of a convolutional layer, batch normalisation layer [13] and ReLU activation layer [21]. The following equations outlines the operation of each convolutional block:

$$\mathbf{y} = \mathbf{W} * \mathbf{x} + \mathbf{b} \tag{1}$$

$$\mathbf{s} = BN(\mathbf{y}) \tag{2}$$

$$\mathbf{h} = ReLU(\mathbf{s}) \tag{3}$$

where $\mathbf{W}$ represents the convolution kernel weights and $\mathbf{b}$ represents the bias. ReLU function returns 0 if s is negative and s otherwise.

This network structure differs from the original FCN structure where it does not contain any pooling layers. This means that all output series from every convolutional block will have the same length as the original time series. The network configurations of the FCN can be found in [28]. At the end of the network, a global average pooling layer [17] and a softmax layer is attached to the back of the network.

2.2 Residual Networks

Residual Networks (ResNet) can be used to extend the original Fully Connected Networks by having shortcut connections across the convolutional blocks. These connects allows gradients to flow through to layers at the back of the network, thus minimizing the vanishing gradient problem [11].

For time series classification, the authors expanded FCN into a deep network. The residual network comprises of 3 residual blocks, each residual block contains 3 convolutional blocks as shown in Eqs. 1–3. The residual connection is added before the activation function near the end of each residual block. The following equations describes the residual block:

$$h_1 = Block_{k_1}(x) \tag{4}$$
$$h_2 = Block_{k_2}(h_1) \tag{5}$$
$$h_3 = Block_{k_3}(h_2) \tag{6}$$
$$y = h_3 + x \tag{7}$$
$$\hat{h} = ReLU(y) \tag{8}$$

where $Block_{k_i}$ refers to the ith convolutional block. ReLU function returns 0 if s is negative and s otherwise.

Like Fully Convolutional Layer, an average pooling layer [17] and softmax layer is attached to the end of the network.

2.3 Random Forests

Random Forests [4] is an ensemble of decision trees which uses concepts of both Bootstrap Aggregating (Bagging) and random subspace to diversity its trees within the ensemble [26]. In Bagging, multiple datasets are generated by randomly picking samples from the original dataset with replacement, which is used to train a decision tree is trained on every generated dataset.

At each node, only a random subsample of features is considered to be selected as the best feature for splitting. Once the best feature is selected from the subsample, a best split is identified and made along that feature axis. The number of features in a subsample can be set using m_{try} parameter.

The base classifier used in the ensemble is Classification And Regression Tree (CART) [4]. To select the best split, we use Gini-impurity as the evaluation criterion, which we will minimise. Gini-impuirity is defined in Eq. 9.

$$Gini = \frac{n_t^l}{n_t}\left[1 - \sum_i^K \left(\frac{n_{w_i}^l}{n_t^l}\right)^2\right] + \frac{n_t^r}{n_t}\left[1 - \sum_i^K \left(\frac{n_{w_i}^r}{n_t^r}\right)^2\right] \tag{9}$$

where n_t, n_t^l, and n_t^r refers to the number of samples that reaches the parent node, the left child node, and the right child node, respectively, K refers to the number of classes, $n_{w_i}^l$ and $n_{w_i}^l$ refers to the number of samples that reaches the left child node and the right child node respectively, that contains class w_i.

2.4 Oblique Random Forests

Oblique Random Forest differs from the original Random Forests that it uses oblique splits instead of axis-parallel splits. This gives decision trees flexibility to produce splits which effectively separate the classes.

One method used to generate hyperplanes which separate the classes is Multisurface Proximal Support Vector Machines (MPSVM). However, the original Support Vector Machine is developed for binary classification only. In order for MPSVM to work on multiclass problems, researchers have to break down the classification problem into smaller ones using methods such as "one-vs-all".

In [30], the authors divide the classes into 2 groups based on Bhattacharyya distance.

In another work, authors propose improvements to Oblique Random Forest [15]. This variant of Oblique Random Forests apply a popular "one-vs-all" concept by creating K partitions. Then, a 'best split' is selected on each of the partitions. Finally, the splits are evaluated based on Gini-impurity and the best split (among all of the 'best splits') is selected.

3 Proposed Solution

In this work, we extract features from ResNet and use the extracted features to train Oblique Random Forests. There are 2 approaches which we can extract features from ResNet. For the first method, we extract the features only from the penultimate layer of ResNet. We first train an original ResNet with the same configurations as [28]. Once the ResNet model is trained, we remove the softmax layer at the end of the trained network. We then pass the inputs through the network, and collect the features at the end of network (or the global poling layer). Here, we extract features from the output of the global average pooling layer as it significantly reduces the number of features in the extracted feature set, which in turn significantly reduces training time complexity of random forests. We should get a feature set containing 128 features.

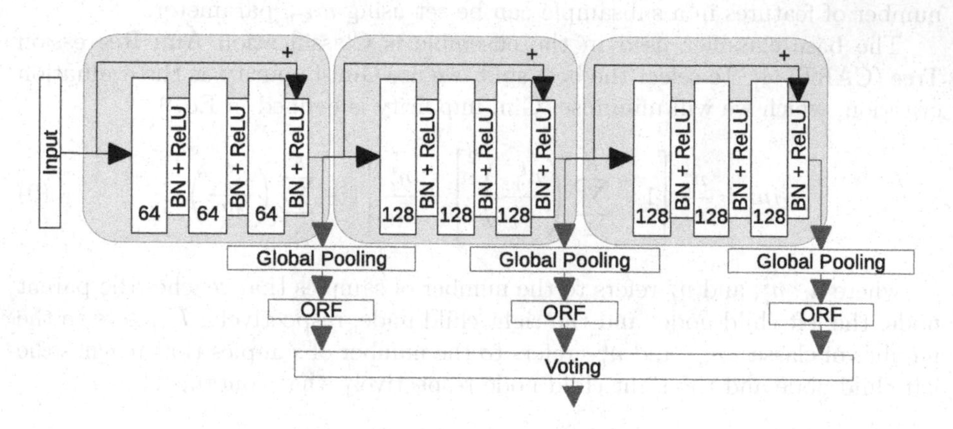

Fig. 1. Feature extraction from ResNet

For the second method, we extract the features at the ends of every residual block in ResNet. In this way, we also extract previously acquired features which can also be useful for classification. Here, the trained ResNet model is identical to the ResNet model in the first method. We attach a global average pooling layer to the outputs of every residual block, and features obtained from 3 global average pooling layers are used as 3 separate feature sets. Global average pooling will reduce the number of features to 64, 128, and 128, which helps to save computational costs. Finally, a classifier is trained on the each feature set, giving us with 3 classifiers. Doing so can encourage more diversification of the trees in ORF. Figure 1 illustrates the process.

For the classification phase, we first perform standardization on the extracted features such that the mean of each feature in the feature set has mean of 0 and standard deviation of 1. Once the feature set is standardized, we train an Oblique Random Forest since Oblique Random Forests performs better than Random Forest in [30]. For both of the feature extraction techniques, we use the improved version of Oblique Random Forests in [15]. For ResNet, we follow the same network configurations as the model proposed in [28].

4 Experiments

4.1 Datasets

For the evaluation of our proposed idea, We use 85 UCR datasets. The length of each sample is between 60 and 2709. The number of samples is between 16 and 8926. The number of classes ranges from 2 (binary) to 60. The datasets comes from a wide range of problems such as electric device, sensor readings, motion captures, spectrographs, ECGs, image outlines and simulated [1]. More details on UCR datasets can be obtained in [5].

4.2 Experiment Setup

We follow the experiment setup in [20]. For each dataset, we use the same training/testing split given by [5]. The classifiers are trained only on the training set and evaluated only on the testing set. For each dataset, We ranked the classifiers based on classification accuracy. The best classifier is ranked 1, the second best classifier is ranked 2, and so on. If there are ties, we average their ranks.

We follow the same network structure used in [28]. ResNet model comprises of 3 residual blocks. Every convolutional layer in all residual blocks has 64, 128, and 128 neurons respectively. In every residual block, the kernel length in layer 1, 2, and 3 is set to 8, 5, and 3 respectively. The models are trained with Adam optimizer [16], with 1500 epochs.

For Oblique Random Forest (ORF), we set the total number of trees in the ensemble to be 500 for each of our methods. For our second method, we set the number of trees in each ORF be 166, bringing the total to $166 \times 3 = 499$ trees. All trees are trained to full size without pruning. We do not tune the parameters since no significant improvements in performance is observed when we attempt to tune the m_{try} parameter. Hence, the m_{try} parameters is set based on the following equation:

$$mtry = floor(\sqrt{M}) \tag{10}$$

where M refers to the number of features.

4.3 Results

To evaluate the performance of 2 different feature extraction methods, we repeat the same experiment on the original dataset and extracted features from 2 different methods. Oblique Random Forest trained on the original dataset is labelled 'ORF', Oblique Random Forest trained on feature extracted using the first method is labelled 'ORF-F1', and Oblique Random Forest trained on feature extracted using the second method is labelled 'ORF-F2'. We experimented with Oblique Random Forests (ORF) used in the works in [15].

In addition, to give a brief idea on how our method compares with other algorithms, we also compare with Euclidean Distance (ED), Dynamic Time Warping (DWT), Derivative DTW (DD_{DTW}), Derivative Transform Distance (DTD_C), Learned Shapelets (LS), Bag of SFA Symbols (BOSS), Time Series Forests (TSF), Time Series Bag of Features (TSBF), Learned Pattern Similarity (LPS), Shapelet Transform (ST), Elastic Ensemble (EE), Collection of Transformation Ensembles (COTE), Multilayer Perceptions (MLP), Fully Connected Networks (FCN), Residual Networks (ResNet) and Echo Memory Networks (EMN). The results are detailed in Table 1, and summarized in Table 2.

4.4 Performance of Our Methods

Here, we compare the performance of our methods. We include ResNet as a baseline classifier, which has average accuracy of 83.85% and average rank of 6.84.

Based on the results obtained, it is clear that Oblique Random Forests requires useful features to be extracted as they performed poorly on the original time series dataset. Residual Networks can serve as one of the useful feature extractor for Oblique Random Forest. Average classification accuracy and rank of ORF-F1 improves by 7.94% and 5.56 respectively.

However, ORF-F1 did not outperform Residual Networks. Average accuracy and rank of ORF-F1 are 0.98% and 0.12 (respectively) worse than ResNet. In addition, we observed that ORF-F1 wins in 34 datasets but loses in 38 datasets (13 are ties). One reason could be that the useful features extracted from ResNet might not be sufficient for Oblique Random Forests to make a reliable prediction.

ORF-F2 scored 0.71% higher accuracy and 0.81 better rank than ORF-F1. ORF-F2 wins ORF-F1 on 43 datasets while loses on 24 other datasets. When compared with ResNet, mean accuracy and rank of ORF-F2 are 0.13% and 0.69 (respectively) better than ResNet. ORF-F2 wins in 45 datasets but loses in 27 datasets.

When compared with other methods, ORF-F1 is ranked sixth, and ORF-F2 is ranked forth. The top 3 classifiers are given in this order: EMN, COTE, FCN. For comparison, Residual Network is ranked fifth.

Our methods performed well among traditional Time Series Classification methods. Both of our methods performs better all other traditional time series methods except COTE. However, when comparing with other deep learning

Table 1. Test accuracy (in %) and average ranks

Dataset	ED	DWT	DD_{DWT}	DTD_C	LS	BOSS	TSF	TSBF	LPS	ST	EE	COTE	MLP	FCN	ResNet	EMN	ORF	ORF-F1	ORF-F2
Adiac	61.1	60.4	70.1	70.1	52.2	76.5	73.1	77.0	77.0	78.3	66.5	79.0	75.2	**85.7**	82.6	82.9	74.7	82.6	82.6
ArrowHead	80.0	70.3	78.9	72.0	84.6	83.4	72.6	75.4	78.3	73.7	81.1	81.1	82.3	**88.0**	81.7	81.7	74.9	84.0	86.9
Beef	66.7	63.3	66.7	66.7	86.7	80.0	76.7	56.7	60.0	**90.0**	63.3	86.7	83.3	75.0	76.7	83.3	80.0	73.3	76.7
BeetleFly	75.0	70.0	65.0	65.0	80.0	90.0	75.0	80.0	80.0	90.0	75.0	80.0	85.0	**95.0**	80.0	80.0	75.0	85.0	85.0
BirdChicken	55.0	75.0	85.0	80.0	80.0	95.0	80.0	90.0	**100.0**	80.0	80.0	90.0	80.0	95.0	90.0	90.0	60.0	90.0	95.0
Car	73.3	73.3	80.0	78.3	76.7	83.3	76.7	78.3	85.0	91.7	83.3	90.0	83.3	91.7	**93.3**	81.7	80.0	**93.3**	**93.3**
CBF	85.2	99.7	99.7	98.0	99.1	99.8	99.4	98.8	99.9	97.4	99.8	99.6	86.0	**100.0**	99.4	**100.0**	90.2	99.7	99.8
Chlorine	65.0	64.8	70.8	71.3	59.2	66.1	72.0	69.2	60.8	70.0	65.6	72.7	**87.2**	84.3	82.8	84.5	80.3	84.6	85.2
CinCECGtorso	89.7	65.1	72.5	85.2	87.0	88.7	98.3	71.2	73.6	95.4	94.2	**99.5**	84.2	81.3	77.1	84.8	77.3	81.7	84.5
Coffee	**100.0**	**100.0**	**100.0**	**100.0**	**100.0**	**100.0**	96.4	**100.0**	**100.0**	96.4	**100.0**	**100.0**	**100.0**	**100.0**	**100.0**	**100.0**	**100.0**	**100.0**	**100.0**
Computers	57.6	70.0	71.6	71.6	58.4	75.6	72.0	75.6	68.0	73.6	70.8	74.0	54.0	**84.8**	82.4	71.6	59.2	80.4	84.0
CricketX	57.7	75.4	75.4	75.4	74.1	73.6	66.4	70.5	69.7	77.2	81.3	80.8	56.9	81.5	**82.1**	78.2	56.7	78.7	80.0
CricketY	56.7	74.4	77.7	77.4	71.8	75.4	67.2	73.6	76.7	77.9	80.5	**82.6**	59.5	79.2	80.5	78.7	60.0	**82.6**	81.3
CricketZ	58.7	75.4	77.4	77.4	74.1	74.6	67.2	71.5	75.4	78.7	78.2	**81.5**	59.2	81.3	81.3	80.8	55.4	80.3	79.2
Diatom	93.5	96.7	96.7	91.5	**98.0**	93.1	93.1	89.9	90.5	92.5	94.4	92.8	96.4	93.0	93.1	97.4	95.1	96.7	91.5
DistPhxAgeGp	62.6	77.0	70.5	66.2	71.9	74.8	74.8	71.2	66.9	77.0	69.1	74.8	82.7	83.5	79.8	84.3	**84.5**	77.7	81.0
DistPhxCorr	71.7	71.7	73.2	72.5	77.9	72.8	77.2	78.3	72.1	77.5	72.8	76.1	81.0	81.2	82.0	**82.2**	80.5	80.3	81.3
DistPhxTW	63.3	59.0	61.2	57.6	62.6	67.6	66.9	67.6	56.8	66.2	64.7	69.8	74.7	79.0	74.0	**79.5**	78.7	75.2	75.0
Earthquakes	71.2	71.9	70.5	70.5	74.1	74.8	74.8	74.8	64.0	74.1	74.1	74.8	79.2	80.1	78.6	81.1	**82.0**	72.7	73.3
ECG200	88.0	77.0	83.0	84.0	88.0	87.0	87.0	84.0	86.0	83.0	88.0	88.0	**92.0**	90.0	87.0	**92.0**	87.0	86.0	88.0
ECG5000	92.5	92.4	92.4	92.4	93.2	94.1	93.9	94.0	91.7	94.4	93.9	**94.6**	93.5	94.1	93.1	94.4	94.0	93.2	93.9
ECGFiveDays	79.7	76.8	76.9	82.2	**100.0**	**100.0**	95.6	87.7	87.9	98.4	82.0	99.9	97.0	98.5	95.5	**100.0**	92.7	**100.0**	99.4
ElectricDevices	55.2	60.2	59.2	59.4	58.7	**79.9**	69.3	70.3	68.1	74.7	66.3	71.3	58.0	72.3	72.8	71.6	64.4	73.9	74.1
FaceAll	71.4	80.8	90.2	89.9	74.9	78.2	75.1	74.4	76.7	77.9	84.9	91.8	88.5	**92.9**	83.4	90.3	73.2	81.5	81.6
FaceFour	78.4	83.0	83.0	81.8	96.6	**100.0**	93.2	**100.0**	94.3	85.2	90.9	89.8	83.0	93.2	93.2	95.5	86.4	95.5	94.3
FacesUCR	76.9	90.5	90.4	90.8	93.9	95.7	88.3	86.7	92.6	90.6	94.5	94.2	81.5	94.8	**95.8**	94.7	76.0	95.0	93.7
FiftyWords	63.1	69.0	75.4	75.4	73.0	70.5	74.1	75.8	81.8	70.5	**82.0**	79.8	71.2	67.9	72.7	75.8	69.0	71.9	69.5
Fish	78.3	82.3	94.3	92.6	96.0	**98.9**	79.4	83.4	94.3	**98.9**	96.6	98.3	87.4	97.1	**98.9**	94.4	83.4	98.3	97.7
FordA	66.5	55.5	72.3	76.5	95.7	93.0	81.5	85.0	87.3	**97.1**	73.8	95.7	76.9	90.6	92.8	93.2	77.2	90.5	93.3
FordB	60.6	62.0	66.7	65.3	91.7	71.1	68.8	59.9	71.1	80.7	66.2	80.4	62.9	88.3	90.0	90.8	67.1	91.7	**92.7**
GunPoint	91.3	90.7	98.0	98.7	**100.0**	**100.0**	97.3	98.7	99.3	**100.0**	99.3	**100.0**	93.3	**100.0**	99.3	99.3	89.3	98.0	98.0
Ham	60.0	46.7	47.6	55.2	66.7	66.7	74.3	76.2	56.2	68.6	57.1	64.8	71.4	76.2	**78.1**	**78.1**	68.6	73.3	68.6
HandOutlines	86.2	88.1	86.8	86.5	48.1	90.3	91.9	85.4	88.1	**93.2**	88.9	91.9	80.7	77.6	86.1	89.1	87.6	82.1	84.5
Haptics	37.0	37.7	39.9	39.9	46.8	46.1	44.5	49.0	43.2	52.3	39.3	52.3	46.1	**55.1**	50.6	51.9	45.1	51.6	52.3
Herring	51.6	53.1	54.7	54.7	62.5	54.7	60.9	64.1	57.8	67.2	57.8	62.5	68.7	**70.3**	59.4	62.5	57.8	53.1	53.1
InlineSkate	34.2	38.4	**56.2**	50.9	43.8	51.6	37.6	38.5	50.0	37.3	46.0	49.5	35.1	41.1	36.5	46.0	31.3	37.1	39.8
InsWngSound	56.2	35.5	35.5	47.3	60.6	52.3	63.3	62.5	55.1	62.7	59.5	**65.3**	63.1	40.2	53.1	64.1	**65.3**	49.2	52.2
ItalyPower	95.5	95.0	95.0	95.1	96.0	90.9	96.0	88.3	92.3	94.8	96.2	96.1	96.6	97.0	96.0	**97.1**	96.7	96.6	96.6
LrgKitApp	49.3	79.5	79.5	79.5	70.1	76.5	57.1	52.8	71.7	85.9	81.1	84.5	48.0	89.6	89.3	90.1	56.3	90.9	**92.0**
Lightning2	75.4	86.9	86.9	86.9	82.0	83.6	80.3	73.8	82.0	73.8	**88.5**	86.9	72.1	80.3	75.4	83.6	73.8	78.7	78.7
Lightning7	57.5	72.6	67.1	65.8	79.5	68.5	75.3	72.6	74.0	72.6	76.7	80.8	64.4	**86.3**	83.6	83.6	68.5	75.3	76.7
Mallat	91.4	93.4	94.9	92.7	95.0	93.8	91.9	96.0	90.8	96.4	94.0	95.4	93.6	**98.0**	97.9	96.2	83.7	96.6	96.5
Meat	93.3	93.3	93.3	93.3	73.3	90.0	93.3	93.3	88.3	85.0	93.3	91.7	93.3	96.7	**100.0**	93.3	93.3	95.0	95.0
MedicalImages	68.4	73.7	73.7	74.5	66.4	71.8	75.5	70.5	74.6	67.0	74.2	75.8	72.9	**79.2**	77.2	77.5	71.7	72.8	78.2
MidPhxAgeGp	51.9	50.0	53.9	50.0	57.1	54.5	57.8	57.8	48.7	64.3	55.8	63.6	73.5	76.8	76.0	**80.0**	78.7	72.0	74.0
MidPhxCorr	76.6	69.8	73.2	74.2	78.0	78.0	**82.8**	81.4	77.3	79.4	78.4	80.4	76.0	79.5	79.3	81.5	67.2	81.3	82.5
MidPhxTW	51.3	50.6	48.7	50.0	50.6	54.5	56.5	59.7	52.6	51.9	51.3	57.1	60.9	61.2	60.7	**63.9**	63.2	58.6	58.6
MoteStrain	87.9	83.5	83.3	76.8	88.3	87.9	86.9	90.3	92.2	89.7	88.3	93.7	86.9	**95.0**	89.5	89.5	88.3	92.9	92.4
NonInv_Thor1	82.9	79.0	80.6	84.1	25.9	83.8	87.6	84.2	81.2	95.0	84.6	93.1	94.2	**96.1**	94.8	93.3	89.8	95.1	95.4
NonInv_Thor2	88.0	86.5	89.3	89.0	77.0	90.1	91.0	86.2	84.1	95.1	91.3	94.6	94.3	**95.5**	95.1	93.9	93.0	95.1	95.1
OliveOil	86.7	83.3	83.3	86.7	16.7	86.7	86.7	83.3	86.7	**90.0**	86.7	**90.0**	40.0	83.3	86.7	86.7	86.7	86.7	86.7
OSULeaf	52.1	59.1	88.0	88.4	77.7	95.5	58.3	76.0	74.0	96.7	80.6	96.7	57.0	98.8	97.9	89.7	52.9	**99.6**	99.2
PhalCorr	76.1	72.8	73.9	76.1	76.5	77.2	80.3	83.0	75.6	76.3	77.3	77.0	83.0	82.6	82.5	**83.2**	82.8	81.8	82.5
Phoneme	10.9	22.8	26.9	26.8	21.8	26.5	21.2	27.6	23.7	32.1	30.5	34.9	9.8	34.5	32.4	23.9	11.6	34.8	**35.4**
Plane	96.2	**100.0**	**100.0**	**100.0**	**100.0**	**100.0**	**100.0**	**100.0**	**100.0**	**100.0**	**100.0**	**100.0**	98.1	**100.0**	**100.0**	**100.0**	98.1	**100.0**	**100.0**
ProxPhxAgeGp	78.5	80.5	80.0	79.5	83.4	83.4	84.9	84.9	79.5	84.4	80.5	**85.4**	82.4	84.9	84.9	**85.4**	84.9	82.4	83.9
ProxPhxCorr	80.8	78.4	79.4	79.4	84.9	84.9	82.8	87.3	84.2	88.3	80.8	86.9	88.7	90.0	91.8	89.0	87.3	**92.1**	90.7
ProxPhxTW	70.7	76.1	76.6	77.1	77.6	80.0	81.5	81.0	73.2	80.5	76.6	78.0	79.7	81.0	80.7	**83.0**	80.2	80.5	82.3
RefDev	39.5	46.4	44.5	44.5	51.5	49.9	**58.9**	47.2	45.9	58.1	43.7	54.7	37.1	53.3	52.8	56.0	36.0	51.5	50.4
ScreenType	36.0	39.7	42.9	43.7	42.9	46.4	45.6	50.9	41.6	52.0	44.5	54.7	40.8	66.7	**70.7**	55.5	41.6	60.0	60.3
ShapeletSim	53.9	65.0	61.1	60.0	95.0	**100.0**	47.8	96.1	86.7	95.6	81.7	96.1	48.3	86.7	**100.0**	99.4	50.6	83.9	**100.0**
ShapesAll	75.2	76.8	85.0	83.8	76.8	90.8	79.2	18.5	87.3	84.2	86.7	89.2	77.5	89.8	91.2	87.3	74.8	**91.5**	89.7
SmlKitApp	34.4	64.3	64.0	64.8	66.4	72.5	81.1	67.2	71.2	79.2	69.6	77.6	38.9	80.3	79.7	69.9	73.1	79.7	**82.4**
SonyAIBOSurf1	69.6	72.5	74.2	71.0	81.0	63.2	78.7	79.5	77.4	84.4	70.4	84.5	72.7	96.8	**98.5**	93.0	80.7	94.3	94.2
SonyAIBOSurf2	85.9	83.1	89.2	89.2	87.5	85.9	81.0	77.8	87.2	93.4	87.8	95.2	83.9	96.2	96.2	92.9	80.6	98.3	**98.4**
StarlightCurves	84.9	90.7	96.2	96.2	94.7	97.8	96.9	97.7	96.3	97.9	92.6	**98.0**	95.7	96.7	97.5	97.8	97.2	96.4	97.8
Strawberry	94.6	94.1	95.4	95.7	91.1	97.6	96.5	95.4	96.2	96.2	94.6	95.1	96.7	96.9	95.8	97.1	**97.7**	97.1	97.1
SwedishLeaf	78.9	79.2	90.1	89.6	90.7	92.2	91.4	91.5	92.0	92.8	91.5	95.5	89.3	96.6	95.8	94.1	90.6	**97.4**	97.1
Symbols	89.9	95.0	95.3	96.3	93.2	**96.7**	91.5	94.6	96.3	88.2	96.0	96.4	85.3	96.2	87.2	95.5	87.8	92.6	91.2
Synth_Cntr	88.0	99.3	99.3	99.7	99.7	96.7	98.7	99.3	98.0	98.3	99.0	**100.0**	95.0	99.0	**100.0**	99.7	97.7	**100.0**	**100.0**
ToeSegmentation1	68.0	77.2	80.7	80.7	93.4	93.9	74.1	78.1	87.7	96.5	82.9	**97.4**	60.1	96.9	96.5	96.5	60.5	96.5	**97.4**
ToeSegmentation2	80.8	83.8	74.6	71.5	91.5	**96.2**	81.5	80.0	86.9	90.8	89.2	91.5	74.6	91.5	86.2	93.1	79.2	91.5	90.8
Trace	76.0	**100.0**	**100.0**	99.0	**100.0**	**100.0**	99.0	98.0	98.0	**100.0**	99.0	**100.0**	82.0	**100.0**	**100.0**	**100.0**	78.0	**100.0**	**100.0**
TeoLeadECG	74.7	90.5	97.8	98.5	99.6	98.1	75.9	86.6	94.8	99.7	97.1	99.3	85.3	**100.0**	**100.0**	99.9	87.3	**100.0**	**100.0**
TwoPatterns	90.7	**100.0**	**100.0**	**100.0**	99.3	99.3	99.1	97.6	98.2	95.5	**100.0**	**100.0**	88.6	89.7	**100.0**	99.9	84.0	**100.0**	99.7
UWavGestAll	94.8	89.2	93.5	93.8	95.3	93.9	95.7	92.6	96.6	94.2	**96.8**	96.4	95.4	82.6	86.8	95.8	94.4	85.0	86.9
UWavGest_X	73.9	72.8	77.9	77.5	79.1	76.2	80.4	**83.1**	82.9	80.3	80.5	82.2	76.8	75.4	78.7	81.3	75.4	77.2	79.1
UWavGest_Y	66.2	63.4	71.6	69.8	70.3	68.5	72.7	73.6	**76.1**	73.0	72.6	75.9	70.3	72.5	66.8	73.6	69.2	67.5	70.6
UWavGest_Z	65.0	65.8	69.6	67.9	74.7	69.5	74.3	**77.2**	76.8	74.8	72.4	75.0	70.5	72.9	75.5	75.5	71.3	75.1	77.1
Wafer	99.5	98.0	98.0	99.3	99.6	99.5	99.6	99.5	99.7	**100.0**	99.7	**100.0**	99.6	99.7	99.7	99.8	99.5	99.9	99.9
Wine	61.1	57.4	57.4	61.1	50.0	74.1	63.0	61.1	63.0	79.6	57.4	64.8	79.6	**88.9**	79.6	81.5	83.3	75.9	81.5
WordSynonyms	61.8	64.9	73.0	73.0	60.7	63.8	64.7	68.8	75.5	57.1	**77.9**	75.7	59.4	58.0	63.2	66.3	57.1	61.9	59.6
Worms	45.5	58.4	58.4	64.9	61.0	55.8	61.0	68.8	70.1	**74.0**	66.2	62.3	34.3	66.9	61.9	58.0	45.9	63.5	61.9
WormsTwoClass	61.0	62.3	64.9	62.3	72.7	**83.1**	62.3	75.3	75.3	**83.1**	68.8	80.5	59.7	72.9	73.5	75.1	61.3	76.2	74.0
Yoga	83.0	83.7	85.6	85.6	83.4	**91.8**	85.9	81.9	86.9	81.8	87.9	87.7	85.5	84.5	85.8	86.6	80.0	85.7	86.6

*Results of ED, DWT, DD_{DWT}, DTD_C, LS, BOSS, TSF, TSBF, LPS, ST, EE, COTE, MLP, FCN, ResNet & EMN are taken from [20].

* Bold values represent the best accuracy/rank.

Table 2. Summary of results

	ED	DWT	DD_{DWT}	DTD_C	LS	BOSS	TSF	TSBF	LPS	ST	EE	COTE	MLP	FCN	ResNet	EMN	ORF	ORF-F1	ORF-F2
Mean Accuracy	70.89	74.04	76.86	76.89	76.91	81.02	77.88	77.80	78.68	82.24	79.23	83.81	75.16	**84.40**	83.85	84.38	75.33	83.27	83.98
Average Rank	16.06	14.64	12.71	12.83	11.11	9.16	10.89	10.91	11.18	8.44	10.11	5.91	12.39	5.93	6.84	**5.25**	12.52	6.96	6.15

methods (MLP, FCN, ResNet and EMN) where competition is more intense, ORF-F1 only wins MLP and ORF-F2 wins both MLP and ResNet.

These results show that the performance of ORF on extract ResNet features leaves much to be desired. There is no free lunch. With 85 dataset having a wide range of characteristics, our best methods performed badly on some datasets, such as Diatom, FiftyWords, HandOutlines, Herring, UWavGestAll, and WordSynonyms.

Table 3. Mean Number of Nodes in a Tree and Mean Training Time across 85 datasets. Mean training times are relative to ResNet. Mean training times is for ORF-F1 and ORF-F2 does not include feature extraction phase.

	ResNet	ORF	ORF-F1	ORF-F2
No. of nodes		144.66	26.17	67.71
Training time	1	0.0692	0.0189	0.0340

In Table 3, we estimate the computational effort to train ORF for all our methods. All our ORF models are trained using the same CPU. Mean training times and number of nodes in each tree are recorded. As a comparison, we also compare our ORF models with ResNet to get a brief idea. ResNet models are trained using a GPU since ResNet trains faster using a GPU.

Training ORF-F2 takes nearly twice as long as ORF-F2. This is evident as trained trees in ORF-F2 have over twice the number of nodes than ORF-F1. This means full-sized trees in ORF-F2 are over twice as complex as those in ORF-F1, with twice as much training parameters. None of our ORF models takes a significant amount of effort when compared to ResNet.

ORF trained without features extraction takes significantly more time than ORF-F1 and ORF-F2. This is due to the large input sizes seen by ORF, leading to even more complex structure of trees in ORF. 63 datasets have input length of more than 128, which is the maximum number of features seen in ORF-F1 and ORF-F2. However, training ORF-F1 and ORF-F2 requires ResNet to be trained, which outweighs their advantages. The total computational effort for ORF-F1 and ORF-F2 will be slightly more than ResNet.

In conclusion, ORF without feature extraction requires significantly less computational effort than ORF-F1 and ORF-F2 but performs significantly worse than both ORF-F1 and ORF-F2.

5 Conclusion

In this paper, we explore using Residual Networks as a feature extractor for Oblique Random Forests. We tried 2 different methods to extract features from Residual networks (ResNet). In our first method, we extract features only from the penultimate layer in ResNet. In our second method, we include intermediate ResNet features by extracting features from every block.

Based on the experiment on 85 UCR datasets, ResNet features help Oblique Random Forests to improve classification performance significantly. In addition, using including intermediate features from Residual Networks significantly improves the performance of Oblique Random Forests.

Some of the possible improvements can include combining more features from other feature extraction techniques or effectively tuning the Oblique Random Forests. More Recurrent Neural Networks can be added into the comparisons. In addition, improvements can be made using other classifiers, such as Oblique Decision Tree Ensemble via Twin Bounded SVM and Ensemble Deep Random Vector Functional Link.

References

1. Bagnall, A., Lines, J., Bostrom, A., Large, J., Keogh, E.: The great time series classification bake off: a review and experimental evaluation of recent algorithmic advances. Data Min. Knowl. Disc. **31**(3), 606–660 (2016). https://doi.org/10.1007/s10618-016-0483-9
2. Baydogan, M.G., Runger, G., Tuv, E.: A bag-of-features framework to classify time series. IEEE Trans. Pattern Anal. Mach. Intell. **35**(11), 2796–2802 (2013)
3. Bostrom, A., Bagnall, A.: Binary shapelet transform for multiclass time series classification. In: Hameurlain, A., Küng, J., Wagner, R., Madria, S., Hara, T. (eds.) Transactions on Large-Scale Data- and Knowledge-Centered Systems XXXII. LNCS, vol. 10420, pp. 24–46. Springer, Heidelberg (2017). https://doi.org/10.1007/978-3-662-55608-5_2
4. Breiman, L.: Random forests. Mach. Learn. **45**(1), 5–32 (2001)
5. Chen, Y., et al.: The UCR time series classification archive, July 2015. www.cs.ucr.edu/~eamonn/time_series_data/
6. Deng, H., Runger, G., Tuv, E., Vladimir, M.: A time series forest for classification and feature extraction. Inf. Sci. **239**, 142–153 (2013). https://doi.org/10.1016/j.ins.2013.02.030
7. Gamboa, J.C.B.: Deep learning for time-series analysis. arXiv preprint arXiv:1701.01887 (2017)
8. Górecki, T., Łuczak, M.: Using derivatives in time series classification. Data Min. Knowl. Disc. **26**(2), 310–331 (2013)
9. Grabocka, J., Schilling, N., Wistuba, M., Schmidt-Thieme, L.: Learning time-series shapelets. In: Proceedings of the 20th ACM SIGKDD International Conference on Knowledge Discovery and Data Mining, KDD 2014, pp. 392–401. Association for Computing Machinery, New York (2014). https://doi.org/10.1145/2623330.2623613

10. Gãrecki, T., Åuczak, M.: Non-isometric transforms in time series classification using DTW. Knowl. Based Syst. **61**, 98–108 (2014). https://doi.org/10.1016/j.knosys.2014.02.011
11. He, K., Zhang, X., Ren, S., Sun, J.: Deep residual learning for image recognition. In: The IEEE Conference on Computer Vision and Pattern Recognition (CVPR), pp. 770–778, June 2016
12. Hills, J., Lines, J., Baranauskas, E., Mapp, J., Bagnall, A.: Classification of time series by shapelet transformation. Data Min. Knowl. Disc. **28**(4), 851–881 (2013). https://doi.org/10.1007/s10618-013-0322-1
13. Ioffe, S., Szegedy, C.: Batch normalization: Accelerating deep network training by reducing internal covariate shift. arXiv preprint arXiv:1502.03167 (2015)
14. Karim, F., Majumdar, S., Darabi, H., Chen, S.: Lstm fully convolutional networks for time series classification. IEEE Access **6**, 1662–1669 (2018)
15. Katuwal, R., Suganthan, P.N.: Enhancing multi-class classification of random forest using random vector functional neural network and oblique decision surfaces. In: 2018 International Joint Conference on Neural Networks (IJCNN), pp. 1–8 (2018)
16. Kingma, D.P., Ba, J.: Adam: a method for stochastic optimization. arXiv preprint arXiv:1412.6980 (2014)
17. Lin, M., Chen, Q., Yan, S.: Network in network. arXiv preprint arXiv:1312.4400 (2013)
18. Lines, J., Bagnall, A.: Time series classification with ensembles of elastic distance measures. Data Min. Knowl. Disc. **29**(3), 565–592 (2014). https://doi.org/10.1007/s10618-014-0361-2
19. Long, J., Shelhamer, E., Darrell, T.: Fully convolutional networks for semantic segmentation. In: The IEEE Conference on Computer Vision and Pattern Recognition (CVPR), pp. 3431–3440, June 2015
20. Ma, Q., Zhuang, W., Shen, L., Cottrell, G.W.: Time series classification with echo memory networks. Neural Networks **117**, 225–239 (2019). https://doi.org/10.1016/j.neunet.2019.05.008
21. Nair, V., Hinton, G.E.: Rectified linear units improve restricted Boltzmann machines. In: Proceedings of the 27th International Conference on Machine Learning (ICML-10), pp. 807–814 (2010)
22. Nweke, H.F., Teh, Y.W., Al-garadi, M.A., Alo, U.R.: Deep learning algorithms for human activity recognition using mobile and wearable sensor networks: state of the art and research challenges. Expert Syst. Appl. **105**, 233–261 (2018). https://doi.org/10.1016/j.eswa.2018.03.056
23. Rajkomar, A., et al.: Scalable and accurate deep learning with electronic health records. NPJ. Digital Med. **1**(1), 18 (2018)
24. Schäfer, P.: The boss is concerned with time series classification in the presence of noise. Data Min. Knowl. Disc. **29**(6), 1505–1530 (2015)
25. Susto, G.A., Cenedese, A., Terzi, M.: Chapter 9 - time-series classification methods: Review and applications to power systems data. In: Arghandeh, R., Zhou, Y. (eds.) Big Data Application in Power Systems, pp. 179–220. Elsevier (2018). https://doi.org/10.1016/B978-0-12-811968-6.00009-7
26. Ho, T.K.: The random subspace method for constructing decision forests. IEEE Trans. Pattern Anal. Mach. Intell. **20**(8), 832–844 (1998)
27. Wang, J., Chen, Y., Hao, S., Peng, X., Hu, L.: Deep learning for sensor-based activity recognition: a survey. Pattern Recogn. Lett. **119**, 3–11 (2019). https://doi.org/10.1016/j.patrec.2018.02.010, deep Learning for Pattern Recognition

28. Wang, Z., Yan, W., Oates, T.: Time series classification from scratch with deep neural networks: a strong baseline. In: 2017 International Joint Conference on Neural Networks (IJCNN), pp. 1578–1585, May 2017. https://doi.org/10.1109/IJCNN.2017.7966039
29. YANG, Q., WU, X.: 10 challenging problems in data mining research. Int. J. Inf. Technol. Decis. Making **05**(04), 597–604 (2006). https://doi.org/10.1142/S0219622006002258
30. Zhang, L., Suganthan, P.N.: Oblique decision tree ensemble via multisurface proximal support vector machine. IEEE Trans. Cybern. **45**(10), 2165–2176 (2015)

P2ExNet: Patch-Based Prototype Explanation Network

Dominique Mercier[1,2](✉), Andreas Dengel[1,2], and Sheraz Ahmed[2]

[1] Computer Science Department, Technische Universität Kaiserslautern, 67663 Kaiserslautern, Germany
[2] Smart Data and Knowledge Services, German Research Center for Artificial Intelligence (DFKI), 67663 Kaiserslautern, Germany
{dominique.mercier,andreas.dengel,sheraz.ahmed}@dfki.de

Abstract. Deep learning methods have shown great success in several domains as they process a large amount of data efficiently, capable of solving difficult classification, forecast, segmentation, and other tasks. However, these networks suffer from their inexplicability that limits their applicability and trustworthiness. Although there exists work addressing this perspective, most of the existing approaches are limited to the image modality due to the intuitive and prominent concepts. Unfortunately, the patterns in the time-series domain are more complex and non-comprehensive, and an explanation for the network decision is pivotal in critical areas like medical, financial, or industry. Addressing the need for an explainable approach, we propose a novel interpretable network scheme, designed to inherently use an explicable reasoning process inspired by the human cognition without the need of additional post-hoc explainability methods. Therefore, the approach uses class-specific patches as they cover local patterns, relevant to the classification, to reveal similarities with samples of the same class. Besides, we introduce a novel loss concerning interpretability and accuracy that constraints P2ExNet to provide viable explanations of the data that include relevant patches, their position, class similarities, and comparison methods without compromising performance. An analysis of the results on eight publicly available time-series datasets reveals that P2ExNet reaches similar performance when compared to its counterparts while inherently providing understandable and traceable decisions.

Keywords: Deep learning · Convolutional neural networks · Time-series analysis · Data analysis · Explainability · Interpretability.

1 Introduction

Nowadays, deep neural networks are popular and used in many different domains comprising image processing, natural language processing, and time-series processing. Though these deep networks have achieved high performance, they are still black boxes in nature. This behavior makes it tough to understand the

H. Yang et al. (Eds.): ICONIP 2020, LNCS 12534, pp. 318–330, 2020.
https://doi.org/10.1007/978-3-030-63836-8_27

reasons behind the decisions. In particular, this black box nature hinders the use of these models in critical domains like medical, autonomous driving, industrial, financial, and raises the need for interpretability methods to provide intuitive and understandable explanations. Only explainable models can are usable in critical domains that require transparency [16].

The existing methods for interpreting decisions of deep learning models are mostly applicable to image modalities. In particular, image concepts are intuitive by default [25]. Besides the image domain, there is only a limited amount of work in the field of time-series as the modalities are more complex and usually not directly interpretable for a human. Nevertheless, these time-series analysis networks and their explanations are pivotal for their industrial and financial use. Therefore, we propose P2ExNet as an approach to deal with time-series data.

Also, existing approaches are mostly post-hoc methods that are applied after the classifier to explain their decisions [7]. Intuitively, these approaches keep the network as it is without any change to the structure, enabling their use on almost every architecture. Usually, this results in an instance-based local explanation that does not explain any global behavior. In contrast to post-hoc methods, the intrinsic methods focus on model design concerning the inference process to provide an understandable global explanation. Ultimately, neither of the two approaches is superior as both have to deal with several limitations regarding the quality, subjectivity [14], the audience, and the domain usage.

To overcome these limitations, we propose a network architecture for time-series analysis based on the standard deep neural network architecture providing a global explanation using representative class-specific prototypes and an instance-based local explanation using patch-based similarities and class-similarities. The inference process of our architecture follows the human-related reasoning process [11] and uses concepts and prototypes [13]. Intuitive class-specific patches explain the network decision. Our approach is superior compared to existing template matching approaches [5] in the manner of generalization and applicability. Our experiments emphasize the use of our network structure by highlighting the comparable performance when compared to a non-interpretable network of the same size over eight publicly available datasets while preserving an intuitive and traceable explanation.

2 Related Work

The field of network interpretability covers post-hoc and intrinsic methods. Based on the use-case, it is not always possible to use both methods as these methods come with restrictions concerning the data and the network. In the following paragraphs, we address the perspectives, their advantages, and drawbacks.

2.1 Post-hoc

Using post-hoc methods to explain the decisions of deep neural networks is a very prominent approach as these methods do not modify the network architecture and can provide an instance base explanation. Furthermore, these methods offer instance-based as well as global explanations resulting in broad applicability.

Instance-Based: A widespread instance-based post-hoc class of approaches in the field of image domain are so-called back-propagation methods [4]. These approaches produce heat-maps highlighting the most relevant and sensitive parts concerning the network decision. There exist enhancements that evolved [27] and take various aspects into account to improve the expressiveness and consistency. Another post-hoc instance-based class of methods are the layer-wise relevance propagation methods [3,10] that produce results that are close to the heat-maps but more stable. In particular, the image domain explored different approaches to visualize the activations [24] or make use of the gradients [18] or saliency [21] to produce heat-maps for instances. However, in the case of the time-series modalities, there exists only a limited amount of work [20].

Global: In contrast to instance-based methods, there exist attempts to compute a global behavior based on the influence of the samples [12,23]. These methods provide an idea of helpful and harmful dataset samples to detect outliers and debug dataset using the sample influence. Another approach is to attach an interpretable architecture to the trained network. As presented in [15], the attachment of an autoencoder before the neural network and a customized loss function for the autoencoder can enhance the interpretability. Siddiqui et al. [19] presented an adoption of this approach for the time-series domain with an adjusted loss function.

2.2 Intrinsic

Intrinsic methods approach the problem from a different perspective by incorporating the interpretability directly. Therefore, they modify the model architecture by introducing interpretable layers [26]. A drawback of these approaches is the restricted learning process that can harm the performance. An intuitive interpretable layer solution are prototype layers to explain model decision [2]. Mainly, two types of prototypes showed to provide reasonable explanations. First, class prototypes that cover the complete input [8,13] and second patch prototypes [6].

2.3 Limitations of Existing Methods

Even though there exists work to explain the network decisions, most of the approaches are limited to image modalities [17]. Furthermore, there is ongoing research investigating the consistency, expressiveness, and subjectivity of these explanations. Some findings prove the inconsistency of saliency-based methods [22] and the expressiveness [1]. Also, methods that use sparsity constraints suffer from the same problems concerning their consistency.

3 P2ExNet: The Proposed Approach

This section provides insights into the proposed approach. It starts with a motivation followed by the general architecture structure, the mathematical background, and the training procedure.

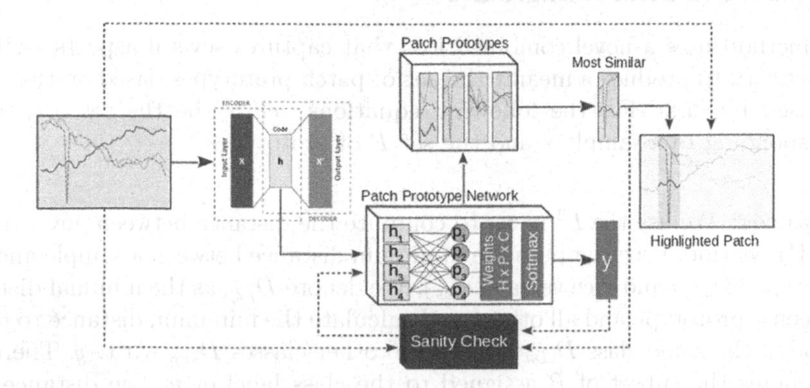

Fig. 1. Inference and testing workflow. Artificially, computed prototypes are evaluated in a similarity-based manner to suggest class-specific patches.

3.1 Motivation: An Understandable Reasoning Behavior

Inspired by human reasoning behavior, we aligned our framework to rely on implicit knowledge about objects and examples already seen before. This approach is similar to the humans' inference process. Precisely, we compare new instances to abstract concepts include class-specific features. The term prototypical knowledge describes the knowledge about these concepts and covers the analogical process to map new to the existing knowledge [9]. Following this process, the proposed method uses shallow representations. These prototypes encode class-specific pattern and provide the decision based on similarity.

3.2 Architecture

Inspired by the work of Gee et al. [8], we combined an autoencoder with a prototype network. The autoencoder consists of several convolutional and max-pooling layers serving as a feature encoding network to provide a latent representation that encodes the relevant features of an input sequence. This representation is fed forward to a custom prototype layer to generate prototypes. Motivated by the work of Chen et al. [6], we use multiple prototypes to represent a sample rather than a single one for the complete input. Precisely, the prototype layer has randomly initialized variables representing patch prototypes of user-defined size. Larger sizes will result in composed concepts, and smaller sizes result in more basic concepts. On top of the prototype layer, we attached a prototype-weight layer to encourage class-specific prototypes and weight their position within the sample to cover the local importance. Finally, a soft-max classification evaluates the similarity scores produced by the prototype layer multiplied with weights of the prototypes, as shown in Fig. 1.

3.3 Mathematical Background

Our method uses a novel combined loss that captures several aspects enabling the network to produce a meaningful set of patch prototypes based on the losses proposed by [6,8]. For the following equations, let S_x be the set of patches corresponding to a sample x and the set P of prototypes.

Distances: We use the L^2 norm to compute the distance between any two vectors. Furthermore, we compute the minimum distance between a sample and any prototype (D_{s2p}) and vice versa (D_{p2s}). We denote D_{p2p} as the minimal distance between a prototype and all others and calculate the minimum distance to a prototype of the same class D_{clst} and to the other classes D_{sep} w.r.t. y. Therefore P_y denotes the subset of P assigned to the class label of y. The distances are shown in Eqs. 1 to 4.

$$D_{s2p}(s) = \min_{p \in P} L^2(s,p) \tag{1}$$

$$D_{p2p}(p) = \min_{p' \in P} L^2(p,p') \tag{2}$$

$$D_{clst}(s,y) = \min_{p \in P_y} L^2(s,p) \tag{3}$$

$$D_{sep}(s,y) = \min_{p \in \{P \setminus P_y\}} D_{s2p}(s,p) \tag{4}$$

Loses: To ensure high-quality prototypes, we introduce our novel patch loss. This loss is a combination of different objectives to achieve good accuracy and an explanation that does not contain duplicates or prototypes that are not class-specific. Our loss combines the following losses:

- **Autoencoder loss:** MSE is used to encourage reconstruction later used for prototype reconstruction.
- **Classification loss:** To produce logits for the softmax cross-entropy we multiply the reciprocal of D_{s2p} and the prototype-weight layer.
- L_{p2s} **and** L_{s2p}: These losses preserve the relation between the input and the prototypes and vice versa as shown in Eq. 5 and 6.
- L_{div}: The diversity among the patch prototypes is computed as shown in Eq. 7.
- L_{clst} **and** L_{sep}: To encourage the network to learn class-specific prototypes we compute L_{clst} and similarly to L_{sep} but with a negative sign. This penalized prototypes close to samples of the wrong class w.r.t. their assigned class.

$$L_{p2s}(x) = \frac{1}{|P|} \sum_{p \in P} D_{p2s}(p) \quad (5)$$

$$L_{s2p}(x) = \frac{1}{|S_x|} \sum_{s \in S_x} D_{s2p}(s) \quad (6)$$

$$L_{div} = \log(1 + \frac{1}{|P|} \sum_{p \in P} D_{p2p}(p))^{-1} \quad (7)$$

$$L_{clst}(x, y) = \frac{1}{|S_x|} \sum_{s \in S_x} D_{clst}(s, y) \quad (8)$$

Our proposed final loss is a linear combination taking into account previously mentioned aspects and ensures meaningful, diverse, and class-specific patch prototypes shown in Eq. 9. By default, we set all lambda values except λ_c to one to find the best compromise between the objectives preserving high accuracy.

$$\begin{aligned} Patch\_Loss(x, y) &= \lambda_c H(x, y) + \lambda_{mse} MSE(x, x) + \lambda_{p2s} L_{p2s}(x) \\ &+ \lambda_{s2p} L_{s2p}(x) + \lambda_{div} L_{div} + \lambda_{clst} L_{clst}(x, y) + \lambda_{sep} L_{sep}(x, y) \quad (9) \end{aligned}$$

3.4 Training Process

The training process of your approach consists of two stages. In the first stage, we fix the weights of the pre-initialized prototype-weight layer to ensure class-specific prototypes. We then train the network until it converges. In the second learning phase, all layers except the prototype-weighting layer are frozen, and the network learns to adjust the prototype weights. The adjustment corrects the prototype class affiliation using the previously trained latent representation.

4 Datasets

We used eight publicly available time-series datasets to emphasize the broad applicability of our approach and examine possible limitations. As a representative set, we used seven different datasets from the UCR Time Series Classification Repository[1] and a point anomaly dataset proposed in [20]. These datasets and their parameters are visualized in Table 1. Note that the Devices dataset corresponds to the 'Electrical devices' dataset taken from the UCR. To have better coverage of different types, we selected the datasets based on the characteristics concerning the number of classes, channels, and time-steps to cover several conditions and show the prototypes. However, we focus on classification datasets.

5 Experiments

In this section, we present our results concerning the performance, applicability, and resource consumption for our proposed approach, highlighting a comparable performance while producing interpretable results.

[1] http://www.timeseriesclassification.com/.

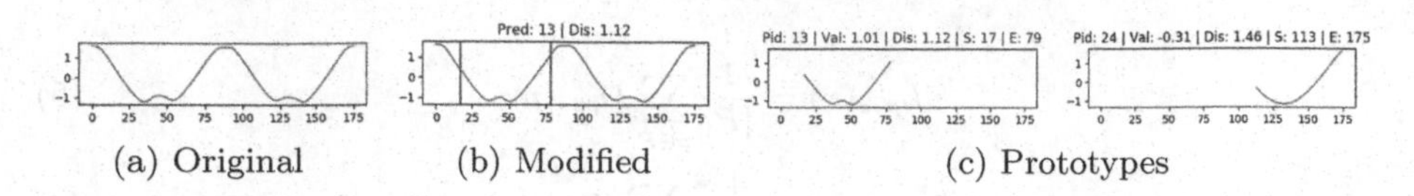

(a) Original (b) Modified (c) Prototypes

Fig. 2. Adiac dataset prototype explanation. a) shows the original series. b) shows the series with the prototype between the red bars. c) shows two prototypes. (Color figure online)

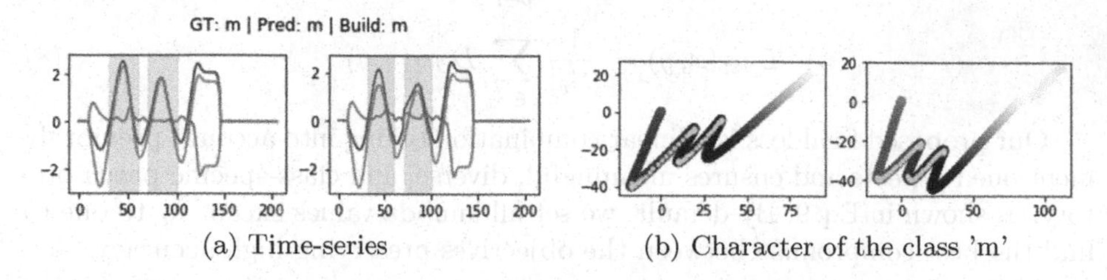

(a) Time-series (b) Character of the class 'm'

Fig. 3. Character dataset prototype explanation. a) shows the original series and the series with the prototypes. b) shows the character output and the modified character.

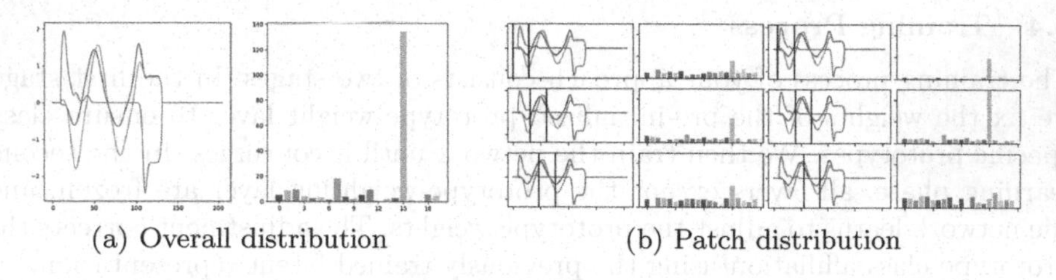

(a) Overall distribution (b) Patch distribution

Fig. 4. Class and prototype distribution. a) shows the class similarities. b) shows some patches and the corresponding class similarities.

5.1 P2ExNet: Instance-Based Evaluation

The proposed method provides the possibility to identify and highlight the parts of the input that were most relevant for the classification. Besides, it provides prototypes along with a sample containing the prototypes to compare it to the original input. Figure 2 shows highlighted regions that were important for the inference on the ADIAC dataset sample. This explanation includes the original sample of the adiac dataset, a modified version, and two prototypes. In the modified version shown in Fig. 2b, we replaced the part between the two red lines with the most important patch prototype to show how close it is to the original part. Figure 2c shows two prototypes. The value of each prototype denoted as 'Val' highlights its contribution towards the classification result. Similarly, Fig. 3 shows a sample from the character trajectories dataset and the mapping of the time-series back to the character. The black value highlights the pressure of the pen, and the yellow part shows the mapping of the prototype back to the

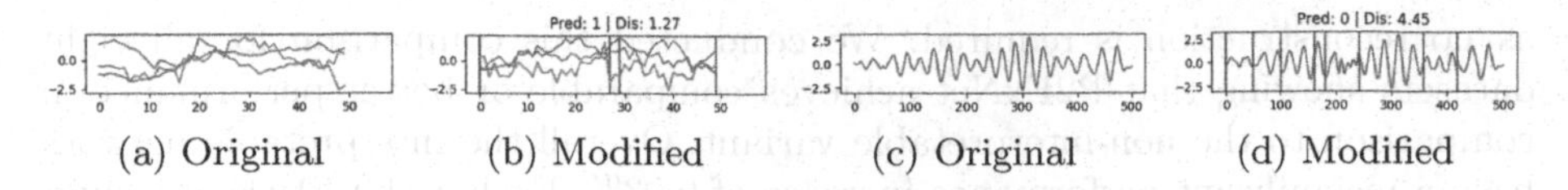

(a) Original (b) Modified (c) Original (d) Modified

Fig. 5. Prototype substitution. a) and c) show original time-series. b) and d) show the corresponding modified samples and their re-classification.

Table 1. Accuracy comparison. A comparison of interpretable and the corresponding non-interpretable counterpart.

Dataset	Classes	Length	Channel	CNN	P2ExNet
Anomaly [20]	2	50	3	**99.79**	93.79
FordA	2	500	1	85.44	**89.32**
Devices	7	96	1	55.42	**62.53**
Adiac	37	176	1	**63.54**	60.15
Crop	24	46	1	68.27	**68.54**
50words	13	270	1	76.84	**81.98**
PenDigits	10	8	2	**94.29**	93.95
Character	20	206	3	**96.53**	91.78

input space. In the case of an incorrect classification, the prototypes have a red caption. Furthermore, in Fig. 4 the class-wise overall and patch-wise distribution provides additional information about similar classes and important patch positions. Especially in Fig. 4b, we show that not all patches have the same importance when it comes to the classification. There are sensitive datasets for which the re-classification can change if the original data gets replaced with a prototype. However, for the classification and the explanation, this is not a problem as it can be solved. A proper re-scaling and adjustment can remove the offset between the prototype and the time-series. In Fig. 5b such a jump in the orange signal is shown and leads to an anomaly. However, the classification of the original signal with the network was correct. Furthermore, some datasets are invariant to small offsets shown in Fig. 5d. That is why re-scaling should be done based on the problem task. In case of a point anomaly task, the patches have to align. In a classification task, it is unlikely that the offset of a single point changes the prediction.

5.2 P2ExNet: Evaluation as a Classifier

Usually, intrinsic interpretability approaches come with an accuracy drop. In Table 1 we present the accuracy trade-off highlighting that our structure is on the same level as the non-interpretable counterpart. To create a network similar to ours without the interpretable part, we replaced the prototype layer with a dense layer and a cross-entropy loss, as suggested by Chen et al. [6]. Furthermore, we removed the decoder as there is no need to restrict the latent representation as no reconstruction is required. We conducted this comparison for all eight datasets showing that P2ExNet achieves comparable or better performance in comparison to the non-interpretable variant. Overall the interpretable network has an insignificant performance increase of 0.03%. Each architecture was superior in four out of the eight datasets. The results show that the accuracy using the interpretable model dropped about 6% on the anomaly dataset but increased 7% on the Electric Devices dataset.

5.3 P2ExNet: Sanity Check

To prove the class-specific and meaningful behavior of the prototypes, we replaced the original time-series once with the most positive and once with the most negative influencing prototypes. In Table 2 we show that the replacement with the most confident prototypes corresponding to the predicted class achieved results close to the default accuracy, whereas the best fit prototype of a different class dramatically decreased the performance as the prediction switched. These results show that our prototypes are class-specific. However, we conducted the second sanity check to investigate the need for the decoder to produce latent representations that are close to the representative prototypes. In Table 3 we show that for the character trajectories, 50words, and the FordA dataset there

Table 2. Replacement of original patch. The second column shows how much data was replaced with the suggested prototypes proposed by P2ExNet. The third column shows whether the prediction was the same as with the original time-series or not. The fourth column shows the P2ExNet accuracy for the original sample and the last column for the sample replacing the original patch with the suggested patch. The first row of each dataset corresponds to replacements with the most similar whereas the second row with the most different prototype.

Dataset	Data replaced	Equal Pred.	P2ExNet Acc.	P2ExNet mod. Acc.
Anomaly	71.99	87.43	93.79	**91.78**
	67.32	19.45		22.72
FordA	51.17	99.92	89.32	**89.40**
	44.95	23.09		32.69
Devices	52.36	81.65	62.53	**60.52**
	65.81	49.81		39.11
Adiac	35.22	85.97	60.15	**55.98**
	69.90	9.11		14.84
Crop	50.50	94.08	68.54	**66.94**
	81.12	22.01		23.28
50words	36.43	93.01	81.98	**77.20**
	52.88	62.50		56.98
PenDigits	69.47	99.31	93.95	**93.54**
	68.65	8.83		11.0
Character	18.15	92.93	91.78	**85.30**
	52.90	31.71		32.87

Table 3. Closeness of prototypes. The difference between representative and generated latent patch prototypes for P2ExNet with and without the use of the decoder are shown.

Dataset	P2ExNet with decoder	P2ExNet without decoder	Improvement
Anomaly	0.6393	**0.4929**	−22.9%
FordA	**0.7018**	1.0315	47.0%
Devices	0.4135	**0.3399**	−17.8%
Adiac	0.538	**0.4993**	−6.2%
Crop	**0.442**	0.4815	8.9%
50words	**0.0413**	0.2086	505.1%
PenDigits	**0.5123**	0.5622	9.7%
Character	**0.0099**	0.5887	5946.5%

is a significant difference if the decoder gets excluded. Also, we compared the representative and decoded prototypes and visualized two prototypes in Fig. 6 highlighting the small difference between the selected representative sample (left) and the decoded one (right). We further provide the latent representation of the character trajectory prototype in Fig. 7. Each plot represents one of the three channels and the blue color encodes the part of the selected sample whereas the orange color decodes the latent representation of the prototype. It is clearly visible that both latent representations share the same pattern and therefore result in a similar decoded prepresentation as shown in Fig. 6b.

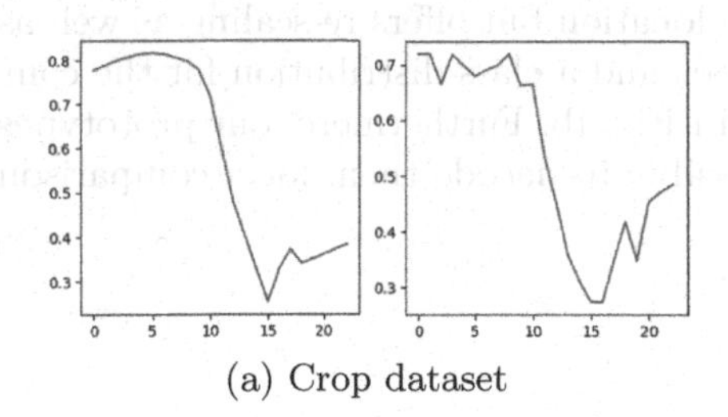

(a) Crop dataset

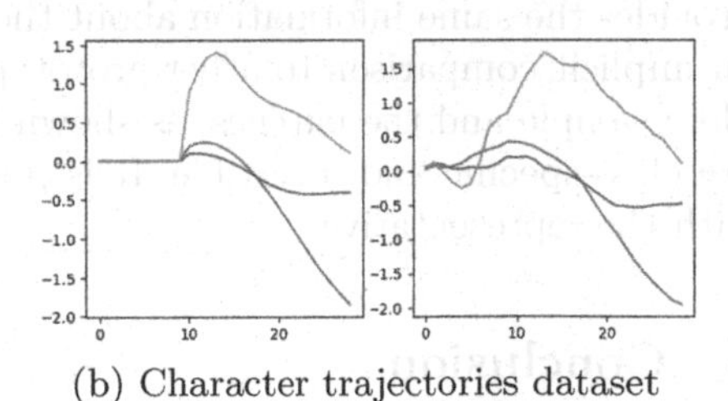

(b) Character trajectories dataset

Fig. 6. Prototype comparison. This figure shows the representative patch based on the distance to the latent prototype and the reconstruction of the latent representation.

5.4 Comparison with Existing Prototype-Based Approaches

We compared the proposed method against existing work [6] and [8]. Precisely, we highlight the explanations and additional outputs. In Fig. 8 we show the explanation of each approach for a character 'a' sample. While [8] explains the class with a prototype providing a single prototype capturing the complete sample, [6] is based on parts of the input leading to a more detailed explanation.

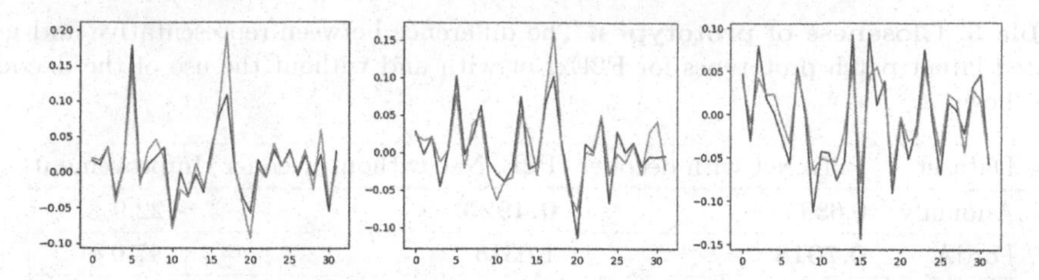

Fig. 7. Latent space difference. The difference between the prototype (orange) and the real sample (blue) in the latent space for each channel are shown. (Color figure online)

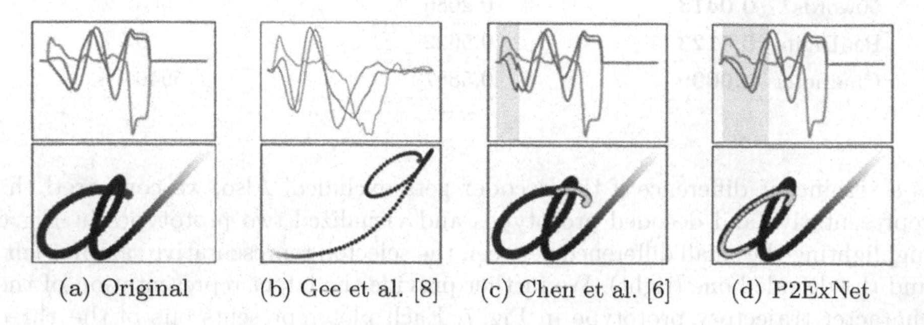

(a) Original (b) Gee et al. [8] (c) Chen et al. [6] (d) P2ExNet

Fig. 8. P2ExNet approaches. Different explanations of the character 'a'.

This method searches a patch for a region in the input image. Precisely, this means additional position information is available. Lastly, our proposed method provides the same information about the location but offers re-scaling as well as an implicit comparison to other prototypes and a class distribution for the complete sample and the patches, as shown in Fig. 4b. Furthermore, our prototypes are class-specific and invertible. It is possible to decode them for a comparison with the representatives.

6 Conclusion

Summarizing our results, we came up with novel network architecture, along with a loss and training procedure aligned to produce interpretable results and an inference process similar to the human reasoning without a significant drop in performance. Further, we proved that the proposed method works for several time-series classification tasks and when excluding the class-specific prototype assignment, our approach is suitable to produce prototypes for regression and forecast tasks. Besides, we compared the proposed method with existing prototype-based methods concerning their interpretable output and time consumption, finding ours superior in both aspects.

Acknowledgements. This work was supported by the BMBF projects DeFuseNN (Grant 01IW17002) and the ExplAINN (BMBF Grant 01IS19074). We thank all members of the Deep Learning Competence Center at the DFKI for their comments and support.

References

1. Alvarez-Melis, D., Jaakkola, T.S.: On the robustness of interpretability methods. arXiv preprint arXiv:1806.08049 (2018)
2. Angelov, P., Soares, E.: Towards explainable deep neural networks (xDNN). arXiv preprint arXiv:1912.02523 (2019)
3. Arras, L., Montavon, G., Müller, K.R., Samek, W.: Explaining recurrent neural network predictions in sentiment analysis. arXiv preprint arXiv:1706.07206 (2017)
4. Bojarski, M., et al.: Visualbackprop: efficient visualization of CNNs. arXiv preprint arXiv:1611.05418 (2016)
5. Brunelli, R.: Template Matching Techniques in Computer Vision: Theory and Practice. Wiley, Chichester (2009)
6. Chen, C., Li, O., Tao, C., Barnett, A.J., Su, J., Rudin, C.: This looks like that: deep learning for interpretable image recognition. arXiv preprint arXiv:1806.10574 (2018)
7. Choo, J., Liu, S.: Visual analytics for explainable deep learning. IEEE Comput. Graphics Appl. **38**(4), 84–92 (2018)
8. Gee, A.H., Garcia-Olano, D., Ghosh, J., Paydarfar, D.: Explaining deep classification of time-series data with learned prototypes. arXiv preprint arXiv:1904.08935 (2019)
9. Gentner, D., Colhoun, J.: Analogical processes in human thinking and learning. In: Glatzeder, B., Goel, V., Müller, A. (eds.) Towards a Theory of Thinking, pp. 35–48. Springer, Heidelberg (2010). https://doi.org/10.1007/978-3-642-03129-8_3
10. Gu, J., Yang, Y., Tresp, V.: Understanding individual decisions of CNNs via contrastive backpropagation. In: Jawahar, C.V., Li, H., Mori, G., Schindler, K. (eds.) ACCV 2018. LNCS, vol. 11363, pp. 119–134. Springer, Cham (2019). https://doi.org/10.1007/978-3-030-20893-6_8
11. Guidoni, P.: On natural thinking. Eur. J. Sci. Educ. **7**(2), 133–140 (1985)
12. Koh, P.W., Liang, P.: Understanding black-box predictions via influence functions. In: Proceedings of the 34th International Conference on Machine Learning, vol. 70, pp. 1885–1894. JMLR. org (2017)
13. Li, O., Liu, H., Chen, C., Rudin, C.: Deep learning for case-based reasoning through prototypes: a neural network that explains its predictions. In: Thirty-Second AAAI Conference on Artificial Intelligence (2018)
14. Lipton, Z.C.: The mythos of model interpretability. arXiv preprint arXiv:1606.03490 (2016)
15. Palacio, S., Folz, J., Hees, J., Raue, F., Borth, D., Dengel, A.: What do deep networks like to see? In: The IEEE Conference on Computer Vision and Pattern Recognition (CVPR), June 2018
16. Samek, W., Wiegand, T., Müller, K.R.: Explainable artificial intelligence: understanding, visualizing and interpreting deep learning models. arXiv preprint arXiv:1708.08296 (2017)
17. Schlegel, U., Arnout, H., El-Assady, M., Oelke, D., Keim, D.A.: Towards a rigorous evaluation of XAI methods on time series. arXiv preprint arXiv:1909.07082 (2019)

18. Selvaraju, R.R., Cogswell, M., Das, A., Vedantam, R., Parikh, D., Batra, D.: Grad-cam: visual explanations from deep networks via gradient-based localization. In: Proceedings of the IEEE International Conference on Computer Vision, pp. 618–626 (2017)
19. Siddiqui, S.A., Mercier, D., Dengel, A., Ahmed, S.: Tsinsight: a local-global attribution framework for interpretability in time-series data. arXiv preprint arXiv:2004.02958 (2020)
20. Siddiqui, S.A., Mercier, D., Munir, M., Dengel, A., Ahmed, S.: Tsviz: demystification of deep learning models for time-series analysis. IEEE Access **7**, 67027–67040 (2019)
21. Simonyan, K., Vedaldi, A., Zisserman, A.: Deep inside convolutional networks: visualising image classification models and saliency maps. arXiv preprint arXiv:1312.6034 (2013)
22. Tomsett, R., Harborne, D., Chakraborty, S., Gurram, P., Preece, A.: Sanity checks for saliency metrics. arXiv preprint arXiv:1912.01451 (2019)
23. Yeh, C.K., Kim, J., Yen, I.E.H., Ravikumar, P.K.: Representer point selection for explaining deep neural networks. In: Advances in Neural Information Processing Systems, pp. 9291–9301 (2018)
24. Yosinski, J., Clune, J., Nguyen, A., Fuchs, T., Lipson, H.: Understanding neural networks through deep visualization. arXiv preprint arXiv:1506.06579 (2015)
25. Zhang, Q.s., Zhu, S.C.: Visual interpretability for deep learning: a survey. Front. Inf. Technol. Electron. Eng. **19**(1), 27–39 (2018)
26. Zhang, Q., Nian Wu, Y., Zhu, S.C.: Interpretable convolutional neural networks. In: Proceedings of the IEEE Conference on Computer Vision and Pattern Recognition, pp. 8827–8836 (2018)
27. Zintgraf, L.M., Cohen, T.S., Adel, T., Welling, M.: Visualizing deep neural network decisions: Prediction difference analysis. arXiv preprint arXiv:1702.04595 (2017)

Prediction of Taxi Demand Based on CNN-BiLSTM-Attention Neural Network

Xudong Guo(✉)

Memory Platform Group, Samsung (China) Semiconductor Co. Ltd., Xi'an, People's Republic of China
guoxudong@bupt.edu.cn

Abstract. As an essential part of the urban public transport system, taxi has been the necessary transport option in the social life of city residents. The research on the analysis and prediction of taxi demands based on the taxi trip records tends to be one of the important topics recently, which is of great importance to optimize the taxi dispatching, minimize the wait-time for passengers and drivers, reduce the time and distances of vacant driving, as well as improve the quality of taxi operation and management. In this paper, we propose the CNN-BiLSTM-Attention model, which consists of Convolutional Neural Networks (CNNs), Bidirectional Long Short Term Memory (BiLSTM) neural networks and the Attention mechanism, to predict the taxi demands at some certain regions. Then we compare the prediction performance of CNN-BiLSTM-Attention model with the baselines. The results show that this model can outperform other models in predicting the taxi demands, which also proves that our CNN-BiLSTM-Attention model is capable of capturing the spatial and temporal features more effectively, and has a better prediction accuracy.

Keywords: CNN-BiLSTM-Attention · Taxi demand · Prediction

1 Introduction

Nowadays, taxi has played an important role in the urban public transport system, as well as the social life of city residents. More and more people tend to choose taxi as the transport option due to its convenience, flexibility and comfort. However, there still exists the imbalance problem between taxi supply and passenger demands. Taxi drivers usually search for the next passenger blindly and randomly, while a lot of passengers often complain the long wait-time to take a taxi. Therefore, the analysis and prediction of taxi demand throughout a region or city is of great importance in improving taxi dispatching effectively, as well as enhancing the satisfaction of taxi passengers. With the aid of big data technologies, more data resources can be available and computable than before. In recent years, taxi trips data has been widely utilized in many tasks, such as taxi dispatching optimization [1], urban traffic condition analysis and prediction [2], and the applications of taxi service strategies [3].

H. Yang et al. (Eds.): ICONIP 2020, LNCS 12534, pp. 331–342, 2020.
https://doi.org/10.1007/978-3-030-63836-8_28

However, there are a few researches focused on the prediction of taxi demand throughout a region or city based on the historical data of taxi trips. Zhao et al. [4] define a maximum predictability method for the taxi demand and prove that the taxi demand can be highly predictable. In addition, three specific algorithms are applied to validate the theory. Qian et al. [5] propose a Gaussian Conditional Random Field (GCRF) model to predict the short-term taxi demand only using the historical taxi data, proving that the model can capture the complex spatio-temporal dependencies to some extent and owns a desirable accuracy in predicting the short-term taxi demand. Yan et al. [6] analyze the data of taxi calls, and propose a Bayesian hierarchical semi-parametric model to predict the taxi demands in the future, besides, this spatio-temporal model is implemented in the cloud computing environment, which can realize the online real-time prediction of taxi demands. The above-mentioned methods mainly focus on predicting the taxi demands by analyzing the historical taxi related data and trying to capture the spatio-temporal features of taxi demands, which can effectively predict taxi passengers in certain regions to some extent. However, this is also a time series forecasting problem, the long-term dependencies exist in the different time periods, which is difficult to be handled by traditional approaches. We need a better method to deal with the long-term dependency and improve the prediction accuracy. With the rapid development of deep learning technologies, Convolutional Neural Network (CNN) and Recurrent Neural Network (RNN) are widely used in a lot of applications. Zhang et al. [7] propose a deep spatio-temporal residual network called ST-Resnet to forecast the inflow and outflow of crowds in each region of a city. The CNNs and residual units in ST-Resnet model can effectively capture the spatio-temporal features of the crowds, but the potential information in the long time series may be lost. Xu et al. [8] propose a real-time method forecasting taxi demands based on LSTM network and achieve an admirable prediction accuracy. As a typical network of Recurrent Neural Network (RNN), Long Short Term Memory (LSTM) [9] network is capable of learning long-term dependencies by utilizing some gating mechanisms to store information for future use, which also has been proved to be powerful in solving some kinds of sequence learning problems. However, it still contains much redundancy for the spatial data. Liu et al. [10] carry out a difficult and challenging task, called taxi origin-destination demand prediction, which mainly aims at predicting the taxi demand between all region pairs in a future time interval. In addition, a novel Contextualized Spatial-Temporal Network (CSTN) is proposed and proved to be effective in predicting taxi demands both in origin and destination. But the spatial and temporal information of taxi demands has not been fully taken into consideration. In this paper, in order to capture the spatio-temporal features comprehensively, and predict the taxi demands more precisely, we propose a new model, called CNN-BiLSTM-Attention, which mainly consists of three essential parts, namely, CNNs (Convolutional Neural Networks), BiLSTM (Bidirectional Long Short-Term Memory) neural networks and the attention mechanism. Then we compare the specific prediction performance of the CNN-BiLSTM-Attention model with other models, such as LSTM, CNN-LSTM and

CNN-LSTM-Attention. The experimental results show that the CNN-BiLSTM-Attention model can outperform these three baselines for the prediction of taxi demands.

2 Models

2.1 CNN-BiLSTM-Attention Model

Here we proposed a model called CNN-BiLSTM-Attention model, which consists of three essential components, namely, Convolutional Neural Networks (CNNs), Bidirectional Long Short Term Memory neural network (BiLSTM) and the Attention mechanism. The structure of CNN-BiLSTM-Attention model is shown in Fig. 1.

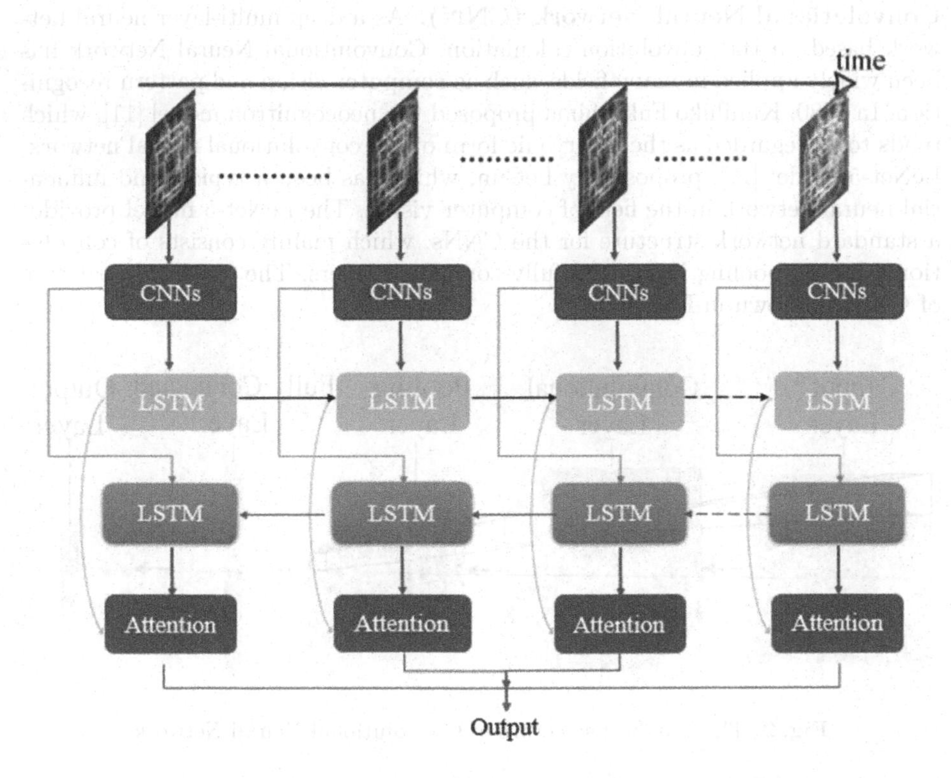

Fig. 1. The structure of CNN-BiLSTM-Attention model

As shown in Fig. 1, the input data is sent to the Convolutional Neural Networks at first, which contains several convolutional layers. The CNNs can be applied to capture the complex spatial information of the taxi demands. Then the BiLSTM neural network will accept the sequence data from the CNNs, due to the fact that the time periods also play an important role in the taxi pick-ups, therefore, the BiLSTM network is used to achieve the potential temporal features in the time series, then the attention mechanism will be adopted in the model, which is able to focus on the relevant elements of the input data, assigning different weights to the elements of the input sequence based on the location and contents of the sequence. More details about each component will be described as below.

2.2 Model Components

As mentioned above, The CNN-BiLSTM-Attention model mainly contains three important components, namely, Convolutional Neural Networks (CNNs), Bidirectional Long Short-Term Memory neural networks (BiLSTM) and Attention mechanism. Each component of this model is described in detail as below.

Convolutional Neural Network (CNN). As a deep multi-layer neural network based on the convolution calculation, Convolutional Neural Network has been widely applied in many fields, such as computer vision and pattern recognition. In 1980, Kunihiko Fukushima proposed the neocognitron model [11], which tends to be regarded as the embryonic form of the convolutional neural network. LeNet-5 model [12], proposed by LeCun, which has been a typical and influential neural network in the field of computer vision. The LeNet-5 model provides a standard network structure for the CNNs, which mainly consists of convolutional layers, pooling layers and fully-connected layers. The standard structure of CNNs is shown in Fig. 2.

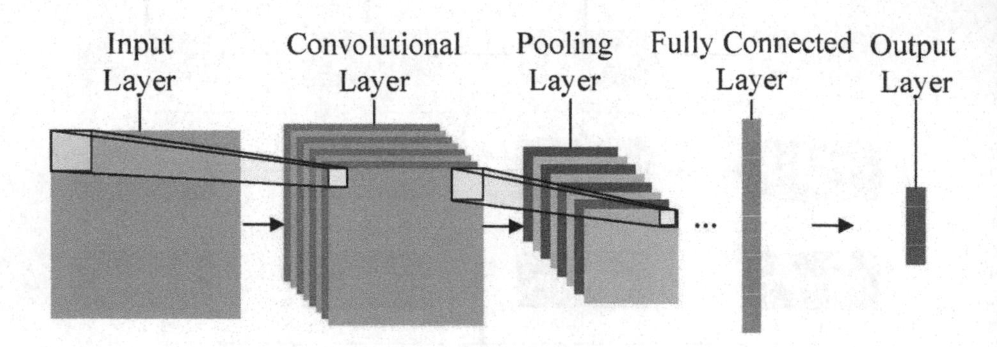

Fig. 2. The standard structure of Convolutional Neural Network

As shown in Fig. 2, the CNNs can effectively capture the spatial information by the convolution operations, as well as its characteristics of weight sharing and sparse connection. The input data usually is the one-dimensional vector or the 2-D matrix. In our experiment, the entire region or area can be treated as an image, while the number of passengers taking a taxi in a location can be seen as the pixels in the image. Therefore, the CNNs in our CNN-BiLSTM-Attention model are applied to capture the underlying spatial features of taxi demands in the target region.

Bidirectional Long Short-Term Memory Neural Network (BiLSTM). BiLSTM neural network is a variant network of the standard LSTM network, which consists of forward LSTM and backward LSTM, providing the access to long range context in both input directions. As a desirable neural network to deal with the long-term dependencies in time series, LSTM is designed to overcome the vanishing gradients through a special gating mechanism.

The standard LSTM architecture is composed of many recurrently connected subnets, known as memory blocks [13]. Each memory block consists of one or several memory cells and three gates, namely, input gate, forget gate and output gate. The input gate aims at selecting the needed new information, and adds it to the cell state. The forget gate tends to remove the information that is no longer required by the memory cell, while the output gate decides what kind of necessary information in the cell should be output. Generally, the gating mechanisms can ensure the cells in LSTM network to store and update the essential information over long periods of time. The equations of these gates (input gate, forget gate and output gate) in LSTM are as follows:

$$i_t = \sigma(W_{xi}x_t + W_{hi}h_{t-1} + W_{ci}c_{t-1} + b_i) \tag{1}$$

$$f_t = \sigma(W_{xf}x_t + W_{hf}h_{t-1} + W_{cf}c_{t-1} + b_f) \tag{2}$$

$$c_t = f_t c_{t-1} + i_t \tanh(W_{xc}x_t + W_{hc}h_{t-1} + b_c) \tag{3}$$

$$o_t = \sigma(W_{xo}x_t + W_{ho}h_{t-1} + W_{co}c_t + b_o) \tag{4}$$

$$h_t = o_t \tanh(c_t) \tag{5}$$

Where i, f and o refer to the input gate, forget gate and output gate respectively, while c stands for the memory cell, W is the weight matrix of hidden state, x_t is the input data at time t, h_{t-1} is the hidden output at time t−1, c_t is the cell state at time t, the activation function of these gates is denoted σ, and b represents the bias value. Given that the event in some time periods in the future tends to exert an influence on the taxi demand at present, for instance, the concert hold in next week may increase the taxi demands in the ticket office this week. Therefore, in our experiment, the BiLSTM network is applied to capture the long-term dependencies in the time series.

Attention Mechanism. The attention mechanism was proposed and applied to solve the image related problems at first, which has been utilized in the field of natural language processing recently. Here we use the attention mechanism to calculate the weights of feature vectors output from the BiLSTM network at different time step, and assign the higher weight to the vital features, making sure that the network can have the better performance. The attention layers proposed by Zhou et al. [14] is used here, the formulas are shown as below.

$$e_t = \tanh(Wh_t + b_t) \tag{6}$$

$$\alpha_t = \frac{\exp(e_t)}{\sum_{j=0}^{t} e_j} \tag{7}$$

$$v = \sum_{t=0}^{n} \alpha_t h_t \tag{8}$$

Here h_t is the hidden state, b_t stands for bias, and v represents the output vector calculated by the weighted sum of the hidden state.

3 Experiment

3.1 Dataset

The dataset used in our experiment is TLC Trip Record Data [15], which includes the historical yellow and green taxi trip records of New York City. This dataset is provided by NYC Taxi and Limousine Commission for research purposes. Each taxi trip in the dataset consists of several records, such as vendor id, pickup datetime, pickup longitude, trip distance, passenger count and so on. Given that the majority of green taxi trip records cannot be available in some years, and also limited by the hardware condition and computational power, here we choose the trip records of the yellow taxi from June 2012 through August 2012 as the training data, and the data from 1st Sept. 2012 to 7th Sept. 2012 is treated as the testing data.

3.2 Preprocessing

The preprocessing for the data is an essential step before the source data is sent to the model, this process mainly includes the treatment of outliers and missing values, as well as the data normalization. In terms of outliers processing, for example, the longitude and latitude values are beyond the boundary of New York City, or the number of passengers exceeds the seating capacity of taxis. Here these kinds of outliers will be removed from the relevant records. When it comes to the missing values, the delete operation is also applied to deal with the related records, ensuring the data integrity. In addition, the Min-Max normalization method is used to scale the input data into the range $[-1, 1]$. In the evaluation phase, the predicted values are re-scaled back to the normal values, then compared with the ground truth.

3.3 Model Hyperparameters

The python libraries, including pandas, matplotlib, Tensorflow and Keras, are utilized to build our models. In our experiment, we choose the Lower Manhattan region, a geographical area with the longitude ranging from $-74.0080°$ to $-73.9740°$ and the latitude ranging from $40.7230°$ to $40.7490°$, as our target area. Then we partition the entire region into equal small ones based on the longitude and latitude, which means that each small area possesses the same size. In order to explore the influence of the geographical partition methods in the process of predicting the taxi demands, the target experimental region are partitioned into 36 (6 * 6) small areas and 144 (12 * 12) small areas respectively. The convolutional neural networks use 32 filters of size 3 * 3, the MaxPooling layer and Flatten layer are also applied. In terms of BiLSTM network, here the time interval is set as one hour, and the time-step length is chosen as 24, which means that the historical taxi trip records in the past 24 time intervals will be used to forecast the number of taxi pickups in the next time period. Besides, to avoid the over-fitting problem, the tackle of dropout is also adopted, the parameter is set as 0.3. The hidden units in the BiLSTM network is set as 64. We compare the performance of the CNN-BiLSTM-Attention model in predicting taxi demands with other three models, namely, LSTM network, CNN-LSTM network and CNN-LSTM-Attention network. Different from other two models, the input shape of LSTM network is a 1 * N vector, where N stands for the number of small areas, while other two models accept a 2-dimensional matrix. Table 1 includes the list of major parameters in our experiments.

Table 1. Important experimental parameters.

Parameter	Value
Number of small areas	36/144
Number of filters	32
Filter size	3 * 3
Time steps	24
Number of units in hidden layer (BiLSTM)	64
Dropout	0.3

3.4 Evaluation Metrics

Due to the fact that multiple areas and time periods are involved in the experiment, here we measure our model by two kinds of prediction error metrics: Root Mean Square Error (RMSE) [16] and Symmetric Mean Absolute Percentage Error (SMAPE) [17]. The SMAPE is an alternative to Mean Absolute Percentage Error (MAPE) when there are zero or near-zero demand for items [18]. Different from the original formula defined by Armstrong in 1985, the widely accepted version of SMAPE without the factor 0.5 in denominator is applied here. The RMSE and SMAPE in regions i over time periods [1−T] are as follows:

$$RMSE_i = \sqrt{\frac{1}{T}\sum_{t=1}^{T}(P_{i,t} - P^*_{i,t})^2} \tag{9}$$

$$SMAPE_i = \frac{1}{T}\sum_{t=1}^{T}\frac{|P_{i,t} - P^*_{i,t}|}{P_{i,t} + P^*_{i,t} + k} \tag{10}$$

Here $P_{i,t}$ represents the real taxi demand in region i at time-step t, while $P^*_{i,t}$ stands for the predicted taxi demand. In order to avoid division by zero when both $P_{i,t}$ and $P^*_{i,t}$ are zero, a small value k is added to Eq. 11 [17]. Similarly, The RMSE and SMAPE of all regions at time-step t would be:

$$RMSE_t = \sqrt{\frac{1}{N}\sum_{i=1}^{N}(P_{i,t} - P^*_{i,t})^2} \tag{11}$$

$$SMAPE_t = \frac{1}{N}\sum_{i=1}^{N}\frac{|P_{i,t} - P^*_{i,t}|}{P_{i,t} + P^*_{i,t} + k} \tag{12}$$

N refers to the number of small regions in the experiment. Based on these metrics, we can get the average RMSE and average SMAPE in regions N as follows:

$$\overline{RMSE\text{-}N} = \frac{1}{N}\sum_{i=1}^{N} RMSE_i \tag{13}$$

$$\overline{SMAPE\text{-}N} = \frac{1}{N}\sum_{i=1}^{N} SMAPE_i \tag{14}$$

Similarly, the average RMSE and average SMAPE in all time periods would be:

$$\overline{RMSE\text{-}T} = \frac{1}{T}\sum_{t=1}^{T} RMSE_t \tag{15}$$

$$\overline{SMAPE\text{-}T} = \frac{1}{T}\sum_{t=1}^{T} SMAPE_t \tag{16}$$

4 Results

4.1 Spatio-temporal Features

The spatio-temporal features of taxi demands play a vital role in the prediction of the number of passengers who would like to take a taxi. Here, a portion of historical taxi trip dataset is used to analyze the spatio-temporal features. In terms of spatial features, in order to ensure the reliability of evaluation results, we randomly choose the records of taxi pickups on 10th May 2012 as the target data to analyze the spatial distribution pattern of taxi demands throughout New York City, which is visualized by the software ArcGIS. The result of taxi demands distribution on 10th May 2012 throughout New York City is shown in Fig. 3. With regard to the temporal features, the taxi records of Metropolitan

Museum of Art on 6th Jun. 2012 and 16th Jun. 2012 are selected to explore the taxi demands at different time periods in both weekday and weekend. The taxi demands during different periods of time at Metropolitan Museum of Art are shown in Fig. 4. As shown in Fig. 3, affected by external multiple factors, such as geography, economic development, road network, and so on, there is an apparent difference about the passenger numbers in different boroughs. As the economic and cultural center of NYC, Manhattan owns the majority of the taxi demands, in addition, due to the location of JFK airport, there still exist many taxi passengers in Queens, while few people would like to request the taxis in Staten Island. As we can see from Fig. 4, in general, the passenger numbers of taxi pickups at Metropolitan Museum of Art on weekend are obviously more than the ones on weekday. Besides, the taxi demands mainly focus on the time period ranging from 10:00 AM to 19:00 PM, which are affected by the opening hours of the museum, as well as the normal work and rest time of citizens.

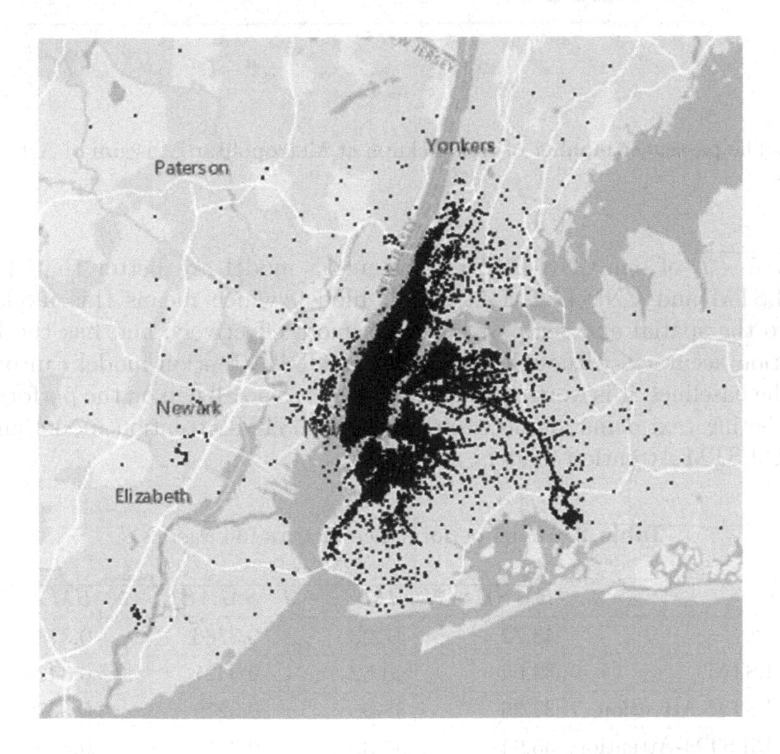

Fig. 3. The distribution of taxi pickups throughout NYC on May 10, 2012

4.2 Prediction Results

In our experiment, we first give the comparison with three other models when the number of divided small areas is set as 36, the results are shown in Table 2 as below. We can find that the $\overline{RMSE-N}$, $\overline{RMSE-T}$, $\overline{SMAPE-N}$ and

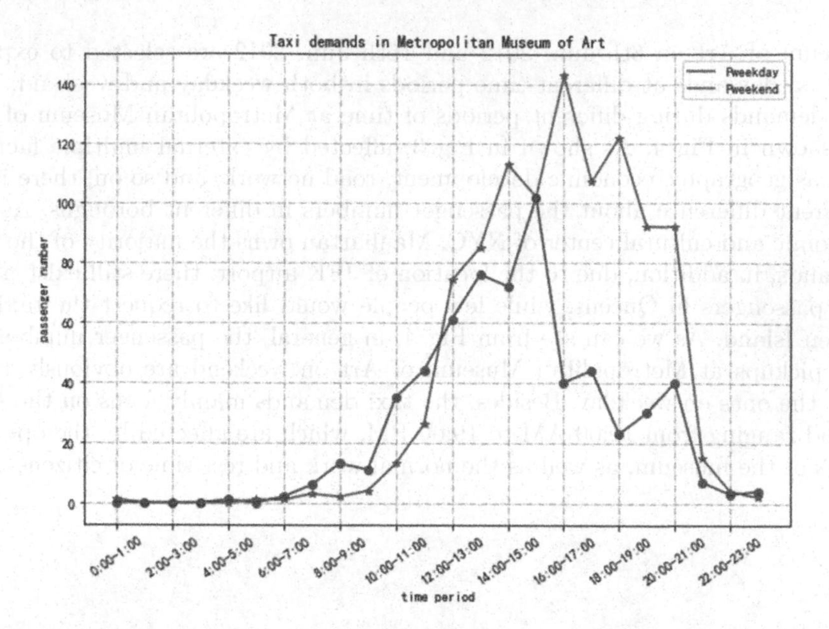

Fig. 4. The passenger number of taxi pickups at Metropolitan Museum of Art at time periods

$\overline{SMAPE-T}$ of our CNN-BiLSTM-Attention model are better than LSTM, CNN-LSTM and CNN-LSTM-Attention model, which means this model can capture the spatial and temporal features more effectively, and has the better prediction accuracy. Although the CNN-BiLSTM-Attention model can outperform the baselines, it is worth noting that there is a small gap on the performance in predicting taxi demands between the CNN-LSTM-Attention model and the CNN-BiLSTM-Attention model.

Table 2. Model performance metrics (36 regions).

Model	$\overline{RMSE-N}$	$\overline{RMSE-T}$	$\overline{SMAPE-N}$	$\overline{SMAPE-T}$
LSTM	45.72	47.23	0.3786	0.3786
CNN-LSTM	39.64	40.52	0.3421	0.3421
CNN-LSTM-Attention	37.26	38.05	0.3279	0.3279
CNN-BiLSTM-Attention	35.94	36.39	0.3158	0.3158

Then in order to evaluate the influence of geographical partition methods in predicting taxi demands, here we give the comparison with three other models when the number of small regions is 144, which is shown in Table 3 as below. As shown in Table 3, it is obvious that the results of $\overline{RMSE-N}$, $\overline{RMSE-T}$, $\overline{SMAPE-N}$ and $\overline{SMAPE-T}$ are all lower than the ones in the condition

Table 3. Model performance metrics (144 regions).

Model	$RMSE-N$	$RMSE-T$	$SMAPE-N$	$SMAPE-T$
LSTM	22.40	22.65	0.1974	0.1974
CNN-LSTM	19.67	19.52	0.1845	0.1845
CNN-LSTM-Attention	18.31	18.49	0.1711	0.1711
CNN-BiLSTM-Attention	18.15	18.23	0.1710	0.1710

where the number of small regions is set as 36, which means that more divided areas can help to reduce the prediction errors and improve the model performance to some extent. In addition, the CNN-BiLSTM-Attention model still has the better prediction performance than other three models, proving that this model owns the desirable prediction accuracy, as well as the reliable stability.

5 Conclusion and Future Work

In this paper, based on the historical taxi trip dataset of New York City, we first analyze the potential spatial and temporal features of taxi demands, then propose a model called CNN-BiLSTM-Attention to predict the number of taxi passengers when the divided small areas is set as 36 or 144. The experimental results show that the prediction performance of the CNN-BiLSTM-Attention model is better than other three baselines, which proves that this model has the better prediction accuracy, as well as the reliable stability. In addition, our experiment also studies the effect of the size of small regions in the prediction problem. The result shows that the prediction accuracy may become better with the smaller region size. For future work, more factors can be taken into consideration, such as weather, traffic condition and so on. We also intend to improve our models with better network structures.

References

1. Yang, Q., Gao, Z., Kong, X., Rahim, A., Wang, J., Xia, F.: Taxi operation optimization based on big traffic data. In: 2015 IEEE 12th International Conference on Ubiquitous Intelligence and Computing and 2015 IEEE 12th International Conference on Autonomic and Trusted Computing and 2015 IEEE 15th International Conference on Scalable Computing and Communications and Its Associated Workshops (UIC-ATC-ScalCom), pp. 127–134. IEEE (2015)
2. Kong, X., Xu, Z., Shen, G., Wang, J., Yang, Q., Zhang, B.: Urban traffic congestion estimation and prediction based on floating car trajectory data. Future Gener. Comput. Syst. **61**, 97–107 (2016)
3. Zhang, D., et al.: Understanding taxi service strategies from taxi GPS traces. IEEE Trans. Intell. Transp. Syst. **16**(1), 123–135 (2015)
4. Zhao, K., Khryashchev, D., Freire, J., Silva, C., Vo, H.: Predicting taxi demand at high spatial resolution: approaching the limit of predictability. In: IEEE International Conference on Big Data (2017)

5. Qian, X., Ukkusuri, S., Yang, C., Yan, F.: Short term taxi demand forecasting using Gaussian conditional random field model. In: Transportation Research Board 2017 Annual Meeting (2017)
6. Yan, H., Zhang, Z., Zou, J.: An online spatio-temporal model for inference and predictions of taxi demand. In: IEEE International Conference on Big Data (2017)
7. Zhang, J., Zheng, Y., Qi, D.: Deep spatio-temporal residual networks for citywide crowd flows prediction (2016)
8. Jun, X., Rahmatizadeh, R., Bölöni, L., Turgut, D.: Real-time prediction of taxi demand using recurrent neural networks. IEEE Trans. Intell. Transp. Syst. **19**(8), 2572–2581 (2018)
9. Hochreiter, S., Schmidhuber, J.: Long short-term memory. Neural Comput. **9**(8), 1735–1780 (1997)
10. Liu, L., Qiu, Z., Li, G., Wang, Q., Ouyang, W., Lin, L.: Contextualized spatial-temporal network for taxi origin-destination demand prediction. IEEE Trans. Intell. Transp. Syst. **PP**(99), 1–13 (2019)
11. Kunihiko and Fukushima: Neocognitron: A self-organizing neural network model for a mechanism of pattern recognition unaffected by shift in position. Biol. Cybern. **36**, 193–202 (1980)
12. Lecun, Y., Bottou, L.: Gradient-based learning applied to document recognition. Proc. IEEE **86**(11), 2278–2324 (1998)
13. Graves, A.: Supervised Sequence Labelling with Recurrent Neural Networks. Springer, Heidelberg (2012). https://doi.org/10.1007/978-3-642-24797-2
14. Zhou, P., Shi, W., Tian, J., Qi, Z., Xu, B.: Attention-based bidirectional long short-term memory networks for relation classification. In: Proceedings of the 54th Annual Meeting of the Association for Computational Linguistics (Volume 2: Short Papers) (2016)
15. NYC taxi & limousine commission. Taxi and limousine commission (TLC) trip record data. https://www1.nyc.gov/site/tlc/about/tlc-trip-record-data.page. Accessed 13 Sept 2020
16. Lv, Y., Duan, Y., Kang, W., Li, Z., Wang, F.Y.: Traffic flow prediction with big data: a deep learning approach. IEEE Trans. Intell. Transp. Syst. **16**(2), 865–873 (2015)
17. Wikipedia. https://en.wikipedia.org/wiki/Symmetric_mean_absolute_percentage_error. Accessed 13 Sept 2020
18. Vanguard software homepage. https://www.vanguardsw.com/business-forecasting-101/forecast-fit/. Accessed 13 Sept 2020

Pruning Long Short Term Memory Networks and Convolutional Neural Networks for Music Emotion Recognition

Madeline Brewer(✉) and Jessica Sharmin Rahman

Australian National University, Canberra, ACT 2601, Australia
{u4673202,jessica.rahman}@anu.edu.au

Abstract. Music can be used as a form of therapy and can reduce symptoms of depression and anxiety. Understanding the relationship between music and physiological reactions could be essential in further developing music therapy. This paper uses machine learning techniques to classify which genre of music is being listen to using physiological responses. Both Long Short Term Memory Networks and Convolutional Neural Networks can be used for making predictions from sequence data. We trained and compared two networks which attempted to classify the genre of music a participant was listening to from their electrodermal activity. An LSTM and a CNN were trained and their accuracy was found to be 69.23% and 72.97% respectively. Pruning of each of the networks was also conducted and it was found that the network structure for both the CNN and the LSTM can be reduced by at least 20% without having a reduction in the accuracy of the model. It was found that the LSTM has only very few important neurons and weights that contribute to the accuracy of the model.

Keywords: Network pruning · Sequence classification · Long short term memory networks · Convolutional neural networks

1 Introduction

The effects of music on humans has long been observed but is still yet to be fully understood. The reason some songs give people chills while others make them want to dance is an interesting and complex problem. It is likely that different genres of music have different effects on people, and understanding how this could be measured is an interesting problem that could lead to breakthroughs in music therapy and music development. If we are able to successfully classify music into genres from physiological signals it would support research into music therapy. To better determine the importance of these physiological responses, this paper looks to identify the effectiveness of model simplification techniques on classification accuracy in two leading neural network architectures.

Time series classification and prediction has been a growing area of focus for deep learning techniques. Time series data includes a multitude of interesting

H. Yang et al. (Eds.): ICONIP 2020, LNCS 12534, pp. 343–352, 2020.
https://doi.org/10.1007/978-3-030-63836-8_29

data, including weather data, stock market forecasts and human physiological signals, which are the focus of this paper. There are many different techniques which can be used to extract the most out of a time series data, including employing the use of Recurrent Neural Networks (RNNs) and Convolutional Neural Networks (CNNs).

Recurrent Neural Networks (RNNs) often suffer from vanishing or exploding gradient problems. Long Short-term Memory (LSTM) networks are an improvement to RNNs by using gating functions as part of their sequence [9]. LSTMs are able to model varying length sequence data. They are able to look at long term dependencies in sequence data [11]. LSTMs are very powerful when dealing with sequence data, as unlike traditional RNNs, they have the ability to remember information from the beginning of the sequence right at the end of the sequence. This means that the network can remember and exploit the most important information from any time point when classifying the sequence. Convolutional Neural Networks (CNNs) are also powerful tools for classifying sequential or spatially related data. CNNs make excellent use of adjacent features, meaning that they excel in spatial classification as adjacent pixels of an image often have related content [7].

Deep learning networks can suffer from the same issues as shallow neural networks, where the optimal size of the model is hard to determine before training occurs and too many neurons in hidden layers can result in the same overfitting problems that occur in shallow neural networks.

The performance of a neural network relies heavily on the number of neurons chosen. Usually, the number of input and output neurons is defined by the problem the network is trying to solve. There may be some ambiguity about the number and type of input neurons to use, however this is a separate area of study. What is far less obvious, however, is how many neurons to use in the hidden layers of the network. There are many rules of thumb when considering how many hidden neurons to include in the network. Some of these include: choosing a number of hidden neurons between the size of the input layer and output layer, choosing a number of hidden neurons which is two thirds the size of the input layer plus the output layer, and choosing the number of hidden neurons which is less than twice the size of the input layer [4].

These rules of thumb give a user an indication of where to start to avoid having too many or too few hidden neurons. However, having too many neurons in the network can lead to inefficient training and cause the model to overfit to the training data, while having too few can result in the network failing to learn to the desired level of performance. There are rules of thumb when adding hidden neurons to a network, including adding one or two hidden units, adding 10% more hidden neurons, or even doubling the amount [12].

Once adequate training has been achieved using these rules of thumb, a question of the optimal network structure arises. If the size of the model was simply doubled to achieve good training performance, some redundant neurons may have been introduced to the network. Reducing the network to the optimal size not only increases the efficiency of the network, but it allows us to determine the appropriate number of hidden neurons for similar problems in the future [5,6].

Many metrics can be used for the assessment of how well a neural network performs. In this paper, we will focus on the percentage of correct classifications the neural network makes, known as the accuracy as a metric of how well the neural network performs. Accuracy is a simple to calculate and easy to understand performance metric which is suitable for the context of this problem.

This paper intends to predict the genre of music a participant is listening to by using the electrodermal activity collected during the song. Two neural networks, an LSTM and a CNN, are trained to attempt to classify the music into the three genres of music. In this paper, we will also attempt to prune the LSTM and CNN, getting rid of unnecessary or redundant neurons without significantly impacting the network performance.

After this introductory section, Sect. 2 of this paper will describe the data set and the neural network features for the LSTM and the CNN. Section 3 will present the results of the neural networks and looks at how the networks performed while pruning. Finally, Sect. 4 of the paper concludes the paper and presents future work.

2 Method

2.1 Data Set

This paper uses the data set developed by Rahman et al. in [13]. The data set contains a set of physiological signals collected from 24 students when exposed to different music stimulus. This paper focuses on the prediction of the genre of the song based on the electrodermal activity of each participant throughout the song. The electrodermal activity of each participant was recorded throughout the song at a sampling rate 4 Hz, meaning that 4 samples were collected each second. The songs were of varying length, but all were about 4 min long. 16 of the 24 participants listened to each song and their electrodermal activity was recorded throughout. 12 different songs were used, with four songs representing three genres, being classical, instrumental, and pop. This single sequential feature will be used for training [13].

2.2 Data Pre-processing

Extensive data pre-processing was required in order to properly train a model able to successfully classify sequential data. Some exploratory data analysis was conducted, and it showed that the values ranged a significant amount between participants. All the participants' data was included in the analysis for each song.

As the data from each participant varied a significant amount, sometimes by an order of magnitude, it was decided that before any analysis the data should be normalised. Min-max normalisation was used on the data of each participant for each song, leaving each participant with a range of values for their electrodermal activity being between zero and one. Min-max normalisation has been shown to improve accuracy of neural nets, and is a simple and easy to implement method of normalisation [8].

In order to have the data in the suitable format for training, an ID for each data point was used. This was generated from the timestamp and participant ID, and song genre. Maintaining the time stamp in this dataset is essential to maintain the sequential nature of the dataset.

The data was split into training and test set in order to train and test our model. The training set contained approximately 80% of the data, with the test set containing the remaining 20%. The data was randomised before being split to ensure that the test set did not contain data from only one or two participants. It was decided that a validation set for this data was not required because of the small data sample size. As there is an even amount of data for each genre, an even sample from class was taken for testing, leaving a balanced set for the training dataset.

Each song was of a slightly different length. In order for the data to be trained using Pytorch, the data was truncated to make each song to the length of the shortest song. This aided with the training of the CNN, even though LSTMs are able to train with sequences of varying lengths. It has been demonstrated that padding can have an effect on the classification of both an LSTM and a CNN, and careful consideration must be made when padding sequences for this reason [2]. It was determined that cutting off the end sequence of the song would not impact the accuracy of the network significantly.

2.3 Network Details

Two networks were trained, an LSTM and a CNN. Both LSTMs and CNNs have shown the ability to classify and predict physiological signals with good accuracy. An LSTM showed accuracy of approximately 85% when predicting emotions from EEG signals [1]. A CNN was used to predict emotions of a participant and achieved a high level of precision and accuracy [15].

The networks were trained, aiming to predict the genre of music the participant listened to based on the sequence of their electrodermal activity. Network parameters were determined based on trial and error results based on the testing accuracy and speed of computation obtained.

It was found that the optimal number of epochs used for both the LSTM and the CNN varied based on other hyperparameters. On many occasions, after several hundred epochs, a dramatic decrease in the testing accuracy was observed. This was deemed to be occurring because the model was overfitting to the training data. In these situations, early stopping of training happened to avoid overfitting the model. Reducing the number of hidden layers also helped to reduced the model overfitting during training.

One of the hyperparameters considered for the LSTM was the sequence length. It was found that long sequence lengths tended to overfit to the training data. In the LSTM model, the learning rate was extremely important to the performance of the training. A high learning rate resulted in the model oscillating between high and low accuracies, with the error of the model increasing and decreasing dramatically. Again, through iterative experimentation a low learning rate of 0.001 was found to produce the most accurate results.

The loss for the LSTM network was calculated using a combination of the softmax function and the negative log likelihood loss function. The optimisation function used was Adam which was chosen because it is an efficient computational method and it performed similarly on accuracy measures as more complex optimisation algorithms [10].

Due to the relatively small size of the dataset, it was decided that a single layer LSTM would be sufficient for achieving the desired accuracy in this model. Salman et al. [14] found that using a multilayer LSTM increased the accuracy of the classification problem from 72% to 80%, however that dataset contained more than 40 000 data points.

Network parameters for the CNN were determined based on trial and error results based on the testing accuracy and speed of computation obtained. Having extremely low learning rates for the model slowed training down a significant amount, so the rate had to be increased to achieve desired results, with a learning rate of 0.01 being the final rate decided upon.

Again, it was found that the optimal number of epochs used varied based on other hyperparameters. On many occasions, after several hundred epochs, a decrease in the testing accuracy was observed. This was deemed to be occurring because the model was overfitting to the training data. In these situations, training was stopped early to avoid overfitting the model.

The loss for the CNN was calculated using the negative log likelihood loss function. The optimisation function used was Stochastic Gradient Descent. A 1 dimensional CNN was used for the sequence data. 1 dimensional CNNs have shown to be comparable to human experts when evaluating pathological voice quality [3].

3 Results and Discussion

3.1 LSTM Sequence Length

Figure 1 shows that when the sequence length is small, the model does not achieve its best accuracy. This may be because there is a pattern in longer sequences. We also see a reduction in accuracy for large sequence lengths. This corresponded to a higher training accuracy, but a lower testing accuracy which leads to the conclusion that the larger sequence length resulted in the model overfitting. It was found that the best accuracy was found at sequence lengths of 32 and 64.

Overall given the size of the data set and the amount of noise present, achieving an accuracy near 70% for the prediction of a genre is pretty good and goes some way of demonstrating the power of the LSTM – using 16 features on the smaller data set only ever yielded about 50% in classification accuracy using a simple feed forward neural network.

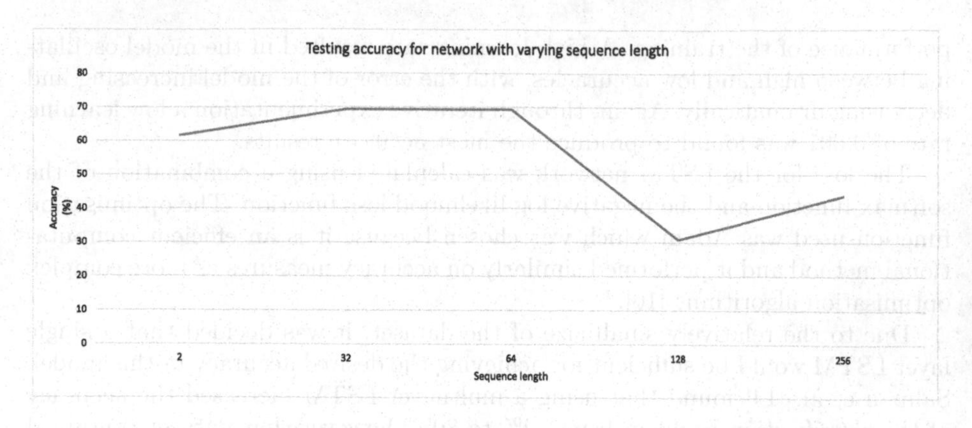

Fig. 1. A graph showing the decrease in accuracy of the LSTM with different sequence lengths.

3.2 Classification Results

We will now look at how the different numbers of epochs changed the accuracy of the LSTM and the CNN.

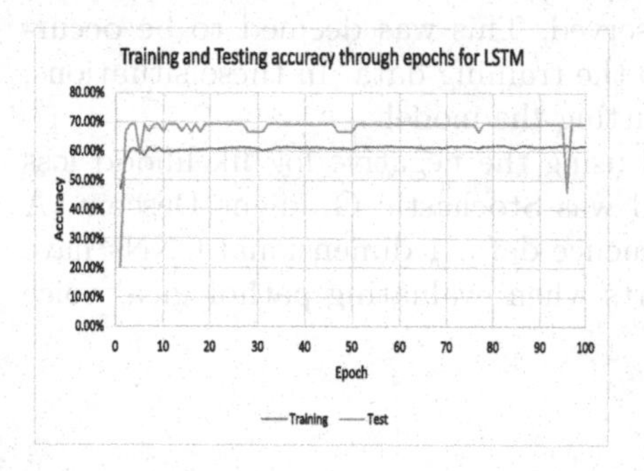

Fig. 2. LSTM classifier accuracy over 100 epochs

Fig. 3. CNN classifier accuracy over 100 epochs

Figure 2 shows us that the LSTM network seems to reach its maximum accuracy fairly early in the training cycle. The testing accuracy sometimes takes sharp dips in accuracy through the epochs, and learning should be stopped when this happens as we are achieving poor test accuracy for no increase in training accuracy.

Unlike the LSTM, the CNN requires more epochs to achieve a higher accuracy, and it would be interesting to see whether this accuracy might become more stable if more epochs were looked at. We see that both the training and testing accuracy has a general upward trend, so it is possible that if the model was left to run for longer that it might obtain a higher accuracy. Overall, the chosen hyperparameters do not appear to indicate any overfitting in the model, however the shape of the curve indicates that the learning rate could potentially be lowered to smooth out some of the jagged decreases in accuracy that are seen in Fig. 3.

3.3 Effects of Network Pruning

Pruning can reduce model complexity, reduce computational time and prevent overfitting of a model. We will now look at the effect of pruning different amounts of neurons in both the LSTM and the CNN.

Due to computing constraints, the model was only run for 10 epochs to test the pruning accuracy, meaning that the model has not always reached its highest accuracy when the pruning accuracy is reported.

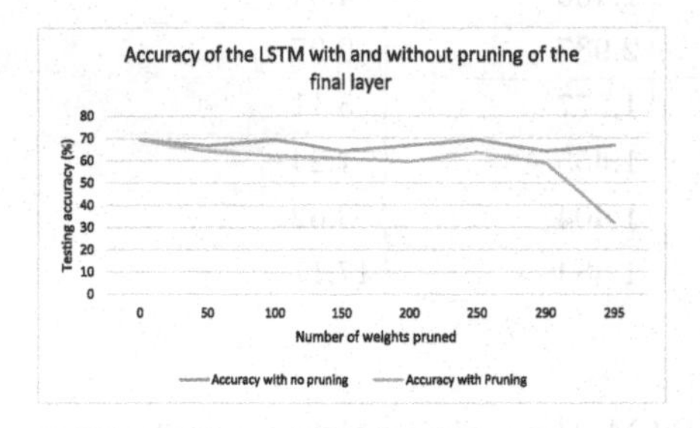

Fig. 4. LSTM accuracy with pruning

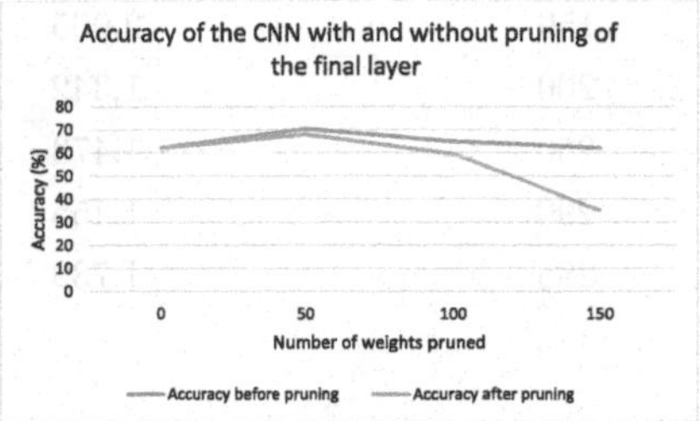

Fig. 5. CNN accuracy with pruning

The LSTM network was trained with 100 hidden neurons, which may be too many for the problem at hand. We have previously discovered that pruning the network can remove several neurons without significantly affecting the accuracy. The network was run with the same hyperparameters, each time pruning a different number of neurons to see what effect this has on the accuracy of the network. The pruning has been done on the final, fully connected layer. As there are three classes for the classification and 100 hidden neurons, there are 300 weights which can be pruned. The results of the pruning can be seen in Fig. 4.

It can be seen from the Fig. 4 that the accuracy removing the number of neurons decreases the accuracy of the network. However, we can also see that there is not a dramatic decrease in the accuracy of the network with pruning until we have removed 295 of the weights. This is somewhat surprising, but indicates that there are only a few neurons and associated weights that strongly contribute to the accuracy of the model.

Similar to the pruning of the LSTM, we see that several neurons can be pruned from the CNN without significantly decreasing the accuracy of the network, presented in Fig. 5. We do, however see a more drastic and earlier decrease in accuracy with pruning which may indicate that there are more neurons associated with the accuracy in the CNN than there were in the LSTM.

The computational advantages of pruning each network was also investigated. Table 1 and Table 2 show the average time taken to train one epoch at each number of neurons pruned for the LSTM and CNN respectively. The computational times can vary significantly, depending on background processes running on the computer used. For this reason, the non-pruning time was calculated at the same time as each of the different pruning metrics.

Table 1. Computational times for LSTM with pruning.

Number of weights pruned	Time before pruning (s)	Time after pruning (s)	Percentage change (%)
0	1.441	1.429	0.88
50	3.097	3.024	2.39
100	1.464	1.400	4.54
150	2.965	2.937	0.97
200	1.342	1.277	5.11
250	1.473	1.455	1.24
290	1.455	1.404	3.62
295	1.739	1.484	17.17

It can be seen in Table 1 that for the LSTM, the training time for each epoch does reduce with an increase in the number of weights pruned. Although this was only run over one epoch on a small data set, it does indicate that pruning an LSTM may be a way to reduce computational time without significantly affecting the accuracy of the network.

Table 2. Computational times for CNN with pruning.

Number of weights pruned	Time before pruning (s)	Time after pruning (s)	Percentage change (%)
0	3.336	3.304	0.966
50	3.295	3.307	–0.362
100	3.281	3.265	0.489
150	3.371	3.395	–0.697

Looking at the computational times for the CNN presented in Table 2, we don't see the same reduction in computing time we saw with the LSTM. This is possibly because there is still a significant amount of computation being done in the CNN, and pruning the final layer is less effective at reducing the computational time.

4 Conclusion and Future Works

Two neural networks were trained and each attempted to classify the genre of music a participant was listening to from their electrodermal activity data collected throughout the participant listening to the track. An LSTM and a CNN were trained and their accuracy was found to be 69.23% and 72.97% respectively. Rahman et al. [10] was able to achieve an accuracy of 96.8% when predicting the genre of the song from the participant's skin response. The paper, unlike this one, used more features in the classification, not just the electrodermal activity in the time-series. Future work may be to consider using more features along with the sequence data to see what accuracy can be obtained from combinations of features. The ability to classify music into genres using electrodermal signals recorded from participants indicates that there are reactions to music that are common across different people.

Pruning of each of the networks was also conducted and it was found that the network structure of each network can be reduced without having a reduction in the accuracy of the model. It also showed that for an LSTM, pruning the network may be an efficient way to reduce computational time without compromising the accuracy of the network.

There were some constraints when training this data, most notably the absence of a GPU to train the data. This means that often training had to be finished early due to time constraints. This analysis could be replicated and more epochs of the data could be run to allow the models to converge more and see the effects of a large number of epochs. In line with this, more tuning of the parameters could be conducted to test the maximum accuracy of the model. Different normalisation techniques could be employed and the effects of this change on the accuracy could be tested. In this analysis, the data was truncated to make the analysis of the sequences more simple. This is not a robust method, as it would not allow sequences of shorter lengths to be included in the analysis. Instead, an analysis with padding of noise or some other kind of padding could be used to allow the whole sequence to be used.

References

1. Alhagry, S., Aly, A., El-Khoribi, R.: Emotion recognition based on EEG using LSTM recurrent neural network. Int. J. Adv. Comput. Sci. Appl. **8**, 10 (2017)
2. Dwarampudi, M., Subba Reddy, N.V.: Effects of padding on LSTMS and CNNs. CoRR, abs/1903.07288 (2019)
3. Fujimura, S., et al.: Classification of voice disorders using a one-dimensional convolutional neural network. J. Voice (2020)
4. Gaurang, P., Ganatra, A., Kosta, Y., Panchal, D.: Behaviour analysis of multilayer perceptrons with multiple hidden neurons and hidden layers. Int. J. Comput. Theor. Eng. **3**, 332–337 (2011)
5. Gedeon, T., Harris, D.: Network reduction techniques. AMSE **1**, 119–126 (1991)
6. Gedeon, T.D.: Indicators of hidden neuron functionality: the weight matrix versus neuron behaviour. In: Proceedings 1995 Second New Zealand International Two-Stream Conference on Artificial Neural Networks and Expert Systems, pp. 26–29 (1995)

7. Hatami, N., Gavet, Y., Debayle, J.: Classification of time-series images using deep convolutional neural networks. CoRR, abs/1710.00886 (2017)
8. Jayalakshmi, T., Santhakumaran, A.: Statistical normalization and back propagation for classification. Int. J. Comput. Theor. Eng. **3**(1), 1793–8201 (2011)
9. Karim, F., Majumdar, S., Darabi, H., Chen, S.: LSTM fully convolutional networks for time series classification. CoRR, abs/1709.05206 (2017)
10. Kingma, D.P., Ba, J.: Adam: a method for stochastic optimization. In: Bengio, Y., LeCun, Y. (eds.) 3rd International Conference on Learning Representations, ICLR 2015, San Diego, CA, USA, 7–9 May 2015, Conference Track Proceedings (2015)
11. Lipton, Z.C. Kale, D.C., Elkan, C., Wetzel, R.: Learning to diagnose with LSTM recurrent neural networks (2015)
12. Ostafe, D.: Neural network hidden layer number determination using pattern recognition techniques. In: 2nd Romanian-Hungarian Joint Symposium on Applied Computational Intelligence, SACI (2005)
13. Rahman, J.S., Gedeon, T., Caldwell, S., Jones, R., Hossain, M.Z., Zhu, X.: Melodious micro-frissons: detecting music genres from skin response. In: 2019 International Joint Conference on Neural Networks (IJCNN), pp. 1–8 (2019)
14. Salman, A.G., Heryadi, Y., Abdurahman, E., Suparta, W.: Single layer and multilayer long short-term memory (LSTM) model with intermediate variables for weather forecasting. Procedia Comput. Sci. **135**, 89–98 (2018). The 3rd International Conference on Computer Science and Computational Intelligence (ICCSCI 2018): Empowering Smart Technology in Digital Era for a Better Life
15. Santamaria-Granados, L., Munoz-Organero, M., Ramirez-González, G., Abdulhay, E., Arunkumar, N.: Using deep convolutional neural network for emotion detection on a physiological signals dataset (AMIGOS). IEEE Access **7**, 57–67 (2019)

Unsupervised Multi-layer Spiking Convolutional Neural Network Using Layer-Wise Sparse Coding

Regina Esi Turkson[1,2(✉)], Hong Qu[1], Yuchen Wang[1], and Moses J. Eghan[2]

[1] University of Electronic Science and Technology of China, Chengdu, China
regina_turkson@yahoo.com, hongqu@uestc.edu.cn, yuchengwang0102@gmail.com
[2] University of Cape Coast, Cape Coast, Ghana
{rturkson,meghan}@ucc.edu.gh

Abstract. Deep learning architecture has shown remarkable performance in machine learning and AI applications. However, training a spiking Deep Convolutional Neural Network (DCNN) while incorporating traditional CNN properties remains an open problem for researchers. This paper explores a novel spiking DCNN consisting of a convolutional/pooling layer followed by a fully connected SNN trained in a greedy layer-wise manner. The feature extraction of images is done by the spiking DCNN component of the proposed architecture. And in achieving the feature extraction, we leveraged on the SAILnet to train the original MNIST data. To serve as input to the convolution layer, we process the raw MNIST data with bilateral filter to get the filtered image. The convolution kernel trained in the previous step is used to calculate the filtered image's feature map, and carry out the maximum pooling operation on the characteristic map. We use BP-STDP to train the fully connected SNN for prediction. To avoid over fitting and to further improve the convergence speed of the network, a dynamic dropout is added when the accuracy of the training sets reaches 97% to prevent co-adaptation of neurons. In addition, the learning rate is automatically adjusted in training, which ensures an effective way to speed up training and slow down the rising speed of the training accuracy at each epoch. Our model is evaluated on the MNIST digit and Cactus3 shape datasets, with the recognition performance on test datasets being 96.16% and 97.92% respectively. The level of performance shows that our model is capable of extracting independent and prominent features in images using spikes.

Keywords: Spiking Deep Convolutional Neural Network (DCNN) · Backpropagation STDP (BP-STDP) · Sparse representation

1 Introduction

Spiking Neural Networks (SNNs) allow learning that is primarily inspired by the brain transformation via discrete action potential in time through adaptive synapses [1]. However, developing a neural network that is efficient and

H. Yang et al. (Eds.): ICONIP 2020, LNCS 12534, pp. 353–365, 2020.
https://doi.org/10.1007/978-3-030-63836-8_30

biologically plausible as SNNs and powerful as deep neural networks in accomplishing different tasks is a prospect in the area of AI and computational neuroscience. Deep learning architecture has many layers of trainable parameters and has shown remarkable performance in machine learning and AI applications [2,3]. Deep Convolutional Neural Networks (DCNNs) have proven outstanding performance in image recognition [4–6], speech recognition [7,8], bioinformatics [9,10] and object detection and segmentation [11,12]. To link biologically plausible learning methods and conventional learning algorithms in neural networks, a number of deep SNNs have recently been developed [13,14]. One of the earliest feed-forward hierarchical CNN of spiking neurons for unsupervised learning was developed in [15]. This networks was extended for larger problems in [16]. A probabilistic STDP rule was later used in [17] to enhance the performance of the model in different object recognition tasks. The work in [18] is one of the deepest and recent STDP-trained convolutional architectures trained by STDP yielding a high accuracy of 98.4% on MNIST. Later, [19] introduced dual accumulator units to address unfavorable interactions of multi-layer training motivated by the design in [18] which used an unsupervised STDP to train all layers concurrently. However, this networks did not use SAILnet and dropout in their implementation. Layer-wise spiking representation learning approaches have been implemented in recent spiking CNNs [20–22] to train convolutional filters. The Foldiak model [23] first presented a set of three learning rules: Hebbian, anti- Hebbian, and homeo-static which attained representation coding in a non-SNN. Zylderberg et al. [24] later modified Foldiak's plasticity rules to develop a sparse representation model with SNNs called SAILnet. However, the lacking features of SAILnet were that the learning rules were not temporally local and the inputs utilized pixel intensity, not spike trains. Without the usage of spike times, the question of training using an approach that is spatio-temporally local and spike-based (as STDP [25]) remains unresolved. Tavanaei and Maida in [21,22] used SAILnet to train orientation selective kernels of the initial layer of a spiking CNN with a feature discovery layer equipped with an STDP variant [17] to extract visual features for classification. A new version of SAILnet was created in [26] to enhance its locality in space and time. However, their implement never included bilateral filters and dropout. Their feature discovery layer also used softmax probability. In the past, experiments of MNIST in SNN used spiking convolution structure to encode the original data into spike trains first, get the feature map, and then use the machine learning classifier to classify the pooled data [18,21]. Although the accuracy of these models are about some percentage points higher than ours, they use machine learning classifier classification to break the biological interpretability of spiking, and we use spiking method to steal the to rail. Since the convolution layer simulates the work of the primary visual cortex of the Primate's eye, our model is better in the biological interpretability. In addition, we use the pooled data to train the classification network, greatly speeding up the fitting speed, and training can be completed in about 20 epochs, which is also to our advantage. We propose a novel spiking DCNN architecture that comprises of a convolutional/pooling layer and a fully connected SNN equipped with an unsupervised learning, all of which undergo bio-inspired learning. We utilized bilateral filter for processing the raw image to obtain the filtered image. To avoid overfitting, we included a dynamic dropout techniques as a mask to randomly disable activation of unit and to shorten the training time. The fully connected SNN is trained using a backpropagation STDP (BP-STDP) rule adopted from [20,27].

2 Network Architecture

This section specifies the details of our spiking DCNN architecture. The network comprises of the SAILnet, a convolutional/pooling layer, and a fully connected SNN. Figure 1 shows the proposed Spiking DCNN architecture.

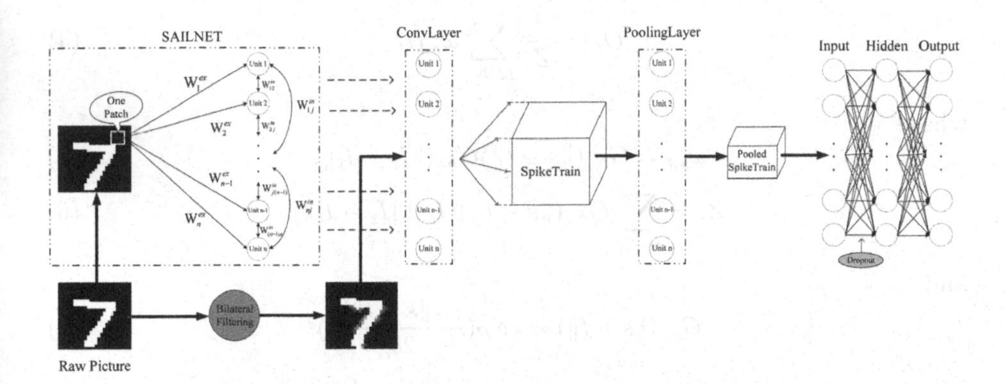

Fig. 1. Architecture of the proposed Spiking DCNN model.

2.1 SAILnet

The SAILnet trained the required filters to learn to represent an image as a set of sparse, distributed features of visual area V1. The SAILnet algorithm uses excitatory weights (W^{ex}), inhibitory weights (W^{inh}) and threshold (θ) as adaptive parameters. It has an input layer and an early representation layer, fully connected laterally by inhibitory weights. SAILnet used spiking neurons for its representation layer and maintained spatial locality in its plasticity rules. However, the lacking features is that the learning rules were not temporally local and the inputs utilized pixel intensity, not spike trains. The training method of our model is consistent with the method in SAILnet. The following formula is used to update the weights:

$$\Delta W_{im}^{inh} = \alpha(n_i n_m - \rho^2) \tag{1}$$

$$\Delta W_{ik}^{ex} = \beta n_i(x_k - n_i W_{ik}^{ex}) \tag{2}$$

$$\Delta\theta_i = \gamma(n_i - \rho) \tag{3}$$

where n_i (or n_m) denotes the number of spikes fired during a stimulus presentation, ρ is the sparsity parameter and x_k is the pixel intensity. Detail description of how stimulus is presented for SAILnet and SAILnet's learning rules is in [22].

2.2 Bilateral Filter

A bilateral filter is a non-linear, edge-preserving, and noise-reducing smoothing filter for images which substitutes the intensity of each pixel with a weighted average of intensity values from neighboring pixels. With bilateral filter, the weighted average of the adjacent pixel is set through the space and the pixel value distance. Critically, the weights depend not only on Euclidean distance of pixels, but also on the radiometric differences and this preserves the structure of the sharp. It is defined and presented as:

$$O_s = \frac{1}{Z_s} \sum_{t \in N_s} \omega_{s,t} I_t \tag{4}$$

where

$$\omega_{s,t} = G_{\delta_\alpha}(\|s - t\|) G_{\delta_\gamma}(\|I_s - I_t\|) \tag{5}$$

$$Z_s = \sum_{t \in N_s} G_{\delta_\alpha}(\|s - t\|) G_{\delta_\gamma}(\|I_s - I_t\|) \tag{6}$$

and

$$G_{\delta_\alpha}(\|s - t\|) = exp(\frac{-(\|s - t\|)^2}{2\delta_\alpha^2}) \tag{7}$$

$$G_{\delta_\gamma}(\|I_s - I_t\|) = exp(\frac{-(\|I_s - I_t\|)^2}{2\delta_\gamma^2}) \tag{8}$$

O_s denotes the output pixel value after bilateral filtering, $\omega_{s,t}$ denotes the weight of any pixel t, Z_s denotes the normalization term, G_{δ_α} is the spatial distance function, G_{δ_γ} is the pixel value distance function, I is the original input image to be filtered, t is the location of any pixel, s being the location of the target pixel and δ_α and δ_γ represent filter radius and filter ambiguity respectively and N_s is the size of the adjacent pixel set of $(2\delta_\alpha + 1) \times (2\delta_\alpha + 1)$.

2.3 Convolution and Pooling

Adopting same strategy in [21,22], we use the SAILnet of spiking neurons for sparse representation of visual features introduced in [24] to train our convolution layer. The trained weight obtained from the sparse coding algorithm is then used as a filter for the convolutional layer. The convolutional layer is represented by a 16-filter, which extracts visual features across the image. A convolutional layer is made up of several neuronal maps and shares weight to lessen the number of parameters as in CNN. Each neuron is discriminative to a visual feature controlled by its input synaptic weights. All the convolutional layers have Leaky Integrate and-Fire (LIF) neurons which gather input from presynaptic units in

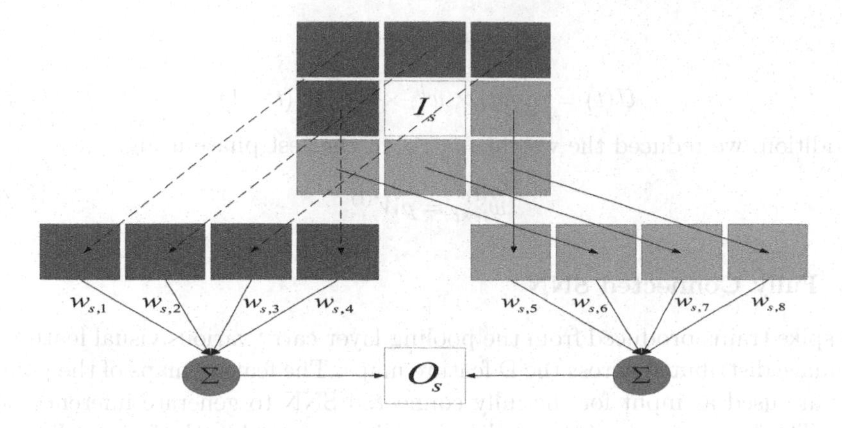

Fig. 2. Bilateral filter distribution of weight average of intensity values of pixelsl.

the form of spikes and fires when their internal potential reach a pre-specified threshold, thereby enabling information in the spiking DCNN to be transferred in the form of spike train instead of real numbers. To generate feature map, the filter is convolved independently with the image over 50ms time step presentation interval, depicting a specific image characteristic. After training, these filter weights are frozen and are no longer trained when used in our model. The purpose of the pooling layers is to reduce the spatial resolution of the feature maps and thus attain spatial invariance to input distortions and translations [2]. Max pooling is use in the layer to choose a neuron with the highest activity in a square neighborhood. The spike train produced from the pooling layer convey varied visual features of the image distributed across the feature maps. The feature maps of the pooling layer are unrolled into a 1-D vector and used as input for the fully connected SNN.

2.4 Dropout

We used dropout to effectively regularize our spiking DCNN. In our experiment, a one-dimensional vector is generated from the output spikes of the pooling layer on which a dynamic dropout is implemented to train the fully connected SNN for final classification. Firstly, the standard model has no dropout. Overfitting normally occurs when training accuracy is above 99%, and at this stage, the testing accuracy neither increase nor decrease. Therefore, adding a dynamic dropout when the accuracy of the training set reaches 97% to prevent co-adaptation of neurons. The whole dropout process is equivalent to averaging many different neural networks. We apply the dynamic dropout rule between the input layer and the hidden layer in the fully connected SNN. Bernoulli function is used to generate 1 or 0 according to a certain probability to decide whether to delete the neuron and it is presented as:

$$r_j^h \sim Bernoulli(p) \tag{9}$$

$$U(t) = input_i \times w_{ij}^h \times r_j^h + U(t-1) \tag{10}$$

In addition, we reduced the weight matrix in the test phase using:

$$w_{test}^{(l)} = pW^{(l)} \tag{11}$$

2.5 Fully Connected SNN

The spike trains produced from the pooling layer carry various visual features of the image distributed across the D feature maps. The feature maps of the pooling layer are used as input for the fully connected SNN to generate inference decisions. The feature maps of the pooling layer is connected to the feature discovery layer which processes these features to recognize and discover more complex features. This layer is implemented in a fully connected feedforward fashion using BP-STDP yielding the predicted classification of the input image. BP-STDP [20,27] shows the network's ability to extract independent and prominent features. BP-STDP uses bio-inspired STDP and high-performance backpropagation (gradient descent) rules. It offers a temporally local learning technique defined by an STDP/anti-STDP rule deduced from the backpropagation weight change rules that can be applied at each time step. We implemented the learning rule for single layer SNN trained with supervision denoted as:

$$\Delta W_{ih}(t) \propto \mu(z_i(t) - r_i(t))s_h(t) \tag{12}$$

The synaptic weights are updated using a teacher signal to switch between STDP and anti-STDP, ensuring the target neuron undergoes STDP and the non-target undergoes anti-STDP. A detailed representation of the BP-STDP is shown in [20, 27]. The various parameters and their associated values used in our experiment are shown in Table 1.

3 Experiment and Results

To evaluate the performance of our proposed spiking DCNN, we run an experiment using the MNIST datasets containing normalized handwritten digits and Cactus3 shape datasets which contain images of shapes of squares, circles, and triangles.

3.1 Datasets Description

The MNIST datasets of handwritten digit was chosen for this investigation due to it ubiquity in machine learning. An MNIST image has 28×28 gray scale images. Each image was divided in an overlapped of 5×5 patches with a stride of 1. Training set of 12,000 individual handwritten digits, each labeled 0–9. ($12{,}000 \times 576 = 6{,}912{,}000$) were used to train the convolution filters. Our test

Table 1. Parameter values of the spiking DCNN used in the simulations

Parameters	Description	Value
α	Hebbian learning rate	0.05
β	Anti-Hebbian rate	0.0001
γ	Threshold adjustment rate	0.01
R	Resistance	1 Ω
C	Capacitance	0.001F
θ^{conv}	Convolutional threshold	5
l_p	Pooling stride	2
a^+	LTP Amplification rate	0.001
a^-	LTD Amplification rate	0.00075
θ^h	Feature discovery neuron's threshold	0.5
T	Time step	50 ms

set consist of 1,800 digits. A patch is presented to a representation neuron to train a 16 convolutional filters. As each picture is input, the weight matrix is updated.

The Cactus3 Shape datasets which is made up of 300 images of squares, circles, and triangles. Each shape has 100 images of pixels. Training datasets of 240 images, 80 images from each shape and testing dataset of 60 images were used.

3.2 Experiments

In the MNIST experiment, we first train the convolution kernel with the raw MNIST data. To serve as input to the convolution, we process the raw MNIST data with bilateral filter to get the filtered image. Then we use the convolution kernel trained in the previous step to calculate the filtered image's feature map, and carry out the maximum pooling operation on the characteristic map. After getting the pooled data, we use BP-STDP to train the fully connected SNN for prediction.

In Fig. 3, the data used are the same, that is, the pooled data obtained by convolution layer and bilateral filter with the same parameters. The difference is the number of neurons used in the hidden layer of BP-STDP which are 50, 100, 500, 1000 and 1500 respectively. The experiments show that BP-STDP of 1000 neurons is the best. In order to conduct more sensitivity test on the number of hidden neurons that gives the best performance of the model, we varied the number of neurons around the 1000 neurons, such as 800 and 1200. It is confirmed that the 800 neurons performed slightly better than 1000 neurons for the proposed model. This is shown in Fig. 4.

To serve as a baseline and for comparative purpose, we use the original MNIST data and the data of MNIST calculated by convolution layer but without bilateral filter, to train and test the BP-STDP network with 800 neurons in hidden layer. The results is shown in Fig. 5. We can conclude that the performance is not as good as compared to our model of combination of convolution layer and bilateral filter.

Also, Fig. 6 shows the output voltage and spike train of neurons in the output layer when different handwritten digits are input into the trained model. The picture on the left shows the membrane voltage of 10 output neurons. The X axis is the time axis and the Y axis is the voltage. The picture on the right is the spike train produced by 10 output neurons, the X axis is the time axis, and the Y axis is the output neuron. The picture of each output spike train corresponds to the voltage map on the left. There are two lines in the figure, which shows the output from the output layer of the input numbers 4 and 7 respectively from top to bottom. In the figure, the output neurons corresponding to the input numbers are represented by triangle symbols, and the others are represented by dots. It can be seen that one output neuron's membrane voltage tends to rise very fast, and the number of spikes far exceeds that of other output neurons, which is exactly the correct output we expect, which ensures the accuracy of prediction. The voltage of other neurons decreased rapidly due to inhibition, and the lowest value was below - 1000 mV. In addition, there are a fewer similar number's output neuron membrane voltage which may rise slowly, for example, the left figure of the second line correctly outputs 7, but the output neuron voltage corresponding to the number 9 rises slowly. However, the figure on the right shows that the output neurons corresponding to number 7 spikes up to 25 times, while the output neurons corresponding to number 9 spikes only once. Correct prediction is still an absolute advantage, which shows that our model is very reliable.

In the Cactus3 shape experiment, we experimented with the Kaggle open-source kernel in order to compare to our model. AlexNet is used as the neural network model for training built using Keras framework. Alexnet has five convolution layers, three pooling layers and two fully connected layers. Because this dataset is simple, we reduce the number of neurons in the five convolutions to 64, 128, 192, 192 and 128 respectively. The number of neurons in the two fully connected layers were 120 and 84 respectively. To achieve better results, data enhancement technique is used to rotate the training set image in three different directions, that is, 90°, 180° and 270° respectively, as shown in Fig. 7. The accuracy of the test set is about 97.92%.

In Fig. 8, the accuracy of AlexNet is compared to our model in 100 training epochs. It is obvious that the overall performance of our model is better than AlexNet. In terms of fitting speed, our model's accuracy can get around 90% at the fifth epoch, while AlexNet needs to be close to 35 epochs to achieve similar accuracy. In terms of accuracy, after the 60th epoch, both models have been fitted. Our model is not only more accurate, but also less volatile.

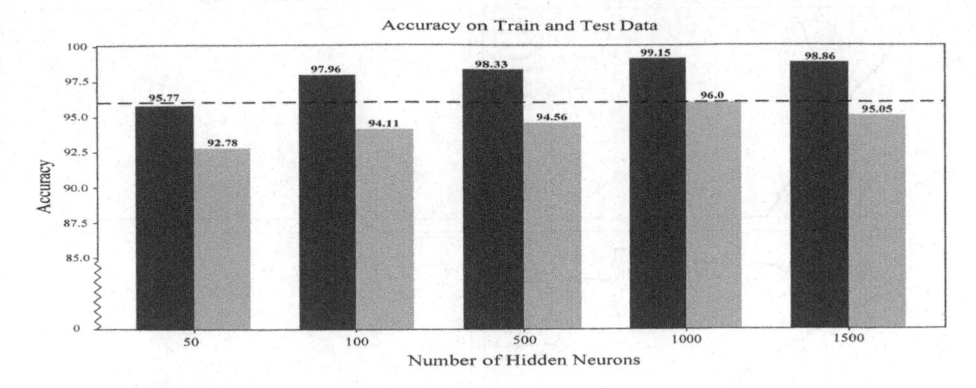

Fig. 3. Accuracy on train and test data for 50, 100, 500, 1000 and 1500 hidden neurons.

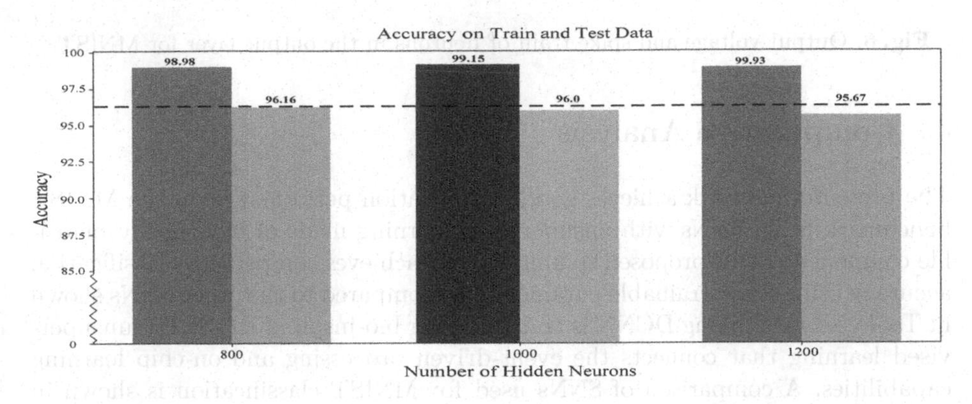

Fig. 4. Accuracy on train and test data for 800, 1000 and 1200 hidden neurons.

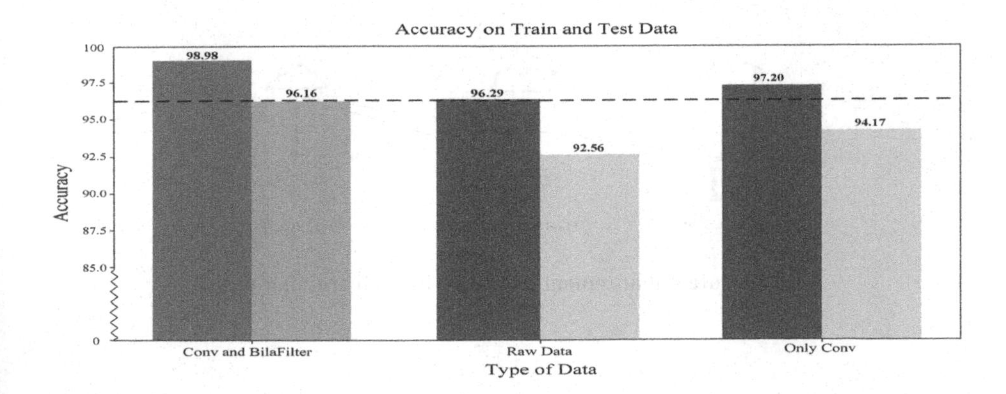

Fig. 5. Accuracy on train and test data for 800 hidden neurons for different data.

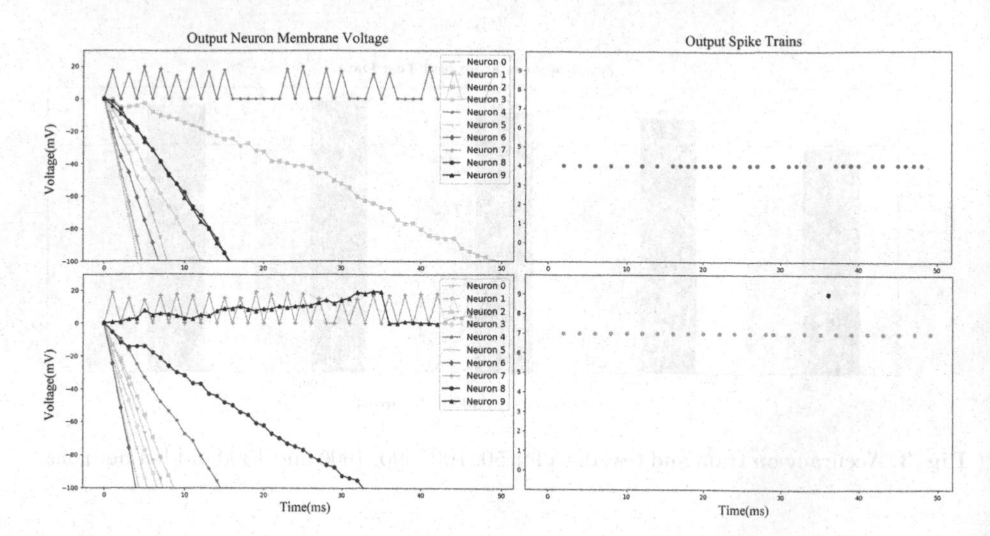

Fig. 6. Output voltage and spike train of neurons in the output layer for MNIST.

4 Comparative Analysis

The presented network achieves good classification performance on the MNIST benchmark using SNNs with unsupervised learning made of biologically plausible components. Our proposed spiking DCNN achieves competitive classification accuracy using fewer trainable parameters as compared to the other SNNs shown in Table 2. Our Spiking DCNN is trained using bio-inspired BP-STDP unsupervised learning that connects the event-driven processing and on-chip learning capabilities. A comparison of SNNs used for MNIST classification is shown in Table 2. From the table, some of the various models had more trainable parameters than our model. However, our spiking DCNN yields a competitive accuracy with fewer synaptic updates.

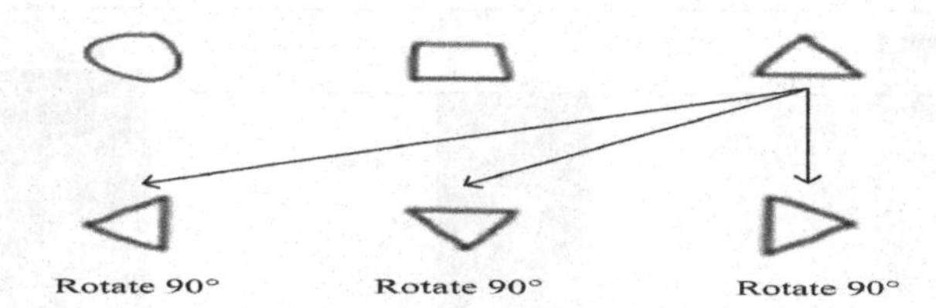

Fig. 7. Data enhancement of shape in each training data.

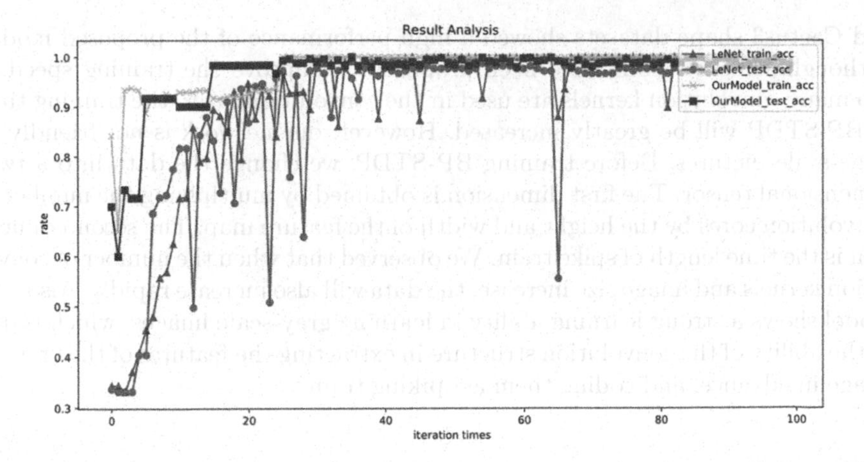

Fig. 8. Result analysis of our model compared to AlexNet in 100 training epoch.

Table 2. Classification accuracy of SNNs for MNIST datasets

SNN topology	Architecture	Spiking encoding scheme	Learning rule	# trainable parameters	Accuracy
Two-layer SNN [28]	Fully connected	Poisson distributed encoding	STDP	5017600	95%
SDNN [18]	Convolutional	Rank-order encoding	STDP+SVM	76500	98.4%
Spiking CNN [22]	Convolutional	Poisson-distributed spike encoding	Sparse Coding+STDP+SVM	590642	98.3%
SpiCNN [29]	Convolutional	Poisson-distributed spike encoding	STDP	25488	91.1%
Spiking DCNN (our model)	**Convolutional**	**SAILNET+ Bilaternal filter**	**BPSTDP**	**356636**	**96.16%**

5 Conclusion

Our work presents a novel spiking DCNN trained in a layer-wise and unsupervised manner. Our model used bilateral filter to process the original images. We extracted the features of the image in advance through a convolution structure, which simulates the function of primary visual cortex in human eyes, making our network more biologically interpretable. Firstly, the standard model has no dropout. Overfitting normally occurs when training accuracy is above 99%, and at this stage, the testing accuracy neither increase nor decrease. Therefore, we added a dynamic dropout when the accuracy of the training set reaches 97% to prevent co-adaptation of neurons. In addition, our learning rate is also adjust automatically in training, which is also an effective way to speed up training and to slow down the rising speed. The experimental results on the MNIST datasets digits and Cactus3 shape datasets showed a high performance of the proposed model. Although some methods have been proposed to improve the training speed, if too many convolution kernels are used in the convolution layer, the training time of BP-STDP will be greatly increased. However, our network is not friendly to large-scale pictures. Before training BP-STDP, we change the data into a two-dimensional tensor. The first dimension is obtained by multiplying the number of convolution cores by the height and width of the feature map. The second dimension is the time length of spike train. We observed that when the number of convolution kernels and image size increase, the data will also increase rapidly. Also, our model shows a strong learning ability in learning gray-scale images, which is due to the ability of the convolution structure in extracting the features of the original image in advance, and coding them as spiking train.

References

1. Tavanaei, A., Ghodrati, M., Kheradpisheh, S.R., Masquelier, T., Maida, A.: Deep learning in spiking neural networks. Neural Netw. **111**, 47–63 (2018)
2. Lecun, Y., Bengio, Y., Hinton, G.: Deep learning. Nature **521**(7553), 436–444 (2015)
3. Schmidhuber, J.: Deep learning in neural networks: an overview. Neural Netw. **61**, 85–117 (2015)
4. Krizhevsky, A., Sutskever, I., Hintom, G.E.: ImageNet classification with deep convolutional neural networks (2012)
5. Rawat, W., Wang, Z.: Deep convolutional neural networks for image classification: a comprehensive review. Neural Comput. **29**(9), 1–98 (2017)
6. Oquab, M., Bottou, L., Laptev, I., Sivic, J.: Learning and transferring mid-level image representations using convolutional neural networks. In: Proceedings IEEE Computer Society Conference on Computer Vision and Pattern Recognition, pp. 1717–1724. IEEE (2014)
7. Abdel-Hamid, O., Deng, L., Yu, D.: Exploring convolutional neural network structures and optimization techniques for speech recognition. In: Proceedings Annual Conference of the International Speech Communication Association, INTERSPEECH, pp. 3366–3370, August 2013
8. Sainath, T.N., Mohamed, A.R., Kingsbury, B., Ramabhadran, B.: Deep convolutional neural networks for LVCSR. In: ICASSP, IEEE International Conference on Acoustics, Speech, and Signal Processing - Proceedings, pp. 8614–8618 (2013)
9. Tavanaei, A., Maida, A.S., Kaniymattam, A., Loganantharaj, R.: Towards recognition of protein function based on its structure using deep convolutional networks. In: Proceedings - 2016 IEEE International Conference on Bioinformatics and Biomedicine, pp. 145–149 (2017)
10. Zeng, H., Edwards, M.D., Liu, G., Gifford, D.K.: Convolutional neural network architectures for predicting DNA-protein binding. Bioinformatics **32**(12), i121–i127 (2016)
11. Zhang, Y., Qiu, Z., Yao, T., Liu, D., Mei, T.: Fully convolutional adaptation networks for semantic segmentation. In: Proceedings IEEE Computer Society Conference on Computer Vision and Pattern Recognition, pp. 6810–6818 (2018)
12. Ronneberger, O., Fischer, P., Brox, T.: U-net: convolutional networks for biomedical image segmentation. In: Navab, N., Hornegger, J., Wells, W.M., Frangi, A.F.

(eds.) MICCAI 2015. LNCS, vol. 9351, pp. 234–241. Springer, Cham (2015). https://doi.org/10.1007/978-3-319-24574-4_28
13. Bengio, Y., Mesnard, T., Fischer, A., Zhang, S., Wu, Y.: STDP-compatible approximation of backpropagation in an energy-based model. Neural Comput. **577**, 555–577 (2017)
14. O'Connor, P., Welling, M.: Deep spiking networks. In: Proceedings of the 33rd International Conference on Machine Learning (2016)
15. Masquelier, T., Thorpe, S.J.: Unsupervised learning of visual features through spike timing dependent plasticity. PLoS Comput. Biol. **3**(2), 0247–0257 (2007)
16. Kheradpisheh, S.R., Ganjtabesh, M., Masquelier, T.: Bio-inspired unsupervised learning of visual features leads to robust invariant object recognition. Neurocomputing **205**, 382–392 (2016)
17. Tavanaei, A., Masquelier, T., Maida, A.S.: Acquisition of visual features through probabilistic spike-timing-dependent plasticity. In: Proceedings International Joint Conference on Neural Networks, pp. 307–314 (2016)
18. Kheradpisheh, S.R., Ganjtabesh, M., Thorpe, S.J., Masquelier, T.: STDP-based spiking deep convolutional neural networks for object recognition. Neural Netw. **99**, 56–67 (2018)
19. Thiele, J.C., Bichler, O., Dupret, A.: Event-based, timescale invariant unsupervised online deep learning with STDP. Front. Comput. Neurosci. **12**, 1–13 (2018)
20. Tavanaei, A., Maida, A.: BP-STDP: approximating backpropagation using spike timing dependent plasticity. Neurocomputing **330**, 39–47 (2019)
21. Tavanaei, A., Maida, A.S.: Multi-layer unsupervised learning in a spiking convolutional neural network. In: Proceedings International Joint Conference on Neural Networks, May 2017, pp. 2023–2030 (2017)
22. Tavanaei, A., Maida, A.S.: Bio-inspired spiking convolutional neural network using layer-wise sparse coding and STDP learning (2016)
23. Foldiak, P.: Forming sparse representation by local anti-Hebbian learning. Biol. Cybern. **170**, 165–170 (1990)
24. Zylberberg, J., Murphy, J.T., DeWeese, M.R.: A sparse coding model with Synaptically local plasticity and spiking neurons can account for the diverse shapes of V1 simple cell receptive fields. PLoS Comput. Biol. **7**(10) (2011)
25. Markram, H., Gerstner, W., Sjöström, P.J.: Spike-timing-dependent plasticity: a comprehensive overview. Front. Synaptic Neurosci. **4**, 2010–2012 (2012)
26. Tavanaei, A., Maida, A.S.: Training a hidden Markov model with a Bayesian spiking neural network. J. Sig. Process. Syst. **90**(2), 211–220 (2016). https://doi.org/10.1007/s11265-016-1153-2
27. Tavanaei, A., Kirby, Z., Maida, A.S.: Training Spiking ConvNets by STDP and gradient descent. In: Proceedings International Joint Conference on Neural Networks, July 2018, pp. 1–8 (2018)
28. Diehl, P.U., Cook, M.: Unsupervised learning of digit recognition using spike-timing-dependent plasticity. Front. Comput. Neurosci. **9**, 1–9 (2015)
29. Lee, C., Srinivasan, G., Panda, P., Roy, K.: Deep spiking convolutional neural network trained with unsupervised spike timing dependent plasticity. IEEE Trans. Cogn. Dev. Syst. **11**(3), 384–394 (2018)

VAEPP: Variational Autoencoder with a Pull-Back Prior

Wenxiao Chen[1,2], Wenda Liu[1], Zhenting Cai[1], Haowen Xu[1,2], and Dan Pei[1,2(✉)]

[1] Department of Computer Science and Technology, Tsinghua University, Beijing, China
{chen-wx17,liuwd17,caizt16,xhw15}@mails.tsinghua.edu.cn, peidan@tsinghua.edu.cn

[2] Beijing National Research Center for Information Science and Technology (BNRist), Beijing, China

Abstract. Many approaches to training generative models by distinct training objectives have been proposed in the past. Variational Autoencoder (VAE) is an outstanding model of them based on log-likelihood. In this paper, we propose a novel learnable prior, Pull-back Prior, for VAEs by adjusting the density of the prior through a discriminator that can assess the quality of data. It involves the discriminator from the theory of GANs to enrich the prior in VAEs. Based on it, we propose a more general framework, VAE with a Pull-back Prior (VAEPP), which uses existing techniques of VAEs and WGANs, to improve the log-likelihood, quality of sampling and stability of training. In MNIST and CIFAR-10, the log-likelihood of VAEPP outperforms models without autoregressive components and is comparable to autoregressive models. In MNIST, Fashion-MNIST, CIFAR-10 and CelebA, the FID of VAEPP is comparable to GANs and SOTA of VAEs.

Keywords: Variational Autoencoder · Deep generative model · Adversarial training

1 Introduction

How to learn deep generative models that are able to capture complex data patterns in high dimension space, *e.g.*, image datasets, is one of the major challenges in machine learning. Many approaches to training generative models by distinct training objectives have been proposed in the past, *e.g.*, Generative Adversarial Network (GAN) [6], flow-based models [11], PixelCNN [20], and Variational Autoencoder (VAE) [10,21]. GANs achieve SOTA in generative models, but likelihood of GANs are poor or incalculable.

The likelihood is important for generative models. VAE uses the variational inference and re-parameterization trick to optimize the evidence lower bound of log-likelihood (ELBO). In the past, researches [12,27] focused on enriching the variational posterior, but recently [26] showed that the standard Gaussian prior

H. Yang et al. (Eds.): ICONIP 2020, LNCS 12534, pp. 366–379, 2020.
https://doi.org/10.1007/978-3-030-63836-8_31

could lead to underfitting. To enrich the prior, several learnable priors have been proposed [2,25,26]. Most of them focus on approximating aggregated posterior which is the integral of the variational posterior and is the optimal prior that maximizes ELBO. However, existing methods based on the aggregated posterior reach limited performance, and the practical meaning of the aggregated posterior is ambiguous. We notice that a discriminator can assess the quality of data and **we argue that it is advisable to adjust the learnable prior by the discriminator, where the discriminator has clear practical meaning.**

We propose a novel learnable prior, Pull-back Prior, based on the discriminator. Firstly, a discriminator $D(x)$ is trained for assessing the quality of images. Then, we define a pull-back discriminator on latent space, by $D(G(z))$, where $G(z)$ is the generator. Finally, we adjust the density of the prior according to the pull-back discriminator.

We propose a training algorithm for VAE with Pull-back Prior (VAEPP), based on SGVB [10] with gradient penalty terms, which mixes the discriminator and the gradient penalty term [7,28] into VAE. Compared to AAE [18], VAEPP uses discriminator to adjust learnable prior while AAE uses discriminator to replace $KL(q(z)||p(z))$. Langevin dynamics, provided by [13] is used in VAEPP to improve the quality of sampling.

The main contributions of this paper are in the following:

- We propose a novel learnable prior, Pull-back Prior, which is adjusted by a discriminator that can assess the quality of data.
- We propose VAEPP framework to use existing techniques of VAE, *e.g.*, flow posterior, WGAN, *e.g.*, gradient penalty strategy, and Langevin dynamics to improve the log-likelihood and quality of sampling.
- In MNIST and CIFAR-10, the log-likelihood of VAEPP outperforms models without autoregressive components and is comparable to autoregressive models. In MNIST, Fashion-MNIST, CIFAR-10, and CelebA, the FID of VAEPP is comparable to GANs and SOTA of VAEs.

2 Background

2.1 VAEs and Learnable Priors

Many generative models aim to minimize the KL-divergence between the empirical distribution $p^*(x)$ and the model distribution $p_\theta(x)$, which leads to maximization likelihood estimation. The vanilla VAE [10] models the joint distribution $p_\theta(x,z)$ and the marginal distribution $p_\theta(x) = \int p_\theta(x,z)\mathrm{d}z$. VAE applies variational inference to obtain the evidence lower bound objective (ELBO):

$$\ln p_\theta(x) \geq \mathbb{E}_{q_\phi(z|x)}[\ln p_\theta(x|z) + \ln p_\theta(z) - \ln q_\phi(z|x)] \triangleq \mathcal{L}(x;\theta,\phi) \qquad (1)$$

where $q_\phi(z|x)$ is the variational encoder and $p_\theta(x|z)$ is the generative decoder. The training objective of VAE is $\mathbb{E}_{p^*(x)}[\mathcal{L}(x;\theta,\phi)]$ and it is optimized by SGVB with the re-parameterization trick. In vanilla VAE, the prior $p_\theta(z)$ is the standard Gaussian.

Recently, [26] showed that the simplistic prior could lead to underfitting. Since then many learnable priors are proposed to enrich the prior. Most of them focused on the aggregated posterior $q_\phi(z)$, which was shown to be the optimal prior that maximizes ELBO according to [26]. The training objective with learnable prior $p_\lambda(z)$ is:

$$\mathcal{L}(\theta, \phi, \lambda) = \mathbb{E}_{p^*(x)}\mathbb{E}_{q_\phi(z|x)} \ln p_\theta(x|z) + \mathbb{E}_{p^*(x)}\mathbb{H}[q_\phi(z|x)] + \mathbb{E}_{q_\phi(z)} \ln p_\lambda(z) \quad (2)$$

$\mathcal{I}, \mathcal{J}, \mathcal{K}$ denote 3 terms in Eq. (2) respectively for short thereafter. Notice that $p_\lambda(z)$ only appears in the last term $\mathcal{K}$ and the optimal solution of $p_\lambda(z)$ is $q_\phi(z)$. [25,26] obtained an approximation of $q_\phi(z)$ with their proposed prior, but reached limited performance.

2.2 GANs and Wasserstein Distance

In vanilla GAN [6], a generator is trained to generate samples for deceiving the discriminator, and a discriminator is trained to distinguish generated samples and real samples. However, vanilla GAN is unstable during the training process. To tackle this problem, Wasserstein distance is introduced by WGAN [1]:

$$W^1(\mu, \nu) = \sup_{Lip(D) \leq 1} \{\mathbb{E}_{\mu(x)} D(x) - \mathbb{E}_{\nu(x)} D(x)\} \quad (3)$$

where $Lip(D) \leq 1$ means that D is 1-Lipschitz, and μ, ν are measures. WGAN is optimized by minimizing $W^1(p^*, p_\theta)$ which can be seen as a min-max optimization.

WGAN makes progress toward stable training but sometimes fails to converge since it uses weight clipping for the Lipschitz constraint. WGAN-GP [7] and WGAN-div [28] improved WGAN by gradient penalty techniques, to achieve a more stable training.

3 Pull-Back Prior

3.1 Intuition of Pull-Back Prior

Definition 1. *The formula of Pull-back Prior is given by:*

$$p_\lambda(z) = \frac{1}{Z} p_\mathcal{N}(z) \cdot e^{-\beta D(G(z))} \quad (4)$$

where $p_\mathcal{N}$ is a simple prior, D is a discriminator, G is a generator, β is a learnable scalar, $f_\lambda(z)$ denotes $p_\mathcal{N}(z)e^{-\beta D(G(z))}$, and $Z = \int_\mathcal{Z} f_\lambda(z)\mathrm{d}z$ is the partition function.

A design proposition of Pull-back Prior is that we increase $p_\lambda(z)$ where z generates better data and decrease $p_\lambda(z)$ where z generates worse data. In Pull-back Prior, D is a discriminator to assess the quality of x, where smaller $D(x)$ indicates x being more similar to real data, as shown in Fig. 1. Such discriminator

$D(x)$ is defined on x, and the pull-back discriminator on z is defined by $D(G(z))$, where $D(G(z))$ represents the ability of z that can generate data with high quality. To increase $p_\lambda(z)$ at the better z and decrease $p_\lambda(z)$ at the worse z, we modify $p_\mathcal{N}(z)$ by $\beta D(G(z))$, and then normalize it by Z. Finally, we obtain the basic formula of Pull-back Prior.

The theoretical derivation for Pull-back Prior is provided in Theorem 1. However, it remains questions about how to obtain D and G, determine β, and calculate Z.

Fig. 1. The discriminators on above images (generated by linear interpolation of two sample from $q_\phi(z)$), are better at both sides and worse at the middle, which validates the intuition that a discriminator can assess the quality of images. Moreover, in VAEPP the density of z which generates better images will increase, and the density of z which generates worse images will decrease.

3.2 How to Obtain D and G

In our model, $G(z) = \mathbb{E}_{p_\theta(x|z)}x$, *i.e.*, the mean of $p_\theta(x|z)$. In our experiments, $p_\theta(x|z)$ is chosen to be a Discretized Logistic [23] or a Bernouli. $G(z)$ is generated by a neural network and it is set as the mean of $p_\theta(x|z)$.

D plays an important role in Pull-back Prior. We shall propose two ways to obtain D in Sect. 4.1 and Sect. 4.2, and compare them later in our experiments.

3.3 How to Determine β

To maximize ELBO, we can obtain the optimal β by (λ contains β and ω, where ω denotes the parameters of D):

$$\beta = \arg\max_\beta \mathcal{L}(\theta, \phi, \lambda) = \arg\max_\beta \mathcal{L}(\theta, \phi, \beta, \omega) \tag{5}$$

When the training coverages, $\partial\mathcal{L}/\partial\beta = 0$. The gradient $\partial\mathcal{L}/\partial\beta$ is:

$$\begin{aligned} \frac{\partial \ln Z}{\partial \beta} &= \frac{1}{Z}\int_{\mathcal{Z}} p_\mathcal{N}(z)e^{-\beta D(G(z))} \cdot (-D(G(z)))\mathrm{d}z = \mathbb{E}_{p_\lambda(z)}[-D(G(z))] \\ \frac{\partial \mathcal{L}}{\partial \beta} &= \mathbb{E}_{q_\phi(z)}[-D(G(z))] - \frac{\partial \ln Z}{\partial \beta} = -\mathbb{E}_{q_\phi(z)}[D(G(z))] + \mathbb{E}_{p_\lambda(z)}[D(G(z))] \end{aligned} \tag{6}$$

The 1st term in Eq. (6) is the mean of the discriminator on reconstructed data (reconstructed data are nearly same as real data in VAE, after few epochs in training). The 2nd term in Eq. (6) is the mean of the discriminator on data generated from p_λ. $\partial\mathcal{L}/\partial\beta = 0$ means that the discriminator can't distinguish reconstructed data and generated data when the training converges. It coincides with the philosophy of GANs that the discriminator can't distinguish real data and generated data when the generator is well-trained.

Noticing that $p_{\mathcal{N}}$ is a special case of p_λ where $\beta = 0$, Pull-back Prior is a general form of the standard Gaussian. We shall compare their performance in experiments.

3.4 The Upper-Bound of Z

It is difficult to calculate the partition function Z exactly. Fortunately for VAEPP, it is acceptable to obtain an upper-bound of Z, denoted by $\hat{Z}$. Using the upper-bound $\hat{Z}$ in training and evaluation, we can obtain lower-bounds of log-likelihood and ELBO (note, $\hat{p}_\theta(x) \leq p_\theta(x)$ indicates $\ln \hat{p}_\theta(x) \leq \ln p_\theta(x)$):

$$\hat{p}_\theta(x) = \int \frac{p_\theta(x|z) f_\lambda(z)}{\hat{Z}} \mathrm{d}z \leq \int \frac{p_\theta(x|z) f_\lambda(z)}{Z} \mathrm{d}z = p_\theta(x)$$

$$\hat{\mathcal{K}} = \mathbb{E}_{q_\phi(z)} \ln \frac{1}{\hat{Z}} f_\lambda(z) \leq \mathbb{E}_{q_\phi(z)} \ln \frac{1}{Z} f_\lambda(z) = \mathcal{K}$$

$$\hat{\mathcal{L}} = \mathcal{I} + \mathcal{J} + \hat{\mathcal{K}} \leq \mathcal{I} + \mathcal{J} + \mathcal{K} = \mathcal{L}$$

The upper-bound $\hat{Z}$ in our model is derived as follows:

$$\hat{Z} = \mathbb{E}_{p^*(x)} \mathbb{E}_{q_\phi(z|x)} \frac{f_\lambda(z)}{\frac{1}{N} q_\phi(z|x)} \geq \mathbb{E}_{p^*(x)} \mathbb{E}_{q_\phi(z|x)} \frac{f_\lambda(z)}{q_\phi(z)} = \mathbb{E}_{q_\phi(z)} \frac{f_\lambda(z)}{q_\phi(z)} = Z \tag{7}$$

The fact that $\hat{Z}$ is an upper-bound of Z comes from:

$$\frac{q_\phi(z|x)}{N} \leq \frac{1}{N} \sum_{i=1}^{N} q_\phi(z|x^{(i)}) \approx \mathbb{E}_{p^*(x)} q_\phi(z|x) = q_\phi(z)$$

In previous VAE literatures [2,10,25] and our paper, it is a common practice to dynamically sample 0/1 binary images (which is exactly the x of our VAE and many other paper's) from real-value grayscale images (whose distribution is denoted by $p^*(e)$). Each pixel value of e is normalized into $[0, 1]$, and then is used as the probability of the corresponding pixel of x being 1 (denoted by $p^*(x|e)$). In such situation, even when the size M of original grayscale image dataset is moderate, the size N of the sampled images dataset is exponentially large. Hence, we shall severely overestimate Z since $\frac{1}{N} q_\phi(z|x) \ll q_\phi(z)$ if directly using Eq. (7). Therefore, we consider to use $p^*(e)$ instead of $p^*(x)$ to estimate a lower bound of $q_\phi(z)$ in such datasets (called Bernouli datasets in our paper). Given that $p^*(x) = \mathbb{E}_{p^*(e)} p^*(x|e)$, we shall have:

$$q_\phi(z) = \mathbb{E}_{p^*(x)} q_\phi(z|x) = \mathbb{E}_{p^*(e)} \mathbb{E}_{p^*(x|e)} q_\phi(z|x) = \mathbb{E}_{p^*(e)} q_\phi(z|e) \tag{8}$$

where $q_\phi(z|e)$ denotes $\mathbb{E}_{p^*(x|e)} q_\phi(z|x)$. Equation (8) suggests that we may train a variational encoder $q_\phi(z|e)$ instead of $q_\phi(z|x)$, along with a generative decoder

$p_\theta(x|z)$, while the log-likelihood estimator is still correct:

$$\mathbb{E}_{p^*(x)} \log p_\theta(x) = \mathbb{E}_{p^*(e)}\mathbb{E}_{p^*(x|e)} \log p_\theta(x) = \mathbb{E}_{p^*(e)}\mathbb{E}_{p^*(x|e)} \log \mathbb{E}_{q_\phi(z|e)} \frac{p_\theta(x,z)}{q_\phi(z|e)}$$

Based on this idea, we then derive $\hat{Z}$ and ELBO as:

$$\hat{Z} = \mathbb{E}_{p^*(e)}\mathbb{E}_{q_\phi(z|e)} \frac{f_\lambda(z)}{\frac{1}{M}q_\phi(z|e)} \geq \mathbb{E}_{p^*(e)}\mathbb{E}_{q_\phi(z|e)} \frac{f_\lambda(z)}{q_\phi(z)} = \mathbb{E}_{q_\phi(z)} \frac{f_\lambda(z)}{q_\phi(z)} = Z$$

$$\begin{aligned}\mathbb{E}_{p^*(x)} \ln p_\theta(x) &= \mathbb{E}_{p^*(e)}\mathbb{E}_{p^*(x|e)} \ln \mathbb{E}_{q_\phi(z|e)} \frac{p_\theta(x|z)p_\lambda(z)}{q_\phi(z|e)} \\ &\geq \mathbb{E}_{p^*(e)}\mathbb{E}_{p^*(x|e)}\mathbb{E}_{q_\phi(z|e)} \ln \frac{p_\theta(x|z)p_\lambda(z)}{q_\phi(z|e)} \\ &= \mathbb{E}_{p^*(x)} \ln p_\theta(x) - \mathbb{E}_{p^*(e)}\mathbb{E}_{p^*(x|e)} KL(q_\phi(z|e), p_\theta(z|x)) \end{aligned} \tag{9}$$

Equation (9) is similar to the original ELBO, and the conclusions in this paper hold for Eq. (9) by repeating derivations for Eq. (9). $\mathcal{L}(\theta, \phi, \lambda)$ denotes Eq. (9) in Bernouli datasets.

Review the estimation of Z. By the theory of importance sampling, p_λ is the optimal choice for the proposal distribution in the estimation of Z. However, it is intractable to sample from p_λ. [2] uses $p_\mathcal{N}$ as the proposal distribution to estimate Z but when $KL(p_\mathcal{N}, p_\lambda)$ is high, the variance of this estimation will be large.

In our experiments, $KL(q_\phi, p_\lambda)$ is much smaller than $KL(p_\mathcal{N}, p_\lambda)$. Therefore, we choose $q_\phi(z)$ as the proposal distribution and use $\frac{1}{N}q_\phi(z|x)$ as a lower bound of $q_\phi(z)$, to obtain $\hat{Z}$ in Eq. (7). The variance of $\hat{Z}$ is acceptable in experiments. In training, $p_\mathcal{N}(z)$ could be used together with $q_\phi(z)$, as the proposal distributions, since $KL(p_\mathcal{N}, p_\lambda)$ is small in the beginning of training.

4 Training and Sampling

In this section, we propose two training methods and a sampling method for VAEPP. The main difference between two trainings method is how to train the discriminator.

4.1 2-Step Training for VAEPP

The discriminator should be obtained by $W^1(p_\theta, p^*)$, suggested by WGAN [1]. However in VAEPP, p_θ is intractable for sampling, since $p_\theta(x) = \mathbb{E}_{p_\lambda(z)} p_\theta(x|z)$ and $p_\lambda(z)$ is intractable for sampling.

When β is small enough, $p_\lambda(z)$ is near to $p_\mathcal{N}(z)$ which is feasible for sampling. Then, $p_\theta(x)$ is near to $p^\dagger(x)$, where $p^\dagger(x) = \mathbb{E}_{p_\mathcal{N}(z)} p_\theta(x|z)$ and $p^\dagger(x)$ is feasible for sampling. Therefore, we try to obtain the discriminator by $W^1(p^\dagger, p^*)$ instead. β is limited by a hyper-parameter. In this way, an discriminator D is trained by:

$$W^1(p^\dagger, p^*) = \sup_{Lip(D) \leq 1} \mathbb{E}_{p^\dagger(x)} D(x) - \mathbb{E}_{p^*(x)} D(x)$$

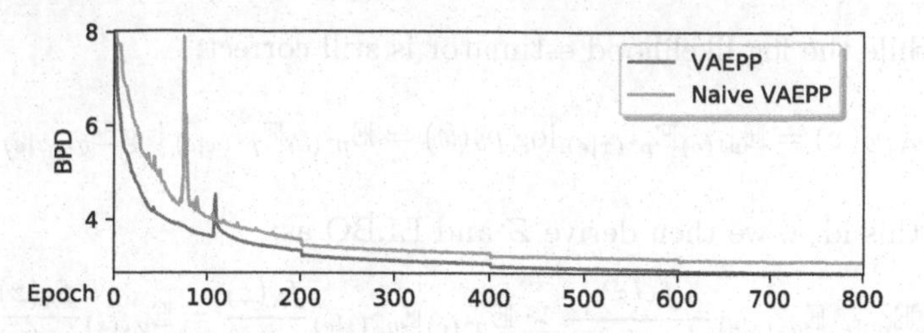

Fig. 2. Training loss of Naive VAEPP and VAEPP on CIFAR-10. Naive VAEPP is more unstable and nearly crashes at 80 epoch while VAEPP has a little acceptable gap. From global view, the training loss of VAEPP is more smooth than Naive VAEPP and is better than Naive VAEPP's over almost all training process, which validates the motivation in Sect. 4.2. There are little gaps at per 200 epoch because learning rate is reduced to half at every 200 epoch.

The other parameters of VAEPP are trained by SGVB:

$$\max_{\theta,\phi,\beta} \mathcal{L}(\theta, \phi, \beta, \omega)$$

Above two optimizations run alternatively, as shown in Algorithm 1. The model trained by 2-step training algorithm is called Naive VAEPP.

4.2 1-Step Training for VAEPP

However, the training process of Algorithm 1 is unstable and inefficient, as shown in Fig. 2. We suspect that the two independent optimizations instead of one whole optimization, may lower the log-likelihood and stability. Therefore, we try to combine the training for $\theta, \phi, \beta, \omega$ into a whole optimization. Our solution is to use SGVB with the gradient penalty term to train VAEPP:

$$\max_{\theta,\phi,\beta} \max_{Lip(D)\leq 1} \mathcal{L}(\theta, \phi, \beta, \omega) \tag{10}$$

Theorem 2 in appendix indicates that it is reasonable to obtain discriminator D during optimizing Eq. (10), and the gradient penalty term should be multiplied by β. Finally, the optimizations for θ, ϕ, β and ω are combined into one, as shown in Algorithm 2. The model trained by 1-step training Algorithm is called VAEPP.

4.3 Sampling from VAEPP

We apply Langevin dynamics to sample z from $p_\lambda(z)$. It could generate natural and sharp images and only requires that $\nabla_z \log p_\lambda(z)$ is computable and $p_\lambda(z_0)$ is high enough where z_0 is the initial point of Langevin dynamics [24]. Moreover, [13] has implemented a Metropolis-Adjusted Langevin Algorithm (MALA) for sampling, where the formula of density also contains a discriminator term. But how to obtain the initial z_0 whose density is high enough is still a problem.

Algorithm 1: 2-step training algorithm for VAEPP

Input: The gradient penalty algorithm R, the batch size b, the number of critic iterations per generator iteration n_c, the parameters for Adam Optimizers, τ.

1 **while** $\theta, \phi, \beta, \omega$ *have not converged* **do**
2   **for** $k = 1, \ldots n_c$ **do**
3   **for** $i = 1, \ldots, b$ **do**
4   Sample $e, x \sim p^*$, $z \sim q_\phi(z|e)$, $\epsilon \sim p_\mathcal{N}$;
5       $Z^{(i)} \leftarrow \frac{1}{2}(e^{-\beta D(G(\epsilon))} + \frac{f_\lambda(z)}{\frac{1}{M} q_\phi(z|e)})$;
6       $\mathcal{L}^{(i)} \leftarrow \ln p_\theta(x|z) + \ln f_\lambda(z) - \ln q_\phi(z|e)$;
7   **end**
8     $\theta, \phi, \beta \leftarrow$ Adam $(\nabla_{\theta,\phi,\beta}(\frac{1}{b}\sum_i^b \mathcal{L}^{(i)} - \ln(\frac{1}{b}\sum_i^b Z^{(i)})), \{\theta, \phi, \beta\}, \tau)$;
9   **end**
10   **for** $i = 1, \ldots, b$ **do**
11   Sample $e, x \sim p^*$, $z \sim p_\mathcal{N}$, $\hat{e} \leftarrow G(z)$;
12   get gradient penalty term $\zeta \leftarrow R(e, \hat{e})$;
13     $L^{(i)} \leftarrow D(\hat{x}) - D(x) + \zeta$;
14   **end**
15   $\omega \leftarrow$ Adam $(\nabla_\omega \frac{1}{b}\sum_i^b L^{(i)}, \omega, \tau)$;
16 **end**

The sampling of VAEPP consists of 3 parts: sample initial z_0 by a GAN modeling $q_\phi(z)$; generate $z \sim p_\lambda(z)$ from initial z_0 by MALA; generate image from z with the decoder. This sampling process is similar to 2-Stage VAE [4]. The main difference between them is that VAEPP applies Langevin dynamics to sample from the explicit prior but 2-Stage VAE doesn't, since the prior of 2-Stage VAE is implicit. In experiments, sampling from the explicit prior may improve the quality of sampling in some datasets.

Accept-Reject Sampling [2] is useless for p_λ because it requires that $p_\lambda(z)/p_\mathcal{N}(z)$ is bounded by a constant T on the support of p_λ, such that a sample could be accepted in expected T times. But it is hard to ensure that there exists a small T in VAEPP.

5 Experiments

5.1 Log-Likelihood Evaluation

We compare our algorithms with other models based on log-likelihood, on MNIST and CIFAR-10 as shown in Table 1, and on Static-MNIST [15], Fashion-MNIST [29], and Omniglot [14], as shown in Table 2. Because the improvement of auto-regressive components is significant, we separate models by whether they use an auto-regressive component. The reason of why VAEPP doesn't use an auto-regressive component is that VAEPP is time-consuming in training, evaluation and sampling due to the huge structure (need additional discriminator)

Algorithm 2: 1-step training algorithm for VAEPP

Input: The gradient penalty method R, the batch size b, the parameters τ for Adam Optimizers.

1 **while** $\theta, \phi, \beta, \omega$ *have not converged* **do**
2   **for** $i = 1, \ldots, b$ **do**
3   Sample $e, x \sim p^*$, $z \sim q_\phi(z|e)$, $\epsilon \sim p_\mathcal{N}$, $\hat{e} \leftarrow G(\epsilon)$, $\zeta \leftarrow R(e, \hat{e})$;
4     $Z^{(i)} \leftarrow \frac{1}{2}(e^{-\beta D(G(\epsilon))} + \frac{f_\lambda(z)}{\frac{1}{M} q_\phi(z|e)})$;
5     $\mathcal{L}^{(i)} \leftarrow \ln p_\theta(x|z) + \ln f_\lambda(z) - \ln q_\phi(z|e) + \beta\zeta$;
6   **end**
7   $\theta, \phi, \beta, \omega \leftarrow$ Adam $(\nabla_{\theta,\phi,\beta}(\frac{1}{b}\sum_i^b \mathcal{L}^{(i)} - \ln(\frac{1}{b}\sum_i^b Z^{(i)})), \{\theta, \phi, \beta, \omega\}, \tau)$
8 **end**

Table 1. Test NLL on MNIST and Bits/dim on CIFAR-10. The data are from [2,3,17, 25,26]. Bits/dim means $-\log p_\theta(x|z)/3072/\ln 2$. VAEPP+Flow means VAEPP with a normalization flow on encoder. The decoder on CIFAR-10 is Discretized Logistic and the decoder on MNIST is Bernouli. Additional, we compare VAE based on $q_\phi(z|x)$ and $q_\phi(z|e)$ on MNIST, whose NLL are 81.10 and 83.30 respectively. Moreover, evaluation using importance sampling based on $q_\phi(z|e)$ has enough small standard deviation (0.01) with 10^8 samples altogether. It validates that $q_\phi(z|e)$ is stable for evaluation and doesn't improve the performance. VAEPP reaches SOTA without autoregressive component, and is comparable to models with autoregressive component.

Model	MNIST	CIFAR	Model	MNIST	CIFAR
With autoregressive			Without autoregressive		
PixelCNN	81.30	3.14	Implicit Optimal Priors	83.21	
DRAW	80.97	3.58	Discrete VAE	81.01	
IAFVAE	79.88	3.11	LARS	80.30	
PixelVAE++	78.00	2.90	VampPrior	79.75	
PixelRNN	79.20	3.00	BIVA	78.59	3.08
VLAE	79.03	2.95	**Naive VAEPP**	76.49	3.15
PixelSNAIL		2.85	**VAEPP**	76.37	2.91
PixelHVAE+VampPrior	78.45		**VAEPP+Flow**	**76.23**	**2.84**

and Langevin dynamics. It is not easy to apply an auto-regressive component on VAEPP since auto-regressive component is also time-consuming. Therefore, how to apply an autoregressive component on VAEPP is a challenging practical work and we leave it for future work. IvOM [19] of VAEPP reaches 0.018, 0.017 on MNIST, CIFAR-10, which shows good data coverage.

We compare Naive VAEPP trained by Algorithm 1 and VAEPP trained by Algorithm 2 on CIFAR-10, as the gradient penalty algorithm is chosen from 3 strategies: WGAN-GP, WGAN-div-1 (sampling the linear interpolation of real data and generated data) and WGAN-div-2 (sampling real data and generated data both) in Table 3.

Table 2. Test NLL on Static MNIST, Fashion-MNIST and Omniglot.

Model	Static MNIST	Fashion	Omniglot
Naive VAEPP	78.06	214.63	90.72
VAEPP	77.73	213.24	89.60
VAEPP+Flow	77.66	213.19	89.24

Table 3. Comparison between Naive VAEPP and VAEPP when gradient penalty strategy varies on CIFAR-10 with dim $\mathcal{Z} = 1024$. For any gradient penalty strategy in the table, VAEPP outperforms Naive VAEPP, which validates the our intuition of Algorithm 2. WGAN-div-1 is chosen as the default gradient penalty strategy since it reaches best performance in VAEPP.

GP Strategy	WGAN-GP	WGAN-div-1	WGAN-div-2
Naive VAEPP	3.15	3.20	4.47
VAEPP	2.95	2.91	2.99

To validate that it is better to use $q_\phi(z)$ to evaluate Z than $p_\mathcal{N}(z)$ in Sect. 3.4, we calculate the $KL(q_\phi(z)||p_\lambda(z))$ and $KL(p_\mathcal{N}(z)||p_\lambda(z))$ on CIFAR-10 and MNIST. The former is smaller than $\mathcal{L} - \mathcal{I}$ [9](180.3 on CIFAR-10 and 12.497 on MNIST), and the latter can be evaluated directly (1011.30 on CIFAR-10 and 57.45 on MNIST). Consequently, $q_\phi(z)$ is much closer to $p_\lambda(z)$ than $p_\mathcal{N}(z)$.

To ensure the variance of estimation $\hat{Z}$ is small enough, the $q_\phi(z|e)$ is chosen as truncated normal distribution (drop the sample whose magnitude is more than 2 standard deviation from the mean) instead of normal distribution, which may reduce the gap between $q_\phi(z)$ and $\frac{1}{M}q_\phi(z|x)$. With 10^9 samples, the variance of $\hat{Z}$ with truncated normal and normal is 0.000967 (truncated normal) and 0.809260 (normal) respectively in MNIST. Therefore, truncated normal is chosen as the default setting.

5.2 Quality of Sampling

As a common sense, the quality of sampling of VAEs is worse than GANs, and it is indeed a reason that we involve the techniques of GAN to improve VAE model: We use the discriminator to adjust learnable prior and a GAN to sample the initial z_0 for Langevin dynamics. These techniques will help VAEPP improve the quality of samples. The samples of VAEPP gets good FID [8], comparable to GANs and 2-Stage VAE (SOTA of VAE in FID), as shown in Table 4. Some generated images of VAEPP are shown in Fig. 3. It is important to notice that the GAN in VAEPP only plays the role that generates z_0 with high $p_\lambda(z_0)$, in latent space with small dimension, instead of image. The ability of VAEPP that generates image from z is totally depend on the decoder.

It is hard to reach best FID, IS [22] and log-likelihood simultaneously with one setting. We observe the fact that when dim $\mathcal{Z}$ (the dimension of latent space)

Fig. 3. Examples of generated images from VAEPP on CelebA [16] and CIFAR-10.

Table 4. FID comparison of GANs and VAEs. Best GAN indicates the best FID on each dataset across all GAN models when trained using settings suggested by original authors [4]. VAEPP uses Bernouli as decoder on MNIST and Discretized Logistic on others. GAN-VAEPP indicates that image is directly sampled from z_0, without Langevin dynamics. In experiments, we found that the FID of VAEPP is usually better than GAN-VAEPP, which means that the explicit prior and Langevin dynamics might be useful for improving the quality of sampling in some datasets.

Model	MNIST	Fashion	CIFAR	CelebA
Best GAN	~ 10	~ 32	~ 70	~ 49
VAE+Flow	54.8	62.1	81.2	65.7
WAE-MMD	115.0	101.7	80.9	62.9
2-StageVAE	12.6	29.3	72.9	44.4
GAN-VAEPP	12.7	26.4	74.1	53.4
VAEPP	12.0	26.4	71.0	53.4

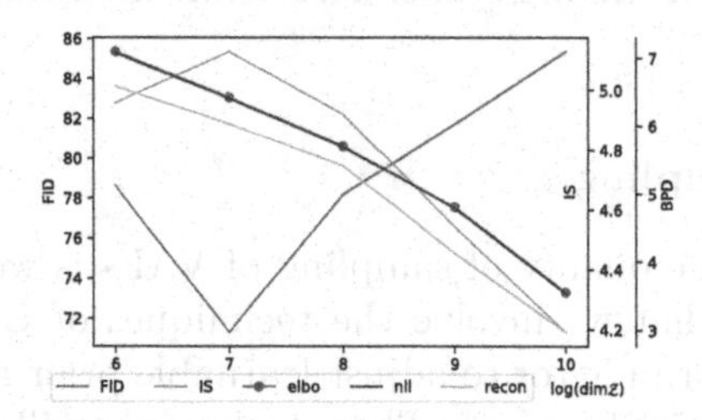

Fig. 4. Comparison of VAEPP with a learnable scalar γ (variance of $p_\theta(x|z)$), as the dimension of latent space varies on CIFAR-10, with metrics BPD, FID and IS. FID and BPD are better when it is smaller and IS is better when it is larger. When $\dim \mathcal{Z}$ is greater than 128, the quality of sampling becomes worse and BPD becomes better as $\dim \mathcal{Z}$ increases. It validates the proposition that $\dim \mathcal{Z}$ should be chosen as a minimal number of active latent dimensions in [4]. It shows an interesting phenomenon that trends of FID and IS, are not same as BPD, maybe greatly different.

increases, the trends of FID and IS are greatly different to log-likelihood's, as shown in Fig. 4. As diagnosis in [4], the variance of $p_\theta(x|z)$ is chosen as a learnable scalar γ, and the $\dim \mathcal{Z}$ is chosen as a number, slightly larger than the dimension of real data manifold. In our experiments, VAEPP reaches best FID when $\dim \mathcal{Z} = 128$.

For better understanding, the values of discriminator on training set are normalized into $\mathcal{N}(0,1)$. To validate the Eq. (6), we calculate $\mathbb{E}_{p_\lambda(z)} D(G(z))$ and $\mathbb{E}_{q_\phi(z)} D(G(z))$. They are 0.092 and 0.015 respectively on CIFAR-10, which means discriminator on generated samples and reconstructed samples are nearly same as on real data. To validate the assumption in Sect. 7 holds in experiment, we calculate $|\mathbb{E}_{p_\theta(x|z)} D(x) - D(G(z))|$, which is an acceptable value (0.019) on CIFAR-10.

6 Conclusion

We propose a novel learnable prior, Pull-back Prior, for VAE, by adjusting the prior through a discriminator assessing the quality of data, with a solid derivation and an intuitive explanation. We propose an efficient and stable training method for VAEPP, by mixing the optimizations of WGAN and VAE into one. VAEPP shows impressive performance in log-likelihood and quality of sampling on common datasets. We believe that VAEPP could lead VAE models into a new stage, with clearer formula, more general framework and better performance.

7 Derivation of Pull-Back Prior

For any given θ, ϕ, search the optimal prior that minimizes the $W^1(p_\theta, p^*)$:

$$\min_\lambda \sup_{Lip(D)\leq 1} \{\mathbb{E}_{p_\lambda(z)} \mathbb{E}_{p_\theta(x|z)} D(x) - \mathbb{E}_{p^*(x)} D(x)\} \tag{11}$$

We use an assumption $\mathbb{E}_{p_\theta(x|z)} D(x) = D(G(z))$ and an approximation D to simplify it. The D in Eq. (11) could be replaced by an approximation D in $W^1(p^\dagger, p^*)$, if p_λ is near $p_\mathcal{N}$, as Sect. 4.1 and Sect. 4.2 does. The simplified optimization is:

$$\min_\lambda \{\mathbb{E}_{p_\lambda(z)} D(G(z)) - \mathbb{E}_{p^*(x)} D(x)\} \quad \textbf{s.t. } KL(p_\lambda, p_\mathcal{N}) = \alpha, \int_\mathcal{Z} p_\lambda(z) \mathrm{d}z = 1$$

Theorem 1. *The optimal solution for this simplified optimization is the Pull-back Prior.*

Proof. It could be solved by Lagrange multiplier method introduced by calculus of variation [5]. The Lagrange function with Lagrange multiplier η, γ is:

$$F(p_\lambda, \eta, \gamma) = \mathbb{E}_{p_\lambda(z)} D(G(z)) - \mathbb{E}_{p^*(x)} D(x) + \eta \int_\mathcal{Z} p_\lambda(z) \mathrm{d}z + \gamma KL(p_\lambda, p_\mathcal{N})$$

By Euler-Lagrange equation, the optimal p_λ satisfies $\frac{\delta F}{\delta p_\lambda} = 0$. Therefore, we obtain

$$\frac{\delta F}{\delta p_\lambda} = D(G(z)) + \eta + \gamma \log \frac{p_\lambda(z)}{p_\mathcal{N}(z)} + \gamma p_\lambda(z) * \frac{1}{p_\lambda(z)} = 0$$

Rewritten it into $\ln p_\lambda(z) = -\frac{1}{\gamma} D(G(z)) + \ln p_\mathcal{N}(z) - (\frac{\eta}{\gamma} + 1)$, which is the Pull-back Prior with $\beta = \frac{1}{\gamma}, \ln Z = (1 + \frac{\eta}{\gamma})$. β is determined by α. In simplified optimization, α is static and need to be searched, *i.e.*, β need to be searched, as Sect. 3.3 does.

Theorem 2. *The optimization process of* $\max_{Lip(D)\leq 1} \mathcal{L}(\theta, \phi, \beta, \omega)$ *is equivalent to the* $\max_{Lip(D)\leq 1} \mathcal{K}$*, which is a lower-bound of* $\beta W^1(p^\dagger, p^*)$.

Proof. $\mathcal{L} = \mathcal{I} + \mathcal{J} + \mathcal{K}$, where $\mathcal{I} + \mathcal{J}$ is independent to D, then $\mathcal{I} + \mathcal{J}$ is constant.

$$\mathcal{K} = -\mathbb{E}_{q_\phi(z)} \beta * D(G(z)) - \ln Z \leq \beta \mathbb{E}_{p_\mathcal{N}(z)} D(G(z)) - E_{q_\phi(z)} D(G(z))$$

where $\ln Z = \ln \mathbb{E}_{p_\mathcal{N}(z)} e^{-\beta * D(G(z))} \geq \mathbb{E}_{p_\mathcal{N}(z)}[-\beta * D(G(z))]$. Then

$$\max_{Lip(D)\leq 1} \mathcal{K} \leq \beta \max_{Lip(D)\leq 1} \{\mathbb{E}_{p^\dagger(x)} D(x) - E_{p_r(x)} D(x)\} = \beta W^1(p^\dagger, p_r)$$

where $p_r(x) = \mathbb{E}_{q_\phi(z)} p_\theta(x|z)$ and $p_r \approx p^*$ is observed in experiments.

Acknowledgments. This work has been supported by National Key R&D Program of China 2019YFB1802504 and the Beijing National Research Center for Information Science and Technology (BNRist) key projects.

References

1. Arjovsky, M., Chintala, S., et al.: Wasserstein generative adversarial networks. In: International Conference on Machine Learning, pp. 214–223 (2017)
2. Bauer, M., Mnih, A.: Resampled priors for variational autoencoders. In: The 22nd International Conference on Artificial Intelligence and Statistics, pp. 66–75 (2019)
3. Chen, X., Mishra, N., et al.: PixelSNAIL: an improved autoregressive generative model. In: International Conference on Machine Learning, pp. 863–871 (2018)
4. Dai, B., Wipf, P.D.: Diagnosing and enhancing VAE models. In: ICLR (2019)
5. Gelfand, I.M., Silverman, R.A., et al.: Calculus of variations. Courier Corporation (2000)
6. Goodfellow, I., Pouget-Abadie, J., et al.: Generative adversarial nets. In: Advances in Neural Information Processing Systems, pp. 2672–2680 (2014)
7. Gulrajani, I., Ahmed, F., et al.: Improved training of Wasserstein GANs. In: NIPS (2017)
8. Heusel, M., Ramsauer, H., et al.: GANs trained by a two time-scale update rule converge to a local Nash equilibrium. In: Advances in Neural Information Processing Systems, pp. 6626–6637 (2017)
9. Hoffman, M.D., Johnson, M.J.: Elbo surgery: yet another way to carve up the variational evidence lower bound. In: Workshop in Advances in Approximate Bayesian Inference, NIPS, vol. 1 (2016)

10. Kingma, D.P., Welling, M.: Auto-encoding variational Bayes. In: ICLR (2014)
11. Kingma, D.P., Dhariwal, P.: Glow: generative flow with invertible 1 × 1 convolutions. In: Advances in Neural Information Processing Systems, pp. 10215–10224 (2018)
12. Kingma, D.P., Salimans, T., et al.: Improved variational inference with inverse autoregressive flow. In: Advances in Neural Information Processing Systems, pp. 4743–4751 (2016)
13. Kumar, R., Goyal, A., et al.: Maximum entropy generators for energy-based models. arXiv preprint arXiv:1901.08508 (2019)
14. Lake, B.M., Salakhutdinov, R., et al.: Human-level concept learning through probabilistic program induction. Science **350**(6266), 1332–1338 (2015)
15. Larochelle, H., Murray, I.: The neural autoregressive distribution estimator. In: AISTATS, pp. 29–37 (2011)
16. Liu, Z., Luo, P., et al.: Deep learning face attributes in the wild. In: Proceedings of the IEEE International Conference on Computer Vision, pp. 3730–3738 (2015)
17. Maaløe, L., Fraccaro, M., et al.: BIVA: a very deep hierarchy of latent variables for generative modeling. In: NeurIPS (2019)
18. Makhzani, A., Shlens, J., Jaitly, N., Goodfellow, J.I.: Adversarial autoencoders. CoRR (2015)
19. Metz, L., Poole, B., et al.: Unrolled generative adversarial networks. In: ICLR (2017)
20. Van den Oord, A., Kalchbrenner, N., et al.: Conditional image generation with PixelCNN decoders. In: Advances in Neural Information Processing Systems, pp. 4790–4798 (2016)
21. Rezende, D.J., Mohamed, S., et al.: Stochastic backpropagation and approximate inference in deep generative models. In: ICML (2014)
22. Salimans, T., Goodfellow, I., et al.: Improved techniques for training GANs. In: Advances in Neural Information Processing Systems, pp. 2234–2242 (2016)
23. Salimans, T., Karpathy, A., Chen, X., Kingma, P.D.: PixelCNN++: improving the PixelCNN with discretized logistic mixture likelihood and other modifications. In: ICLR (2017)
24. Song, Y., Ermon, S.: Generative modeling by estimating gradients of the data distribution. In: Advances in Neural Information Processing Systems, pp. 11895–11907 (2019)
25. Takahashi, H., Iwata, T., et al.: Variational autoencoder with implicit optimal priors. In: Proceedings of the AAAI Conference on Artificial Intelligence, vol. 33, pp. 5066–5073 (2019)
26. Tomczak, J., Welling, M.: VAE with a VampPrior. In: International Conference on Artificial Intelligence and Statistics, pp. 1214–1223 (2018)
27. Tomczak, J.M., Welling, M.: Improving variational auto-encoders using householder flow. arXiv preprint arXiv:1611.09630 (2016)
28. Wu, J., Huang, Z., Thoma, J., Acharya, D., Van Gool, L.: Wasserstein divergence for GANs. In: Ferrari, V., Hebert, M., Sminchisescu, C., Weiss, Y. (eds.) ECCV 2018. LNCS, vol. 11209, pp. 673–688. Springer, Cham (2018). https://doi.org/10.1007/978-3-030-01228-1_40
29. Xiao, H., Rasul, K., et al.: Fashion-MNIST: a novel image dataset for benchmarking machine learning algorithms (2017)

Why Do Deep Neural Networks with Skip Connections and Concatenated Hidden Representations Work?

Oyebade K. Oyedotun(✉) and Djamila Aouada

Interdisciplinary Centre for Security, Reliability and Trust (SnT), University of Luxembourg, 1855 Luxembourg City, Luxembourg
{oyebade.oyedotun,djamila.aouada}@uni.lu
http://wwwen.uni.lu/snt

Abstract. Training the classical-vanilla deep neural networks (DNNs) with several layers is problematic due to optimization problems. Interestingly, skip connections of various forms (e.g. that perform the summation or concatenation of hidden representations or layer outputs) have been shown to allow the successful training of very DNNs. Although there are ongoing theoretical works to understand very DNNs that employ the summation of the outputs of different layers (e.g. as in the residual network), there is none to the best of our knowledge that has studied why DNNs that concatenate of the outputs of different layers (e.g. as seen in Inception, FractalNet and DenseNet) works. As such, we present in this paper, the first theoretical analysis of very DNNs with concatenated hidden representations based on a general framework that can be extended to specific cases. Our results reveal that DNNs with concatenated hidden representations circumnavigate the singularity of hidden representation, which is catastrophic for optimization. For substantiating the theoretical results, extensive experiments are reported on standard datasets such as the MNIST and CIFAR-10.

Keywords: Deep networks · Skip connection · Optimization · Generalization

1 Introduction

Classical deep neural networks (DNNs) have only one path from the input layer to the output layer for information flow. This type of DNNs are commonly referred to as 'PlainNets', and many works [1,2] have reported improved results on various tasks by simply extending the depth of previous state-of-the-art PlainNets. Following this observation, theoretical studies that characterize the role of model depth for the compact representation of complicated functions can be found in [3,4]. Interestingly, training PlainNets with many layers of feature representations (i.e. very deep PlainNets) is difficult, since optimization typically becomes problematic when the number of model layers exceeds fifteen [5,6]. However, it is known that the optimization problem of very deep PlainNets can be mitigated by employing skip connections of various forms. A popular form of skip connection in the literature is based on the summation of the output of any given layer with the outputs of earlier layers. Some very DNNs that adopt

H. Yang et al. (Eds.): ICONIP 2020, LNCS 12534, pp. 380–392, 2020.
https://doi.org/10.1007/978-3-030-63836-8_32

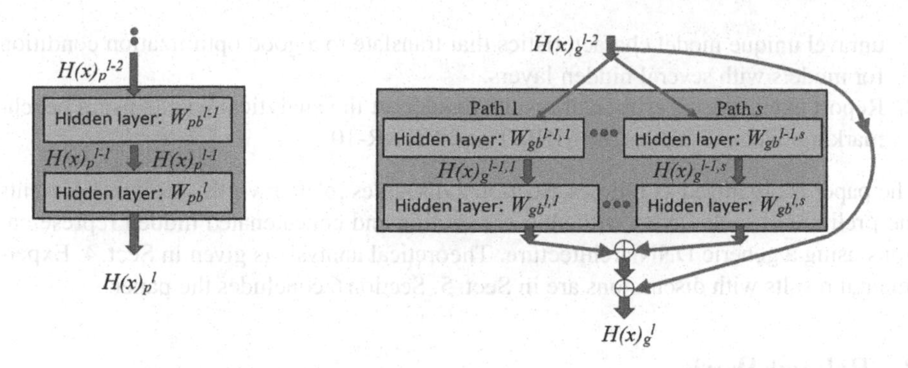

Fig. 1. Generic DNN block used for analysis, where ⊕ denotes the vertical concatenation operation. Left: PlainNet block. Right: ConcNeXt block.

this type of construction include the ResNet [7], ResNeXt [8] and S-ResNet [9]. While very impressive results for different tasks have been reported in several works using skip connections, the literature is still lacking of a concrete theoretical account of their operation. We note that analytical studies have been limited to the ResNet, since using skip connections that sum the hidden representation of different layers generally results in complicated DNN models. Notwithstanding, controversies such as the useful number of layers [10], how optimization problem is circumnavigated [10–12], and the source of improved generalization capacity [10] continue to trail the ResNet. Hence, it is not surprising that the theoretical analysis of similar but more complicated models such as the ResNeXt [8] and S-ResNet [9] are outrightly missing in the literature.

The other popular and successful form of skip connection in the literature entails the concatenation of the output of different layers. DNNs that employ this type of construction are the Inception [13], Inception V3 [14], FractalNet [15] and DenseNet [16]. Despite the success of these DNN models, to the best of our knowledge, there are no theoretical works to understand their operation. This is not surprising, given the complicated architectural construction of this type of DNNs in comparison to the DNNs that employ skip connections and the summation of hidden representations [7–9]. In this paper, we provide a theoretical framework for understanding the complicated operation of very DNN models that employ skip connections with concatenated hidden representations. We acknowledge the proliferation of DNN architectures that employ skip connections and concatenated hidden representations in the literature [13–16]. As such, a generic DNN architecture that can be adapted to specific cases is employed for analysis in this paper. Specifically, our generic architecture is similar to the ResNeXt, where the summation operation is simply replaced by the concatenation operation. We borrow the nomenclature of the ResNeXt so that the DNN architecture used in this paper is referred to as 'ConcNeXt'. We note that the theoretical results for the generic DNN architecture used in this paper are relatable to similar DNNs such as the Inception [13] and DenseNet [16]. Namely, our contributions are as follows.

1. Present the first theoretical study of DNNs with concatenated hidden representations that relies on elements of linear algebra and random matrix theory. Our results

unravel unique model characteristics that translate to a good optimization condition for models with several hidden layers.
2. Report extensive experimental results to support the analytical results using benchmarking datasets such as the MNIST and CIFAR-10.

The paper is organized as follows. Section 2 discusses related works. Section 3 presents the preliminaries on DNNs with skip connection and concatenated hidden representations using a generic DNN architecture. Theoretical analysis is given in Sect. 4. Experimental results with discussions are in Sect. 5. Section 6 concludes the paper.

2 Related Work

Deep Neural Networks with Skip Connections: Considering the proliferation of different DNN architectures, there is a growing concern in recent times of their interpretability. Particularly, we are interested in how the architectures of the DNNs impact their training characteristics and performances. Interestingly, the unconventional construction of the state-of-the-art DNNs, which employ various forms of skip connections obfuscate their operation. Nevertheless, we note that considerable progress has been made in understanding the operation of DNNs that employ the summation of the outputs of hidden layers with the outputs of the previous layers; particularly, the ResNet [7]. For instance, the work [10] argued that the effective depth of the ResNet is significantly smaller than the architectural depth. Namely, the ResNet with 110 layers was shown to have an effective depth of 17 layers [10]. The gradients shattering problem was studied in [11], where skip connections were found to be helpful for fostering well-structured gradients that make the training of very DNNs successful. The unrolled iterative estimation concept was proposed in [12,17], where it was argued that groups of ResNet blocks iteratively refine the representations computed in a particular stage, and new representations are computed in other stages [12]. Another work from the dynamical systems perspective is in [18]. Importantly, a unanimous account of the operation of the ResNet remains a challenge in the deep learning community. Subsequently, the literature is lacking of the theoretical study of DNNs that employ skip connections and the concatenation of the outputs of hidden layers with the outputs of previous layers. Some DNNs that use this type of construction include the Inception [13,14], FractalNet [15] and DenseNet [16]. We presume that the main reason for the lack of analytical study for this type of DNNs is their more complicated constructions and operations.

Studying Deep Linear Neural Networks: The strict theoretical study of DNNs with all its 'bells and whistles' often results in mathematical intractability. Among other simplifications that make DNN amenable to theoretical analysis is the linear activation function assumption [19–21]. In fact, stricter assumptions such as the number of hidden units, data points and convexity are in the literature [20,22]. Interestingly, it is known that the theoretical results obtained using linear DNNs are mostly applicable for non-linear DNNs [23,24]. This observation is not confounding, since a linear DNN with two or more layers is still a non-convex optimization problem in the parameter space.

3 Preliminaries

We discuss as preliminaries the simplified and generic DNN architecture used for the theoretical analysis in the paper. Specifically, the building blocks for the PlainNet and ConcNeXt that the analyses are based on are shown in Fig. 1, where we use $\oplus$ to denote the vertical concatenation operation. Note that we assume the linear activation function for analysis. However, we show via experiments (in Sect. 5) that theoretical results are valid for practical models that use the non-linear activation function such as the rectified linear function. First, we give the following axiom that is important for expressing the hidden representations of the ConcNeXt in interesting forms.

Axiom 1. *Let a matrix $\boldsymbol{C} \in \mathbb{R}^{n \times N}$ be the vertical concatenation of matrices $\boldsymbol{A} \in \mathbb{R}^{n_1 \times N}$ and $\boldsymbol{B} \in \mathbb{R}^{n_2 \times N}$ as in $\boldsymbol{C} = (\boldsymbol{A} \oplus \boldsymbol{B})$; where $n = n_1 + n_2$. Assuming $\boldsymbol{A} = \boldsymbol{D}\boldsymbol{B}$ with $\boldsymbol{D} \in \mathbb{R}^{n_1 \times n_2}$, so that $\boldsymbol{C} = (\boldsymbol{D}\boldsymbol{B} \oplus \boldsymbol{B})$. Then, we can write $\boldsymbol{C} = (\boldsymbol{D} \oplus \mathbf{I})\boldsymbol{B}$, where $\mathbf{I} \in \mathbb{R}^{n_2 \times n_2}$ is the identity matrix.*

3.1 Plain Network (PlainNet)

The output of the PlainNet block is strictly hierarchical as in Fig. 1 (left); there is only one connecting input. Let the input data to the PlainNet DNN model be $\boldsymbol{X} \in \mathbb{R}^{n \times N}$. Subsequently, let the input to a hypothetical PlainNet block composed of two hidden layers be $\boldsymbol{H}(\boldsymbol{X})_p^{l-2} \in \mathbb{R}^{n \times N}$ as in Fig. 1 (left); where $\boldsymbol{W}_{pb}^{l} \in \mathbb{R}^{n \times n}$ and $\boldsymbol{W}_{pb}^{l-1} \in \mathbb{R}^{n \times n}$ are the layers' parameters initialized as Gaussian random matrices, and pb indicates the weights in a PlainNet block. Thus, the output of the PlainNet block is

$$\boldsymbol{H}(\boldsymbol{X})_p^{l} = \boldsymbol{W}_{pb}^{l} \boldsymbol{W}_{pb}^{l-1} \boldsymbol{H}(\boldsymbol{X})_p^{l-2}. \tag{1}$$

For the sake of simplicity, the transformation $\boldsymbol{W}_{pb}^{l} \boldsymbol{W}_{pb}^{l-1}$ given in Eq. (1) is lumped as $\boldsymbol{W}_{p}^{l} = \boldsymbol{W}_{pb}^{l} \boldsymbol{W}_{pb}^{l-1}$, where $\boldsymbol{W}_{p}^{l} \in \mathbb{R}^{n \times n}$. Hence, the output of the PlainNet block can be simply expressed as

$$\boldsymbol{H}(\boldsymbol{X})_p^{l} = \boldsymbol{W}_{p}^{l} \boldsymbol{H}(\boldsymbol{X})_p^{l-2}. \tag{2}$$

3.2 Concatened Network of Hidden Representations (ConcNeXt)

The ConcNeXt employs skip connections that concatenate the output of the previous block with the outputs of the s parallel paths in the current block in Fig. 1 (right); where $s \in \mathbb{N}^{+}$ is referred to as the cardinality of the ConcNeXt with $1 \leq k \leq s$. As such, let the weight matrix at layer l and path k be $\boldsymbol{W}_{gb}^{l,k} \in \mathbb{R}^{q \times q}$, where gb indicates the weights in the ConcNeXt block. Similar to the ResNeXt [8], the parameterization of the ConcNeXt is such that $q \leftarrow q/s$. That is, the dimensions of the weight matrix is reduced with an increase in cardinality. For compactness, we use $\Upsilon_{k=1}^{s} \boldsymbol{\Gamma}^{k}$ to denote successive vertical concatenations of the variable $\boldsymbol{\Gamma}^{k}$, where the dimension of $\boldsymbol{\Gamma}^{k}$ permits the concatenation operations. That is, $\Upsilon_{k=1}^{s} \boldsymbol{\Gamma}^{k} = (\boldsymbol{\Gamma}^{1} \oplus \cdots \oplus \boldsymbol{\Gamma}^{k} \oplus \cdots \oplus \boldsymbol{\Gamma}^{s})$. Given the block's input, $\boldsymbol{H}(\boldsymbol{X})_g^{l-2} \in \mathbb{R}^{q \times N}$, the output of the ConcNeXt block, $\boldsymbol{H}(\boldsymbol{X})_g^{l}$, with the parameters $\boldsymbol{W}_{gb}^{l,k}, \boldsymbol{W}_{gb}^{l-1,k} \in \mathbb{R}^{q \times q}$ initialized as Gaussian random matrices is

$$\boldsymbol{H}(\boldsymbol{X})_g^{l} = (\Upsilon_{k=1}^{s} \boldsymbol{W}_{gb}^{l,k} \boldsymbol{W}_{gb}^{l-1,k} \boldsymbol{H}(\boldsymbol{X})_g^{l-2} \oplus \boldsymbol{H}(\boldsymbol{X})_g^{l-2}), \tag{3}$$

Again, for simplicity, the transformation $\boldsymbol{W}_{gb}^{l,k}\boldsymbol{W}_{gb}^{l-1,k}$ in Eq. (3) is lumped as $\boldsymbol{W}_{g}^{l,k} = \boldsymbol{W}_{gb}^{l,k}\boldsymbol{W}_{gb}^{l-1,k}$, where $\boldsymbol{W}_{g}^{l,k} \in \mathbb{R}^{q\times q}$. Finally, factorizing Eq. (3) using Axiom 1 gives

$$\boldsymbol{H}(\boldsymbol{X})_g^l = (\Upsilon_{k=1}^{s}\boldsymbol{W}_g^{l,k} \oplus \mathbf{I})\boldsymbol{H}(\boldsymbol{X})_g^{l-2}. \quad (4)$$

4 Theoretical Study of DNNs with Skip Connections and Concatenated Hidden Represenations

This section theoretically investigates why DNNs with skip connections and concatenated hidden representations alleviate the training problems of very DNNs. The analysis of the ConcNeXt is positioned relative to the optimization problem of very deep PlainNets based on the singularity of hidden representations. First, the definitions, propositions and lemmas that are crucial for the theoretical study are given as follows. Note that all proofs are in the supplementary material.

Definition 1. *The condition number of a matrix $\boldsymbol{A} \in \mathbb{R}^{n\times N}$ denoted $\kappa(\boldsymbol{A})$ is given as*

$$\kappa(\boldsymbol{A}) = \sigma_{max}(\boldsymbol{A})/\sigma_{min}(\boldsymbol{A}), \quad (5)$$

where $\sigma_{max}(\boldsymbol{A})$ and $\sigma_{min}(\boldsymbol{A})$ are the maximum and minimum singular values of $\boldsymbol{A}$, respectively. From Definition 1, singularity implies that $\sigma_{min}(\boldsymbol{A}) = 0$, so that $\kappa(\boldsymbol{A}) = \infty$. Generally, the optimization of ill-posed problems are very difficult [25,26].

Proposition 1. *Given a matrix $\boldsymbol{W} \in \mathbb{R}^{n\times n}$ whose column vectors, $\boldsymbol{w}_i$, are drawn from a Gaussian or uniform distribution, the probability that $\boldsymbol{W}$ is non-singular, $P(\boldsymbol{w}_i \notin W_{-i}^{span})$, is*

$$P(\boldsymbol{w}_i \notin W_{-i}^{span}) = 1 : 1 \leq i \leq n \quad (6)$$

Corollary 1. *The initialization [27,28] of an m-layer DNN weight matrices, $\{\boldsymbol{W}^l \in \mathbb{R}^{n\times n}\}_{l=1}^{m}$, follows Proposition 1, and hence are all non-singular. That is, $\sigma_{min}(\boldsymbol{W}^l) \neq 0 : 1 \leq l \leq m$.*

Corollary 2. *For an m-layer DNN, Corollary 1 gives $0 < \sigma_{min}(\boldsymbol{W}^l) < \infty : 1 \leq l \leq m$. Subsequently, popular weights initialization schemes [27,28] yield $P(\sigma_{min}(\boldsymbol{W}^l) < 1) = 1 : 1 \leq l \leq m$.*

Assumption 1. *The input to an m-layer DNN, $\boldsymbol{X} \in \mathbb{R}^{n\times N}$, is non-singular. This is an important assumption required for the validity of many machine learning algorithms.*

Lemma 1. *Let $\boldsymbol{X} \in \mathbb{R}^{m\times n}$ and $\boldsymbol{V} \in \mathbb{R}^{n\times p}$ be matrices. If $\boldsymbol{V}$ is singular, then their product, $\boldsymbol{Y} = \boldsymbol{XV}$, is also singular.*

Lemma 2. *Let $\boldsymbol{X} \in \mathbb{R}^{m\times n}$ and $\boldsymbol{V} \in \mathbb{R}^{n\times p}$ be two matrices. If $\boldsymbol{X}$ and $\boldsymbol{V}$ are non-singular, then their product, $\boldsymbol{Y} = \boldsymbol{XV}$, is also non-singular.*

Lemma 3 (Solymosi [29]). *Let $\boldsymbol{Z} \in \mathbb{R}^{n\times n}$ be the set of non-singular matrices. Thus, for many pairs $\boldsymbol{X}_1$, $\boldsymbol{X}_2 \in \boldsymbol{Z}$, the sum $\boldsymbol{Y} = \boldsymbol{X}_1 + \boldsymbol{X}_2$ is non-singular. i.e. $\sigma_{min}(\boldsymbol{Y}) \neq 0$.*

Proposition 2. *Assuming that the optimal solution for a DNN is $\boldsymbol{\theta}$. A relative change of the hidden representation at layer l, $\Delta \boldsymbol{H}(\boldsymbol{X})^l$, results into a relative solution change, $\Delta\theta$, which can be expressed as follows*

$$\frac{\| \Delta\boldsymbol{\theta} \|}{\| \boldsymbol{\theta} \|} \leq \kappa(\boldsymbol{H}(\boldsymbol{X})^l)\frac{\| \Delta\boldsymbol{H}(\boldsymbol{X})^l \|}{\| \boldsymbol{H}(\boldsymbol{X})^l \|} : 0 \leq l \leq m, \quad (7)$$

where $\boldsymbol{\theta} = \{\boldsymbol{W}^l\}_{l=1}^m$, and $\boldsymbol{H}(\boldsymbol{X})^0 = \boldsymbol{X}$ for $l = 0$ is the input to the DNN. We note that changes in the hidden representation $\| \Delta\boldsymbol{H}(\boldsymbol{X})^l \|$ can result from small intentional variations (or inevitable noise) that are captured in the input data batch, $\boldsymbol{X}$.

Definition 2. *For an m-layer DNN, let the input to layer l be $\boldsymbol{H}(\boldsymbol{X})^{l-1}$, and the weight and error gradient at iteration t be $\boldsymbol{W}^l(t)$ and $\boldsymbol{\Delta}^l(t)$, respectively. The weight update for $\boldsymbol{W}^l$ at iteration $t+1$ denoted $\boldsymbol{W}^l(t+1)$ is*

$$\boldsymbol{W}^l(t+1) = \boldsymbol{W}^l(t) + \eta\boldsymbol{\Delta}^l(t)\boldsymbol{H}(\boldsymbol{X})^{l-1} : 1 \leq l \leq m, \quad (8)$$

where η is the learning rate.

4.1 PlainNet (PlainNet)

Let the input of a linear m-layer PlainNet be $\boldsymbol{X} \in \mathbb{R}^{n\times N}$, and the hidden layers parameterized by $\boldsymbol{\theta}_p = \{\boldsymbol{W}_p^l \in \mathbb{R}^{n\times n}\}_{l=1}^m$. The output of the PlainNet in the last layer m, $\boldsymbol{H}(\boldsymbol{X})_p^m$, can be written as

$$\boldsymbol{H}(\boldsymbol{X})_p^m = \prod_{l=1}^{m}\boldsymbol{W}_p^l\boldsymbol{W}_p^{l-1}\cdots\boldsymbol{W}_p^2\boldsymbol{W}_p^1\boldsymbol{X}. \quad (9)$$

Let $\gamma_{p_{min}}^m = \prod_{l=1}^m \sigma_{min}(\boldsymbol{W}_p^l)$ characterize the cummulative outcome of the product of the minimum singular values of different layer weights, $\sigma_{min}(\boldsymbol{W}_p^l)$. Particularly, $\gamma_{p_{min}}^m$ allows the characterization of the singularity or near-singularity of the hidden representation $\boldsymbol{H}(\boldsymbol{X})_p^m$ in a DNN using Eq. (5). Considering that $m \gg 1$ for very deep PlainNets, Corollary 2 where $P(\sigma_{min}(\boldsymbol{W}_p^l) < 1) = 1 : 1 \leq l \leq m$ yields $\gamma_{p_{min}}^m \ll 1$. Importantly, for $m \gg 1$, it is observed that $\gamma_{p_{min}}^m$ can become extremely small such that insufficient machine floating point precision results in a rounding-off to zero. i.e. $\gamma_{p_{min}}^m = 0$. As such, the transformation caused by $\prod_{l=1}^m \boldsymbol{W}_p^l$ collapses space, and is thus singular. Subsequently, using Lemma 1, the result of the transformation of $\boldsymbol{X}$ based on $\prod_{l=1}^m \boldsymbol{W}_p^l$ in Eq. (9) is singular; that is, $\boldsymbol{H}(\boldsymbol{X})_p^m$ is singular, and $\kappa(\boldsymbol{H}(\boldsymbol{X})_p^m) = \infty$.

Remark 1. From Proposition 2, $\| \Delta\boldsymbol{\theta}_p \| / \| \boldsymbol{\theta}_p \|$ is unbounded given $\kappa(\boldsymbol{H}(\boldsymbol{X})_p^m) = \infty$ for the PlainNet. Consequently, small changes in the hidden representation, $\Delta\boldsymbol{H}(\boldsymbol{X})^l$, in Proposition 2 are so magnified that optimization is haphazard in the solution space defined by $\boldsymbol{\theta}_p$. The constant and extreme fluctuation of $\boldsymbol{\theta}_p$ means that optimization cannot progress, and convergence to any decent local minima is impossible.

Since $\boldsymbol{H}(\boldsymbol{X})_p^m$ is singular, Lemma 1 shows that the term $\eta\boldsymbol{\Delta}^m(t)\boldsymbol{H}(\boldsymbol{X})^{m-1}$ in Definition 2 is singular too so that the weight update, $\boldsymbol{W}^m(t+1)$ *can become badly conditioned*; this observation is confirmed via experiments in Sect. 5. In constrast, $\gamma_{p_{min}}^m$ does not become extremely small for shallow PlainNets where typically $m < 20$ such that $\kappa(\boldsymbol{H}(\boldsymbol{X})_p^m) < a : a \in \mathbb{R}$. Hence, $\| \Delta\boldsymbol{\theta}_p \| / \| \boldsymbol{\theta}_p \|$ in Proposition 2 for the shallow PlainNet is bounded by a reasonable value, and the fluctuation of $\boldsymbol{\theta}_p$ due to $\Delta\boldsymbol{H}(\boldsymbol{X})^l$ is moderate so that shallow PlainNets can be successfully optimized. i.e. model optimization converges.

4.2 Concatened Network of Aggregated Hidden Representations (ConcNeXt)

The theoretical analysis of the hidden representations of the ConcNeXt show that they are never singular. However, we first give the following important lemma and axiom that allow the characterization of the minimum singular value of the outcome of the concatenation of any matrix and the identity matrix.

Lemma 4. *Let a matrix $\boldsymbol{B} \in \mathbb{R}^{n\times n_2}$ be the vertical concatenation of any matrix $\boldsymbol{A} \in \mathbb{R}^{n_1\times n_2}$ with $n_1 \geq n_2$ and the identity $\mathbf{I} \in \mathbb{R}^{n_2\times n_2}$ as in $\boldsymbol{B} = [\boldsymbol{A} \oplus \mathbf{I}]$, so that $n = n_1 + n_2$. Let the singular values of $\boldsymbol{B}$ and $\boldsymbol{A}$ be $\{\sigma(\boldsymbol{B})_i\}_{i=1}^{n_2}$ and $\{\sigma(\boldsymbol{A})_i\}_{i=1}^{n_2}$, respectively. Then, it can be shown that $\sigma(\boldsymbol{B})_i = \sqrt{(\sigma(\boldsymbol{A})_i)^2 + 1} : 1 \leq i \leq n_2$.*

The implication of Lemma 4 is that $\sigma_{min}(B) \geq 1 \because \sigma_{min}(A) \geq 0$.

Axiom 2. *Let a matrix $\boldsymbol{B} \in \mathbb{R}^{q\times z}$ be the vertical concatenation of s random matrices $\boldsymbol{A}_k \in \mathbb{R}^{q_k\times z} : 1 \leq k \leq s$ with $q = \sum_{k=1}^{s} q_k$, as in $\boldsymbol{B} = \Upsilon_{k=1}^{s}\boldsymbol{A}_k = [\boldsymbol{A}_1 \oplus \cdots \oplus \boldsymbol{A}_k \oplus \cdots \oplus \boldsymbol{A}_s]$. Then, we conclude that the aggregated matrix, $\boldsymbol{B}$, is also a random matrix.*

Let the input of a linear m-layer ConcNeXt be $\boldsymbol{X} \in \mathbb{R}^{n\times N}$, and the hidden layers parameterized by $\boldsymbol{W}_g^1 \in \mathbb{R}^{q\times n}$ and $\{\boldsymbol{W}_g^l \in \mathbb{R}^{q\times q}\}_{l=2}^m$, so that $\boldsymbol{\theta}_g = \{\boldsymbol{W}_g^l\}_{l=1}^m$. Following Eq. (4), the output of the ConcNeXt in the last layer m, $\boldsymbol{H}(\boldsymbol{X})_g^m$, can be written as

$$\boldsymbol{H}(\boldsymbol{X})_g^m = \prod_{l=1}^{m}(\Upsilon_{k=1}^{s}\boldsymbol{W}_g^{l,k} \oplus \mathbf{I})\boldsymbol{X}. \tag{10}$$

Using Axiom 2, $\Upsilon_{k=1}^{s}\boldsymbol{W}_g^{l,k}$ in Eq. (10) can be aggregated in a compact form so that we have $\boldsymbol{W}_a^l = \Upsilon_{k=1}^{s}\boldsymbol{W}_g^{l,k}$. Therefore, Eq. (10) becomes

$$\boldsymbol{H}(\boldsymbol{X})_g^m = \prod_{l=1}^{m}(\boldsymbol{W}_a^l \oplus \mathbf{I})\boldsymbol{X}. \tag{11}$$

Let $\gamma_{g_{min}}^m = \prod_{l=1}^{m}\sigma_{min}(\boldsymbol{W}_a^l \oplus \mathbf{I})$. In Eq. (11), we have $\sigma_{min}(\boldsymbol{W}_a^l \oplus \mathbf{I}) \geq 1 : 1 \leq l \leq m$ from Lemma 4. As such, $\gamma_{g_{min}}^m \geq 1$ so that the overall transformation effect of $\prod_{l=1}^{m}[\boldsymbol{W}_a^l \oplus \mathbf{I}]$ in Eq. (11) does not collapse space, and thus is non-singular. Since $\prod_{l=1}^{m}(\boldsymbol{W}_a^l \oplus \mathbf{I})$ is non-singular and $\boldsymbol{X}$ is non-singular from Assumption 1, we can conclude using Lemma 2 that $\boldsymbol{H}(\boldsymbol{x})_g^m$ in Eq. (11) is non-singular too. i.e $\kappa(\boldsymbol{H}(\boldsymbol{x})_g^m) < a : a \in \mathbb{R}$.

Remark 2. Considering $\kappa(\boldsymbol{H}(\boldsymbol{X})_g^m) < a : a \in \mathbb{R}$, $\| \Delta\boldsymbol{\theta}_g \| / \| \boldsymbol{\theta}_g \|$ in Proposition 2 for the ConcNeXt is bounded. Thus, small changes in the hidden representation, $\Delta\boldsymbol{H}(\boldsymbol{X})^m$, in Proposition 2 results in moderate changes in the solution, $\boldsymbol{\theta}_g$. The stability of $\boldsymbol{\theta}_g$ means that optimization can progress successfully, and decent local minima can be reached. Again, considering that $\boldsymbol{H}(\boldsymbol{X})_g^m$ is non-singular, the term $\eta\boldsymbol{\Delta}^m(t)\boldsymbol{H}(\boldsymbol{X})^{m-1}$ in Definition 2 is non-singular by applying Lemma 2. Finally, noting that $\boldsymbol{W}_g^m$ is non-singular from Proposition 1, the weight update $\boldsymbol{W}_g^m(t+1)$ in Eqn. (8) is non-singular with a high probablity using Lemma 3. Our experiments indeed confirm these interesting observations.

5 Experiments

5.1 Experimental Settings

This section reports the experimental results that validate the theoretical analysis given in Sect. 4 using 110 layers PlainNet and ConcNeXt models, which are multilayer perceptrons (MLPs) trained on the MNIST [30] dataset. All model parameters are initialized from Gaussian distributions using the He method [27]. In addition, all models are trained using gradient descent, learning rate = [0.0001–0.1], momentum rate = 0.9, batch normalization (BN), weight decay of 10^{-4}, batch size of 128, and for 60 epochs. Table 1 shows the model architectures and number of parameters for the different models. Note that for showing the agreement between our theoretical analysis and practical models, the experiments reported herein use the rectified linear activation. For additional experiments using the linear activation function, see Section A7.2 in the supplementary material. Furthermore, Section A8 in the supplementary material contains the experiments for convolutional neural network (CNN) models using CIFAR-10 dataset [31]. In the following section, important results for the MLP models are reported.

5.2 Results and Discussion

Training results are given in Table 1, where the PlainNet clearly reflects poor optimization. In contrast, ConcNeXt-110 ($s = 1$) and ConcNeXt-110 ($s = 2$) are both well optimized. For testing, ConcNeXt-110 ($s = 2$) slightly outperforms ConcNeXt-110 ($s = 1$); similar results for $s = 2$ on the harder CIFAR-10 dataset are provided in Section A8 of the supplementary material. The hidden representations for PlainNet-110 are shown in Fig. 2, where singularity (i.e. units respond in similar fashion for different input data) is seen. In contrast, ConcNeXt-110 ($s = 1$) in Fig. 2 shows no singularity problems (i.e. units respond in different manners for different input data); the hidden representations for ConcNeXt-110 ($s = 2$) is similar to that of ConcNeXt-110 ($s = 1$) so are not shown; Section A7.1 of the supplementary material contains results for ConcNeXt-110 ($s = 2$). Figure 3 and Fig. 4 show the weight values and hidden units' outputs in the different layers of the MLPs, respectively. It is seen that PlainNet-110 has bizzarely high weight and units' output values, while ConcNeXt-110 ($s = 1$) has reasonable weight and units' output values; ConcNeXt-110 (s = 2) has values that are similar to ConcNeXt-110 ($s = 1$), and thus not shown.

Figure 5 and Fig. 6 show the conditions of the weights and hidden representations in the MLPs, respectively. From Fig. 5 (left), the weights of PlainNet-110 have extremely high conditions numbers, while the weights of ConcNeXt-110 ($s = 1$) and ConcNeXt-110 ($s = 2$) both have small condition numbers. From Fig. 5 (right), the weights of PlainNet-110 have zero singular values starting from the 25th layer, while the weights of ConcNeXt-110 ($s = 1$) and ConcNeXt-110 ($s = 2$) both have a minimum singular value of one for all the layers. Figure 6 (left) shows that the hidden representations of PlainNet-110 have extremely high and even infinite condition numbers that plague optimization, while ConcNeXt-110 ($s = 1$) and ConcNeXt-110 ($s = 2$) both have small condition numbers that foster successful optimization. For clarity, Fig. 6 (right) shows that the hidden representations of ConcNeXt-110 ($s = 2$) operates with smaller

Table 1. 110 layers MLP details and results on the MNIST dataset.

MLP model	PlainNet-110	ConcNeXt-110 ($s = 1$)	ConcNeXt-110 ($s = 2$)
Hidden units per layer	180	50	25
Parameters	3.66M	3.92M	3.77M
Training accuracy	13.22%	100%	100%
Testing accuracy	12.45%	98.42%	98.83%

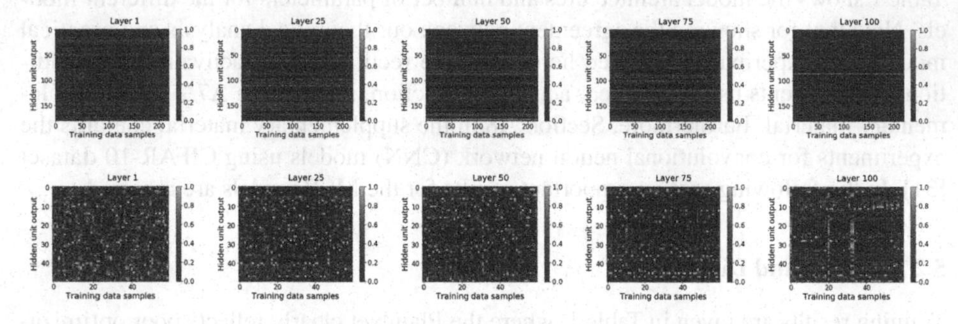

Fig. 2. Hidden representations for the 110 layers MLPs using randomly chosen batch of input data from the MNIST dataset. Top row: PlainNet-110. Bottom row: ConcNeXt-110 ($s = 1$).

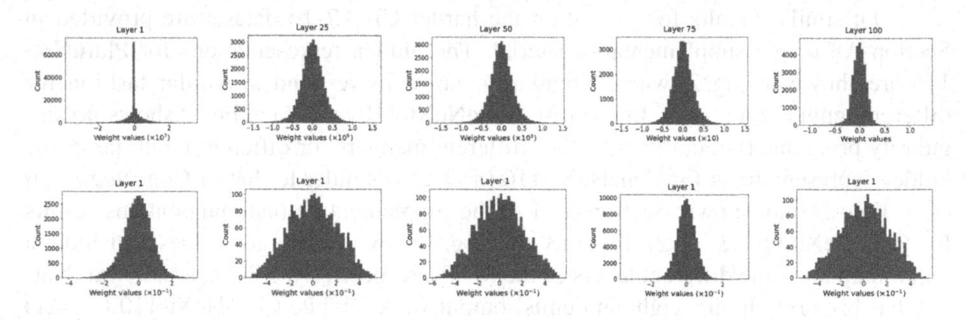

Fig. 3. Weights distribution for the 110 layers MLPs using randomly chosen batch of input data from the MNIST dataset. Top row: PlainNet-110. Bottom row: ConcNeXt-110 ($s = 1$).

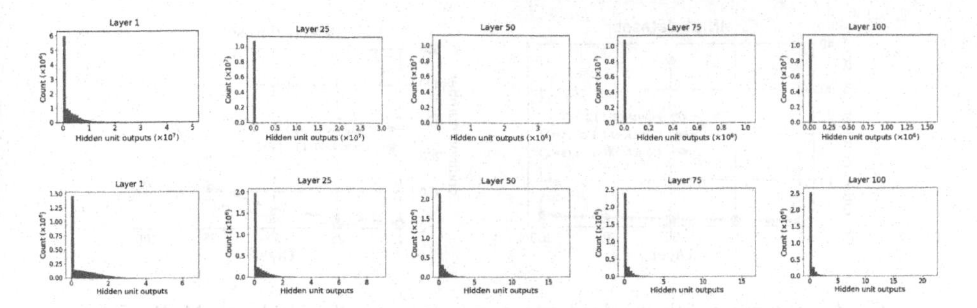

Fig. 4. Units' outputs distribution for the 110 layers MLPs using randomly chosen batch of input data from the MNIST dataset. Top row: PlainNet-110. Bottom row: ConcNeXt-110 ($s = 1$).

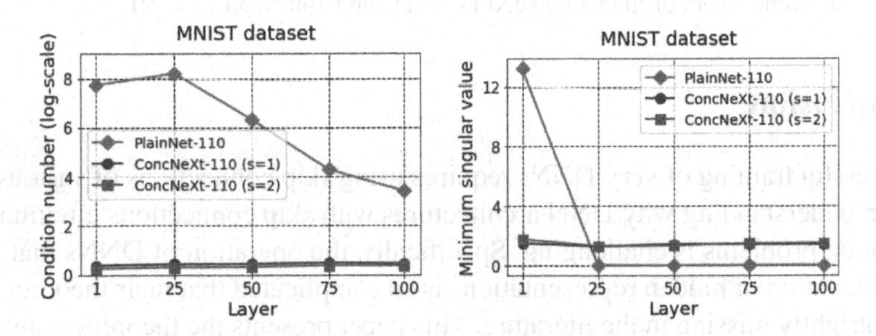

Fig. 5. Condition of the layer weight in the MLPs. Left: condition number of the weights in the different layers *plotted to log-scale*. Right: smallest singular value of the weights in the different layers.

condition numbers as compared to the ConcNeXt-110 ($s = 1$). Subsequently, using this observation in 2 explains the improved generalization performance of ConcNeXt-110 ($s = 2$) over ConcNeXt-110 ($s = 1$).

5.3 Main Insights into Models Concatenated Hidden Representations

The main insights from the theoretical and empirical results are summarized as follows.

1. The minimum singular values for the weights in the PlainNet are invariably less than one. i.e. $\sigma_{min}(\boldsymbol{W}_p^l) < 1 : 1 \leq l \leq m$. As such, for very DNNs where $m \gg 1$, the repeated multiplication of the input data, $\boldsymbol{X}$, by the model weights $\{\boldsymbol{W}_p^1, \cdots, \boldsymbol{W}_p^m\}$ causes some components of $\boldsymbol{X}$ to vanish so that collinearity and thus singularity occur in the resulting output $\boldsymbol{H}(\boldsymbol{X})_p^m$. Finally, singularity plagues model optimization.

2. Skip connections that concatenate hidden representations operate such that the minimum singular values for the compactly expressed model weights are $\sigma_{min}(\boldsymbol{W}_a^l \oplus \mathbf{I}) \geq 1 : 1 \leq l \leq m$. Thus, for very DNNs where $m \gg 1$, the repeated multiplication of the input data, $\boldsymbol{X}$, by the model weights $\{(\boldsymbol{W}_a^1 \oplus \mathbf{I}), \cdots, (\boldsymbol{W}_a^m \oplus \mathbf{I})\}$ does not cause any component of $\boldsymbol{X}$ to vanish, and thus collinearity and singularity are mitigated. Consequently, such models with several layers can be successfully optimized.

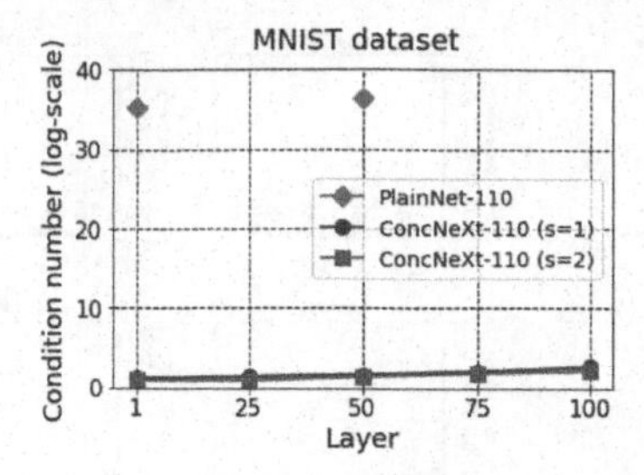

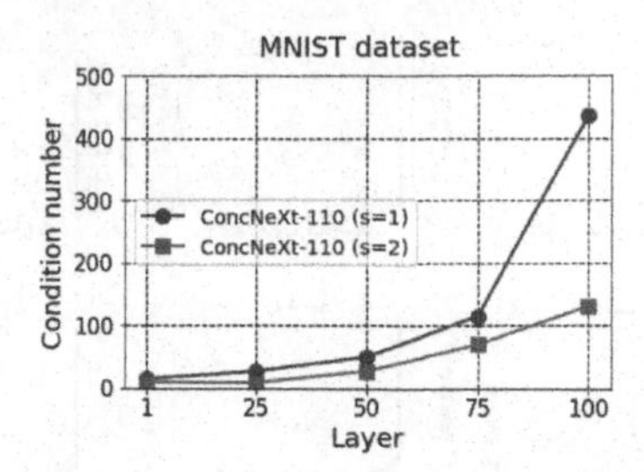

Fig. 6. Condition number of the hidden layer representations in the 110 layers MLPs. The layers for which values are not shown have infinite condition numbers. Left: condition number of the hidden representations in the different layers. Right: condition number of the hidden representations in the different layers of the ConcNeXt ($s = 1$) and ConcNeXt ($s = 2$).

6 Conclusion

The successful training of very DNNs requires using skip connections of various forms. However, understanding why DNN architectures with skip connections circumnavigate optimization problems is challenging. Specifically, the operation of DNNs that employ the concatenation of hidden representations is so complicated that their theoretical analysis is outrightly missing in the literature. This paper presents the theoretical analysis of DNNs with skip connections and concatenated hidden representations, which to the best of our knowledge is the first in the literature. Our theoretical results that are confirmed via extensive experiments show that concatenating hidden representations improve the condition of the hidden representations by mitigating singularity problems in hidden representations that ensue from the repeated multiplication of model weights.

Acknowledgments. This work was funded by the National Research Fund (FNR), Luxembourg, under the project reference R-AGR-0424-05-D/Björn Ottersten and CPPP17/IS/11643091/IDform/Aouada.

References

1. Lee, C.-Y., Xie, S., Gallagher, P., Zhang, Z., Tu, Z.: Deeply-supervised nets. In: Artificial Intelligence and Statistics, pp. 562–570 (2015)
2. Simonyan, K., Zisserman, A.: Very deep convolutional networks for large-scale image recognition. In: International Conference on Learning Representations (2015)
3. Eldan, R., Shamir, O.: The power of depth for feedforward neural networks. In: Conference on Learning Theory, pp. 907–940 (2016)
4. Safran, I., Shamir, O.: Depth-width tradeoffs in approximating natural functions with neural networks. In: Proceedings of the 34th International Conference on Machine Learning, vol. 70, pp. 2979–2987. JMLR. org (2017)
5. Oyedotun, O. K., El Rahman Shabayek, A., Aouada, D., Ottersten, B.: Highway network block with gates constraints for training very deep networks. In: Proceedings of the IEEE Conference on Computer Vision and Pattern Recognition Workshops, pp. 1658–1667 (2018)
6. Srivastava, R. K., Greff, K., Schmidhuber, J.: Training very deep networks. In: Advances in Neural Information Processing Systems, pp. 2377–2385 (2015)

7. He, K., Zhang, X., Ren, S., Sun, J.: Deep residual learning for image recognition. In: Proceedings of the IEEE Conference on Computer Vision and Pattern Recognition, pp. 770–778 (2016)
8. Xie, S., Girshick, R., Dollár, P., Tu, Z., He, K.: Aggregated residual transformations for deep neural networks. In: Proceedings of the IEEE Conference on Computer Vision and Pattern Recognition, pp. 1492–1500 (2017)
9. Oyedotun, O.K., Aouada, D., Ottersten, B., et al.: Training very deep networks via residual learning with stochastic input shortcut connections. In: Liu, D., Xie, S., Li, Y., Zhao, D., El-Alfy, E.S. (eds.) International Conference on Neural Information Processing, vol. 10635, pp. 23–33. Springer, Cham (2017). https://doi.org/10.1007/978-3-319-70096-0_3
10. Veit, A., Wilber, M. J., Belongie, S.: Residual networks behave like ensembles of relatively shallow networks. In: Advances in Neural Information Processing Systems, pp. 550–558 (2016)
11. Balduzzi, D., Frean, M., Leary, L., Lewis, J.P., Ma, K.W.D., McWilliams, B.: The shattered gradients problem: if resnets are the answer, then what is the question? In: International Conference on Machine Learning, pp. 342–350 (2017)
12. Greff, K., Srivastava, R.K., Schmidhuber, J.: Highway and residual networks learn unrolled iterative estimation. In: International Conference Learning Representations (2017)
13. Szegedy, C.,et al.: Going deeper with convolutions. In: Proceedings of the IEEE Conference on Computer Vision and Pattern Recognition, pp. 1–9 (2015)
14. Szegedy, C., Vanhoucke, V., Ioffe, S., Shlens, J., Wojna, Z.: Rethinking the inception architecture for computer vision. In: Proceedings of the IEEE Conference on Computer Vision and Pattern Recognition, pp. 2818–2826 (2016)
15. Larsson, G., Maire, M., Shakhnarovich, G.: FractalNet: ultra-deep neural networks without residuals. In: International Conference on Learning Representations (2017)
16. Huang, G., Liu, Z., Van Der Maaten, L., Weinberger, K. Q.: Densely connected convolutional networks. In: Proceedings of the IEEE Conference on Computer Vision and Pattern Recognition, pp. 4700–4708 (2017)
17. Jastrzebski, S., Arpit, D., Ballas, N., Verma, V., Che, T., Bengio, Y.: Residual connections encourage iterative inference. In: International Conference on Learning Representations (2018)
18. Chang, B., Meng, L., Haber, E., Tung, F., Begert, D.: Multi-level residual networks from dynamical systems view. In: International Conference on Learning Representations (2018)
19. Zhou, Y., Liang, Y.: Critical points of neural networks: analytical forms and landscape properties. In: International Conference on Learning Representations (2017)
20. Sonoda, S., Murata, N.: Transport analysis of infinitely deep neural network. J. Mach. Learn. Res. **20**(1), 31–82 (2019)
21. Nguyen, Q., Hein, M.: Optimization landscape and expressivity of deep cnns. In: International Conference on Machine Learning, pp. 3730–3739 (2018)
22. Laurent, T., Brecht, J.: Deep linear networks with arbitrary loss: all local minima are global. In: International Conference on Machine Learning, pp. 2902–2907 (2018)
23. Kawaguchi, K.: Deep learning without poor local minima. In: Advances in Neural Information Processing Systems, pp. 586–594 (2016)
24. Saxe, A.M., McClelland, J.L., Ganguli, S.: Exact solutions to the nonlinear dynamics of learning in deep linear neural networks. In: International Conference on Learning Representations (2014)
25. Neubauer, A.: A new gradient method for ill-posed problems. Numer. Funct. Anal. Optim. **39**(6), 737–762 (2018)
26. Neubauer, A., Scherzer, O.: A convergence rate result for a steepest descent method and a minimal error method for the solution of nonlinear ill-posed problems. Zeitschrift für Analysis und ihre Anwendungen **14**(2), 369–377 (1995)

27. He, K., Zhang, X., Ren, S., Sun, J.: Delving deep into rectifiers: surpassing human-level performance on imagenet classification. In: Proceedings of the IEEE International Conference on Computer Vision, pp. 1026–1034 (2015)
28. Glorot, X., Bengio, Y.: Understanding the difficulty of training deep feedforward neural networks. In: Proceedings of the Thirteenth International Conference on Artificial Intelligence and Statistics, pp. 249–256 (2010)
29. Solymosi, J.: The sum of nonsingular matrices is often nonsingular. Linear Algebra Appl. **552**, 159–165 (2018)
30. LeCun, Y., Cortes, C.: MNIST handwritten digit database. http://yann.lecun.com/exdb/mnist/. Accessed Oct 2019
31. Krizhevsky, A., Nair, V., Hinton, G.: CIFAR-10, CIFAR-100 (Canadian institute for advanced research). http://www.cs.toronto.edu/~kriz/cifar.html. Accessed Oct 2019

Recommender Systems

AMBR: Boosting the Performance of Personalized Recommendation via Learning from Multi-behavior Data

Chen Wang[1,2,5], Shilu Lin[1,2], Zhicong Zhong[1,2], Yipeng Zhou[3], and Di Wu[1,2,4](✉)

[1] School of Data and Computer Science, Sun Yat-Sen University, Guangzhou, China
{wangch287,linshl7,zhongzhc3}@mail2.sysu.edu.cn, wudi27@mail.sysu.edu.cn

[2] Guangdong Key Laboratory of Big Data Analysis and Processing, Guangzhou, China

[3] Department of Computing, Faculty of Science and Engineering, Macquarie University, Sydney, NSW 2109, Australia
yipeng.zhou@mq.edu.au

[4] The Key Laboratory of Machine Intelligence and Advanced Computing, Sun Yat-sen University, Ministry of Education, Beijing, China

[5] PCL Research Center of Networks and Communications, Peng Cheng Laboratory, Shenzhen, China

Abstract. The performance of personalized recommendation can be further improved by exploiting multiple user behaviors (*e.g.*, browsing, adding-to-cart, product purchasing) to predict items of user interests. However, the challenge lies in how to accurately model the relations among multiple user behaviors. The commonly adopted cascade relation over-simplifies the problem and cannot model the real user behavior patterns. In this paper, we propose a novel multi-behavior recommendation algorithm called AMBR (Attentive Multi-Behavior Recommendation), which can well capture the complicated relations among multiple behaviors. AMBR integrates the representation learning module and the matching function learning module into one framework. By utilizing the modern neural network techniques, AMBR is more flexible in modeling the relations of multiple behaviors without presuming a fixed cascade relation. Finally, we also conduct a set of experiments based on two real-world datasets, and the results show that our AMBR algorithm significantly outperforms other state-of-the-art algorithms by over 8.6%, 9.3% in terms of HR and NDCG.

Keywords: Personalized recommendation · Multi-behavior data · Neural networks

1 Introduction

In the past decade, personalized recommendation has become an essential functionality of many Internet services (*e.g.*, online shopping, online video), which

H. Yang et al. (Eds.): ICONIP 2020, LNCS 12534, pp. 395–406, 2020.
https://doi.org/10.1007/978-3-030-63836-8_33

helps end users discover items of their interests when confronting with a plethora of items in the system. Traditional recommender systems commonly utilize only one type of user behavior (*e.g.*, product purchasing) to make recommendation decisions, which fails to explore more valuable information contained in multiple user behaviors (*e.g.*, website browsing, adding-to-cart) [3].

Recent studies show that the performance of recommender systems can be improved by taking multiple user behaviors into account [4,9]. Thereafter, the problem of multi-behavior recommendation has gained significant attention from both academic researchers and industrial practitioners. Collective Matrix Factorization (CMF) [9], the extension of MF (Matrix Factorization), was proposed to cope with multi-behavior data by optimizing the likelihood of each type of user behaviors separately. Nevertheless, CMF only resorts to the dot-product interaction between latent vectors, which cannot seize the complicated relations between multiple behaviors very well. Other sophisticated recommendation algorithms were designed though they were based on some strong assumptions. For instance, the work [4] assumed that browsing, adding-to-cart, and purchasing should occur sequentially.

However, it is challenging to accurately model the impact of multiple user behaviors on personalized recommendation. Existing works have a few drawbacks: *First*, it is not reasonable to assume a simple sequential relation among behaviors. For example, it is assumed that purchasing behavior should occur before adding-to-cart, and adding-to-cart happens before browsing in Gao *et al.* [4]. But it is common that users may purchase an item recommended via online social networks from their friends without adding-to-cart behavior [6]; *Second*, latent semantics can reflect the relations among multiple behaviors better, but existing approaches have not taken them well into consideration [7,9].

In this paper, we propose a novel multi-behavior recommendation algorithm called AMBR (Attentive Multi-Behavior Recommendation) to address the aforementioned drawbacks. To model the impact of multiple user behaviors more accurately, we enhance the multi-behavior recommendation framework with representation learning and matching function learning, where the knowledge among different behavior channels are learned, transferred and fused in a seamless model. To learn the latent semantics of multiple behaviors, we further propose to adopt the attention mechanism to capture the relations among multiple user behaviors. Specifically, we apply the attention mechanism to the widely used Neural Collaborative Filtering (NCF) model via deep neural networks. In this way, our model can better model the complex relations among multiple behaviors. We also define a new personalized objective function, which can be efficiently solved by existing optimization methods.

In a nutshell, the main contributions of this work are listed as follows.

- We design the AMBR algorithm by integrating representation learning and matching function learning into one framework. To learn the representation of multiple behaviors, we propose to utilize the attention mechanism to capture the relations of behaviors in the latent space.

- We propose a new user-specified multi-behavior matching score function, which can better depict the relations between multiple user behaviors than the previous cascade matching function. In this way, our model can easily handle more complex behavior contexts.
- We conduct extensive experiments on two real-world datasets. The experiment results reveal that our AMBR algorithm can improve the recommendation performance by at least 8.6%, 9.3% in terms of HR and NDCG compared with state-of-the-art algorithms.

The rest of this paper is organized as follows. In Sect. 2, we review the latest related works. In Sect. 3, we present our proposed algorithm in detail. The experiment settings and results are shown in Sect. 4. Finally, we conclude our work in Sect. 5.

2 Related Work

Recently, due to the decent capacity to fit non-linear functions, neural networks were widely used to digest auxiliary information for recommendations, such as audio information, textual description of items, and visual information [1]. Although deep-learning based methods have been paid much attention, the aforementioned methods still resorted to the MF-based models to depict the interaction between users and items. To learn a more complex user-item interaction function, a neural network architecture, named Neural Collaborative Filtering (NCF), was proposed by He *et al.* [5]. They designed several user-item matching functions through different deep neural networks. Our work extends the architecture of NCF for multi-behavior recommendation through the perspectives of both representation learning and matching function learning.

On the other hand, previous works on multi-behavior recommendation learned the recommendation model in two different ways, either from the perspective of representation learning or matching function learning. A typical method of the first type is the Collective Matrix Factorization (CMF) model [9], which factorized multiple user-side matrices while sharing the item-side matrix. Further, Zhao *et al.* [11] extended the CMF to build user profiles by using a variety of consumption and publishing behaviors in social media. They employed matrix factorization techniques to model each user's behaviors as a separate entry in the input user-by-topic matrix. The main limitations of these works lie on that they utilized the same optimization objective for different behaviors and did not analyze the relations between different behavior types, which should contain more useful information for recommendations.

Some other works dealt with the multi-behavior recommendation from the perspective of matching function learning. Loni *et al.* [7] proposed the Multi-Feedback Bayesian Personalized Ranking (MF-BPR), a pairwise method which extended the Bayesian Personalized Ranking (BPR) [8] by assigning different types of user behaviors with different levels of feedback. The work in Gao *et al.* [4] extended the architecture of NCF via a multi-task learning framework, which is named as Neural Multi-Task Recommendation (NMTR). NMTR accounted for the cascade relation among different types of behaviors (*e.g.*, a user must click on a product before purchasing it) and performed a joint optimization using multi-task learning method. However, in many circumstances, different behaviors do not have a strict cascade relation. To solve this problem, our work proposes a new user-specified optimization objective which does not rely on the cascade relation between behaviors.

3 Design of AMBR Algorithm

Assume that there are n types of behaviors in a multi-behavior recommendation system, and $\mathcal{N}$ denotes the set of all behavior types, where $\mathcal{N} = \{1, 2, ..., n\}$. We denote the n-th behavior as the *key behavior* such as the buying behavior in E-commerce systems or viewing behavior in online video platforms. Given a triple record $(u, i, \mathbf{y})$ from the dataset, u indicates a particular user and i represents a particular item respectively. Let $\mathbf{y} = \{y^1, y^2, ..., y^n\}$ denote the labels of each type of behavior (either observed in the history or not), and we aim to estimate the likelihood of the key behavior $\hat{y}^n$ as precisely as possible.

3.1 Framework Overview

As shown in Fig. 1, the overall framework of our AMBR algorithm can be summarized into the following blocks:

- **Input Layer.** Without loss of generality, we feed the user one-hot vector and item one-hot vector into the model.
- **Attention Based Representation Learning.** We resort to the attention mechanism to exploit the relations between multiple behaviors in the latent space.
- **Multi-behavior Matching Function Learning.** We utilize the non-linear NCF units as the sub-module to predict the matching scores for each type of behaviors and we fuse the matching scores to make personalized recommendation.
- **Multi-task Optimization Module.** We optimize the objective function for each observed behavior simultaneously to make the model more robust.

3.2 Attention Based Representation Learning

Let $\mathbf{v}_u^U$ and $\mathbf{v}_i^I$ denote the one-hot vectors for the current input of user u and item i respectively. Then we project each type of user behaviors associated with the input item into the latent space via linear embedding:

$$\begin{cases} \mathbf{e}_u^1 = \mathbf{W}^1 \mathbf{v}_u^U, \\ \mathbf{e}_u^2 = \mathbf{W}^2 \mathbf{v}_u^U, \\ \quad \cdots \\ \mathbf{e}_u^n = \mathbf{W}^n \mathbf{v}_u^U; \end{cases} \tag{1}$$

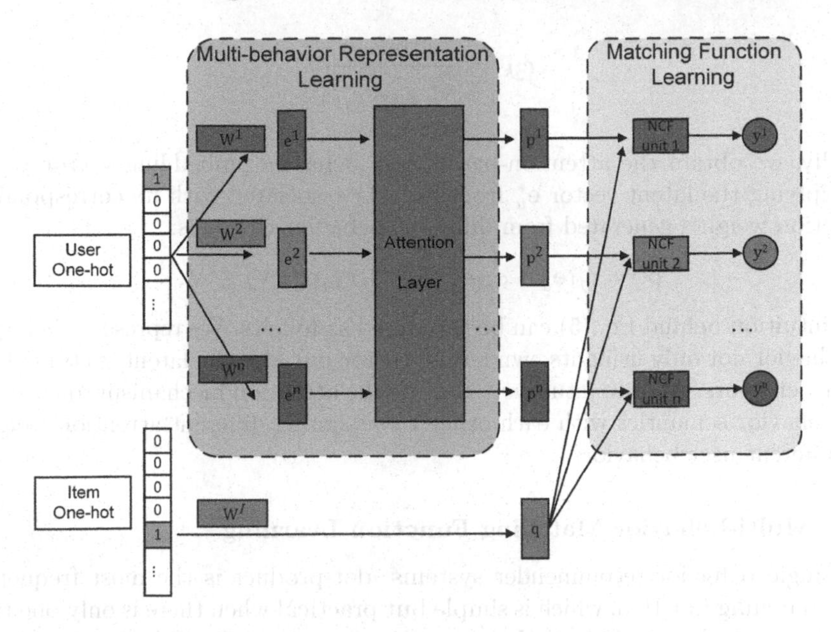

Fig. 1. The overall framework of our AMBR algorithm.

and

$$\mathbf{q}_i = \mathbf{W}^I \mathbf{v}_i^I, \tag{2}$$

where $\mathbf{W}^*$ denotes the parameter matrix used in linear embedding layer. $\mathbf{e}_u^*$ and $\mathbf{q}_i$ denote the behavior embedding vector and item embedding vector respectively.

After having assigned each type of behavior a separate latent vector, we now take the inter-behavior relations into consideration. Actually, it is non-trivial to exploit the behavior relations in the latent space since there is no fixed ordinal order in real-life scenarios as we have stated in Sect. 1. Thankfully, we can resort to the attention mechanism, which reflects the context influence of each word for the sentence in the machine translation domain [10]. For the scenario of recommendation where multiple user behaviors are available, some behavior implies a positive feedback to the key behavior while other behaviors may neutralize. We reflect these relations using the attention mechanism with some customized settings for multi-behavior recommendation.

Let $\mathbf{C} \in \mathbb{R}^{N \times N}$ denote the attention weight matrix, where c_{jk} denotes the attention weight between behavior j and k. Then we calculate each element of matrix $\mathbf{C}$ as:

$$c_{jk} = (\mathbf{e}_u^j)^T \mathbf{M}\, \mathbf{e}_u^k,\ \forall j, k \in \mathcal{N}; \tag{3}$$

where $\mathbf{M}$ denotes the parameters in the attention layer. Note, it is not necessary that c_{jk} is equal to c_{kj}. After obtaining the attention score matrix, we also normalize it by applying the softmax function along its row axis to guarantee that each row vector of $\mathbf{C}$ sums to 1:

$$c_{jk} = \frac{exp(c_{jk})}{\sum_{k=1}^{n} exp(c_{jk})}. \tag{4}$$

Finally, we obtain the attention-based user behavior embedding vector $\mathbf{p}_u^*$ by multiplying the latent vector $\mathbf{e}_u^*$ from Eq. (1) associated with its corresponding attention weights generated from different behavior channels:

$$\mathbf{p}_u^j = c_{j1}\mathbf{e}_u^1 + c_{j2}\mathbf{e}_u^2 + ... + c_{jn}\mathbf{e}_u^n, \ \forall j \in \mathcal{N}. \tag{5}$$

The intuition behind Eq. (5) can be explained as follows. We represent each type of behavior not only using its own latent vector but also the latent vectors of the other behaviors. And we intuitively exploit the attention mechanism to co-relate the behavior semantics with each other by assigning different attention weights for different user behaviors.

3.3 Multi-behavior Matching Function Learning

For single behavior recommender systems, dot-product is the most frequently used matching function, which is simple but practical when there is only one type of behavior to be considered. However, the circumstance will be more complex when we design a recommendation algorithm based on multiple behaviors. That is to say, given a certain commodity, the interaction scores of different behaviors may vary a lot. For example, when we choose a mobile phone, it is more likely for us to view and compare relevant ones several times. In this case, the score of viewing should be much higher than that of the purchasing behavior. To learn the interaction scores for multi-behavior recommendation, we assign different matching functions between each specialized behavior latent vector and item latent vector via:

$$\begin{aligned} \hat{y}_{ui}^1 &= f^1(\mathbf{p}_u^1, \mathbf{q}_i | \theta^1), \\ \hat{y}_{ui}^2 &= f^1(\mathbf{p}_u^2, \mathbf{q}_i | \theta^2), \\ &... \\ \hat{y}_{ui}^n &= f^n(\mathbf{p}_u^n, \mathbf{q}_i | \theta^n). \end{aligned} \tag{6}$$

Here f^* denotes different matching score functions and θ^* denotes the parameters used in the function.

Recently, the NCF unit [5] has been well exploited to generate matching scores. In our model, we also resort to the NCF units to generate matching scores for multiple behaviors and f can be implemented as:

$$\begin{cases} f_{GMF}^j(\mathbf{p}_u^j, \mathbf{q}_i) = \sigma(\mathbf{s}^T(\mathbf{p}_u^j \otimes \mathbf{q}_i)) \\ f_{MLP}^j(\mathbf{p}_u^j, \mathbf{q}_i) = \sigma(\mathbf{s}^T \mathbf{z}_L) \\ f_{NeuMF}^j(\mathbf{p}_u^j, \mathbf{q}_i) = \sigma(\mathbf{s}^T \begin{bmatrix} \mathbf{p}_u^j \otimes \mathbf{q}_i \\ \mathbf{z}_L \end{bmatrix}) \end{cases} \forall j \in \mathcal{N},$$

where $\mathbf{s}^T$ and $\mathbf{z}_L$ denote the layer-specific settings in the original paper [5] and σ denotes the sigmoid function. Generally, the choice for a specific NCF function used for our algorithm should depend on the dataset, and we leave this discussion in Sect. 4.

Now we focus our attention on the fusion of matching scores generated from different behavior channels. Previous works relate the matching scores between different behaviors by setting an ordinal order uniformly, which is unreasonable in real-life scenarios. Considering two users whose behavior patterns vary a lot: user A browses commodity many times every day. He/She just regards Internet suffering as a daily entertainment but never purchasing. While user B only interacts with the E-commerce system when necessary. It is obvious that the way to fuse matching scores on the above two circumstances should be different. Thus we introduce another user-specific variable $\mathbf{h}_u = \{h_u^1, h_u^2, ..., h_u^n\} \in \mathbb{R}^N$ to reflect the heterogeneity. Then we update our estimated matching scores in Eq. (6) as:

$$\hat{y}_{ui}^j = \begin{cases} h_u^1 * \hat{y}_{ui}^1 + h_u^2 * \hat{y}_{ui}^2 + ... + h_u^n * \hat{y}_{ui}^n & , if\ j = n \\ \hat{y}_{ui}^j & , if\ j \neq n \end{cases} \tag{7}$$

Here, the user-specific vector $\mathbf{h}_u$ is another trainable variable in our model, which reflects the behavior diversities among different users. In our method, we calculate the value of $\mathbf{h}_u$ via:

$$h_u^j = \frac{N_u^j}{\sum_{t=1}^{n} N_u^t}, \ \forall j \in \mathcal{N}, \tag{8}$$

where N_u^j denotes the times of behavior j observed in user u's history. Through Eq. (8), we bound the value of h_u^j in the range of $[0, 1]$ to avoid the problem that the prediction of the key behavior is over-influenced by the other behaviors. Since our final objective is to exploit the relations between the key behavior and the other behaviors, thus it is a vector of dimension n. If the number of behaviors to be predicted is more than one, then it can be set as a matrix.

3.4 Model Learning

After having generated the matching scores of different user behaviors from Eq. (7), we resort to the Multi-Task Learning (MTL) paradigm to learn different optimization objectives simultaneously in a shared model. For a single behavior $j \in \mathcal{N}$, the cross entropy loss can be written as:

$$L^j = -\Big(\sum_{(u,i)\in\mathcal{Y}_+^j} log\, \hat{y}_{ui}^j + \sum_{(u,i)\in\mathcal{Y}_-^j} log(1 - \hat{y}_{ui}^j)\Big). \tag{9}$$

Taking all behaviors into the MTL paradigm, our final optimization objective can be written as:

$$L = -\sum_{j=1}^{n} \lambda_j \Big(\sum_{(u,i)\in\mathcal{Y}_+^j} log\, \hat{y}_{ui}^j + \sum_{(u,i)\in\mathcal{Y}_-^j} log(1 - \hat{y}_{ui}^j)\Big). \tag{10}$$

Here, λ_* denotes a weight representing the extent of the importance of every single behavior. The advantage of our model is that we use a more general approach to optimize multiple objectives simultaneously. For the case that we only aim to optimize the key behavior, we can simply set λ_n as 1 and other λ_* as 0.

4 Experiments

4.1 Experimental Settings

Datasets. We evaluate the performance of our algorithm using two real-world E-commerce datasets. The statistics of the two datasets are summarized in Table 1.

- **Tmall Dataset.** This dataset is released in IJCAI-15 challenge, which is collected from Tmall, one of the largest E-commerce systems in China. Three types of user behaviors including viewing, adding-to-cart and purchasing are available within the time period from 01/05/2014 to 30/11/2014.
- **Beibei Dataset.** Beibei is another E-commerce website providing maternal and infant products in China. Since the original dataset is not publicly available, we use the data sampled by Ding *et al.* [3].

Table 1. The statistics of the datasets used in our experiment

Dataset	User#	Item#	View#	Add-to-cart#	Purchase#
Tmall	12,921	22,570	531,640	24,681	160,840
Beibei	10,000	49,488	952,791	–	156,883

Evaluation Metrics. We adopt the leave-one-out method to split the training set and the test set. Following the experimental settings in previous works, we use **HR@K** and **NDCG@K** to evaluate the performance of all the algorithms. HR measures whether the test item is contained in the top-K list. While, NDCG accounts for the position influence by assigning higher scores if the hits are at top ranks.

Baselines. We compare our methods with several baselines, which can be further divided into two groups depending on whether the algorithms are designed for multi-behavior recommendation.

The compared **single-behavior methods** are as follows.

- **ItemkNN** [2]: This is a typical memory-based algorithm, which calculates the similarity between item-vectors. And it has been widely used in previous works for comparison.

- **BPR** [8]: BPR is an optimization criterion derived from the maximum posterior estimator for personalized ranking. We implement it to optimize the matrix factorization model.
- **NCF** [5]: NCF is a neural-based framework for collaborative filtering. It has three instances (*i.e.*, **GMF**, **MLP** and **NeuMF**) which model user-item interactions in different ways. We evaluate all three optional models for comparison.

We also compare our methods with the following **multi-behavior methods**.

- **CMF** [11]: It extends the MF-based methods through factorizing multiple user-side behavior matrices simultaneously, while sharing the item-side matrix.
- **MF-BPR** [7]: MF-BPR utilizes an extended sampling method which reflects different types of behaviors with different levels.
- **NMTR** [4]: This is the state-of-the-art algorithm for multi-behavior recommendation, which optimizes the cascade relation between multiple behaviors using neural networks.

Parameter Settings. We implement our algorithm and all the other baseline models using Tensorflow. For neural-based models, we initialize the parameters in the same way as He *et al.* [5]. For models that have multiple hidden layers, *i.e.*, NCF, NMTR and AMBR, we tune the number of hidden-layers from 1 to 4. We set the embedding size as 64 and the negative sampling ratio as 4 uniformly, which can achieve good performance. The batch size is selected through a grid search in [512, 1024, 2048]. Similarly, the learning rate is in [0.001, 0.002, 0.005, 0.01] and the epoch size is in [30, 40, 50]. Three optimizers including SGD, Adam and Adagrad are implemented. Finally, we also apply L_2 regularization for all methods to prevent over-fitting. The best parameters of our algorithm are presented in Table 2.

Table 2. The best parameters of AMBR algorithm

Dataset	Embedding size	Learning rate	Epoch size	Batch size	L2 Regularization	Optimizer
Tmall	64	0.002	30	2048	10^{-5}	Adam
Beibei	64	0.001	30	1024	10^{-6}	Adam

4.2 Performance Comparison

Table 3 shows the performance comparison between our AMBR algorithm and other baseline algorithms. We set different random seeds for several times and report the best performance score for each algorithm.

Table 3. The overall performance comparison for different algorithms

Method	HR@50	NDCG@50	HR@100	NDCG@100	HR@200	NDCG@200
Tmall Dataset						
ItemKNN	0.0198	0.0059	0.0248	0.0068	0.0284	0.0073
BPR	0.0302	0.0087	0.0484	0.0115	0.0793	0.0158
GMF	0.0202	0.0053	0.0351	0.0080	0.0606	0.0115
MLP	0.0192	0.0050	0.0336	0.0073	0.0577	0.0106
NeuMF	0.0193	0.0050	0.0337	0.0073	0.0576	0.0106
CMF	0.0164	0.0045	0.0376	0.0081	0.0620	0.0114
MF-BPR	0.0317	0.0093	0.0532	0.0127	0.0861	0.0174
NMTR	0.0336	0.0086	0.0573	0.0130	0.0937	0.0178
AMBR	**0.0386**	**0.0103**	**0.0657**	**0.0146**	**0.1077**	**0.0205**
Beibei Dataset						
ItemKNN	0.0198	0.0059	0.0248	0.0068	0.0284	0.0073
BPR	0.0384	0.0101	0.0652	0.0141	0.1041	0.0196
GMF	0.0517	0.0155	0.1148	0.0281	0.1738	0.0365
MLP	0.0499	0.0137	0.1083	0.0236	0.1760	0.0329
NeuMF	0.0514	0.0139	0.1053	0.0231	0.179	0.0328
CMF	0.0579	0.0166	0.1286	0.0280	0.1998	0.0379
MF-BPR	0.0375	0.0102	0.0653	0.0145	0.1069	0.0203
NMTR	0.0663	0.0173	0.1304	0.0276	0.2020	0.0364
AMBR	**0.0737**	**0.0191**	**0.1416**	**0.0307**	**0.2196**	**0.0398**

Table 4. The ablation study for AMBR algorithm on Tmall dataset

Method	HR@50	NDCG@50	HR@100	NDCG@100	HR@200	NDCG@200
AMBR/WtAtt	0.0337	0.0087	0.0632	0.0134	0.1039	0.0191
AMBR/WtFuse	0.0304	0.0082	0.0538	0.0120	0.0882	0.0168
AMBR	0.0386	0.0128	0.0657	0.0146	0.1077	0.0205

We can observe that AMBR achieves the best performance over the other baseline algorithms in terms of both HR and NDCG. Overall speaking, our algorithm can achieve the relative improvements of 27.8%/11.9% in comparison with the best single-behavior/multi-behavior baseline algorithms (*i.e.*, BPR/NMTR) in terms of HR on Tmall dataset. The performance gains can be explained from two perspectives. We not only propose an elaborately designed method for behavior representation learning but also optimize a more reasonable user-specific matching function than previous works. We also find that the algorithms which utilize multiple behaviors achieve better performance than that of single behavior algorithms. This demonstrates the effectiveness to exploit the information in multiple behaviors.

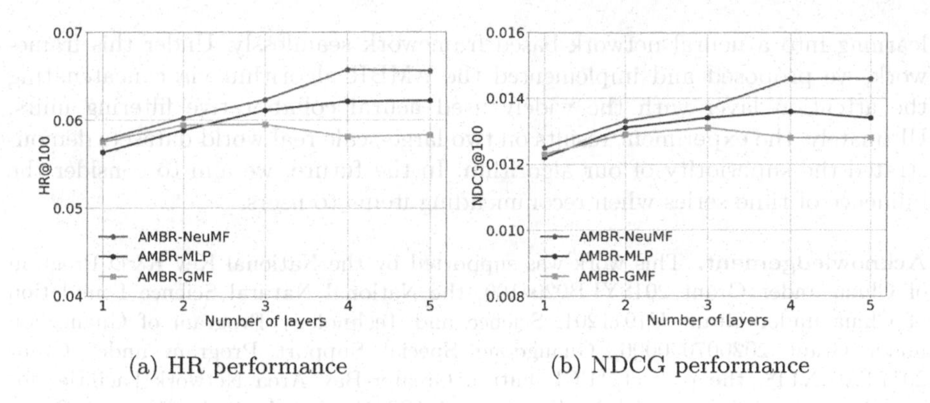

(a) HR performance (b) NDCG performance

Fig. 2. The performance of the AMBR algorithm with different layer settings on Tmall dataset.

4.3 Ablation Study

To investigate which component is the main contributing component in our algorithm, we conduct extensive experiments with two variations of AMBR. We denote the ablation of the attention mechanism in Eq. (5) as AMBR/WtAtt, which means we remove the attention layer from our architecture. And the ablation of the variable $\mathbf{h_u}$ in Eq. (7) is named as AMBR/WtFuse. Table 4 shows the experiment results on Tmall dataset, from which we find both of the two components play a significant role in our design. They learn different knowledge when working at multi-behavior recommendation.

4.4 Layer Settings

To choose an appropriate NCF function and investigate how the number of hidden layers influences our algorithm, we conduct extensive experiments with all three optional NCF units as we have mentioned in Sect. 3.3. And we present the results by varying the number of layers from 1 to 5 on Tmall dataset in Fig. 2. From Fig. 2 we find that, the recommendation performance gains as we increase the layer number for AMBR algorithms. This demonstrates the efficiency of the deep-learning based methods for multi-behavior recommendation. On the other hand, when we increase the layer number continuously, the performance stagnates. In general, it is unnecessary to set the layer number greater than 4, since too many layer parameters may lead to over-fitting.

5 Conclusion

In this paper, we firstly analysed the drawbacks of the previous multi-behavior recommendation algorithms. To overcome these shortcomings, we proposed a unified framework by integrating representation learning and matching function learning into a neural network based framework seamlessly. Under this framework, we proposed and implemented the AMBR algorithm via concatenating the attention layer with the widely used neural collaborative filtering units. Ultimately, the experiment results on two large-scale real world datasets demonstrated the superiority of our algorithm. In the future, we aim to consider the influence of time series when recommending items to users.

Acknowledgement. This work was supported by the National Key R&D Program of China under Grant 2018YFB0204100, the National Natural Science Foundation of China under Grant U1911201, Science and Technology Program of Guangzhou under Grant 202007040006, Guangdong Special Support Program under Grant 2017TX04X148, the project "PCL Future Greater-Bay Area Network Facilities for Large-scale Experiments and Applications" (LZC0019), the Australia Research Council under Grant DE180100950.

References

1. Chen, C., et al.: An efficient adaptive transfer neural network for social-aware recommendation. In: Proceedings of the 42nd International ACM SIGIR Conference on Research and Development in Information Retrieval (2019)
2. Deshpande, M., Karypis, G.: Item-based top-N recommendation algorithms. ACM Trans. Inf. Syst. (TOIS) **22**(1), 143–177 (2004)
3. Ding, J., Yu, G., He, X., Feng, F., Li, Y., Jin, D.: Sampler design for Bayesian personalized ranking by leveraging view data. IEEE Trans. Knowl. Data Eng. (2019). https://doi.org/10.1109/TKDE.2019.2931327
4. Gao, C., et al.: Learning to recommend with multiple cascading behaviors. IEEE Trans. Knowl. Data Eng. (2019). https://doi.org/10.1109/TKDE.2019.2958808
5. He, X., Liao, L., Zhang, H., Nie, L., Hu, X., Chua, T.S.: Neural collaborative filtering. In: Proceedings of the 26th International Conference on World Wide Web, pp. 173–182. International World Wide Web Conferences Steering Committee (2017)
6. Lin, T.H., Gao, C., Li, Y.: CROSS: cross-platform recommendation for social e-commerce. In: Proceedings of the 42nd International ACM SIGIR Conference on Research and Development in Information Retrieval, pp. 515–524. ACM (2019)
7. Loni, B., Pagano, R., Larson, M., Hanjalic, A.: Bayesian personalized ranking with multi-channel user feedback. In: Proceedings of the 10th ACM Conference on Recommender Systems, pp. 361–364. ACM (2016)
8. Rendle, S., Freudenthaler, C., Gantner, Z., Schmidt-Thieme, L.: BPR: Bayesian personalized ranking from implicit feedback. In: Proceedings of the Twenty-Fifth Conference on Uncertainty in Artificial Intelligence, pp. 452–461. AUAI Press (2009)
9. Singh, A.P., Gordon, G.J.: Relational learning via collective matrix factorization. In: Proceedings of the 14th ACM SIGKDD International Conference on Knowledge Discovery and Data Mining, pp. 650–658. ACM (2008)
10. Vaswani, A., et al.: Attention is all you need. In: Advances in Neural Information Processing Systems, pp. 5998–6008 (2017)
11. Zhao, Z., Cheng, Z., Hong, L., Chi, E.H.: Improving user topic interest profiles by behavior factorization. In: Proceedings of the 24th International Conference on World Wide Web, pp. 1406–1416. International World Wide Web Conferences Steering Committee (2015)

Asymmetric Pairwise Preference Learning for Heterogeneous One-Class Collaborative Filtering

Yongxin Ni, Zhuoxin Zhan, Weike Pan(✉), and Zhong Ming(✉)

National Engineering Laboratory for Big Data System Computing Technology, College of Computer Science and Software Engineering, Shenzhen University, Shenzhen, China
{niyongxin2016,zhanzhuoxin2018}@email.szu.edu.cn, {panweike,mingz}@szu.edu.cn

Abstract. Heterogeneous one-class collaborative filtering (HOCCF) is a recent and important recommendation problem which involves two different types of one-class feedback such as purchases and examinations. In this paper, we propose a generic asymmetric pairwise preference assumption and a novel like-minded user-group construction strategy for the HOCCF problem. Specifically, our generic assumption contains six different pairwise preference relations derived from the heterogeneous feedback, where we introduce a series of weighting strategies to make our assumption more reasonable. Our group construction strategy introduces richer interactions within user-groups, which is expected to learn the users' preference more accurately. We then design a novel recommendation model called asymmetric pairwise preference learning (APPLE). Extensive empirical studies show that our APPLE can recommend items significantly more accurately than the closely related state-of-the-art methods on three real-world datasets.

Keywords: Asymmetric pairwise preference learning · Heterogeneous one-class collaborative filtering · Implicit feedback · User-group

1 Introduction

The main purpose of a recommender system is to deliver a personalized ranked list of items for each user accurately, which helps a user discover some items that he/she is interested in. In the community of recommender systems, there have been lots of works [8,13] for the one-class collaborative filtering (OCCF) [9] problem, where the input data only contains one single type of users' behaviors such as purchases.

However, in real-world applications, users' behaviors are usually in heterogeneous forms, including more than one type of one-class feedback such as both purchases and examinations, which is usually called heterogeneous OCCF (HOCCF) [11]. In this paper, we study the HOCCF problem and propose a

H. Yang et al. (Eds.): ICONIP 2020, LNCS 12534, pp. 407–419, 2020.
https://doi.org/10.1007/978-3-030-63836-8_34

generic asymmetric pairwise preference assumption, a weighting strategy and a group construction strategy, and finally obtain our recommendation model called asymmetric pairwise preference learning (APPLE).

Firstly, on the basis of the asymmetric pairwise preference assumption of a very recent work [8], we propose a generic assumption by extending (i) "a user's preference to a purchased item and an un-purchased item" to "a user's preference to a purchased item, an examined item and an un-interacted item", and (ii) "the preference of a purchase user-group and an un-purchase user-group to an item" to "the preference of a purchase user-group, an examination user-group and an un-interaction user-group to an item". Our generic assumption introduces more pairwise relations so as to exploit the users' purchases and examinations more completely.

Secondly, we find that it may be not the best to assign different pairwise preference relations the same weight since each user (or user-group) may have his/her (or their) own criterion. We therefore introduce some dynamic weighting factors to adjust the weight appropriately for different users. In particular, we design some weighting factors for two of four relations involving examinations, because we find that this type of user behaviors are the key difference between OCCF and HOCCF, which may cause over-learning without interference.

Thirdly, we propose a strategy for the construction of a user-group, which is the first like-minded user-group construction strategy for a recommendation algorithm as far as we know. Specifically, for each user-group who purchased or examined an item i, we sample the users by our similarity-priority strategy rather than by a random selection strategy in most previous works. Moreover, our strategy for user-group construction can be readily applied to other algorithms or real-world applications.

We then conduct extensive empirical studies and show the effectiveness of our APPLE in comparison with the very competitive state-of-the-art methods.

2 Related Work

Before introducing our solution to the HOCCF problem, we first discuss some closely related works on addressing the one-class collaborative filtering (OCCF) problem [9] and the heterogeneous OCCF problem [11], respectively.

2.1 One-Class Collaborative Filtering

In a real-world recommender system such as an e-commerce platform, users' one-class feedback such as purchases are easier to be collected than users' categorical ratings to items. For this reason, recommendation with users' one-class feedback or OCCF [4,6,9] becomes more and more important compared with the counterpart of multi-class collaborative filtering with users' explicit categorical ratings [14].

For modeling users' one-class feedback in OCCF, there are two main branches of methods, including neighborhood-based methods [2] and factorization-based methods [7,13]. In neighborhood-based methods, we often first mine similar users (or items) to a target user (or a candidate item), and then make recommendation based on the assumption that users with similar taste in the past will have similar taste in the future (or a user's taste in the past is similar to that in the future). In factorization-based methods, we usually learn some latent representation of each user and each item by factorizing the (user, item) interaction matrix, with which we can then make recommendation by the inner product of each (user, item) pair or other forms.

Among the factorization-based methods, different methods adopt different preference assumptions, including pointwise preference assumption [6,9] and pairwise preference assumption [13], where the former assumes that a user likes (and dislikes) an interacted (and an un-interacted) item, and the latter assumes that the hidden preference score between an interacted (user, item) pair is larger than that of an un-interacted (user, item) pair. The pairwise preference assumption has well been recognized to be a better one for it is more relaxed and more likely to be satisfied in real situations, based on which some important works have been developed such as Bayesian personalized ranking (BPR) [13] and its extensions [10]. Very recently, a new asymmetric pairwise preference assumption is proposed with improved performance, which involves two types of pairwise preference assumptions, one defined on the original (user, item) interaction matrix and the other defined on the transposed (user, item) interaction matrix.

However, most factorization-based methods based on the pairwise preference assumption are designed for modeling one single type of one-class feedback, which may not be able to capture users' preferences sufficiently for the case with two different types of one-class feedback such as both purchases and examinations in HOCCF.

2.2 Heterogeneous One-Class Collaborative Filtering

In order to model the heterogeneous one-class feedback such as purchases and examinations in HOCCF (instead of the homogeneous feedback in OCCF), we have to answer some fundamental questions, e.g., "how to deal with the difference between two different types of feedback", "how to address the uncertainty of the examinations", etc.

In transfer via joint similarity learning (TJSL) [11], the authors first extend the factored item similarity model [7] by designing an expanded prediction rule to involve two different types of feedback, and then design an iterative algorithm to adaptively identify some likely to be purchased items from the examinations for each user so as to address the uncertainty issue. The expanded prediction rule and the iterative learning procedure may not be very efficient. In view-enhanced alternative least square (VALS) [3], the authors design more than one loss functions for the two types of one-class feedback, and then assign different weight for different losses, which is usually difficult to be determined. In BPR for heterogeneous implicit feedback (BPRH) [12], the authors generalize the concept of item-set from homogeneous one-class feedback [10] to heterogeneous one-class feedback, and use two types of pairwise preference relations for the two types of feedback. However, they do not consider the pairwise relations from the auxiliary perspective of the transposed (user, item) interaction matrix, which thus may not be sufficient in modeling users' preferences. In efficient heterogeneous collaborative filtering (EHCF) [1], the authors develop an efficient non-sampling optimization method by leveraging the sparsity of the positive-only data and also incorporate the context of each level's behaviors. However, it may not learn the users' preferences well since the impact from the neighboring users' behaviors are not modeled.

In this paper, we focus on how to generalize the very recent and effective asymmetric pairwise preference learning approach [8] to the HOCCF problem in order to capture the users' preferences more sufficiently.

3 Our Solution

3.1 Problem Definition

In HOCCF, we have two types of one-class feedback, i.e., purchases $\mathcal{R}^{\mathcal{P}} = \{(u,i)\}$ and examinations $\mathcal{R}^{\mathcal{E}} = \{(u,k)\}$. We use $\mathcal{I}_u^{\mathcal{P}} = \{i|(u,i) \in \mathcal{R}^{\mathcal{P}}\}$ and $\mathcal{I}_u^{\mathcal{E}} = \{k|(u,k) \in \mathcal{R}^{\mathcal{E}}\}$ to denote the set of purchased items and the set of examined items by user u, respectively. Our goal is to exploit the (user, item) pairs in $\mathcal{R}^{\mathcal{P}}$ and $\mathcal{R}^{\mathcal{E}}$ so as to recommend a personalized ranked list of items for each user $u \in \mathcal{U}$ from the set of not-yet purchased items $\mathcal{I} \backslash \mathcal{I}_u^{\mathcal{P}}$. We illustrate the studied problem in Fig. 1 and list some commonly used notations in Table 1.

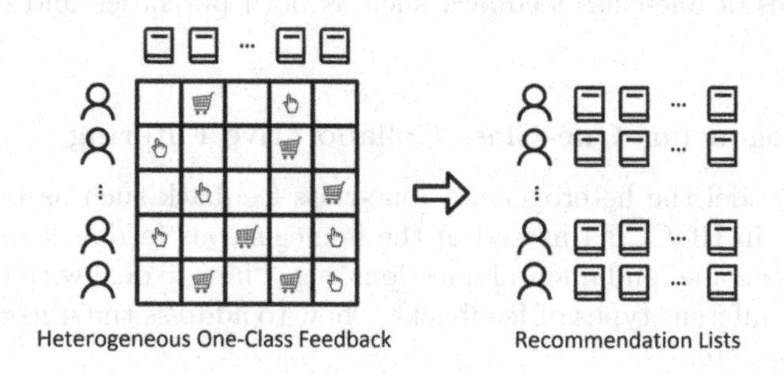

Fig. 1. Illustration of ranking-oriented recommendation with heterogeneous one-class feedback in HOCCF.

3.2 A Generic Preference Assumption

In modeling users' homogeneous one-class feedback (e.g., purchases), the well-known pairwise preference assumption used in Bayesian personalized ranking (BPR) [13] is as follows,

$$\hat{r}_{ui} > \hat{r}_{uj}, \tag{1}$$

Table 1. Notations and descriptions.

Notation	Description
n	Number of users
m	Number of items
$\mathcal{U} = \{u\}$	The whole set of users
$\mathcal{I} = \{i\}$	The whole set of items
u, v, w	User ID
i, k, j	Item ID
$\mathcal{R}^{\mathcal{P}} = \{(u, i)\}$	Purchases
$\mathcal{I}_u^{\mathcal{P}} = \{i \mid (u, i) \in \mathcal{R}^{\mathcal{P}}\}$	Items purchased by user u
$\mathcal{U}_i^{\mathcal{P}} = \{u \mid (u, i) \in \mathcal{R}^{\mathcal{P}}\}$	Users who purchased item i
$\mathcal{P} \subseteq \mathcal{U}_i^{\mathcal{P}}$	A set of users
$\mathcal{R}^{\mathcal{E}} = \{(v, k)\}$	Examinations
$\mathcal{I}_v^{\mathcal{E}} = \{k \mid (v, k) \in \mathcal{R}^{\mathcal{E}}\}$	Items examined by user v
$\mathcal{U}_k^{\mathcal{E}} = \{v \mid (v, k) \in \mathcal{R}^{\mathcal{E}}\}$	Users who examined item k
$\mathcal{E} \subseteq \mathcal{U}_k^{\mathcal{E}}$	A set of users
$\mathcal{R}^{\mathcal{N}} = \{(w, j)\}$	Un-interacted pairs
$\mathcal{I}_u^{\mathcal{N}} = \{i \mid (u, i) \in \mathcal{R}^{\mathcal{N}}\}$	Items un-interacted with by user u
$\mathcal{U}_i^{\mathcal{N}} = \{w \mid (w, j) \in \mathcal{R}^{\mathcal{N}}\}$	Users who un-interacted with item i
$\mathcal{N} \subseteq \mathcal{U}_i^{\mathcal{N}}$	A set of users
$\hat{r}_{ui}$	Predicted preference
d	Number of latent dimensions
$U_{u\cdot} \in \mathbb{R}^{1 \times d}$	User u's latent feature vector
$V_{i\cdot} \in \mathbb{R}^{1 \times d}$	Item i's latent feature vector
$b_i \in \mathbb{R}$	Item i's bias

where $(u, i) \in \mathcal{R}^{\mathcal{P}}$ and $j \in \mathcal{I}_u^{\mathcal{N}}$ denote that user u has purchased the item i and has not interacted with the item j, respectively. The pairwise relation in Eq. (1) means that the preference of the user u to the item i is larger, which is reasonable because a user usually prefers a purchased item to an un-interacted one.

Very recently, a novel asymmetric pairwise preference assumption is used in asymmetric BPR (ABPR) [8], which involves a horizontal relation in Eq. (1) and a vertical one,

$$\hat{r}_{ui} > \hat{r}_{uj}, \; \hat{r}_{\mathcal{P}i} > \hat{r}_{\mathcal{N}i}, \tag{2}$$

where $\mathcal{P} \subseteq \mathcal{U}_i^{\mathcal{P}}$ is a set of users who purchased item i and $\mathcal{N} \subseteq \mathcal{U}_i^{\mathcal{N}}$ is a set of users who have not interacted with item i. The newly introduced vertical relation $\hat{r}_{\mathcal{P}i} > \hat{r}_{\mathcal{N}i}$ in Eq. (2) is defined on groupwise preference, i.e., the preference of a group of users $\mathcal{P}$ to item i is larger than that of a group of users $\mathcal{N}$ to the same item, which makes the relation more comparable in comparison with the relation defined on two single users [8].

However, the asymmetric assumption in Eq. (2) is designed only for one single type of one-class feedback instead of both purchases and examinations in HOCCF. As a response, we propose a novel and generic preference assumption,

$$\hat{r}_{ui} > \hat{r}_{uk} > \hat{r}_{uj},\ \hat{r}_{\mathcal{P}i} > \hat{r}_{\mathcal{E}i} > \hat{r}_{\mathcal{N}i}, \tag{3}$$

where $(u, i) \in \mathcal{R}^{\mathcal{P}}$ is a purchasing record, $k \in \mathcal{I}_u^{\mathcal{E}}$ and $j \in \mathcal{I}_u^{\mathcal{N}}$ are items that have been examined and have not been interacted with by user u, respectively, and $\mathcal{P} \subseteq \mathcal{U}_i^{\mathcal{P}}$, $\mathcal{E} \subseteq \mathcal{U}_i^{\mathcal{E}}$ and $\mathcal{N} \subseteq \mathcal{U}_i^{\mathcal{N}}$ are sets of users who have purchased, examined and un-interacted with item i, respectively.

3.3 Weighting Strategy

In our generic preference assumption shown in Eq. (3), we have six pairwise relations, i.e., $\hat{r}_{ui} > \hat{r}_{uj}$, $\hat{r}_{ui} > \hat{r}_{uk}$, $\hat{r}_{uk} > \hat{r}_{uj}$, $\hat{r}_{\mathcal{P}i} > \hat{r}_{\mathcal{N}i}$, $\hat{r}_{\mathcal{P}i} > \hat{r}_{\mathcal{E}i}$ and $\hat{r}_{\mathcal{E}i} > \hat{r}_{\mathcal{N}i}$. We realize that the examined item k in the relation $\hat{r}_{ui} > \hat{r}_{uk}$ is very different from the item j in the relation $\hat{r}_{ui} > \hat{r}_{uj}$ or the relation $\hat{r}_{uk} > \hat{r}_{uj}$, because $|\mathcal{I}_u^{\mathcal{E}}| << |\mathcal{I}_u^{\mathcal{N}}|$ results in the consequence that the item $k \in \mathcal{I}^{\mathcal{E}}$ is very likely to be over exploited. This issue also exists for the relation $\hat{r}_{\mathcal{P}i} > \hat{r}_{\mathcal{E}i}$.

The above observation motivates us to design a weighting strategy for the two relations $\hat{r}_{ui} > \hat{r}_{uk}$ and $\hat{r}_{\mathcal{P}i} > \hat{r}_{\mathcal{E}i}$. To obtain some reasonable weights, we take advantage of a user's/user-group's behaviors and covert them into some kind of ratios.

Specifically, for each user u, we exploit his/her purchase and examination information, and introduce an individual purchase-to-examination (iP2E) ratio,

$$\alpha_u = \frac{|\mathcal{I}_u^{\mathcal{P}}|}{|\mathcal{I}_u^{\mathcal{P}}| + |\mathcal{I}_u^{\mathcal{E}}|}. \tag{4}$$

The ratio α_u is derived from one single user u, which may be insufficient to represent the weight for the relation $\hat{r}_{\mathcal{P}i} > \hat{r}_{\mathcal{E}i}$, because we should consider the overall effect of both user-group $\mathcal{P}$ and user-group $\mathcal{E}$. We thus introduce a conventional purchase-to-examination (cP2E) ratio,

$$\mathcal{A}_c = \frac{\sum_{u \in \mathcal{P} \cup \mathcal{E}} \alpha_u}{|\mathcal{P} \cup \mathcal{E}|}. \tag{5}$$

However, the weighting factor for $\hat{r}_{\mathcal{P}i} > \hat{r}_{\mathcal{E}i}$, i.e., cP2E, is still unable to accurately reflect the overall standard of user-groups $\mathcal{P}$ and $\mathcal{E}$ in $\hat{r}_{\mathcal{P}i} > \hat{r}_{\mathcal{E}i}$ since $\mathcal{A}_c$ here should be a group-oriented factor. Hence, we regard the group $\mathcal{P} \cup \mathcal{E}$ as a whole and further design a groupwise purchase-to-examination (gP2E) ratio,

$$\mathcal{A}_g = \frac{|\bigcup_{u \in \mathcal{P}} \mathcal{I}_u^{\mathcal{P}}|}{|\bigcup_{u \in \mathcal{P}} \mathcal{I}_u^{\mathcal{P}}| + |\bigcup_{v \in \mathcal{E}} \mathcal{I}_v^{\mathcal{E}}|}. \tag{6}$$

Since gP2E reflects the overall standard of user-groups $\mathcal{P}$ and $\mathcal{E}$ more accurately than cP2E, we expect that the relation $r_{\mathcal{P}i} > r_{\mathcal{E}i}$ weighted with gP2E will perform better than that with cP2E, which can also be found in our empirical studies.

Finally, we further introduce a tunable parameter β $(0 < \beta \leq 1)$ to iP2E, cP2E and gP2E in order to have a more flexible strategy, i.e., $\frac{\alpha_u}{\beta}$, $\frac{\mathcal{A}_c}{\beta}$ and $\frac{\mathcal{A}_g}{\beta}$. Notice that our weighting factors α_u, $\mathcal{A}_c$ and $\mathcal{A}_g$ are different for different users and user-groups, which makes the corresponding relations more likely to hold. For example, some user may purchase an item once the item meets his/her requirement, while others may be very cautious when shopping.

3.4 Like-Mined User-Group Construction

For each user u, we can learn the influence of different users of $\mathcal{P}$, $\mathcal{E}$ and $\mathcal{N}$ with an equal probability under the traditional random sampling strategy. But it should be noted that this widely adopted strategy can be further improved by taking the similarities among the users into consideration.

We hope to find a sampling strategy for $\mathcal{P}$ and $\mathcal{E}$, which not only preserves the advantage of the random sampling strategy but also makes the selected users of the two groups as similar to user u as possible. This similarity-priority strategy can also eliminate the effect of occasionality when constructing a group. For example, a user u likes fruits and thus he/she examines an avocado k, while another user u' does not like fruits but also examines the same item just out of curiosity. In this case, the previous random sampling strategy will unreasonably add u' to $\mathcal{U}_i^{\mathcal{E}}$ though u and u' are not related (they both just happened to examine the same item k). Such occasionality is detrimental to the preference learning of user u, because it brings deficiency to the previous random sampling strategy.

Specifically, for the construction of a user-group $\mathcal{P}$, we randomly divide $\mathcal{U}_i^{\mathcal{P}}\backslash u$ into $|\mathcal{U}_i^{\mathcal{P}}\backslash u|/(|\mathcal{P}|-1)$ groups, where each group contains $|\mathcal{P}|-1$ users. We then compute the similarity between each group-user u' and user u by the purchase information. After that, we pick up the group with the highest overall similarity, and obtain our $\mathcal{P}$ by adding u into the selected group. The purchase similarity is defined as follows,

$$S_{\mathcal{P}} = \sum_{u' \in \mathcal{P}\backslash\{u\}} \mathrm{JI}_{u,u'}^{\mathcal{P}} + \delta \times \sum_{u_1,u_2 \in \mathcal{P}\backslash\{u\}, u_1 \neq u_2} \mathrm{JI}_{u_1,u_2}^{\mathcal{P}}, \tag{7}$$

where $\mathrm{JI}_{u,u'}^{\mathcal{P}} = |\mathcal{I}_u^{\mathcal{P}} \cap \mathcal{I}_{u'}^{\mathcal{P}}|/|\mathcal{I}_u^{\mathcal{P}} \cup \mathcal{I}_{u'}^{\mathcal{P}}|$ is the Jaccard index. As we can see, the weight of the similarity between users within the group is reduced by $\delta < 1$. The reason is that the group we finally pick up may still have a large difference among its internal users (e.g., the similarity between one group-user and user u is very low, while those between other group-users and user u are very high), resulting in unreasonable partial similarities within the group. We adopt a similar way to construct $\mathcal{E}$, where the examination similarity is defined as follows,

$$S_{\mathcal{E}} = \sum_{u' \in \mathcal{E}} \text{JI}_{u,u'}^{\mathcal{E}} + \delta \times \sum_{u_1, u_2 \in \mathcal{E}, u_1 \neq u_2} \text{JI}_{u_1,u_2}^{\mathcal{E}}, \quad (8)$$

where $\text{JI}_{u,u'}^{\mathcal{E}} = |\mathcal{I}_u^{\mathcal{E}} \cap \mathcal{I}_{u'}^{\mathcal{E}}| / |\mathcal{I}_u^{\mathcal{E}} \cup \mathcal{I}_{u'}^{\mathcal{E}}|$ and $u \notin \mathcal{E}$. Notice that there may be insufficient candidate users in $\mathcal{U}_i^{\mathcal{P}}$ or $\mathcal{U}_i^{\mathcal{E}}$ for an item i, for which we will then use all the available $|\mathcal{U}_i^{\mathcal{P}}|$ and/or $|\mathcal{U}_i^{\mathcal{E}}|$ users instead in the training process.

Algorithm 1. The algorithm of APPLE.

1: **for** $t = 1, 2, .., T$ **do**
2: **for** $t_2 = 1, 2, .., |\mathcal{R}^{\mathcal{E}} \backslash \sum_{(u,i) \in \mathcal{R}^{\mathcal{P}}} \{(u,k) \in \mathcal{R}^{\mathcal{E}}\}|$ **do**
3: Randomly pick a (user, item) pair (u, k) from $\mathcal{R}^{\mathcal{E}} \backslash \sum_{(u,i) \in \mathcal{R}^{\mathcal{P}}} \{(u,k) \in \mathcal{R}^{\mathcal{E}}\}$.
4: Randomly pick an item j from $\mathcal{I}_u^{\mathcal{N}}$.
5: Calculate the gradients w.r.t. the tentative objection function $-\ln \sigma(\hat{r}_{uk} - \hat{r}_{uj}) + \frac{\alpha}{2}||V_{k\cdot}||^2 + \frac{\alpha}{2}||V_{j\cdot}||^2 + \frac{\alpha}{2}||U_{u\cdot}||^2$.
6: Update the corresponding model parameters.
7: **end for**
8: **for** $t_3 = 1, 2, .., |\mathcal{R}^{\mathcal{P}} \backslash \sum_{(u,k) \in \mathcal{R}^{\mathcal{E}}} \{(u,i) \in \mathcal{R}^{\mathcal{P}}\}|$ **do**
9: Randomly pick a (user, item) pair (u, i) from $\mathcal{R}^{\mathcal{P}} \backslash \sum_{(u,k) \in \mathcal{R}^{\mathcal{E}}} \{(u,i) \in \mathcal{R}^{\mathcal{P}}\}$.
10: Randomly pick an item j from $\mathcal{I}_u^{\mathcal{N}}$.
11: Pick highest-similarity $|\mathcal{P}| - 1$ users from $\mathcal{U}_i^{\mathcal{P}} \backslash \{u\}$.
12: Randomly pick $|\mathcal{E}|$ users from $\mathcal{U}_i^{\mathcal{E}}$.
13: Randomly pick $|\mathcal{N}|$ users from $\mathcal{U}_i^{\mathcal{N}}$.
14: Calculate the gradients w.r.t. the tentative objection function $-\ln \sigma(\hat{r}_{ui} - \hat{r}_{uj}) - \frac{1}{|\mathcal{N}|} \sum_{w \in \mathcal{N}} \ln \sigma(\hat{r}_{\mathcal{P}i} - \hat{r}_{wi}) - \frac{\mathcal{A}}{\beta} \frac{1}{|\mathcal{E}|} \sum_{v \in \mathcal{E}} \ln \sigma(\hat{r}_{\mathcal{P}i} - \hat{r}_{vi}) - \frac{1}{|\mathcal{N}|} \sum_{w \in \mathcal{N}} \ln \sigma(\hat{r}_{\mathcal{E}i} - \hat{r}_{wi}) + \frac{\alpha}{2}||V_{i\cdot}||^2 + \frac{\alpha}{2}||V_{j\cdot}||^2 + \frac{\alpha}{2}||b_i||^2 + \frac{\alpha}{2}||b_j||^2 + \sum_{u' \in \mathcal{P}} [\frac{\alpha}{2}||U_{u'\cdot}||^2 + \frac{\alpha}{2}||b_{u'}||^2] + \sum_{v \in \mathcal{E}} [\frac{\alpha}{2}||U_{v\cdot}||^2 + \frac{\alpha}{2}||b_v||^2] + \sum_{w \in \mathcal{N}} [\frac{\alpha}{2}||U_{w\cdot}||^2 + \frac{\alpha}{2}||b_w||^2]$.
15: Update the corresponding model parameters.
16: **end for**
17: **for** $t_4 = 1, 2, .., |\mathcal{R}^{\mathcal{P}}|$ **do**
18: Randomly pick a (user, item) pair (u, i) from $\mathcal{R}^{\mathcal{P}}$.
19: Randomly pick an item k from $\mathcal{I}_u^{\mathcal{E}}$.
20: Randomly pick an item j from $\mathcal{I}_u^{\mathcal{N}}$.
21: Pick highest-similarity $|\mathcal{E}|$ users from $\mathcal{U}_i^{\mathcal{E}}$.
22: Pick highest-similarity $|\mathcal{P}| - 1$ users from $\mathcal{U}_i^{\mathcal{P}} \backslash \{u\}$.
23: Randomly pick $|\mathcal{N}|$ users from $\mathcal{U}_i^{\mathcal{N}}$.
24: Calculate the gradients w.r.t. the tentative objection function in Eq.(9).
25: Update the corresponding model parameters.
26: **end for**
27: **end for**

Table 2. Results of our APPLE and two closely related baseline methods, i.e., BPR and ABPR. Notice that we fix $\delta = 0.5$, and use the best value of β from $\{0.1, 0.2, \ldots, 1.0\}$. The significantly best results (p-value is smaller than 0.05) are marked in bold.

Dataset	Method	Prec@5	Recall@5	F1@5	NDCG@5	1-call@5
ML100K	BPR	0.0552±0.0006	0.1032±0.0019	0.0673±0.0007	0.0874±0.0020	0.2425±0.0034
	ABPR	0.0606±0.0012	0.1173±0.0049	0.0744±0.0020	0.0956±0.0021	0.2655±0.0070
	APPLE	**0.0679**±0.0003	**0.1291**±0.0028	**0.0828**±0.0007	**0.1089**±0.0022	**0.2880**±0.0048
ML1M	BPR	0.0928±0.0008	0.0829±0.0002	0.0717±0.0003	0.1121±0.0010	0.3609±0.0018
	ABPR	0.0931±0.0014	0.0809±0.0009	0.0706±0.0010	0.1150±0.0019	0.3596±0.0042
	APPLE	**0.1079**±0.0017	**0.0974**±0.0014	**0.0840**±0.0011	**0.1314**±0.0028	**0.4109**±0.0044
Alibaba	BPR	0.0050±0.0006	0.0193±0.0026	0.0077±0.0009	0.0138±0.0017	0.0246±0.0031
	ABPR	0.0050±0.0004	0.0194±0.0015	0.0077±0.0006	0.0145±0.0007	0.0246±0.0018
	APPLE	**0.0079**±0.0002	**0.0317**±0.0006	**0.0124**±0.0003	**0.0224**±0.0003	**0.0391**±0.0012

3.5 Objective Function and Algorithm

Based on our preference assumption and weighting strategy, we reach the objective function of our model for each $(u, i, k, j, \mathcal{P}, \mathcal{E}, \mathcal{N})$,

$$\min_{\Theta} \ell(u, i, k, j, \mathcal{P}, \mathcal{E}, \mathcal{N}) + \text{reg}(u, i, k, j, \mathcal{P}, \mathcal{E}, \mathcal{N}), \tag{9}$$

where $\ell(u, i, k, j, \mathcal{P}, \mathcal{E}, \mathcal{N}) = -\ln\sigma(\hat{r}_{ui} - \hat{r}_{uj}) - \frac{\alpha_u}{\beta}\ln\sigma(\hat{r}_{ui} - \hat{r}_{uk}) - \ln\sigma(\hat{r}_{uk} - \hat{r}_{uj}) - \frac{1}{|\mathcal{N}|}\sum_{w\in\mathcal{N}}\ln\sigma(\hat{r}_{\mathcal{P}i} - \hat{r}_{wi}) - \frac{A}{\beta}\frac{1}{|\mathcal{E}|}\sum_{v\in\mathcal{E}}\ln\sigma(\hat{r}_{\mathcal{P}i} - \hat{r}_{vi}) - \frac{1}{|\mathcal{N}|}\sum_{w\in\mathcal{N}}\ln\sigma(\hat{r}_{\mathcal{E}i} - \hat{r}_{wi})$ is the loss function, and $\text{reg}(u, i, k, j, \mathcal{P}, \mathcal{E}, \mathcal{N}) = \frac{\alpha}{2}||V_{i\cdot}||^2 + \frac{\alpha}{2}||V_{k\cdot}||^2 + \frac{\alpha}{2}||V_{j\cdot}||^2 + \frac{\alpha}{2}||b_i||^2 + \frac{\alpha}{2}||b_k||^2 + \frac{\alpha}{2}||b_j||^2 + \sum_{u'\in\mathcal{P}}[\frac{\alpha}{2}||U_{u'\cdot}||^2 + \frac{\alpha}{2}||b_{u'}||^2] + \sum_{v\in\mathcal{E}}[\frac{\alpha}{2}||U_{v\cdot}||^2 + \frac{\alpha}{2}||b_v||^2] + \sum_{w\in\mathcal{N}}[\frac{\alpha}{2}||U_{w\cdot}||^2 + \frac{\alpha}{2}||b_w||^2]$ is the regularization term used to avoid overfitting, and $\Theta = \{U_{u\cdot}, V_{i\cdot}, b_u, b_i, u \in \mathcal{U}, i \in \mathcal{I}\}$ is a set of model parameters to be learned. We then have the gradients of the parameters w.r.t. the objective function, and obtain a stochastic gradient descent based algorithm enhanced with our group construction strategy, which is shown in Algorithm 1.

The time complexity of our APPLE mainly includes two parts, i.e., the pairwise preference learning and the construction of like-minded user groups, where the former is comparable to that of the closely related method ABPR, and the latter is dependent on the group size, which is usually a small constant such as 3 in our experiments.

4 Experiments

4.1 Datasets and Evaluation Metrics

We employ three real-world datasets [11] in our empirical studies, including MovieLens 100K (ML100K), MovieLens 1M (ML1M), and Alibaba2015

Table 3. Results of our APPLE with cP2E and gP2E ($\beta = 1$ and $\delta = 0.5$).

Dataset	Strategy	Prec@5	NDCG@5
ML100K	cP2E	0.0600	0.0939
	gP2E	**0.0635**	**0.1025**
ML1M	cP2E	0.1021	0.1244
	gP2E	**0.1030**	**0.1261**
Alibaba	cP2E	0.0078	0.0217
	gP2E	**0.0079**	**0.0224**

(Alibaba). ML100K and ML1M are both users' 5-star ratings on movies, where the former contains 100,000 ratings assigned by 943 users to 1,682 movies, and the latter contains 1,000,209 ratings assigned by 6,040 users to 3,952 movies. The pre-processing steps of ML100K and ML1M for the HOCCF problem can be found in [11]. Alibaba is a real data with users' purchases and clicks. We follow previous works and adopt five ranking-oriented evaluation metrics, including Precision@5 (Prec@5), Recall@5, F1@5, NDCG@5 and 1-call@5.

4.2 Baselines and Configurations

We compare our APPLE[1] with two closely related methods, i.e., BPR [13] and ABPR [8]. We fix the number of latent dimensions $d = 20$, the learning rate $\gamma = 0.01$, the group size $|\mathcal{P}| = |\mathcal{E}| = |\mathcal{N}| = 3$, and search the best value of the iteration number $T \in \{100, 500, 1000\}$ and the best value of the tradeoff parameter $\alpha \in \{0.001, 0.01, 0.1\}$ on the regularization terms for each method on the three datasets via NDCG@5.

For both the baseline methods and our APPLE, we implement them in the same stochastic gradient descent (SGD) based algorithm. We use $\beta = 1$ and $\delta = 0.5$ as default values, and adjust their values in the range of $\{0.1, 0.2, \ldots, 1.0\}$ when specified. We also compare the effect of our proposed cP2E and gP2E weighting strategy on the construction of user-groups in our empirical studies, and compare the recommendation performance of similarity-priority user-group

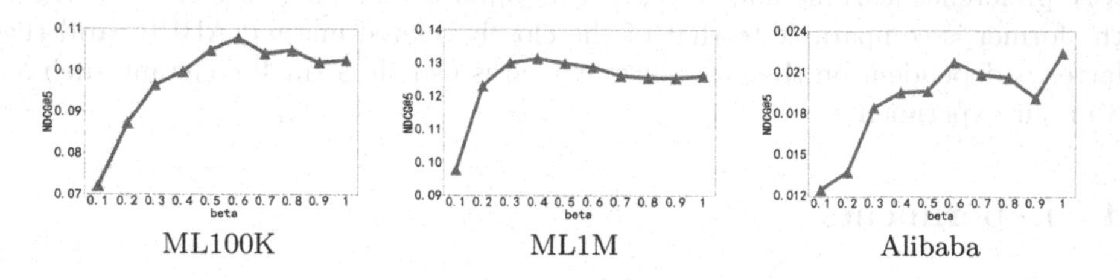

Fig. 2. Results of our APPLE with different values of β ($\delta = 0.5$).

[1] http://csse.szu.edu.cn/staff/panwk/publications/APPLE/.

Table 4. Results of our APPLE with two different user-group construction strategies ($\beta = 1$ and $\delta = 0.5$).

Dataset	Strategy	Prec@5	NDCG@5
ML100K	Random	0.0598	0.0951
	Similarity-priority	**0.0635**	**0.1025**
ML1M	Random	0.1002	0.1228
	Similarity-priority	**0.1030**	**0.1261**
Alibaba	Random	0.0069	0.0200
	Similarity-priority	**0.0079**	**0.0224**

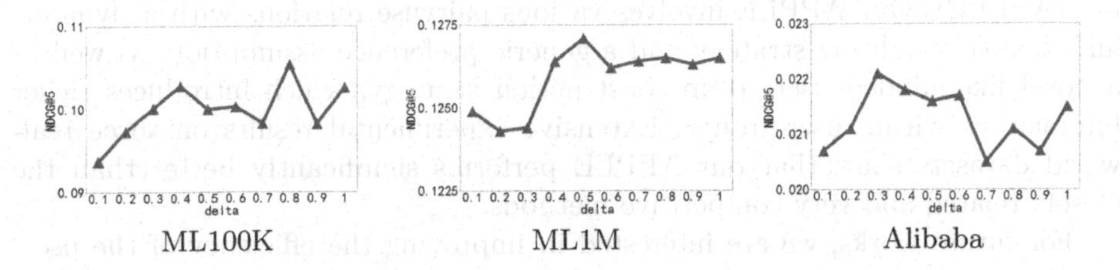

Fig. 3. Results of our APPLE with different values of δ ($\beta = 1$).

construction strategy with that of random sampling user-group construction strategy. Moreover, we conduct empirical studies to support our weighting strategy to only two relations and our hypothesis that similarities within groups should be reduced.

4.3 Results

We fix $\delta = 0.5$, tune $\beta \in \{0.1, 0.2, \ldots, 1.0\}$ and report the results in Table 2. We can see that our APPLE performs significantly better than BPR and ABPR, demonstrating its effectiveness in modeling heterogeneous one-class feedback.

We then fix $\beta = 1$, $\delta = 0.5$ and study the impact of the two different weighting factors, i.e., cP2E and gP2E, and report the results in Table 3. We can see that our groupwise weighting strategy is always better, which shows that it can better reflect the overall standard of user-groups $\mathcal{P}$ and $\mathcal{E}$ in $\hat{r}_{\mathcal{P}i} > \hat{r}_{\mathcal{E}i}$.

In order to study the performance of β with different values in $\{0.1, 0.2, \ldots, 1.0\}$, we fix $\delta = 0.5$ and report the trends in Fig. 2. We can see that a medium or large value of β is usually better, which also provides us some guidance in deploying our solution in a real-world application.

In order to study the effectiveness of our like-minded user-group construction strategy, we fix $\beta = 1$, $\delta = 0.5$ and report the results of the random strategy and ours in Table 4. We can see that our strategy is always better, which shows that it can indeed eliminate the effect of occasionality to some extent. Notice that our strategy is generic and can be readily applied to other algorithms.

Finally, we fix $\beta = 1$ and study the impact of δ in our similarity-priority strategy, and report the results of $\delta \in \{0.1, 0.2, \ldots, 1.0\}$ in Fig. 3. We can see that a medium value of δ is usually better, i.e., 0.8, 0.5 and 0.3 on ML100K, ML1M, and Alibaba, respectively, which shows the effectiveness of reducing the similarity weight between users within a group. In real deployment, we may safely fix $\delta = 0.5$ for simplicity.

5 Conclusions and Future Work

In this paper, we study the HOCCF problem and propose a generic asymmetric pairwise preference learning model called asymmetric pairwise preference learning (APPLE). Our APPLE involves various pairwise relations with a dynamic and flexible weighting strategy and a generic preference assumption, as well as a novel like-minded user-group construction strategy, which introduces richer interactions within user-groups. Extensive experimental results on three real-world datasets show that our APPLE performs significantly better than the closely related and very competitive methods.

For future works, we are interested in improving the efficiency of the user-group construction strategy and generalizing our APPLE to some deep learning architectures [5].

Acknowledgement. We thank the support of National Natural Science Foundation of China Nos. 61872249, 61836005 and 61672358. Weike Pan and Zhong Ming are the corresponding authors for this work.

References

1. Chen, C., Zhang, M., Zhang, Y., Ma, W., Liu, Y., Ma, S.: Efficient heterogeneous collaborative filtering without negative sampling for recommendation. In: Proceedings of the 34th AAAI Conference on Artificial Intelligence, pp. 19–26 (2020)
2. Deshpande, M., Karypis, G.: Item-based top-N recommendation algorithms. ACM Trans. Inf. Syst. **22**(1), 143–177 (2004)
3. Ding, J., et al.: Improving implicit recommender systems with view data. In: Proceedings of the 27th International Joint Conference on Artificial Intelligence. IJCAI 2018, pp. 3343–3349 (2018)
4. He, X., He, Z., Song, J., Liu, Z., Jiang, Y.G., Chua, T.S.: NAIS: neural attentive item similarity model for recommendation. IEEE Trans. Knowl. Data Eng. **30**(12), 2354–2366 (2018)
5. He, X., Liao, L., Zhang, H., Nie, L., Hu, X., Chua, T.S.: Neural collaborative filtering. In: Proceedings of the 26th International Conference on World Wide Web. WWW 2017, pp. 173–182 (2017)
6. Hu, Y., Yehuda, K., Chris, V.: Collaborative filtering for implicit feedback datasets. In: Proceedings of the 8th IEEE International Conference on Data Mining. ICDM 2008, pp. 263–272 (2008)
7. Kabbur, S., Ning, X., Karypis, G.: FISM: factored item similarity models for top-N recommender systems. In: Proceedings of the 19th ACM SIGKDD International Conference on Knowledge Discovery and Data Mining. KDD 2013, pp. 659–667 (2013)

8. Ouyang, S., Li, L., Pan, W., Ming, Z.: Asymmetric Bayesian personalized ranking for one-class collaborative filtering. In: Proceedings of the 13th ACM Conference on Recommender Systems. RecSys 2019, pp. 373–377 (2019)
9. Pan, R., et al.: One-class collaborative filtering. In: Proceedings of the 8th IEEE International Conference on Data Mining. ICDM 2008, pp. 502–511 (2008)
10. Pan, W., Chen, L., Ming, Z.: Personalized recommendation with implicit feedback via learning pairwise preferences over item-sets. Knowl. Inf. Syst. **58**(2), 295–318 (2019). https://doi.org/10.1007/s10115-018-1154-5
11. Pan, W., Liu, M., Ming, Z.: Transfer learning for heterogeneous one-class collaborative filtering. IEEE Intell. Syst. **31**(4), 43–49 (2016)
12. Qiu, H., Liu, Y., Guo, G., Sun, Z., Zhang, J., Nguyen, H.T.: BPRH: Bayesian personalized ranking for heterogeneous implicit feedback. Inf. Sci. **453**, 80–98 (2018)
13. Rendle, S., Freudenthaler, C., Gantner, Z., Schmidt-Thieme, L.: BPR: Bayesian personalized ranking from implicit feedback. In: Proceedings of the 25th Conference on Uncertainty in Artificial Intelligence. UAI 2009, pp. 452–461 (2009)
14. Salakhutdinov, R., Mnih, A.: Probabilistic matrix factorization. In: Annual Conference on Neural Information Processing Systems. NeurIPS 2007, pp. 1257–1264 (2007)

DPR-Geo: A POI Recommendation Model Using Deep Neural Network and Geographical Influence

Jun Zeng(✉), Haoran Tang, and Junhao Wen

School of Big Data and Software Engineering, Chongqing University, Chongqing, China
{zengjun,tanghaoran,jhwen}@cqu.edu.cn

Abstract. In the wake of developments in artificial intelligence, deep learning technology has been used in location-based social networks (LBSNs) to provide web services that meet the needs of users. Point of interest (POI) recommendation, as one of the most important mobile services, aims to recommend new satisfactory POIs to users according to their historical records. However, existing models that uses original high-dimension user vector or location vector cannot capture useful information from historical records effectively. Meanwhile, most of them complete recommendation service only in terms of user's perspective or location's perspective. Hence, in this paper, we propose a novel deep learning framework for POI recommendation. Firstly, we use a multi-layer neural network to reduce the dimension of user vector and location vector. Then, we construct a union neural network by concatenating and multiplying vectors to obtain the preferences of users. Finally, considering the unique geographical characteristic of location, we model the distance probability to enhance recommendation. Experimental results on real-world dataset demonstrate our model outperforms some popular recommendation algorithms and achieves our expected goal.

Keywords: Recommendation system · Point of interest · Deep learning · Neural network · Geographical influence

1 Introduction

With the rapid prevalence of artificial intelligence, location-based social networks (LBSNs) that link the physical and virtual world have grown in popularity, such as Facebook, Twitter and Foursquare [18]. Point of interest (POI) recommendation, one of the basic mobile services in LBSNs, aims at recommending new locations to users who haven't visited them before by mining their historical check-ins on locations [3]. This facilitates people's outdoor activities. Meanwhile, large-scale check-in data that describes users' behavior on visiting locations provides us with opportunities to design absorbing service [4].

H. Yang et al. (Eds.): ICONIP 2020, LNCS 12534, pp. 420–431, 2020.
https://doi.org/10.1007/978-3-030-63836-8_35

Existing POI recommendation models based on classical machine learning use the original high-dimension user vector and location vector to complete service, such as matrix factorization [7]. However, in the real world, users may have visited a few locations while most of locations are unvisited, which causes the challenge of learning users' complex preferences [6]. Moreover, there is always some hidden information between users and locations [2]. Most models ignore the union between users and location and they provide recommendation just from user's perspective or location's perspective. Deep learning, emerging as powerful tools for relation extraction [9], is able to solve the problems mentioned above effectively. By applying deep learning technology, it's possible to mine potential preferences and mobile patterns of users by dimension reduction and union computation. The user preferences indicate the POIs users prefer to visit, which helps us design recommendation services. Different from other types of recommendation, one of the most prominent features for POI recommendation is the geographical distance of location [11]. For instance, in most cases, users tend to visit the locations that are closer to them.

In this paper, we focus on how to use deep learning technology to deeply mine users' preferences for POIs and how geographic distance plays its role. Therefore, we propose a deep POI recommendation framework which also takes into account the distance probability of location. There are two parts of our deep model. One is for dimension reduction and the other is for union computation. For simplicity, we also call POI location.

The main contributions of this paper can be summarized as follows.

- In order to solve the problem of high-dimension vectors of user and location, we construct two different multi-layer neural networks to reduce the dimensions of user vector and location vector respectively. This extracts the key information of users and locations and alleviates the data sparsity to a certain extent.
- For the sake of computing the recommendation prediction from the union of user and location, we design an innovative union neural network by concatenating and multiplying the results of the first step respectively to obtain the potential preference of user. This is our core which mines the correlation between user and location effectively.
- For making our recommendation more explainable in real-world, we model the geographical distance by probability distribution to further enhance recommendation and improve the performance.
- The experimental results on two real-world datasets demonstrate our proposed model outperforms some popular recommendation algorithms.

The rest of our paper is organized as follows. Section 2 introduces the related work of POI recommendation. Section 3 presents our proposed deep model. Section 4 discusses and analyzes the experimental results. Finally, Sect. 5 concludes the paper.

2 Related Work

Nowadays, deep learning technology has been widely used in industry, including the field of recommendation system. In our previous work, we propose a deep recommendation framework based on restricted Boltzmann machine and non-negative matrix factorization [17]. We also propose a recurrent neural network model with self-attention to predict the next locations of users [15]. In addition, we integrate geographical and social influence into collaborative filtering to make recommendation more reasonable [16]. Xue proposes a novel matrix factorization model based on neural network to learn a common low dimensional space for both users and locations [12]. Yang proposes a general and principled framework to solve the sparsity problem by smoothing among users and locations [13]. Our proposed model in this paper is inspired by the work of Xue and Yang. Moreover, Chang proposes a content-aware POI embedding model that consists check-in layer and text layer [1].

Geographical distance is one of the most prominent features for POI recommendation. Ye argues that geographical influence among locations plays an import role in users' behaviors and model it by power law distribution [14]. Gao proposes a joint model that utilizes social network information and geographical distance [5]. Wang exploits the geographical influence to improve POI recommendation by three different factors [11]. Liu proposes an adversarial learning model based on geographical information by fusing geographical features and generative adversarial networks [8]. It can be seen that the using of geographical distance will improve the performance of POI recommendation. Hence, we decide to apply geographical distance into our proposed model.

3 DPR-Geo: POI Recommendation Using Deep Neural Network and Geographical Influence

POI recommendation aims to recommend new locations to users by mining their historical check-in records. In this section, we will introduce our recommendation model DPR-Geo on the whole. The framework is shown in Fig. 1. First, for the sake of extracting useful information of users and locations, we construct a multi-layer neural network to reduce the dimensions of user vector and location vector respectively. Second, we design a union neural network by concatenating and multiplying the user vector and location vector after reducing dimensions and then get the potential preferences. Third, considering that geographical distance plays an important role in POI recommendation, we model distance by power law distribution to enhance final recommendation. The details of each step will be presented in following subsections.

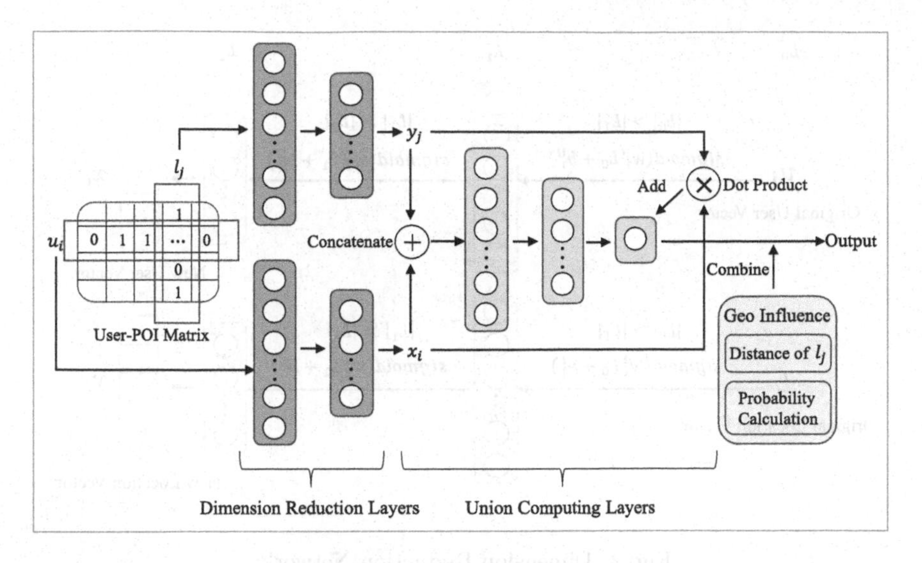

Fig. 1. The framework of DPR-Geo.

3.1 Problem Formulation

Suppose we have a set L of locations $\{l_1, l_2, \cdots, l_n\}$, a set U of users $\{u_1, u_2, \cdots, u_m\}$. The user-POI matrix is R: $U \times L$ and R_{ij} denotes the correlation between user u_i and location l_j, which is defined as follows.

$$R_{ij} = \begin{cases} 1, if\ u_i\ has\ visited\ l_j \\ 0, if\ l_j\ is\ unvisited\ by\ u_i \end{cases} \tag{1}$$

Most research takes the number of times that u_i visits l_j as the value of R_{ij}. However, we focus on whether u_i will visit a new location in the future. So we adopt 0–1 value. Our goal is to predict the unknown $\widehat{R}_{ij}$ and produce a list Rec_{u_i} of recommendations for u_i, which is defined as follows:

$$\widehat{R}_{ij} = F\left(u_i, l_j | \theta\right) \tag{2}$$

$$Rec_{u_i} = \left\{ l_j | sorted\ by\ \widehat{R}_{ij}, K \right\} \tag{3}$$

where F is our model, θ is parameter set and K is the length of Rec_{u_i}. Our task is to recommend satisfactory POIs to users and make them enjoy our service.

3.2 Dimension Reduction Network

The original vectors of users and locations are high dimensional, so we need to find a common space of them. Followed by Xue [12], in order to extract the key information of users and locations, we construct a multi-layer neural network for

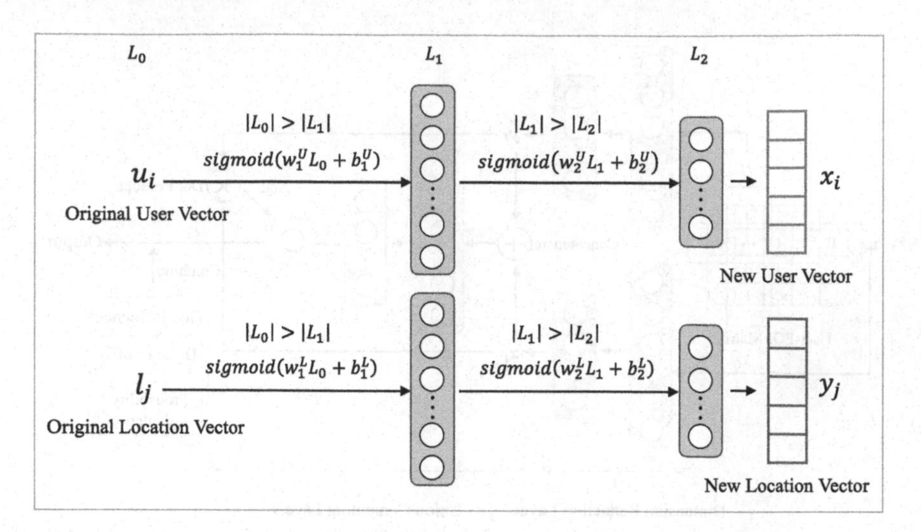

Fig. 2. Dimension Reduction Network.

dimension reduction that is shown in Fig. 2. The neural network is computed as follows:

$$L_0 = u_i \; or \; l_j \tag{4}$$

$$L_k = f\left(w_k L_{k-1} + b_k\right), k = 1, 2, \ldots, N-1 \tag{5}$$

where L_k is the layer k, w_k and b_k are weight matrix and bias of L_k corresponding. The activation function f is chosen as sigmoid function that many deep models adopt. We reduce the dimension of user and location separately to find a same-dimension space for them. Therefore, after reducing dimension, the user vector and location vector are defined as follows:

$$x_i = f\left(w_N^U f\left(\cdots f\left(w_2^U f\left(w_1^U u_i + b_1^U\right) + b_2^U\right)\cdots\right) + b_N^U\right) \tag{6}$$

$$y_j = f\left(w_N^L f\left(\ldots f\left(w_2^L f\left(w_1^L l_j + b_1^L\right) + b_2^L\right)\ldots\right) + b_N^L\right) \tag{7}$$

where w_N^U and b_N^U are weight matrix and bias of the dimension reduction network for users, w_N^L and b_N^L are weight matrix and bias of dimension reduction network for locations. x_i and y_j are the results of reducing dimension.

3.3 Union Network

Most POI recommendation systems calculate preference only from the perspective of user or location. However, we argue that the union of user and location contains more hints of user's preference. Therefore, we construct our union neural network that is shown in Fig. 3. This is the core of DPR-Geo because it extracts the correlation between user and location. The input in Fig. 3 is the

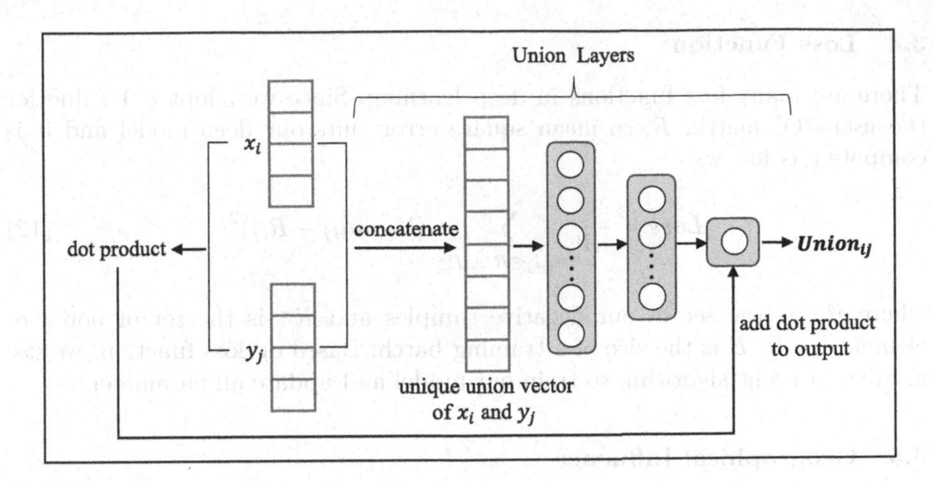

Fig. 3. Union neural network.

low-dimension user vector and location vector after dimension reduction and the output is the preference whose value is from 0 to 1.

Since we have the vectors of user and location, now we need to calculate the potential preference of user. Inspired by Yang [13], we use two different methods to construct union neural network. On one hand, we concatenate the vectors of user and location. The concatenation is a unique vector and it will help us mine the preference from the view of both user and location. We input the concatenation of x_i and y_j to union neural network, which is computed as follows:

$$\tilde{L}_0 = x_i \oplus y_j \tag{8}$$

$$\tilde{L}_k = f\left(\tilde{w}_k \tilde{L}_{k-1} + \tilde{b}_k\right), k = 1, 2, \ldots, M-1 \tag{9}$$

where $\tilde{L}_k$ is the layer k, $\tilde{w}_k$ and $\tilde{b}_k$ are weight matrix and bias of $\tilde{L}_k$. Same as dimension reduction network, the activation function is sigmoid function. Note that it's vector concatenation, not vector addition. On other hand, considering that x_i and y_j can be regarded as the results of matrix factorization, we calculate the dot product of x_i and y_j. Compared with the original vectors of user and location, our x_i and y_j are low dimensional and have key information of them. So x_i and y_j can perform better in matrix factorization. Next, we calculate the potential union preference, which is defined as follows:

$$z_{ij} = f\left(w_M^Z f\left(\ldots f\left(w_1^Z\left(x_i \oplus y_j\right) + b_1^Z\right)\ldots\right) + b_M^Z\right) \tag{10}$$

$$Union_{ij} = \frac{1}{2}\left(z_{ij} + x_i \cdot y_j\right) \tag{11}$$

where w_M^Z and b_M^Z are parameters of the union neural network. The reason why we average the two methods we mentioned above is we assume that they both have equal effects. Meanwhile, it will simplify the training process. In our future work, we will try other measures to combine them.

3.4 Loss Function

There are many loss functions in deep learning. Since we adopt 0–1 value for the user-POI matrix R, so mean square error suits our deep model and it is computed as follows:

$$Loss = \frac{1}{|B|} \sum_{u_i, l_j \in R^+ \cup R^-} (Union_{ij} - R_{ij})^2 \tag{12}$$

where R^- is the set of our negative samples and R^+ is the set of non-zero elements in R. B is the size of a training batch. Based on loss function, we use gradient descent algorithm to train our model and update all parameters.

3.5 Geographical Influence

Distance between two locations plays an important role when recommending [14]. If a location is far away from user's current location, there is little chance that user will visit it. Hence, in order to model the geographical influence, we adopt a power law distribution that is shown as follows:

$$Pro\,(l_j|l_p) = a \times dis(l_j, l_p)^b \tag{13}$$

where l_p is user's current location and $Pro(l_j|l_p)$ represents the probability of visiting the new location l_j after l_p. a and b are parameters of power law distribution.

It's difficult to use the geographical model directly. For obtaining the unknown parameters, we need to convert our power law distribution to a linear model by using logarithmic representation, which is defined as follows:

$$\log Pro = \log a + b \log dis\,(l_j, l_p) \tag{14}$$

$$\delta\,(C, \omega) = a' + b \log \omega \tag{15}$$

where ω is parameter set, C represents $dis(l_j, l_p)$ and a' is equal to $\log a$. Based on least squares method, the objective function that needs to be minimized is defined as follows:

$$\min \frac{1}{2} \sum_{C \in D} (\delta\,(C, \omega) - t\,(C))^2 + \frac{\varphi}{2} \|\omega\|^2 \tag{16}$$

where D is a real-world dataset and $t(C)$ is the logarithmic value of the true distance probability derived from D. The last term is a regularization and φ is the weight. Now we get our distance model. Therefore, the locations that user has visited is denoted by L_u and the probability of unvisited location l_j for L_u is calculated as follows:

$$Pro\,(l_j|L_u) = \prod_{l_p \in L_u} Pro\,(l_j|l_p) \tag{17}$$

3.6 POI Recommendation

We have introduced our deep learning model and geographical influence respectively. For the sake of improving the performance of our deep model, now we integrate it with the geographical influence. Hence, the prediction of unknown $\widehat{R}_{ij}$ is computed as follows:

$$\widehat{R}_{ij} = \alpha \times \frac{Pro\left(l_j|L_u\right)}{\max\limits_{l_p \in L-L_u} Pro\left(l_p|L_u\right)} + (1-\alpha) \times Union_{ij} \tag{18}$$

where α is the weight and $\alpha \in [0,1]$. Finally, we rank all predictions for user u_i according to $\widehat{R}_{ij}$ and produce a recommendation list with Top-K POIs to user.

4 Experiments

In this section, we evaluate our proposed model with some popular recommendation algorithms on two real-world datasets. In order to avoid over-fitting, N and M in our model are both set to 3.

4.1 Datasets

We employ two real-world datasets collected from two cities on Foursquare. One is Honolulu, Hawaii. The other is Atlanta, Georgia. There are 33884 check-ins made with Honolulu and they are produced by 768 users on 4716 locations. The average check-in of per user is 44. For Atlanta dataset, there are 43987 check-ins and they are made by 3238 users on 4853 locations. The average check-in of each user is 13. Obviously, the dataset of Atlanta is more sparse than that of Honolulu. We randomly select 80% of the locations of each user as training data and the remaining 20% as test data. Moreover, for the sake of effectiveness of experiments, the users who have visited less than 6 locations and the locations that have been visited by less than 6 users are removed from datasets.

4.2 Evaluation Metrics

Precision, recall and F1-score are most used evaluation metrics in POI recommendation system. F1-score is the combination of precision and recall. We adopt F1-score to find the optimal α. Precision and recall are used in comparing our model with other algorithms. All evaluation metrics are defined as follows:

$$Precision@K = \frac{1}{|U|} \sum_{u \in U} \frac{|Rec_u \cap Test_u|}{|Rec_u|} \tag{19}$$

$$Recall@K = \frac{1}{|U|} \sum_{u \in U} \frac{|Rec_u \cap Test_u|}{|Test_u|} \tag{20}$$

$$F1-score@K = 2\frac{Precision@K \times Recall@K}{Precision@K + Recall@K} \tag{21}$$

where Rec_u is the recommendation list for user u and $Test_u$ is the test data of user u. K is the length of recommendation list, which is set to 5, 10 and 15.

4.3 Optimal α

In our proposed model DPR-Geo, there is a parameter α that controls the influence of graphical distance. Since we need to determine the best α, we use F1-score that combines precision and recall. The result of α is shown in Fig. 4. In Honolulu dataset, it's obvious that all curves first go up and then go down after reaching their peaks $\alpha = 0.4$ basically. It indicates that appropriate geographical influence can enhance the output of our neural networks. The curve of $K = 5$ goes down sharply and is even lower than other curves after $\alpha = 0.7$, which tells us it is more sensitive to distance when K is small. In Atlanta dataset, all curves keep the same trend of Honolulu dataset while there are some fluctuations. Their peaks are around $\alpha = 0.6$. The curve of $K = 5$ also goes down sharply. The possible reason is that geographical distance has more significant influence on the curves with small K.

In a word, geographical distance will improve our POI recommendation results and its influence is significantly different in terms of different datasets.

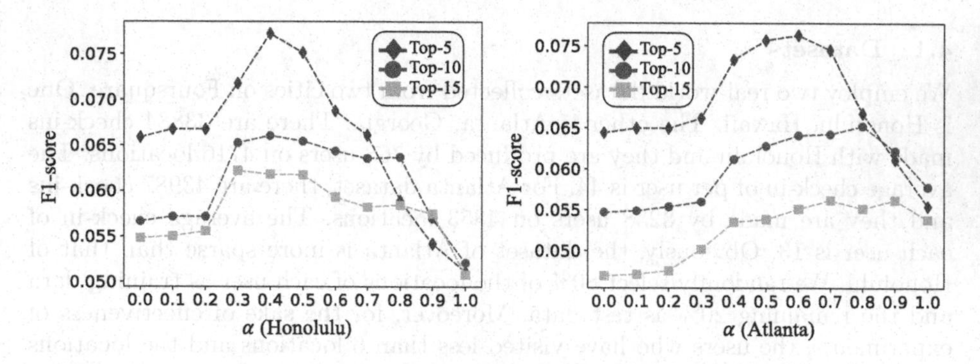

Fig. 4. F1-score under different α.

4.4 Performance Comparison

To comprehensively demonstrate the effectiveness of our proposed model DPR-Geo, we compare it with following popular recommendation algorithms:

POP: A standard model that recommends popular POIs to users.

NMF: A classical non-negative matrix factorization.

BPR [10]**:** Bayesian personalized ranking via optimizing the ordering relationship of users and POIs.

DMF [12]**:** A novel deep matrix factorization for recommendation system, which learns a common low dimensional space for both users and recommendations.

DPR: Our proposed deep model that dose not consider geographical influence.

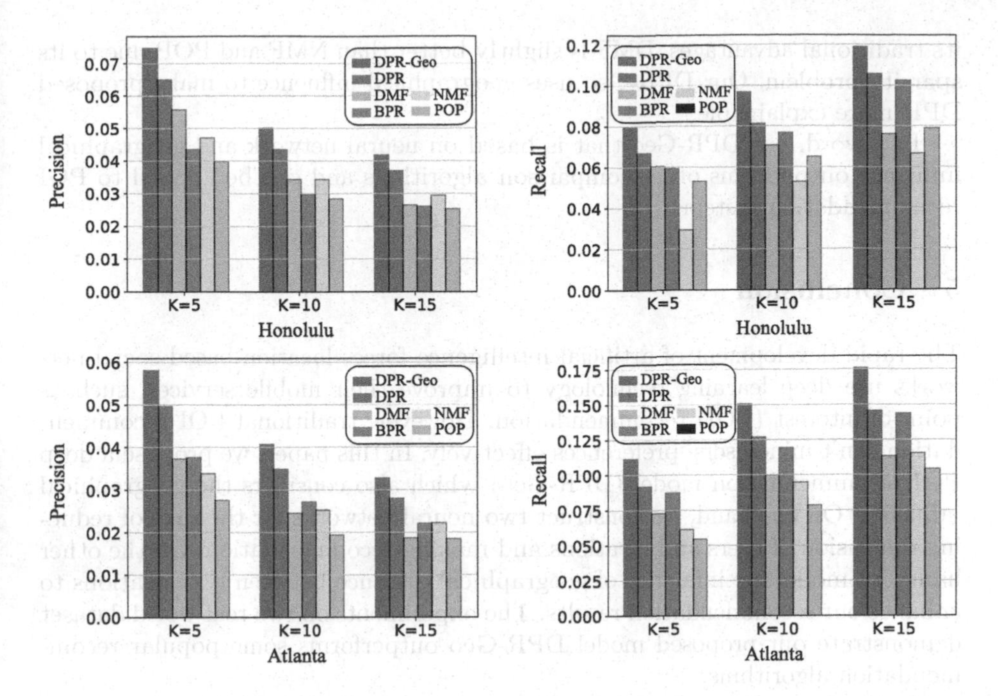

Fig. 5. Precision and Recall.

Our final model which integrates DPR and geographical influence is called **DPR-Geo**. We set $\alpha = 0.4$ and $\alpha = 0.6$ for Honolulu dataset and Atlanta dataset respectively. The performance comparison is shown in Fig. 5 and we summarize the following observations.

For Honolulu dataset, DPR-Geo outperforms other algorithms in terms of both precision and recall. In precision, NMF and DMF are better than POP and BPR, which indicates that matrix factorization still has some advantages. However, matrix factorization can't mine users' preferences deeply. Hence, our DPR-Geo extract the potential correlation between users and locations based on neural networks. In recall, BPR and POP are superior to DMF and NMF. The possible reason is that recommending POIs according to POI-pair method or popularity is more suitable since Honolulu is a famous tourist city. DPR-Geo performs better than DPR, which demonstrates integrating geographical influence into deep model can improve performance effectively.

For Atlanta dataset, DPR-Geo is still superior to other algorithms in terms of both precision and recall. In precision, DMF is not as good as BPR and even inferior to NMF when the length of recommendation list increases. This is caused by the data sparsity because the average check-in is 13 in Atlanta dataset while the average check-in of Honolulu dataset is 44. Therefore, DMF can't capture more useful information of users. POP is the worst model, which shows recommendation based on popularity can't achieve excellent result. In recall, BPR is only second to DPR-Geo and DPR. It indicates that BPR keeps its traditional advantages. DMF is slightly better than NMF and POP due to its sparsity problem. Our DPR-Geo uses geographical influence to make proposed DPR more explainable.

In a word, our DPR-Geo that is based on neural network and geographical influence outperforms other comparison algorithms and can be applied to POI recommendation system.

5 Conclusion

The rapid development of artificial intelligence forces location-based social networks use deep learning technology to improve their mobile services, such as point-of-interest (POI) recommendation. However, traditional POI recommendation can't mine users' preferences effectively. In this paper, we propose a deep POI recommendation model (DPR-Geo) which also considers the geographical influence. On one hand, we construct two neural networks for the sake of reducing dimension of users and locations and making recommendation. On the other hand, we model the influence of geographical distance between two locations to enhance our recommendation results. The experiments on two real-world dataset demonstrate our proposed model DPR-Geo outperforms some popular recommendation algorithms.

In the future, first of all, we will improve the neural networks of our deep model by trying different structures and layers. In our union network, we adopt a simple combination of concatenating and multiplying vectors. We will find a more suitable measure to finish that. Secondly, we will take into account more factors that may affect the performance of our model, such as the categories of POIs. Category can be regarded as the explicit semantic expression of POI itself. So we are able to capture the explicit preferences of users from categories.

Acknowledgments. This research is sponsored by Natural Science Foundation of Chongqing, China (No. cstc2020jcyj-msxmX0900) and the Fundamental Research Funds for the Central Universities (No. 2020CDJ-LHZZ-040).

References

1. Chang, B., Park, Y., Park, D., Kim, S., Kang, J.: Content-aware hierarchical point-of-interest embedding model for successive POI recommendation. In: IJCAI 2018, pp. 3301–3307 (2018)
2. Chen, J., Lian, D., Zheng, K.: Improving one-class collaborative filtering via ranking-based implicit regularizer. In: Proceedings of the AAAI Conference on Artificial Intelligence, vol. 33, pp. 37–44 (2019)
3. Chen, J., Zhang, W., Zhang, P., Ying, P., Niu, K., Zou, M.: Exploiting spatial and temporal for point of interest recommendation. Complexity **2018** (2018)
4. Gao, H., Liu, H.: Data analysis on location-based social networks. In: Chin, A., Zhang, D. (eds.) Mobile Social Networking. CSS, pp. 165–194. Springer, New York (2014). https://doi.org/10.1007/978-1-4614-8579-7_8

5. Gao, H., Tang, J., Liu, H.: gSCorr: modeling geo-social correlations for new check-ins on location-based social networks. In: Proceedings of the 21st ACM International Conference on Information and Knowledge Management, pp. 1582–1586 (2012)
6. He, X., He, Z., Du, X., Chua, T.S.: Adversarial personalized ranking for recommendation. In: The 41st International ACM SIGIR Conference on Research & Development in Information Retrieval, pp. 355–364 (2018)
7. Lian, D., Zhao, C., Xie, X., Sun, G., Chen, E., Rui, Y.: GeoMF: joint geographical modeling and matrix factorization for point-of-interest recommendation. In: Proceedings of the 20th ACM SIGKDD International Conference on Knowledge Discovery and Data Mining, pp. 831–840 (2014)
8. Liu, W., Wang, Z.J., Yao, B., Yin, J.: Geo-ALM: poi recommendation by fusing geographical information and adversarial learning mechanism. In: IJCAI 2019, pp. 1807–1813 (2019)
9. Ouyang, L., Tang, H., Xiao, G.: Chinese text relation extraction with multi-instance multi-label BLSTM neural networks. In: The 31st International Conference on Software Engineering & Knowledge Engineering, pp. 337–448 (2019)
10. Rendle, S., Freudenthaler, C., Gantner, Z., Schmidt-Thieme, L.: BPR: Bayesian personalized ranking from implicit feedback. In: Proceedings of the Twenty-Fifth Conference on Uncertainty in Artificial Intelligence (UAI 2009), pp. 452–461 (2012)
11. Wang, H., Shen, H., Ouyang, W., Cheng, X.: Exploiting POI-specific geographical influence for point-of-interest recommendation. In: IJCAI 2018, pp. 3877–3883 (2018)
12. Xue, H.J., Dai, X., Zhang, J., Huang, S., Chen, J.: Deep matrix factorization models for recommender systems. In: IJCAI 2017, Melbourne, Australia, vol. 17, pp. 3203–3209 (2017)
13. Yang, C., Bai, L., Zhang, C., Yuan, Q., Han, J.: Bridging collaborative filtering and semi-supervised learning: a neural approach for POI recommendation. In: Proceedings of the 23rd ACM SIGKDD International Conference on Knowledge Discovery and Data Mining, pp. 1245–1254 (2017)
14. Ye, M., Yin, P., Lee, W.C., Lee, D.L.: Exploiting geographical influence for collaborative point-of-interest recommendation. In: Proceedings of the 34th International ACM SIGIR Conference on Research and Development in Information Retrieval, pp. 325–334 (2011)
15. Zeng, J., He, X., Tang, H., Wen, J.: A next location predicting approach based on a recurrent neural network and self-attention. In: Wang, X., Gao, H., Iqbal, M., Min, G. (eds.) CollaborateCom 2019. LNICST, vol. 292, pp. 309–322. Springer, Cham (2019). https://doi.org/10.1007/978-3-030-30146-0_21
16. Zeng, J., Li, F., He, X., Wen, J.: Fused collaborative filtering with user preference, geographical and social influence for point of interest recommendation. Int. J. Web Serv. Res. (IJWSR) **16**(4), 40–52 (2019)
17. Zeng, J., Tang, H., Li, Y., He, X.: A deep learning model based on sparse matrix for point-of-interest recommendation. In: The 31st International Conference on Software Engineering & Knowledge Engineering, pp. 379–492 (2019)
18. Zhou, X., Mascolo, C., Zhao, Z.: Topic-enhanced memory networks for personalised point-of-interest recommendation. In: Proceedings of the 25th ACM SIGKDD International Conference on Knowledge Discovery & Data Mining, pp. 3018–3028 (2019)

Feature Aware and Bilinear Feature Equal Interaction Network for Click-Through Rate Prediction

Lang Luo, Yufei Chen(✉), Xianhui Liu, and Qiujun Deng

College of Electronic and Information Engineering, Tongji University, Shanghai 201804, China

yufeichen@tongji.edu.cn

Abstract. Advertising recommendation is crucial for many Internet companies because it largely affects their business income, and click-through rate (CTR) plays a key role in it. Most of the current CTR prediction models pay less attention to the feature importance before feature interaction. Besides, during bilinear feature interaction (BI), these models simply use hadamard product or inner product and implicitly introduce unnecessary feature order noise. In this paper, we propose a model called Feature Aware and Bilinear Feature Equal Interaction Network (FaBeNET). On the one hand, it can be aware of the feature importance and keep original feature as many as possible through the Squeeze-and-Excitation Residual Network (SE-ResNet); On the other hand, it assigns an interaction matrix to each feature, so the BI can be equally and effectively learned by the combination of hadamard product and inner product. On this basis, a deep neural network is used to learn higher-order feature interaction. Experiments show that FaBeNet achieves performance 0.7919 AUC and 0.4581 Logloss, which is better than other models, such as the DCN, xDeeepFM, and FiBiNET.

Keywords: Click-through rate · Squeeze-excitation residual network · Bilinear Feature Equal Interaction · Recommender systems

1 Introduction

Recommendation system is indispensable for many Internet companies because their quality directly affects the business income of the platforms [1–3]. The core problem of it is CTR prediction, which aims to predict the probability of users clicking on advertisements or items [4,5]. In the literature, numerous methods have been applied to predict CTR, which can be divided into two types: traditional models and deep learning models.

For traditional models, Logistic regression (LR)model [6] is the forerunner and foundation of many popular CTR prediction models. Despite the strong interpretability, it is difficult to deal with the problem of increasingly sparse and high-dimensional data. The Factorization Machine (FM) model [7] decomposes

H. Yang et al. (Eds.): ICONIP 2020, LNCS 12534, pp. 432–443, 2020.
https://doi.org/10.1007/978-3-030-63836-8_36

the weight into the inner product of two vectors to carry out the interaction of variables, which can deal with the problem of data sparsity better. The Field-aware Factorization Machine (FFM) model [8] adds the concept of "field" based on the FM model and uses field-aware embedding vector to complete feature interaction. The Attentional Factorization Machine (AFM) model [9] uses the popular attention mechanism recently to make different feature interactions contribute differently to the prediction. However, these traditional methods can only learn the second-order interaction of feature, their linear expression ability limits the final prediction effect.

For deep learning models, Factorization machine supported neural network (FNN) model [10] introduces the DNN structure and uses the pre-trained factorization machine for field embedding. Wide&Deep (W&D) model [1] uses a linear model of "wide part" and a DNN model of "deep part" to model low-order and high-order feature interactions simultaneously. However, expertise feature engineering is still required as input for "wide part". DeepFM [11], the Factorization-Machine based neural network, can be trained end-to-end without any feature engineering because it replaces the "wide part" with the FM model and the two parts share the feature embedding vector. Neural Factorization Machine [12] stacks deep neural networks on top of the output of the second-order feature interactions to model higher-order features. Deep & Cross Network [13] introduces a special cross-network module for feature interaction. But the output of CrossNet is limited in a special form. To solve this problem, the eXtreme Deep Factorization Machine (xDeepFM) [14] designs a Compressed Interaction Network (CIN) module to learn explicit high-order feature interactions better. FiBiNET [15] introduces the SENet [16] mechanism to learn the feature importance and three types of BI layer to learn fine-grained feature interaction.

Many challenges need to be faced for CTR prediction. Different features should be given different weights. It is important to dynamically learn the feature importance while ensuring that the information of the original features is not lost. Besides, the interaction between features should be simple and effective and the interaction result should has nothing to do with the order of features. However, few models can achieve the above two points. To solve these problems, we propose a model called Feature Aware and Bilinear Feature Equal Interaction Network (FaBeNET). The contributions of our work are as follows.

* We introduce SE-ResNet [16,17] to dynamically learn the feature importance and keep original feature information. Thanks to the shortcut connection of residual network, the original feature embeddings don't need to enter the BI layer, which reduces a large number of parameters.
* In BI layer, we propose a new type of feature interaction called "Bilinear Feature Equal Interaction". In this interaction, each feature corresponds to an interaction matrix, so the bilinear feature interaction can then be equally and effectively learned by the combination of hadamard product and inner product. It successfully solves the problem of introducing feature order noise during feature interaction.

2 Feature Aware and Bilinear Feature Equal Interaction Network

In this section, we will describe the structure of our FaBeNET model (Fig. 1). For the sake of simplicity, we omit the logistic regression part of the model, because it is relatively simple and common. Our model mainly includes the following six parts: sparse input and embedding layer, SE-ResNet layer, bilinear feature equal interaction layer, concatenation layer, DNN network layer, and output layer.

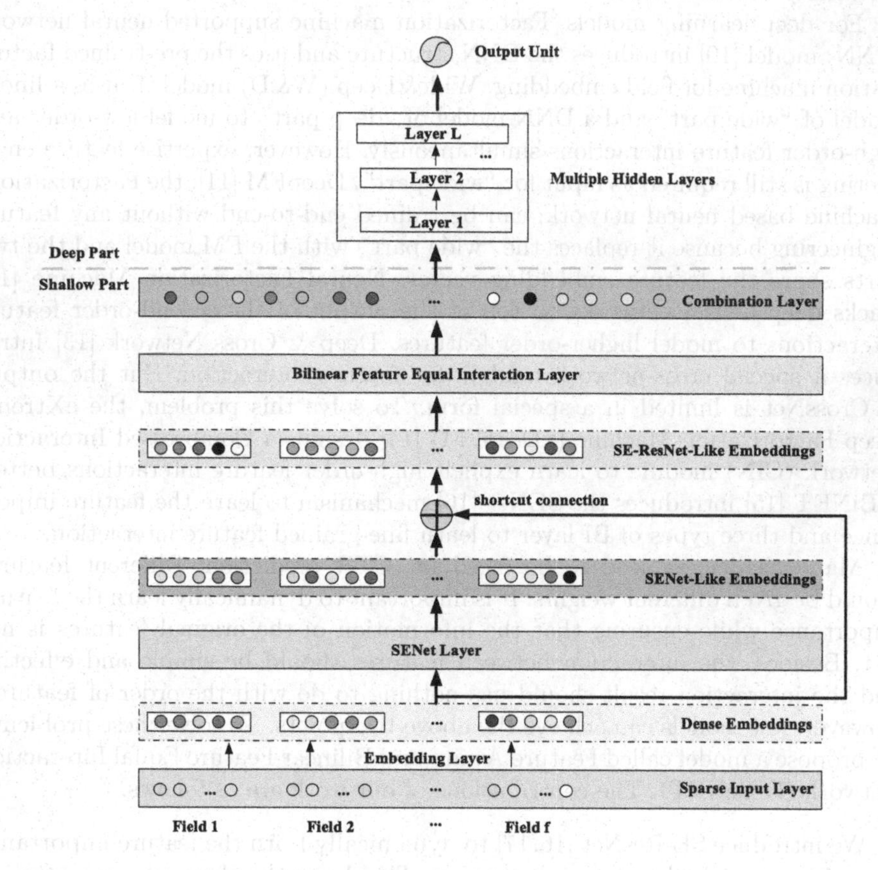

Fig. 1. FaBeNet model structure.

2.1 Sparse Input and Embedding Layer

In CTR prediction, input features are usually categoric. Sparse inputs are vectors formed by one-hot encoding of them. To reduce the dimension of input features, we use embedding technology to transform these sparse inputs into dense embeddings [10]:

$$e_i = W_{embed,i} x_i \quad (1)$$

where $x_i \in R^{n_v^i}$ is the one-hot encoding vector of the i-th categorical feature, $W_{embed,i} \in R^{n_e \times n_v^i}$ is the corresponding embedding matrix, and n_e is the embedding size, n_v^i is the vocabulary size of the i-th feature. It will be optimized together with other parameters in the network. $e_i \in R_e^n$ is the i-th embedding. Finally, we stack all the embeddings together to get the output of this layer:

$$E = [e_1^T, e_2^T, \cdots, e_{n_f}^T] \quad (2)$$

where n_f represents the number of features.

2.2 SE-ResNet Layer

SENet has the ability to automatically be aware of the importance of the channel. The shortcut connection structure of residual network can keep original feature information and enhance the diversity of features. SE-ResNet is a combination of the above two models and has achieved success in many fields [18]. In this section, We introduce the SE-ResNet (Fig. 2) to make FaBeNet dynamically be aware of and learn the feature importance. Besides, the jump connection structure can alleviate the subsequent network degradation of FaBeNet.

As shown in Fig. 2, SE-ResNet can be divided into four parts: squeeze, execution, rescale and shortcut connection. We will introduce them in detail as follows.

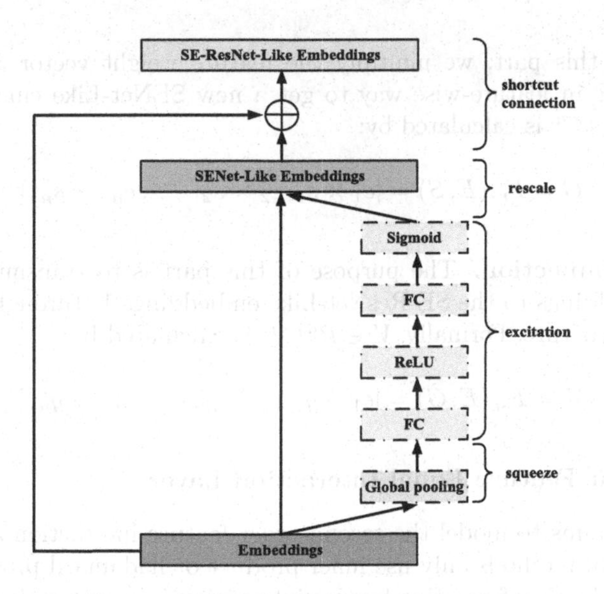

Fig. 2. SE-ResNet structure.

Squeeze. The purpose of this part is to squeeze the global information of each feature embedding into a feature-aware static by using global pooling:

$$z_i = F_{sq}(e_i) = \frac{1}{n_e}\sum_{i=1}^{n_e} e_i^{(t)} \quad (3)$$

Then, we can get a statistic vector Z which is generated by shrinking E through its dimensions n_f:

$$Z = [z_1, z_2, \cdots z_{n_f}] \quad (4)$$

As the output of the transformation E, $Z \in R^{n_f}$ can be interpreted as a collection of the local descriptors whose statistics are expressive for the whole feature embeddings.

Excitation. The purpose of this part is to fully capture feature-wise dependencies. We opt to employ a simple gating mechanism with two fully connected (FC) layers to learn the importance of the feature. Formally, the weights of feature embeddings S is calculated by:

$$S = F_{ex}(Z) = \sigma(W_2\delta(W_1 Z)) \quad (5)$$

where $S \in R^{n_f}$ refers to the weight vector, and σ, δ is Sigmoid function and ReLU function respectively. The parameters of this layer are $W_1 \in R^{n_f \times \frac{n_f}{r}}$, $W_2 \in R^{\frac{n_f}{r} \times n_f}$, and r is reduction ratio which is a hyper-parameter of FaBeNet.

Rescale. In this part, we multiply the feature weight vector S and feature embeddings E in feature-wise way to get a new SENet-Like embeddings. Formally, $G \in R^{n_f \times n_e}$ is calculated by:

$$G = F_{rs}(E, S) = [e_1 \times s_1, e_2 \times s_2, \cdots, e_{n_f} \times s_{n_f}] \quad (6)$$

Shortcut Connection. The purpose of this part is to transmit the original feature embeddings to the SE-ResNet-Like embeddings V through the shortcut connection structure. Formally, $V \in R^{n_f \times n_e}$ is calculated by:

$$V = F_{sc}(E, G) = [e_1 + g_1, e_2 + g_2, \cdots, e_{n_f} + g_{n_f}] \quad (7)$$

2.3 Bilinear Feature Equal Interaction Layer

The BI layer aims to model the second-order feature interaction automatically. The traditional methods only use inner product or hadamard product, which is too simple to learn informative feature interaction in a sparse dataset. FiBiNet combines inner product and hadamard product and proposes three types of feature interaction methods:

Field-All Type.

$$p_{i,j} = v_i * W \odot v_j \tag{8}$$

where $*$ denotes the inner product and $\odot$ denotes the hadamard product. p_{ij} is the result of feature interaction between v_i and v_j. W is the interaction matrix that is shared among all (v_i, v_j). This feature interaction method is too simple to express all the feature interaction information with only one interaction matrix, which limits the expression ability of the model.

Field-Each Type.

$$p_{i,j} = v_i * W_i \odot v_j \tag{9}$$

where W_i is the corresponding parameter matrix of the i-th feature. The total number of interaction matrices is $n_f - 1$. This feature interaction method is asymmetric in feature interaction because $v_i * W_i \odot v_j \neq v_j * W_j \odot v_i$. Prediction results will change as the order of input features changes. This method is unreasonable because it implicitly introduces feature order noise.

Field-Interaction Type.

$$p_{i,j} = v_i * W_{ij} \odot v_j \tag{10}$$

where W_{ij} is the corresponding parameter matrix of interaction between feature i and feature j. In this feature interaction method, the total number of interaction matrices is $\frac{n_f \times (n_f - 1)}{2}$. On the one hand, too many parameters lead to long training time; On the other hand, it is easy to cause overfitting problem.

Field-Equal Type. As also shown in Fig. 3, we propose a new type of interaction method named "Bilinear Feature Equal Interaction". For the convenience of subsequent comparison experiments, we can also call it "Field-Equal Type". Taking the i-th embedding v_i and j-th embedding v_j as an example, the result of feature interaction p_{ij} is calculated by:

$$p_{i,j} = (v_i * W_i) \odot (v_j * W_j) \tag{11}$$

where $R_{ij} = \{(i,j)_{i \in \{1,2,\cdots,n_f\}, j \in \{1,2,\cdots,n_f\}, j > i}\}$, $(i,j) \in R_{ij}$, $p_{ij} \in R^{n_e}$, $x_i \in R^{n_e}$, $x_j \in R^{n_e}$. $W_i \in R^{n_e \times n_e}$, $W_j \in R^{n_e \times n_e}$ are the corresponding parameter matrix of x_i and x_j, respectively. In this interaction mode, the position of features is equal because they are symmetrical in feature interaction, which successfully solves the problem of introducing feature order noise in "Field-Each Type". The total number of interaction matrices is n_f. Compared to "Field-Each Type", the cost it pays is just one more parameter matrix to learn and one more inner product to perform. Besides, it has stronger expression ability than "Field-All Type" and fewer parameters than "Field-Interaction Type".

In this layer, the SE-ResNet-Like embeddings V is transformed into interaction vectors:

$$P = [p_1, p_2, \cdots, p_{n_p}] \tag{12}$$

where $P \in R^{n_p \times n_e}$, $n_p = \frac{n_f \times (n_f - 1)}{2}$.

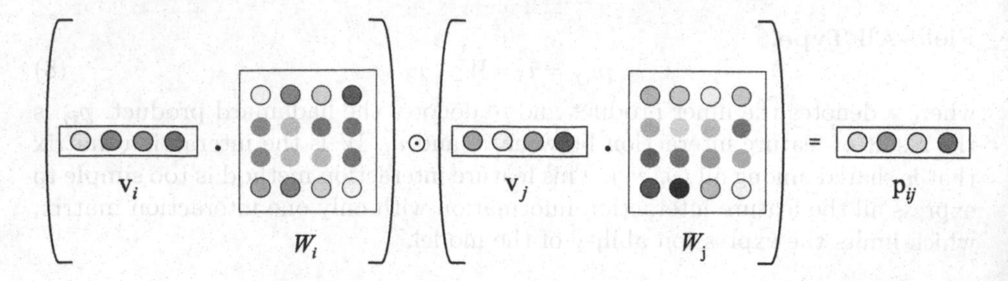

Fig. 3. Field-equal type bilinear feature interaction.

2.4 Concatenation Layer

In the task of CTR estimation, the input features also contain some dense numerical features. Let the normalized dense features be $D \in R^{n_d}$, where n_d is the number of dense features. We concatenate dense features D and interaction vectors P in this layer. The concatenate vector $C \in R^{n_d+n_p \times n_e}$ is calculated by:

$$C = F_{concat}(D, P) = [d_1, \cdots, d_{n_d}, p_1, \cdots, p_{n_p}] \tag{13}$$

* Shallow Type: When we sum C element by element, and then use sigmoid function to get CTR prediction results, we call it shallow-FaBeNet.
* Deep Type: If we input C to the subsequent DNN layer to calculate the final CTR prediction result, we call it deep-FaBeNet.

2.5 DNN Network Layer

This part is a fully-connected feed-forward neural network with each MLP layer having the following formula:

$$h^{l+1} = f(W^l h^l + b^l) \tag{14}$$

Where $h^{l+1} \in R^{n_l+1}$, $h^l \in R^{n_l}$ are the $(l+1)$-th and l-th hidden layer output, respectively. $W^l \in R^{n_{l+1} \times n_l}$, $b \in R^{n_l+1}$ are the model weight and bias which need to be learned. Finally, this layer outputs an real number y_d, which can be calculated by:

$$y_d = f(W^{L+1} h^L + b^L) \tag{15}$$

where L is the number of deep layers.

2.6 Output Layer

In this part, we give the final CTR prediction formulation of out proposed model:

$$\hat{y} = \sigma(w_0 + \sum_{i=0}^{n_x} w_i x_i + y_d) \tag{16}$$

where $\hat{y} \in (0, 1)$ is the predicted value of CTR, n_x is the feature size, x is the original input and w is the corresponding weights in the linear part.

Our Loss function is Logloss:

$$\mathcal{L} = -\frac{1}{N}\sum_{i=0}^{N}(Ny_i log(\hat{y}_i) + (1 - y_i)log(1 - \hat{y}_i)) \tag{17}$$

where N is the size of training set, y_i is the true label of the i-th sample x_i.

3 Experiments

In this section, we compare our proposed modes FaBeNet with existing models in CTR prediction. Moreover, we analyze the impact of our two improvements: SE-ResNet and "Field-Equal Type" through ablation study.

3.1 Dataset

The dataset of our experiment is the Criteo Display ADs[1]. Due to the limitation of experimental hardware, we randomly sample 10% of them as our experimental dataset. And then we split the experimental dataset randomly into two parts: 90% is for training, while the rest is for testing.

3.2 Experiments Setting

For all shallow models, we set the mini-batch size, embedding size to 1000, 10, respectively. For the optimization method, we use the AdaGrad [19] with learning rate of 0.001 as optimization method.

For all deep models, we set mini-batch size, embedding size, DNN structure to 1000, 6, (600, 600, 600) and use Adam [20] with learning rate of 0.0002 as optimization method. For FiBiNet and FaBeNet, We set reduction ratio to 6. For FaBeNet, we set dropout [21] to 0.3. For other models, we set dropout to 0.5.

3.3 Model Comparison

Firstly, from the Table 1 we can observe that the results of deep models are better than shallow models, which also proves the effectiveness of modeling high-order feature interaction.

Secondly, LR performs poorly in shallow model comparison experiments because it has no effective measures to solve the data sparseness problem. Although FM alleviates data sparseness through factorization, it does not perform well because it does not model the importance of features. Shallow-FiNiNet-interaction is the best performer in shallow-FiBiNet, both in and shallow-FaBeNet has a greater improvement than AFM and FFM. Shallow-FaBeNet is not as effective as shallow-FiNiNet-interaction, probably because we did not concatenate the original embeddings with Se-ResNET like embeddings after passing

[1] http://labs.criteo.com/downloads/download-terabyte-click-logs/.

Table 1. Comparison results of different models

Model Class	Model	AUC	Logloss	Runtime Per Epoch(s)
Shallow	LR [6]	0.7548	0.4874	2170
	FM [7]	0.7665	0.4784	2202
	AFM [9]	0.7697	0.4757	4316
	FFM [8]	0.7715	0.4743	4416
	shallow-FiBiNet-all [15]	0.7739	0.4725	3550
	shallow-FiBiNet-each [15]	0.7754	0.4714	3880
	shallow-FiBiNet-interaction [15]	**0.7760**	**0.4708**	4330
	shallow-FaBeNet(Ours)	0.7757	0.4712	3891
Deep	FNN [10]	0.7873	0.4634	4814
	W& D [1]	0.7783	0.4705	6050
	DeepFM [11]	0.7878	0.4632	4770
	NFM [12]	0.7876	0.4633	7894
	DCN [13]	0.7872	0.4636	7702
	xDeepFM [14]	0.7883	0.4630	4984
	deep-FiBiNet-all [15]	0.7913	0.4589	5360
	deep-FiBiNet-each [15]	0.7915	0.4586	6052
	deep-FiBiNet-interaction [15]	0.7912	0.4591	6950
	deep-FaBeNet(Ours)	**0.7919**	**0.4581**	5372

through the BI layer, which makes the shallow models without DNNs too simple. However, the parameters of shallow-FaBeNet are much less than that of shallow-FiNiNet-interaction and can be a good solution when time performance is required.

Thirdly, in the comparative experiment of the deep models, the deep-FabeNet proposed by us achieves the best performance overall baseline methods on the Criteo dataset. The performance of W&D model is the worst because there are no artificial features in this experiment. The best performance of deep-FiBiNet is the deep-FiBiNet-each. Both it and our proposed deep-FaBeNet are better than DeepFM, xDeepFM, and other models in terms of evaluation metrics, which shows the effectiveness of introducing mechanisms such as SENet to learn feature importance before DNN. Compared with deep-FiBiNet-each, deep-FaBeNet replaces SENet with SE-ResNet and improves the feature interaction mode. The performance of deep-FaBeNet shows that the SE-ResNet mechanism and "Field-Equal Type" bilinear interaction are effective. On the one hand, the SE-ResNet mechanism automatically learns the importance of features while preserving the information of the original embeddings, which enhances the diversity of features. At the same time, the original embedding vector does not need to enter the Bi layer for interaction, which reduces the number of parameters. On the other hand, the feature interaction mode of “Field-Equal Type” makes the features equal in the interaction and solves the problem of introducing feature sequence noise.

3.4 Ablation Study

Ablation Study of SE-ResNet. In order to further illustrate the effectiveness of SE-ResNet, we carry out the ablation experiments over FaBeNet in this section to explore the role of it. The comparison results are shown in the Table 2.

Table 2. Ablation study result of SE-ResNet

Model Class	Model	AUC	Logloss
Shallow	BASE (Ours)	**0.7757**	**0.4712**
	NO-RES	0.7748	0.4718
	NO-SE-RES	0.7710	0.4747
Deep	Deep-BASE	**0.7919**	**0.4581**
	NO-RES	0.7716	0.4588
	NO-SE-RES	0.7910	0.4593

We set shallow-FaBeNet as the base model, and NO-RES means replacing SE-ResNet with SENet, NO-SE-RES means removing the SE-ResNet. From the table, we can observe that the SE-ResNet structure is indispensable to FaBeNet, especially in the shallow model. When we remove the SE-ResNet structure, the performance of the model declines apparently, which illustrates the effectiveness of modeling the importance of features. At the same time, no matter in the shallow model or the deep model, the SE-ResNet structure performs better than the SENet structure, which shows the effectiveness of retaining the original feature information through the shortcut connection.

Ablation Study of “FILE-EQUAL TYPE”. In order to further illustrate the effectiveness of our proposed field-equal type bilinear-interaction, we carry out the ablation experiments in this section to explore the role of field-equal type bilinear-interaction in FaBeNet. The comparison results are shown in the Table 3:

We set shallow-FaBeNet as the base model, and NO-BI means removing the bilinear interaction layer, Base-Each means that replacing the ‘Field-equal type’ with ‘Field-each type’ and so on. From the table, we can observe that whether the shallow model or the deep model, the performance of the model has dropped to a large extent after removing the BI layer, which shows that the BI layer is essential too. In the deep model, the “Field-equal type” interaction method achieves the best performance among the four feature interaction methods, which shows that our improvement is effective. Although the effect of the “field-equal

Table 3. Ablation study result of "Field-equal Type"

Model Class	Model	AUC	Logloss
Shallow	BASE (Ours)	0.7757	0.4712
	NO-BI	0.7708	0.4749
	BASE-all [15]	0.7741	0.4723
	BASE-each [15]	0.7755	0.4713
	BASE-interaction [15]	**0.7763**	**0.4706**
Deep	Deep-BASE (Ours)	**0.7919**	**0.4581**
	NO-BI	0.7906	0.4596
	Deep-BASE-all [15]	0.7912	0.4591
	Deep-BASE-each [15]	0.7914	0.4589
	Deep-BASE-interaction [15]	0.7910	0.4593

type" is not as good as the "Field-interaction type" in the shallow model, it has fewer parameters and is better in effect than the remaining feature interaction methods.

4 Conclusion

In the task of CTR prediction, being aware of the feature importance and designing a reasonable and effective feature interaction mode is the key factor. We propose a model that can dynamically learn the feature importance and perform equal type bilinear feature interaction: FaBeNet. On the one hand, it can learn the feature importance and keep original feature information through the SE-ResNet. On the other hand, we propose a new bilinear feature interaction mode by integrating hadamard product and inner product to perform equal interaction. Experimental results show that our proposed model FaBeNet achieves state-of-art performance in AUC and Logloss. The ablation studys in the experiment verify the effectiveness of the SE-ResNet mechanism and the "Field-Equal Type" bilinear feature interaction, which provided experimental support for our improvements.

Acknowledgment. This work was supported by the National High Technology Research, Development Program of China (No. 2018YFB1703500), the Shanghai Innovation Action Project of Science and Technology (No. 19511105502), and the Fundamental Research Funds for the Central Universities.

References

1. Cheng, H.T., Koc, L., Harmsen, J., et al.: Wide & deep learning for recommender systems. In: Proceedings of the 1st Workshop on Deep Learning for Recommender Systems, pp. 7–10 (2016)

2. Graepel, T., Candela, J.Q., Borchert, T., et al.: Web-scale Bayesian click-through rate prediction for sponsored search advertising in Microsoft's Bing search engine. Omnipress (2010)
3. He, X., Pan, J., Jin, O., et al.: Practical lessons from predicting clicks on ads at Facebook. In: Proceedings of the Eighth International Workshop on Data Mining for Online Advertising, pp. 1–9 (2014)
4. Davidson, J., Liebald, B., Liu, J., et al.: The YouTube video recommendation system. In: Proceedings of the Fourth ACM Conference on Recommender Systems, pp. 293–296 (2010)
5. Beel, J., Gipp, B., Langer, S., et al.: Research-paper recommender systems: a literature survey. Int. J. Digit. Libr. **17**(4), 305–338 (2016). https://doi.org/10.1007/s00799-015-0156-0
6. Kleinbaum, D.G., Dietz, K., Gail, M., et al.: Logistic Regression. Springer, New York (2002)
7. Rendle, S.: Factorization machines. In: 2010 IEEE International Conference on Data Mining, pp. 995–1000. IEEE (2010)
8. Juan, Y., Zhuang, Y., Chin, W.S., et al.: Field-aware factorization machines for CTR prediction. In: Proceedings of the 10th ACM Conference on Recommender Systems, pp. 43–50 (2016)
9. Xiao, J., Ye, H, He, X., et al.: Attentional factorization machines: learning the weight of feature interactions via attention networks. arXiv preprint arXiv:1708.04617 (2017)
10. Zhang, W., Du, T., Wang, J.: Deep learning over multi-field categorical data. In: Ferro, N., et al. (eds.) ECIR 2016. LNCS, vol. 9626, pp. 45–57. Springer, Cham (2016). https://doi.org/10.1007/978-3-319-30671-1_4
11. Guo, H., Tang, R., Ye, Y., et al.: DeepFM: a factorization-machine based neural network for CTR prediction. arXiv preprint arXiv:1703.04247 (2017)
12. He, X., Chua, T.S.: Neural factorization machines for sparse predictive analytics. In: Proceedings of the 40th International ACM SIGIR Conference on Research and Development in Information Retrieval, pp. 355–364 (2017)
13. Wang, R., Fu, B., Fu, G., et al.: Deep & cross network for ad click predictions. In: Proceedings of the ADKDD 2017, pp. 1–7 (2017)
14. Lian, J., Zhou, X., Zhang, F., et al.: xDeepFM: combining explicit and implicit feature interactions for recommender systems. In: Proceedings of the 24th ACM SIGKDD International Conference on Knowledge Discovery & Data Mining, pp. 1754–1763 (2018)
15. Huang, T., Zhang, Z., Zhang, J.: FiBiNET: combining feature importance and bilinear feature interaction for click-through rate prediction. In: Proceedings of the 13th ACM Conference on Recommender Systems, pp. 169–177 (2019)
16. Hu, J., Shen, L., Sun, G.: Squeeze-and-excitation networks. In: Proceedings of the IEEE Conference on Computer Vision and Pattern Recognition, pp. 7132–7141 (2018)
17. He, K., Zhang, X., Ren, S., et al.: Deep residual learning for image recognition. In: Proceedings of the IEEE Conference on Computer Vision and Pattern Recognition, pp. 770–778 (2016)
18. Linsley, D., Scheibler, D., Eberhardt, S., et al.: Global-and-local attention networks for visual recognition. arXiv preprint arXiv:1805.08819 (2018)
19. Duchi, J., Hazan, E., Singer, Y.: Adaptive subgradient methods for online learning and stochastic optimization. J. Mach. Learn. Res. **12**(Jul), 2121–2159 (2011)
20. Kingma, D.P., Ba, J.: Adam: a method for stochastic optimization. arXiv preprint arXiv:1412.6980 (2014)
21. Srivastava, N., Hinton, G., Krizhevsky, A., et al.: Dropout: a simple way to prevent neural networks from overfitting. J. Mach. Learn. Res. **15**(1), 1929–1958 (2014)

GFEN: Graph Feature Extract Network for Click-Through Rate Prediction

Mei Yu[1,2,3], Chengchang Zhen[1,2,3], Ruiguo Yu[1,2,3], Xuewei Li[1,2,3], Tianyi Xu[1,2,3], Mankun Zhao[1,2,3], Hongwei Liu[4], Jian Yu[1,2,3], and Xuyuan Dong[5](✉)

[1] College of Intelligence and Computing, Tianjin University, Tianjin, China
{yumei,chengchangzhen,rgyu,xuewei,tianyi.xu, zmk,yujian}@tju.edu.cn

[2] Tianjin Key Laboratory of Cognitive Computing and Application, Tianjin, China

[3] Tianjin Key Laboratory of Advanced Networking (TANK Lab), Tianjin, China

[4] Foreign Language, Literature and Culture Studies Center, Tianjin Foreign Studies University, Tianjin, China
liuhongwei@tjfsu.edu.cn

[5] Information and Network Center, Tianjin University, Tianjin, China
dongxuyuan@tju.edu.cn

Abstract. Click-Through Rate (CTR) is a fundamental task in personalized advertising and recommender systems. It is vital to model different orders feature interactions in click-through rate. There are many proposed methods in this field such as FM and its variants. However, many current methods can not adequately and precisely extract feature interactions. In this paper, we improve an effective light weight method called the Graph Feature Extract Network (GFEN) to further explicitly or implicitly model low-order and high-order feature interactions information via Graph Convolutional Network (GCN) and Global Recombination Network (GRN). GCN explicitly models local high-order feature interactions. GRN automatically captures global high-order feature interactions via the diversity pooling layer and recombines global and local feature interactions via fully connection layer. We conduct extensive experiments on there real-world datasets and show that our model achieves the best performance compared to the existing state-of-the-art methods such as CKE, DKN, Wide&Deep, RippleNet etc. Our proposed GFEN is a very light weight model, which can be applied to other complicate model such as Ripplenet based knowledge graph and deep learning models based CTR. The whole model can be efficiently fit on large-scale raw input feature.

Keywords: Recommender system · Click-through rate · Graph Convolution Network · Deep neural network

1 Introduction

Click-Through Rate prediction is a critical problem for many applications such as online advertising and recommender systems which predicting the probabilities

H. Yang et al. (Eds.): ICONIP 2020, LNCS 12534, pp. 444–454, 2020.
https://doi.org/10.1007/978-3-030-63836-8_37

of users clicking on ads or items. Now it is a big challenge for users to obtain meaningful information in so much information. Recommender system aims to address this problem and make personalized recommendations for users. In click-through rate task, it is vital to capture feature interactions of users. But in current internet companies like Amazon, the user's representation vector is high-dimensional and sparse because of a lot of categorial features. As a result, the combinatorial features are more sparse. With such sparse and high-dimensional input features, some models such as Linear Regression are easily overfitted, which can not accurately learn users' features.

In order to alleviate dimension explosion, most models need to convert the one-hot encoding vector of user and item to latent low dimension embedding vector. And, due to the more structure information in knowledge graph, many researchers begin to make use of knowledge graph to improve the recommendation performance and explainable [10,18,19]. However, many current models can not precisely capture feature interactions.

In this paper, we propose GFEN, which consists of Graph Convolution Network (GCN) and Global Recombination Network (GRN) to precisely extract high-order feature interactions. Graph Convolutional Network (GCN) is able to explicitly model the local high-order feature interactions. And Global Recombination Network (GRN) combines the diversity pooling layer with fully connection layer, which can automatically capture and recombine global and local high-order feature interactions. In Global Recombination Network (GRN), we can get the global information of multi perspectives via diversity pooling layer. Then the fully connection layer recombines global and local feature interactions. Our results are substantially more than state-of-the-art models. Empirically, we conduct many experiments in three real-world datasets. The experiments show that GFEN has achieved remarkable performance.

Our main contributions are listed as follows:

- We introduce a light weight module called Graph Feature Extract Network (GFEN) which can be used to enhance the performance of complicated models based knowledge graph.
- We combine Graph Convolutional Network (GCN) and Global Recombination Network (GRN) to capture high-order feature interactions via different perspective information.
- We conduct experiments on three real-world datasets, and the results prove the effectiveness of GFEN over state-of-the-art baselines.

2 Related Work

2.1 Factorization Machine

Factorization Machine (FM) [15] is the most successful model for click-through rate. FM converts a high dimensional sparse vector to a low dimensional dense real-value vector to solve the dimensional disaster problem. Meanwhile, FM models all feature interactions between variables using factorized parameters. These

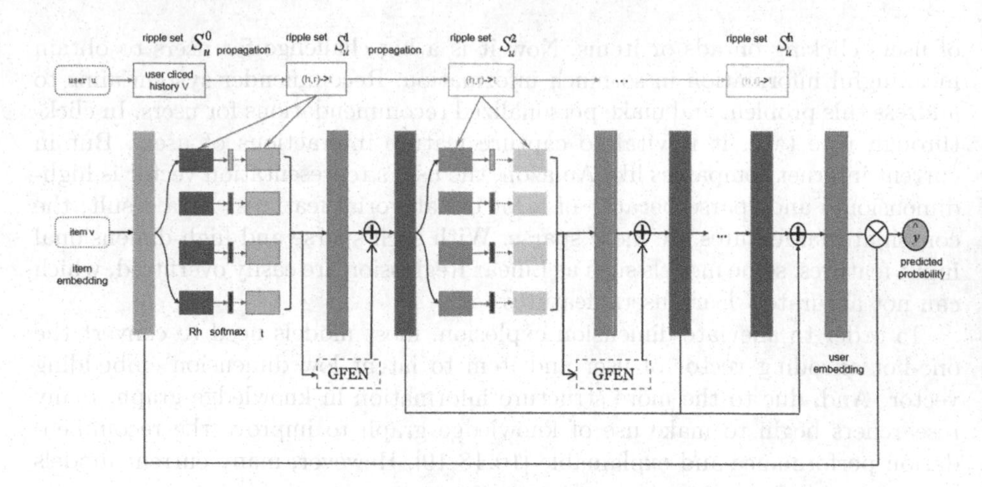

Fig. 1. Overview of the Enhanced Ripplenet.

interaction information is vital to predict users preferences in click-through rate tasks. Thus, some researchers have made a lot of improvements on the basis of FM, like AFM [6], NFM [8], DeepFM [6] and xDeepFM [12] etc.

2.2 Feature Generation by Convolutional Neural Networks

Recently, due to its ability to extract potential features information, CNN [1, 3,21] has achieved great success in many fields such as computer vision and natural language processing. As a result, many neural network models [4,6,12, 24,25] based CTR have also been proposed in recent years. How to effectively model the feature interactions is the key factor for most of these neural network based models. Liu [13] proposed "Feature Generation by Convolutional Neural Network (FGCNN)". It uses Convolutional Neural Network (CNN) and Multi-Layer Perception(MLP), which complements each other, to learn global and local feature interactions for feature generation. Be inspired of it, our model adopts Graph Convolutional Network (GCN) [11,14] that is able to more capture users' local feature interactions in knowledge graph.

3 Our Method

We aim to dynamically learn a user's feature. To this end, we propose the Graph Feature Extract Network(GFEN) for click-through rate prediction. We apply our model on Ripplenet to improve its effective performance. So in this section, we will describe some definitions in Ripplenet as depicted in Fig. 1. Our proposed GFEN can enhance the performance of captureing feature interactions in knowledge graph.

3.1 Basic Framework

We regard Ripplenet as our the basic net to enhance it. Input and embedding layer are applied upon the raw feature to convert it to a low dimensional, dense real-value vector. A users historical preferences are used to find more interactive interests in knowledge graph. In each interest propagation, we can infer a interest vector T from a user vector H and relation vector R. These related tail embeddings can be used as the latent features of the user embedding. Finally, the user embedding and item embedding are used to predict the probability of the user's clicking.

We will introduce GFEN into Ripplenet to improve its performance. One problem in Ripplenet is that the tail feature embedding in each hop are only calculated by weighted sum. It can not precisely capture a users feature interactions in each hop. And it can not been take into account that high-order features interaction information between features in each tail feature embedding. The tail embedding from different relationships paths in every hop has a lot of meaningful feature information. And these feature interactions information [13] is helpful to the representation of a user's embedding. So we propose Graph Feature Extract Network(GFEN) to extract and recombine user's linear features and high-order feature interactions.

3.2 Propagation Process in Ripple Network

Given the item v embedding and the user's 1-hop ripple set S_u^1, we can get the similarity of each triple and item v.

$$p_i = soft\max\left(v^T R_i h_i\right) = \frac{\exp\left(v^T R_i h_i\right)}{\sum_{(h,r,t)\in S_u^1} \exp\left(v^T R_i h_i\right)} \tag{1}$$

After the similarity weights are obtained, we take sum of tails in S_u^1 weighted by corresponding weights.

$$o_u^1 = \sum\nolimits_{(h,r,t)\in S_u^1} p_i t_i \tag{2}$$

o_u^1 is also used to update item v's vector, so that the same operation can be repeated in order to get o_u^2. Then we can get the final feature representation vector of user.

$$u = o_u^1 + o_u^2 + \cdots + o_u^H \tag{3}$$

Lastly, the final feature interaction vector of user u and the embedding of the item v inner product to output the probability of clicking.

$$\hat{y} = \sigma\left(u^T v\right) \tag{4}$$

Here, σ is the Sigmoid function.

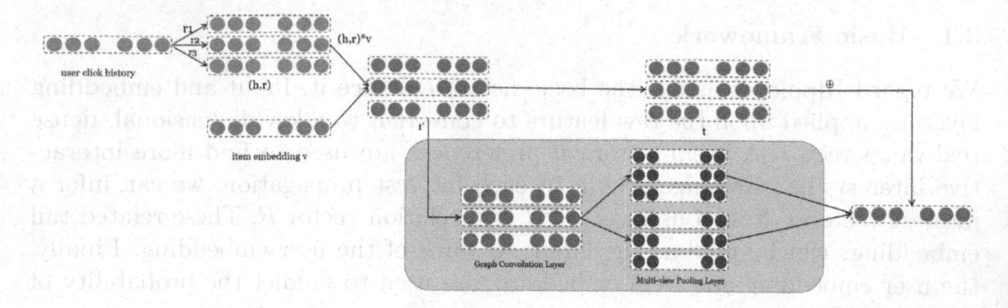

Fig. 2. Graph feature extract network (GFEN).

3.3 Graph Feature Extract Network

Since we aim to dynamically learn global and local feature interactions, we proposed Graph Feature Extract Network (GFEN). As depicted in Fig. 2, the model GFEN consists of Graph Convolutional Network (GCN) and Global Recombination Network (GRN). In Graph Convolutional Network (GCN) [5], we model explicitly local high-order feature interactions. In order to model multi perspectives local feature interactions, we adopt GCN module to dynamically extract feature interactions via multi scaled convolution kernel such as $f^{2\times2}$ and $f^{3\times3}$ and multi perspective pooling. The following is operation steps in GFEN.

Assuming that the output of the first Graph Convolutional Layer is O_{c1}, we can formulate the operation convolution part:

$$\begin{aligned} O_{c1} &= \varepsilon_u^k * f \\ O_{c1} &= pooling\left(O_{c1}\right) \\ O_{c1} &= ReLU\left(O_{c1}\right) \end{aligned} \tag{5}$$

where O_{c1} is the concatenation of O_{c1}^{mean} and O_{c1}^{max}. As we all know, CNN [9] only models the local feature information of the receptive field area, so it will lose some meaningful information. Thus, we adopt diversity pooling layers [14] to capture different perspective feature interactions. And we concatenate the feature interactions of mean and max pooling. It contains different perspective feature interactions [20].

In Global Recombination Network (GRN), we automatically recombine high-order feature interactions via fully connection layer. We formulate recombination layer operation:

$$\begin{aligned} O_{f1} &= O_{c1} \times W \\ O_{f1} &= ReLU\left(O_{f1}\right) \end{aligned} \tag{6}$$

where W is the weight in fully connection layers, O_{f1} is the output of GFEN.

All offset values bias are omitted for the simplification of the formulations. So we can get the user feature embedding after H hops. In this way, we capture the linear feature information o_u^i and the high-order feature interactions of a user. We introduce high-order feature interactions to enhance Ripplenet.

$$u = \left(o_u^1 + O_{GFEN}^1\right) + \left(o_u^2 + O_{GFEN}^2\right) + \cdots + \left(o_u^H + O_{GFEN}^H\right) \tag{7}$$

3.4 Objective Function

In the Enhanced Ripplenet, we still choose cross-loss entropy as the objective function. In order to avoid the overfitted problem, the L2 regularization term of the parameters is also added.

$$\begin{aligned} \min L = & \sum_{(u,v)\in Y} -\left(y_{uv}\log\sigma\left(u^T v\right) + \left(1 - y_{uv}\right)\log\left(1 - \sigma\left(u^T v\right)\right)\right) \\ & + \tfrac{\lambda_1}{2}\left(\|V\|_2^2 + \|E\|_2^2 + \sum_{r\in R}\|R\|_2^2\right) + \tfrac{\lambda_2}{2}\sum_{r\in R}\left\|I_r - E^T R E\right\|_2^2 \\ & + \tfrac{\lambda_3}{2}\left(\|W_c\|_2^2 + \|W_f\|_2^2\right) \end{aligned} \tag{8}$$

Here, y_{uv} is the ground truth value and $u^T v$ is the predicted value.

Our model chooses Adam's method to optimize the loss function. By calculating the gradient of the loss function with respect to the model parameters, we use back-propagation to update all the trainable parameters to obtain the best model. We will carry out some experiments in Sect. 4.

4 Experiments

In order to verify the effectiveness of our model, we conduct experiments on three different popular datasets: MovieLens-1M[1], Book-crossing[2] and Last.FM[3]. We arrange our experiments with 2 T K80 GPUs. The results of our proposed method are compared with several state-of-the-art CTR techniques based knowledge graph.

4.1 Datasets and Metrics

Movielens-1M [7] is a widely used benchmark dataset in movie recommendations, which consists of approximately 1 million explicit ratings (from 1–5) on the Movie-Lens website. Book-Crossing contains 114,9780 explicit ratings(from 0 to 10) of books in the Book-Crossing community. Last.FM contains musician listening information from a set of 2K users from Last.fm online music system. Since MovieLens-1M, Book-Crossing are explicit information, we need transform explicit feedback data into implicit feedback data, where 1 indicates that the user rated the item, otherwise 0. We split all datasets into two parts:80% is for training, while the rest is for testing.

In our experiments, we adopt two metrics. AUC: Area Under the ROC Curve measures the probability that a CTR predictor will assign a higher score to a

[1] https://grouplens.org/datasets/movielens/1m/.

[2] http://www2.informatik.uni-freiburg.de/~cziegler/BX/.

[3] https://grouplens.org/datasets/hetrec-2011/.

randomly chosen positive item than a randomly chosen negative item. A higher AUC score indicates a better performance. ACC: Accuracy reflects the rate at which the CTR predictor accurately identifies true positives and negatives. A higher ACC score indicates a better performance (Table 1).

Table 1. Dataset statistics.

Datasets	MovieLens-lM	Book-Crossing	Last.FM
Users	6036	17860	1872
Items	2445	14967	3846
Interactions	753772	139746	42346
1-hop triples	19098	15682	11546
2-hop triples	133002	54340	1169

Table 2. The overall performance of proposed model.

Model	MovieLens-1M		Book-Crossing		Last.FM	
	AUC	ACC	AUC	ACC	AUC	ACC
PER	71.2%	66.7%	62.3%	58.8%	63.3%	59.6%
CKE	79.6%	73.9%	67.4%	63.5%	74.4%	67.3%
DKN	65.5%	58.9%	62.1%	59.8%	60.2%	58.1%
LibFM	89.2%	81.2%	68.5%	63.9%	77.7%	70.9%
Wide&Deep	90.3%	82.2%	71.1%	62.3%	75.6%	68.8%
Ripplenet	92.1%	84.6%	72.2%	65.1%	76.8%	69.1%
Ours	**92.3%**	**84.8%**	**73.1%**	**68.9%**	**80.5%**	**74.3%**

4.2 Baselines Methods

We compare our proposed GFEN base Ripplenet with the following baseline models. PER [22] considers the knowledge graph as a heterogeneous information network and extract meta-paths feature-based to represent the connections between users and items. CKE [23] combines the structure information in knowledge graph, item comment texts, visual information of the item with the collaborative filtering method to realize recommendation as a whole framework. DKN [18], a model for news recommendation, treats entity embedding, entity context embedding and word embedding as multi-channels to input the recommendation module for click-through rate prediction. LibFM [16] is a general feature-based method. This method uniformly uses the attributes of users and items as input to the recommendation algorithm. Wide&Deep [2] joints trained wide linear models for linear feature information and deep neural networks for non-linear feature information for recommendation. Ripplenet [17] is a memory network, spreading user preferences in knowledge graph to predict next click for user.

4.3 Results and Discussion

Overall Performance. In this subsection, we summarize the overall performance on MovieLens-1M, Book-Crossing and Last.FM in Table 2. It shows the results of our model and the existing methods on three datasets. Obviously, our model consistently outperforms other models such as PER, CKE, DKN etc. The results indicate that GFEN is an effective method on many real-world datasets.

Among all compared methods, our proposed GFEN achieves the best performance. The best AUC and ACC on MovieLens-1M is 92.1% and 84.6%. The best AUC and ACC is 73.1% and 68.9% on Book-Crossing dataset. The best AUC and ACC is 80.5% and 74.3% on Last.FM dataset. The proposed model has achieved a significant improvement when compared to existing state-of-the-art models. It proves that GFEN is effective to model feature interactions in knowledge graph. The GCN layer learns local feature interactions, which dynamically captures feature interactions. The GRN layer recombines local and global feature interactions via pooling and fully connection layer.

Table 3. Performance of different reception sizes in GCN

Reception size	MovieLens-1M	Book-Crossing	Last.FM
	AUC	AUC	AUC
2*2	92.35%	72.58%	81.26%
3*3	92.14%	73.47%	81.33%
4*4	92.13%	72.25%	80.22%

Table 4. Performance of different pooling types in GRN

Pooling type	MovieLens-1M	Book-Crossing	Last.FM
	AUC	AUC	AUC
Mean Pooling	92.24%	72.45%	80.96%
Max Pooling	92.35%	74.25%	81.33%
Both	92.3%	73.1%	80.5%

Ablation Analysis. To further validate and gain deep insights into our model, we conduct some ablation studies. As shown in Table 3 and Table 4, we conduct ablation experiments about the parameter of reception sizes in GCN. On Book-Crossing and Last.FM dataset, we can observe that increasing reception size improves model performance at the beginning. However, the performance is degraded if the reception size keeps increasing. When the size reaches 3*3, model gets best performance. And the size of 2*2 achieves the maximum AUC on MovieLens-1M.

The Table 4 shows that the pooling operation in GRN is how to impact the performance of our model. We can observe that the performance of max pooling on AUC metric outperforms mean pooling on the three datasets. Therfore, its better to adopt max pooling in experiments.

5 Conclusion

In this paper, we proposed GFEN, which is a light weight module. It consists of two parts: Graph Convolutional Network (GCN) and Global Recombination Network (GRN). Graph Convolutional Network (GCN) is able to explicitly model the local high-order feature interactions. And Global Recombination Network (GRN) combines the diversity pooling layers with fully connection layers. In GRN, we can get the global information of multi perspectives via diversity pooling layer. Then the fully connection layer recombines global and local feature interactions. We conduct extensive experiments in three real-world datasets. The results demonstrate the significant superiority of GFEN over strong baselines.

Acknowledgment. This work is jointly supported by National Natural Science Foundation of China (No.61877043 and 61877044).

References

1. Chan, P.P.K., Hu, X., Zhao, L., Yeung, D.S., Liu, D., Xiao, L.: Convolutional neural networks based click-through rate prediction with multiple feature sequences. In: Proceedings of the Twenty-Seventh International Joint Conference on Artificial Intelligence, IJCAI 2018, Stockholm, Sweden, 13–19 July 2018, pp. 2007–2013 (2018). https://doi.org/10.24963/ijcai.2018/277
2. Cheng, H., et al.: Wide & deep learning for recommender systems. In: Proceedings of the 1st Workshop on Deep Learning for Recommender Systems, DLRS@RecSys 2016, Boston, MA, USA, 15 September 2016, pp. 7–10 (2016)
3. Conneau, A., Schwenk, H., Barrault, L., LeCun, Y.: Very deep convolutional networks for natural language processing. CoRR abs/1606.01781 (2016). http://arxiv.org/abs/1606.01781
4. Covington, P., Adams, J., Sargin, E.: Deep neural networks for YouTube recommendations. In: Proceedings of the 10th ACM Conference on Recommender Systems, Boston, MA, USA, 15–19 September 2016, pp. 191–198 (2016). https://doi.org/10.1145/2959100.2959190
5. Fout, A., Byrd, J., Shariat, B., Ben-Hur, A.: Protein interface prediction using graph convolutional networks. In: 2017 Advances in Neural Information Processing Systems 30: Annual Conference on Neural Information Processing Systems, Long Beach, CA, USA, 4–9 December 2017, pp. 6530–6539 (2017). http://papers.nips.cc/paper/7231-protein-interface-prediction-using-graph-convolutional-networks
6. Guo, H., Tang, R., Ye, Y., Li, Z., He, X.: DeepFM: a factorization-machine based neural network for CTR prediction. In: Proceedings of the Twenty-Sixth IJCAI 2017, Melbourne, Australia, 19–25 August 2017, pp. 1725–1731 (2017)
7. Harper, F.M., Konstan, J.A.: The movielens datasets: history and context. ACM Trans. Interact. Intell. Syst. **5**(4), 19:1–19:19 (2016). https://doi.org/10.1145/2827872

8. He, X., Chua, T.: Neural factorization machines for sparse predictive analytics. In: Proceedings of the 40th International ACM SIGIR Tokyo, Japan, 7–11 August 2017, pp. 355–364 (2017)
9. Huang, G., Liu, Z., Weinberger, K.Q.: Densely connected convolutional networks. CoRR abs/1608.06993 (2016). http://arxiv.org/abs/1608.06993
10. Li, Z., Cui, Z., Wu, S., Zhang, X., Wang, L.: Fi-GNN: modeling feature interactions via graph neural networks for CTR prediction. In: Proceedings of the 28th ACM International Conference on Information and Knowledge Management, CIKM 2019, Beijing, China, 3–7 November 2019, pp. 539–548 (2019). https://doi.org/10.1145/3357384.3357951
11. Li, Z., Chen, Q., Koltun, V.: Combinatorial optimization with graph convolutional networks and guided tree search. In: Advances in Neural Information Processing Systems 31: Annual Conference on Neural Information Processing Systems 2018, NeurIPS 2018, Montréal, Canada, 3–8 December 2018, pp. 537–546 (2018). http://papers.nips.cc/paper/7335-combinatorial-optimization-with-graph-convolutional-networks-and-guided-tree-search
12. Lian, J., Zhou, X., Zhang, F., Chen, Z., Xie, X., Sun, G.: xDeepFM: combining explicit and implicit feature interactions for recommender systems. In: KDD 2018, London, UK, 19–23 August 2018, pp. 1754–1763 (2018)
13. Liu, B., Tang, R., Chen, Y., Yu, J., Guo, H., Zhang, Y.: Feature generation by convolutional neural network for click-through rate prediction. In: World Wide Web Conference, WWW 2019, San Francisco, CA, USA, 13–17 May 2019, pp. 1119–1129 (2019)
14. Nguyen, T.H., Grishman, R.: Graph convolutional networks with argument-aware pooling for event detection. In: Proceedings of the Thirty-Second AAAI Conference on Artificial Intelligence, (AAAI-18), the 30th Innovative Applications of Artificial Intelligence (IAAI-18), and the 8th AAAI Symposium on Educational Advances in Artificial Intelligence (EAAI-18), New Orleans, Louisiana, USA, 2–7 February 2018, pp. 5900–5907 (2018). https://www.aaai.org/ocs/index.php/AAAI/AAAI18/paper/view/16329
15. Rendle, S.: Factorization machines. In: ICDM 2010, pp. 995–1000 (2010)
16. Rendle, S.: Factorization machines with libFM. ACM TIST **3**(3), 57:1–57:22 (2012)
17. Wang, H., et al.: RippleNet: propagating user preferences on the knowledge graph for recommender systems, Torino, Italy, 22–26 October 2018, pp. 417–426 (2018)
18. Wang, H., Zhang, F., Xie, X., Guo, M.: DKN: deep knowledge-aware network for news recommendation. In: WWW 2018, Lyon, France, 23–27 April 2018, pp. 1835–1844 (2018)
19. Wang, H., Zhang, F., Zhao, M., Li, W., Xie, X., Guo, M.: Multi-task feature learning for knowledge graph enhanced recommendation. In: Conference, WWW 2019, San Francisco, CA, USA, 13–17 May 2019, pp. 2000–2010 (2019)
20. Yao, H., et al.: Deep multi-view spatial-temporal network for taxi demand prediction. In: Proceedings of the Thirty-Second AAAI Conference on Artificial Intelligence, (AAAI-18), the 30th innovative Applications of Artificial Intelligence (IAAI-18), and the 8th AAAI Symposium on Educational Advances in Artificial Intelligence (EAAI-18), New Orleans, Louisiana, USA, 2–7 February 2018, pp. 2588–2595 (2018). https://www.aaai.org/ocs/index.php/AAAI/AAAI18/paper/view/16069
21. Yin, W., Schütze, H.: Multichannel variable-size convolution for sentence classification. CoRR abs/1603.04513 (2016). http://arxiv.org/abs/1603.04513
22. Yu, X., et al.: Personalized entity recommendation: a heterogeneous information network approach. In: WSDM 2014, New York, NY, USA, 24–28 February 2014, pp. 283–292 (2014)

23. Zhang, F., Yuan, N.J., Lian, D., Xie, X., Ma, W.: Collaborative knowledge base embedding for recommender systems. In: Proceedings of the 22nd ACM SIGKDD Francisco, CA, USA, 13–17 August 2016, pp. 353–362 (2016)
24. Zhou, G., et al.: Deep interest evolution network for click-through rate prediction. In: The Thirty-Third AAAI Conference on Artificial Intelligence, AAAI 2019, The Thirty-First Innovative Applications of Artificial Intelligence Conference, IAAI 2019, The Ninth AAAI Symposium on Educational Advances in Artificial Intelligence, EAAI 2019, Honolulu, Hawaii, USA, 27 January–1 February 2019, pp. 5941–5948 (2019). https://doi.org/10.1609/aaai.v33i01.33015941
25. Zhou, G., et al.: Deep interest network for click-through rate prediction. In: Proceedings of the 24th ACM SIGKDD International Conference on Knowledge Discovery & Data Mining, KDD 2018, London, UK, 19–23 August 2018, pp. 1059–1068 (2018). https://doi.org/10.1145/3219819.3219823

JUST-BPR: Identify Implicit Friends with Jump and Stay for Social Recommendation

Runsheng Wang[1], Min Gao[1(✉)], Junwei Zhang[1], and Quanwu Zhao[2,3]

[1] School of Big Data and Software Engineering, Chongqing University, Chongqing, China
{wangruns,gaomin,jw.zhang}@cqu.edu.cn

[2] School of Economics and Business Administration, Chongqing University, Chongqing, China
zhaoquanwumx@cqu.edu.cn

[3] Chongqing Key Laboratory of Logistics, Chongqing University, Chongqing, China

Abstract. Recommender constantly suffers from the problems of data sparsity and cold-start. As suggested by social theories, people often alter their ways of behaving and thinking to cater to social environments, especially their friends. For this reason, prior studies have integrated social relations into recommender systems to help infer user preference when there are few available data, which is known as social recommendation. However, explicit social relations are also sparse and meanwhile are usually noisy. To enhance social recommendation, a few studies identify more reliable implicit relations for each user over the user-item and user-user networks. Among these research efforts, meta-paths guided search shows state-of-the-art performance. However, designing meta-paths requires prior knowledge from domain experts, which may hinder the applicability of this line of research. In this work, we propose a novel social recommendation model (**JUST-BPR**) with a meta-path-free strategy to search for implicit friends. Concretely, We adopt the idea of 'Jump and Stay', which is a heterogeneous random walk technique, to social recommendation. Based on this idea, we manage to bypass the design of meta-paths and obtain high-quality implicit relations in a more efficient way. Then we integrate these implicit relations into an augmented social Bayesian Personalized Ranking model for top-N recommendation. Experiments on two real-world datasets show the superiority of the proposed method and demonstrate the differences between implicit friends discovered by meta-paths and JUST-BPR, respectively.

Keywords: Recommender systems · Social networks · Heterogeneous networks · Random walk

1 Introduction

Internet users are often overwhelmed by a mass amount of information online. To cope with this issue, recommender systems have emerged to capture the

H. Yang et al. (Eds.): ICONIP 2020, LNCS 12534, pp. 455–466, 2020.
https://doi.org/10.1007/978-3-030-63836-8_38

user's preferences. The task of the recommender system is to predict a rating of user u on item i or to generate a personalized recommendation list for the given user u [4]. Despite the effectiveness, traditional recommender systems that only leverage user-item interaction records suffer from the data sparsity and cold-start issues because users can only consume a tiny proportion of millions of products (i.e., the density of the user-item rating matrix in commercial recommender systems is often less than 1% [7]). Therefore, there is a need to go beyond user-item ratings and incorporate more auxiliary information to improve recommender systems.

As suggested by social science theories [1], people often alter their ways of behaving and thinking to cater to social environments, especially their friends. That means, users' friends may influence and shape their personal tastes. For this reason, prior studies incorporate social relations into recommender systems [4,6,7,14,20] to alleviate the above two issues, which is known as social recommendation. As the complementary information of user-item interaction behaviors, social relations can improve the performance of recommender systems. However, new issues are along with this new paradigm of recommender system. Those are, people often seek their friends' advice before making decisions in real-life and in the social network. The difference between the two cases is that we know which friends we can trust in real-life, while it is difficult for the social recommender system to know who is trusted more by the target user in the social network since social relations are often noisy because of the openness nature of the social network. Besides, a few studies reveal that the cold-start users are 'cold' in both user-item interactions and social networks [19]. As a result, these user can hardly benefit from the explicit social relations. To make better use of social relations, some meta-paths guided methods [16,17] are proposed to discover implicit and reliable friends to enhance social recommendation. However, in the context of a heterogeneous information network composed of user-item interactions and user-user social relations, using meta-paths guided random walks requires strong prior knowledge from domain experts [2] to design a set of meaningful meta-path, which hinders the applicability of this line of research.

Inspired by the idea of 'Jump and Stay' [3], in this paper, we develop a social Bayesian Personalized Ranking model for social recommendation, named **JUST-BPR**, with a meta-path-free strategy to discover implicit and reliable friends for each user. For a better social recommendation performance, we face the following challenges: (1) how to utilize the idea of 'Jump and Stay' to discover implicit and reliable social relations; (2) how to integrate implicit and reliable friends discovered without meta-paths into the recommendation process. To cope with these two challenges, JUST-BPR is designed with two stages: searching for implicit relations and generating social recommendations, and mainly includes the following five components: (1) a component which constructs user-item and user-user heterogeneous networks with positive feedback and negative feedback respectively; (2) a component which obtains unfixed sequences of node types over the positive heterogeneous network and negative heterogeneous network using 'Jump and Stay'; (3) a component which learns the user/item embedding based on a heterogeneous skip-gram model; (4) a component to get the user's implicit and reliable positive friends and negative friends using user similarity; and (5) a component which generates personalized recommendation list with an item ranking model. Among these components, components (1) and (2) deal with challenge (1), while the rest deal with challenge (2).

Overall, the main contributions of this paper are summarized as follows.

- We develop a meta-path-free strategy to discover implicit relations based on the idea of 'Jump and Stay'.
- We develop a social Baysian Personalized Ranking model for social recommendation named JUST-BPR, which utilizes positive feedback and negative feedback to discover implicit and reliable friends without meta-paths guided random walks.
- Extensive experiments on real-world datasets demonstrate the feasibility of decoupling meta-paths from the process of discovering implicit and reliable friends. The results verify the effectiveness of our meta-path-free idea.

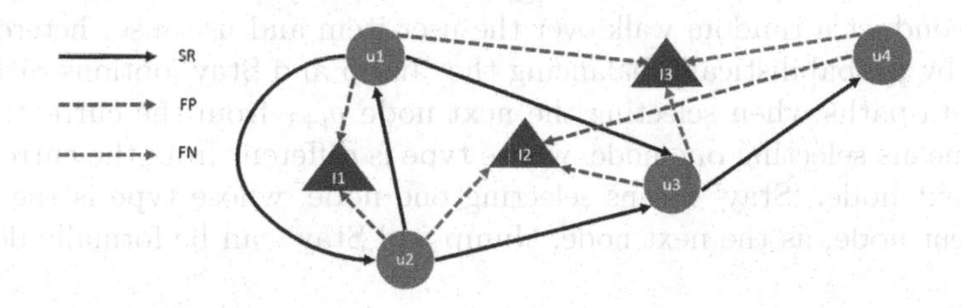

Fig. 1. Heterogeneous information network used in this paper.

2 Discover Implicit Friends with Jump and Stay for Recommendation

2.1 Preliminaries

Heterogeneous Information Network. A heterogeneous information network is defined as $G = (V, E, T)$, where V denotes the set of nodes and E denotes the set of edges. Each node $v \in V$ and each edge $e \in E$ are associated with their mapping functions $\phi(v) : v \rightarrow T_V$ and $\phi(e) : e \rightarrow T_E$, respectively. T_V and T_E denote the node types and relation types, where $|T_V| + |T_E| > 2$. For an edge $e \in E$, if it connects two nodes of different types, it is called heterogeneous edge. Otherwise, it is called homogeneous edge.

Positive Feedback and Negative Feedback. Positive feedback $P_u = \{(u, i)\}$ refers to a set of user-item pairs in which user u shows affection to item i. Negative feedback $N_u = \{(u, j)\}$ refers to a set of user-item pairs in which user u dislikes item j (e.g. giving low ratings).

Implicit Friends. Implicit friends refer to a pair of users who have similar tastes or preferences but are not necessarily connected to each other in the social networks.

Let U denote the user set, and I denote the item set. Figure 1 is an example of a heterogeneous information network composed of user-item interaction records with positive feedback FP and negative feedback FN and user-user social relations SR, which includes two types of nodes and three types of edges. In other words, $|T_V| = 2$ and $|T_E| = 3$.

2.2 Generating Node Corpora with Jump and Stay

Meta-paths [12] are extensively used in heterogeneous mining tasks. As the data in social recommender systems can be organized as a heterogeneous network, meta-paths have also been introduced to social recommendation to discover implicit relations [16,17]. However, designing meta-paths requires prior knowledge from domain experts, which hinder the applicability of meta-paths-based work. Inspired by heterogeneous graph embedding with the 'Jump and Stay' [3], we propose a meta-paths-free strategy to discover implicit friends. Concretely, we first conduct a random walk over the user-item and user-user heterogeneous network by probabilistically balancing the 'Jump and Stay' options rather than using meta-paths when selecting the next node v_{i+1} from the current node v_i. 'Jump' means selecting one node, whose type is different from the current node, as the next node. 'Stay' means selecting one node, whose type is the same as the current node, as the next node. 'Jump and Stay' can be formally defined as follows.

- **Jump** to a target domain $q \in T_V$: choosing one of those nodes in the target domain q, which is connected to the current node v_i with a heterogeneous edge, as the next node v_{i+1}.
- **Stay** in the current domain: choosing one of those nodes in the current domain, which is connected to the current node v_i with a homogeneous edge, as the next node v_{i+1}.

After defining 'Jump and Stay', we can probabilistically control the random walk using two steps rather than meta-paths. The first step is to determine whether to jump or stay. If the jump decision is made, the second step is to decide which domain to jump. Otherwise, we will stay in the current domain. Given a node v_i, the probability of staying in the current domain is listed as follows.

$$P_{stay}(v_i) = \begin{cases} 1, & \text{if } V_{jump}(v_i) = \varnothing \\ 0, & \text{if } V_{stay}(v_i) = \varnothing \\ \alpha^l, & \text{otherwise ,} \end{cases} \tag{1}$$

where $V_{jump}(v_i)$ and $V_{stay}(v_i)$ denote a set of nodes, which are connected to the current node v_i with heterogeneous edges and homogeneous edges, respectively. α^l denotes an exponential decay function, where $\alpha \in [0, 1]$ is the initial stay probability, and l denotes the number of nodes that consecutively visited in the same domain of the current node v_i. Using this method, the probability function P_{stay} can not only help us to balance the number of heterogeneous edges and homogeneous edges but also help us to avoid staying too long in the same domain [3]. However, according to [10,16], if the next node v_{i+1} has more shared neighbors with the current node v_i, the connection strength will be more reliable. To this end, we define the transition probability from the current node v_i to the node v_j^q in the next domain q as follows:

$$P_{trans}(v_{i+1}^q|v_i) = \frac{N(v_i) \cap N(v_{i+1}^q)}{N(v_i)}, \tag{2}$$

where $N(v_i)$ denotes the neighbors connected to the current node v_i. Using this method, we can not only balance the number of heterogeneous edges and homogeneous edges, but also take personalized jump of nodes into account. The next step is how to balance the variety of nodes, i.e., the node distribution over different node types.

As mentioned in [3], we also use a fixed-length queue Q_{hist} to memorize the node types, which have been visited recently. When a jump decision is made at the current node v_i, the next target domain q will be selected as follows:

$$q = \begin{cases} q|q \in T_V \wedge q \neq T_{v_i}, & \text{if } Q_{hist} = \varnothing \\ q|q \in T_V \wedge q \notin Q_{hist}, & \text{otherwise} \end{cases}, \tag{3}$$

where T_{v_i} denotes the type of the current node v_i. Eventually, utilizing the idea of 'Jump and Stay', positive node corpora C_p and negative node corpora C_n will be generated from positive heterogeneous information network and negative heterogeneous information network, respectively. The node corpora will be used to generate the representation of user preferences in the next section.

2.3 Node Embedding to Discover Implicit Friends

After the node corpora is generated, i.e., C_p, we learn the user/item embedding based on a heterogeneous Skip-Gram model [2], which is a variant of word2vec [8,9]. The model tries to maximize the network probability in terms of local structure in a random walk. Formally, given a network $G = (V, E)$, node v_i, the objective function is

$$\arg\max_{\Theta} \sum_{v_i \in V} \sum_{t \in T_V} \sum_{v_j \in C_t(v_i)} p(v_j|v_i; \theta), \tag{4}$$

where $C_t(v_i)$ denotes the context of v_i with window size w, and $p(v_j|v_i;\theta)$ is the conditional probability of having a context node v_j for a given node v_i. The probability is commonly defined as a softmax function:

$$p(v_j|v_i; \theta) = \frac{e^{X_{v_i} \cdot X_{v_j}}}{\sum_{u \in V} e^{X_{v_i} \cdot X_{v_u}}}, \tag{5}$$

where X_{v_i} represents the embedding vector of node v_i. Moreover, to achieve efficient optimization, negative sampling [9] is adopted. Given a node v_i and a small set of negative nodes with size M, Eq. 4 is updated by maximize the following objective function:

$$\log \sigma\left(X_{v_i} \cdot X_{v_j}\right) + \sum_{m=1}^{M} \mathbb{E}_{v_m \sim P(v)} \left[\log \sigma\left(-X_{v_m} \cdot X_{v_i}\right)\right], \tag{6}$$

where $\sigma(x) = \frac{1}{1+e^{-x}}$, and $P(v)$ is the pre-defined sampling distribution determined by the node degree. Using this method, the user embedding can be learned, and then we identify top-K implicit friends for each user by computing the cosine similarity of user embeddings over C_p and C_n, respectively. Then we can get K positive friends $PF(u)$ and K negative friends $NF(u)$.

2.4 JUST-BPR: BPR with Jump and Stay

After obtaining $PF(u)$ and $NF(u)$ for each user, we find that these two sets share an overlapped part. We name the users in the joint set *perfect friends* since they have some shared preferences on both postive and negative feedback. Then we augment the assumption of social Bayesian Personalized Ranking [20], which assumes that the items consumed by the current user u should be ranked higher than items consumed by u's friends, and higher than items neither consumed by u nor u's friends, to that

$$x_{ui} \succeq x_{up} \succeq x_{uf} \succeq x_{uu} \succeq x_{un}, \tag{7}$$

where x_{ui} denotes items which are consumed and given a positive feedback by u, x_{up} denotes items which are consumed and given a positive feedback by u's perfect friends, x_{uf} denotes items which are consumed and given a positive feedback by u's positive friends who share similar preferences over C_p, x_{uu} denotes items have not been consumed by u, and x_{un} denotes items which are consumed and given a negative feedback by either u or u's negative friends who share similar preferences over C_n. $x_{ui} \cup x_{up} \cup x_{uf} \cup x_{uu} \cup x_{un} = I$, and they are disjoint with each other. Therefore, the likelihood function of our recommendation model JUST-BPR can be formulated as follows:

$$\begin{aligned} \mathcal{L}(\Theta) = \prod_{u \in U} \Big(& \prod_{i \in x_{ui}, j \in x_{up}} Pr[i \succeq j] \prod_{j \in x_{up}, k \in x_{uf}} Pr[j \succeq k] \\ & \prod_{k \in x_{uf}, l \in x_{uu}} Pr[k \succeq l] \prod_{l \in x_{uu}, m \in x_{un}} Pr[l \succeq m] \Big). \end{aligned} \tag{8}$$

Finally, a local minimum can be obtained through the gradient descent technique.

3 Experiments

In this section, we conduct extensive experiments on two real-world datasets to evaluate the effectiveness of the proposed method JUST-BPR.

3.1 Experimental Setup

Datasets. Table 1 summarizes the main statistics of the two datasets used in our experiments. The LastFM[1] is the collection of listening records. The songs, which interacted with the current user only once, are treated as negative feedback. For Douban[2] with a rating scale from 1 to 5, positive feedback and negative feedback contain the ratings of 4 and 5 and the ratings of 1 and 2, respectively. It should be noted that we use 80% of the data as the training set and the remaining 20% as the testing set. The experiments are conducted with five-fold cross-validation for ten times, and the average results are reported.

Table 1. Statistics of the datasets.

Datasets	#Users	#Items	#Feedbacks	Density	#Relations
LastFM	1,892	17,632	92,834	0.278%	25,434
Douban	2,831	15,918	636,436	1.412%	35,624

Baselines. To demonstrate the superiority of the JUST-BPR, we compare our approach with the following methods:

- **BPR:** It is a classic Bayesian personalized ranking method proposed in [11]. It only uses the user-item ratings for recommendation with the assumption that users tend to assign higher ranks to items that they have observed.
- **SBPR:** This model is a social Bayesian personalized ranking method proposed in [20]. It assumes that users are more likely to prefer items that their friends have consumed.
- **IF-BPR:** It is a implicit relation-based social recommendation method proposed in [16]. It uses meta-paths guided random walks over the heterogeneous network to identify users' implicit friends.
- **RSGAN:** It is an adversarial social recommendation method proposed in [17]. It relies on meta-paths guided walks to obtain seeded friends as well.

Metrics. We use one relevance-based metric (*Recall*@10) and two ranking-based metrics (*MAP*@10 and *NDCG*@10) to evaluate the recommendation results of JUST-BRP and the baselines.

3.2 Recommendation Performance

Table 2 demonstrates the performance of different recommendation methods on two real-world datasets. We can draw the following conclusions from the table.

[1] https://www.last.fm.
[2] https://www.douban.com.

First, our proposed method outperforms all the compared baselines on both datasets. Second, we can observe that the proposed method achieves better performance than SBPR and BPR, which only leverages explicit social relations and user-item interactions, respectively. It shows the importance of taking implicit social information into account. Third, according to IF-BPR and RSGAN, we can find that the proposed method can achieve similar, even better, recommendation results in most cases without using meta-paths guided random walks. Namely, utilizing the idea of 'Jump and Stay' to discover implicit and reliable friends for social recommender system is useful.

Table 2. Statistics of the datasets.

Dataset	Metric	BPR	SBPR	IF-BPR	RSGAN	JUST-BPR	Improv
LastFM	Recall@10	0.1270	0.1322	0.1441	0.1364	0.1543	7.064%
	MAP@10	0.0559	0.0594	0.0653	0.0605	0.0710	8.710%
	NDCG@10	0.1220	0.1269	0.1394	0.1307	0.1496	7.329%
Douban	Recall@10	0.0494	0.0536	0.0568	0.0543	0.0622	9.398%
	MAP@10	0.0870	0.0908	0.1056	0.0969	0.1074	1.630%
	NDCG@10	0.1658	0.1721	0.1923	0.1805	0.1959	1.895%

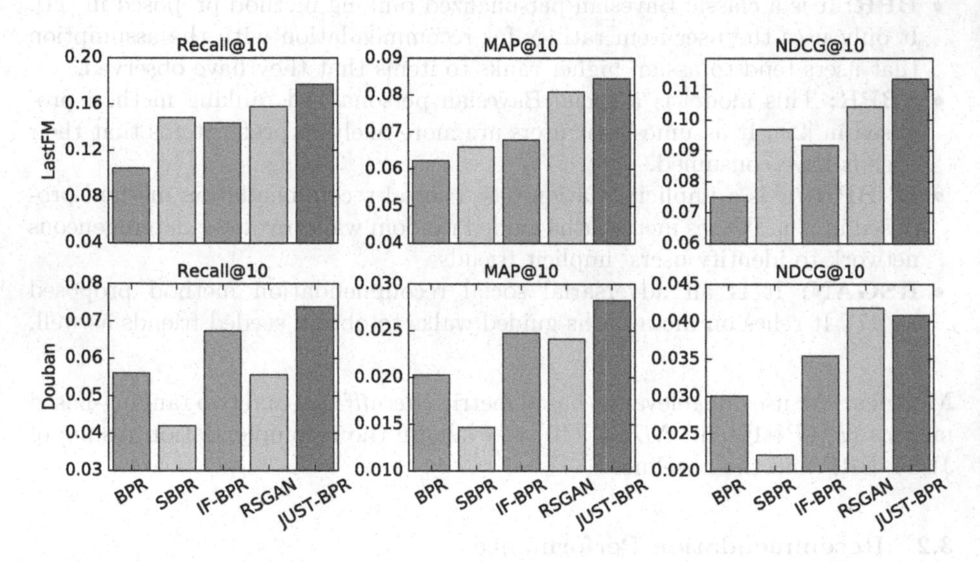

Fig. 2. Performance evaluations on cold-start users.

As mentioned in the beginning, one of the biggest challenges in the recommender system is to recommend items for cold-start users. They have very few

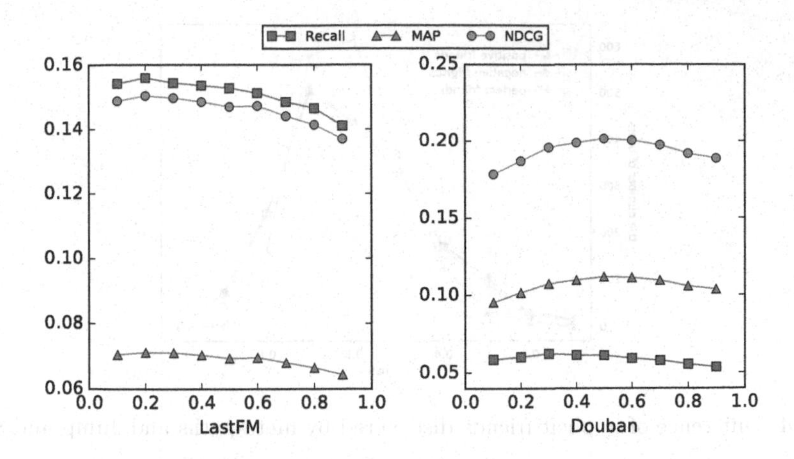

Fig. 3. Performance impact of different stay probability α.

interaction records with the items. To compare our method with the other methods, we first select users whose interaction records are less than ten as cold-start users and then conduct experiments on them. As shown in Fig. 2, we can see that JUST-BPR performs better than other methods. Interestingly, we notice that SBPR shows better results than BPR on LastFM dataset, but is inferior to BPR on Douban dataset. This confirms the issue that using explicit social relations is affected by noise again.

3.3 Parameter Effect Analysis

In this part, we investigate the impact of the stay probability α for the recommendation. It helps us to balance the number of heterogeneous edges and homogeneous edges in the process of identifying implicit friends. We show the impact of α on the recommendation performance in Fig. 3. In this experiment, we tune α in the range of [0.1, 0.9] with a step of 0.1 to report the corresponding performance. We can observe that the proposed method JUST-BPR performs best when α lies in the range [0.2, 0.4] on LastFM and [0.3, 0.5] on Douban. We can draw that heterogeneous edges play the primary role in the process of discovering implicit friends. When α is too large, the random walk will stay in the current domain with high probability, and a homogeneous edge will be selected. To this end, α is set to 0.3 in this experiment.

3.4 Implicit Friends Identified by Jump and Stay vs. Meta-Paths

In this part, we will explore why the model is useful. Specifically, we explore the difference between implicit friends found by meta-paths and those discovered by JUST-BPR. It is natural for us to ask whether the implicit friends discovered by meta-paths and those discovered by 'Jump and Stay' overlap with each other. To explore this, we extract implicit and reliable friends with meta-paths used

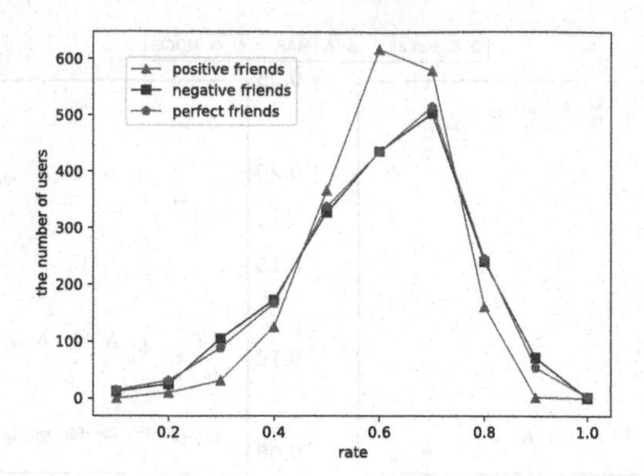

Fig. 4. Difference of implicit friends discovered by meta-paths and Jump and Stay.

in [16,17] and 'Jump and Stay' on LastFM, respectively. As expected, we observe that more than 56% of implicit friends discovered by 'Jump and Stay' appear in those discovered by meta-paths. Figure 4 illustrates that more than 56% of implicit positive friends discovered by Jump and Stay appear in those discovered by meta-paths, more than 51% of implicit negative friends discovered by 'Jump and Stay' appear in those discovered by meta-paths, and more than 55% of implicit perfect friends discovered by 'Jump and Stay' appear in those discovered by meta-paths. It shows that using the idea of 'Jump and Stay' can also identify implicit reliable friends without meta-paths guided random walk.

4 Related Work

The early exploration of social recommendation mainly focuses on the incorporation of explicit social relations. Owing to the flexibility of integrating prior knowledge, matrix factorization (MF) [4] is the most used model during this period. The representative work of MF-based social recommendation models can be categorized into three groups: Co-factorization methods [7], Ensemble methods [6], and regularization methods [4]. Besides, some studies also investigate the effect of social relations in item ranking [20] and user exposure [13]. With the boom of deep learning, deep neural networks have also been introduced to social recommendation. A few models which are based on deep forward networks [17] and graph neural networks [15,18] were proposed successively.

After the negative findings of directly using explicit social relations were reported, researchers then shift attention to exploring implicit social relations. Ma *et al.* [5] are the first to investigate the effect of implicit relations. In their work, rating information is used to identify similar users as implicit friends, but if two users do not share common items, the work is inapplicable. Zhang *et al.* [19] use network embedding to deal with the problem that neighborhood-based methods cannot connect users without common neighbors, and enable distant similar users identification. Yu *et al.* [16] adopt meta-paths to social recommendation and use specific meta-paths to guide the random walk-based search of implicit relations, which shows state-of-the-art performance. However, manually designing meta-paths requires expert knowledge and hence their work cannot be generalized to new situations. In this paper, we fill this gap and propose 'Jump and Stay' to conduct random walk, which has a better applicability.

5 Conclusion

This paper aims to utilize both positive and negative feedback to find implicit and reliable friends for social recommender system using the idea of 'Jump and Stay' instead of a set of meaningful meta-paths. In our method, we first generate a sequence of nodes from the heterogeneous network composed of user-item interactions and user-user social relations. Secondly, we map each user node into the latent feature space by word embedding technology and find top-k implicit friends according to the similarity between each node. Finally, the implicit relations are integrated into the social Bayesian personalized ranking method. The experimental results on two real-world datasets show the effectiveness of the proposed method. Moreover, we compare the difference between implicit and reliable friends, identified by meta-paths guided random walks and JUST-BPR.

Acknowledgements. This study was supported by the National Key Research and Development Program of China (2018YFF0214706), the Natural Science Foundation of Chongqing, China (cstc2020jcyj-msxmX0690), the Graduate Student Research Innovation Project (CYS19052), the Fundamental Research Funds for the Central Universities of Chongqing University (2020CDJ-LHZZ-039), the Key Research Program of Chongqing Technology Innovation and Application Development (cstc2019jscx-fxydX0012), and the Key Research Program of Chongqing Science & Technology Commission(cstc2019jscx-zdztzxX0031).

References

1. Cialdini, R.B., Goldstein, N.J.: Social influence: compliance and conformity. Annu. Rev. Psychol. **55**, 591–621 (2004)
2. Dong, Y., Chawla, N.V., Swami, A.: metapath2vec: scalable representation learning for heterogeneous networks. In: Proceedings of the 23rd ACM SIGKDD International Conference on Knowledge Discovery and Data Mining, pp. 135–144. ACM (2017)
3. Hussein, R., Yang, D., Cudré-Mauroux, P.: Are meta-paths necessary?: Revisiting heterogeneous graph embeddings. In: Proceedings of the 27th ACM International Conference on Information and Knowledge Management, pp. 437–446. ACM (2018)
4. Jamali, M., Ester, M.: A matrix factorization technique with trust propagation for recommendation in social networks. In: Proceedings of the Fourth ACM Conference on Recommender Systems, pp. 135–142. ACM (2010)
5. Ma, H.: An experimental study on implicit social recommendation. In: International ACM SIGIR Conference on Research and Development in Information Retrieval, pp. 73–82 (2013)
6. Ma, H., King, I., Lyu, M.R.: Learning to recommend with social trust ensemble. In: Proceedings of the 32nd International ACM SIGIR Conference on Research and Development in Information Retrieval, pp. 203–210. ACM (2009)

7. Ma, H., Yang, H., Lyu, M.R., King, I.: SoRec: social recommendation using probabilistic matrix factorization. In: Proceedings of the 17th ACM Conference on Information and Knowledge Management, pp. 931–940. ACM (2008)
8. Mikolov, T., Chen, K., Corrado, G., Dean, J.: Efficient estimation of word representations in vector space. arXiv preprint arXiv:1301.3781 (2013)
9. Mikolov, T., Sutskever, I., Chen, K., Corrado, G.S., Dean, J.: Distributed representations of words and phrases and their compositionality. In: Advances in Neural Information Processing Systems, pp. 3111–3119 (2013)
10. Onnela, J.P., et al.: Structure and tie strengths in mobile communication networks. Proc. Nat. Acad. Sci. **104**(18), 7332–7336 (2007)
11. Rendle, S., Freudenthaler, C., Gantner, Z., Schmidt-Thieme, L.: BPR: Bayesian personalized ranking from implicit feedback. In: Proceedings of the Twenty-Fifth Conference on Uncertainty in Artificial Intelligence, pp. 452–461. AUAI Press (2009)
12. Sun, Y., Han, J., Yan, X., Yu, P.S., Wu, T.: PathSim: meta path-based top-K similarity search in heterogeneous information networks. Proc. VLDB Endowment **4**(11), 992–1003 (2011)
13. Wang, M., Zheng, X., Yang, Y., Zhang, K.: Collaborative filtering with social exposure: a modular approach to social recommendation. arXiv preprint arXiv:1711.11458 (2017)
14. Wang, X., Hoi, S.C., Ester, M., Bu, J., Chen, C.: Learning personalized preference of strong and weak ties for social recommendation. In: Proceedings of the 26th International Conference on World Wide Web, International World Wide Web Conferences Steering Committee, pp. 1601–1610 (2017)
15. Wu, L., Sun, P., Hong, R., Fu, Y., Wang, X., Wang, M.: SocialGCN: an efficient graph convolutional network based model for social recommendation. arXiv preprint arXiv:1811.02815 (2018)
16. Yu, J., Gao, M., Li, J., Yin, H., Liu, H.: Adaptive implicit friends identification over heterogeneous network for social recommendation. In: Proceedings of the 27th ACM International Conference on Information and Knowledge Management, pp. 357–366. ACM (2018)
17. Yu, J., Gao, M., Yin, H., Li, J., Gao, C., Wang, Q.: Generating reliable friends via adversarial training to improve social recommendation. In: 2019 IEEE International Conference on Data Mining (ICDM), pp. 768–777. IEEE (2019)
18. Yu, J., Yin, H., Li, J., Gao, M., Huang, Z., Cui, L.: Enhance social recommendation with adversarial graph convolutional networks. arXiv preprint arXiv:2004.02340 (2020)
19. Zhang, C., Yu, L., Wang, Y., Shah, C., Zhang, X.: Collaborative user network embedding for social recommender systems. In: Proceedings of the 2017 SIAM International Conference on Data Mining, pp. 381–389. SIAM (2017)
20. Zhao, T., McAuley, J., King, I.: Leveraging social connections to improve personalized ranking for collaborative filtering. In: Proceedings of the 23rd ACM International Conference on Conference on Information and Knowledge Management, pp. 261–270. ACM (2014)

Leveraging Knowledge Context Information to Enhance Personalized Recommendation

Jiong Wang[1,2], Yingshuai Kou[1,2], Yifei Zhang[1,2], Neng Gao[1(✉)], and ChenYang Tu[1]

[1] State Key of Laboratory of Information Security, Institute of Information Engineering, Chinese Academy of Sciences, Beijing, China
{wangjiong,kouyingshuai,zhangyifei,gaoneng,tuchenyang}@iie.ac.cn
[2] School of Cyber Security, University of Chinese Academy of Sciences, Beijing, China

Abstract. Knowledge graphs (KGs) have proven to be effective to improve the performance of recommendation. However, with the tremendous increase of users and items, existing methods still face several challenging problems: (1) path-based methods rely heavily on manually designed meta-path; (2) embedding-based methods lack sufficient considerations of user personality. To overcome the shortcomings of previous works, we propose a novel model, named **KCER**, short for leveraging **K**nowledge **C**ontext to **E**nhance **R**ecommendation. Firstly, KCER generates the representation of knowledge context associating with specific user-item pairs. Then to obtain enriched user representations, we leverage a gated attention network to extracted meaningful information from the associated knowledge context and user dedicated ID embedding. We conduct extensive experiments on three real-world datasets to evaluate the model. The experimental results show the superiority of KCER compared with other state-of-the-art methods.

Keywords: Recommender system · Collaborative filtering · Knowledge graph · Gated attention network

1 Introduction

With the development of the internet, information overload is one of the dilemmas we confront with [8]. Recommender system (RS) is an effective means to solve this problem as it can help select the appropriate data to satisfy the users' needs. However, with the tremendous increase of users and items, traditional methods are difficult to cope with the highly sparse data [17]. Recently, many efforts have been devoted to taking advantage of the knowledge graph (KG) to improve the performance of RS [18]. In the KG, entities (nodes) are connected by various relationships (edges) to form a heterogeneous network [11]. In this way, KG can strengthen the connection between items, which helps to explore users'

H. Yang et al. (Eds.): ICONIP 2020, LNCS 12534, pp. 467–478, 2020.
https://doi.org/10.1007/978-3-030-63836-8_39

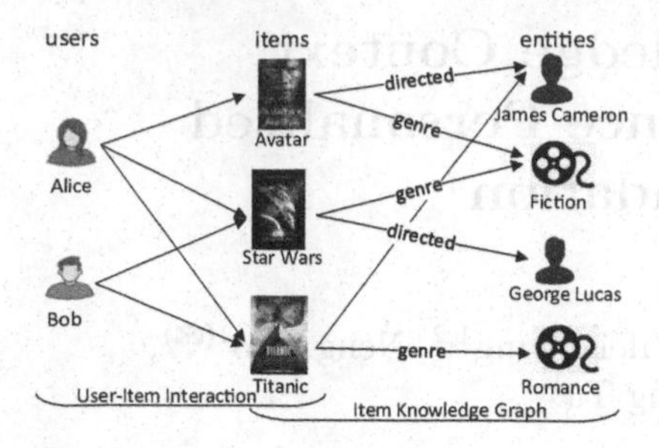

Fig. 1. Illustration of knowledge graph enhanced movie recommendation system.

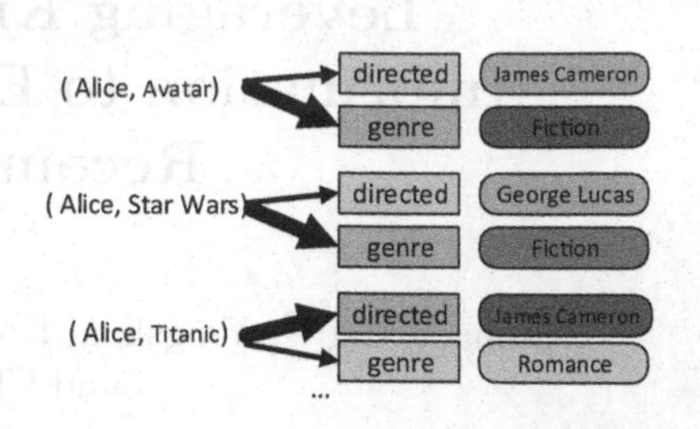

Fig. 2. A possible illustration of knowledge context effect on specific user-item pairs.

potential preferences and alleviate the cold-start problem [1]. Figure 1 shows a toy example of the knowledge graph enhanced movie recommendation system.

In the past years, many approaches have been studied [2]. The mainstream methods can be divided into two categories. The first is path-based methods, which adopt hand-craft meta-paths to model complex relations between users and items. This approach often requires a lot of manual experience. The other category is embedding-based methods, which aim at incorporating the knowledge graph embedding into the traditional recommendation framework. For example, Deep Knowledge-Aware Network (DKN) [12] is a convolutional neural network (CNN) based framework to make news recommendations in which the information of news is enriched by associated entities in the knowledge graph. RippleNet [11] aims at item recommendation in which users' interests are enriched by entities they have interacted with. Although embedding-based method is more efficient, simply introducing knowledge representation methods may bring some irrelevant information, that users are not interested in.

According to the above discussion, we can see that it is necessary for us to find a more effective way to use knowledge context. Figure 2 shows our key idea. We argue that users have various preferences for the auxiliary information on different items. For example, the user *Alice* pays more attention to the genre when interacting with the item *Avatar*, but pays more attention to the director when interacting with the item *Titanic*. Out of such considerations, we propose a novel recommendation model named *KCER*, short for leveraging *K*nowledge *C*ontext to *E*nhance *R*ecommendation. KCER first generates the representation of user-specific knowledge context, which can be used to infer the user's potential interests. Then, to capture the meaningful information from the knowledge context, a gated attention network is utilized to adaptively fuse the knowledge context and users' dedicated information.

In sum, our main contributions are outlined as follows:

– We propose a novel deep collaborative filtering model named KCER for item recommendation. It takes both the user-item interaction and item-based

knowledge information into account and is able to improve the performance of recommendation.
- KCER explicitly generates the representation of knowledge context associated with the specific user-item pairs and further adopts it to enhance user final representation via a gated attention network.
- We conduct extensive experiments on three real-world datasets to demonstrate the effectiveness of our proposed model.

2 Related Work

There is a rich line of research on item recommendation. Traditional recommender system approaches mainly focused on user-item interaction, including explicit and implicit feedback. Most recently, the exploration of deep neural network on recommendation has attracted the attention of researchers. There are two main ways to use the knowledge graph to enhance recommendation.

Path-based methods explores various types of paths to describe the associations between users and items. HeteRec [15] first introduces meta-paths to leverage the heterogeneous information for personalized recommendation. Limited to the expressive power of a single path, FMG [20] first fuses the heterogeneous information via meta-graph to generate recommendation results. Recently, RKGE [9] learns semantic representations of both entities and paths between entities for characterizing user preferences towards items. Compared with path-based method, our model is more flexible, since it does not need to design handcrafted meta-paths.

Embedding-based methods utilize KGE algorithms to introduce structural information in the knowledge graph. CKE [16] leverages structure, textual and visual information from the knowledge base for recommendation. Deep knowledge-aware networks (DKN) [12] incorporates knowledge entity embeddings into a CNN framework for news recommendation. RippleNet [11] explores users' potential interests by propagating the preferences on the item-based knowledge graph. AKUPM [10] figures out the most related part of the incorporated entities based on depicting intra-entity-interaction and inter-entity-interaction among entities. Recently, graph neural networks (GNN) [4] are used to explore the complex interaction between users and items. For instance, KGCN [13] adopts GNN to capture inter-item relatedness effectively, i.e., mining their associated attributes on the KG. The major difference between these method and our KCER method is that KCER utilizes the combination information of different user-item pairs to extract related information from the knowledge context and followed with a deep fusion network, which is beneficial for making final recommendation.

3 Task Formulation

In order to introduce our recommendation method more clearly, we will give a detailed explanation of the notations and target problem.

Definition 1 (User-item Interaction Graph). Given a recommendation scenario, let $\mathcal{U} = \{u_1, u_2, ..., u_M\}$ denote a set of M users and $\mathcal{V} = \{v_1, v_2, ..., v_N\}$ denote a set of N items. Here, we can represent the interaction between users and items as graph $\mathcal{G}_1$, which is defined as $\{(u_i, v_j, y_{ij}) | u_i \in \mathcal{U}, v_j \in \mathcal{V}\}$. Since we focus on the user's implicit feedback on items, the link $y_{ij} = 1$ indicates there is an observed interaction between user i and item j.

Definition 2 (Item Knowledge Graph). In addition to the interaction data, side information such as item attributes, can help distinguish different relations between items. We represent the auxiliary information in the form of knowledge graph $\mathcal{G}_2$, which consists of triples of entities and relationships. Formally, it is defined as $\{(h, r, t) | h, t \in \mathcal{E}, r \in \mathcal{R}\}$, where $\mathcal{E}, \mathcal{R}$ denote the set of entities and relations, respectively.

Task Description. We now formulate the recommendation task to be addressed in this paper:

Input: user item interaction graph $\mathcal{G}_1$ and item knowledge graph $\mathcal{G}_2$.

Output: a prediction function $\mathcal{F}(\cdot)$ that predicts the probability $\hat{y}_{ij}$ that user u_i would interact with item v_j.

4 Our Proposed Model

In this section, we will introduce the proposed KCER model, which exploits knowledge context in an end-to-end fashion. We will first present the overall architecture of KCER and then introduce the details of each layer of it. Finally, we offer the optimization procedures.

4.1 Model Overview

We present the overall architecture for the proposed model in Fig. 3. As we can see, KCER consist of four main components: (1) embedding layer, (2) knowledge context generating layer, (3) knowledge fusion layer and (4) prediction layer. We explicitly incorporate item-based auxiliary information as the knowledge context to capture user preference. To enrich the representation of users and items, a neural fusion layer is adopted. Due to the adaptive incorporation of knowledge context information, our model is able to yield better performance.

4.2 Input and Embedding Layer

Following previous work [3], we set up a lookup layer to transform the one-hot representation of users and items into low-dimensional dense vectors. Formally, given a user-item pair (u_i, v_j), we first convert each of them into a low-dimensional dense vector $\mathbf{u}_i$ and $\mathbf{v}_j$. For the associated entities and relations in the knowledge graph, we also adopt a lookup layer to transform them into low-dimensional dense vectors.

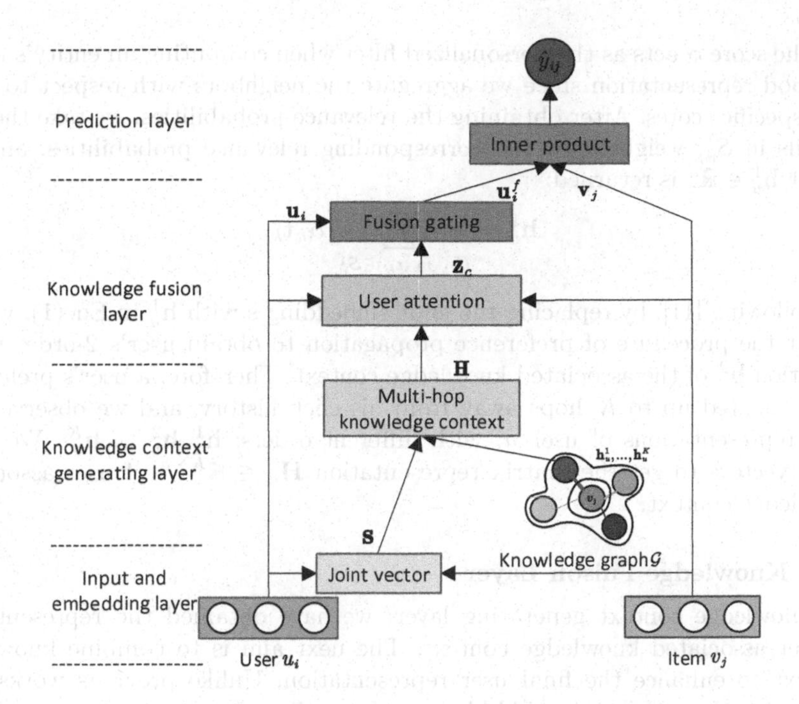

Fig. 3. The architecture of our proposed model KCER.

4.3 Knowledge Context Generating Layer

In order to model the interactions between users and items in a more fine-grained way, we first generate knowledge context representation based on the specific user-item pair. Consider a candidate pair of user u_i and item(entity) v_j. Following [11], we use $\mathcal{S}_{u_i}^1$ to denote the set of triples directly connected to user u_i's clicked history, and r_k to denote the relation between head entity h_k and tail entity t_k. We argue that users will consider relations and relation values simultaneously when interacting with the corresponding items. Therefore, inspired by [19], we implement attention score α as follows:

$$\alpha_k^* = tanh(\mathbf{s} + \mathbf{s} \odot \mathbf{r}_k + \mathbf{b})\mathbf{t}_k^T,$$
$$\alpha_k = \frac{\exp(\alpha_k^*)}{\sum_{(h_k, r_k, t_k) \in \mathcal{S}_{u_i}^1} \exp(\alpha_k^*)}, \tag{1}$$

where $\mathbf{b} \in \mathbb{R}^d$ is a global bias vector and $\odot$ denotes the element-wise product of vectors. Note that $\mathbf{s}$ is the joint embedding of specific user-item pair (u_i, v_j). We apply the following operation to transform:

$$\mathbf{s} = \frac{\mathbf{u}_i \odot \mathbf{v}_j}{\|\mathbf{u}_i\| \, \|\mathbf{v}_j\|}. \tag{2}$$

The denominator added in Eq. (2) is used to normalize and make the generated vectors have the same scale.

The score α acts as the personalized filter when computing an entity's neighborhood representation since we aggregate the neighbors with respect to these user-specific scores. After obtaining the relevance probabilities, we take the sum of tails in $\mathcal{S}^1_{u_i}$ weighted by the corresponding relevance probabilities, and the vector $\mathbf{h}^1_u \in \mathbb{R}^d$ is returned:

$$\mathbf{h}^1_u = \sum_{(h_k, r_k, t_k) \in \mathcal{S}^1_{u_i}} \alpha_k \mathbf{t}_k. \tag{3}$$

Following [11], by replacing the joint embedding $\mathbf{s}$ with $\mathbf{h}^1_u$ in Eq. (1), we can repeat the procedure of preference propagation to obtain user's 2-order representation $\mathbf{h}^2_u$ of the associated knowledge context. Therefore, a user's preference is propagated up to K hops away from his click history, and we observe multiple representations of user u_i with different orders: $\mathbf{h}^1_u, \mathbf{h}^2_u, ..., \mathbf{h}^K_u$. We stack these vectors to get the matrix representation $\mathbf{H}_u \in \mathbb{R}^{K \times d}$ of user associated knowledge context.

4.4 Knowledge Fuison Layer

By knowledge context generating layer, we have obtained the representation of user associated knowledge context. The next aim is to combine knowledge context to enhance the final user representation. Unlike previous works concatenating these two kinds of hidden representations, we propose a neural fusion layer to adaptively merge them. We first adopt the multi-dimensional attentional layer [6] to get the final vectorized representation, since the user focuses on multiple aspects of the knowledge context. The aggregation operation is computed as follows:

$$\begin{aligned} \mathbf{A} &= softmax(\mathbf{W}^1 tanh(\mathbf{W}^2 \mathbf{u}_i + \mathbf{W}^3 \mathbf{v}_j + \mathbf{W}^4 \mathbf{H}^T_u + \mathbf{b}^1) + \mathbf{b}^2), \\ \mathbf{Z}_c &= \mathbf{A}\mathbf{H}_u, \\ \mathbf{z}_c &= avg\_pooling(\mathbf{Z}_c), \end{aligned} \tag{4}$$

where $\mathbf{W}^1 \in \mathbb{R}^{d_a \times d}, \mathbf{W}^2, \mathbf{W}^3, \mathbf{W}^4 \in \mathbb{R}^{d \times d}$ is the weight matrix, $\mathbf{b}^1 \in \mathbb{R}^d, \mathbf{b}^2 \in \mathbb{R}^{d_a}$ is the bias term, and the $softmax$ is performed along the second dimension of its input. We adopt the average pooling to aggregate the context matrix representation $\mathbf{Z}_c \in \mathbb{R}^{d_a \times d}$ into a vector form, i.e., $\mathbf{z}_c \in \mathbb{R}^d$.

Then, to get the final user representation, we adopt the gating mechanism to fuse knowledge context representation $\mathbf{z}_c$ and user dedicated ID embedding $\mathbf{u}_i$. The computation is as follows:

$$\begin{aligned} \mathbf{g} &= \sigma(\mathbf{W}^{g1} \mathbf{u}_i + \mathbf{W}^{g2} \mathbf{z}_c + \mathbf{b}^g), \\ \mathbf{u}^f_i &= \mathbf{g} \odot \mathbf{u}_i + (\mathbf{1} - \mathbf{g}) \odot \mathbf{z}_c, \end{aligned} \tag{5}$$

where $\mathbf{W}^{g1}, \mathbf{W}^{g2} \in \mathbb{R}^{d \times d}$, $\mathbf{b}^g \in \mathbb{R}^d$ are the parameters in the gating layer. $\sigma(\cdot)$ denotes the *sigmoid* function. By using a gating layer, we can extract useful information from the two representations and combine them smoothly. So far, we have improved the user preference model by considering the knowledge context, and next the model (i.e., $\mathbf{u}^f_i$,$\mathbf{v}_j$) is utilized to recommend items.

4.5 Prediction Layer

In this layer, the estimated preference score $\hat{y}_{ij}$ is calculated given the user and item latent representations $\mathbf{u}_i^f, \mathbf{v}_j$. Similar to [3], we conduct the inner product operation on user and item representations to predict their matching score:

$$\hat{y}_{ij} = \sigma(\mathbf{u}_i^{f^T} \mathbf{v}_j), \tag{6}$$

where $\sigma(\cdot)$ denotes the *sigmoid* function. For each user, all items are ranked by their corresponding preference scores and the top-k lists are recommended to the user.

4.6 Learning Algorithm

The entire KCER model is trained in an end-to-end manner. Similar to [3], we adopt cross-entropy as the loss of the recommendation task. The final loss function of KCER is represented as follows:

$$\mathcal{L} = - \sum_{(i,j) \in \mathcal{Y}^+ \cup \mathcal{Y}^-} (y_{ij} \log(\hat{y}_{ij}) + (1 - y_{ij}) \log(1 - \hat{y}_{ij})) + \lambda ||\Theta||^2, \tag{7}$$

where the $\mathcal{Y}^+$ denotes all the observed interactions and $\mathcal{Y}^-$ denotes the unobserved interactions. λ is the balancing parameters. Θ denotes all the trainable parameters in the proposed KCER model. The last term is the L_2 term. To make computation more efficient, we use a negative sampling strategy during training process and n is the number of negative samples for user u_i. We use mini-batch stochastic gradient descent (SGD) algorithm to minimize this loss function.

5 Experiment and Evaluation

In this section, we conduct experiments with the aim of answering the following research questions: **RQ1:** Does our proposed model KCER outperform the state-of-the-art knowledge-based recommender methods? **RQ2:** What is the design choices of KCER on the quality of item recommendation? **RQ3:** How sensitive is our model to hyper-parameters? In what follows, we first present the experimental settings, followed by answering the above research questions.

5.1 Experimental Settings

Datasets. We evaluate our models on three public available datasets. Movielens-1M[1] contains approximately 1 million ratings and is a widely used benchmark dataset in movie recommender systems. Book-Crossing[2] collects user ratings for books from Book-Crossing community. Last.FM[3] contains listening information of approximately 2 thousand users from Last.FM online music system. The basic statistics of the three dataset are summarized in Table 1.

[1] https://grouplens.org/datasets/movielens/.
[2] http://www2.informatik.uni-freiburg.de/~cziegler/BX/.
[3] https://grouplens.org/datasets/hetrec-2011/.

Table 1. Basic statistics of the datasets.

Datasets	#Users	#Items	#Interactions	Density	#Triples
MovieLens-1M	6,036	2,347	753,772	5.3%	1,241,995
Book-Crossing	17,860	14,910	139,746	0.05%	151,500
Last.FM	1,872	3,846	42,346	0.59%	15,518

Baselines. In this subsection, we compare the proposed model KCER with the following state-of-the-art methods:

- LibFM [7]. It is a widely used feature-based factorization model in CTR scenarios. Following [11], we concatenate user ID, item ID, and the corresponding averaged entity embeddings learned from TransR [5] as input for LibFM.
- PER [14]. It is a path-based method, which generates latent feature for users and items by exploring meta-paths.
- CKE [16]. It is an embedding-based method, which leverages structural, textual and visual information from the knowledge base for recommendation. We implement CKE as CF plus a structural knowledge module in this paper.
- RippleNet [11]. It is a memory-network-like approach that propagates users' preferences on the KG for recommendation.
- KGCN [13]. It is a GCN-based model to aggregating neighbors' information, which reflects users' personalized and potential interests.

Evaluation Metrics. We randomly split each dataset into training, validation and test set in a ratio of 6:2:2 and evaluate our method in two experiment scenarios: (1) in click-through rate prediction, we use *Accuracy (ACC)* and AUC to evaluate the performance prediction; (2) in top-K recommendation, we adopt *Precision@K*, *Recall@K* to evaluate the recommendation results.

5.2 Performance Comparison (RQ1)

CTR Prediction Results. Table 2 summarizes the results of all these methods in CTR prediction. From the evaluation results, we observe that the proposed KCER model consistently outperforms all the state-of-the-art baselines. Compared with the best results of baselines, KCER achieves an improvement of 1.5%, 1.6%, 2.3% w.r.t. AUC on three datasets respectively.

We find that PER is inferior to other baselines. This is probably because the paths discovered by hand-crafted meta-paths introduce noise information, which leads to degraded model performance. CKE performs poorly on three datasets compared with other embedding-based methods. One possible reason is that CKE's introduction of KGE is too simple and lacks considerations of sufficient user personalization. KGCN and RippleNet are the most competitive ones, since they can adaptively mine higher-order information in the knowledge graph.

Table 2. Evaluation of CTR prediction.

Methods	Movielens-1M		Book-Crossing		Last.FM	
	AUC	*ACC*	*AUC*	*ACC*	*AUC*	*ACC*
LibFM	0.906	0.828	0.691	0.713	0.785	0.729
PER	0.711	0.665	0.634	0.589	0.645	0.604
CKE	0.801	0.739	0.678	0.640	0.743	0.672
RippleNet	0.913	0.831	0.729	0.661	0.767	0.694
KGCN	0.910	0.828	0.727	0.660	0.783	0.715
KCER	**0.927**	**0.839**	**0.741**	**0.693**	**0.803**	**0.744**

Top-K Item Recommendation Results. We test the top-K (K = 2 to 20) item recommendations on Movielens-1M and Book-Crossing dataset in Fig. 4. Several observations stand out:

- Path-based method, such as PER, performs comparably poorly than other baselines. One possible reason is that PER depends heavily on the quality of meta-paths, which require extensive domain knowledge to define. Unreasonable paths may easily introduce noise.
- FM method outperforms CKE by a large margin. This is because the cross features in FM actually serve as the second-order connectivity between users and entities. However, CKE models connectivity on the granularity of triples, which neglects the high-order connectivity.
- KCER achieves the best performance in general and obtains high improvements over the state-of-the-art methods in both metrics of *Precision@K* and *Recall@K*. In particular, KCER improves over the strongest baseline w.r.t. *Recall@20* by 1.3% and 7.9% in Movielens-1M and Book-Crossing. Moreover, compared to the recent method RippleNet, KCER adopts attention mechanism to distinguish the importance of different orders in KGs, rather than simple concatenation operation used in RippleNet.

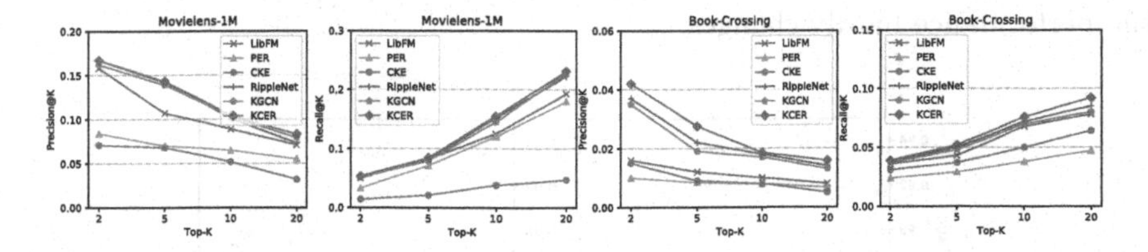

Fig. 4. Evaluation of top-K item recommendation, where K ranges from 2 to 20 on Movielens-1M and Book-Crossing datasets.

5.3 Ablation Analysis (RQ2)

Since KCER is composed of several important design decisions, we next present an ablation analysis to study the impacts of these decisions on recommendation quality. Table 3 shows the ablation analysis results. We use KCER-p to present that we generate the knowledge context representation without considering user-specific preference, i.e., replacing **s** with the embedding of the item. KCER-f indicates we replace the gated fusion with concatenation operator. From the results shown in Table 3, we have some observations. KCER performs better than KCER-p, because the joint embedding can help to extract information associated with user-specific preference from knowledge context representation. KCER performs better than KCER-f, because the gated attention network can adaptively fuse the knowledge context and user's intrinsic information.

Table 3. The ablation analysis on Movielens-1M and Book-Crossing datasets.

Methods	Movielens-1M		Book-Crossing	
	AUC	ACC	AUC	ACC
KCER	**0.927**	**0.839**	**0.740**	**0.693**
KCER-p	0.919	0.830	0.728	0.683
KCER-f	0.912	0.824	0.733	0.688

5.4 Sensitivity Analysis of Hyper-parameters (RQ3)

In this section, we investigate the influence of parameters d and λ in KCER. We vary d from 10 to 150 and search the λ from $\{10^{-7}, 10^{-6}, 10^{-5}, 10^{-4}, 10^{-3}\}$, respectively, while keeping other parameters fixed. The results of AUC on Movielens-1m are presented in Fig. 5. We observe that, with the increase of d, the performance is boosted at first since embeddings with a large dimension can encoder more useful information. Small λ can improve our model and KCER reaches its best when λ is set to 10^{-6}. When the value of λ continues to decrease, the performance hardly changes.

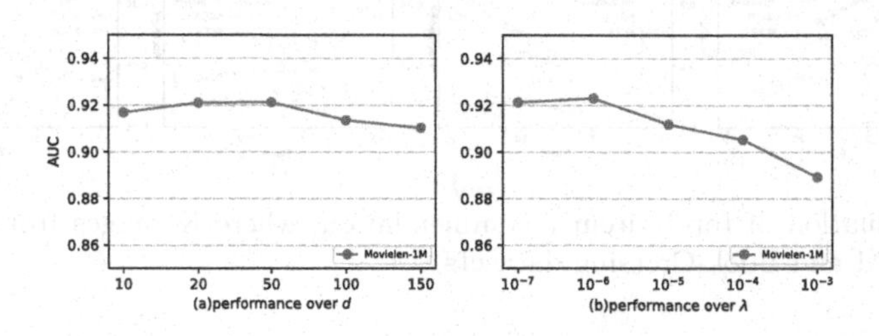

Fig. 5. Parameter sensitivity of KCER on the Movielens-1M dataset.

6 Conclusion

In this paper, we propose a novel deep knowledge-enhanced model KCER to learn the complex interactions between users and items with the help of an item-based knowledge graph. The two main factors of the proposed model are (i) generation of user-item pairs' specific knowledge context representation; (ii) deep fusion process to improve final user representation. More specifically, we generate the multi-hop knowledge context representation based on the joint embedding of user-item pairs. Then we design a knowledge fusion layer to filter out the related parts in the multi-hop knowledge context and obtain the final enriched user representation. Experiment results on several public datasets demonstrate the effectiveness of our proposed model. For future work, we plan to extend KCER by considering the fine-grained composition of entities and relations in KGs.

Acknowledgments. This work is supported by the National Key Research and Development Program of China.

References

1. Chen, W.H., Hsu, C.C., Lai, Y.A., Liu, V., Yeh, M.Y., Lin, S.D.: Attribute-aware collaborative filtering: survey and classification. arXiv preprint arXiv:1810.08765 (2018)
2. Guo, Q., et al.: A survey on knowledge graph-based recommender systems. arXiv preprint arXiv:2003.00911 (2020)
3. He, X., Liao, L., Zhang, H., Nie, L., Hu, X., Chua, T.S.: Neural collaborative filtering. In: Proceedings of the 26th International Conference on World Wide Web, pp. 173–182. International World Wide Web Conferences Steering Committee (2017)
4. Kipf, T.N., Welling, M.: Semi-supervised classification with graph convolutional networks. arXiv preprint arXiv:1609.02907 (2016)
5. Lin, Y., Liu, Z., Sun, M., Liu, Y., Zhu, X.: Learning entity and relation embeddings for knowledge graph completion. In: Twenty-Ninth AAAI Conference on Artificial Intelligence (2015)
6. Lin, Z., et al.: A structured self-attentive sentence embedding. arXiv preprint arXiv:1703.03130 (2017)
7. Rendle, S.: Factorization machines with libFM. ACM Trans. Intell. Syst. Technol. (TIST) **3**(3), 1–22 (2012)
8. Srivastava, R., Palshikar, G.K., Chaurasia, S., Dixit, A.: What's next? A recommendation system for industrial training. Data Sci. Eng. **3**(3), 232–247 (2018)
9. Sun, Z., Yang, J., Zhang, J., Bozzon, A., Huang, L.K., Xu, C.: Recurrent knowledge graph embedding for effective recommendation. In: Proceedings of the 12th ACM Conference on Recommender Systems, pp. 297–305 (2018)
10. Tang, X., Wang, T., Yang, H., Song, H.: AKUPM: attention-enhanced knowledge-aware user preference model for recommendation. In: Proceedings of the 25th ACM SIGKDD International Conference on Knowledge Discovery & Data Mining, pp. 1891–1899 (2019)
11. Wang, H., et al.: RippleNet: propagating user preferences on the knowledge graph for recommender systems. In: Proceedings of the 27th ACM International Conference on Information and Knowledge Management, pp. 417–426 (2018)

12. Wang, H., Zhang, F., Xie, X., Guo, M.: DKN: deep knowledge-aware network for news recommendation. In: Proceedings of the 2018 World Wide Web Conference, pp. 1835–1844 (2018)
13. Wang, H., Zhao, M., Xie, X., Li, W., Guo, M.: Knowledge graph convolutional networks for recommender systems. In: WWW, pp. 3307–3313. ACM (2019)
14. Yu, X., et al.: Personalized entity recommendation: a heterogeneous information network approach. In: Proceedings of the 7th ACM International Conference on Web Search and Data Mining, pp. 283–292 (2014)
15. Yu, X., et al.: Recommendation in heterogeneous information networks with implicit user feedback. In: Proceedings of the 7th ACM Conference on Recommender Systems, pp. 347–350 (2013)
16. Zhang, F., Yuan, N.J., Lian, D., Xie, X., Ma, W.Y.: Collaborative knowledge base embedding for recommender systems. In: Proceedings of the 22nd ACM SIGKDD International Conference on Knowledge Discovery and Data Mining, pp. 353–362 (2016)
17. Zhang, S., Yao, L., Sun, A.: Deep learning based recommender system: a survey and new perspectives. arXiv preprint arXiv:1707.07435 (2017)
18. Zhang, W., Cao, Y., Xu, C.: SARC: split-and-recombine networks for knowledge-based recommendation. In: 2019 IEEE 31st International Conference on Tools with Artificial Intelligence (ICTAI), pp. 652–659. IEEE (2019)
19. Zhang, W., Paudel, B., Zhang, W., Bernstein, A., Chen, H.: Interaction embeddings for prediction and explanation in knowledge graphs. In: Proceedings of the Twelfth ACM International Conference on Web Search and Data Mining, pp. 96–104 (2019)
20. Zhao, H., Yao, Q., Li, J., Song, Y., Lee, D.L.: Meta-graph based recommendation fusion over heterogeneous information networks. In: Proceedings of the 23rd ACM SIGKDD International Conference on Knowledge Discovery and Data Mining, pp. 635–644 (2017)

LHRM: A LBS Based Heterogeneous Relations Model for User Cold Start Recommendation in Online Travel Platform

Ziyi Wang[1(✉)], Wendong Xiao[1], Yu Li[2], Zulong Chen[1], and Zhi Jiang[1]

[1] Alibaba Group, Hangzhou, China
{jianghu.wzy,xunxiao.xwd,zulong.cz,jz105915}@alibaba-inc.com
[2] Department of Computer Science and Technology, Hangzhou Dianzi University, Hangzhou, China
liyucomp@hdu.edu.cn

Abstract. Most current recommender systems used the historical behaviour data of user to predict user' preference. However, it is difficult to recommend items to new users accurately. To alleviate this problem, existing user cold start methods either apply deep learning to build a cross-domain recommender system or map user attributes into the space of user behaviour. These methods are more challenging when applied to online travel platform (e.g., Fliggy), because it is hard to find a cross-domain that user has similar behaviour with travel scenarios and the Location Based Services (LBS) information of users have not been paid sufficient attention. In this work, we propose a LBS-based Heterogeneous Relations Model (LHRM) for user cold start recommendation, which utilizes user's LBS information and behaviour information in related domains (e.g., Taobao) and user's behaviour information in travel platforms (e.g., Fliggy) to construct the heterogeneous relations between users and items. Moreover, an attention-based multi-layer perceptron is applied to extract latent factors of users and items. Through this way, LHRM has better generalization performance than existing methods. Experimental results on real data from Fliggy's offline log illustrate the effectiveness of LHRM.

Keywords: Recommender system · Cold start · Cross domain

1 Introduction

Recommender Systems (RSs) aim to improving the Click-Through Rate (CTR), post-Click conVersion Rate (CVR) and stay time in the application. Most current RSs are based on the intuition that users' interests can be inferred from their historical behaviours or other users with similar preference [21]. Unfortunately,

Supported by National Natural Science Foundation of China (No. 61802098).

H. Yang et al. (Eds.): ICONIP 2020, LNCS 12534, pp. 479–490, 2020.
https://doi.org/10.1007/978-3-030-63836-8_40

recommendation algorithms are generally faced with data sparsity and cold start problems so that RSs cannot guarantee high recommendation accuracies [6,7].

Cold start problem refers to making recommendations when there are no prior interactions available for a user or an item [10,12,14], which falls into two forms: (1) new user cold start problem (2) new item cold start problem [7]. In new user cold start problem, a new user has just registered to the system and RS has no behaviour information about the user except some basic attributes [15]. In new item cold start problem, a new item is presently added to the online recommendation platform and RS has no ratings on it [15]. Compared with new item cold start problem, the new user cold start problem is more difficult and has been attracting greater interest [1]. In this paper, we focus on user cold start problem. Existing methods, including cross-domain recommendation algorithms [6,9,17,19,20], Lowrank Linear Auto-Encoder (LLAE) [11] have been proposed and achieved great success for user cold start problem.

User cold start recommendation over online travel platforms (e.g., Fliggy) are more challenging, thus existing methods cannot work well. LLAE [11] can reconstruct user behavior from user attributes, but even for active user, travel is a low-frequency demand and user behaviour is quite sparse. Therefore, the generalization performance of LLAE is limited by the sparse behaviour of users. Cross-domain algorithms try to utilize explicit or implicit feedbacks from multiple auxiliary domains to improve the recommendation performance in the target domain [6]. Unfortunately, it is hard to find a cross-domain that user has similar behaviour with travel scenarios and the LBS information of users have not been paid sufficient attention. Unconditional fuse the user behaviour information from other domains may introduce much noise. More importantly, user's travel intention is strongly related to user' LBS information. The intuition is that, users who are geographical situation closer may have similar travel intention.

To alleviate the user cold start problem in travel scenarios, we propose a LBS based Heterogeneous Relations Model (LHRM) for user cold start recommendation in online travel platform. LHRM firstly constructs heterogeneous relations between users and items and then apply an attention-based multi-layer perceptron to learn the latent factors of users and items. Heterogeneous relations is proposed in [2], which include user-user couplings, item-item couplings, and user-item couplings. It is increasingly recognized that modeling such multiple heterogeneous relations is essential for understanding the non-IID nature and characteristics of RSs [2,3]. In order to relieve the problem of data-sparse, user behaviour information in a specific category of items in related domains (e.g., Taobao) is used to learn the embedding representation of user. The background is that, more than 80% of Fliggy users have Taobao platform account[1], and most of them are cold start users in Fliggy, but they have rich behaviours on Taobao. Then LBS information and user behaviour information in a specific category of items in Taobao domain are concatenated to construct the heterogeneous relations between users. User behaviour information in Fliggy domain is used to

[1] Fliggy and Taobao jointly use Taobao platform account, and relevant data sharing has been informed to users and obtained user's consent.

construct the heterogeneous relations between items. Meanwhile, user attributes are mapped into the space of user behaviour in Fliggy domain. After obtaining the side information and the embedding representation of user and items, an attention-based multi-layer perceptron is applied to extract higher level features and make the recommendation results more accurately for cold start user in Fliggy.

In summary, the contributions of this paper are multi-fold:

- We propose a LBS-based Heterogeneous Relations Model (LHRM) for user cold start recommendation in travel scenarios, which utilize the LBS information and fuses user behaviour information in specific category in Taobao domain to improve the recommendation performance.
- A new heterogeneous relations between users and items is proposed in LHRM, which can represent the relationship between users who with similar preference better.
- Comprehensive experimental results on real data demonstrate the effectiveness of the proposed LHRM model.

The rest of this paper is organized as follows: Sect. 2 review the related work. Section 3 describe the proposed model in detail. Section 4 focus on the experimental results about the proposed model, including performance evaluation on real data from Fliggy's offline log. At last, we conclude the paper in Sect. 5.

2 Related Work

The main issue of the cold start problem is that, there is no available information can be required for making recommendations [7]. There has been extensive research on cold start problem in recommender systems. In the section, we mainly review the related work about user cold start problem.

Cross domain [5,6,8,9,17,19] recommendation algorithms have attracted much attention in recent years, which utilize explicit or implicit feedbacks from multiple auxiliary domains to improve the recommendation performance in the target domain. [6] proposed a Review and Content based Deep Fusion Model (RC-DFM), which contains four major steps: vectorization of reviews and item contents, generation of latent factors, mapping of user latent factors and cross-domain recommendation. Through this way, the learned user and item latent factors can preserve more semantic information. [8] proposed the collaborative cross networks (CoNet), which can learn complex user-item interaction relationships and enable dual knowledge transfer across domains by introducing cross connections from one base network to another and vice versa. [19] combine an online shopping domain with information from an ad platform, and then apply deep learning to build a cross-domain recommender system based on shared users of these two domains, to alleviate the user cold start problem.

Servel recent works model the relationship between user attributes and user behaviour. With the assumption that people with the similar preferences would have the similar consuming behavior, [11] proposed a Zero-Shot Learning (ZSL) method for user cold start recommendation. Low-rank Linear Auto-Encoder (LLAE) consists of two parts, a low-rank encoder maps user behavior into user attributes and a symmetric decoder reconstructs user behavior from user attributes. LLAE takes the efficiency into account, so that it suits large-scale problem.

A non-personalized recommendation algorithm is proposed in [16]. The authors hypothesize that combining distinct non-personalized RSs can be better to conquer the most first-time users than traditional ones. [16] proposed two RSs to balance the recommendations along the profile-oriented dimensions. Max-Coverage and Category-Exploration aims to explore user coverage to diversify the items recommended and conquer more first-time users.

3 The Proposed Approach

3.1 Problem Statement

Most current Recommender Systems (RSs) based on the intuition that users' interests can be inferred from their historical behaviours (such as purchase and click history) or other users with similar preference. However, it is difficult to recommend items to new users accurately. User cold start problem is a long-standing problem in recommender systems. In this work, we define the cold start user as the user who have not any behaviours on Fliggy in the past one month. Specifically, more than 20% of users are cold start users in Fliggy everyday and the optimization task of user cold start is becoming very important in Fliggy. The problem can be summarized as follows:

Problem: Given a target domain D_t, and a source domain D_s, user u is new for D_t, but it has interactions in D_s, recommend top k items for u in D_t.

3.2 Notations

In this paper, we use lowercase and uppercase letters to represent vector and matrix, respectively. We denote active users in two domain intersection as the target users $\{u_t\}$. For every target u_t, we denote the basic attributes as $x_{u_t} = \{x_1, x_2, x_3, ..., x_k\}$, the heterogeneous relation of user-user as $E_{u_g} = \{e_{u_1}, e_{u_2}, e_{u_3}, ..., e_{u_n}; e_{u_t}\}$, each e_{u_n} is the representation vector of u_n. We denote the items that target users have interacted as the target items $\{i_t\}$. For every target i_t, we denote the basic attributes as $x_{i_t} = \{x_1, x_2, x_3, ..., x_j\}$, heterogeneous relation of item-item as $E_{i_g} = \{e_{i_1}, e_{i_2}, e_{i_3}, ..., e_{i_m}; e_{i_t}\}$, each e_{i_m} is the representation vector of i_m. We use U to denote the target user matrix, and I to denote the target item matrix, $Y \in \{0,1\}^{|U|*|I|}$ be the relationship matrix between U and I, where $Y_{u,i} = 1$ means u_n clicked i_m.

3.3 Geohash Algorithm

In order to map user' LBS information (such as latitude and longitude) to a range, we use the **Geohash** algorithm [18]. Geohash is a public domain geocode system invented in 2008 by Gustavo Niemeyer and G.M. Morton, which encodes a geographic location into a short string of letters and digits (e.g., geohash6 (31.1932993, 121.4396019) = *wtw37q*). Length of different Geohash strings represent different area of region, for example, geohash5 means a square area of about 10 km^2.

3.4 LBS Based Heterogeneous Relations Model

It is worth noting that users' interests can be inferred from historical behaviours or other users with similar preference and benefited from heterogeneous relations. Moreover, user's travel intention is strongly related to user' LBS information (such as latitude and longitude), which based on the intuition that users who are geographical situation closer may have similar travel intention. To active this, we propose LBS based Heterogeneous Relations Model (LHRM), in which LBS information is used to construct the heterogeneous relation between users. The framework of LHRM illustrated in Fig. 1. LHRM contains two modules: heterogeneous relations construction module and representation learning module.

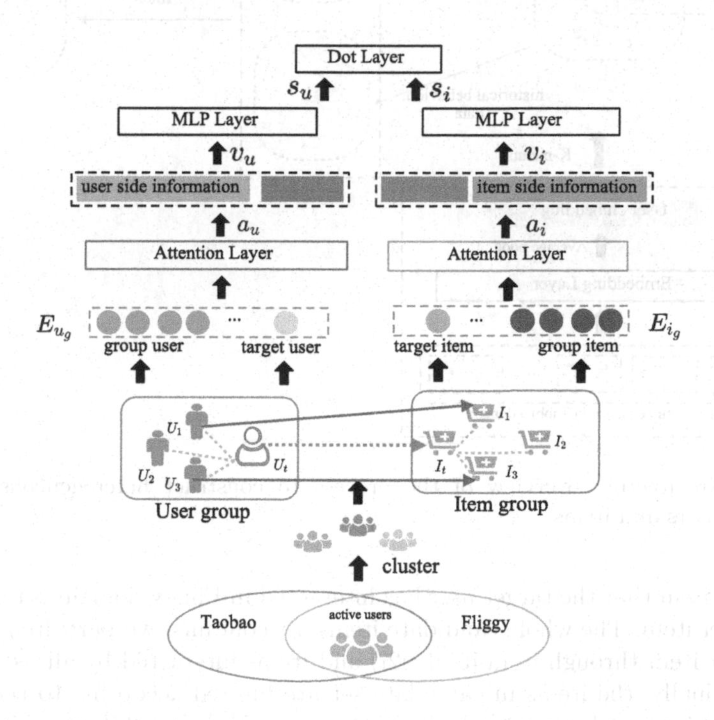

Fig. 1. Framework of the proposed LHRM

Heterogeneous Relations Construction Module. The detailed process of constructing the heterogeneous relations between users and items is shown in

Fig. 2. We can see that user's historical behaviours sequence and LBS sequence in Taobao domain are concatenated and input into a embedding layer, which pre-trained by skip-gram algorithm [13]. Specifically, items not related to travel are filtered out and user's latitude and longitude information is mapped to a string with the length of 5 by the geohash5 algorithm. After the embedding layer, we adopt average-pooling to generate the corresponding vector representation of user. In order to generate different user groups, we utilize K-means algorithm to cluster users according to their representation vectors. For each user group, any user can be regarded as the target user, and the other users are regarded as the friends of the target user.

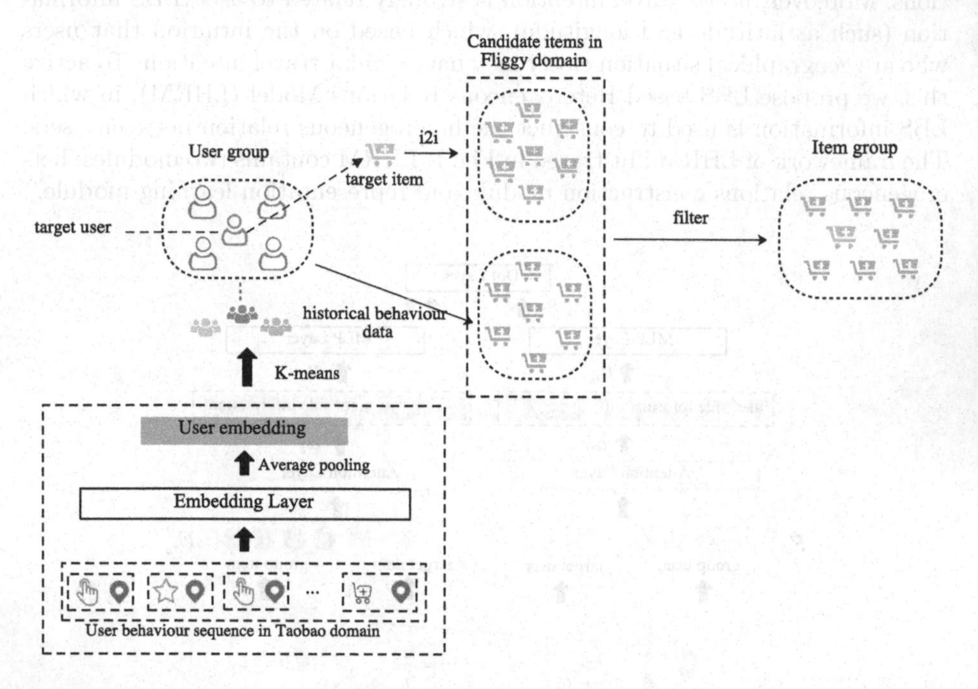

Fig. 2. Architecture overview of the process to construct heterogeneous relations between users and items

Each item that the target user has interacted in Fliggy domain is regarded as the target item. The whole candidate items set contains two part: items recalled by target item through item-item (i2i) and items interacted by all users in user group. Finally, the items in candidate set are filtered according to the topic of the target item and generate the item group. In this way, all items in item group are more related, and which can be represented by a pre-trained item embedding vector in Fliggy domain.

Representation Learning Module. An attention-based multi-layer perceptron is used to learn the latent factors of users and items. After the process of heterogeneous relations construction, E_{u_g} and E_{i_g} are generated. Then we adopt an attention layer to focus on the relevant parts of E_{u_g} and E_{i_g}. The implementation of attention for sequence-to-one networks on E_{u_g} is shown in Eq. (1) and Eq. (2):

$$\alpha_{ti} = \frac{exp(score(e_{u_t}, e_{u_i}))}{\sum_{i'=1}^{n+1} exp(score(e_{u_t}, e_{u_{i'}}))} \tag{1}$$

where:

$$score(e_{u_t}, e_{u_{i'}}) = e_{u_t}\mathbf{W}_a e_{u_{i'}} \tag{2}$$

$\mathbf{W}_a$ is the learnable parameters in attention layer, and the output of attention layer is computed as Eq. (3):

$$\mathbf{a}_u = \sum_{i=1}^{n+1} \alpha_{ti} \times e_{u_i} \tag{3}$$

Similarly, attention layer is implemented on E_{i_g}, and the output is a_i. The MultiLayer Perceptron (MLP) layer is a feed-forward neural network, which can generalize better to unseen feature combinations through low-dimensional dense embeddings learned for the sparse features [4]. We denote the input of MLP layer as v_u and v_i, the output of MLP layer as s_u and s_i. $v_u = [a_u, x_{u_t}]$, $v_i = [a_i, x_{i_t}]$. The output of LHRM is the preference score of u_t for i_t, we denote $\hat{y}$ as the output of dot layer. $\hat{y}$ is computed as Eq. (4):

$$\hat{y} = \frac{1}{1 + exp(-s_u \cdot s_i)} \tag{4}$$

y is binary labels with $y = 1$ or $y = 0$ indicating whether click or not. The logistic loss of LHRM is shown in Eq. (5):

$$L(y, \hat{y}) = -ylog(\hat{y}) - (1 - y)log(1 - \hat{y}) \tag{5}$$

For clarity, we show the key steps of our algorithm in Algorithm 1.

4 Experiment

In this section, we conduct experiments on Fliggy and Taobao's offline log dataset to evaluate the performance of LHRM and some baseline models.

4.1 Compared Methods

In the experiments, we compare the following methods.

- **Hot:** Hot is a non-personalized recommendation algorithm, which recommends items to new users according to the popularity score of item in Fliggy domain.

Algorithm 1. LBS based Heterogeneous Relations Model

Input: User's behaviour sequence in Taobao domain S_T, user's LBS sequence in Taobao domain L_T, user's behaviour sequence in Fliggy domain S_F, user attributes X_u, item attributes X_i
Output: Latent factors of users S_u, latent factors of items S_i

1: S_T and L_T are used to construct heterogeneous relation between users by K-means, and output E_{u_g}
2: S_F is used to construct heterogeneous relation between items, and output E_{i_g}
3: E_{u_g} and E_{i_g} are input into attention layer, and output a_u, a_i
4: the concatenation vector of $[a_u, x_{u_t}]$ and $[a_i, x_{i_t}]$ are input into MLP layer, and output s_u, s_i
5: s_u, s_i are input into dot layer, and output $\hat{y}$
6: update all parameters according to $L(y, \widehat{y}) = -ylog(\widehat{y}) - (1-y)log(1-\widehat{y})$
7: **Cold start:**
8: $\hat{y}_{new} = s_{u_{new}} \cdot s_i$
9: **Recommendation:**
10: Computing the similarity score between the new user and all candidate items, and recommend the top-k items

- **HERS:** Heterogeneous relations-Embedded Recommender System (HERS) is proposed in [9], which based on ICAUs to model and interpret the underlying motivation of user-item interactions by considering user-user and item-item influences and can handle the cold start problem effectively.
- **MaxCov:** Max-Coverage (MaxCov) [16] is a non-personalized recommendation algorithm, which aims to explore user coverage to diversify the items recommended and conquer more first-time users.
- **LHRM:** Lbs based heterogeneous relations model proposed in this paper.

Popularity score of items is very important factor in cold start recommendation, therefore, when implementing LHRM and HERS in experiments, we fuse the popularity score of items with the score of LHRM and HERS's output. Then, final preference score of u_t for i_t calculated in Eq. (6):

$$\widehat{y} = Score_{LHRM/HERS} \times Pop\_Score_{item} \tag{6}$$

4.2 Implementation Details

We set the maximum length of user group and item group to 10. The number of cluster center is set to 1000. The dimension of latent factors of user and item is a hyper parameter, we set it to 32, 64, 128 and 256 in the experiments. In order to evaluate the performance of proposed methods, we adopt two evaluation metrics, i.e., Hit Rate (HR@30,@50,@100,@200), Normalized Discounted Cumulative Gain (NDCG@30,@50,@100,@200). HR is a metric of shotting accurately at target items. NDCG is a cumulative measure of ranking quality, which is more sensitive to the relevance of higher ranked items.

4.3 Datasets

During our survey, no public datasets for user cold start recommendation in travel scenarios. To evaluate the proposed approach, we collect the offline log data from Fliggy and Taobao domain in the past one month as the dataset. Generally, impression and click sample as positive samples, impression but not click samples as negative samples. The statics of dataset is illustrated in Table 1.

Table 1. Statistics of dataset. (pos - positive, neg - negative, M - Million)

	Training	Validation	Testing
# of samples	7.68M	1.56M	1.34M
# of pos samples	3.64M	0.35M	0.062M
# of neg samples	4.04M	1.21M	1.28M
# of users	1.6M	0.437M	0.012M
# of items	0.15M	0.086M	0.2M

4.4 Results

We show the experimental results of different models in Table 2. Among all methods, LHRM achieves the best performance in terms of all metrics. Specifically, when the dimension of latent factors of user and item is set to 32, the HR and NDCG are the highest.

Table 2. Comparison of different models on dataset.

	HR				NDCG			
	@30	@50	@100	@200	@30	@50	@100	@200
Hot	0.034	0.065	0.128	0.169	0.008	0.014	0.023	0.029
HERS	0.039	0.088	0.142	0.245	0.011	0.02	0.028	0.041
MaxCov	0.035	0.075	0.0993	0.1989	0.002	0.007	0.024	0.036
LHRM-32	**0.0754**	**0.109**	**0.184**	**0.266**	**0.022**	**0.028**	**0.039**	**0.05**
LHRM-64	0.056	0.097	0.152	0.254	0.016	0.023	0.031	0.044
LHRM-128	0.0728	0.101	0.149	0.255	0.02	0.025	0.032	0.044
LHRM-256	0.052	0.089	0.15	0.253	0.014	0.02	0.03	0.043

Table 2 shows the HR and NDCG on target items, and all existing methods did not work well on cold start users. Generally, in practical applications, we not only care whether the recommended items will be clicked, but also whether the recommended items are related to the target items. Therefore, we evaluate the hit rate of different models under different degree of relevance with target items. Experimental results are shown in Fig. 3. We can see that, LHRM-32 is very competitive, MaxCov performs best when calculating hit rate according to whether the destination same as the target items.

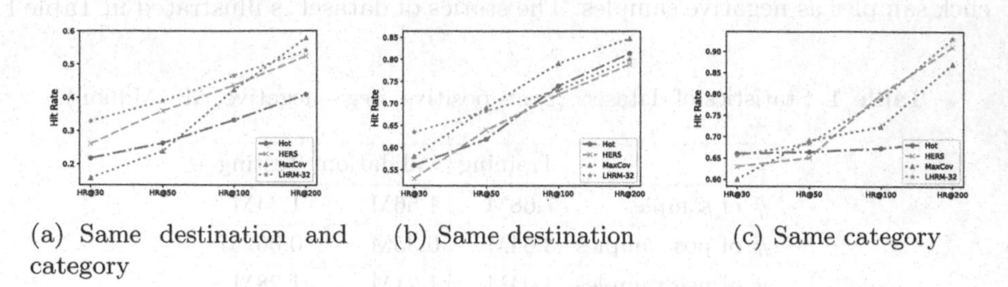

(a) Same destination and category (b) Same destination (c) Same category

Fig. 3. Comparison of different models w.r.t different degree of relevance with target items

5 Conclusion

In this paper, we point out two challenges of user cold start recommendation in travel platform: i) it is hard to find a cross-domain that user has similar behaviour with travel scenarios ii) LBS information of users have not been paid sufficient attention. To address this problem, we propose LBS based heterogeneous relations model. LHRM utilizes user's LBS information and behaviour information in Taobao domain and user's behaviour information in Fliggy domain to construct the heterogeneous relations between users and items. Moreover, an attention-based multi-layer perceptron is applied to extract latent factors of users and items. Experimental results on real data from Fliggy's offline log illustrate the effectiveness of LHRM.

References

1. Bobadilla, J., Ortega, F., Hernando, A., Bernal, J.: A collaborative filtering approach to mitigate the new user cold start problem. Knowl. Based Syst. **26**, 225–238 (2012). https://doi.org/10.1016/j.knosys.2011.07.021
2. Cao, L.: Coupling learning of complex interactions. Inf. Process. Manag. **51**(2), 167–186 (2015). https://doi.org/10.1016/j.ipm.2014.08.007
3. Cao, L.: Non-IID recommender systems: a review and framework of recommendation paradigm shifting. Engineering **2**(2), 212–224 (2016)
4. Cheng, H., et al.: Wide & deep learning for recommender systems. In: Proceedings of the 1st Workshop on Deep Learning for Recommender Systems, pp. 7–10 (2016). https://doi.org/10.1145/2988450.2988454

5. Elkahky, A.M., Song, Y., He, X.: A multi-view deep learning approach for cross domain user modeling in recommendation systems. In: Proceedings of the 24th International Conference on World Wide Web, WWW 2015, Florence, Italy, 18–22 May 2015, pp. 278–288 (2015). https://doi.org/10.1145/2736277.2741667
6. Fu, W., Peng, Z., Wang, S., Xu, Y., Li, J.: Deeply fusing reviews and contents for cold start users in cross-domain recommendation systems. In: The Thirty-Third AAAI Conference on Artificial Intelligence, AAAI 2019, The Thirty-First Innovative Applications of Artificial Intelligence Conference, IAAI 2019, The Ninth AAAI Symposium on Educational Advances in Artificial Intelligence, EAAI 2019, Honolulu, Hawaii, USA, 27 January–1 February 2019, pp. 94–101 (2019). https://doi.org/10.1609/aaai.v33i01.330194
7. Gope, J., Jain, S.K.: A survey on solving cold start problem in recommender systems. In: 2017 International Conference on Computing, Communication and Automation (ICCCA), pp. 133–138 (2017)
8. Hu, G., Zhang, Y., Yang, Q.: CoNet: collaborative cross networks for cross-domain recommendation. In: Proceedings of the 27th ACM International Conference on Information and Knowledge Management, CIKM 2018, Torino, Italy, 22–26 October 2018, pp. 667–676 (2018). https://doi.org/10.1145/3269206.3271684
9. Hu, L., Jian, S., Cao, L., Gu, Z., Chen, Q., Amirbekyan, A.: HERS: modeling influential contexts with heterogeneous relations for sparse and cold-start recommendation. In: The Thirty-Third AAAI Conference on Artificial Intelligence, AAAI 2019, The Thirty-First Innovative Applications of Artificial Intelligence Conference, IAAI 2019, The Ninth AAAI Symposium on Educational Advances in Artificial Intelligence, EAAI 2019, Honolulu, Hawaii, USA, 27 January–1 February 2019, pp. 3830–3837 (2019). https://doi.org/10.1609/aaai.v33i01.33013830
10. Lam, X.N., Vu, T., Le, T.D., Duong, A.D.: Addressing cold-start problem in recommendation systems. In: Proceedings of the 2nd International Conference on Ubiquitous Information Management and Communication, ICUIMC 2008, Suwon, Korea, 31 January–1 February 2008, pp. 208–211 (2008). https://doi.org/10.1145/1352793.1352837
11. Li, J., Jing, M., Lu, K., Zhu, L., Yang, Y., Huang, Z.: From zero-shot learning to cold-start recommendation. In: The Thirty-Third AAAI Conference on Artificial Intelligence, AAAI 2019, The Thirty-First Innovative Applications of Artificial Intelligence Conference, IAAI 2019, The Ninth AAAI Symposium on Educational Advances in Artificial Intelligence, EAAI 2019, Honolulu, Hawaii, USA, 27 January–1 February 2019, pp. 4189–4196 (2019). https://doi.org/10.1609/aaai.v33i01.33014189
12. Lika, B., Kolomvatsos, K., Hadjiefthymiades, S.: Facing the cold start problem in recommender systems. Expert Syst. Appl. **41**(4), 2065–2073 (2014). https://doi.org/10.1016/j.eswa.2013.09.005
13. Mikolov, T., Chen, K., Corrado, G., Dean, J.: Efficient estimation of word representations in vector space. In: 1st International Conference on Learning Representations, ICLR 2013, Scottsdale, Arizona, USA, 2–4 May 2013, Workshop Track Proceedings (2013). http://arxiv.org/abs/1301.3781
14. Nadimi-Shahraki, M., Bahadorpour, M.: Cold-start problem in collaborative recommender systems: efficient methods based on ask-to-rate technique. CIT **22**(2), 105–113 (2014). http://cit.srce.unizg.hr/index.php/CIT/article/view/2223
15. Saraswathi, K., Mohanraj, V., Saravanan, B., Suresh, Y., Senthilkumar, J.: Survey: a hybrid approach to solve cold-start problem in online recommendation system. Social Science Electronic Publishing

16. Silva, N., Carvalho, D., Pereira, A.C.M., Mourão, F., da Rocha, L.C.: The pure cold-start problem: a deep study about how to conquer first-time users in recommendations domains. Inf. Syst. **80**, 1–12 (2019). https://doi.org/10.1016/j.is.2018.09.001
17. Song, T., Peng, Z., Wang, S., Fu, W., Hong, X., Yu, P.S.: Review-based cross-domain recommendation through joint tensor factorization. In: Database Systems for Advanced Applications - 22nd International Conference, DASFAA 2017, Suzhou, China, 27–30 March 2017, Proceedings, Part I, pp. 525–540 (2017). https://doi.org/10.1007/978-3-319-55753-3_33
18. Vukovic, T.: Hilbert-Geohash - hashing geographical point data using the Hilbert space-filling curve (2016)
19. Wang, H., et al.: Preliminary investigation of alleviating user cold-start problem in e-commerce with deep cross-domain recommender system. In: Companion of The 2019 World Wide Web Conference, WWW 2019, San Francisco, CA, USA, 13–17 May 2019, pp. 398–403 (2019). https://doi.org/10.1145/3308560.3316596
20. Wang, X., Peng, Z., Wang, S., Yu, P.S., Fu, W., Hong, X.: Cross-domain recommendation for cold-start users via neighborhood based feature mapping. In: Database Systems for Advanced Applications - 23rd International Conference, DASFAA 2018, Gold Coast, QLD, Australia, 21–24 May 2018, Proceedings, Part I, pp. 158–165 (2018). https://doi.org/10.1007/978-3-319-91452-7_11
21. Zhu, H., et al.: Learning tree-based deep model for recommender systems. In: Proceedings of the 24th ACM SIGKDD International Conference on Knowledge Discovery & Data Mining, pp. 1079–1088 (2018). https://doi.org/10.1145/3219819.3219826

Match4Rec: A Novel Recommendation Algorithm Based on Bidirectional Encoder Representation with the Matching Task

Lingxiao Zhang(✉), Jiangpeng Yan, Yujiu Yang, and Li Xiu(✉)

Tsinghua Shenzhen International Graduate School, Shenzhen 518055, China
zhang-lx18@mails.tsinghua.edu.cn, li.xiu@sz.tsinghua.edu.cn

Abstract. Characterizing users' interests accurately plays a significant role in an effective recommender system. The sequential recommender system can learn powerful hidden representations of users from successive user-item interactions and dynamic users' preferences. To analyze such sequential data, the use of self-attention mechanisms and bidirectional neural networks have gained much attention recently. However, there exists a common limitation in previous works that they only model the user's main purposes in the behavioral sequences separately and locally, lacking the global representation of the user's whole sequential behavior. To address this limitation, we propose a novel bidirectional sequential recommendation algorithm that integrates the user's local purposes with the global preference by additive supervision of the matching task. Particularly, we combine the mask task with the matching task in the training process of the bidirectional encoder. A new sample production method is also introduced to alleviate the effect of mask noise. Our proposed model can not only learn bidirectional semantics from users' behavioral sequences but also explicitly produces user representations to capture user's global preference. Extensive empirical studies demonstrate our approach considerably outperforms various baseline models.

Keywords: Recommendation · Sequential recommendation · Matching task

1 Introduction

Recommender Systems can help users obtain a more customized and personalized recommendation experience by characterizing users exhaustively and mine their interests precisely. A widely used approach to building quality recommender systems in real applications is collaborative filtering (CF) [1]. But such a method takes users' shopping behaviors as isolated manners, while these behaviors usually happen successively in a sequence. Recently, sequential recommendations based on users' historical interactions have attracted increasing attention. They

H. Yang et al. (Eds.): ICONIP 2020, LNCS 12534, pp. 491–503, 2020.
https://doi.org/10.1007/978-3-030-63836-8_41

model the sequential dependencies over the user-item interactions (e.g., like or purchase) in sequences to capture user interests [2]. Two basic paradigms of the pattern have proliferated: unidirectional (left-to-right) and bidirectional sequential model. The former, including Markov Chains (MC) [3], Recurrent Neural Networks (RNNs) [4] and self-attentive sequential model [5], is more close to the order of interactions between users and items in many real-world applications, yet it is not sufficient to learn optimal representations for user behavior sequences. The latter, like BERT4Rec [6], premeditates various unobservable external factors and does not follow a rigid order assumption, which is beneficial to incorporate context from both sides for sequence representation learning. However, the aforementioned bidirectional sequential model only relays on capturing the user's main purposes, which are reflected by relatively important items distributed in different local areas of the whole sequence. Therefore, there exited a limitation that these models cannot always conjecture the user's main purposes without the global knowledge, especially, when the sequence is quite short or the user just clicks something aimlessly.

In this paper, we consider the user's entire sequential behavior as the supplement of the local purposes. We integrate the mask task with the matching task by the novel mask setting of the unambiguous user sequential representation during the training processing of the bidirectional model. The matching task which usually directly build the mapping between the user's whole behavior sequence and the targeted items, treats the recommendation problems as the matching problem and can measure the user's global preference on items [7]. The mask task [6,8] is adopted to substitute the objective in unidirectional models for the bidirectional models. Some items in the users' behavioral sequences are masked in certain probability (e.g., replace them with a special token [mask]). Then, the recommender model predicts the ids of those masked items based on their surrounding context, which is a mixture of both the left and right context. To integrate the matching task in such a mask task, we use a special token "[UID]" to explicitly represent individual users, inspired by doc2vec [9]. Then we concatenate the user token with several item tokens from a sequence to train a bidirectional encoder model. Thus, our model can determine whether or not each users' semantic vector (i.e., the output of the user token) and items' semantic vectors (including positive samples and negative samples) are well matched. Because the output of the user token has merged various correlations among items in each sequence, this method can also be applied to variable-length pieces of sequences and expressly form the user representation. To alleviate the effect of mask noise in the training, we produce instances that only compute the loss function of the matching task between the original user behavioral sequence and items. Extensive experiments on four datasets show that our model outperforms various state-of-the-art sequential models consistently.

In conclusion, the contributions of this paper are listed as follows: a) We integrate the matching task with the bidirectional recommender model by the novel mask setting during the training. In this way, user's local purposes and the global preference in the behavioral sequence can be combined to boost the performance of the recommender system. b) We propose a novel sample production method to alleviate the effect of mask noise in the training for matching task. c) Extensive experiments show that our model outperforms state-of-the-art methods on four benchmark datasets.

2 Related Works

In this section, we briefly introduce several works closely related to ours. We first discuss general recommendation, followed by sequential recommendation and the process of the matching task in the recommendation.

As mentioned above, Collaborative Filtering (CF) [1] is one of the most widely used general recommendations that takes users' shopping behaviors as isolated manners. Recently, deep learning techniques have been introduced for general recommendation. Some researchers tried to use more auxiliary information (e.g., text [10], images [11], acoustic [12]) into recommendation systems. Some works focused on replacing conventional matrix factorization (NCF [13]) with neural networks (e.g., AutoRec [14] and CDAE [15]).

Different from the above methods, sequential recommendation systems consider orders in users' behaviors. Early works adopted the Markov chain to model the transition matrices over user-item interactions in a sequence [3]. Then recurrent neural networks (RNN) are widely used for sequential recommendation [4]. Apart from RNNs, other deep learning models are also applied in sequential recommendation systems. For example, Caser [16] learns sequential patterns through both horizontal and vertical convolutional filters. Recently, the use of attention mechanisms in recommendation has got the substantial performance. SASRec [5] applies a two-layer Transformer decoder to capture the user's sequential behaviors in left-to-right order (i.e., Transformer language model). BERT4Rec [6] uses a two-layer Transformer decoder with the help of the Cloze task to achieve bidirectional information mining, which is closely related to our work.

Matching tasks in the recommendation is used to capture the user's global preference on items. The fundamental problem of matching tasks is the semantic gap because users and items are heterogeneous objects, and there may not be any overlap between the features [7]. To address the problem, matching tasks usually are performed at the semantic level. Thus, the strong representation ability of the models is the key to improving recommendation performance. Deep learning methods are widely used in the matching task because of their great potentials of abstracting representations for data objects [13–15]. In this paper, we use Transformer to unambiguously represent individual users and perform the matching task, aiming to model user's global preference in the sequence to get better recommendation performance.

3 Methodology

In this section, we introduce our model architecture and several detailed modules. Firstly, some important variables are defined as the following. Considering a user's interaction sequence $S^{u_i} = [v_1^{u_i}, v_2^{u_i}, v_3^{u_i}, ..., v_{|V|}^{u_i}]$, the next item $v_{n+1}^{u_i}$ needs to be predicted by the sequential recommendation algorithm, where $v_i^u \in V$, item set $V = \{v_1, v_2, v_3, ..., v_{|V|}\}$, and $u_i \in U$, item set $U = \{u_1, u_2, u_3, ..., u_{|U|}\}$. Predicted probability can be formalized as $p(v_{n+1}^{u_i} = v|S^{u_i})$.

3.1 Model Architecture

Our model architecture is shown in Fig. 1, which is made up of the embedding layer, transformer layers, and the output layer.

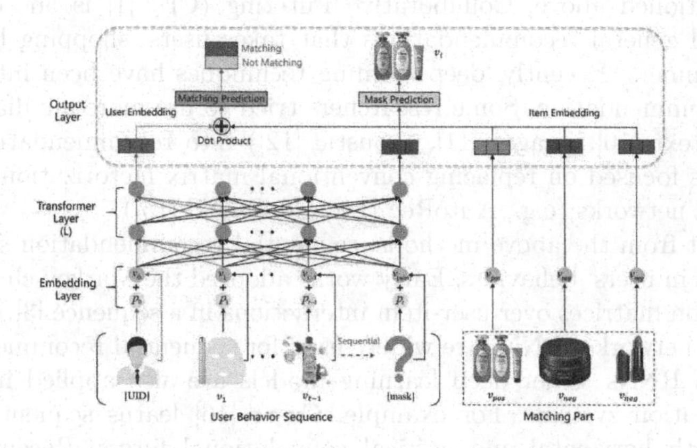

Fig. 1. Our proposed model architecture.

In the embedding layer, the input sequential items are mapped into item embedding and position embedding. Note that to expressly build the user representation, we add a special token "[UID]" at the beginning of a sequence and share the weights from the item embedding with positive and negative items that are used in following matching task. After the embedding layer, we stack L Transformer layers to catch dependencies of items in each sequence. Different from other sequential models such as RNN, the self-attention mechanism directly computes dependencies of tokens in sequences rather than through accumulative dependencies in the last time. In the output layer, the model needs to predict masked items and determine if user-item pairs are well matched.

The embedding layer and the output layer is specially designed for our proposed model, the transformer layers share the same structure of BERT [8], which is used to process neural language sequence, and is also used in BERT4Rec [6] for recommendation problem.

3.2 Embedding Layer

In our model, given a user's interaction sequence S^{u_i}, we set a restriction on the maximum sequence length N to make sure our model can handle. In other words, we only consider the most recent N-1 actions (except special token "[UID]" at first of the sequence). The input embedding has two types: user embedding and item embedding. User embedding $E \in \mathbf{R}^{N \times d}$ is made up by summing item embedding $E_v \in \mathbf{R}^{N \times d}$ and position embedding $E_n \in \mathbf{R}^{N \times d}$:

$$E = E_v + E_p. \tag{1}$$

Additionally, we use shared weights from item embedding E_v to map one positive item (i.e., the last item) and n random sampled negative items (non-interaction items) to item space. A visualization of this construction can be seen in Fig. 2.

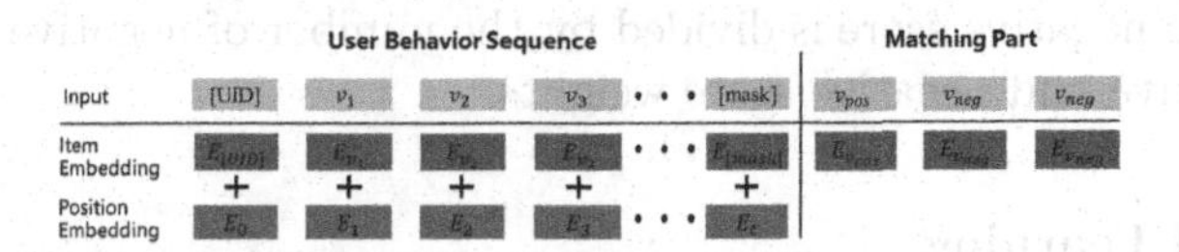

Fig. 2. Input representation of our model. The input embedding has two types: user embedding and item embedding.

Take the case shown in Fig. 1 to help readers to understand. The consumer finally buys a shampoo, while we build a triple [shampoo, cream, lipstick] for the matching task. The shampoo is positive, the latter two items are negative.

3.3 Transformer Layer

As mentioned above, Transformer Layer is first proposed in [8] to build the bidirectional semantic representation for language understanding. Every Transformer layer is mainly constructed by a multi-head self-attention sub-layer and a feed-forward network [17]. Based on [17], we employ a residual connection, layer normalization and dropout around each of the two sublayers to avoid overfitting model and vanishing gradient. The process is formulated as follow:

$$g(x) = x + \text{Dropout}(g(\text{LayerNorm}(\text{x}))), \tag{2}$$

where $g(x)$ represents the self-attention layer or the feed-forward network.

3.4 Output Layer

In the output layer, we need to deal with two tasks: masked items prediction and matching prediction. For masked items, we get the final output H_t^L after L Transformer layers, where t means the masked item v_t is at time step t. Softmax is employed as the activation function. The process is formulated as follow:

$$P_{Mask}(v_t) = \text{Softmax}(H_t^L W_p + b_p), \tag{3}$$

where b_p is a learnable projection bias, W_p is the projection matrix. In order to alleviate overfitting and reducing the model size, we make W_p share weights from the item embedding matrix in the embedding layer.

For matching prediction, we extract the first final output H_1^L (i.e., the final output of "[UID]"), and a positive item E_v^{pos} and n negative items E_v^{neg} that have been mapped into item semantic space in the embedding layer. We calculate matching scores of a positive one and negative ones. The process is defined by:

$$Score_{pos} = E_v^{pos} \bullet H_1^L, \tag{4}$$

$$Score_{neg} = (E_v^{neg} \bullet H_1^L)/n. \tag{5}$$

Note that the negative score is divided by the number of negative sampling n to balance positive and negative score weights.

3.5 Model Learning

For unidirectional sequential recommendation, the task of predicting the next item tends to be adopted in their models. For example, these models create N-1 samples (like $([v_1], v_2)$ and $([v_1, v_2], v_3)$ from the original length N behavioral sequence. But for the bidirectional sequential recommendation, if we also adopt this strategy to train model, these models create (N-1)! samples, which is time-consuming and infeasible. Thus we employ the mask task (same as [6,8]) to efficiently train our model. Different from [6], we add special tokens "[UID]" at the first of the sequence, which is used in the matching task. Here is a mask example in our model:

Input: $[\text{[UID]}, v_1, v_2, v_3, v_4, v_5] \rightarrow [\text{[UID]}, v_1, \text{[mask]}_2, v_3, \text{[mask]}_4, v_5]$,
Labels: $\text{[mask]}_2 = v_2, \text{[mask]}_4 = v_4$.

where we randomly mask the proportion ρ of all items in the input sequence (i.e., replace with special token "[mask]"), and we always mask all of the successive same items at once to prevent the information leakage as far as possible. Our model needs to predict these masked items' ids based on their surrounding items. We define the negative log-likelihood loss for each masked input $S^{u'}$:

$$Loss_{mask} = \frac{1}{|S^u_{mask}|} \sum_{v_{mask} \in S^u_{mask}} - \log P_{mask}(v_{mask} = v^*_{mask} | S^{u'}), \tag{6}$$

where $S^{u'}$ is the masked version for user behavior history S^u, S^u_{mask} is the random masked items in it, v^*_{mask} is the label for the masked item v_{mask}, and the probability $P_{mask}(\bullet)$ is defined in Eq. (3). In multiple epochs, we produce different masked samples to train a more powerful model.

Simultaneously, we add a new matching task into our model to capture user's global preference. We adopt the binary cross-entropy loss for each user:

$$Loss_{matching} = -(\log(\sigma(score_{pos} \bullet c)) + \log(1 - \sigma(score_{neg} \bullet c)), \tag{7}$$

where $score_{pos}$ and $score_{neg}$ are defined in Eq. (4) and (5) separately, c is a scaling coefficient, which is assigned to 10 by us. In multiple epochs, we randomly generate n negative items for each user sequence. And the total loss is the sum of the mask loss and the matching loss, shown as the following equation:

$$Loss = Loss_{mask} + Loss_{matching}. \tag{8}$$

We propose a new sample production method including three types. Firstly, we create samples used in the computation of the total loss. To address the mismatch between training and prediction, we create another type of sample that only masks the last item in the input user behavior sequences (same as [12]). To enhance our model's power of representations and alleviate the effect of mask noise, we also produce samples that are made of original sequences and the matching part. Here, we mix up these samples to train. Three types of samples are listed as follows:

Mask+Matching: [[[UID], v_1, [mask]$_2$, v_3, [mask]$_4$, v_5], [v_{pos}, v_{neg_1}, v_{neg_2}]],
ThelastMask: [[[UID], v_1, v_2, v_3, v_4, v_5, [mask]$_6$], []],
Matching: [[[UID], v_1, v_2, v_3, v_4, v_5], [v_{pos}, v_{neg_1}, v_{neg_2}]].

In the prediction stage, we adopt a conventional strategy: sequential prediction (i.e., predicting the last item based on the final hidden representation of the sequence).

4 Experiments and Discussions

4.1 Datasets and Baselines

We evaluate the proposed model on four representative datasets from three real-world applications, which vary significantly in domains and sparsity. **Amazon** [18] datasets contain product reviews and metadata from Amazon online shopping platform. They are separated into 24 categories according to the top level. In this work, we employ the small review subsets of "Beauty" and "Video Games" category. **Steam** [5] datasets contain reviews from the Steam video game platform. **MovieLens** [19] is a popular benchmark dataset, including several million movie ratings, reviews, etc. We employ the "MovieLens-1M" version.

For the preprocessing procedure, we use a common strategy from [4–6,13]. For all datasets, we transfer all ratings or reviews to implicit feedbacks (i.e., representing as numeric 1). Then, we group the interaction records by users and arrange them into sequences ordered by timestamp. We leave out users and items with fewer than five feedbacks. The statistics of the processed datasets are shown in Table 1. It needs to emphasize that we employ review datasets of Amazon rather than rating datasets, which is different from [5,6].

To verify the effectiveness of our method, we choose the following baselines: **POP** is a simple baseline that ranks items according to their popularity.

Table 1. Statistics of processed datasets.

Dataset	#users	#items	#actions	Avg. length
Beauty	22363	12101	0.23M	6.88
Video Games	24303	10672	0.26M	7.54
Steam	334730	13047	5.3M	10.59
ML-1M	6040	3416	1.0M	163.5

NCF [13] is a general framework that replaces an inner product with a neural network to learn the matching function. **FPMC** [3] combines an MF term with first-order MCs to capture long-term preferences and short-term transitions respectively. **GRU4Rec+** [4] uses GRU with a new cross-entropy loss functions and sampling strategy to achieve session-based recommendation. **Caser** [16] employs CNN in both horizontal and vertical ways to model high-order MCs for the sequential recommendation. **SASRec** [5] uses a left-to-right Transformer language model to capture users' sequential behaviors. **BERT4Rec** [6] uses a two-layer Transformer decoder with the help of the Cloze task to mine bidirectional sequential information.

For GRU4Rec+, Caser, SASRec, we use codes provided by the corresponding authors. For NCF, FPMC, and BERT4Rec, our model, we implement them by using *TensorFlow*. All parameters are initialized by using truncated normal distribution in the range [−0.02, 0.02]. We consider the $\ell 2$ weight decay from $\{1, 0.1, 0.01, 0.001\}$, and dropout rate from $\{0, 0.1, 0.2, ..., 0.9\}$, learning rate from {1e−3, 1e−4, 1e−5}, $\beta 1 = 0.9$, $\beta 2 = 0.999$, the hidden dimension size d from $\{16, 32, 64\}$. For our model, we set the layer number $L = 2$ and head number $h = 2$ and set the maximum sequence length $N = 200$ for ML-1m, $N = 50$ for Beauty, Video Games, and Steam datasets, and we employ the same mask proportion ρ with [6] (i.e., $\rho = 0.6$ for Beauty and Video Games, $\rho = 0.4$ for Steam, $\rho = 0.2$ for MovieLens-1M). We consider the number negative sampling n from $\{5, 10, 20\}$. All hyper-parameters are tuned on the validation sets. All results are under their optimal hyper-parameter settings.

4.2 Evaluation Metrics

To evaluate the performances of the recommendation models, we adopt the leave-one-out evaluation (i.e., next item recommendation) task, which is widely used in [4–6,16]. For each user, we select the most recent action of his/her behavioral sequence as the test set, treat the second most recent action as the validation set, and utilize the remainder as the train set. Note that during testing, the input sequence is a combination of the train set and the validation set.

We adopt Normalized Discounted Cumulative Gain (NDCG), Hit Ratio (HR) and Mean Reciprocal Rank (MRR) metrics as evaluation metrics. In this work, we report HR and NDCG with k = 10. The higher value means better performance for all metrics. To avoid computing heavily on all item predictions, we randomly sample 100 negative items and rank these negative items with the ground-truth item for each user. Based on the rankings of these 101 items, the evaluation metrics can be evaluated.

4.3 Recommendation Performance

Table 2. Recommendation performance. The best performing method in each row is boldfaced, and the second-best method in each row is underlined.

Datasets	Metric	POP	NCF	FPMC	GRU4 Rec+	Caser	SAS Rec	Bert4 Rec	Ours	Improv.
Beauty	NDCG @10	0.1793	0.2567	0.2937	0.2354	0.2705	0.3019	0.3298	**0.3370**	2.1%
	HR@10	0.3363	0.4217	0.4064	0.3943	0.4223	0.4654	0.4943	**0.5009**	1.3%
	MRR	0.1553	0.2229	0.2773	0.1105	0.2424	0.1865	0.2431	**0.2957**	21.6%
Video Games	NDCG @10	0.2512	0.3778	0.3225	0.4634	0.4137	0.4738	0.4947	**0.5163**	4.1%
	HR@10	0.4385	0.6031	0.5211	0.7137	0.6307	**0.7320**	0.6861	0.7217	−1.4%
	MRR	0.2151	0.3241	0.2793	0.2379	0.3616	0.3134	0.4119	**0.4396**	6.7%
Steam	NDCG @10	0.4927	0.4996	0.5768	0.5465	0.5950	0.6171	0.6316	**0.6460**	2.2%
	HR@10	0.7556	0.7629	0.8216	0.7986	0.8310	0.8440	0.8633	**0.8760**	1.4%
	MRR	0.4225	0.4295	0.5087	0.5247	0.5292	0.4177	0.5488	**0.5835**	6.3%
ML-1M	NDCG @10	0.2455	0.4094	0.5258	0.5456	0.5408	0.5354	0.5483	**0.5669**	3.3%
	HR@10	0.4458	0.6856	0.7439	0.7514	0.7769	0.7889	0.7546	**0.8096**	2.6%
	MRR	0.2070	0.3398	0.3600	0.4039	0.4517	0.3039	0.4728	**0.4973**	5.1%

Table 2 illustrates the results of all methods on the four datasets. The last column is the improvements of our method relative to the best baseline. In our re-implementation of BERT4Rec, we reported different results compared to the original paper. The following three reasons need to be considered: firstly, we employ different datasets; secondly, we adopt a uniform negative sampling method instead of sampling according to items' popularity during the evaluation; thirdly, we reproduce it according to the published paper without the configuration file. From the results, we can summarize that:

The non-personalized POP method gets the worst performance on all datasets because of just considering the number of interactions. In general, the sequential methods outperform traditional non-sequential methods such as NCF due to successive sequential information. This observation explains sequential information is beneficial to the improvement of recommendation performance. Particularly, on sparse dataset Video Games, FPMC performs worse than NCF only based collaborative filtering. This means that these datasets only have little additional sequential information and the neural model having more parameters is magnificent to recommendation performance. Among sequential recommendation baselines, on dense dataset ML-1m, Caser gets better performance than FPMC, which suggests that high-order interactions are useful for long input sequences. Furthermore, SASRec outperforms RNN (GRU4Rec+), CNN (Caser) sequential model, on the whole, meaning that the self-attention mechanism is more powerful for sequential feature extraction. BERT4Rec basically gets better performance than SASRec, suggesting that bidirectional sequential information is beneficial for the recommendation system.

According to the results, our method improved the best baseline on all four datasets w.r.t. the three metrics, especially, gaining 9.93% MRR improvements (on average) against the strongest baselines. Compared with BERT4Rec, an additional matching task and more abundant samples make our model outperform by a large margin w.r.t. the three metrics, which means the matching task is an important auxiliary tool to improve recommendation performance.

Meanwhile, Fig. 3 visualizes average correlation coefficients of output sequences on Beauty of the first 10 items to qualitatively reveal the model's behavior. From the result, some tendencies can be concluded as follows: a) The users' representations are more affected by recent behavior, which is consistent with our common sense. Because recent items usually play a more important role in predicting the future. b) Items in our model tend to highlight the items on both sides, especially the surrounding items. This indicates bidirectional information has been mined successfully.

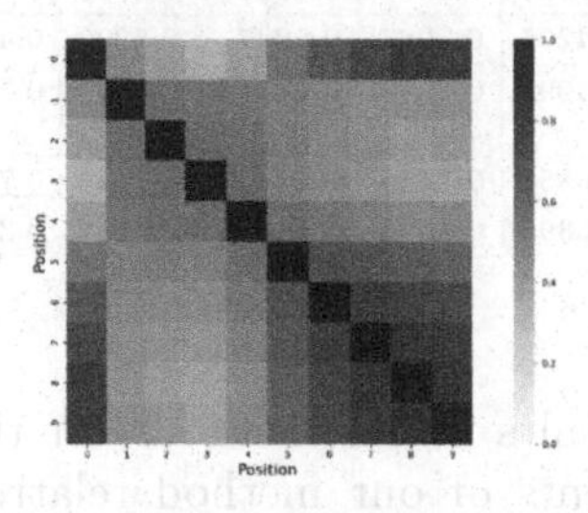

Fig. 3. Heat-map of average correlation coefficients of output sequences on Beauty at different positions. The first position "0" denotes "[UID]".

4.4 Ablation Study

Finally, we use the ablation study to analyze numerous key components of our model to better understand their impacts. Table 3 shows the results of our default version and its variants on all four datasets (with d = 32).

w/o PE. Without the position embedding, the sequential model becomes the sequential model based on isolated actions. The attention weight on each item depends only on item embedding, which leads to the rapid decline of recommendation performance, especially on dense datasets, because long sequences have more noise actions.

Table 3. Ablation analysis (MRR). The bold score indicates performance better than the default version, while "↓" indicates performance drop more than 10%.

Architecture	Beauty	V-Games	Steam	ML-1M
Default	0.2957	0.4396	0.5835	0.4973
w/o PE	0.2777	0.3911↓	0.5591	0.2955↓
w/o Matching Task	0.2406↓	0.4101	0.5462	0.4513
1 head ($h = 1$)	0.2818	0.4078	0.5763	0.4611
4 heads ($h = 4$)	0.2826	0.4216	**0.5841**	**0.5012**
1 layer ($L = 1$)	0.2756	0.4273	0.5713	0.4656
3 layers ($L = 3$)	**0.2981**	**0.4435**	**0.5907**	**0.5076**

w/o Matching Task. The variant only adopts the mask task as an objective task (like BERT4Rec [6]). The recommendation performance witnesses a noticeable decrease on sparse datasets (e.g., Beauty) because the phenomenon of the mask dilemma is more common for short sequences. (i.e., more masked items mean less available context information and vice versa.)

Head Number h**.** Multi-headed attention can expand the model's ability to focus on different positions. We observe that long sequence datasets benefit from a larger h (e.g., ML-1M), which means users' multiple interests are mined.

Layer Number L**.** The results demonstrate that hierarchical Transformer layers can help model learn more complicated item transition patterns. This confirms the validity of the self-attention mechanism.

4.5 Space and Time Complexity Analysis

A theoretical analysis of the time and space complexity is presented as follows:

Space Complexity. The learned parameters in our model are from the embedding layer, the transformer layers and the output layer. The total number of parameters is $O(|V|d + Nd + d^2)$, where $|V|$ means the number of the item set, d is the size of hidden dimension, N means the maximum sequence length. BERT4Rec [6] also has $O(|V|d + Nd + d^2)$.

Time Complexity. The time complexity of our model is mainly due to transformer layers, which is $O(dN^2 + Nd^2)$. BERT4Rec has $O(|V|d + Nd + d^2)$. With GPU acceleration and $N = 6$ in experiments, the difference is minor.

5 Conclusion

Recently, deep bidirectional sequential architecture proposed for neural language processing has brought impressive progress in recommender systems. In this paper, we optimize the bidirectional encoder representation recommender system via the additive matching task by the special token "[UID]" representing users. This method explicitly provides representations of users and captures user's global preference and main attentions in the sequence. Extensive experimental results on four real-world datasets indicate that our model outperforms state-of-the-art baselines. In the future, we will try to fuse heterogeneous interactions (e.g., purchase, review, clicks, etc.) in our model to achieve better performance.

Acknowledgements. This work was partial supported by National Natural Science Foundation of China (Grant No. 41876098)

References

1. Koren, Y., Bell, R.: Advances in collaborative filtering. In: Ricci, F., Rokach, L., Shapira, B. (eds.) Recommender Systems Handbook, pp. 77–118. Springer, Boston (2015). https://doi.org/10.1007/978-1-4899-7637-6_3
2. Wang, S., Hu, L., et al.: Sequential recommender systems: challenges, progress and prospects. In: IJCAI, pp. 6332–6338. Morgan Kaufmann, Macao (2019)
3. Rendle, S., Freudenthaler, C., Thieme, L.S.: Factorizing personalized Markov chains for next-basket recommendation. In: WWW, pp. 811–820. ACM, New York (2010)
4. Hidasi, B., Karatzoglou, A.: Recurrent neural networks with top-k gains for session-based recommendations. In: CIKM, pp. 843–852. ACM, New York (2018)
5. Kang, W.C., Julian, M.: Self-attentive sequential recommendation. In: ICDM, pp. 197–206. IEEE, Singapore (2018)
6. Sun, F., Liu, J., et al.: BERT4Rec: sequential recommendation with bidirectional encoder representations from transformer. In: CIKM, Beijing, China, pp. 1441–1450. ACM (2019)
7. Xu, J., He, X., Li, H.: Deep learning for matching in search and recommendation. In: SIGIR, Ann Arbor, USA, pp. 1365–1368. ACM (2018)
8. Devlin, J., Chang, M.W., et al.: BERT: pre-training of deep bidirectional transformers for language understanding. In: NAACL, NAACL, New Orleans, USA (2018)
9. Le, Q., Mikolov, T.: Distributed representations of sentences and documents. In: ICML, Beijing, China. ACM (2014)
10. Kim, D., Park, C., et al.: Convolutional matrix factorization for document context-aware recommendation. In: RecSys, pp. 233–240. ACM, New York (2016)
11. Kang, W.C., Fang, C., et al.: Visually-aware fashion recommendation and design with generative image models. In: ICDM, New Orleans, USA, pp. 207–216. IEEE (2017)
12. Oord, A.v.d., Dieleman, S., Schrauwen, B.: Deep content-based music recommendation. In: NIPS, pp. 2643–2651. MIT Press, Lake Tahoe (2013)
13. He, X., Liao, L., et al.: Neural collaborative filtering. In: WWW, Perth, Australia, pp. 173–182. ACM (2017)
14. Sedhain, S., Menon, A.K., et al.: AutoRec: autoencoders meet collaborative filtering. In: WWW, pp. 111–112. ACM, New York (2015)
15. Wu, Y., DuBois, C., et al.: Collaborative denoising auto-encoders for top-N recommender systems. In: WSDM, pp. 153–162. ACM, New York (2016)

16. Tang, J., Wang, K.: Personalized top-n sequential recommendation via convolutional sequence embedding. In: WSDM, Marina Del Rey, USA, pp. 565–573. ACM (2018)
17. Vaswani, A., Shazeer, N., et al.: Attention is all you need. In: NIPS, pp. 5998–6008. MIT Press, Long Beach (2017)
18. McAuley, J., Targett, C., et al.: Image-based recommendations on styles and substitutes. In: SIGIR, pp. 43–52. ACM, New York (2015)
19. Harper, F.M., Konstan, J.A.: The MovieLens datasets: history and context. ACM Trans. Interact. Intell. Syst. **5**(4), 19:1–19:19 (2015)

Multi-level Feature Extraction in Time-Weighted Graphical Session-Based Recommendation

Mei Yu[1,2,3], Suiwu Li[1,2,3], Ruiguo Yu[1,2,3], Xuewei Li[1,2,3], Tianyi Xu[1,2,3], Mankun Zhao[1,2,3], Hongwei Liu[4], and Jian Yu[1,2,3](✉)

[1] College of Intelligence and Computing, Tianjin University, Tianjin, China
{yumei,suiwuli2018,rgyu,lixuewei,tianyi.xu,zmk,yujian}@tju.edu.cn
[2] Tianjin Key Laboratory of Cognitive Computing and Application, Tianjin, China
[3] Tianjin Key Laboratory of Advanced Networking (TANK Lab), Tianjin, China
[4] Foreign Language, Literature and Culture Studies Center, Tianjin Foreign Studies University, Tianjin, China
liuhongwei@tjfsu.edu.cn

Abstract. Session-based recommendation aims to simulate users' behavior through a series of anonymous sessions. Recent research work mainly introduces deep learning into the recommender systems, and has achieved relatively good results. Previous research only focused on the clicked item thus ignoring the time information, that is dwell time for each item. It is undeniable that the length of dwell time on an item can reflect the user's preferences to a certain extent. And they lack the mining latent features of items. In this paper, we propose to explore multi-level feature extraction in time-weighted graphical session-based recommendation, abbreviated as F-TGNN. In F-TGNN, we first construct graphs for session sequences, in which the dwell time between two items is used as the weight of the corresponding edge. Then we use gated Graph Neural Network (GNN) to learn the transitions of items in the session sequence and obtain the embedding of each item. After that, we propose a Feature Extraction Module (FEM) to mine sequential patterns from item-level and contextual information between items from sequence-level. Finally, the predicted score for each item to be the next click is calculated. Extensive experiments conducted on two real datasets show that F-TGNN evidently outperforms the state-of-the-art session-based recommendation methods consistently.

Keywords: Session-based recommendation · Dwell time · Graph neural network · Feature Extraction

1 Introduction

Session-based recommender system (SRS) [8,10,14] takes into account the information embedded from one session to another and takes a session as the basic unit for recommendation. With the development of deep learning, more and more

H. Yang et al. (Eds.): ICONIP 2020, LNCS 12534, pp. 504–515, 2020.
https://doi.org/10.1007/978-3-030-63836-8_42

people have applied it to the SRS and achieved better results than traditional methods. [1] is first to use Recurrent Neural Network (RNN) in session-based recommendation and then improved by [11], who use data enhancement techniques and take user's time-shifting behavior into account. [3] design a global and local RNN recommender NARM to capture users' sequential behavior and main purposes simultaneously. Similar to NARM, STAMP [7] propose s short memory model to combines RNN and attention mechanism. But RNN-based methods focus on the final hidden state, so the general interest features of long-distance items may be omitted. At the same time, this one-way relationship is not applicable to some sequences with complex transitions. To solve these problems, GNN has been introduced to construct session graphs to learn the transitions between items. [16] first propose SR-GNN to use GNN with attention mechanism in SRS and take the result to a new level. [17] then further improves the result by applying the self-attention mechanism, which better considers the impact of long-term interest on recommended performance.

Above methods still have some limitations. First, most research work ignored the impact of dwell time on recommender performance but only regarded the positions of items in a session as a time click sequence. Second, most previous research is still insufficiency in mining user's sequential patterns and extracting the latent feature of items in the session.

In light of these problems, we introduce F-TGNN. The main contributions of this work are as follows:

- We add dwell time as the weight of edges in the process of building the session graph, which explicitly expresses the user's preference for items.
- We propose Feature Extraction Module to capture important features and mine sequential patterns at item-level and selecting contextual information between items at sequence-level. The model learn the feature interactions among the session embedding from local to global in a hierarchical manner.
- We conduct experiments on two real datasets which demonstrate the superior performance of the proposed model with other state-of-the-art methods.

2 Related Work

In this chapter, we introduce the deep learning-based recommendation method and the related work on Graph Neural Network.

Deep Learning-Based Approach. When it comes to session-based recommendations, [1] proposes RNN model (GRU4REC) to simulate sequential mode in SRS. On the base of [1,11] enhances the GRU4REC model by using data enhancement techniques and considering temporal changes in users' behavior. In addition to RNN, a neighborhood-based method to capture co-occurrence signals is proposed by [2]. In terms of using Convolutional Neural Network (CNN), [13] combine session clicks with content features such as item descriptions and item categories to generate recommendations by using a three-dimensional CNN. [12]

model recent items as an "image" and use horizontal/vertical convolutional for top-N sequential recommendation. [15] proposed a list-wise deep neural network model to train a ranking model. With the development of attention mechanism in deep learning, [3] introduces NARM, a neural attention recommender with encoder-decoder architecture, to capture users' sequential patterns and global purpose. After that, [7] proposes STAMP that utilizes MLP networks and an attention network to effectively capture the users' long-term interests and short-term interests.

Graph Neural Network Based Method. Recently, Graph neural network (GNN) is a type of method for processing graph domain information based on deep learning. Due to its better performance and interpretability, GNN has recently become a widely used graph analysis method, such as script event prediction [6], situation recognition [4], and image classification [9]. As for session-based recommendations, the SR-GNN model proposed by [16] constructs session sequences into graphs and uses gated GNN [5] for embedded representation. [17] then enhances SR-GNN model by adding a self-attention network to better capture global preference of users.

3 Method

In this section, we first introduce the definitions and notations of session-based recommendation, and then introduce our model F-TGNN in detail. Figure 1 shows our model F-TGNN. Specifically, our model F-TGNN is divided into the following steps. First, we dynamically construct graphs for session sequences with dwell time consideration and learn the transitions of items through gated GNN. Then, Feature Extraction Module is used to extract feature interactions and mine sequential patterns in the specific session from item-level and extract contextual information between items from sequence-level. Finally, the predicted score for each item to be the next click is calculated.

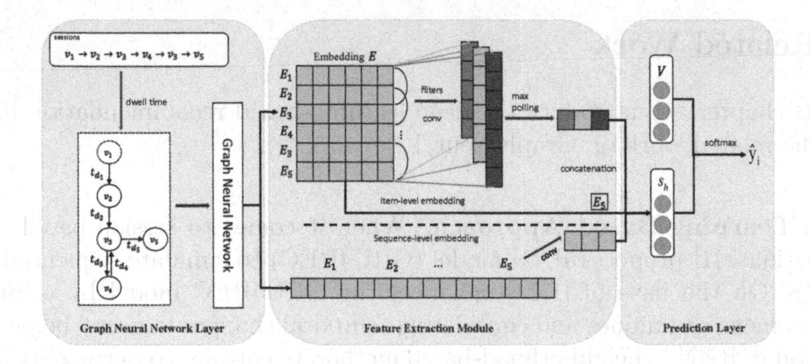

Fig. 1. The workflow of the proposed F-TGNN method.

3.1 Notations

In session-based recommendation, let $V = \{v_1, v_2, \cdots, v_{|V|}\}$ represents a set of unique items in all sessions. For an anonymous session $s = [v_{s_1}, v_{s_2}, \cdots v_{s_n}]$, can be ordered by its corresponding timestamp $t = [t_1, t_2, \cdots t_n]$, where $v_{s_i} \in V$ indicates an item that a user clicks in a session. Except for the last item, the dwell time corresponding to other items is $t_{dwell} = [t_{d_1}, t_{d_2}, \cdots t_{d_{(n-1)}}]$, where $t_{d_i} = t_{i+1} - t_i$. The goal of session-based recommendation is to predict the user's next click, which is $v_{s_{n+1}} \in V$ in s. In such a model, for a particular session, the probability $\hat{y}$ of all items indicating the score of the corresponding item. Finally, the items with top-k highest score will be the recommended candidate items.

3.2 Build a Session Graph

Each session can be constructed into a directed graph $G_s = (\mathcal{V}_s, \mathcal{E}_s)$, each node represents an item $v_{s_i} \in \mathcal{V}_s$ and each edge $(v_{s_{i-1}}, v_{s_i}) \in \mathcal{E}_s$ means that the user clicks v_{s_i} after clicking $v_{s_{i-1}}$.

A phenomenon commonly encountered in real life is that the more a user likes an item, the more time he will spend on this item, thus the corresponding dwell time will be longer. Based on this, we use the dwell time as the weight of each edge in the graph. Next, we introduce the process of constructing the adjacency matrix, which determines how nodes in the subgraph interact with each other. Adjacency matrix is represented by $A \in \mathbb{R}^{n \times 2n}$, where n is the number of nodes. Since multiple items may be repeated in sequence, a normalized weight value to each edge has been assigned, which is calculated as the dwell time of a certain edge divided by the sum of the dwell time of the starting node of the edge. Figure 2 is an example of a directed session graph and its adjacency matrix A.

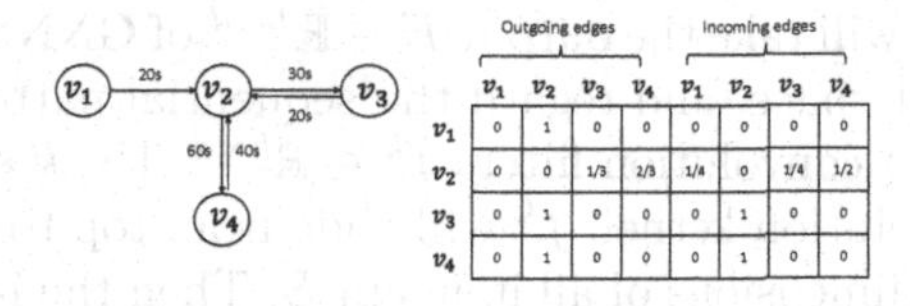

	v_1	v_2	v_3	v_4	v_1	v_2	v_3	v_4
v_1	0	1	0	0	0	0	0	0
v_2	0	0	1/3	2/3	1/4	0	1/4	1/2
v_3	0	1	0	0	0	1	0	0
v_4	0	1	0	0	0	1	0	0

Fig. 2. An example of a session graph and its adjacency matrix A.

3.3 Graph Neural Network Layer

After building the session graphs, the next thing we need to do is to use gated GNN to learn the embedding of items and the transitions between items in the session. The basic recurrence of gated GNN is as follows:

$$\mathrm{a}^{(\mathrm{t})} = A^T \mathrm{v}^{(t-1)} + b \quad (1)$$
$$\mathrm{z}^{(\mathrm{t})} = \sigma(W_{\mathrm{z}}\mathrm{a}^{(\mathrm{t})} + U_{\mathrm{z}}\mathrm{v}^{(t-1)}) \quad (2)$$
$$\mathrm{r}^{(\mathrm{t})} = \sigma(W_{\mathrm{r}}\mathrm{a}^{(\mathrm{t})} + U_{\mathrm{r}}\mathrm{v}^{(t-1)}) \quad (3)$$
$$\mathrm{c}^{(\mathrm{t})} = \tanh(W_{\mathrm{o}}\mathrm{a}^{(\mathrm{t})} + U_{\mathrm{o}}(\mathrm{r}^{(\mathrm{t})} \odot \mathrm{v}^{(t-1)})) \quad (4)$$
$$\mathrm{v}^{(\mathrm{t})} = (1 - \mathrm{z}^{(\mathrm{t})}) \odot \mathrm{v}^{(\mathrm{t}-1)} + \mathrm{z}^{(\mathrm{t})} \odot \mathrm{c}^{(\mathrm{t})} \quad (5)$$

The whole process is like a typical GRU-based update, which integrates information from other nodes and previous states to update the current hidden state of the target node . Where r and z are reset gate and update gate respectively. $\mathrm{v}^{(t-1)} = [\mathrm{v}_1^{t-1}, ..., \mathrm{v}_n^{t-1}]$ is the node embedding in the session, and each node $\mathrm{v_i} \in \mathbb{R}^d$, d is the dimension of the latent vector. σ is the logistic sigmoid function, and $\odot$ is element-wise multiplication. The final state is the combination of the previous hidden state and the candidate state, under the control of the update gate.

3.4 Feature Extraction Module (FEM)

In FEM, we use CNN to extract latent feature and mine sequential patterns from item-level and select contextual information between items from sequence-level.

Utilize CNN at Item-Level. [12,18,19] has used CNN for sequence recommendation, which shows that convolution facilitates the mining of sequential patterns within the serialized items. For example, when a user has clicked the pencil and eraser, his sequential pattern maybe "(pencil, eraser)→ pencil case". So when a convolution filter slides on the line of items a user clicked, it will select the sequential pattern by having larger values in the latent dimensions where pencil and eraser have larger values.

In our model, we will take the output $E \in \mathbb{R}^{L \times d}$ of GNN as the "image" of the L items in the latent space, and regard the sequential patterns as local features of "image". We use n convolution filters $f^k \in \mathbb{R}^{h \times d}$, $1 \leq k \leq n$, $h \in \{1, \cdots, L\}$ is the height of a convolution kernel. f^k will slide from top to bottom and interact with the horizontal dimensions of all items in E. Then the i-th convolution value is:

$$c_i^k = \Phi_c(E_{i:i+h-1} \cdot f^k + b_c) \quad (6)$$

where i is in the range of $[1, L - h + 1]$, $E_{i:i+h-1}$ denotes the i-th row to the row $i + h - 1$. $\cdot$ indicating the inner product of f^k and $E_{i:i+h-1}$. Φ_c is the activation function of the convolutional layer, b_c is a bias term. The convolution output of f^k is:

$$c^k = [c_1^k \, c_2^k \cdots c_{L-h+1}^k] \quad (7)$$

Then, to select the most important features extracted by the filters, we use the max pooling operation. The final output of the n convolution filters will be:

$$o = \{\max(c^1), \max(c^2), \cdots, \max(c^n)\} \quad (8)$$

where $o \in \mathbb{R}^n$, which is considered to be the specific feature extracted by the n convolution filters, i.e., the sequential patterns at item-level in the current session.

Utilize CNN at Sequence-Level. Different from that utilizing CNN at item-level, we horizontally connect sequential items to form an overall representation of the session sequence, denoted by $E_s \in \mathbb{R}^{1\times(L\times d)}$. We think that the overall embedding E_s keeps the timing information and the context information of each item. Then we use convolution filter $f_s^k \in \mathbb{R}^{1\times(L\times d)}$ to extract this information at sentence-level (the process is similar to that at item-level), which makes the extracted features more continuous. The final output of the n convolution filters is denoted by $\widetilde{o} \in \mathbb{R}^n$, which selects the important contextual information from a global perspective.

3.5 Prediction Layer

After the convolution, we cascade the output of the FEM with the last vector of the GNN output E_L, and then perform a linear transformation:

$$s_h = W' \begin{bmatrix} o \\ \widetilde{o} \\ E_L \end{bmatrix} + b' \tag{9}$$

where $W' \in \mathbb{R}^{d\times(d+2n)}$ and $b' \in \mathbb{R}^d$ are the weight matrix and the bias term respectively, E_L is the last item embedding of current session. Similar to [7,16], adding the final state E_L can better represent sequential patterns because the item that the user will click on is largely related to the item that was clicked recently. o intend to capture user's sequential patterns at item-level and $\widetilde{o}$ is proposed to capture important context information at sentence-level.

Then, the score s_i is computed:

$$\widehat{s}_i = s_h^T \cdot \mathrm{v_i} \tag{10}$$

Finally, a softmax function is utilized to get the output vector of the model:

$$\widehat{y} = soft\max(\widehat{s}) \tag{11}$$

where $\widehat{s} \in \mathbb{R}^{|V|}$ denotes the score of all items in the item set, and $\widehat{y} \in \mathbb{R}^{|V|}$ indicates the probability of appearing in the ranking list.

3.6 Objective Function

The objective function can be defined as the cross-entropy of the prediction score $\widehat{y}$:

$$L(y, \widehat{y}) = -\sum_{i=1}^{|V|} y_i \cdot \log(\widehat{y}_i) + \lambda\|\theta\|^2 \tag{12}$$

where y is the true probability distribution, $\widehat{y}$ is the predicted rough distribution, θ represents all the parameters that can be learned, and λ is the regularization weight.

4 Experiments

The structure of the experimental part is as follows: (1) Introduce the datasets, evaluation metrics and comparison methods. (2) Comparison with other baseline methods. (3) Ablation analysis and show the experimental results under different experimental parameters. The code is available at[1].

4.1 Datasets

Two public datasets i.e., Yoochoose[2] and Diginetica[3] are used in our experiment. As the same setting of [3,7,16], we filter out the session sequence with length 1 and the items that occur less than 5 times in the datasets. In addition, same as [11,16], according to the time order of a sequence, we generate the corresponding subsequences. More specifically, for a session sequence $s = [v_{s_1}, v_{s_2}, \cdots, v_{s_n}]$, we generate a series of sequences and labels $([v_{s_1}], v_{s_2}), ([v_{s_1}, v_{s_2}], v_{s_3}), \cdots, ([v_{s_1}, v_{s_2}, \cdots, v_{s_{n-1}}], v_{s_n})$. Since the data volume of Yoochoose dataset is very large, similar to [3,7,16], we use its most recent 1/64 fragment. The detailed description of the two datasets is shown in Table 1.

Table 1. Statistics of the datasets used in our experiments

Datasets	Yoochoose 1/64	Diginetica
All the clicks	557248	982961
Train sessions	369859	719470
Test sessions	55898	60858
All the items	16766	43097
Average length	6.16	5.12

4.2 Evaluation Metrics

Like [16], we use P@20 and MRR@20 as evaluation metrics.

P@20: Precision@20 is used to measure the accuracy of recommenders. It is the proportion of desired items in the top 20 in all test cases.

$$P@k = \frac{n_{hit}}{N} \tag{13}$$

where N is the number of test cases, n_{hit} means the number of cases when the desired item is in the top-k ranking list.

---

[1] https://github.com/DebonairLi/F-TGNN.
[2] http://2015.recsyschallenge.com/challege.html.
[3] http://cikm2016.cs.iupui.edu/cikm-cup.

MRR@20: Average reciprocal rank is the average of the reciprocal ranks of the desired items in all test cases. The reciprocal rank is set to zero if the rank is larger than 20. The larger the value of MRR, the better the recommendation effect.

$$MRR@k = \frac{1}{N}\sum_{t \in R} \frac{1}{rank(t)} \tag{14}$$

where R denotes the top-k ranking list.

4.3 Parameter Setup

The dimension of the embedded vector used in our model is d = 50 in Diginetica and 100 in Yoochoose. All parameters are initialized using a Gaussian distribution with a mean of 0 variance of 0.01. All parameters are optimized through, in which the initial learning rate is 0.001 and will decay by 0.1 after every 3 epochs. In addition, the value of batch size is 100, and the L2 penalty is 10^{-5}. The number of convolution filters are 32 in Diginetica and 64 in Yoochoose1/64 respectively.

4.4 Baseline Methods

To prove the validity of proposed method F-TGNN, we compare it with the latest deep learning-based methods:

GRU4REC [1]: It is the first method that utilize RNN to build users' sequential patterns in session-based recommendation.

NARM [3]: Based on RNN, they propose to use attention mechanism for session-based recommendation to capture users' general purpose and short-term interest.

STAMP [7]: A model that learns long and short interest with a short-term attention/memory priority module.

SR-GNN [16]: Use GNN to learn the transitions of items in a session, and then utilize soft-attention to learn global preference and the current interest.

GC-SAN [17]: Based on SR-GNN, add self-attention mechanism to better consider long-term interests.

4.5 Comparison with Baseline Methods

Table 2 shows the performance of the baseline methods and our method F-TGNN on two metrics, in which the best results are indicated in bold type. And the best results proves the validity of our model.

GRU4REC is a first method that uses GRU in SRS, which proves the deep learning-based methods have good result in SRS. With the development of attention mechanism, NARM and STAMP use it with neural networks and improves

Table 2. The performance of FTSR with other baseline methods over two datasets

Method	Yoochoose 1/64		Diginetica	
	P@20	MRR@20	P@20	MRR@20
GRU4REC	60.64	22.89	29.45	8.33
NARM	68.32	28.63	49.70	16.17
STAMP	68.74	29.67	45.64	14.32
SR-GNN	70.57	30.94	50.73	17.59
GC-SAN	70.77	30.78	50.99	17.45
F-TGNN	**71.30**	**31.48**	**52.78**	**18.48**

the results further. Recently, SR-GNN first propose to use Gated GNN for learning session graphs, which is used to learn the contextual transitions instead of one-way relations. On the base of SR-GNN, GC-SAN add self-attention blocks to consider long-term interest more accurately.

However, compared to above methods, our model F-TGNN first add dwell into the process of constructing session graph to express a user's preference for certain items. Then we take session as an image and use Feature Extraction Module to mine sequential patterns at item-level and extract contextual information at sequence-level. The results show correctness of our method.

4.6 Ablation Analysis

For in-depth analysis, we propose related variants:
(1) SR-GNN: The baseline SR-GNN proposed in 2019.
(2) SR-GNN+Time: The SR-GNN model adds dwell time in the process of constructing session graph.
(3) SR-GNN+FEM: The SR-GNN model adds Feature Extraction Module to select features and mine sequential patterns, and we remove the soft-attention from original SR-GNN model.
(4) F-TGNN: The overall F-TGNN model we propose.

We keep other parameter settings stay the same and the result of SR-GNN, F-TGNN and its related variants is shown in Fig. 3 and Fig. 4.

First of all, from the figure we can see that the experimental results of F-TGNN and its variants are superior to SR-GNN using only GNN, which shows that our method is effective in session-based recommendation. Then, it can be seen from (1), (2) that when we add dwell time into the process of constructing session graphs, the result improves, proving that considering time attribute helps sequential modeling. From (1) and (3), we can conclude that the use of FEM has an obvious effect on the improvement of the model effect and is the core component of our F-TGNN. At item-level, we use CNN to extract features and mine sequential patterns and at sequence-level, we use CNN to select important contextual information among items. Finally, as can be seen from (2), (3) and (4), If we combine the first two, the effect is the best.

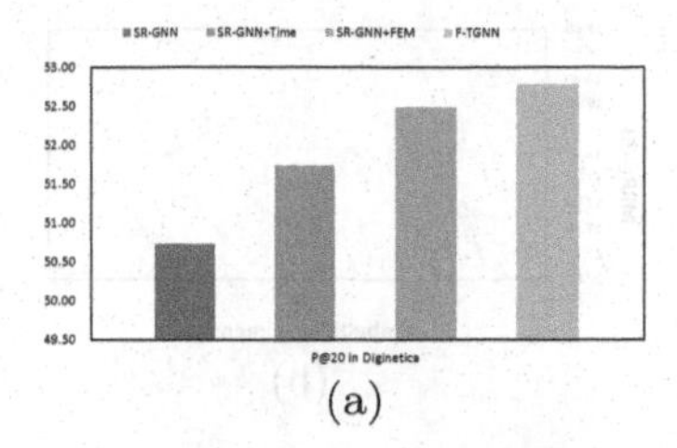

(a)

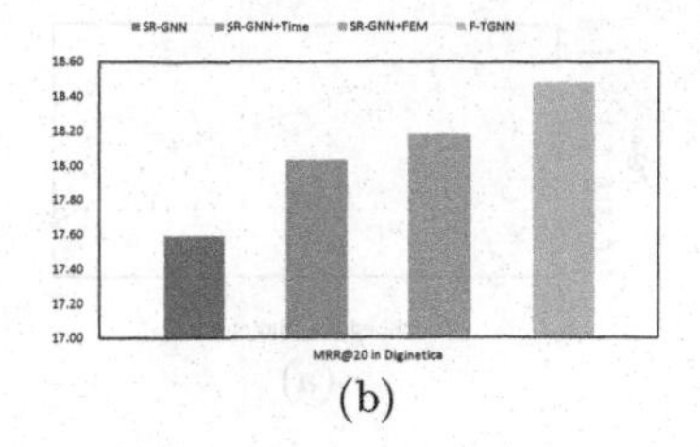

(b)

Fig. 3. The performance of SR-GNN, F-TGNN and it's related variants in Diginetica. This figure should be viewed in colour. (Color figure online)

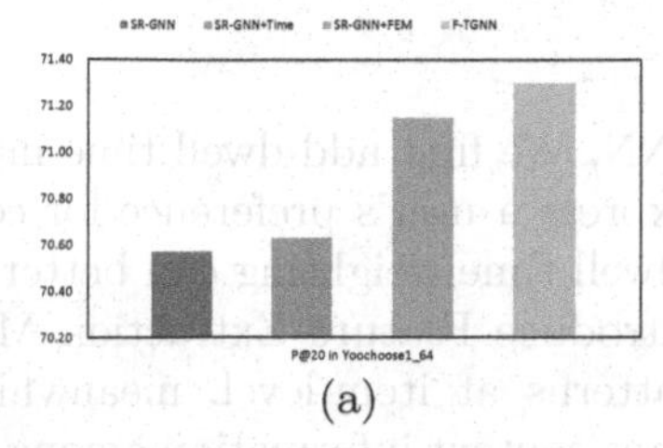

(a)

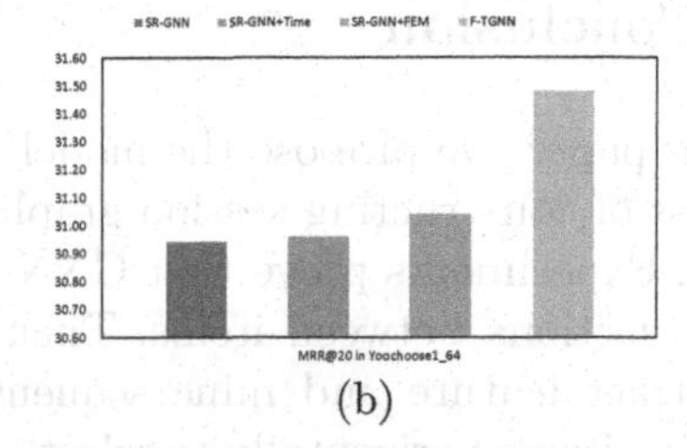

(b)

Fig. 4. The performance of SR-GNN, F-TGNN and it's related variants in Yoochoose1/64. This figure should be viewed in colour. (Color figure online)

4.7 The Sensitivity of Hyper-parameters

Due to the space limit, we only analyse the effects on Diginetica.

Table 3. Influence of different number of convolution filters on experimental results in Diginetica.

Number of filters (n)	4	8	16	32	64
P@20	51.37	51.60	52.36	**52.78**	52.46
MRR@20	17.78	17.85	18.17	**18.48**	18.09

Firstly, different convolution filters on experimental results are compared and the results are show in Table 3. It can be seen that as the number of convolution filters increases, the value of both metrics first increase and then decrease, and reach the maximum when $n = 32$. Therefore, the number of convolution filters should be selected appropriately. If the number is too small, important features cannot be fully extracted. If the number is too large, the extracted features may be contrary to the prediction, and will increase the model training time.

Secondly, we analyze the effect of different embedding dimensions on the experimental results. From Fig. 5, we can see that as the embedding dimension increases, both metrics increase first and then decrease. So we choose the embedding dimension 50 as the best choice for the experimental parameters.

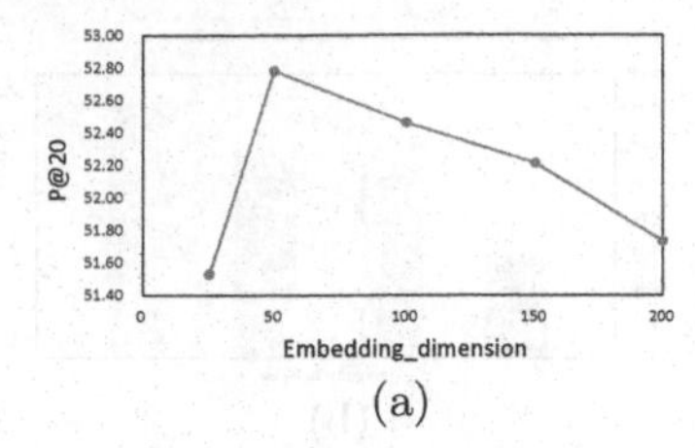

(a)

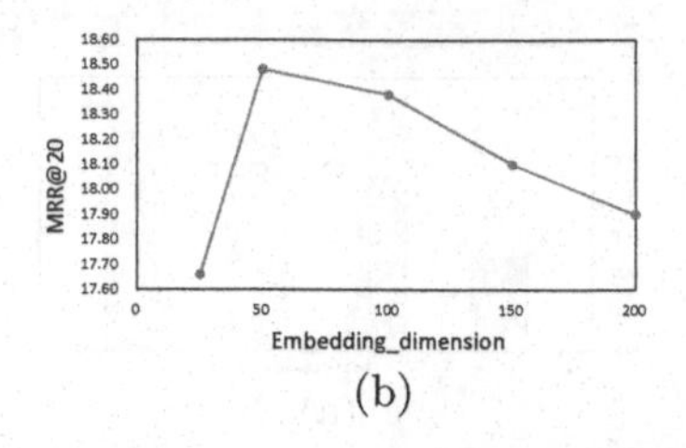

(b)

Fig. 5. The performance of embedding dimension on dataset Diginetica.

5 Conclusion

In this paper, we propose the model F-TGNN. We first add dwell time into the process of constructing session graphs to express a user's preference for certain items. Experiments prove that GNN with dwell time weighting can better learn the transitions between items. Then we introduce Feature Extraction Module to extract feature and mine sequential patterns at item-level, meanwhile we combine items horizontally to select important context information among items at sequence-level, which improves the deficiencies of previous work on feature modeling of session items. Experimental results from two public datasets show the superiority of F-TGNN over state-of-the-art methods.

Acknowledgments. This work is jointly supported by National Natural Science Foundation of China (61877043) and National Natural Science Foundation of China (61877044).

References

1. Hidasi, B., Karatzoglou, A., Baltrunas, L., Tikk, D.: Session-based recommendations with recurrent neural networks. In: ICLR (2016). https://iclr.cc/archive/www/doku.php%3Fid=iclr2016:accepted-main.html
2. Jannach, D., Ludewig, M.: When recurrent neural networks meet the neighborhood for session-based recommendation. In: RecSys 2017, pp. 306–310 (2017). https://doi.org/10.1145/3109859.3109872
3. Li, J., Ren, P., Chen, Z., Ren, Z., Lian, T., Ma, J.: Neural attentive session-based recommendation. In: CIKM, pp. 1419–1428 (2017). https://doi.org/10.1145/3132847.3132926
4. Li, R., Tapaswi, M., Liao, R., Jia, J., Urtasun, R., Fidler, S.: Situation recognition with graph neural networks. In: ICCV, pp. 4183–4192 (2017). https://doi.org/10.1109/ICCV.2017.448
5. Li, Y., Tarlow, D., Brockschmidt, M., Zemel, R.S.: Gated graph sequence neural networks. In: ICLR 2016 (2016). https://iclr.cc/archive/www/doku.php%3Fid=iclr2016:accepted-main.html
6. Li, Z., Ding, X., Liu, T.: Constructing narrative event evolutionary graph for script event prediction. In: IJCAI 2018, pp. 4201–4207 (2018). https://doi.org/10.24963/ijcai.2018/584

7. Liu, Q., Zeng, Y., Mokhosi, R., Zhang, H.: STAMP: short-term attention/memory priority model for session-based recommendation. In: SIGKDD, pp. 1831–1839 (2018). https://doi.org/10.1145/3219819.3219950
8. Ludewig, M., Jannach, D.: Evaluation of session-based recommendation algorithms. User Model. User Adap. Inter. **28**, 331–390 (2018). https://doi.org/10.1007/s11257-018-9209-6
9. Marino, K., Salakhutdinov, R., Gupta, A.: The more you know: using knowledge graphs for image classification. In: CVPR, pp. 20–28 (2017). https://doi.org/10.1109/CVPR.2017.10
10. Quadrana, M., Jannach, D., Cremonesi, P.: Tutorial: sequence-aware recommender systems. In: WWW 2019, p. 1316 (2019). https://doi.org/10.1145/3308560.3320091
11. Tan, Y.K., Xu, X., Liu, Y.: Improved recurrent neural networks for session-based recommendations. In: DLRS, pp. 17–22 (2016). https://doi.org/10.1145/2988450.2988452
12. Tang, J., Wang, K.: Personalized top-n sequential recommendation via convolutional sequence embedding. In: WSDM, pp. 565–573 (2018). https://doi.org/10.1145/3159652.3159656
13. Tuan, T.X., Phuong, T.M.: 3D convolutional networks for session-based recommendation with content features. In: RecSys, pp. 138–146 (2017). https://doi.org/10.1145/3109859.3109900
14. Wang, S., Cao, L., Wang, Y.: A survey on session-based recommender systems. CoRR (2019). http://arxiv.org/abs/1902.04864
15. Wu, C., Yan, M.: Session-aware information embedding for e-commerce product recommendation. In: CIKM 2017, pp. 2379–2382 (2017). https://doi.org/10.1145/3132847.3133163
16. Wu, S., Tang, Y., Zhu, Y., Wang, L., Xie, X., Tan, T.: Session-based recommendation with graph neural networks. In: AAAI, pp. 346–353 (2019). https://doi.org/10.1609/aaai.v33i01.3301346
17. Xu, C., et al.: Graph contextualized self-attention network for session-based recommendation. In: IJCAI, pp. 3940–3946 (2019)
18. Yan, A., Cheng, S., Kang, W., Wan, M., McAuley, J.J.: CosRec: 2D convolutional neural networks for sequential recommendation. In: CIKM 2019, pp. 2173–2176 (2019). https://doi.org/10.1145/3357384.3358113
19. Yuan, F., Karatzoglou, A., Arapakis, I., Jose, J.M., He, X.: A simple convolutional generative network for next item recommendation. In: WSDM 2019, pp. 582–590 (2019). https://doi.org/10.1145/3289600.3290975

Time Series Analysis

3ETS+RD-LSTM: A New Hybrid Model for Electrical Energy Consumption Forecasting

Grzegorz Dudek[1(✉)], Paweł Pełka[1], and Slawek Smyl[2]

[1] Electrical Engineering Faculty, Częstochowa University of Technology, Częstochowa, Poland
{dudek,p.pelka}@el.pcz.czest.pl

[2] Uber Technologies, 555 Market Street, San Francisco, CA 94104, USA
slaweks@hotmail.co.uk

Abstract. This work presents an extended hybrid and hierarchical deep learning model for electrical energy consumption forecasting. The model combines initial time series (TS) decomposition, exponential smoothing (ETS) for forecasting trend and dispersion components, ETS for deseasonalization, advanced long short-term memory (LSTM), and ensembling. Multi-layer LSTM is equipped with dilated recurrent skip connections and a spatial shortcut path from lower layers to allow the model to better capture long-term seasonal relationships and ensure more efficient training. Deseasonalization and LSTM are combined in a simultaneous learning process using stochastic gradient descent (SGD) which leads to learning TS representations and mapping at the same time. To deal with a forecast bias, an asymmetric pinball loss function was applied. Three-level ensembling provides a powerful regularization reducing the model variance. A simulation study performed on the monthly electricity demand TS for 35 European countries demonstrates a high performance of the proposed model. It generates more accurate forecasts than its predecessor (ETS+RD-LSTM [1]), statistical models such as ARIMA and ETS as well as state-of-the-art models based on machine learning (ML).

Keywords: Exponential smoothing · Long short-term memory · Mid-term load forecasting

1 Introduction

The power system load is a nonlinear and nonstationary process that can change rapidly due to many factors such as macroeconomic variations, weather, electricity prices, consumer types and habits, etc. Therefore, electricity demand forecasting, which is essential for the power system operation and planning, is a big

The project financed under the program of the Polish Minister of Science and Higher Education titled "Regional Initiative of Excellence", 2019–2022. Project no. 020/RID/2018/19, the amount of financing 12,000,000.00 PLN.

H. Yang et al. (Eds.): ICONIP 2020, LNCS 12534, pp. 519–531, 2020.
https://doi.org/10.1007/978-3-030-63836-8_43

challenge. In this study we consider mid-term electrical load forecasting (MTLF) focusing on monthly electricity demand forecasting over 12 months horizon.

MTLF methods can be roughly classified into statistical/econometrics methods or ML methods [2]. The former include ARIMA, ETS and linear regression (LR). ARIMA and ETS can deal with seasonal TS but LR requires additional operations such as decomposition or extension of the model with periodic components [3]. Limited adaptability of the statistical MTLF models and problems with nonlinear relationship modeling have increased researchers' interest in ML and AI tools [4]. Of these, neural networks (NNs) are the most popular because of their attractive features including learning capabilities, universal approximation property, nonlinear modeling and massive parallelism. Some examples of using NNs for MLTF are: [5] where NN uses historical loads and weather variables to predict monthly demand and is trained by heuristic algorithms to improve performance, [6] where Kohonen NN is used, [7] where NNs are supported by fuzzy logic, [8] where generalized regression NN is used, [9] where weighted evolving fuzzy NN is used, and [10] where NNs, LR and AdaBoost are combined.

Recent trends in ML such as deep recurrent NNs (RNNs), are very attractive for TS forecasting [11]. RNNs are able to exhibit temporal dynamic behavior using their internal state to process sequences of inputs. Recent works have reported that RNNs, such as the LSTM, provide high accuracy in forecasting and outperform most of the traditional statistical and ML methods [12]. Some application examples of LSTMs to load forecasting can be found in [13–15].

In [1] we proposed a hybrid residual dilated LSTM and ETS model (ETS+RD-LSTM) for MTLF. This model was based on the winning submission to the M4 forecasting competition 2018 [16], developed by Slawek Smyl [17]. A simulation study confirmed the high performance of the model and its competitiveness with classical models such as ARIMA and ETS as well as state-of-the-art ML models. In this work we extend ETS+RD-LSTM by introducing initial TS normalization, i.e. detrending and unifying the variance. This method of TS preprocessing we used in our previous works achieving very good results [18]. Recently we used it for LSTM model obtaining a 15% reduction in error [19]. We expect that the TS initial normalization, which simplifies the relationship between input and output data, allows ETS+RD-LSTM to improve its performance.

2 Forecasting Model

The proposed model is a modified version of ETS+RD-LSTM which we described in [1]. We extend ETS+RD-LSTM by introducing initial TS normalization. A normalization procedure removes a trend and unifies variance of the TS. The normalized TS exhibit yearly patterns which are further removed using deseasonalization as an integral part of ETS+RD-LSTM. The normalized and deseasonalized TS are forecasted using RD-LSTM. Then the forecasts are reseasonalized using seasonal components extracted by ETS. To reduce the model variance, we use ensembling at three levels which aggregates individual forecasts.

The resulting aggregated forecasts are finally denormalized. For denormalization, the forecasts of the mean yearly demand and yearly dispersion are needed. They are produced by additional two ETS modules.

2.1 Architecture and Features

An architecture of the proposed forecasting system is shown in Fig. 1. The system components are as follows (symbols in italics denote the sets of TS or forecasts):

- Normalization – each original monthly electricity demand TS is normalized. This procedure removes a trend from the TS and unifies its variance. Normalization module loads a set of TS (Z), calculates the series of yearly mean demands ($\overline{Z}$) and yearly dispersions (Σ) for each TS, and determines normalized series (Y).
- ETS – exponential smoothing modules for forecasting the yearly mean demands and their dispersions. These values are necessary for denormalization.
- Deseasonalization – each normalized TS is deseasonalized. This procedure extracts the seasonal components, S, individually for each series using ETS (ETSd module), and determines deseasonalized TS, X.
- RD-LSTM – residual dilated LSTM for forecasting the normalized and deseasonalized TS, X.
- Reseasonalization – each TS forecast produced by RD-LSTM is reseasonalized using inverse operations to deseasonalization.
- Ensembling – the reseasonalized forecasts produced by RD-LSTM are averaged. The ensembling module receives the sets of individual forecasts, $\hat{Y}_k^r$, and returns an aggregated forecast for each TS, $\hat{Y}_{\text{avg}}$.
- Denormalization – the averaged forecasts $\hat{Y}_{\text{avg}}$ are denormalized using forecasted values of the yearly means, $\hat{\overline{Z}}$, and dispersions, $\hat{\Sigma}$.

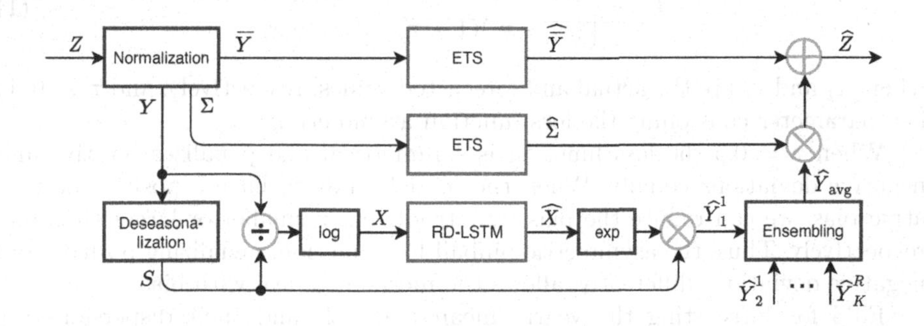

Fig. 1. The proposed forecasting system architecture

The proposed system has a hybrid and hierarchical structure. It combines statistical modeling (ETS), advanced ML (RD-LSTM), and ensembling. ETS is used as a forecasting model for yearly means, $\overline{Z}$, and dispersions, Σ, as well as for extraction of seasonal components (ETSd). The preprocessed TS, without trend and seasonal variations, are forecasted using RD-LSTM. Details of data preprocessing and flow are described in Subsect. 2.2.

The TS are exploited in a hierarchical manner, meaning that both local and global components are utilized in order to extract and combine information at either a series or a dataset level, thus enhancing the forecasting accuracy. The global features are learned by RD-LSTM across many TS (cross-learning). The specific features of each individual TS, such as trend, variance, and seasonality, are extracted by normalization and ETSd modules. Thus, each series has a partially unique and partially shared model.

The strength of RD-LSTM, which revealed in M4 competition, is cross-learning, i.e., using many series to train a single model. This is unlike standard statistical TS algorithms, where a separate model is developed for each series. Another important ingredient in the success of the proposed method precursor in the M4 competition was the on-the-fly preprocessing that was an inherent part of the training process. Crucially, the parameters of this preprocessing (in the proposed model these are twelve initial seasonal components and smoothing coefficient β, see Subsect. 2.2) were being updated by the same overall optimization procedure (SGD) as weights of RD-LSTM, with the overarching goal of minimizing forecasting errors. This enables the model to simultaneous optimization of data representation, i.e. searching for the most suitable representations of input and output data for RD-LSTM, and forecasting performance.

ETSd is used as the preprocessing tool. It extracts a seasonal component which is used for deseasonalization of the normalized TS. ETSd was inspired by the Holt-Winters multiplicative seasonal model. However, it has been simplified by removing trend and level components (see Subsect. 2.2). This is because the input TS are normalized, i.e. they have no trend and their level is one. ETSd is optimized simultaneously with RD-LSTM using pinball loss function [17]:

$$L_t = \begin{cases} (x_t - \hat{x}_t)\tau & \text{if } x_t \geq \hat{x}_t \\ (\hat{x}_t - x_t)(1-\tau) & \text{if } \hat{x}_t > x_t \end{cases} \tag{1}$$

where x_t and $\hat{x}_t$ are the actual and forecasted values, respectively, and $\tau \in (0, 1)$ is a parameter controlling the loss function asymmetry.

When $\tau = 0.5$ the loss function is symmetrical and penalizes positive and negative deviations equally. When the model tends to have a positive or negative bias, we can reduce the bias by introducing τ smaller or larger than 0.5, respectively. Thus, the asymmetric pinball loss function, penalizing positive and negative deviations differently, allows the method to deal with bias.

ETS for forecasting the yearly mean demands and their dispersions are defined as innovations state space models [20]. They combine the seasonal, trend and error components in different ways (additively or multiplicatively). For each TS the optimal ETS model is selected using Akaike information criterion (AIC).

Ensembling is used for reduction the model variance related to the stochastic nature of SGD, and also related to data and parameter uncertainty. Ensembling is seen as a much more powerful regularization technique than more popular alternatives, e.g. dropout or L2-norm penalty [21]. In our case, ensembling combines individual forecasts at three levels: stage of training level, data subset level and model level. At the stage of training level, the forecasts produced by L most recent training epochs are averaged. This can reduce the effect of stochastic searching, i.e. calming down the noisy SGD optimization process. At the data subset level, we use K models which learn on the subsets of the training set, $\Psi_1, \Psi_2, ..., \Psi_K$. Each k-th model produces forecasts for TS included in its own training subset Ψ_k. Then the forecasts produced by the pool of K models are averaged individually for each TS. The third level of ensembling simply averages the forecasts for each TS generated in R independent runs of a pool of K models. In each run, the training subsets Ψ_k are created anew (see [1] for details).

2.2 Time Series Processing

A monthly electricity demand TS exhibits a trend, yearly seasonality and random component (see Fig. 2(a)). To simplify the forecasting problem, the TS is preprocessed as follows. Let $\{z_t\}_{t=1}^N$ be a monthly electricity demand TS starting from January and ending in December. This TS is divided into yearly subsequences $\{z_t^i\}_{t=12(i-1)+1}^{12(i-1)+12}$, $i = 1, ..., N/12$. Each i-th subsequence is expressed by a vector $\mathbf{z}_i = [z_{i,1} z_{i,2} \dots z_{i,12}]^T$. The normalized version of $\mathbf{z}_i$, $\mathbf{y}_i = [y_{i,1} y_{i,2} \dots y_{i,12}]^T$, is determined as follows:

$$y_{i,j} = \frac{z_{i,j} - \overline{z}_i}{\sigma_i} + 1 \tag{2}$$

where $j = 1, ..., 12$, $\overline{z}_i$ is a mean of subsequence $\{z_t^i\}$, and $\sigma_i = \sqrt{\sum_{j=1}^{12}(z_{i,j} - \overline{z}_i)^2}$ is a measure of its dispersion.

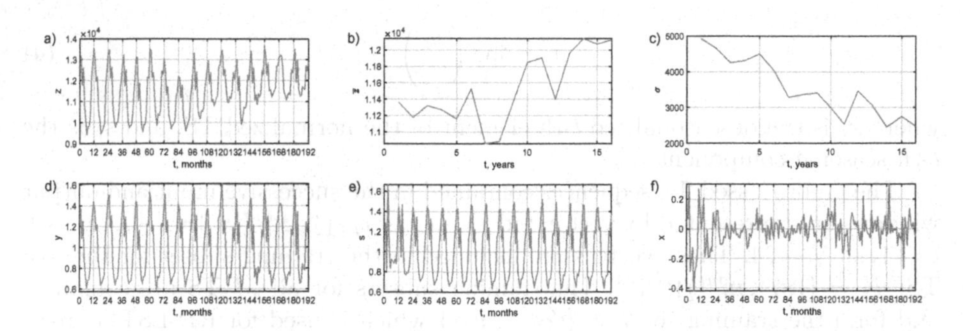

Fig. 2. TS preprocessing: (a) original TS $\{z_t\}$, (b) yearly mean demand TS $\{\overline{z}_t\}$, (c) yearly dispersion TS $\{\sigma_t\}$, (d) normalized TS $\{y_t\}$, (e) seasonal component TS $\{s_t\}$, and (f) normalized and deseasonalized TS $\{x_t\}$

Note that yearly subsequences $\{z_t^i\}$ have different means and dispersions (see Fig. 2(a)). After normalization they are unified, i.e. all yearly subsequences have an average of one, the same variance and also a unity of length. They carry information about the shapes of the yearly sequences. Now we create a new TS composed of normalized subsequences representing successive yearly periods: $\{y_t\} = \{\mathbf{y}_i^T\}_{i=1}^{N/12} = \{y_{1,1}, y_{1,2}, ..., y_{N/12,12}\}$. This TS is shown in Fig. 2(d). Note its regular character and stationarity.

The TS of the mean yearly demand, $\{\overline{z}_i\}_{i=1}^{N/12}$, and yearly dispersion, $\{\sigma_i\}_{i=1}^{N/12}$, are shown in Fig. 2(b) and (c), respectively. They are forecasted by ETS one step ahead (for the next year) and used for denormalization.

The normalized TS, $\{y_t\}$, is further deseasonalized. To do so, we use a simplified Holt-Winters multiplicative seasonal model with only one component:

$$s_{t+12} = \beta y_t + (1 - \beta)s_t \tag{3}$$

where s_t is the seasonal component at timepoint t and $\beta \in [0, 1]$ is a smoothing coefficient.

The seasonal component is shown in Fig. 2(e). It is used for deseasonalization during the on-the-fly preprocessing. The TS $\{y_t\}$ is deseasonalized in each training epoch using the updated values of seasonal components. These updated values are calculated from (3), where parameters, 12 initial seasonal components and β, are increasingly fine tuned for each TS in each epoch by SGD.

The TS is deseasonalized using rolling windows: input and output ones. The input window contains twelve consecutive elements of the TS which after deseasonalization will be the RD-LSTM inputs. The corresponding output window contains the next twelve consecutive elements, which after deseasonalization will be the RD-LSTM outputs. The TS fragments inside both windows are deseasonalized by dividing them by the relevant seasonal component. Then, to limit the destructive impact of outliers on the forecasts, a squashing function, log(.), is applied. The resulting deseasonalization can be expressed as follows:

$$x_t = \log\left(\frac{y_t}{s_t}\right) \tag{4}$$

where x_t is the deseasonalized t-th element of the normalized TS, and s_t is the t-th seasonal component.

The preprocessed TS sequences contained in the successive input and output windows are represented by vectors: $\mathbf{x}_t^{in} = [x_t \dots x_{t+12}]$, $\mathbf{x}_t^{out} = [x_{t+13} \dots x_{t+24}]$, $t = 1, ..., N - 24$. These vectors are included in the training subset for the i-th TS: $\Phi_i = \{(\mathbf{x}_t^{in}, \mathbf{x}_t^{out})\}_{t=1}^{N-24}$. The training subsets for all M TS are combined and form the training set $\Psi = \{\Phi_1, ..., \Phi_M\}$ which is used for RD-LSTM cross-learning. Note the dynamic character of the training set. It is updated in each epoch because the seasonal components in (4) are updated by SGD.

The forecasts produced by RD-LSTM, $\hat{x}_t$, are reseasonalized as follows:

$$\hat{y}_t = s_t \exp(\hat{x}_t) \tag{5}$$

where s_t is determined from (3) on the basis of the TS history.

Finally, the TS is denormalized using transformed Eq. (2):

$$\hat{z}_{i,j} = (\hat{y}_{i,j} - 1)\hat{\sigma}_i + \hat{\bar{z}}_i \tag{6}$$

where i refers to the forecasted yearly period, $j = 1, ..., 12$, $\hat{\bar{z}}_i$ and $\hat{\sigma}_i$ are the forecasted yearly mean and dispersion for period i.

2.3 Residual Delated LSTM

The RD-LSTM architecture used in this study is shown in Fig. 3(a) [1]. It is composed of four recurrent layers and a linear unit LU. The first layer consists of the standard LSTM block shown in Fig. 3(b). The subsequent three layers consist of RD-LSTM blocks, i.e. blocks equipped with dilated recurrent skip connections and a spatial shortcut path from lower layers (Fig. 3(c)).

A standard LSTM block consists of hidden state $\mathbf{h}_t$ and cell state $\mathbf{c}_t$. The cell state contains information learned from the previous time steps which can be added to or removed from the cell state using the gates: input gate (i), forget gate (f) and output gate (o). At each time step t, the block uses the past state, $\mathbf{c}_{t-1}$ and $\mathbf{h}_{t-1}$, and input $\mathbf{x}_t$ to compute output $\mathbf{h}_t$ and updated cell state $\mathbf{c}_t$. The hidden and cell states are recurrently connected back to the block input. All of the gates are controlled by the hidden state of the past cycle and input $\mathbf{x}_t$. The equations for a standard LSTM block are shown in Table 1.

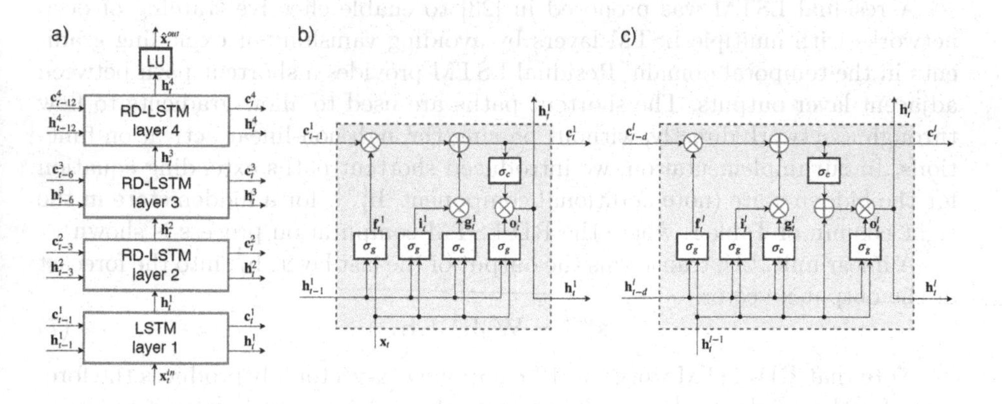

Fig. 3. RD-LSTM architecture (a), LSTM block (b), and RD-LSTM block (c)

The RD-LSTM blocks employ dilation mechanism proposed in [22]. It is to solve three main problems related to RNN learning on long sequences: complex dependencies, vanishing and exploding gradients, and efficient parallelization. It is characterized by multi-resolution dilated recurrent skip connections. To compute the current states of the LSTM block, the last $d - 1$ states are skipped, i.e. a dilated LSTM block receives as input states $\mathbf{c}_{t-d}$ and $\mathbf{h}_{t-d}$, where $d > 1$ is a dilation. Usually multiple dilated recurrent layers are stacked with hierarchical dilations to construct a system, which learns the temporal dependencies

Table 1. Equations for the forward pass of LSTM blocks

Standard LSTM block	RD-LSTM block
$\mathbf{f}_t^1 = \sigma_g(\mathbf{W}_f^1\mathbf{x}_t + \mathbf{V}_f^1\mathbf{h}_{t-1}^1 + \mathbf{b}_f^1)$	$\mathbf{f}_t^l = \sigma_g(\mathbf{W}_f^l\mathbf{h}_t^{l-1} + \mathbf{V}_f^l\mathbf{h}_{t-d}^l + \mathbf{b}_f^l)$
$\mathbf{i}_t^1 = \sigma_g(\mathbf{W}_i^1\mathbf{x}_t + \mathbf{V}_i^1\mathbf{h}_{t-1}^1 + \mathbf{b}_i^1)$	$\mathbf{i}_t^l = \sigma_g(\mathbf{W}_i^l\mathbf{h}_t^{l-1} + \mathbf{V}_i^l\mathbf{h}_{t-d}^l + \mathbf{b}_i^l)$
$\mathbf{g}_t^1 = \sigma_c(\mathbf{W}_g^1\mathbf{x}_t + \mathbf{V}_g^1\mathbf{h}_{t-1}^1 + \mathbf{b}_g^1)$	$\mathbf{g}_t^l = \sigma_c(\mathbf{W}_g^l\mathbf{h}_t^{l-1} + \mathbf{V}_g^l\mathbf{h}_{t-d}^l + \mathbf{b}_g^l)$
$\mathbf{o}_t^1 = \sigma_g(\mathbf{W}_o^1\mathbf{x}_t + \mathbf{V}_o^1\mathbf{h}_{t-1}^1 + \mathbf{b}_o^1)$	$\mathbf{o}_t^l = \sigma_g(\mathbf{W}_o^l\mathbf{h}_t^{l-1} + \mathbf{V}_o^l\mathbf{h}_{t-d}^l + \mathbf{b}_o^l)$
$\mathbf{c}_t^1 = \mathbf{f}_t^1 \otimes \mathbf{c}_{t-1}^1 + \mathbf{i}_t^1 \otimes \mathbf{g}_t^1$	$\mathbf{c}_t^l = \mathbf{f}_t^l \otimes \mathbf{c}_{t-d}^l + \mathbf{i}_t^l \otimes \mathbf{g}_t^l$
$\mathbf{h}_t^1 = \mathbf{o}_t^1 \otimes \sigma_c(\mathbf{c}_t^1)$	$\mathbf{h}_t^l = \mathbf{o}_t^l \otimes (\sigma_c(\mathbf{c}_t^l) + \mathbf{h}_t^{l-1})$

where $\mathbf{W}$, $\mathbf{V}$ and $\mathbf{b}$ are input weights, recurrent weights and biases, respectively, σ_c is a hyperbolic tangent function, σ_g is a sigmoid activation function $(1 + e^{-x})^{-1}$, $\otimes$ denotes the Hadamard product, superscript 1 refers to the first layer of RD-LSTM network, where we use the standard LSTM block, superscript l indicates the layer number for RD-LSTM blocks (from 2 to 4 in our case), and d is a dilation (3, 6 or 12 in our case).

of different scales at different layers. In [22], it was shown that this solution can reliably improve the ability of recurrent models to learn long-term dependency. Dilated RNN can be particularly useful for seasonal TS. In this case dilations can be related to seasonality. In our case we use d = 3, 6 and 12.

A residual LSTM was proposed in [23] to enable effective training of deep networks with multiple LSTM layers by avoiding vanishing or exploding gradients in the temporal domain. Residual LSTM provides a shortcut path between adjacent layer outputs. The shortcut paths are used to allow gradients to flow through a network directly, without passing through non-linear activation functions. In our implementation, we introduced shortcut paths extending equation for the hidden state (note additional component, $\mathbf{h}_t^{l-1}$, for a hidden state in the right column of Table 1, where the RD-LSTM computation process is shown).

A linear unit, LU, transforms the output of the last layer, $\mathbf{h}_t^4$, into the forecast of the output x-vector:

$$\hat{\mathbf{x}}_t^{out} = \mathbf{W}_x\mathbf{h}_t^4 + \mathbf{b}_x \tag{7}$$

Note that RD-LSTM works on 12-component x-vectors. It produces the forecasts for the whole yearly period receiving the previous yearly period as input. The parameters of RD-LSTM, i.e. input weights $\mathbf{W}$, recurrent weights $\mathbf{V}$, and biases $\mathbf{b}$, are learned using SGD in the cross-learning mode simultaneously with the ETSd parameters. The length of the cell and hidden states, m, the same for all layers, was selected on the training set to ensure the highest performance.

3 Results

The proposed forecasting model is applied for monthly electricity demand forecasting for 35 European countries. The real-world data are taken from the ENTSO-E repository (www.entsoe.eu). The TS lengths vary from 5 to 24 years.

The forecasting problem is to produce the forecasts for the twelve months of 2014 (last year of data) using data from the previous period for training. For hyperparameter selection the model learned on the TS fragments up to 2012, and then it was validated on 2013. The selected hyperparameters were used to build the model for 2014: number of epochs 10, learning rate 10^{-3}, length of the cell and hidden states $m = 40$, asymmetry parameter in pinball loss $\tau = 0.4$, ensembling parameters: $L = 5$, $K = 4$, $R = 3$. RD-LSTM was implemented in C++ relying on the DyNet library and run in parallel on an 8-core CPU. We employ R implementation of ETS (function `ets` from package `forecast`).

The proposed model was compared with its predecessor, ETS+RD-LSTM [1], and other state-of-the-art models based on ML as well as classical statistical models. They include: k-nearest neighbor weighted regression model, k-NNw, fuzzy neighborhood model, FNM, general regression NN model, GRNN, multilayer perceptron, MLP, adaptive neuro-fuzzy inference system, ANFIS, LSTM model, ARIMA model, and ETS model. All ML models were used also in +ETS versions, where the TS were initially normalized and the yearly mean and dispersion were forecasted using ETS (just like in this study). Details of the comparative models can be found in [18,19,24,25].

Table 2 shows the forecast results averaged over 35 countries, i.e. median of absolute percentage error (APE), mean APE (MAPE), interquartile range of APE as a measure of the forecast dispersion, root mean square error (RMSE),

Table 2. Results comparison among proposed and comparative models

Model	Median *APE*	*MAPE*	*IQR*	*RMSE*	*MPE*
k-NNw	2.89	4.99	3.85	368.79	−1.87
FNM	2.88	4.88	4.26	354.33	−2.03
N-WE	2.84	5.00	3.97	352.01	−1.91
GRNN	2.87	5.01	4.02	350.61	−1.87
k-NNw+ETS	2.71	4.47	3.52	327.94	−1.25
FNM+ETS	2.64	4.40	3.46	321.98	−1.26
N-WE+ETS	2.68	4.37	3.36	320.51	−1.26
GRNN+ETS	2.64	4.38	3.51	324.91	−1.26
MLP	2.97	5.27	3.84	378.81	−1.37
MLP+ETS	3.11	4.80	4.12	358.07	−1.71
ANFIS	3.56	6.18	4.87	488.75	−2.51
ANFIS+ETS	3.54	6.32	4.26	464.29	−1.30
LSTM	3.73	6.11	4.50	431.83	−3.12
LSTM+ETS	3.08	5.19	4.54	366.45	−1.41
ARIMA	3.32	5.65	5.24	463.07	−2.35
ETS	3.50	5.05	4.80	374.52	−1.04
ETS+RD-LSTM	2.74	4.48	3.55	347.24	−1.11
3ETS+RD-LSTM	2.64	4.09	3.13	314.01	−0.32

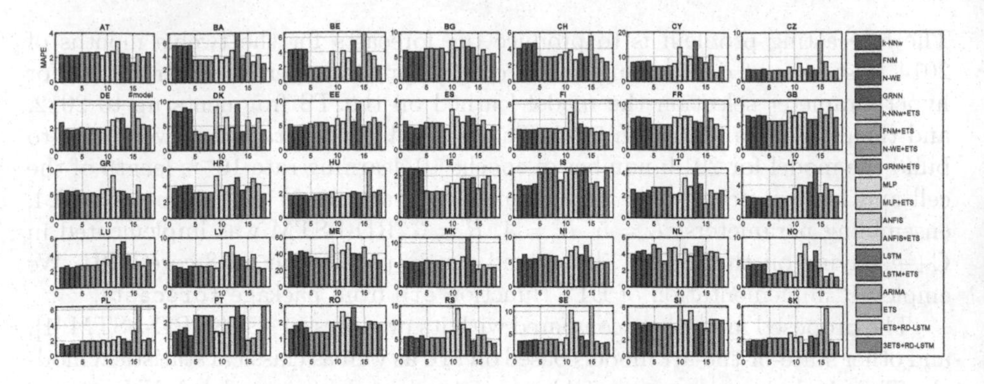

Fig. 4. MAPE for each country

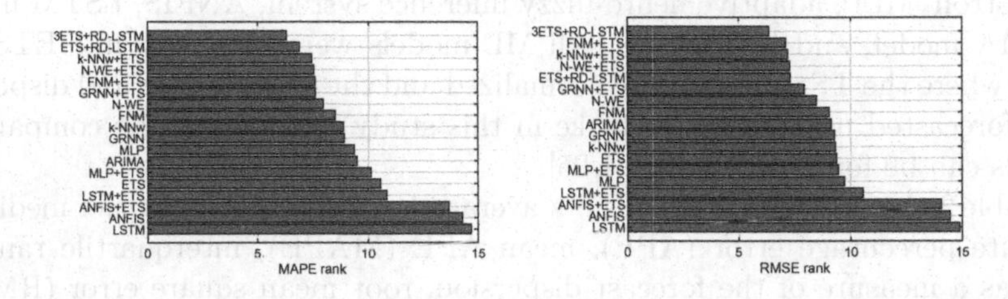

Fig. 5. Rankings of the models

and mean PE (MPE). The proposed model is denoted by 3ETS+RD-LSTM. As can be seen from this table, all error measures indicate that 3ETS+RD-LSTM is the most accurate model comparing with its competitors. It outperforms its predecessor, ETS+RD-LSTM, by 8.7% in MAPE and 9.5% in RMSE.

MPE provides information on potential forecast bias. All the models produced negatively biased forecasts, i.e. overpredicted. But for 3ETS+RD-LSTM, the t-test did not reject the null hypothesis that PE comes from a normal distribution with mean equal to zero (p-value $= 0.44$). All other models did not pass this test. So it can be concluded that 3ETS+RD-LSTM, as the only model, produced unbiased forecasts. Note that 3ETS+RD-LSTM has the mechanism to deal with bias. The loss function (1) asymmetry is controlled by parameter τ. It was selected as 0.4, which allowed the model to reduce the negative bias.

Figure 4 depicts more detailed results, MAPE for each country. As can be seen, in most cases 3ETS+RD-LSTM is one of the most accurate models. Figure 5 depicts the model rankings based on MAPE and RMSE. They show average ranks of the models in the rankings for individual countries. Note the first position of 3ETS+RD-LSTM in both rankings.

Examples of forecasts produced by the selected models are shown in Fig. 6. For PL and DE data, MAPE is on the low level around 2% while for GB data the forecasts are strongly underestimated, over 5%. This is because the demand

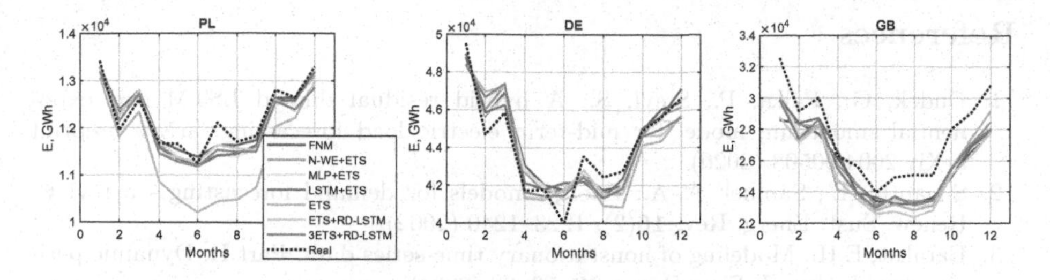

Fig. 6. Examples of forecasts produced by selected models

for GB went up unexpectedly in 2014 despite the downward trend observed in the previous period.

Summarizing experimental research, it should be noted that the forecasting model performance depends significantly on the appropriate TS preprocessing. Although LSTM deals with raw data, without preprocessing [19], introducing initial normalization and dynamic deseasonalization in 3ETS+RD-LSTM improved significantly LSTM performance.

It should be noted that LSTM based models are more complex than other comparative models. Due to the huge number of parameters and complicated learning procedure using backpropagation through time, the learning time of LSTM is much longer than for other comparative models.

4 Conclusion

In this work, we proposed an extended hybrid RD-LSTM and ETS model for MTLF. It combines initial TS decomposition into three components (normalized TS, trend, and dispersion), ETS modules for trend and dispersion forecasting, ETS for deseasonalization, advanced LSTM, and ensembling. The model has a hierarchical structure composed of a global part learned across many TS (LSTM) and a TS specific part (normalization and deseasonalization). Deseasonalization and LSTM are combined in a simultaneous learning process using SGD which leads to learning TS representations and mapping at the same time.

We used residual dilated LSTM, which can capture better long-term seasonal relationships and ensure more efficient training. This is because of dilated recurrent skip connections and a spatial shortcut path from lower layers. To deal with a forecast bias, an asymmetric pinball loss function was applied. Three-level ensembling provides regularization reducing the model variance, which has sources in the stochastic nature of SGD, and also in data and parameter uncertainty.

An experimental study, monthly electricity demand forecasting for 35 European countries, demonstrated the state-of-the-art performance of the proposed model. It generated more accurate forecasts than its predecessor (ETS+RD-LSTM), classical models such as ARIMA and ETS as well as state-of-the-art models based on ML.

References

1. Dudek, G., Pełka, P., Smyl, S.: A hybrid residual dilated LSTM end exponential smoothing model for mid-term electric load forecasting. arXiv preprint arXiv:2004.00508 (2020)
2. Suganthi, L., Samuel, A.-A.: Energy models for demand forecasting - a review. Renew. Sust. Energ. Rev. **16**(2), 1223–1240 (2002)
3. Barakat, E.H.: Modeling of nonstationary time-series data. Part II. Dynamic periodic trends. Int. J. Elec. Power **23**, 63–68 (2001)
4. González-Romera, E., Jaramillo-Morán, M.-A., Carmona-Fernández, D.: Monthly electric energy demand forecasting with neural networks and Fourier series. Energ. Convers. Manage. **49**, 3135–3142 (2008)
5. Chen, J.F., Lo, S.K., Do, Q.H.: Forecasting monthly electricity demands: an application of neural networks trained by heuristic algorithms. Information **8**(1), 31 (2017)
6. Gavrilas, M, Ciutea, I, Tanasa, C.: Medium-term load forecasting with artificial neural network models. In: IEEE Conference on Electricity Distribution Publication, vol. 6 (2001)
7. Doveh, E., Feigin, P., Hyams, L.: Experience with FNN models for medium term power demand predictions. IEEE Trans. Power Syst. **14**(2), 538–546 (1999)
8. Pełka, P., Dudek, G.: Medium-term electric energy demand forecasting using generalized regression neural network. In: Świątek, J., Borzemski, L., Wilimowska, Z. (eds.) ISAT 2018. AISC, vol. 853, pp. 218–227. Springer, Cham (2019). https://doi.org/10.1007/978-3-319-99996-8_20
9. Pei-Chann, C., Chin-Yuan, F., Jyun-Jie, L.: Monthly electricity demand forecasting based on a weighted evolving fuzzy neural network approach. Int. J. Elec. Power **33**, 17–27 (2011)
10. Ahmad, T., Chen, H.: Potential of three variant machine-learning models for forecasting district level medium-term and long-term energy demand in smart grid environment. Energy **160**, 1008–1020 (2018)
11. Hewamalage, H., Bergmeir, C., Bandara, K.: Recurrent neural networks for time series forecasting: current status and future directions. arXiv preprint arXiv:1909.00590v3 (2019)
12. Yan, K., Wang, X., Du, Y., Jin, N., Huang, H., Zhou, H.: Multi-step short-term power consumption forecasting with a hybrid deep learning strategy. Energies **11**(11), 3089 (2018)
13. Bedi, J., Toshniwal, D.: Empirical mode decomposition based deep learning for electricity demand forecasting. IEEE Access **6**, 49144–49156 (2018)
14. Zheng, H., Yuan, J., Chen, L.: Short-term load forecasting using EMD-LSTM neural networks with a XGBboost algorithm for feature importance evaluation. Energies **10**(8), 1168 (2017)
15. Narayan, A., Hipel, K.-W.: Long short term memory networks for short-term electric load forecasting. In: IEEE International Conference on Systems, Man, and Cybernetics (SMC), pp. 2573–2578 (2017)
16. Makridakis, S., Spiliotis, E., Assimakopoulos, V.: The M4 competition: results, findings, conclusion and way forward. Int. J. Forecast. **34**(4), 802–808 (2018)
17. Smyl, S.: A hybrid method of exponential smoothing and recurrent neural networks for time series forecasting. Int. J. Forecast. **36**(1), 75–85 (2020)
18. Dudek, G., Pełka, P.: Pattern similarity-based machine learning methods for mid-term load forecasting: a comparative study. arXiv preprint arXiv:2003.01475 (2020)

19. Pelka, P., Dudek, G.: Pattern-based long short-term memory for mid-term electrical load forecasting. In: 2020 International Joint Conference on Neural Networks (IJCNN), Glasgow, United Kingdom, pp. 1–8 (2020). https://doi.org/10.1109/IJCNN48605.2020.9206895
20. Hyndman, R.-J., Koehler, A.-B., Ord, J.-K., Snyder, R.-D.: Forecasting with Exponential Smoothing: The State Space Approach. Springer, Heidelberg (2008). https://doi.org/10.1007/978-3-540-71918-2
21. Oreshkin, B.-N., Carpov, D., Chapados, N., Bengio, Y.: N-BEATS: neural basis expansion analysis for interpretable time series forecasting. arXiv preprint arXiv:1905.10437v4 (2020)
22. Chang, S., Zhang, Y., Han, W., Yu, M., Guo, X., Tan, W., et al.: Dilated recurrent neural networks. arXiv preprint arXiv:1710.02224 (2017)
23. Kim, J., El-Khamy, M., Lee, J.: Residual LSTM: design of a deep recurrent architecture for distant speech recognition. arXiv preprint arXiv:1701.03360 (2017)
24. Pełka, P., Dudek, G.: Pattern-based forecasting monthly electricity demand using multilayer perceptron. In: Rutkowski, L., Scherer, R., Korytkowski, M., Pedrycz, W., Tadeusiewicz, R., Zurada, J.M. (eds.) ICAISC 2019. LNCS (LNAI), vol. 11508, pp. 663–672. Springer, Cham (2019). https://doi.org/10.1007/978-3-030-20912-4_60
25. Pełka, P., Dudek, G.: Neuro-fuzzy system for medium-term electric energy demand forecasting. In: Borzemski, L., Świątek, J., Wilimowska, Z. (eds.) ISAT 2017. AISC, vol. 655, pp. 38–47. Springer, Cham (2018). https://doi.org/10.1007/978-3-319-67220-5_4

A Deep Time Series Forecasting Method Integrated with Local-Context Sensitive Features

Tianyi Chen[1], Canghong Jin[1(✉)], Tengran Dong[1], and Dongkai Chen[2]

[1] Zhejiang University City College, Hangzhou 310015, China
{31701007,31701009}@stu.zucc.edu.cn, jinch@zucc.edu.cn
[2] Dartmouth College, Hanover, NH 03755, USA
Dongkai.Chen.GR@dartmouth.edu

Abstract. Time series forecasting predicts values in future timestamps based on historically observed series information. Algorithms based on deep neural networks such as Temporal Convolution Network(TCN) have outperformed traditional methods such as Autoregressive Integrated Moving Average Model(ARIMA). However, most existing deep learning approaches suffer from the insufficient ability to capture the seasonality features in series adequately since the network structure ignores the fact that the importance of points the series varies a lot. The local context that reflects a sub-segment of seasonality can indicate potential patterns of the series based on the periodicity. Therefore, we tend to exploit local information from historical records. To this end, we develop a novel strategy to extract local context sensitivity information and integrate them into the current state-of-the-art TCN model, namely LS-TCN. This information enables an improvement in capturing the series pattern and fluctuation, as well as providing transferable guidance for forecasting in the next steps. Experiments conducted on three different real-world series datasets demonstrate that our method significantly outperforms the state-of-the-art models, especially in autocorrelation series corpus.

Keywords: Time series forecasting · Temporal convolution network · Point context similarity

1 Introduction

Time series forecasting task involves using historical information predicting feature timestamps values which applies to many real-world scenarios from traffic speed forecasting in the intelligent transportation systems to stock price predicting in the market decision and long-term energy demand forecasting in resource management. Moreover, we also have witnessed the success of deep learning in time series forecasting task [2], especially in big-data temporal series due to their capacity to extract complex patterns automatically without laborious hand-crafted features [16].

H. Yang et al. (Eds.): ICONIP 2020, LNCS 12534, pp. 532–543, 2020.
https://doi.org/10.1007/978-3-030-63836-8_44

Previous empirical studies have demonstrated that deep learning-based algorithms outperform traditional algorithms such as ARIMA model [15] and convolutional architectures such as TCN can outperform recurrent canonical networks, e.g., LSTM in multiple time series datasets [6,11]. Moreover, several hybrid TCN approaches, e.g., Multi-Stage TCN (MS-TCN), Ensemble Empirical Mode Decomposition-Temporal Convolutional Network (EEMD-TCN) and Temporal Graph Convolutional Network (T-GCN) are better than TCN and other deep recurrent models cross a diverse range of tasks and datasets [4,22].

Most deep learning approaches suffer from the insufficient ability to capture the local information and seasonality features in series adequately. As the trend and seasonal patterns are ingredients that frequently occur in time series, which can impact future outcomes, periodic features extracted from trend and seasonality can be in close reflection with the corresponding series. Although TCN architecture filters larger area by enlarging the number of layers to get more receptive fields to achieve significantly longer for training, their abilities to pick up on seasonality and trends and to adapt series-specific features are insufficient due to the unbiased sampling.

In particular, the influence of points in various time periods is different. Therefore, the intuitive idea is to give the points in the series different weights that could be integrated during the learning process to discover the seasonality and trends. To comprehensively take the seasonal information into consideration, we develop LS-TCN, in which the sub-segment of seasonality is represented as a local context. The network can extract periodic knowledge for every point and fuse it with the original series.

The key contributions of this paper include:

- We propose a mechanism to fetch the importance features of every point under its micro-context condition, which considers both seasonality of global series and its local neighbors. Various and sweeping methods are adopted to compute the correlation between the sample and target area in corpora.
- We propose a novel model LS-TCN to leverage both micro-context sensitivity and global longer periodic dependencies which does not need additional features but only utilizes generated decomposition features by the series itself.
- We conduct experiments on multiple real-world datasets to demonstrate that the model has a better capability of capturing patterns of the series as well as making more robust forecasting.

2 Related Work

2.1 Time Series Forecasting

Time Series forecasting is a significant task in machine learning. Furthermore, time series exists in every aspect of life—for instance, traffic flow, company revenue, stock index, and so on. An extensive amount of data can be regarded as a time series.

As deep learning is gaining popularity in recent years, lots of sophisticated neural networks are applied to time series forecasting, including LSTM [5] and WaveNet [1] which was initially designed for audio generation [13]. Meanwhile, the attention mechanism is applied to time series forecasting tasks as well [9].

Most methods based on TCN simply rely on the dilation convolution with a growing receptive field to capture longer temporal dependencies, which ignores the local periodic characteristics of the convolution features [21]. Therefore a method that can capture the locality at a point level to have better performance is required.

2.2 Series Significance Representation

Dynamic time warping was a widely used technique for time series similarity measure [3], even though it has excellent performance and can get high accuracy, the time consumed may be computationally expensive, which leads to a method that segment the series first and extract features in order to balance between the computational cost and accuracy [10]. Discrete wavelet transformation (DWT) [19] compresses the time series [8] and projects the signal into a tiling of the time-frequency plane for feature extraction [14].

The detection of periodic patterns in time series contributes to forecasting tasks. Using suffix trees as the underlying data structure can discover the periodicity [7]. Meanwhile, date-compensated discrete Fourier transform (DCDFT) is shown as a powerful tool for identifying periodic components [20]. Other network-based methods can capture the periodicity as well as processing the noise [18].

3 Framework

To address the short and long term time series forecasting problem, We propose a novel framework that is based on fusing seasonal features with local context to help better capture semantic features. Figure 1 shows the architecture of our neural network model.

Specifically, we first determine the size of the sliding window in order to sample the contexts properly. Then we sample the series into segment levels to obtain contexts. The micro-context represents a sub-segment characteristic of the seasonality of the series. Then we can extract local sensitivity features by estimating micro-contexts pair by pair. We use causal convolution to make sure the dimension of the series after each layer consistent. We stack dilated convolution layers to capture longer periodic local sensitivity information and learn valid periodic features. The whole framework of LS-TCN is trained in an end-to-end manner. Next, we introduce each module of LS-TCN in detail.

3.1 Optimal Sampling for Contexts Generation

Given a time series $\mathcal{X} = \{x_1, x_2, x_3, ..., x_{n-1}, x_n, x_{n+1}\}$, where $n+1$ denotes the total length of the series and $x_{i+1} \in \mathcal{X}$ denotes the next timestamp value of the

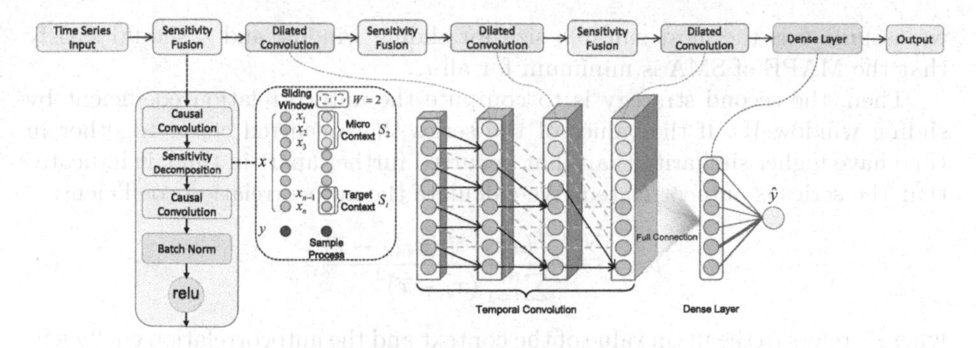

Fig. 1. The architecture of our model. In this case, the size of the sliding window W is 2. Here y denotes the ground-truth value of the next point. In the sampling process, The closest W points form the target context $\mathcal{X}_t$ and $\mathcal{X}_2$ is the second micro-context. Sensitivity fusion captures the periodic local dependencies and fuses the prior knowledge with the original series. The detail of the sensitivity fusion layer will be illustrated in Sect. 3.3. The receptive fields grow exponentially as the layer goes deeper, enabling the dilated convolution to uncover local patterns. After the propagation through a dense layer, the network forecasts a value $\hat{y}$. The last full connection reflects that the point weights vary.

series. We define the *micro context*:

$$\mathcal{X}_i = \{x_{i-\frac{W}{2}}, ..., x_{i-1}, x_{i+1}, ..., x_{i+\frac{W}{2}}\}, \tag{1}$$

where $i \in [1, n+1]$. For each point $x_i \in \mathcal{X}$, we treat its neighbors with window W as a context to represent local features. If the length of a micro context is less than $|W|$, we use zero-padding to maintain the same input length. Meanwhile, we define the *target context*:

$$\mathcal{X}_t = \{x_{n-W+1}, ..., x_{n-1}, x_n\}, \tag{2}$$

where we denote the closest $|W|$ points to x_{i+1} since they may have a more substantial influence on the forecasting result.

In order to compute the importance of points in the series, the first step is to select proper contexts. Here $|W|$ denotes the size of the sliding window of a context that represents the local features. We use moving average and autocorrelation coefficient methods to calculate the optimal window W, thus enhance the ability of the network to capture periodic trend features.

First, we determine W by the moving average method, which is a low-pass filter and could filter out the high-frequency disturbance in time series. For each point $x_i \in \mathcal{X}$, to measure the percentage deviation between $\mathcal{X}_t$ and $\mathcal{X}_i$, we compute the mean absolute percentage error(MAPE) as a metrics by simple moving average model(SMA) under various window size, ranged from 1 to n. The SMA in computed by:

$$SMA(\mathcal{X}, k) = \frac{\sum_{j=1}^{k} x_{i-j+1}}{k}, \tag{3}$$

we assume k is the most suitable size for sliding window and k should satisfy that the MAPE of SMA is minimum for all i.

Then, the second strategy is to compute the autocorrelation coefficient by sliding window W. If the values of the series that occurred closer together in time have higher similarity than that occurred further apart in time, it indicates that the series is autocorrelated. We calculate the autocorrelation coefficient:

$$\rho_t = \frac{\sum_{i=1}^{n-t}(x_i - \overline{x})(x_{i+t} - \overline{x})}{\sum_{i=1}^{n}(x_i - \overline{x})^2}, \tag{4}$$

where $\overline{x}$ refers to the mean value of the context and the autocorrelation coefficient indicates the degree to which two different contexts interact with each other. Then we select t when ρ_t is maximum. Since k and t are different, the final size of window w is token into account both suitable values, such as weighted ranking algorithms.

Eventually, there will be 1 target context and n micro contexts. The target context can be represented as a target vector $V_t \in \mathbb{R}^{1\times W}$, and all micro contexts can be represented as a vector $V_s \in \mathbb{R}^{n\times W}$.

3.2 Integration of Context Significance Features

After obtaining the micro contexts that represent the semantics, the next step is to integrate the significance features of each context by comparing the context $\mathcal{X}_i$ and target context $\mathcal{X}_t$ by the following three methods:

Value Square Deviation(VSD):

$$VSD(\mathcal{X}_i, \mathcal{X}_t) = \frac{\sum_{j=1}^{|W|}(\mathcal{X}_{ij} - \mathcal{X}_{tj})^2}{|W|}, \tag{5}$$

where $|W|$ represents the sliding window size; $\mathcal{X}_{ij}$ refers to the j-th value of micro context $\mathcal{X}_i$; $\mathcal{X}_{tj}$ corresponds to the j-th value of target context $\mathcal{X}_t$. VSD measures the average square deviation between two contexts.

Value Mean Deviation(VMD):

$$VMD(\mathcal{X}_i, \mathcal{X}_t) = \frac{\sum_{j=1}^{|W|}|\mathcal{X}_{ij} - \mathcal{X}_{tj}|}{|W|}, \tag{6}$$

VMD measures the average mean deviation between two contexts.

Dot Product Ratio (DPR):

$$DPR(\mathcal{X}_i, \mathcal{X}_j) = \frac{\sum_{j=1}^{|W|}\mathcal{X}_{ij} \times \mathcal{X}_{tj}}{\sum_{j=1}^{|W|}\mathcal{X}_{ij}^2}, \tag{7}$$

DPR is used to measures the ratio of the dot product between two contexts, and the value range is $[\frac{n-2}{2n-1}, 1]$.

Then, all three computed importance features are integrated by concatenation:

$$\mathbb{X} = \mathcal{X} \oplus \mathcal{X}_{VSD} \oplus \mathcal{X}_{VMD} \oplus \mathcal{X}_{DPR}, \tag{8}$$

where the symbol $\oplus$ denotes concatenation in channel dimension, $\mathbb{X}$ corresponds to the time series after integration, $\mathcal{X}_{VSD}$ refers to the VSD feature series, $\mathcal{X}_{VMD}$ is the VMD feature series, and $\mathcal{X}_{DPR}$ represents the DPR feature series.

Therefore, the original series now contains additional seasonality and local sensitivity information for every point in multiple dimension space $\mathcal{X}' \in \mathbb{R}^{4 \times n}$.

3.3 Local-Context Sensitive Convolution

After integrating the local sensitivity information with the time series, we can extract features by the dilated convolution layers.

Step 1: Sensitivity Fusion. At this step, we aim to employ sensitivity decomposition to transform the time series to a tensor that contains the prior periodic knowledge of the original series. Since the number of channels of the input series may not be constant, we first compress the channels of the series through a preprocess causal convolution layer before sensitivity decomposition. We define the convolution computation:

$$X' = Conv(\mathcal{X}, K_1), \tag{9}$$

where $\mathcal{X} \in \mathbb{R}^{n_c \times n}$ refers to the series with n_c channels, K_1 denotes the convolution kernel with kernel size $K_1 \in \mathbb{R}^{n_c \times 1 \times 1}$.

Then we can compute context importance features of each point by the methods mentioned in Sect. 3.2 and concatenate the original series and feature series. Now that the context features are distributed in multiple channels, we apply another causal convolution layer to enable these features to interact with each other. Meanwhile, it can keep the channel dimension consistent.

$$\mathcal{X}_1' = Conv(X', K_2), \tag{10}$$

where $\mathcal{X}_1'$ represents the local sensitive series of the 1-st layer and K_2 corresponds to the convolution kernel with kernel size $K_1 \in \mathbb{R}^{4 \times d_{c1} \times 1}$ where d_{c1} denotes the hidden dimension of the 1-st layer.

This process of sensitivity fusion is shown in Fig. 2.

Step 2: Temporal Convolution. At this step, we stack three dilated convolution layers with dilation equal to 1, 2, 4, respectively, to enlarge the receptive field, enabling the network to capture longer periodic local sensitivity information. Before each dilated convolution, we fuse the sensitivity information with the current series. Therefore, the context importance is enhanced through each layer, guiding the network to predict based on different point weights.

$$\mathcal{X}_2 = Conv(\mathcal{X}_1', K_3, dilation = 1), K_3 \in \mathbb{R}^{d_{c1} \times d_{c2} \times 1} \tag{11}$$

$$\mathcal{X}_2' = SensitivityFusion(\mathcal{X}_2), \mathcal{X}_2' \in \mathbb{R}^{d_{c2} \times n-1} \tag{12}$$

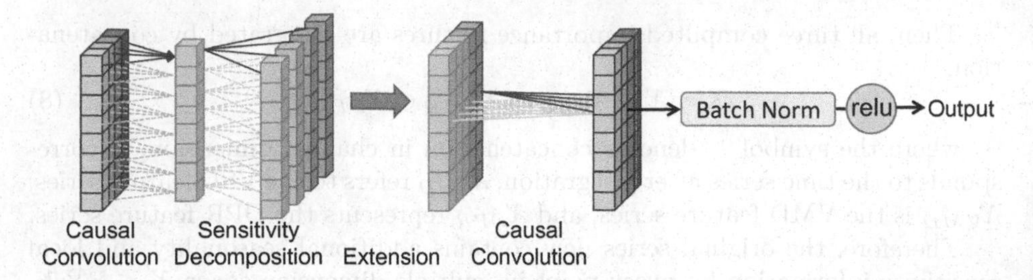

Fig. 2. Structure of the Sensitivity Fusion layer. The causal convolution is used for channel transformation as well as interaction. After decomposition, we extend the original series with the sensitivity vectors in channel direction for fusing. Note that the dimension of the series before and after sensitivity fusion remains the same, only except that the local periodic pattern is included.

$$\mathcal{X}_3 = Conv(\mathcal{X}_2^{'}, K_4, dilation = 2), K_4 \in \mathbb{R}^{d_{c2} \times d_{c3} \times 1} \tag{13}$$

$$\mathcal{X}_3^{'} = SensitivityFusion(\mathcal{X}_3), \mathcal{X}_3^{'} \in \mathbb{R}^{d_{c3} \times n-3} \tag{14}$$

$$\mathcal{X}_4 = Conv(\mathcal{X}_3^{'}, K_5, dilation = 4), K_5 \in \mathbb{R}^{d_{c3} \times d_{c4} \times 1}, \tag{15}$$

where $\mathcal{X}_i^{'}$ corresponds to the local sensitive series activated by the i-th sensitivity fusion layer; K_i denotes the convolution kernel in the i-th layer, concretely, $\mathcal{X}_1 \in \mathbb{R}^{d_{c1} \times n}$ and $\mathcal{X}_4 \in \mathbb{R}^{d_{c4} \times n-7}$.

Step 3: Forecasting. Eventually, we propagate the learned features through a dense layer to predict the next value of the series. Since the original series is fused with prior local sensitive knowledge through the temporal convolution, the points in the dense layer have different weights, which guides the network to adjust the forecasting result to get closer to reality automatically.

4 Experiments

In this section, we first describe the experimental setup and then assess the performance of LS-TCN the time series forecasting task comparing with several deep learning models as baselines.

4.1 Experimental Setup

Dataset Descriptions. To compare our model and baselines in the task of next value forecasting, we supply three public real-world sequential datasets, including:

PEMS-BAY[1]**:** A stable dataset in the field of transportation with few noises. The mean autocorrelation coefficient of the dataset is 0.31. The dataset contains

[1] https://github.com/liyaguang/DCRNN.

traffic speed collected by 325 sensors in the Bay Area of California, starting from Jan 1, 2017, through May 31, 2017. We aggregate traffic speed into 5 min interval with size 1000×92.

SML2010[2]**:** Dataset collected from a monitor system that corresponds to approximately 40 days of monitoring data. The dataset has strong periodicity, and the autocorrelation coefficient is 0.68. We aggregate the temperature data into 15 min intervals. There are 4048 points in total, and we use the first 2800 for training, the rest for validation.

NASDAQ100[3]**:** We use a subset MSFT of the full NASDAQ100 stock dataset, which includes 105 days' stock data starting from July 26. The dataset has a regular periodicity with the autocorrelation coefficient being 0.31, and the fluctuation is small.

All datasets have been applied Z-Score normalization and split into training set (60%), validation set (20%) and test set (20%) in chronological order.

Comparison with the Baseline Methods. To prove the validity of the proposed approach, we compare the following forecasting methods:

LSTM [12]**:** A commonly used model implements a gated system that controls neural information processing. We set hidden dimension $d^h = 10$ with one layer stacked;

WaveNet [13]**:** A seminal CNN model that has been generalized for broader sequence modeling problems. We set residual channel 32; skip channel 128 with layer $K = 4$ for each block; three blocks stacked in total;

Transformer [17]**:** A model that comprises a sequence of encoders with the self-attention mechanism. We set eight layers in total; query size $q = 32$; value size $v = 32$; hidden dimension $d^h = 256$; attention window size 32; dropout rate $\beta = 0.3$. The rest of parameter settings remained the same as in the original paper;

TCN [11]**:** A convolutional architecture which is autoregressive, able to process sequences of arbitrary length. Three dilated convolution layers stacked; each layer has the kernel size 2 and stride 1.

LS-TCN: Parameters of our model are provided in Table 1. The source code of our implementation is available at Github[4].

Evaluation Metrics. To measure the effectiveness of different time series forecasting methods, we utilize the following evaluation metrics:

[2] https://archive.ics.uci.edu/ml/datasets/SML2010.
[3] http://cseweb.ucsd.edu/~yaq007/NASDAQ100_stock_data.html.
[4] https://github.com/NewWesternCEO/LS-TCN/.

Table 1. LS-TCN parameter settings

Learning rate λ	0.001	Weight decay	0.0001	Batch size	256
Dilation factors	[1,2,4]	Sliding window size W	4	Extend dimension	4
Dense layers	2	Units in dense layers	[73, 12]	Kernal size	2

Mean Absolute Percentage Error(MAPE):

$$MAPE(y, \hat{y}) = \frac{\sum_{i=1}^{n} \left| \frac{y_i - \hat{y}_i}{y_i} \right|}{n} \times 100\%, \tag{16}$$

where y refers to the ground-truth value, and $\hat{y}$ denotes the predicted value. MAPE is used to measure the relative errors, often reported as a percentage.

Mean Absolute Error(MAE):

$$MAE(y, \hat{y}) = \frac{\sum_{i=1}^{n} |y_i - \hat{y}_i|}{n}, \tag{17}$$

MAE is used to measure the average absolute error between the predicted value and the ground-truth value.

Root Mean Square Error(RMSE):

$$RMSE(y, \hat{y}) = \sqrt{\frac{\sum_{i=1}^{n} (y_i - \hat{y}_i)^2}{n}}, \tag{18}$$

RMSE is applied to measure the deviation between the predicted value and the ground-truth value. RMSE is more sensitive to outliers.

All the compared models are trained and tested multiple times to eliminate outliers, and the results are averaged to reduce random errors.

4.2 Forecasting Performance Evaluation

Table 2 provides the average forecasting error averaged over 10 runs on the PEMS-BAY traffic dataset, where the **best** results are highlighted and the second-best results are underlined.

Table 2. Performance evaluation of PEMS-BAY traffic dataset

Methods	15 min			30 min			1 hour		
	MAPE	MAE	RMSE	MAPE	MAE	RMSE	MAPE	MAE	RMSE
LSTM(Malhotra et al. 2015)	7.54%	3.13	6.00	7.60%	3.23	6.06	8.34%	3.62	6.61
WaveNet(van den Oord et al. 2016)	17.90%	7.26	11.52	17.88%	7.25	11.47	18.05%	7.24	11.54
Transformer(Vaswani et al. 2017)	10.98%	6.34	10.98	14.70%	6.16	10.48	13.73%	6.01	9.98
TCN(Liu et al. 2019)	7.11%	3.97	5.01	7.56%	4.17	5.36	9.05%	4.72	6.31
LS-TCN	**3.81%**	**1.84**	**3.37**	**4.42%**	**2.07**	**4.42**	**5.58%**	**2.55**	**5.09**

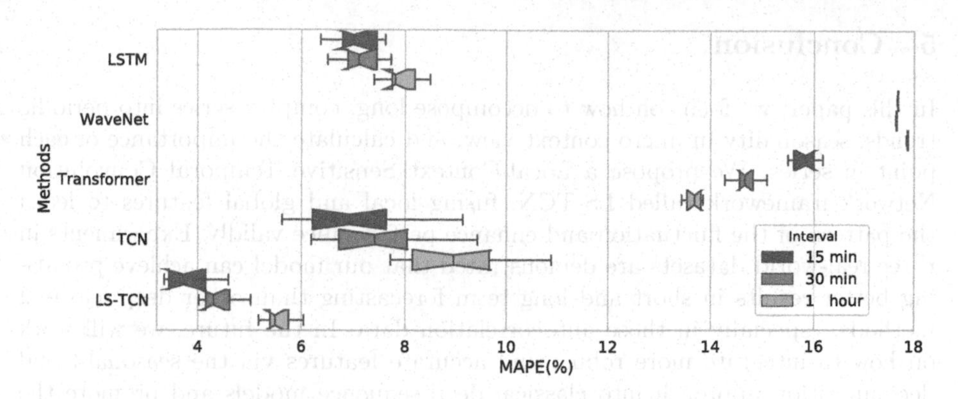

Fig. 3. Variance estimation of 10 runs on the PEMS-BAY dataset. For each box in the figure, the vertical line in the middle denotes the mean value, the vertical line on the left indicates the minimum value. In contrast, the one on the right denotes the maximum value. And the notch denotes the confidence interval.

The 1-h forecasting results of the mean and confidence interval of MAPE of 10 runs on the PEMS-BAY dataset is shown in Fig. 3. The LS-TCN not only has achieved lower forecasting error than all the compared methods but also has lower variance and more excellent stability than the original TCN. The short-term forecasting of LS-TCN outperforms the long-term forecasting since it can capture the local features better in the former case.

Table 3 provides the average 12-point forecasting error averaged over 10 runs on all three series datasets, where the **best** results are highlighted and the second-best results are underlined.

In order to observe the influence between seasonality of series and effectiveness of our module, we calculate the autocorrelation value of these three datasets. The results are 0.31 for PEMS-BAY, 0.68 for SML2010 and 0.31 for NASDAQ100. According to the results in Table 3, We note that LS-TCN has better performance than other baselines in PEMS-BAY and SML2010, where observations occur at different points are more similar. By contrast, LS-TCN is not so good in RMSE value than LSTM in those data whose seasonality is not so related as NASDAQ100. Therefore, our LS-TCN is more suitable in these seasonal and serial correlation series.

Table 3. Performance evaluation of 12-point forecasting on three datasets

Methods	PEMS-BAY			SML2010			NASDAQ100		
	MAPE	MAE	RMSE	MAPE	MAE	RMSE	MAPE	MAE	RMSE
LSTM(Malhotra et al. 2015)	8.34%	3.62	6.61	9.28%	1.64	2.07	0.99%	0.58	**0.74**
WaveNet(van den Oord et al. 2016)	18.05%	7.24	11.54	12.47%	2.59	2.99	6.52%	3.83	4.09
Transformer(Vaswani et al. 2017)	13.73%	6.01	9.98	15.43%	2.57	3.46	2.28%	1.34	1.64
TCN(Liu et al. 2019)	9.05%	4.72	6.31	7.75%	1.61	1.84	1.16%	0.74	0.95
LS-TCN	**5.58%**	**2.55**	**5.09**	**5.79%**	**1.21**	**1.54**	**0.67%**	**0.39**	1.01

5 Conclusion

In this paper, we focus on how to decompose long, complex series into periodic trends, seasonality in micro context view, and calculate the importance of each point in series. We propose a Local-Context Sensitive Temporal Convolution Network framework called LS-TCN, fusing local and global features to learn the pattern of the fluctuation and enhance performance validly. Experiments in three real-world datasets are demonstrated that our model can achieve promising better results in short and long term forecasting than other deep learning methods, especially in those autocorrelation data. In the future, we will work on how to integrate more robust and accurate features via the seasonal-trend decomposition approach into classical deep sequence models and promote the neural network architecture as well.

Acknowledgments. Funding support for this research was in part provided by Zhejiang Department of Education (No.Y20194137).

References

1. Cho, C.H., Lee, G.Y., Tsai, Y.L., Lan, K.C.: Toward stock price prediction using deep learning. In: Proceedings of UCC, pp. 133–135 (2019)
2. Cirstea, R.G., Micu, D.V., Muresan, G.M., Guo, C., Yang, B.: Correlated time series forecasting using multi-task deep neural networks. In: Proceedings of CIKM, pp. 1527–1530 (2018)
3. Dau, H.A., Begum, N., Keogh, E.: Semi-supervision dramatically improves time series clustering under dynamic time warping. In: Proceedings of CIKM, pp. 999–1008 (2016)
4. Farha, Y.A., Gall, J.: Ms-tcn: multi-stage temporal convolutional network for action segmentation. In: Proceedings of ICPR, pp. 3575–3584 (2019)
5. Fu, R., Zhang, Z., Li, L.: Using lstm and gru neural network methods for traffic flow prediction. In: Proceedings of YAC, pp. 324–328 (2016)
6. Lea, C., Flynn, M.D., Vidal, R., Reiter, A., Hager, G.D.: Temporal convolutional networks for action segmentation and detection. In: proceedings of ICPR, pp. 156–165 (2017)
7. Li, C.E., Chang, Y.I.: A time-position join method for periodicity mining in time series databases. In: Proceedings of ICS, pp. 294–299 (2016)
8. Li, D., Bissyande, T.F.D.A., Klein, J., Le Traon, Y.: Time series classification with discrete wavelet transformed data: insights from an empirical study. In: Proceedings of SEKE (2016)
9. Li, S., et al.: Enhancing the locality and breaking the memory bottleneck of transformer on time series forecasting. In: Proceedings of NIPS, pp. 5243–5253 (2019)
10. Li, Z., Zhang, H., Wu, S., Zhao, Y.: Similarity measure of time series based on feature extraction. In: Proceedings of ICCCBDA, pp. 13–16 (2020)
11. Liu, Y., Dong, H., Wang, X., Han, S.: Time series prediction based on temporal convolutional network. In: Proceedings of ICIS, pp. 300–305 (2019)
12. Malhotra, P., Vig, L., Shroff, G., Agarwal, P.: Long short term memory networks for anomaly detection in time series. Proc. ESANN **89**, 89–94 (2015)

13. van den Oord, A., et al: A generative model for raw audio. In: Proceedings of ISCASSW, pp. 125–125 (2016)
14. Rodrigues, M.W., Zárate, L.E.: Time series analysis using synthetic data for monitoring the temporal behavior of sensor signals. In: Proceedings of SMC, pp. 453–458 (2019)
15. Siami-Namini, S., Tavakoli, N., Siami Namin, A.: A comparison of arima and lstm in forecasting time series. In: Proceedings of ICMLA, pp. 1394–1401 (2018)
16. Torres, J.F., Troncoso, A., Koprinska, I., Wang, Z., Martínez-Álvarez, F.: Deep learning for big data time series forecasting applied to solar power. In: Grana, M., et al. (eds.) SOCO'18-CISIS'18-ICEUTE'18 2018. AISC, vol. 771, pp. 123–133. Springer, Cham (2019). https://doi.org/10.1007/978-3-319-94120-2_12
17. Vaswani, A., et al.: Attention is all you need. In: Proceedings of NIPS, pp. 5998–6008 (2017)
18. Vijesh, V., Drisya, G.V., Kumar, K.S.: Network based periodicity detection of a time series. In: Proceedings of RAICS, pp. 199–203 (2018)
19. Wang, J., Wang, Z., Li, J., Wu, J.: Multilevel wavelet decomposition network for interpretable time series analysis. In: Proceedings of CIKM, pp. 2437–2446 (2018)
20. Wang, L., Zhou, L., Xu, G., Li, X.: Date-compensated discrete fourier transform for periodicity detection in unevenly astronomical time series. In: Proceedings of AIPC, vol. 2073 (2019)
21. Wang, Y., Liu, Z., Hu, D., Zhang, M.: Multivariate time series prediction based on optimized temporal convolutional networks with stacked auto-encoders. In: Proceedings of ACML, pp. 157–172 (2019)
22. Yan, S., Xiong, Y., Lin, D.: Spatial temporal graph convolutional networks for skeleton-based action recognition. In: Proceedings of AAAI (2018)

Benchmarking Adversarial Attacks and Defenses for Time-Series Data

Shoaib Ahmed Siddiqui[1,2](✉), Andreas Dengel[1,2], and Sheraz Ahmed[1]

[1] German Research Center for Artificial Intelligence (DFKI), 67663 Kaiserslautern, Germany
{andreas.dengel,sheraz.ahmed}@dfki.de

[2] TU Kaiserslautern, 67663 Kaiserslautern, Germany
shoaib_ahmed.siddiqui@dfki.de

Abstract. The adversarial vulnerability of deep networks has spurred the interest of researchers worldwide. Unsurprisingly, like images, adversarial examples also translate to time-series data as they are an inherent weakness of the model itself rather than the modality. Several attempts have been made to defend against these adversarial attacks, particularly for the visual modality. In this paper, we perform detailed benchmarking of well-proven adversarial defense methodologies on time-series data. We restrict ourselves to the L_∞ threat model. We also explore the trade-off between smoothness and clean accuracy for regularization-based defenses to better understand the trade-offs that they offer. Our analysis shows that the explored adversarial defenses offer robustness against both strong white-box as well as black-box attacks. This paves the way for future research in the direction of adversarial attacks and defenses, particularly for time-series data.

Keywords: Time-series · Deep learning · Adversarial attacks · Adversarial defenses

1 Introduction

Time-series data is ubiquitous in this era of internet-of-things (IoT) and industry 4.0 where millions of sensors are generating data at an extremely high frequency [5,8,9,11]. With this increasing amount of data, there has been a wide-scale deployment of deep models for time-series analysis. Deep learning models have proven to be susceptible to changes in the input which can significantly alter the predictions of the classifier [12]. This raises a serious concern over the real-world deployment of these models. This vulnerability has been particularly explored in the context of images [2,6,10,12].

A similar vulnerability exists also in the context of other modalities as this is primarily a weakness of the current learning paradigm rather than the modality [12]. Since many of the time-series models are deployed security-critical scenarios, there has been an increasing interest regarding the robustness of these

H. Yang et al. (Eds.): ICONIP 2020, LNCS 12534, pp. 544–554, 2020.
https://doi.org/10.1007/978-3-030-63836-8_45

models [5,8,9,11]. Therefore, gradient-based attacks prevalent in the computer-vision community have also been translated to time-series data [5,9,11].

Various defenses have been proposed to circumvent this adversarial vulnerability [10,14,15]. Some of these defenses are specific to the visual modality, while a wide range of literature has been focused on either training on these adversarial examples or adding an additional regularization term that forces the predictions to be consistent within a specific neighborhood of the example. In this paper, we employ some of the most well-recognized defense methodologies tested on images and evaluate their robustness for time-series data to establish a proper benchmark.

2 Related Work

Adversarial examples were first discovered by Szegedy et al. (2013) [12] as a consequence of trying to solve an inverse optimization problem. Since then, a wide range of literature has focused on this security aspect including both development of more sophisticated defenses as well as advances in attacks in order to break these defenses.

Adversarial attacks can be mainly categorized into two different categories namely white-box attacks and black-box attacks. White-box attacks assume access to model architecture and parameters, therefore, they can effectively and efficiently attack the model using the gradient information. Black-box attacks, on the other hand, require access to either the output probabilities or even just the label, making them more applicable in real-world settings. However, black-box attacks usually require thousands or even millions of queries to the model to compute just a single adversarial example.

Szegedy et al. (2013) [12] presented the first adversarial attack based on box-constrained L-BFGS to mine adversarial examples. Goodfellow et al. (2014) [6] proposed a fast version of the attack by assuming the linearity of classifiers around the input. With this assumption, they were able to use a single step attack based on the gradient which they named Fast Gradient-Sign Method (FGSM). Madry et al. (2017) [10] proposed an iterative version of FGSM with a random restart which they named Projected Gradient Descent (PGD) and claimed it to be an optimal first-order adversary. Another famous attack is Carlini-Wagner [4] attack. However, this is mainly designed for L_2 norm-based attacks while we only focus on L_∞ norm-based attacks in this paper. Boundary attack [2] introduced by Brendel et al. (2017) formed the basis for decision-based attacks where the attacker assumes access only to the output label. SIMple Black-box Attack (SIMBA) [7] proposed by Guo et al. (2019) greatly simplified the attack pipeline by assuming access to the output probabilities of the model. The attack mines adversarial examples by just randomly perturbing pixels if they have a negative impact on the output probability.

As attacks have progressed, more and more sophisticated methods have been developed to defend against these attacks. However, most of these attacks were either shown to be masking the gradient or poorly tested [1]. One of the most effective methods to defend against adversarial attacks is PGD-based adversarial training [10]. The robust model is trained on the generated adversarial examples rather than the original inputs in this case. As PGD is one of the most powerful white-box attacks, PGD-based adversarially trained models were shown to be significantly superior in terms of robustness as compared to other techniques [1,10]. Zhang et al. (2019) [15] proposed TRADES which uses an additional regularization term along with the conventional cross-entropy loss that minimizes the discrepancy between clean and adversarial predictions. Feature denoising [14] proposed by Zie et al. (2019) introduced additional denoising operators in the network. The whole network was then trained using adversarial training [10]. The idea was based on minimizing the discrepancy between the feature maps of a clean and adversarial example. A large fraction of adversarial defense literature has been focused on provable defenses that provide formal guarantees against the worst-case adversary. However, they are usually prohibitively slow and unable to scale to large datasets [13]. We don't include these methods in our comparison and leave it as future work.

Minor efforts have also been made in terms of extending these attacks for time-series data. Siddiqui et al. (2019) [11] showed that gradient-based adversarial attacks were effective for both time-series classification as well as time-series regression networks. Fawaz et al. (2019) [5] analyzed a range of different time-series datasets and showed that deep models trained on time-series data are vulnerable to adversarial attacks. Both these papers only explore the vulnerability of networks, which is not surprising given that adversarial attacks exploit the machine learning optimization framework, rather than a specific modality [6]. Karim et al. (2020) [9] and Harford et al. (2020) [8] employed Gradient Adversarial Transformation Network (GATN) for attacking models. Since they also considered classical time-series models that are non-differentiable, they used a knowledge distillation approach to train a student network mimicking the predictions of the original classifier. Therefore, what they explored were just transfer attacks which are a rather weak form of black-box attacks. On the other hand, we evaluate using both the strongest white-box as well as black-box attacks to truly establish the robustness of the evaluated defenses. Our work specifies a proper threat model when evaluating against attacks as well as considers both strong white-box and black-box attacks which haven't been explored in the context of time-series data. Therefore, we not only benchmark adversarial defenses but also establish a benchmark for strong adversarial attacks to be considered for future work which has been missing in the prior work [3].

3 Method

3.1 Threat Model

The threat model specifies the conditions under which the considered defense is designed to be secure [3]. We consider L_∞ threat model and use an epsilon of 0.3 for training robust models. There are no box-constraints for time-series data due to their variable input range in contrast to the visual modality where the pixels take on a discrete value from [0, 255]. Therefore, in our case, the data was normalized with zero-mean and unit standard deviation which justified the choice of 0.3 as the epsilon value.

3.2 Adversarially Robust Models (using Adversarial Defenses)

A robust model is a model which is robust against these minor corruptions of the input signal. A robust model is usually obtained by training a model with a particular adversarial defense methodology. There is a very wide range of literature on the topic of adversarial defenses. However, most of these defenses were shown to be broken by a stronger attack [1]. Therefore, we only explored defenses that withstood these attacks when evaluated on images. We will now discuss each of the evaluated defenses in detail.

Adversarial Training. Madry et al. (2017) [10] proposed a robust optimization algorithm which they named as adversarial training. Adversarial training is one of the most simple and widely accepted adversarial defenses in the literature. The idea is to just train a classifier on the attacked examples rather than the original ones. As the model inherently learns to be robust to these attacks during training, this naturally re-configures the decision boundaries of the network.

$$\mathbf{w}^* = \underset{\mathbf{w}}{\arg\min} \quad \underset{\mathbf{x}' \in B_p(\mathbf{x}, \epsilon)}{\arg\max} \; \mathcal{L}(\Phi(\mathbf{x}'; \mathbf{w}), y')$$

where y' indicates the model's prediction on the input $\mathbf{x}$. The maximization problem is approximately solved by generating an adversarial example using the PGD attack which is considered to be the optimal first-order adversary [10]. The biggest advantage of adversarial training is that there are no additional hyperparameters making model training very easy and convenient.

TRADES. Zhang et al. (2019) [15] introduced TRADES which smoothens the predictions of the classifier around the input by employing an additional term in the final objective alongside the conventional cross-entropy.

$$\mathbf{w}^* = \underset{\mathbf{w}}{\arg\min} \; \left\{ \mathcal{L}_{CE}(\Phi(\mathbf{x}; \mathbf{w}), y) + \underset{\mathbf{x}' \in B_p(\mathbf{x}, \epsilon)}{\arg\max} \; \mathcal{L}_{KL}(\Phi(\mathbf{x}; \mathbf{w}), \Phi(\mathbf{x}'; \mathbf{w}))/\lambda \right\}$$

where $\mathcal{L}_{CE}$ represents the conventional cross-entropy loss on clean data while $\mathcal{L}_{KL}$ computes the KL-divergence between the logits obtained from the original example and the computed adversarial example. The maximization problem is again approximately solved by generating an adversarial example using the PGD attack. TRADES introduces an additional hyperparameter λ which controls the trade-off between clean accuracy and adversarial robustness.

Feature Denoising. Xie et al. (2019) [14] introduced the idea of feature denoising based on their observation that the feature maps computed for an adversarial example are significantly noiser than the ones computed for the original image. Therefore, in order to circumvent this problem, they used denoising operators before max-pooling layers in their network. They compared different denoising operators, and found Gaussian Non-Local Means (GNLM) to be the most effective one. This can be represented as:

$$y_i = \frac{1}{\sum_{\forall j \in \mathcal{N}} f(x_i, x_j)} \sum_{\forall j \in \mathcal{N}} f(x_i, x_j) \times x_j$$

where y_i denotes the ith output, $\mathcal{N}$ denotes all the spatial locations on the feature map and $f(x_i, x_j)$ captures the similarity between x_i and x_j. Since we use the Gaussian version of non-local means, the similarity function is given by $f(x_i, x_j) = e^{\frac{1}{\sqrt{d}}\theta(x_i)^t \phi(x_j)}$ where $\theta(x_i) \in \mathbb{R}^{64}$ and $\phi(x_j) \in \mathbb{R}^{64}$ represents two embedded versions of the input implemented via 1×1 convolution, and d denotes the number of channels. Based on their findings, we also introduce an additional GLNM denoising layer before every pooling layer in all of our networks. Since this denoising layer is also learned during adversarial training, we compare it's impact when training the model using different adversarial defense techniques.

3.3 Robust Evaluation (using Adversarial Attacks)

The aim of robust evaluation is to precisely identify the robustness of the trained robust model. Carlini et al. [3] provided comprehensive guidelines to evaluate robust models in order to avoid pitfalls which prior defense methods could not avoid, providing a false sense of security. For this reason, we included two major black-box as well as two major white-box attacks to compare. In total, we used 5 different attacks, where one black-box attack i.e. noise attack is a rather weak attack, but serves as a trivial baseline.

<u>**Evaluation Metric**</u>

There are many different choices when evaluating robust models. This includes the input examples to consider as well as the target to use for computing the adversarial example. In our case, we compute robust accuracy on all examples regardless of whether they were correctly classified or not. We always conduct attack using the original labels instead of the model's prediction in order to ensure that the attack does not mistakenly move an incorrectly classified example to a correct class, providing a false sense of robustness. However, regardless of the choice of this metric, the key takeaways from our experiments still remain the same. We evaluate robustness using untargeted attacks as they can be considered worst-case adversaries. Targeted attacks are usually much more relevant in practice but harder to find as compared to untargeted attacks.

<u>**White-box Attacks**</u>

FGSM [6]: Goodfellow et al. (2014) [6] posited that the lack of robustness of deep models is due to their linear nature. Therefore, they used this linear approximation to develop a Fast Gradient-Sign based attack Method (FGSM) which takes just a single step in the direction pointed by the gradient.

$$\mathbf{x}_a = \mathbf{x} + \epsilon\ sign(\frac{\partial}{\partial \mathbf{x}}\mathcal{L}(\Phi(\mathbf{x};\mathbf{w}), y))$$

where y can either be the original label or the model's prediction, and *sign* returns the sign of the gradient. We used 100 random restarts for FGSM-100 and picked the best adversarial example from these 100 restarts.

PGD [10]: Projected-Gradient Descent (PGD) is an iterative version of FGSM with an additional random restart.

$$\mathbf{x}_a^{(t+1)} = Clip_{\mathbf{x},\epsilon}\Big\{\mathbf{x}_a^{(t)} + \alpha\ sign(\frac{\partial}{\partial \mathbf{x}_a^{(t)}}\mathcal{L}(\Phi(\mathbf{x}_a^{(t)};\mathbf{w}), y))\Big\}$$

where α is per-step update size, $Clip_{\mathbf{x},\epsilon}$ binds the L_∞ norm of the perturbation to be ϵ, and $\mathbf{x}_a^{(t)}$ indicates the adversarial example obtained after the tth optimization step. In the case of PGD, $\mathbf{x}_a^{(0)} = \mathbf{x} + \delta$ where δ is a random perturbation within the L_∞ norm-ball. We used 10 random restarts for PGD-10 and picked the best adversarial example from these 10 restarts. Each every restart, we use 100 PGD steps with $\alpha = 2 \times \epsilon/T$ where T is the number of PGD steps.

Black-box Attacks

Noise Attack: One of the most simple and preliminary attacks is the random noise attack. The idea is to generate a set of random vectors and pick the best one out of them. We used a set of 100 random vectors for NOISE-100.

Boundary Attack [2]: Boundary attack was the first decision-based attack. It starts with a random input that is not classified as the given label y and walks towards the original input until it hits the decision boundary. At this point, the attack starts moving orthogonal to the decision boundary until it encounters the closest attainable point to the given example $\mathbf{x}$. Boundary attack minimizes the L_2 norm of perturbation, therefore, it is not particularly optimized for the L_∞ models that we consider in our case. In order to compute the robust accuracy metric that we report, we compute the L_∞ norm of the computed perturbation and then discard any perturbations which exceed this budget. Therefore, in many cases, although being a powerful attack when considering euclidean distances, it does not provide high success rates when evaluated on L_∞ norm.

Simple Black-box Attack (SIMBA) [7]: SIMBA is one of the most simple black-box attacks which uses the predicted probabilities from the classifier to decide the perturbation vector. SIMBA attack considers all of the input points in the sequence and computes their impact on the probability of the predicted class if perturbed by either $\{-\epsilon, \epsilon\}$. It then chooses perturbation, whichever maximally reduces the probability of the predicted class. If the probability of the predicted class is not impacted by the chosen input point, the attack leaves it intact, hence, it also minimizes the L_0 norm of the perturbation. The number of input points queried to compute the adversarial example is restricted when considering high-dimensional images in order to make the attack time efficient. However, we did not impose any such restrictions as the number of points in a sequence is usually smaller in contrast to the number of pixels in an image.

3.4 Dataset

Due to lack of space, we only present results on the famous character trajectories dataset[1]. We also experimented with several other datasets and found the results to be consistent between these different datasets. The character trajectories dataset contains hand-written characters using a Wacom tablet. Only three dimensions are kept for the final dataset which includes x, y and pen-tip force. The sampling rate was set to be 200 Hz. The data was numerically differentiated and Gaussian smoothen with $\sigma = 2$. The task is to classify the characters into 20 different classes. This dataset is comprised of 2858 character samples divided into 1383 training, 606 validation and 869 test sequences. Each sequence is comprised of 206 time-steps with three channels. Since we need to constrain the input range for precisely defining the epsilon norm-ball to consider within our attack and defense framework, we normalize the data to have zero mean and unit standard deviation.

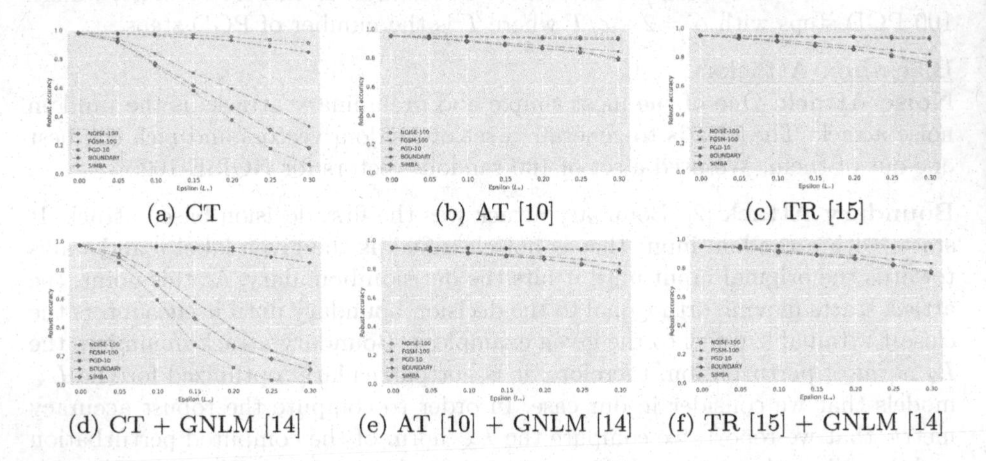

(a) CT (b) AT [10] (c) TR [15]

(d) CT + GNLM [14] (e) AT [10] + GNLM [14] (f) TR [15] + GNLM [14]

Fig. 1. Adversarial robustness curves computed for the different robust models using five different attacks. CT stands for conventional training, AT stands for adversarial training, TR stands for TRADES, and GNLM stands for Gaussian Non-Local Means.

4 Results

Our main results are presented in Fig. 1. We report the robust accuracy of the classifier considering different attack methods on a range of different epsilon values starting from 0.05 to 0.3 with an increment of 0.05. These plots precisely

[1] https://archive.ics.uci.edu/ml/datasets/Character+Trajectories.

capture the robustness of the model against adversarial attacks of different magnitudes. Since we trained a range of different TRADES models using different values of $1/\lambda$, we only report the best one out here, and explore the impact of these hyperparameters later in Sect. 4.2. It is evident from the plot that conventional training results in poor robustness against these attacks, almost reducing the classifier's accuracy to 0% when considering the worst-case adversary. However, when using defense methodologies such as adversarial training or TRADES, the model gains robustness against both white-box as well as black-box attacks with a slightly detrimental effect on the clean accuracy of the model.

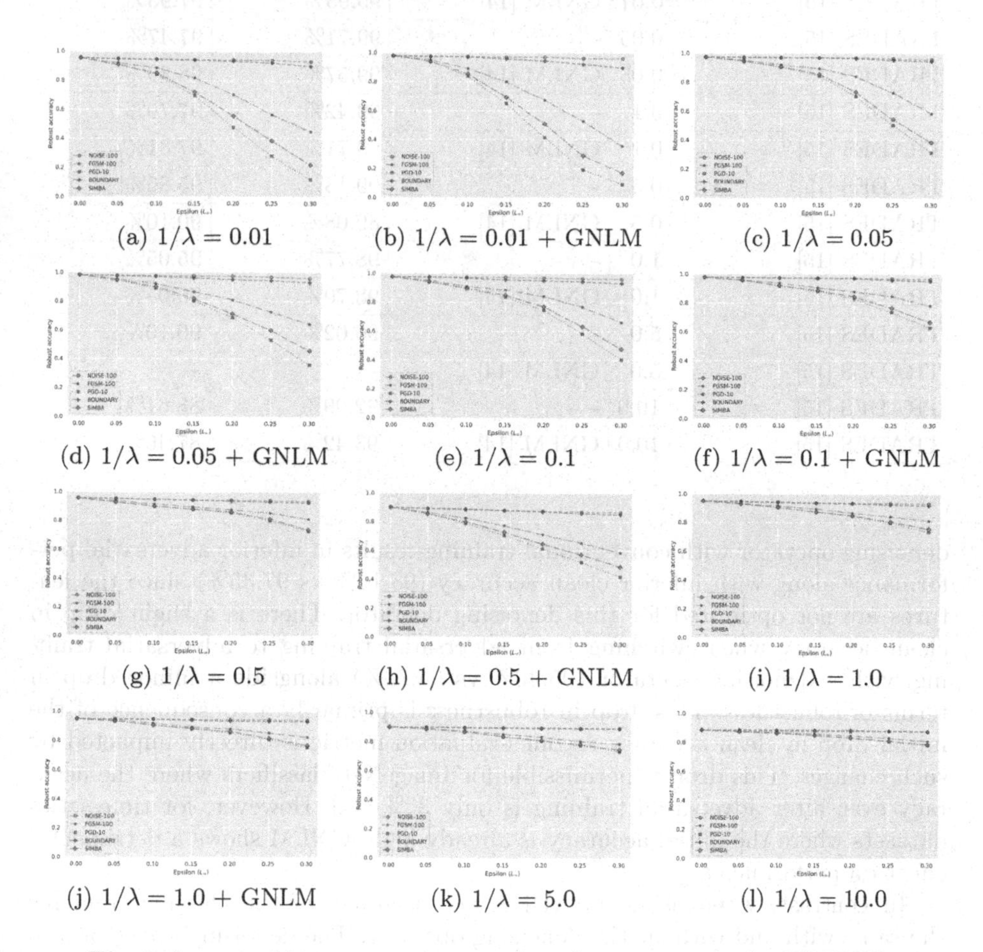

(a) $1/\lambda = 0.01$ (b) $1/\lambda = 0.01$ + GNLM (c) $1/\lambda = 0.05$

(d) $1/\lambda = 0.05$ + GNLM (e) $1/\lambda = 0.1$ (f) $1/\lambda = 0.1$ + GNLM

(g) $1/\lambda = 0.5$ (h) $1/\lambda = 0.5$ + GNLM (i) $1/\lambda = 1.0$

(j) $1/\lambda = 1.0$ + GNLM (k) $1/\lambda = 5.0$ (l) $1/\lambda = 10.0$

Fig. 2. Impact of hyperparameters on the robustness curves for TRADES.

4.1 Quantifying the Impact of Denoising Operators

Figure 1 also presents a comparison of using Gaussian Non-Local Means (GNLM) as a denoising operator alongside the use of different training schemes. Using

Table 1. Clean accuracy of the different models

Defense	$1/\lambda$	Denoising operator	Train accuracy	Test accuracy
-	-	-	100.00%	98.16%
-	-	GNLM [14]	100.00%	97.35%
Adversarial Training [10]	-	-	99.13%	95.40%
Adversarial Training [10]	-	GNLM [14]	98.70%	94.71%
TRADES [15]	0.01	-	99.49%	95.63%
TRADES [15]	0.01	GNLM [14]	99.93%	97.93%
TRADES [15]	0.05	-	99.71%	97.47%
TRADES [15]	0.05	GNLM [14]	99.57%	98.39%
TRADES [15]	0.1	-	99.42%	97.70%
TRADES [15]	0.1	GNLM [14]	99.71%	97.81%
TRADES [15]	0.5	-	99.13%	95.86%
TRADES [15]	0.5	GNLM [14]	89.08%	90.10%
TRADES [15]	1.0	-	98.77%	95.05%
TRADES [15]	1.0	GNLM [14]	98.70%	95.05%
TRADES [15]	5.0	-	96.02%	90.10%
TRADES [15]	5.0	GNLM [14]	-	-
TRADES [15]	10.0	-	92.99%	88.61%
TRADES [15]	10.0	GNLM [14]	93.42%	87.46%

denoising operator with conventional training results in inferior adversarial performance along with inferior clean accuracy (98.16% vs 97.35%) since the features are not optimized for this denoising operator. There is a slight drop in clean accuracy when switching from adversarial training to adversarial training with a denoising operator (95.40% vs 94.71%) alongside a minor drop in terms of robustness. This drop in robustness is primarily a consequence of the initial drop in clean accuracy as our evaluation metric is directly impacted by such changes. This drop is permissible for ImageNet classifiers where the accuracy even after adversarial training is only 35% [14]. However, for time-series datasets where the initial accuracy is already high, GNLM shows a detrimental effect on performance.

In contrast, when using TRADES, the accuracy of the classifier remains the same with and without the denoising operator. The denoising operator also positively impacts the robustness of the model. In comparison to adversarial training, there is no impact on clean accuracy when using TRADES with $1/\lambda = 1.0$. Table 1 summarizes the clean accuracies of the model under different settings for a direct comparison.

4.2 Sensitivity to Regularization Hyperparameters

Figure 2 visualizes the robustness curves for different values of the hyperparameter $1/\lambda$ used when training the robust model using TRADES. We also list the clean accuracies of the models in Table 1 for a direct comparison. It is evident from the table and the figure that higher regularization leads to lower clean accuracy as expected alongside higher robustness against adversarial attacks. It is important to note that the network failed to converge in many cases when using $1/\lambda > 1.0$ and GNLM denoising operator.

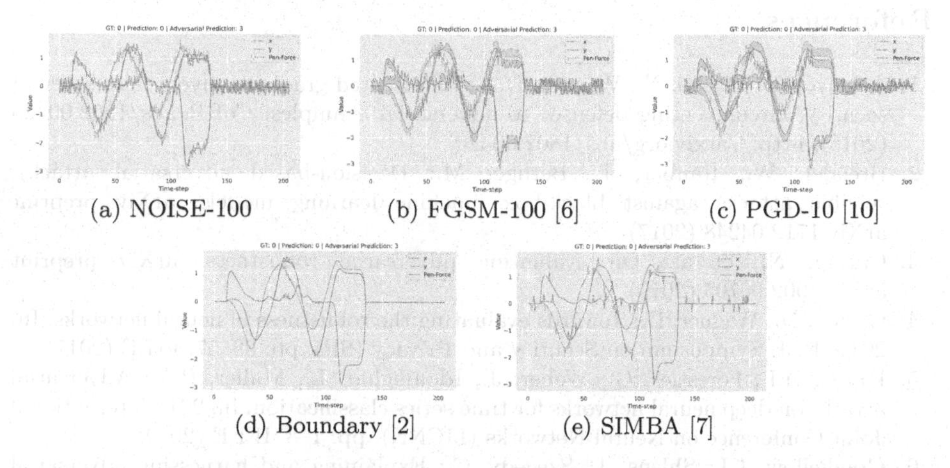

(a) NOISE-100 (b) FGSM-100 [6] (c) PGD-10 [10]

(d) Boundary [2] (e) SIMBA [7]

Fig. 3. Generated adversarial examples using an ϵ of 0.3. The original signal is highlight using a solid line while the attacked signal is represented using different line styles. The shaded area highlights the difference between the two signals.

4.3 Attacked Examples

Figure 3 presents a particular example from the character trajectories dataset on the undefended model. We visualized examples generated from a rather high value of epsilon i.e. 0.3. This is to ensure that the differences between different attacks are properly highlighted. It is interesting to note that all attacks changed the label to the same target class 3 indicating that the two classes are similar in the feature space. Almost all attacks exhausted the L_∞ perturbation budget except for boundary and SIMBA attack as boundary attack minimizes the L_2 norm of the perturbation while SIMBA additionally minimizes the L_0 norm of the perturbation alongside the L_∞ norm.

5 Conclusion

This paper establishes an important benchmark regarding the robustness of time-series classification models trained using different adversarial defense techniques.

Our analysis shows that the defenses evaluated for visual modality provide similar robustness against adversarial attacks on time-series data.

Future work should be mainly targeted towards the evaluation of these adversarial attacks for regression networks. While it is easy to quantify the impact in terms of success rate for classification networks, this is much harder to report when considering real-valued outputs. Another important direction is to compare provable robustness methods on time-series data and evaluate their efficacy as compared to the defenses considered here.

References

1. Athalye, A., Carlini, N., Wagner, D.A.: Obfuscated gradients give a false sense of security: circumventing defenses to adversarial examples. CoRR abs/1802.00420 (2018). http://arxiv.org/abs/1802.00420
2. Brendel, W., Rauber, J., Bethge, M.: Decision-based adversarial attacks: reliable attacks against black-box machine learning models. arXiv preprint arXiv:1712.04248 (2017)
3. Carlini, N., et al.: On evaluating adversarial robustness. arXiv preprint arXiv:1902.06705 (2019)
4. Carlini, N., Wagner, D.: Towards evaluating the robustness of neural networks. In: 2017 IEEE Symposium on Security and Privacy (SP), pp. 39–57. IEEE (2017)
5. Fawaz, H.I., Forestier, G., Weber, J., Idoumghar, L., Muller, P.A.: Adversarial attacks on deep neural networks for time series classification. In: 2019 International Joint Conference on Neural Networks (IJCNN), pp. 1–8. IEEE (2019)
6. Goodfellow, I.J., Shlens, J., Szegedy, C.: Explaining and harnessing adversarial examples. arXiv preprint arXiv:1412.6572 (2014)
7. Guo, C., Gardner, J.R., You, Y., Wilson, A.G., Weinberger, K.Q.: Simple black-box adversarial attacks. arXiv preprint arXiv:1905.07121 (2019)
8. Harford, S., Karim, F., Darabi, H.: Adversarial attacks on multivariate time series. arXiv preprint arXiv:2004.00410 (2020)
9. Karim, F., Majumdar, S., Darabi, H.: Adversarial attacks on time series. IEEE Trans. Pattern Anal. Mach. Intell. (2020)
10. Madry, A., Makelov, A., Schmidt, L., Tsipras, D., Vladu, A.: Towards deep learning models resistant to adversarial attacks. arXiv preprint arXiv:1706.06083 (2017)
11. Siddiqui, S.A., Mercier, D., Munir, M., Dengel, A., Ahmed, S.: Tsviz: demystification of deep learning models for time-series analysis. IEEE Access **7**, 67027–67040 (2019)
12. Szegedy, C., et al.: Intriguing properties of neural networks. arXiv preprint arXiv:1312.6199 (2013)
13. Wong, E., Kolter, Z.: Provable defenses against adversarial examples via the convex outer adversarial polytope. In: International Conference on Machine Learning, pp. 5286–5295 (2018)
14. Xie, C., Wu, Y., Maaten, L.V.D., Yuille, A.L., He, K.: Feature denoising for improving adversarial robustness. In: Proceedings of the IEEE Conference on Computer Vision and Pattern Recognition, pp. 501–509 (2019)
15. Zhang, H., Yu, Y., Jiao, J., Xing, E.P., Ghaoui, L.E., Jordan, M.I.: Theoretically principled trade-off between robustness and accuracy. arXiv preprint arXiv:1901.08573 (2019)

Correlation-Aware Change-Point Detection via Graph Neural Networks

Ruohong Zhang(✉), Yu Hao, Donghan Yu, Wei-Cheng Chang, Guokun Lai, and Yiming Yang

Carnegie Mellon University, 5000 Forbes Ave, Pittsburgh, PA 15213, USA
{ruohongz,yuhao2,dyu2,wchang2,guokun,yiming}@cs.cmu.edu

Abstract. Change-point detection (CPD) aims to detect abrupt changes over time series data. Intuitively, effective CPD over multivariate time series should require explicit modeling of the dependencies across input variables. However, existing CPD methods either ignore the dependency structures entirely or rely on the (unrealistic) assumption that the correlation structures are static over time. In this paper, we propose a Correlation-aware Dynamics Model for CPD, which explicitly models the correlation structure and dynamics of variables by incorporating graph neural networks into an encoder-decoder framework. Extensive experiments on synthetic and real-world datasets demonstrate the advantageous performance of the proposed model on CPD tasks over strong baselines, as well as its ability to classify the change-points as correlation changes or independent changes.

Keywords: Multivariate time series · Change-point detection · Graph neural networks

1 Introduction

Change-point detection (CPD) aims to detect abrupt property changes over time series data. In this study, change-points are detected through the changes of *dynamics* and *correlation* of variables. Dynamics refers to the physical property that determines a variable's modus operandi and correlation describes the interactions between variables. Previous CPD methods [1,2] model dynamics by parametric distributions like Hidden Markov Models (HMM), but they don't explicitly capture the correlation information. Other works capture static correlation structures in the multivariate time series [3], but they can't detect any correlation changes. We propose a **Co**rrelation-aware **D**ynamics Model for **C**hange-**p**oint **D**etection (CoRD-CPD) which incorporates graph neural networks into an encoder-decoder framework to explicitly model both changeable correlation structure and variable dynamics. We refer to the changes of correlation structure as **correlation changes** and the changes of variable dynamics as **independent changes**, as shown in Fig. 1.

H. Yang et al. (Eds.): ICONIP 2020, LNCS 12534, pp. 555–567, 2020.
https://doi.org/10.1007/978-3-030-63836-8_46

Our model is capable of distinguishing the two types of changes, which could have a broader impact on decision-making. In financial markets, traders use pair trading strategy to profit from correlated stocks, such as Apple and Samsung (both are phone sellers), which share similar dips and highs. News about Apple expanding markets may independently raise its price without breaking its correlation with Samsung. However, news about Apple building self-driving cars will break its correlation with Samsung, and establish new correlations with automobile companies. While both of them are change-points, the former is an independent change of variables and the latter is a correlation change between variables. Knowing the type of change can guide financial experts to choose trading strategies properly.

Our contributions can be summarized as follows:

- We propose CoRD-CPD to capture both changeable correlation structure and variable dynamics.
- Our CoRD-CPD classifies the change-points as correlation changes or independent changes, and ensembles them for robust CPD.
- Experiment on synthetic and real datasets demonstrates that our model can bring enhanced interpretability and improved performance in CPD tasks.

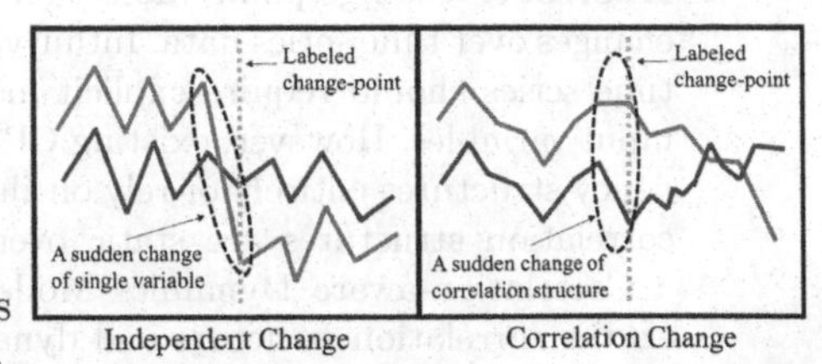

Fig. 1. (Left) an independent change of one variable and (Right) a correlation change between two variables. The red vertical line is the labeled change-point.

2 Method for CPD

A multivariate time series is denoted by $\mathbf{x} \in \mathbf{R}^{T \times N \times M}$, where T is the time steps, N is the number of variables and M is the number of features for each variable. We study the CPD problem in a retrospective setting and assume there is one change-point per $\mathbf{x} = \{\mathbf{x}^j\}_{j=1}^{T}$. The change-point at time step t satisfies:

$$\{\mathbf{x}^1, \mathbf{x}^2, \ldots, \mathbf{x}^{t-1}\} \sim \mathbb{P}$$
$$\{\mathbf{x}^t, \mathbf{x}^{t+1}, \ldots, \mathbf{x}^T\} \sim \mathbb{Q}$$

where $\mathbb{P}$ and $\mathbb{Q}$ denotes two different distributions. We attribute this difference to a correlation change (of the correlation structure), an independent change (of variable dynamics), or a mixture of both.

Correlation Change corresponds to the change of the correlation structure of multivariate time series, which is modeled by correlation matrices $\mathbf{A} \in \mathbb{R}^{T \times N \times N}$. At each time step, the pairwise interaction between variables (A_{ij}^t) is represented as a continuous value between 0 and 1, indicating how much they are correlated.

The correlation change score $\mathbf{s}_r$ is calculated by the L_1 distance between two neighboring correlation matrices:

$$s_r^t = \|\mathbf{A}^t - \mathbf{A}^{t-1}\|_1, t > 1 \tag{1}$$

Independent Change corresponds to the change of the variable dynamics. Given the current values of time series (and the extracted correlation matrices), if the dynamics rule is followed, the expected values of the future time steps predicted by our model will be close to the observed values; Otherwise, the difference will be large. This difference is used as the independent change score $\mathbf{s}_d$. Formally, we use the Mean Squared Error (MSE) as a metric to compare the expected values $\hat{\mathbf{w}}^{t+1} = \{\hat{\mathbf{x}}^i\}_{i=t+1}^{t+k}$ with the observed values $\mathbf{w}^{t+1} = \{\mathbf{x}^i\}_{i=t+1}^{t+k}$ over a window of size k.

$$s_d^t = \text{MSE}(\hat{\mathbf{w}}^t, \mathbf{w}^t), t > 1 \tag{2}$$

Note that if only a correlation change takes place, the expected value $\hat{\mathbf{w}}^t$ should not be different from the observed value $\mathbf{w}^t$, since we model a conditional probability $P(\mathbf{x}^t|\mathbf{x}^{<t}, \mathbf{A})$ and any correlation change will be factored in.

Ensemble of Change-point Scores aims to combine the correlation change with the independent change, because in real world applications, change-points could be resulted from a mixture of both. A simple way to ensemble them (for $\mathbf{s}_{en}$) is to sum the normalized scores of $\mathbf{s}_r$ and $\mathbf{s}_d$:

$$\mathbf{s}_{en} = \text{Norm}(\mathbf{s}_r) + \text{Norm}(\mathbf{s}_d) \tag{3}$$

$$\text{Norm}(\mathbf{s}) = \frac{\mathbf{s} - u_s}{\sigma_s} \tag{4}$$

where u_s and σ_s are mean and standard deviation of score $\mathbf{s}$.

In order to use our CPD methods above, we need to model correlation matrices and to be able to predict a future window of time steps based on the extracted correlation. We will introduce our CoRD-CPD in the next section.

3 Correlation-Aware Dynamics Model

The CoRD-CPD has an encoder for correlation extraction and a decoder for variable dynamics. Given a time series $\mathbf{x}$, the encoder models a distribution of correlation matrix $q_\phi(\mathbf{A}^t|\mathbf{x})$ for each time step t, and by factorization,

$$q_\phi(\mathbf{A}|\mathbf{x}) = \prod_{t=1}^{T} q_\phi(\mathbf{A}^t|\mathbf{x}) \tag{5}$$

The decoder models a distribution of time steps $p_\theta(\mathbf{x}|\mathbf{A})$ auto-regressively,

$$p_\theta(\mathbf{x}|\mathbf{A}) = \prod_{t=1}^{T} p_\theta(\mathbf{x}^t|\mathbf{x}^{<t}, \mathbf{A}^{t-1}) \tag{6}$$

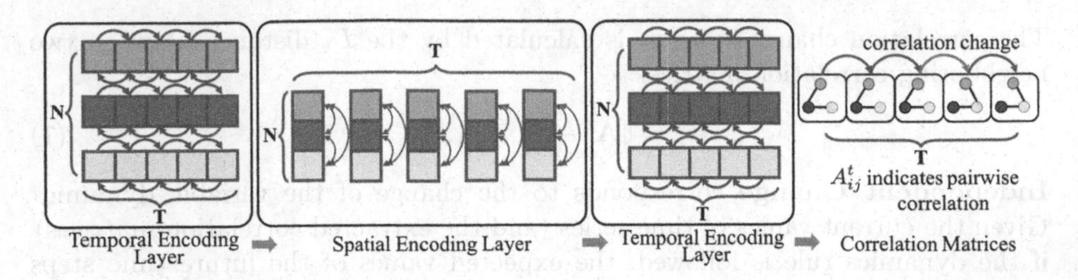

Fig. 2. CorD-CPD Encoder: the encoder extracts correlation matrices from multivariate time series. The temporal encoding layer captures time dependent features, and the spatial encoding layer models relational features between variables.

The objective function maximizes the log likelihood,

$$\mathcal{L}_{obj} = \mathbb{E}_{q_\phi(\mathbf{A}|\mathbf{X})}[\log p_\theta(\mathbf{x}|\mathbf{A})] \tag{7}$$

3.1 Correlation Encoder

The encoder infers a correlation matrix $\mathbf{A}^t$ at each time step, which depends on both temporal features and variable interactions. To leverage both sources, we propose Temporal Encoding Layers (TEL) to extract features across time steps and Spatial Encoding Layers (SEL) to extract features from variable interactions. As shown in Fig. 2, the two types of layers are alternatively applied to progressively incorporate temporal and correlation features into latent embeddings. Practically, we found 2 TEL and 1 SEL is enough for our tasks.

For each layer, let $\mathbf{h} \in \mathbb{R}^{T\times N\times K}$ denote the input and let $\tilde{\mathbf{h}} \in \mathbb{R}^{T\times N\times K'}$ denote the output, where T is the time steps, N is the number of variables, and K, K' are the number of input and output features respectively. The input to the initial layer is the multivariate time series data $\mathbf{x} \in \mathbb{R}^{T\times N\times M}$. The posterior distribution of the correlation matrix is modeled by

$$q_\phi(A^t_{ij}|\mathbf{x}) = \text{Softmax}(\text{Linear}([\tilde{\mathbf{h}}^t_{(f)i}; \tilde{\mathbf{h}}^t_{(f)j}])) \tag{8}$$

$$\tilde{\mathbf{h}}_{(f)} = \text{TEL}_2(\text{SEL}(\text{TEL}_1(\mathbf{x}))) \tag{9}$$

where $[\cdot;\cdot]$ is the concatenation operator and $\tilde{\mathbf{h}}_{(f)}$ is the embedding of the final layer. As an additional trick, we apply Gumbel-Softmax [4] to enforce sparse connections in correlation matrices in order to reduce noise.

Temporal Encoding Layer (TEL). Leverages information across T time steps (independently for each variable). For a fixed variable i, let $\mathbf{h}_i = \{\mathbf{h}^t_i\}^{t=N}_{t=1}$ denote the embeddings of that variable at all time steps. We offer two implementations of TEL with different neural architectures: RNN_{TEL} and $\text{Trans}_{\text{TEL}}$.

RNN_{TEL} is a bidirectional GRU network [5]:

$$\overrightarrow{\mathbf{h}_i}^t = \overrightarrow{\mathrm{GRU}}(\overrightarrow{\mathbf{h}_i}^{t-1}, \mathbf{h}_i^t) \tag{10}$$

$$\overleftarrow{\mathbf{h}_i}^t = \overleftarrow{\mathrm{GRU}}(\overleftarrow{\mathbf{h}_i}^{t+1}, \mathbf{h}_i^t) \tag{11}$$

$$\tilde{\mathbf{h}}_i^t = [\overrightarrow{\mathbf{h}_i}^t, \overleftarrow{\mathbf{h}_i}^t] \tag{12}$$

where $\overrightarrow{\mathbf{h}_i}^t, \overleftarrow{\mathbf{h}_i}^t$ are intermediate representation from forward and backward GRU. The output $\tilde{\mathbf{h}}^t$ is a concatenation of embeddings from both directions.

$Trans_{TEL}$ uses the Transformer model [6] with self-attention to capture temporal dependencies. For the self-attention layer, the input is transformed into query matrices $\mathbf{Q}_i^t = \mathbf{h}_i^t \mathbf{W}_Q$, key matrices $\mathbf{K}_i^t = \mathbf{h}_i^t \mathbf{W}_K$ and value matrices $\mathbf{V}_i^t = \mathbf{h}_i^t \mathbf{W}_V$. Here $\mathbf{W}_Q, \mathbf{W}_K, \mathbf{W}_V$ are learnable parameters. Finally, the dot-product attention is a weighted sum of value vectors:

$$\tilde{\mathbf{h}}_i^t = \mathrm{softmax}\left(\frac{\mathbf{Q}\mathbf{K}^T}{\sqrt{d_k}}\right) \cdot \mathbf{V} \tag{13}$$

where d_k is the size of hidden dimension. Similar to [6], we use residual connection, layer normalization and positional encoding for Trans_{TEL}.

Spatial Encoding Layer (SEL). Leverages the information between the N variables (independently at each time step) via graph neural networks (GNN) [7]. For a fixed time step t, let $\mathbf{h}^t = \{\mathbf{h}_i^t\}_{i=1}^N$ denote the embeddings all variables at time t. The output is obtained by

$$\tilde{\mathbf{h}}^t = \mathrm{GNN}(\{\mathbf{h}_i^t\}_{i=1}^N) \tag{14}$$

where a GNN module is implemented by the feature aggregation and combination operations:

$$\mathbf{e}_{ij} = f_e([\mathbf{h}_i^t; \mathbf{h}_j^t]) \tag{15}$$

$$\tilde{\mathbf{h}}_j = f_v(\mathbf{h}_j + \sum_{i \neq j} \mathbf{e}_{ij}) \tag{16}$$

where Eq. 15 aggregates features between neighboring nodes and Eq. 16 combines those features by a summation. $f_e(\cdot)$ and $f_v(\cdot)$ are non-linear neural networks for which we provide two implementations: $\mathrm{GNN}_{\mathrm{SEL}}$ and $\mathrm{Trans}_{\mathrm{SEL}}$.

$\mathrm{GNN}_{\mathrm{SEL}}$ is implemented by a multilayer perceptron (MLP) and Trans_{SEL} is implemented by the Transformer model. Compared with MLP, Transformer has could be advantageous for spatial encoding because of well-designed self-attention, residual connection and layer normalization. The positional encoding layer is removed from Transformer because the variables are order invariant.

3.2 Dynamics Decoder

At a high level, the decoder learns the dynamics of variables by predicting the future time steps to be as close as the observed values. Instead of predicting the value of $\hat{\mathbf{x}}_i^{t+1}$ directly, we predict the change $\Delta\hat{\mathbf{x}}_i^t = \hat{\mathbf{x}}_i^{t+1} - \mathbf{x}_i^t$ as shown in Fig. 3.

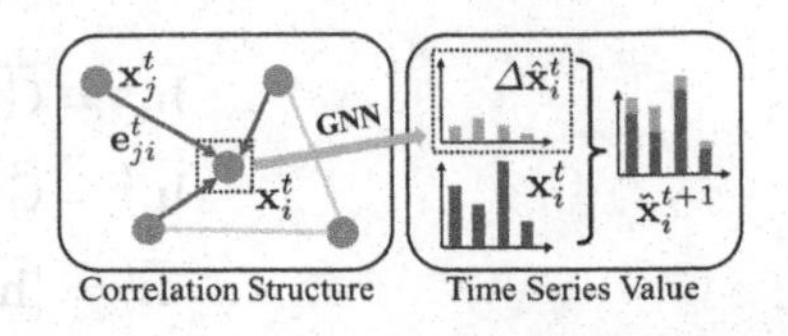

Fig. 3. CorD-CPD Decoder: Given a correlation matrix, the decoder predicts the change of future steps.

Since the prediction has to factor in the correlation between variables, we also need GNN to incorporate correlation matrices into feature embeddings. Again, the feature aggregation and combination operations are performed on the input $\mathbf{x}^t$,

$$\mathbf{e}_{ji}^t = \mathbf{A}_{ji}^t g_e([\mathbf{x}_j^t; \mathbf{x}_i^t]) \tag{17}$$

$$\tilde{\mathbf{h}}_i^t = g_v(\mathbf{x}_i^t + \sum_{j \neq i} \mathbf{e}_{ji}^t) \tag{18}$$

where the functions $g_e(.)$ and $g_v(.)$ are MLPs. We model $\Delta\hat{\mathbf{x}}_i^t = g_{\text{out}}(\tilde{\mathbf{h}}_i^{\leq t})$, where $g_{\text{out}}(\tilde{\mathbf{h}}_i^{\leq t})$ can be $\text{MLP}(\tilde{\mathbf{h}}_i^t)$ or $\text{RNN}(\tilde{\mathbf{h}}_i^{\leq t})$ depending on the application. Together, $\hat{\mathbf{x}}_i^{t+1} = \mathbf{x}_i^t + \Delta\hat{\mathbf{x}}_i^t = \mathbf{x}_i^t + g_{\text{out}}(\tilde{\mathbf{h}}_i^{\leq t})$.

The log likelihood of density $p_\theta(\mathbf{x}|\mathbf{A})$ can be expressed as:

$$\log p_\theta(\mathbf{x}|\mathbf{A}) = \sum_{t=1}^{T} \log p_\theta(\mathbf{x}^t|\mathbf{x}^{<t}, \mathbf{A}^{t-1}) \tag{19}$$

$$= \sum_i \sum_{t=1}^{T} \log \mathcal{N}(\mathbf{x}_i^t|\hat{\mathbf{x}}_i^t, \sigma^2\mathbf{I}) \tag{20}$$

$$\propto -\sum_i \sum_{t=2}^{T} \frac{\|\mathbf{x}_i^t - \hat{\mathbf{x}}_i^t\|_2^2}{2\sigma^2} \tag{21}$$

Maximizing Eq. 21 is equivalent to minimizing $\mathcal{L}_{obj} = \sum_i \sum_{t=2}^{T} \frac{\|\mathbf{x}_i^t - \hat{\mathbf{x}}_i^t\|_2^2}{2\sigma^2}$.

Since change-points are sparse in time series data, we introduce an additional regularization to ensure the smoothness of correlation matrix:

$$\mathcal{L}_{smooth} = \frac{1}{T-1} \sum_{t=2}^{T} \|\mathbf{A}^t - \mathbf{A}^{t-1}\|_2^2 \tag{22}$$

Finally, the loss function is $\mathcal{L} = \mathcal{L}_{obj} + \lambda\mathcal{L}_{smooth}$, where λ controls the relative strength of smoothness regularization.

4 Experiment with Physics Simulations

4.1 Particle-Spring Change-Point Dataset

We developed a dataset with a simulated physical particle-spring system. The system contains $N = 5$ particles that move in a rectangular space. Some randomly selected pairs (out of the 10 pairs in total) of particles are connected by invisible springs. The motion of particles are determined by the laws of physics such as Newton's law, Hooke's law, and Markov property. The trajectories of length $T = 100$ of the particles are recorded as the multivariate time series data. Each variable has $M = 4$ features: location l_x, l_y and speed v_x, v_y.

While the physical system is similar to the one in [3], we additionally design 3 types of change-points by perturbing the location, speed, and connection at a random time step between $[25, 75]$:

- **location**: A perturbation to the current location sampled from $\mathcal{N}(0, 0.1)$, where the range of the location is $[-5, 5]$.
- **speed**: A perturbation to the current speed by sampled from $\mathcal{N}(0, 0.02)$, where range of the speed is $[-1, 1]$.
- **connection**: re-sample connections and ensure that at least 5 out of 10 pairs of connections are changed.

The change of location or speed (both are dynamics) belongs to the independent change, and the change of connection (a type of correlation) belongs to the correlation change. Since the change-point is either a correlation change or an independent change, we are able to test the ability of our model to classify them.

We generate 500 time series for each type of change and mix them together (totally 1500 time series) as training data. For validation and testing data, we generate 100 time series for each type of change and evaluate on them separately. Our model is unsupervised, so the validation set is only used for hyperparameter tuning. In real world datasets, human labeled change-points are scarce in quantity, which usually results in large variance in evaluation. As a remedy, our synthetic data can be generated in a large amount to reduce such a variance in testing.

4.2 Evaluation Metric and Baselines

For quantitative evaluation of CPD performance, we consider two metrics:

Area-Under-the-Curve (AUC) of the receiver operating characteristic (ROC) is a metric commonly used in the CPD literature [8].

Triangle Utility (TRI) is a hinge-loss-based metric: $\max(0, 1 - \frac{\|y-l\|}{w})$, where $w = 15$ is the margin, l and y are the labeled and predicted change-points.
Both of the metrics range from $[0, 1]$ and higher values indicate better predictions. However, AUC treats the change-point scores at each time step independently, without considering any temporal patterns. TRI considers the distance between the label and the predicted change-point (the one with highest change-point score), but it doesn't measure the quality of predictions at the other time steps. We use both metrics because they complement with each other.

Next, we introduce 6 baselines of the state-of-the-art statistical and deep learning models:

- **ARGP-BODPD** [9] is Bayesian change-point model that uses auto-regressive Gaussian Process as underlying predictive model.
- **RDR-KCPD** [10] uses relative density ratio technique that considers f-divergence as the dissimilarity measure.
- **Mstats-KCPD** [11] uses kernel maximum mean discrepancy (MMD) as dissimilarity measure on data space.
- **KL-CPD** [8] uses deep neural models for kernel learning and generative method to learn pseudo anomaly distribution.
- **RNN** [5] is a recurrent neural network baseline to learn variable dynamics from multivariate time series (without modeling correlations).
- **LSTNet** [12] combines CNN and RNN to learn variable dynamics from long and short-term temporal data (without modeling correlations).

Table 1. AUC and TRI metrics on synthetic datasets for the prediction of location, speed and connection change. Our CoRD-CPD (evaluated with $\mathbf{s}_{en}$) has the best performance on both metrics among all the baselines.

Model	Location		Speed		Connection	
	AUC	TRI	AUC	TRI	AUC	TRI
ARGP-BOCPD	0.5244	0.0880	0.5231	0.0660	0.5442	0.1287
RDR-KCPD	0.5095	0.0680	0.5279	0.1093	0.5234	0.0860
Mstats-KCPD	0.5380	0.0730	0.5369	0.0727	0.5508	0.0833
RNN	0.5413	0.2567	0.5381	0.2660	0.5446	0.3047
LSTNet	0.5817	0.3487	0.5817	0.3460	0.5337	0.2193
KL-CPD	0.5247	0.1053	0.5378	0.1352	0.5574	0.3127
GNN_{SEL}+RNN_{TEL}	0.9864	0.9740	0.9700	**0.9320**	**0.9681**	**0.9153**
$\text{Trans}_{\text{SEL}}$+$\text{RNN}_{\text{TEL}}$	**0.9885**	**0.9773**	**0.9755**	0.9080	0.9469	0.9040
GNN_{SEL}+$\text{Trans}_{\text{TEL}}$	0.9692	0.9333	0.9609	0.8473	0.8840	0.8527

4.3 Main Results

Table 1 shows the performance of the statistical baselines (first panel), the deep learning baselines (second panel) and our proposed CoRD-CPD (third panel).

Statistical Baselines are not as competitive as the other deep learning models among all the types of changes. One explanation is that those models have strong assumption on the parameterization of probability distributions, which may hurt the performance on datasets that demonstrate complicated interactions of variables. The dynamics rule of the physics system can be hardly captured by those methods.

Deep Learning Baselines are slightly better than the statistical models, in which the LSTNet has the best performance on location and speed changes.

Table 2. Our CoRD-CPD separately computes the scores for correlation change (cor) and independent change (ind). The correlation change score is high on the connection data, while the correlation change score is high on the location and speed data.

Model	Type	Location		Speed		Connection	
		AUC	TRI	AUC	TRI	AUC	TRI
GNN_{SEL}+RNN_{TEL}	Cor	0.5145	0.3153	0.5590	0.3553	**0.9649**	**0.9073**
	Ind	**0.9835**	**0.9727**	**0.9587**	**0.9493**	0.8093	0.7320
$\text{Trans}_{\text{SEL}}$+ RNN_{TEL}	Cor	0.4944	0.2626	0.5463	0.3266	**0.9755**	**0.9273**
	Ind	**0.9859**	**0.9720**	**0.9685**	**0.9233**	0.7774	0.6460
GNN_{SEL}+$\text{Trans}_{\text{TEL}}$	Cor	0.5544	0.3467	0.5832	0.4266	**0.9098**	**0.8787**
	Ind	**0.9855**	**0.9693**	**0.9623**	**0.9133**	0.7912	0.7620

Since LSTNet has a powerful feature extractor for long and short-term temporal data, it is better at learning variable dynamics. However, as correlation plays an important role in the synthetic data, ignoring it will hurt performance in general.

CoRD-CPD is evaluated on the test data by the ensemble score $\mathbf{s}_{en}$. It has the best performance on both metrics among all the baselines. We didn't include the result of $\text{Trans}_{\text{SEL}}$+$\text{Trans}_{\text{TEL}}$, because empirically it is harder to converge. $\text{Trans}_{\text{SEL}}$+$\text{RNN}_{\text{TEL}}$ is the best at detecting the independent changes, while GNN_{SEL}+RNN_{TEL} is the best at detecting the correlation changes. The reason could be that the Transformer models are better at identifying local patterns, while RNNs are more stable at combining features with long term dependencies. Among the three types of changes, the score of connection change is lower than that of the other two, indicating the detection of correlation changes is harder than independent changes.

4.4 Change-Point Type Classification

Our CoRD-CPD separately computes change-point scores for correlation change ($\mathbf{s}_r$) and independent change ($\mathbf{s}_d$). We show the ability of our model to separate the two types of changes based on the scores.

Correlation vs. Independent Change. In Table 2, the correlation change (cor) and independent change (ind) are separately evaluated. The correlation change scores ($\mathbf{s}_r$) are high on the connection data, while the independent change scores ($\mathbf{s}_d$) are high on location and speed change. This result shows that our system can indeed distinguish the two types of changes.

Location&speed. The independent changes can be successfully distinguished. For location and speed data, AUC of the independent changes is over 0.97, close to a perfect detection; AUC of the correlation change is close to 0.5, nearly a random guess. Therefore, our system doesn't signal a correlation change for location and speed data, but it gives a strong signal of an independent change.

Connection. The correlation changes are harder to be detected, but CorD-CPD gives a good estimation. In the connection data, AUC of correlation change are higher than independent change, but the gap was smaller than that in location and speed data. The reason could be that the errors made by encoder are propagated into the decoder, and thus made the forecasting of time series values inaccurate.

Classification Method. While our model shows a potential to distinguish the two types of changes, we want it to be able to classify them. We propose to use the difference between normalized correlation change score $\mathbf{s}_r$ and independent change score $\mathbf{s}_d$ as an indicator of change-point type, at time t:

$$\mathrm{Norm}(\mathbf{s}_r)^t - \alpha\,\mathrm{Norm}(\mathbf{s}_d)^t \begin{cases} \geq \tau, & \text{correlation change} \\ < \tau, & \text{independent change} \end{cases}$$

where $\alpha = 0.75$ is our design choice, and τ is a threshold to separate the correlation change and the independent change. Moving the value of τ controls the type I error (False Positive) and the type II error (False Negative). To measure the classification quality by leveraging the error, ROC AUC is a typical solution.

We classify the change-point types under two settings: with label and without label, according to whether the labeled change-point is provided.

With Label: When a labeled change-point is provided by human experts, our model classifies it as either a correlation change or an independent change, whichever dominates.

Without Label: When the label information is unavailable, our model performs classification from the predicted change-point with the highest $\mathbf{s}_{en}$ score.

The results are shown in Table 3. Our best model Trans$_{\mathrm{SEL}}$+RNN$_{\mathrm{TEL}}$ achieves an ROC AUC of 0.979 (with label) and 0.973 (without label). This indicates that our model has a strong ability to discriminate the two types of change-points under both settings. GNN$_{\mathrm{SEL}}$+Trans$_{\mathrm{TEL}}$ has the worst classification performance, which is consistent to the observation in Table 2 that it is not good at capturing correlation changes.

In the next experiment, we set $\tau = 0$ and report the classification accuracy on the three data types. As shown in right part of Table 3, a high accuracy of 98% on identifying the location and speed change demonstrates that our model can predict the independent changes well. For correlation changes, the Trans$_{\mathrm{SEL}}$+RNN$_{\mathrm{TEL}}$ shows the best performance by achieving 93% on supervised setting and 84% on unsupervised setting.

When labeled change-points are not provided, the classification task could be more difficult, because the it relies on the predicted change-points. If a predicted change-point is far from the ground truth, the classification is prone to errors.

Table 3. ROC AUC metric demonstrates the ability of our model to separate the two types of change-points. When $\tau = 0$, we report the change-point classification accuracy on the 3 types of data.

Model	ROC AUC	Location	Speed	Connection
With Label				
GNN_{SEL}+RNN_{TEL}	0.972	**98%**	**97%**	68%
$Trans_{SEL}$+RNN_{TEL}	**0.979**	**98%**	96%	**93%**
GNN_{SEL}+$Trans_{TEL}$	0.916	**98%**	96%	87%
Without Label				
GNN_{SEL}+ RNN_{TEL}	0.969	**98%**	**96%**	73%
$Trans_{SEL}$+RNN_{TEL}	**0.973**	96%	92%	**84%**
GNN_{SEL}+$Trans_{TEL}$	0.929	91%	83%	75%

5 Experiments with Physical Activity Monitoring

In addition to our synthetic dataset, we test our CorD-CPD on real-world data: the PAMAP2 Physical Activity Monitoring dataset [13]. The dataset contains sensor data collected from 9 subjects performing 18 different physical activities, such as walking, cycling, playing soccer, etc. Specifically, the variables we consider are $N = 3$ Inertial Measurement Units (IMU) on wrist, chest and ankle respectively, measuring $M = 10$ features including temperature, 3D acceleration, gyroscope and magnetometer. The change-points are labeled as the transitions between activities.

To account for the transitions between activities, the independent changes could possibly include the rising of temperature and the correlation changes could be from the switch of different moving patterns between wrist, chest and ankle.

Table 4. We report the performance of our CorD-CPD on a real-world multivariate time series dataset (PAMAP2). The variables are sensors and the features includes temporatures and 3-D motions. The change-points are transitions between activities.

Model	AUC	TRI
ARGP-BOCPD	0.5079	0.1773
RDR-KCPD	0.5633	0.1933
Mstats-KCPD	0.5112	0.1480
RNN	0.5540	0.2393
LSTNet	0.5688	0.3145
KL-CPD	0.5326	0.2102
GNN_{SEL}+RNN_{TEL}	0.7868	0.7574
$Trans_{SEL}$+ RNN_{TEL}	0.7903	0.7750
GNN_{SEL}+ $Trans_{TEL}$	**0.8277**	**0.8020**

The data was sample every 0.01 second over totally 10 hours. In the pre-processing, we down-sample the time-series by 20 time steps and then slice them into windows of a fixed length $T = 100$ steps. Each window contains exactly one transition from range $[25, 75]$. There are totally 184 multivariate time series with change-points: 150 of them are used as training, 14 are used as validation and 20 are used as testing.

The results are shown in Table 4. Our CoRD-CPD achieves the best performance among the 6 statistical and deep learning baselines. We attribute the enhanced performance to the ability of CoRD-CPD to better model the two types of changes and to successfully ensemble them. In real life scenarios, a change-point could arise from a mixture of independent change and correlation change. The experiment results show that explicitly modeling both types of changes injects a positive inductive bias during learning, and thus enhances the performance of CPD tasks.

6 Conclusion

In this paper, we study the CPD problem on multivariate time series data under the retrospective setting. We propose CoRD-CPD to explicitly model the correlation structure by incorporating graph neural networks into an encoder-decoder framework. CoRD-CPD can classify change-points into two types: the correlation change and the independent change. We conduct extensive experiments on physics simulation dataset to demonstrate that CoRD-CPD can distinguish the two types of change-points. We also test it on the real-word PAMAP2 dataset to show the enhanced performance on CPD over competitive statistical and deep learning baselines.

References

1. Zhang, N.R., Siegmund, D.O., Ji, H., Li, J.Z.: Detecting simultaneous changepoints in multiple sequences. Biometrika **97**(3), 631–645 (2010)
2. Montanez, G.D., Amizadeh, S., Laptev, N.: Inertial hidden markov models: modeling change in multivariate time series. In: Twenty-Ninth AAAI Conference on Artificial Intelligence (2015)
3. Kipf, T., Fetaya, E., Wang, K.C., Welling, M., Zemel, R.: Neural relational inference for interacting systems. In: International Conference on Machine Learning, pp. 2693–2702 (2018)
4. Jang, E., Gu, S., Poole, B.: Categorical reparameterization with gumbel-softmax. In: International Conference on Learning Representations (ICLR) (2017)
5. Cho, K., et al.: Learning phrase representations using RNN encoder-decoder for statistical machine translation. arXiv preprint arXiv:1406.1078 (2014)
6. Vaswani, A., et al.: Attention is all you need. In: Advances in Neural Information Processing Systems, pp. 5998–6008 (2017)
7. Kipf, T.N., Welling, M.: Semi-supervised classification with graph convolutional networks. arXiv preprint arXiv:1609.02907 (2016)

8. Chang, W.C., Li, C.L., Yang, Y., Póczos, B.: Kernel change-point detection with auxiliary deep generative models. arXiv preprint arXiv:1901.06077 (2019)
9. Saatçi, Y., Turner, R.D., Rasmussen, C.E.: Gaussian process change point models. In: ICML, pp. 927–934 (2010)
10. Liu, S., Yamada, M., Collier, N., Sugiyama, M.: Change-point detection in time-series data by relative density-ratio estimation. Neural Netw. **43**, 72–83 (2013)
11. Li, S., Xie, Y., Dai, H., Song, L.: M-statistic for kernel change-point detection. In: Advances in Neural Information Processing Systems, pp. 3366–3374 (2015)
12. Lai, G., Chang, W.C., Yang, Y., Liu, H.: Modeling long-and short-term temporal patterns with deep neural networks. In: The 41st International ACM SIGIR Conference on Research & Development in Information Retrieval, pp. 95–104. ACM (2018)
13. Reiss, A., Stricker, D.: Introducing a new benchmarked dataset for activity monitoring. In: 2012 16th International Symposium on Wearable Computers, pp. 108–109. IEEE (2012)

DPAST-RNN: A Dual-Phase Attention-Based Recurrent Neural Network Using Spatiotemporal LSTMs for Time Series Prediction

Shajia Shan[1,2], Ziyu Shen[1,2], Bin Xia[1,2], Zheng Liu[1,2], and Yun Li[1,2](✉)

[1] Jiangsu Key Laboratory of Big Data Security and Intelligent Processing, Nanjing University of Posts and Telecommunications, Nanjing, China
liyun@njupt.edu.cn

[2] School of Computer Science, Nanjing University of Posts and Telecommunications, Nanjing, China

Abstract. For time series forecasting, the weight distribution among multivariables and the long-short-term time dependence are always very important and challenging. Traditional machine forecasting can't automatically select the effective features of multivariable input and can't capture the time dependence of sequences. The key to solve this problem is to capture the spatial correlations at the same time, the spatiotemporal relationships at different times and the long-term dependence of the temporal relationships between different series. In this paper, inspired by human attention mechanism including encoder-decoder model, we propose DPAST-based RNN (DPAST-RNN) for long-term time series prediction. Specifically, in the first phase we use attention mechanism to extract relevant features at each time adaptively then we use stacked LSTM units to extract hidden information of time series both from time and space dimensions. In the second phase, we use another attention mechanism to select the related hidden state in encoder to the hidden state of the decoder at the current time to make context vector which is embed into recurrent neural network in decoder. Thorough empirical studies based upon the VM-Power dataset we collected on OpenStack and the NASDAQ 100 Stock dataset demonstrate that the DPAST-RNN can outperform state-of-the-art methods for time series prediction.

Keywords: Time series prediction · Spatiotemporal LSTM · Attention mechanism · Encoder-decoder model

1 Introduction

Time series prediction algorithm has a wide range of applications, e.g., fine-grained photovoltaic output prediction [3], financial prediction [20], environmental forecasting [21], heart and brain signal analysis [7] and prediction of geo-sensor over future hours [13]. Generally, time series prediction can be divided

H. Yang et al. (Eds.): ICONIP 2020, LNCS 12534, pp. 568–578, 2020.
https://doi.org/10.1007/978-3-030-63836-8_47

into single variable problem and multivariable problem. However, in most cases, multivariable time series prediction problem is more in line with the needs of practical modeling. Different from the single variable time series prediction with strong periodicity, the problem of the multivariable prediction is mainly reflected in the following aspects: the correlation between the multivariable features at the same time, the correlation between the multivariable features at different times and the correlation between the multivariable features and the time of the target sequence. For some classical methods in time series prediction, ARIMA [1] assumes that the sequence variation is stable, so it is not suitable for non-stationary and multivariate time prediction. Support vector regression (SVR) [14], as a traditional regression method is used for time series prediction where feature sequences are mapped into high dimensional space, which pays more attention to the spatial correlations of these exogenous series at the same time, but ignores the time dependence. With the development of neural network, recurrent neural network (RNN) [18] especially Long short-term memory units (LSTM) [10] and gated recurrent unit (GRU) [5] are widely used in time series prediction. The encoder-decoder network structure was first proposed by Sutskever et al. [19] to solve the sequence to sequence machine translation problem. RNN based encoder-decoder network [5] was initially applied to machine translation. However, with the increasing length of vector representation, the performance of the encoder-decoder network deteriorated rapidly. Therefore, Bahdanau et al. [2] proposed the attention mechanism based on encoder-decoder structure. Attention mechanism has been widely used in machine translation [6], image caption [4], exogenous time series prediction [9], etc. Due to the success of attention-based encoder-decoder networks in sequence learning, Qin et al. [17] employ two-stage attention mechanism based on encoder-decoder structure to forecast multivariate time series. To capture the spatial dependency between sensors, Liang et al. [13] added global attention in GeoMAN. However, the decoder part of the models mentioned above does not fully consider the cyclic relationship between the target information and the encoded data in time.

In this paper, we use spatiotemporal LSTMs in the encoder network to obtain more accurate spatiotemporal relationship of the input data, and then embed the context information into the LSTM in the decoder network to enhance the attention of the target sequence to the encoding information in time. In addition, we build OpenStack virtual environment to collect VM power dataset and use DTW to preprocess and filter the data. The contributions of our work are three-fold:

- In the stage of data preprocessing, we use DTW [15] to analysis the original multivariate data and extract the effective feature variables in our dataset.
- In addition, considering that the single-layer LSTM can not transfer the effective information of multivariate input data, we use the spatiotemporal LSTMs to encode time series information as the input of decoder after the input attention mechanism.

– In the decoder, we embed the context vector generated by the temporal attention mechanism into recurrent neural network, so as to obtain a more accurate spatiotemporal relationship.

2 Model

The framework of the proposed forecasting model is shown in Fig. 1, which consists of encoder and decoder. The two phases attention modules are contained in the encoder and decoder respectively. The first phase in Encoder can adaptively select the most relevant input features while the second phase in Decoder uses categorical information to decode the stimulus. The Encoder encodes the time series conditioned on the input attention through the spatiotemporal LSTMs. In the decoder, the temporal attention is used to generated context vector c_t which represents a weighted sum of previous encoder hidden state across all the time steps. Then we combine the c_t with the hidden state in LSTM unit as the new hidden state fed to LSTM.

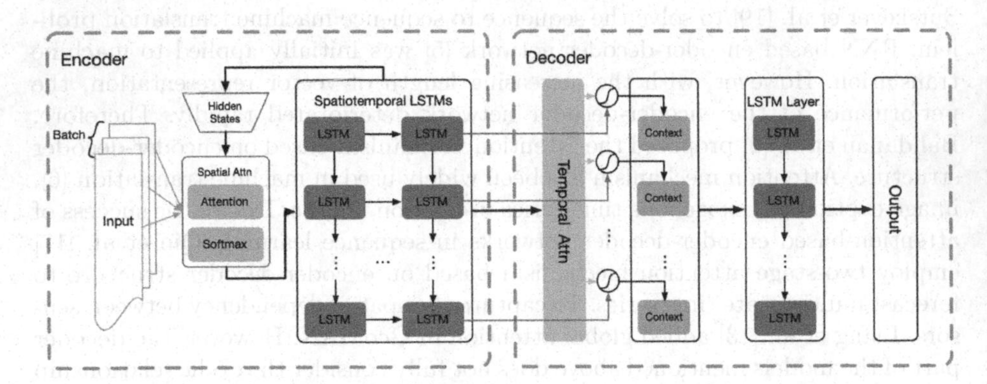

Fig. 1. Graphical illustration of the Dual-Phase Attention-based Recurrent Neural Network using Spatiotemporal LSTMs model.

2.1 Encoder

The encoder is used to encode the input sequence in time window T into the feature representation through RNN. Inspired by the DSTP [11] model which can select elementary stimulus features in the early stages of processing and input attention mechanism in DA-RNN [17], we use spatial attention to select the relevant driving series adaptively.

For time series prediction, given the input sequence $X = (\mathbf{x^1}, \mathbf{x^2} ..., \mathbf{x^n})^\top$ where n is the number of driving (exogenous) series, it can be divided into a series of time windows with T. Given the k-th input driving (exogenous) series $x^k = (x_1^k, x_2^k, ..., x_T^k)^\top$, we can construct an input attention mechanism by referring to the previous hidden state h_{t-1} and the cell state s_{t-1} in the encoder LSTM unit with:

$$e_t^k = \mathbf{v_e}^\top tanh(W_e[h_{t-1}; s_{t-1}] + U_e x^k) \tag{1}$$

and

$$\alpha_t^k = \frac{exp(e_t^k)}{\sum_{i=1}^{n} exp(e_t^i)} \tag{2}$$

where $v_e \in R^T$, $W_e \in R^{T\times 2m}$, $U_e \in R^{T\times T}$ are parameters to learn. After that, we employ a softmax function to ensure all the attention weights at per time step sum to one. With these attention weights, we can adaptively extract the driving time series with:

$$\tilde{x} = (\alpha_t^1 x_t^1, \alpha_t^2 x_t^2, ..., \alpha_t^n x_t^n) \tag{3}$$

Then the encoder is applied to learn a mapping from x_t to h_t (at time step t) with $h_t = f_e(h_{t-1}, x_t)$ can be updated as $h_t = f_e(h_{t-1}, \tilde{x}_t)$ where f_e is a spatiotemporal LSTM architecture based on LSTM units can be summarized as follows:

$$f_t = \sigma(W_f[h_{t-1}; x_t] + b_f) \tag{4}$$

$$i_t = \sigma(W_i[h_{t-1}; x_t] + b_i) \tag{5}$$

$$o_t = \sigma(W_o[h_{t-1}; x_t] + b_o) \tag{6}$$

$$s_t = f_t \odot s_{t-1} + i_t \odot tanh(W_s[h_{t-1}; x_t] + b_s) \tag{7}$$

$$h_t = o_t \odot tanh(s_t) \tag{8}$$

where $[h_{t-1}; x_t] \in R^{m+n}$ is a concatenation of the previous hidden state h_{t-1} and the current input x_t, W_f, W_i, W_o, $W_s \in R^{m\times(m+n)}$, and b_f, b_i, b_o, $b_s \in R^m$ are parameters to learn.

In order to enhance the ability of LSTM to capture long-term memory, we use two layers of stacked LSTM to transmit information in space and time. At every time step t, the first layer of LSTM is $h_t^l = f_e^l(h_{t-1}^l, \tilde{x}_t)$ where $l = 1$. Given the current level of LSTM layer l where $l \geqslant 2$, the output can be updated with:

$$h_t^l = f_e^l(h_{t-1}^l, h_t^{l-1}) \tag{9}$$

then the output is a concatenation of the previous T hidden state of the LSTM units as the encoded input driving series.

2.2 Decoder

In order to predict the output $\tilde{y}_t$, we use another LSTM to decode the input infomation. In the decoder, the attention weight of the decoder hidden state at time t is calculated based upon the previous decoder hidden state d_{t-1} and the cell state of the LSTM unit $s_{t-1}^{'}$ with:

$$l_i^t = \mathbf{v_d}^\top tanh(W_d[d_{t-1}; s_{t-1}^{'}] + U_d h_i), 1 \leqslant i \leqslant T \tag{10}$$

$$\beta_t^i = \frac{exp(l_t^i)}{\sum_{j=1}^{T} exp(l_t^j)} \tag{11}$$

where $[d_{t-1}; s'_{t-1}] \in R^{2p}$ is a concatenation of the previous hidden state and cell state of the LSTM unit in the decoder and h_i is concatenation of the hidden state in last time window T. $v_d \in R^m$, $W_d \in R^{m\times 2p}$, $U_d \in R^{m\times m}$ are parameters to learn. The weights of the i-th encoder hidden states β_t^i represent the importance it take at time t^i. Since each encoder hidden state h_i is mapped to a temporal component of the input, the context vector c_t can be computed as a weighted sum of all encoder hidden states $\{h_1, h_2, ..., h_T\}$,

$$c_t = \sum_{i=1}^{T} \beta_t^i h_i \tag{12}$$

Then the updated history target value can be combined with c_{t-1} and the given target series $y_{t-1} = \{y_{t-1}, y_{t-1}, ..., y_{t-1}\}$:

$$\tilde{y}_{t-1} = \mathbf{y_{t-1}}^\top \cdot c_{t-1} \tag{13}$$

where $\mathbf{y_{t-1}}^\top \cdot c_{t-1}$ is the point product of the decoder input y_{t-1} and the computed context vector c_{t-1}.

In order to enhance the influence of context vector on decoder, we combine the context vector with the hidden state of the decoder at every moment, the new hidden state can be updated after a linear layer as,

$$\tilde{d}_t = \mathbf{v_c}^\top tanh(W_c[c_t; d_{t-1}]) \tag{14}$$

where $[c_t; d_{t-1}] \in R^{m\times p}$ is a concatenation of the previous hidden state in LSTM unit of decoder and the current context vector c_t. We choose the nonlinear function f_d as a LSTM unit [10] to model long-term dependencies. Then the hidden state d_t can be updated as:

$$d_t = f_d(\tilde{d}_{t-1}, \tilde{y}_{t-1}) \tag{15}$$

and the final prediction can be computed as:

$$\tilde{y_T} = \mathbf{v_y}^\top (W_y[d_T; c_T] + b_w) + b_v \tag{16}$$

where $[d_T; c_T] \in R^{p+m}$ is a concatenation of the decoder hidden state and the context vector and the W_y, b_w, b_v are the parameters to learn.

2.3 Training Procedure

The model is based on encoder-decoder structure and parameters can be learned by standard back propagation with mean squared error as the objective function:

$$L(y_T, \tilde{y}_T) = \frac{1}{N}\sum_{i=1}^{N}(y_T^i - \tilde{y}_T^i)^2 \tag{17}$$

where N is the number of training samples. We choose Adam optimizer [12] to train the model and the size of the minibatch is 128. The learning rate is 0.001. Specifically, the proposed DPAST-RNN can make the loss function converge quickly.

3 Experiments

In this section, we first introduce the two datasets for this experiment. In addition, we introduce the collection process of VM-Power dataset. Then we discuss the parameter settings for DPAST-RNN and the evaluation metrics. Finally, we compare the DPAST-RNN with three different baseline methods.

3.1 Data Acquisition

In order to verify the performance of our DPAST-RNN model on more time series data, we configured the OpenStack environment to collect the indicators and real power of the virtual machine to make VM-Power dataset. There is an OpenStack controller node, an OpenStack compute node, and a monitor node for collecting the data from the compute node. These nodes are connected to the same (Local Area Network) LAN. The power of IT equipment can be measured by the Power Distribution Unit (PDU). The architecture is shown in Fig. 2.

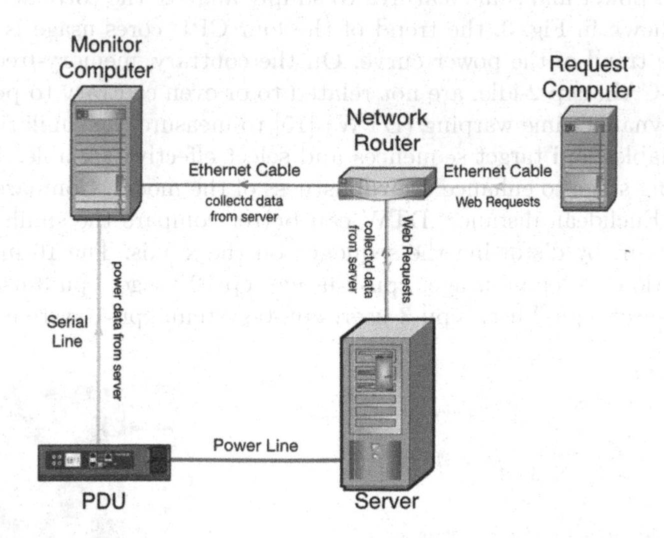

Fig. 2. The architecture of data collection procedure for VM-Power.

We deployed a collector called collectd [8] on the compute node to collect metrics of the compute node. The sampling frequency of collectd is set to 1 Hz, the same as the sampling frequency of PDU. Specifically, we use a client machine with a Quad-core CPU to request web resource and collect virtual machine metrics per seconds with real power in PDU.

3.2 Datasets and Setup

In this experiment, we used two datasets NASDAQ 100 Stock and VM-Power as shown in Table 1 where the size of encoder hidden states m and decoder hidden states p are set as $m = p = 64$ and 128 to test the performance of different methods for time series prediction.

Table 1. The statistics of two datasets.

Dataset	Driving series	Target series	Size		
			Train	Valid	Test
VM-Power	10	1	2636	263	528
NASDAQ 100 Stock	80	1	40551	4055	8111

The NASDAQ 100 Stock is a public dataset which contains the stock prices of 81 major corporations under NASDAQ 100. In this dataset, we use the share price of NDX as the target sequence and the share price of the remaining 80 companies as the driving time sequence.

From over 100 metrics we collected in origin VM-Power dataset, we draw a line chart of power and some features to simply analyze the correlation between them. As shown in Fig. 3, the trend of the four CPU cores usage is roughly as same as the trend of the power curve. On the contrary, memory-free, memory-cached, irq-CAL, cup-2-idle, are not related to or even contrary to power trend, so we use dynamic time warping (DTW) [15] to measure the similarity between feature variables and target sequences and select effective variables in the data preprocessing stage to enhance the robustness of the model. Compared with the traditional Euclidean distance, DTW can better compare the similarity of two time waveforms by distorting the sequence on the x-axis. The 10 metrics from DTW selection are cpu-0-usage, cpu-1-usage, cpu-2-usage, cpu-3-usage, cpu-0-user, cpu-1-user, cpu-2-user, cpu-3-user, cpu-0-system, cpu-1-system.

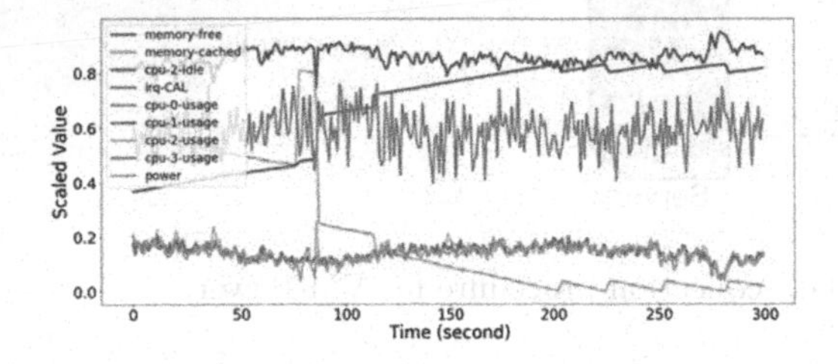

Fig. 3. The curves of power and the features selected of VM-Power dataset.

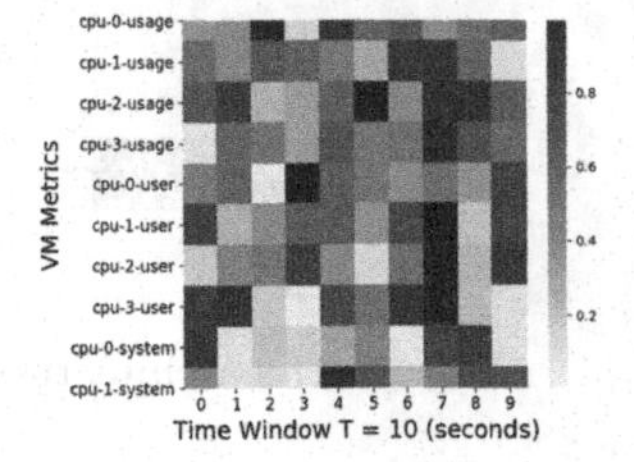

Fig. 4. Plot of input spatial attention weights in one time window $T = 10$ for 10 virtual machine energy consumption index variables in VM-Power dataset.

3.3 Parameter Settings

We initialized the size of hidden states 128 both in encoder and decoder and choose the window size T=10 where T $\in$ {5, 10, 15, 20, 25} that achieve the best performance over the validation set are used for evaluation. To measure the effectiveness of various methods for time series prediction, we consider two different evaluation metrics, root mean squared error (RMSE) [16] and mean absolute error (MAE). Given y_t is the target at time t and $\hat{y_t}$ is the predicted value at time t, RMSE is defined as $RMSE = \sqrt{\frac{1}{N}\sum_{i=1}^{N}(y_i^t - \hat{y_i^t})^2}$ and MAE is defined as $MAE = \frac{1}{N}\sum_{i=1}^{N}|y_i^t - \hat{y_i^t}|$.

3.4 Results: Time Series Prediction

We compared our DPAST-RNN with three baseline methods in two datasets and proved its effectiveness. The results of prediction in two datasets are shown in Fig. 5 and 6. Among these baselines, LSTM [10] is a basic method to address time series prediction in RNN. From the prediction results in Fig. 5, the model based on RNN can better predict the time series data with more severe fluctuations. For the rising part of continuous oscillation, our model can better reduce the time delay. We also show the visual attention distribution in Fig. 4. We observe that the different characteristic variables get different weights in time window T which indicates that input attention mechanism can effectively extract the relevant driving sequence.

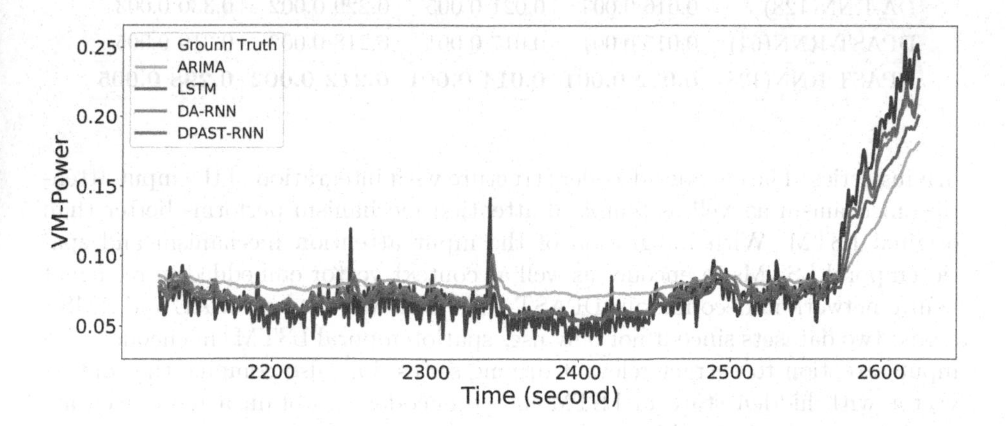

Fig. 5. VM-Power prediction result.

The time series prediction results of DPAST-RNN and baseline methods over the two datasets are shown in Table 2. In Table 2, the results of the *RMSE* of ARIMA is generally worse than the RNN based methods. This is because ARIMA only consider the target series rather than the relationship between

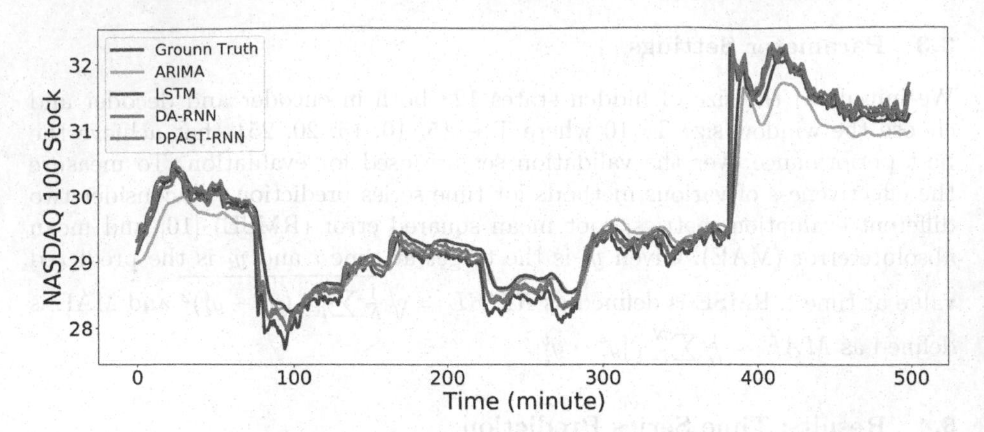

Fig. 6. NASDAQ 100 Index prediction result.

Table 2. Time series prediction results over the Vm-Power dataset and NASDAQ 100 Stock dataset (best performance displayed in boldface).

Models	VM-Power dataset		NASDAQ 100 Stock dataset	
	MAE	RMSE	MAE	RMSE
ARIMA	1.97	2.66	0.92	1.47
LSTM(64)	0.282 0.003	0.362 0.003	0.262 0.005	0.390 0.003
LSTM(128)	0.270 0.003	0.347 0.003	0.251 0.005	0.380 0.003
DA-RNN(64)	0.014 0.003	0.019 0.001	0.216 0.002	0.310 0.003
DA-RNN(128)	0.016 0.004	0.021 0.005	0.229 0.002	0.330 0.003
DPAST-RNN(64)	0.015 0.001	0.017 0.001	0.218 0.002	0.319 0.005
DPAST-RNN(128)	**0.012 0.001**	**0.014 0.001**	**0.212 0.002**	**0.298 0.005**

driving series. The encoder-decoder structure with integration of the input attention mechanism as well as temporal attention mechanism performs better than original LSTM. With integration of the input attention mechanism and spatiotemporal LSTMs in encoder as well as context vector embedded in recurrent neural network in decoder, our DPAST-RNN achieves the best MAE and RMSE across two datasets since it not only uses spatiotemporal LSTMs in encoder with input attention to extract relevant driving series, but also combine the context vector with hidden state in LSTM in the encoder to obtain a more accurate spatiotemporal relationship across all time steps.

4 Conclusion and Future Work

In this paper, we propose a DPAST-RNN model based on spatiotemporal LSTM network for time series prediction, which consists of two phases attention mechanism. In the proposed model, we use DTW to remove the noise of multivariate input time series. In the encoder part of DPAST-RNN, the spatiotemporal

LSTMs can accurately encode the driving series after input attention mechanism. In the decoder part of DPAST-RNN, the updated hidden state in LSTM with context vector can naturally capture the long-range temporal information of the encoded inputs. The experimental results on two datasets demonstrate a higher performance than other baseline methods.

In the future, we will explore time series prediction based on attention mechanism without RNN structure. Moreover, we will extend our method to solve the problem of long-term prediction.

Acknowledgments. This work was supported by National Key Research and Development Program of China (2018YFB1003702) and Jiangsu Scientific Research Innovation Practice Project (KYCX20_0760).

References

1. Amini, M.H., Kargarian, A., Karabasoglu, O.: ARIMA-based decoupled time series forecasting of electric vehicle charging demand for stochastic power system operation. Electr. Power Syst. Res. **140**, 378–390 (2016)
2. Bahdanau, D., Cho, K., Bengio, Y.: Neural machine translation by jointly learning to align and translate. arXiv preprint arXiv:1409.0473 (2014)
3. Chakraborty, P., Marwah, M., Arlitt, M., Ramakrishnan, N.: Fine-grained photovoltaic output prediction using a Bayesian ensemble. In: Twenty-Sixth AAAI Conference on Artificial Intelligence (2012)
4. Cheng, Y., Huang, F., Zhou, L., Jin, C., Zhang, Y., Zhang, T.: A hierarchical multimodal attention-based neural network for image captioning. In: Proceedings of the 40th International ACM SIGIR Conference on Research and Development in Information Retrieval, pp. 889–892 (2017)
5. Cho, K., et al.: Learning phrase representations using RNN encoder-decoder for statistical machine translation. arXiv preprint arXiv:1406.1078 (2014)
6. Di Gangi, M.A., Federico, M.: Deep neural machine translation with weakly-recurrent units. arXiv preprint arXiv:1805.04185 (2018)
7. Fernandez-Fraga, S., Aceves-Fernandez, M., Pedraza-Ortega, J., Ramos-Arreguin, J.: Screen task experiments for EEG signals based on SSVEP brain computer interface. Int. J. Adv. Res. **6**(2), 1718–1732 (2018)
8. Forster, F., Harl, S.: Collectd - the system statistics collection daemon (2012). https://collectd.org
9. Guo, T., Lin, T.: Multi-variable LSTM neural network for autoregressive exogenous model. arXiv preprint arXiv:1806.06384 (2018)
10. Hochreiter, S., Schmidhuber, J.: Long short-term memory. Neural Comput. **9**(8), 1735–1780 (1997)
11. Hübner, R., Steinhauser, M., Lehle, C.: A dual-stage two-phase model of selective attention. Psychol. Rev. **117**(3), 759 (2010)
12. Kingma, D.P., Ba, J.: Adam: a method for stochastic optimization. arXiv preprint arXiv:1412.6980 (2014)
13. Liang, Y., Ke, S., Zhang, J., Yi, X., Zheng, Y.: GeoMAN: multi-level attention networks for geo-sensory time series prediction. In: IJCAI, pp. 3428–3434 (2018)
14. Liu, J., Zio, E.: SVM hyperparameters tuning for recursive multi-step-ahead prediction. Neural Comput. Appl. **28**(12), 3749–3763 (2017). https://doi.org/10.1007/s00521-016-2272-1

15. Müller, M.: Dynamic time warping. In: Müller, M. (ed.) Information Retrieval for Music and Motion, pp. 69–84. Springer, Heidelberg (2007). https://doi.org/10.1007/978-3-540-74048-3_4
16. Plutowski, M., Cottrell, G., White, H.: Experience with selecting exemplars from clean data. Neural Netw. **9**(2), 273–294 (1996)
17. Qin, Y., Song, D., Chen, H., Cheng, W., Jiang, G., Cottrell, G.: A dual-stage attention-based recurrent neural network for time series prediction. arXiv preprint arXiv:1704.02971 (2017)
18. Rumelhart, D.E., Hinton, G.E., Williams, R.J.: Learning representations by back-propagating errors. Nature **323**(6088), 533–536 (1986)
19. Sutskever, I., Vinyals, O., Le, Q.V.: Sequence to sequence learning with neural networks. In: Advances in Neural Information Processing Systems, pp. 3104–3112 (2014)
20. Wu, Y., Hernández-Lobato, J.M., Ghahramani, Z.: Dynamic covariance models for multivariate financial time series. arXiv preprint arXiv:1305.4268 (2013)
21. Zamora-Martinez, F., Romeu, P., Botella-Rocamora, P., Pardo, J.: On-line learning of indoor temperature forecasting models towards energy efficiency. Energy Build. **83**, 162–172 (2014)

ForecastNet: A Time-Variant Deep Feed-Forward Neural Network Architecture for Multi-step-Ahead Time-Series Forecasting

Joel Janek Dabrowski[1(✉)], YiFan Zhang[2], and Ashfaqur Rahman[3]

[1] Data61, CSIRO, Brisbane, QLD 4067, Australia
joel.dabrowski@data61.csiro.au
[2] Agriculture and Food, CSIRO, Brisbane, QLD 4067, Australia
Yi-Fan.Zhang@csiro.au
[3] Data61, CSIRO, Sandy Bay, TAS 7005, Australia
Ashfaqur.Rahman@data61.csiro.au

Abstract. Recurrent and convolutional neural networks are the most common architectures used for time-series forecasting in deep learning literature. Owing to parameter sharing and repeating architecture, these models are *time-invariant* (shift-invariant in the spatial domain). We demonstrate how time-invariance in such models can reduce the capacity to model time-varying dynamics in the data. We propose ForecastNet which uses a deep feed-forward architecture and interleaved outputs to provide a *time-variant* model. ForecastNet is demonstrated to model time varying dynamics in data and outperform statistical and deep learning benchmark models on several seasonal time-series datasets.

Keywords: Time-invariance · ForecastNet · Forecasting · Deep learning

1 Introduction

Multi-step-ahead forecasting involves the prediction of some variable(s), several time-steps into the future, given past and present data. Over the set of time-steps, various time-series components such as complex trends, seasonality, and noise may be observed at a range of scales or resolutions. Increasing the number of steps ahead that are forecast increases the range of scales that need to be modelled. An accurate forecasting method is required to model all these components over the complete range of scales.

Deep sequence models based on the recurrent neural network are popular for time-series forecasting [6,9,10]. The recurrence in the RNN produces a replicated set of cells over time. This replication creates a time-invariant model. The same is true for the convolutional neural network (CNN), though the property

H. Yang et al. (Eds.): ICONIP 2020, LNCS 12534, pp. 579–591, 2020.
https://doi.org/10.1007/978-3-030-63836-8_48

is more commonly known as shift-invariance in the spatial domain. Time invariance means that the model is not able to vary in time. Intuitively, this restricts the model's ability to adapt to changes. Such changes may be in the form of variations in dynamics and long-range dependencies.

To address this, we propose ForecastNet; a time-variant deep feed-forward architecture with interleaved outputs. Between the interleaved outputs, are flexible network structures (hidden blocks) whose parameters and (optionally) architecture vary over time. We demonstrate four variations of ForecastNet to highlight its flexibility and compare these with state-of-the-art benchmark models. ForecastNet is shown to be accurate and robust in terms of performance variation.

The contributions of this study are: (1) our novel ForecastNet model[1]; (2) we provide the first study (to our knowledge) which specifically targets time-invariance in deep learning models, and show how time-invariance can cause popular deep learning models to fail – which we believe has significant implications to the field as a whole; and (3) we provide a comparison with several benchmark models on seasonal time-series datasets.

2 Motivations and Related Work

The RNN comprises a set of cell structures with parameters that are replicated over time. The replication has key benefits such as parameter sharing and an ability to handle varying sequence lengths. However, the time invariant properties can reduce its capacity to model complex dependencies over time. In comparison, ForecastNet is not time-invariant. Another challenge with RNNs is that learning long sequences can be difficult due to complex dependencies over time and vanishing gradients [4]. ForecastNet however, mitigates vanishing gradient problems and advocates a deep architecture by using shortcut-connections [8,14] and by interleaving outputs between hidden blocks.

State-of-the-art deep sequence models include multiple RNNs linked in various configurations. The sequence-to-sequence (encoder-decoder) model [15] is a prominent configuration. This model sequentially links two RNNs (an encoder and a decoder) through a fixed size vector, such as the last encoder cell state. This vector can form a bottleneck between the encoder and decoder and earlier inputs have to pass through several layers to reach the decoder. The attention model [1] addresses these problems by adding an *attention mechanism* between the encoder and decoder to ascribe relevance to each encoder cell. Unlike the sequence-to-sequence and attention models, ForecastNet does not have a separate encoder and decoder. Challenges in linking these entities are thus removed.

The convolutional neural network (CNN) has also been applied to modelling sequential data. WaveNet [11] is a seminal CNN model which uses multiple layers of dilated causal convolutions for raw audio synthesis. The Temporal Convolutional Network (TCN) [2] generalises WaveNet for broader sequence modelling

[1] Code is available at https://github.com/jjdabr/forecastNet.

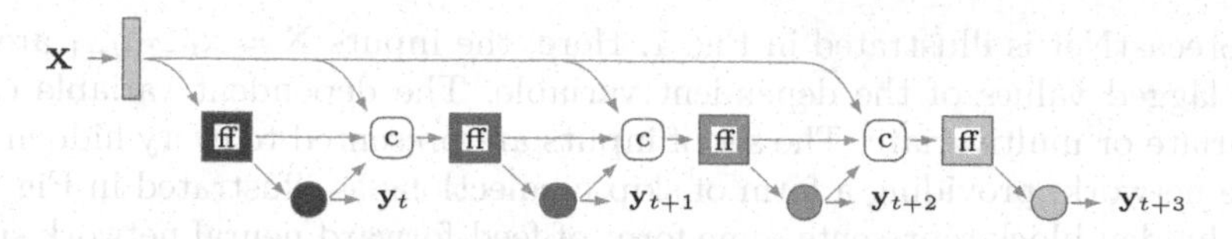

Fig. 1. ForecastNet structure to provide a forecast $\mathbf{y}_{t:t+3}$ given $\mathbf{X} = \mathbf{x}_{t-T_i:t-1}$ as inputs (rectangle). A hidden block (squares labelled with ff) comprises some form of feed-forward neural network structure. The first hidden block takes $\mathbf{X}$ as an input. The remaining blocks take a concatenation (rounded square blocks labelled with c) of the input $\mathbf{X}$, the previous hidden block, and the previous output. Outputs (circles) can take various forms such as mixture densities or linear layers. Each hidden block and output are illustrated with a different shade to indicate that they are unique in terms of parameters (they are not shared) and (optionally) architecture.

problems. The CNN is shift-invariant (or translation-invariant). This is beneficial in image processing, however, it directly translates to time-invariance in time-series applications. We however demonstrate how ForecastNet is able to accommodate convolutional layers but still provide a time-variant model.

A model that has successfully departed from the RNN and CNN architectures is the transformer model [16]. Though the transformer has been highly successful in natural language processing, in its canonical form, it has limitations in time-series analysis. The first limitation is that it does not assume any sequential structure of the inputs [16]. Positional encoding in the form of a sinusoid is injected into the inputs to provide information on the sequence order. Temporal structure is key in time-series modelling and is what time-series models are usually designed to model. The second limitation is that the majority of the processing operates over the dimension of the input embedding. The Transformer is thus *not* designed to operate on low dimensional time-series signals such as univariate time-series. ForecastNet is specifically designed to model the temporal structure of the data and it is not limited to high-dimensional inputs.

3 ForecastNet Architecture

Definition 1. *ForecastNet is a feed-forward neural network comprising a set of T_i inputs, T_o outputs (forecasts), and a set of sequentially connected hidden blocks. Each hidden blocks comprises some form of feed-forward neural network. ForecastNet's output at time t is given by*

$$\mathbf{y}_t = f_t(\mathbf{h}_t) \quad and \quad \mathbf{h}_t = g_t(\mathbf{X}, \mathbf{h}_{t-1}, \mathbf{y}_{t-1}) \tag{1}$$

where $\mathbf{X} = \mathbf{x}_{\tau-T_i:\tau-1}$ are the inputs comprising the T_i observations up to the first forecast time τ, $\mathbf{y}_t$ is the forecast at time $t = [\tau, \ldots, \tau + T_o - 1]$, $\mathbf{h}_t$ is the hidden block output, f_t is the output layer, and g_t is a function describing the hidden block. The functions f_t and g_t are indexed by t as their parameters and (optionally) architecture vary over time. Initially $\mathbf{y}_{\tau-1} = \mathbf{h}_{\tau-1} = \emptyset$ such that $\mathbf{h}_\tau = g_\tau(\mathbf{X})$.

ForecastNet is illustrated in Fig. 1. Here, the inputs $\mathbf{X} = \mathbf{x}_{t-T_i:t-1}$ are a set of T_i lagged values of the dependent variable. The dependent variable can be univariate or multivariate. The set of inputs are presented to every hidden block in the network, providing a form of skip connections as illustrated in Fig. 1.

A hidden block represents some form of feed-forward neural network such as a multi-layered perceptron (MLP), a CNN, or self-attention. Each hidden block can be heterogeneous in terms of architecture. However, even if the architecture of each hidden block is identical (as used in this study), each block is provided with its own *unique* set of parameters (weights). The hidden blocks are intended to model the time-series dynamics. Links between hidden blocks model local dynamics and sets of hidden blocks model longer-term dynamics.

Outputs follow hidden blocks. Each output in ForecastNet provides a forecast one-step into the future. The deeper the network, the more outputs there are. ForecastNet thus naturally scales in complexity with increased forecast reach. To provide forecasts with uncertainty, a mixture density network [3,13] output with a single Gaussian is used. A linear and softplus layer predict the mean and the variance respectively. The model is trained with gradient descent to optimise the Gaussian log-likelihood function. We also demonstrate ForecastNet with a linear output layer that is optimised using the mean squared error.

4 ForecastNet Properties

4.1 Time-Variance

A time-invariant system is defined as *a system for which a time shift of the input sequence causes a corresponding shift in the output sequence* [12]. Proving a system is time-invariant involves showing that, passing a delayed input through a system is the same as delaying the output of the system by the same amount.

Theorem 1. *If the hidden blocks and output layers of ForecastNet are time-varying, ForecastNet is not time-invariant.*

Proof. Let $\mathbf{h'}_t = \mathbf{h}_{t-m}$ and $\mathbf{X}$ the T_i inputs up to the first forecast time τ. Time-invariance requires that $\mathbf{y}_{t-m} = \mathbf{y'}_t$. From (1), $\mathbf{y}_{t-m}$ is given by

$$\mathbf{y}_{t-m} = f_{t-m}(g_{t-m}(\mathbf{X}, \mathbf{h}_{t-m-1}, \mathbf{y}_{t-m-1}))$$

and $\mathbf{y'}_t$ is given by

$$\mathbf{y'}_t = f_t(g_t(\mathbf{X}, \mathbf{h'}_{t-1}, \mathbf{y'}_{t-1})) = f_t(g_t(\mathbf{X}, \mathbf{h}_{t-m-1}, \mathbf{y'}_{t-1})).$$

As $f_t \neq f_{t-m}$ & $g_t \neq g_{t-m} \Rightarrow \mathbf{y}_{t-m} \neq \mathbf{y'}_t$, ForecastNet is *not* time-invariant. □

Owing to no parameter sharing in ForecastNet, its parameters (and optionally architecture) vary over time (over the sequence of inputs and forecast horizon) which cause $f_t \neq f_{t-m}$ and $g_t \neq g_{t-m}$. This results in a time-variant model. In comparison, the RNN, is time-invariant according to the following:

Theorem 2. *A RNN with inputs* $\mathbf{x}_t$ *and hidden states* $\mathbf{h}_t$*, given by*

$$\mathbf{h}_t = f(W\mathbf{x}_t, V\mathbf{h}_{t-1}) \tag{2}$$

is time invariant if f is a function that is constant in time; W and V are parameters and weights that are constant in time; and when $\mathbf{h}_t$ *is initialised to some initial value* $\mathbf{h}_0$ *immediately before the first input is received.*

Proof. The RNN in (2) can be represented in a telescopic form given by

$$\mathbf{h}_t = f(W\mathbf{x}_t, [Vf(W\mathbf{x}_{t-1}, [\dots[Vf(W\mathbf{x}_1, V\mathbf{h}_0)]])]) \tag{3}$$

where $\mathbf{h}_0$ is a constant. Given $\mathbf{x}'_t = \mathbf{x}_{t-m}$, show that $\mathbf{h}_{t-m} = \mathbf{h}'_t$. For $\mathbf{h}_{t-m}$:

$$\mathbf{h}_{t-m} = f(W\mathbf{x}_{t-m}, [Vf(W\mathbf{x}_{t-m-1}, [\dots[Vf(Wx_1, V\mathbf{h}_0)]])])$$

For $\mathbf{h}'_t$

$$\begin{aligned}\mathbf{h}'_t &= f(W\mathbf{x}'_t, [Vf(W\mathbf{x}'_{t-1}, [\dots[Vf(W\mathbf{x}'_1, V\mathbf{h}'_0)]])]) \\ &= f(W\mathbf{x}_{t-m}, [Vf(W\mathbf{x}_{t-m-1}, [\dots[Vf(W\mathbf{x}_{1-m}, V\mathbf{h}'_0)]])])\end{aligned}$$

However, $\mathbf{x}_{t-m} = 0 \; \forall \; t \le m$ and $\mathbf{h}_{t-m} = 0 \; \forall \; t \le m$. Furthermore, assume that the system is always initialised with the constant $\mathbf{h}_0$. Thus

$$\mathbf{h}_t = f(W\mathbf{x}_t, [Vf(W\mathbf{x}_{t-1}, [\dots[Vf(W\mathbf{x}_1, V\mathbf{h}_0)]])])$$

and $\mathbf{h}_{t-m} = \mathbf{h}'_t$ proving the RNN is time-variant. □

4.2 Interleaved Outputs

The vanishing/exploding gradient problem stems from the repeated application of the chain rule of calculus producing a long chain of *factors*. By *interleaving outputs* between hidden blocks in ForecastNet, the chain is broken into a *sum of terms* which is more stable than a product of factors.

To show this, consider ForecastNet with L hidden blocks each containing a single fully connected layer, with a neuron's linear combination $z^{[l]}$ (before the activation), an output $a^{[l]}$ (after the activation), layer weights $W^{[l]}$, and cost function $\mathcal{L}$. Application of the chain rule produces

$$\frac{\partial \mathcal{L}}{\partial W^{[l]}} = \sum_{k=0}^{\frac{L-1-l}{2}} \frac{\partial \mathcal{L}}{\partial z^{[l+2k+1]}} \frac{\partial z^{[l+2k+1]}}{\partial a^{[l+2k]}} \Psi_k \frac{\partial a^{[l]}}{\partial W^{[l]}} \tag{4}$$

where $\Psi_k = 1$ if $k = 0$ and $\Psi_k = \prod_{j=1}^{k} \frac{\partial z^{[l+2j]}}{\partial a^{[l+2(j-1)]}}$ for $k > 0$.

Similarly, consider a multi-layered perceptron (MLP) with L-layers. Repeated application of the chain rule of calculus results in the following product

$$\frac{\partial \mathcal{L}}{\partial W^{[l]}} = \frac{\partial \mathcal{L}}{\partial a^{[L]}} \frac{\partial a^{[L]}}{\partial z^{[L]}} \frac{\partial z^{[L]}}{\partial a^{[L-1]}} \frac{\partial a^{[L-1]}}{\partial z^{[L-1]}} \cdots \frac{\partial z^{[l+2]}}{\partial a^{[l+1]}} \frac{\partial a^{[l+1]}}{\partial z^{[l+1]}} \frac{\partial z^{[l+1]}}{\partial a^{[l]}} \frac{\partial a^{[l]}}{\partial z^{[l]}} \frac{\partial z^{[l]}}{\partial W^{[l]}} \tag{5}$$

This product of factors can be expressed in the form λ^L. If $|\lambda| < 1$, the product vanishes towards 0 as L grows. In comparison, (4) contains a sum of terms, which does not tend to zero (vanish) as the product of factors does[2].

The interleaved outputs mitigate, but do not eliminate the vanishing gradient problem. The term Ψ_k contains a product of derivatives. The number of factors in this product grows proportional to k. For a distant layer where k is large, Ψ_k will have many factors. Gradients back-propagated from distant layers are thus still susceptible to vanishing gradient problems. However, for nearby outputs where k is small, Ψ_k will have few factors and so gradients from *nearby* outputs are less likely to experience vanishing gradient problems. Thus, nearby outputs can provide guidance to local parameters during training, resulting in improved convergence as the effective depth of the network is reduced.

4.3 Memory Requirements

Parameter sharing is an effective way of reducing memory requirements in a neural network. With the repeating architecture in the RNN and CNN, a network size remains constant irrespective of the sequence length. As ForecastNet has different parameters associated with each sequence step, its memory requirements increase with increased forecast reach and input sequence length. For problems with extremely large forecast horizons, or where computing resources have minimal memory capacity, the hidden block architecture must be constrained.

Parameter sharing additionally provides a means to handle varying sequence lengths. ForecastNet is designed for fixed length sequences. In seasonal time-series forecasting, ForecastNet is sized according to the seasonal period of the data. This has the benefit of incorporating prior knowledge of the data's temporal structure into the network.

5 Methods and Datasets

5.1 Datasets

A set of models are compared on a synthetic dataset and nine real-world datasets sourced from various domains. These include weather, environmental, energy, aquaculture, and meteorological domains. Datasets are hand-picked to ensure that they provide a sufficiently challenging problem and have a seasonal component. Properties such as varying seasonal amplitude, seasonal shape, trend, and noise were sought. For example, the shape of the seasonal cycle changes over time in most datasets. The dataset's properties are summarised in Table 1.

The synthetic dataset is used to provide a baseline. The data is generated according to $x_t = 2\sin(5\omega t) + 1/3\sin(\omega t)$, where ω is the angular frequency and t denotes time. The low frequency sinusoid emulates a long-term time-varying trend, whereas the high frequency sinusoid emulates seasonality.

[2] Note that RNNs can have a derivative with a form similar to (4) if each cell has a target. However, if targets are not provided, such as in the encoder of the sequence-to-sequence model, the derivative reduces to a form similar to (5).

Table 1. Dataset properties.

Dataset	Abbrev	Resolution	Period ($\mathcal{T}$)	Length	Minimum	Maximum	Mean	Std. Dev
Synthetic	Synth	–	20	4320	−2.33	2.33	−0.00	1.43
England temperature[1]	Weath	Monthly	12	3261	0.10	18.80	9.27	4.75
River flow[1]	River	Monthly	12	1492	3290	66500	23157.60	13087.40
Electricity[2]	Elect	Hourly	24	19224	5514	14580	8709.79	1360.29
Traffic Volume[3]	Traff	Hourly	24	8776	125	7217	3269.26	2021.57
Lake levels[1]	Lake	Monthly	12	648	10	20	15.08	2.00
Dissolved Oxygen [5]	DO	Hourly	24	2422	5.66	7.94	6.50	0.53
pH [5]	pH	Hourly	24	2422	8.07	11.15	8.56	0.21
Pond temperature [5]	Temp	Hourly	24	2422	24.38	31.97	27.74	1.85
Ozone[1]	Ozone	Monthly	12	516	266	430	338.00	38.30

[1] https://robjhyndman.com/tsdl/
[2] https://www.aemo.com.au/Electricity/National-Electricity-Market-NEM/Data-dashboard
[3] https://archive.ics.uci.edu/ml/index.php

Table 2. Model configuration. $\mathcal{T}$ is the seasonal period of the dataset.

Model	Configuration
FN	A hidden block comprises two He initialised [7] fully connected hidden layers, each with 24 ReLU neurons
FN2	Identical to FN2 but with a linear output layer instead of a mixture density layer
cFN	A hidden block comprises a convolutional layer with 24 filters, each with a kernel size of 2, followed by a average pooling layer with a pool size of 2. The convolutional and pooling layers are duplicated and followed by a fully connected layer with 24 ReLU neurons
cFN2	Identical to cFN2 but with a linear output layer instead of a mixture density layer
DeepAR	The sequence-to-sequence architecture with single layered LSTMs are used in the encoder and decoder. The mixture density output of the network is a Gaussian distribution
TCN	The TCN contains a convolutional layer with 32 filters, each with a kernel size of 2 for the Synthetic, Weather, Elect., and River datasets. For the remaining datasets, the TCN contains a convolutional layer with 64 filters, each with a kernel size of 3. The output contains a dense layer with $\mathcal{T}$ linear units
Attention	Encoder has a bidirectional LSTM and the decoder has a single layered LSTM. The LSTM cells are configured with 24 ReLU units
Seq2Seq	Encoder and decoder use a single layered LSTM with 24 ReLU units
MLP	Feed forward MLP with a single hidden layer comprising $4\mathcal{T}$ ReLU hidden neurons. A set of $2\mathcal{T}$ inputs are provided and a set of $\mathcal{T}$ linear outputs are used
DLM	The DLM used is the free-form seasonal model with a zero order trend component [17]. Model fitting is performed using a modified Kalman filter[1]
SARIMA	Standard form (p,d,q)(P,D,Q)s with: **Synthetic**: (1,1,1)(1,1,0)20, **Weather**: (2,0,3)(0,1,0)12, **Elect.**: (3,1,4)(0,1,0)24, **River**: (2,0,4)(0,1,0)12, **Traff.**: (3,1,1)(0,1,0)24, **Lake**: (2,0,8)(0,1,0)12, **DO**: (2,0,6)(0,1,0)24, **pH**: (4,1,3)(0,1,0)24, **Temp.**: (2,1,4)(0,1,0)24, and **Ozone**: (3,0,4)(0,1,0)12,

[1] See https://pydlm.github.io/index.html

5.2 Models

Four variations of ForecastNet are tested, and are denoted by FN, FN2, cFN, and cFN2. Two have fully connected blocks and two have convolutional blocks. These are compared with four deep-learning based benchmark models: deepAR [13], the TCN [2], the sequence-to-sequence (Seq2Seq) model [15], and the attention model [1] (the Transformer is not considered for reasons discussed in Sect. 2). For completeness, a (traditional) MLP, a free-form seasonal Dynamic Linear Model (DLM) [17], and a seasonal autoregressive moving average (SARIMA) model are additionally included. The DLM (a state space model) and the SARIMA model are well-known statistical models used for time-series forecasting. All models except the ForecastNet and the MLP models are time-invariant. The benchmark models are configured in their default form (as close as possible to how their authors present them). Configuration details of the models are provided in Table 2.

The models are tested on datasets that have a seasonal component with period denoted by $\mathcal{T}$. The number of inputs in all models is set to $2\mathcal{T}$ and the number of outputs (forecast-steps) is set to $\mathcal{T}$. Thus, the models are trained to forecast one seasonal cycle ahead in time, given the two previous cycles of data.

5.3 Training and Testing

The datasets are scaled to the range of $[0, 1]$. For each dataset, the first 90% of the dataset sequence is used as the training set and the last 10% is used as the test set. The training and test sets each comprise a long sequence of values. These sequences are converted into a set of samples that the models can process. A sample is extracted using a sliding window of length $3\mathcal{T}$. The first $2\mathcal{T}$ elements in this window form the input sequence to the model, and the last $\mathcal{T}$ elements form the forecast target values. The sliding window is slid element-by-element across the dataset sequence to produce a set of samples. The set of training samples are shuffled prior to training.

The ADAM algorithm is used in all machine learning models. The learning rate is selected using a grid-search over the range $10^{-i}, i = [2, \ldots, 6]$. Early stopping is used to address overfitting and defines the number of epochs. The Mean Absolute Scaled Error (MASE) is used to evaluate forecasting performance.

6 Results and Discussion

6.1 Time-Invariance Demonstration

To demonstrate the effect of time-invariance in a model, we test the models on a time-varying dataset. The synthetic dataset is adapted to include amplitude modulation, which is a time-varying operation. The modified synthetic dataset is given by $x_t = \frac{1}{2}\sin\left(\frac{1}{6}\omega t\right)\left(\frac{3}{5}\sin\left(\omega t\right) + \frac{1}{5}\sin\left(\frac{1}{5}\omega t\right)\right)$. The first sinusoid performs amplitude modulation, the second sinusoid emulates a seasonal cycle, and the third sinusoid emulates a time-varying trend.

The amplitude modulation repeats every 6 cycles of the seasonal period. A model is only presented with two seasonal cycles as inputs and thus does not observe the complete amplitude modulation pattern. A time-variant model's parameters are able to change over time, which enables the model to adapt to the

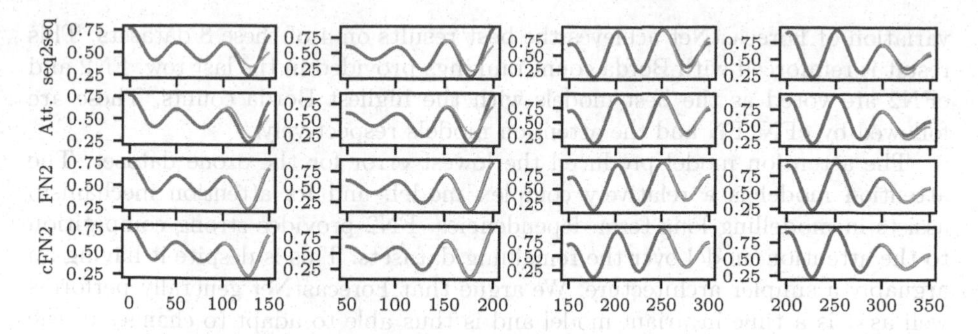

Fig. 2. Time-variant synthetic dataset forecasts for the seq2seq, Att., FN2, and cFN2 models at starting time indices 0, 50, 150 and 200.

variations in amplitude. As amplitude modulation is a time-variant operation, a time-invariant model is expected to perform poorly in forecasting.

The seq2seq, attention, FN2, and cFN2 models are trained on the dataset. Results of forecasts from inputs starting at time indices 0, 50, 150 and 200 are presented in Fig. 2. As expected, the time-invariant seq2seq model is not able to provide accurate forecasts when there are large variations in the amplitude. The attention model performs slightly better given its attention mechanism. However, the ForecastNet models are able to adapt to the large variations in the amplitude demonstrating the effectiveness of a time-variant model.

Table 3. Average MASE of the models results over the test datasets. The last row indicates the sum of Borda counts of the models over the datasets (a higher value indicates more points in the voting score). Boldface numbers highlight top results.

	FN	cFN	FN2	cFN2	deepAR	Seq2Seq	Attention	TCN	MLP	DLM	SARIMA
Synth	0.0039	0.0089	**0.0004**	0.0032	0.0284	0.0106	0.0359	0.0476	0.0082	0.6400	0.2853
Weath	0.4630	0.4030	0.4556	**0.3142**	0.4615	0.4258	0.3703	0.4729	0.4655	0.5221	0.6087
Elect	1.1173	1.0430	0.8946	**0.5410**	1.7747	1.0045	1.3858	1.0924	1.3389	1.7267	1.2580
River	0.7137	0.6630	0.6644	**0.3936**	0.8602	0.5735	0.5282	0.8510	0.8514	0.7655	0.8655
Traff	2.2343	1.9470	1.4356	**0.8208**	2.0066	1.7806	1.9400	2.1968	2.3575	2.3997	2.3163
Lake	**1.4239**	1.6127	1.5815	1.6858	1.6111	1.7250	1.5575	2.0268	1.5659	1.9521	1.6929
DO	**0.5354**	0.6197	0.6162	0.5380	0.7128	0.7335	2.1052	0.7776	0.7654	0.7746	0.6449
pH	1.4099	1.2276	1.2623	**1.0101**	2.6355	1.7021	1.4157	1.3456	1.9020	1.2930	1.6850
Temp	1.9008	1.9020	1.6624	1.9951	1.9503	2.1274	2.2314	3.1774	2.1010	3.6660	**1.6449**
Ozone	0.7235	0.7849	0.6903	0.7933	1.0266	1.4955	**0.5816**	0.6921	0.6603	0.7552	0.8871
Borda	74	76	**90**	**90**	43	57	65	42	51	31	41

6.2 Model Comparison Error Results

The average MASE over all forecasts on each dataset's test set is provided in Table 3. ForecastNet produces the best results on 8 of the 10 datasets. The cFN2 variation of ForecastNet achieves the best results on 4 of these 8 datasets. This result is reinforced with Borda count rankings provided in the last row. FN2 and cFN2 are voted as the best models with the highest Borda counts. These are followed by cFN, FN and the attention models respectively.

The attention model produced the lowest error for the ozone dataset. The attention model is a relatively complex model, and its attention mechanism assists in modelling long-term dependencies. FN2 provides strong competition to the attention model over the remaining datasets. This is despite it having an arguably a simpler architecture. We argue that ForecastNet generally performs well as it is a time-invariant model and is thus able to adapt to changes in the dynamics such variations in amplitude, trend, and seasonal cycle shape.

Increasing the model complexity generally resulted in improved forecast performance. For example, cFN generally outperforms FN, and attention model outperforms seq2seq. However, simpler models do not fail on the datasets. For example, the MLP provided comparably accurate forecasts despite its simplicity. Note however, that the MLP is a time-variant model.

The DLM and SARIMA statistical models performed well despite being linear models. For example, the SARIMA model achieved the lowest error on the pond temperature dataset. This suggests that the dynamics of this dataset are more linear. However, with the non-linear trends, amplitudes, and cyclic shapes in the other datasets, the DLM and SARIMA models did not perform as well the non-linear neural network-based models.

Of the deep neural network-based models, the TCN performed the worst on several datasets. However, the authors suggest that the model is in a simplified form and improved results may be possible by using a more advanced architecture [2]. Furthermore, the TCN is designed to perform dilated convolutions over many samples. Of the datasets used in this study, the maximum number of input samples was 48 for datasets with a period of 24 h. This may be too few to demonstrate the effectiveness of the TCN.

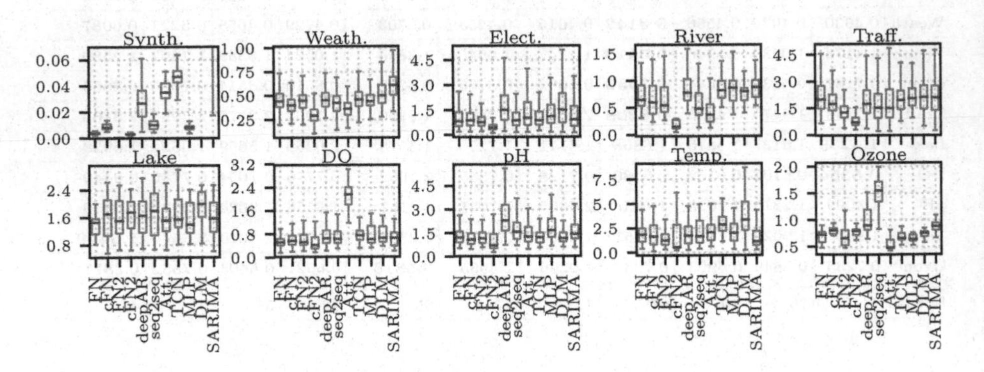

Fig. 3. Box plot of the MASE over the set of forecasts produced for each training dataset. The DLM and SARIMA boxes are outside of the plot range for the synthetic dataset. The labels on the horizontal axis are common for all plots.

6.3 Model Comparison Box-Whisker Plots

Box-whisker plots of the results over all the forecasts in each dataset's test set are provided in Fig. 3. ForecastNet consistently produced small boxes with low median values. The small boxes indicate that there is little variation in the accuracy over the set of forecasts. This indicates some form of robustness in the ForecastNet model. The low median values indicate a high level of accuracy. We argue that the time-variance properties of ForecastNet allow it to generalise better over the variations in the data dynamics. This results in the smaller boxes and lower median values.

There was significant variation over the different models in the synthetic dataset box-whisker plots. For this dataset, the fully connected networks such as FN, FN2 and MLP have small boxes. DeepAR had a large box which indicates a high variation in the forecast accuracy.

The models generally perform well on the weather dataset. This may be due to a more consistent seasonal amplitude in this dataset compared with the other real-world datasets. The lake and pH datasets have varying trends, amplitudes, and seasonal shape resulting in higher errors. ForecastNet and the attention model seem to model these variations better given their lower errors.

In datasets such as electricity, traffic, and pH, ForecastNet produced low errors with small boxes indicating reliable and accurate forecasts. Especially in the electricity and traffic datasets, it is evident that increased model complexity and removing the mixture density output results in lower errors and a more robust model. The mixture density outputs can reduce accuracy because the learning algorithm seeks to *simultaneously* minimise two variables in the normal distribution's log likelihood function. This is opposed to minimising a single variable in mean squared error loss function used for a linear output layer.

7 Summary and Conclusion

In this study, ForecastNet is proposed as time-variant deep neural architecture for multi-step-ahead time-series forecasting. Its architecture breaks from convention of structuring a model around the RNN or CNN. The result is a model that is time-variant compared with the RNN and CNN, which are time-invariant.

We demonstrate how time-invariant models can fail on a time-variant dataset. Additionally, we provide a comparison of several deep learning and statistical models on a range 10 of seasonal time-series datasets, selected from various domains. We demonstrate that ForecastNet is both accurate and robust on all datasets. It outperforms other models in terms of MASE on 8 of the 10 datasets and is ranked as the best performing model overall with Borda counts.

ForecastNet is a flexible architecture that lends itself well to various layer structures and is easily implemented in modern deep learning platforms. To improve accuracy, we suggest experimenting with various approaches such as using batch normalisation, shortcut-connections within and between hidden blocks, and augmenting the inputs with temporal information such as position. Integrating self-attention into the model will provide benefits relating to model interpretability. In future work, investigating ways of handling variable sequence lengths could be investigated to make the model more widely applicable to problems such as natural language processing.

References

1. Bahdanau, D., Cho, K., Bengio, Y.: Neural machine translation by jointly learning to align and translate. In: Proceedings of the International Conference on Learning Representations (2015). http://arxiv.org/abs/1409.0473
2. Bai, S., Kolter, J.Z., Koltun, V.: An empirical evaluation of generic convolutional and recurrent networks for sequence modeling. arXiv preprint arXiv:1803.01271 (2018)
3. Bishop, C.M.: Mixture density networks. Technical report NCRG94004, Aston University (1994)
4. Chang, S., et al.: Dilated recurrent neural networks. In: Guyon, I., et al. (eds.) Advances in Neural Information Processing Systems 30, pp. 77–87. Curran Associates, Inc. (2017). http://papers.nips.cc/paper/6613-dilated-recurrent-neural-networks.pdf
5. Dabrowski, J.J., Rahman, A., George, A., Arnold, S., McCulloch, J.: State space models for forecasting water quality variables: an application in aquaculture prawn farming. In: Proceedings of the 24th ACM SIGKDD International Conference on Knowledge Discovery and #38; Data Mining. (KDD 2018), pp. 177–185, ACM, New York (2018). https://doi.org/10.1145/3219819.3219841
6. Du, S., Li, T., Yang, Y., Horng, S.J.: Multivariate time series forecasting via attention-based encoder-decoder framework. Neurocomputing (2020). https://doi.org/10.1016/j.neucom.2019.12.118, http://www.sciencedirect.com/science/article/pii/S0925231220300606
7. He, K., Zhang, X., Ren, S., Sun, J.: Delving deep into rectifiers: surpassing human-level performance on imagenet classification. In: 2015 IEEE International Conference on Computer Vision (ICCV), pp. 1026–1034, December 2015. https://doi.org/10.1109/ICCV.2015.123
8. He, K., Zhang, X., Ren, S., Sun, J.: Deep residual learning for image recognition. arXiv preprint arXiv:1512.03385 (2015)
9. Hewamalage, H., Bergmeir, C., Bandara, K.: Recurrent neural networks for time series forecasting: current status and future directions. arXiv preprint arXiv:1909.00590 (2019)
10. Mariet, Z., Kuznetsov, V.: Foundations of sequence-to-sequence modeling for time series. In: Chaudhuri, K., Sugiyama, M. (eds.) Proceedings of Machine Learning Research. Proceedings of Machine Learning Research, vol. 89, pp. 408–417. PMLR, 16–18 April 2019. http://proceedings.mlr.press/v89/mariet19a.html
11. Oord, A.V.D., et al.: WaveNet: a generative model for raw audio. arXiv preprint arXiv:1609.03499 (2016)
12. Oppenheim, A.V., Schafer, R.W.: Discrete-Time Signal Processing. Pearson education signal processing series, 3rd edn. Pearson, London (2009)
13. Salinas, D., Flunkert, V., Gasthaus, J., Januschowski, T.: DeepAR: probabilistic forecasting with autoregressive recurrent networks. Int. J. Forecast. (2019). https://doi.org/10.1016/j.ijforecast.2019.07.001, http://www.sciencedirect.com/science/article/pii/S0169207019301888

14. Srivastava, R.K., Greff, K., Schmidhuber, J.: Training very deep networks. In: Cortes, C., Lawrence, N.D., Lee, D.D., Sugiyama, M., Garnett, R. (eds.) Advances in Neural Information Processing Systems 28, pp. 2377–2385. Curran Associates, Inc. (2015). http://papers.nips.cc/paper/5850-training-very-deep-networks.pdf
15. Sutskever, I., Vinyals, O., Le, Q.V.: Sequence to sequence learning with neural networks. In: Ghahramani, Z., Welling, M., Cortes, C., Lawrence, N.D., Weinberger, K.Q. (eds.) Advances in Neural Information Processing Systems 27, pp. 3104–3112. Curran Associates, Inc. (2014). http://papers.nips.cc/paper/5346-sequence-to-sequence-learning-with-neural-networks.pdf
16. Vaswani, A., et al.: Attention is all you need. In: Guyon, I., Luxburg, U.V., Bengio, S., Wallach, H., Fergus, R., Vishwanathan, S., Garnett, R. (eds.) Advances in Neural Information Processing Systems 30, pp. 5998–6008. Curran Associates, Inc. (2017). http://papers.nips.cc/paper/7181-attention-is-all-you-need.pdf
17. West, M., Harrison, J.: Bayesian Forecasting and Dynamic Models. Springer Series in Statistics, Springer, New York (1997). https://doi.org/10.1007/0-387-22777-6

Memetic Genetic Algorithms for Time Series Compression by Piecewise Linear Approximation

Tobias Friedrich, Martin S. Krejca(✉), J. A. Gregor Lagodzinski, Manuel Rizzo, and Arthur Zahn

Hasso Plattner Institute, University of Potsdam, Potsdam, Germany
{tobias.friedrich,martin.krejca,gregor.lagodzinski}@hpi.de, arthur.zahn@student.hpi.de

Abstract. *Time series* are sequences of data indexed by time. Such data are collected in various domains, often in massive amounts, such that storing them proves challenging. Thus, time series are commonly stored in a compressed format. An important compression approach is *piecewise linear approximation* (PLA), which only keeps a small set of time points and interpolates the remainder linearly. Picking a subset of time points such that the PLA minimizes the mean squared error to the original time series is a challenging task, naturally lending itself to heuristics. We propose the *piecewise linear approximation genetic algorithm* (PLA-GA) for compressing time series by PLA. The PLA-GA is a memetic $(\mu + \lambda)$ GA that makes use of two distinct operators tailored to time series compression. First, we add special individuals to the initial population that are derived using established PLA heuristics. Second, we propose a novel local search operator that greedily improves a compressed time series. We compare the PLA-GA empirically with existing evolutionary approaches and with a deterministic PLA algorithm, known as Bellman's algorithm, that is optimal for the restricted setting of sampling. In both cases, the PLA-GA approximates the original time series better and quicker. Further, it drastically outperforms Bellman's algorithm with increasing instance size with respect to run time until finding a solution of equal or better quality – we observe speed-up factors between 7 and 100 for instances of 90,000 to 100,000 data points.

Keywords: Genetic algorithms · Time series compression · Piecewise linear approximation · Hybridization · Experimental study

1 Introduction

In the modern age of Industry 4.0 and the Internet of Things, sensors are used in abundance in smart devices, cars, or to monitor production facilities. This leads to a huge accumulation of data over time, whose collection speed is only going to increase in the nearby future [9]. A prevalent type of such sensory data

H. Yang et al. (Eds.): ICONIP 2020, LNCS 12534, pp. 592–604, 2020.
https://doi.org/10.1007/978-3-030-63836-8_49

are time series – sequences of measurements indexed by time. In order to store these vast amounts of data in the long term and to allow scalable data analysis techniques, it is often necessary to compress time series [17,22]. To this end, various approaches exist that construct a time series of reduced dimensionality as an approximation of the original, known as *time series representation* [15,27].

In this setting, the use of *lossy compression* methods [18,19,37], adopted from the research on multimedia data, has been often advocated under several terminologies. These methods trade precision for a higher compression factor so that the reconstructed time series is only an approximation of the original, frequently omitting noise, outliers, and other information deemed not worthy of their memory cost. While some of these methods represent the time series in a transformed domain (e.g., involving discrete Fourier [1] or Wavelet transforms [31]), in certain application fields, it is important to maintain the original time domain and the related time stamps information, such as in GPS or accelerometer data [20].

In this work, we study *piecewise linear approximation* (PLA [21,24]), which is among the most important time series representation procedures for lossy compression and has been shown to be efficient in terms of memory and transmission cost compared to other similar methods [37]. PLA compresses a time series by representing it via a sequence of linear segments whose quality is assessed by an error measure between the original and the reconstructed series. This leads to a combinatorial problem similar to NP-hard problems like cluster analysis or subset selection in regression [35]. Consequently, heuristics are applied. We analyze to what extent evolutionary algorithms (EAs) can be used for this problem.

Related Work. Using EAs as heuristics for time series analysis is not a novel approach [3]. However, most research is focused on detecting break points in time series, that is, certain points in the series that have interesting or important structural properties [3,5,7]. To the best of our knowledge, the only prior research concerning EAs for compression by PLA was conducted by Duràn-Rosal et al. [10–12]. The authors consider the restricted setting of *sampling,* for which the compressed series is restricted to consist only of points contained in the original time series. Their works introduce a memetic genetic algorithm [12], particle swarm optimization algorithm [10], and coral reef optimization algorithm [11], all augmented with a local search operator to improve intermediate solutions.

Outside the domain of EAs, other heuristics have been considered for compression by PLA [34]. Especially, in the restricted setting of sampling, Bellman's algorithm [4] is optimal and deterministic. However, compression by PLA usually concerns the slightly different setting where one is not given a number of points m to pick but an error bound ε instead [8,14,26]. The goal is to find a compressed series with a mean squared error (MSE) of at most ε to the original time series. Thus, the number of points to pick can be chosen freely by the compression algorithm. Recent advances were made in this area [23,36]. The benefit of this approach is that the result is guaranteed to be within the specified error distance (if a result is returned). On the downside, the user has no control over the memory that the compressed time series occupies. However, note that the approaches of a given error bound ε and a given compression size m can be roughly converted into one another by performing a binary search on the parameter not provided.

Our Results. We propose the *piecewise linear approximation genetic algorithm* (PLA-GA, Alg. 1), a memetic $(\mu + \lambda)$ genetic algorithm (GA) variant for general time series compression via PLA. It features a seeding and a local search operator, both specific to this domain. Seeding adds special individuals to the initial population, which are computed via established PLA heuristics. The local search operator is a novel contribution that greedily improves a compressed time series.

We empirically analyze the impact of these two memetic operators on the PLA-GA's performance (see 4.2). We find that both operators are favorable, as they help the PLA-GA improve its MSE as well as its run time (see Fig. 1). We then evaluate the performance of the PLA-GA against competing approaches (see Sect. 4.3 and 4.4), which all operate under the sampling restriction. First, we compare the PLA-GA to the two latest and best-performing EAs [10,11] (see Sect. 4.3), which are the only EAs for compression by PLA, to the best of our knowledge. We observe that the PLA-GA outperforms them both with respect to the MSE as well as run time (see Table 1). Second, we compare the PLA-GA to Bellman's algorithm [4] (Sect. 4.4), a deterministic optimal algorithm for the sampling restriction, which the PLA-GA is not restricted to. First, we provide the PLA-GA with a run time budget equivalent to that of Bellman's algorithm. We observe a large variance of the MSE in the results, with the PLA-GA usually outperforming Bellman's – reducing the error of Bellman's up to 55 % (Table 2). Then, we analyze how long the PLA-GA takes to find a solution of Bellman's quality or better. Our results show that the PLA-GA drastically outperforms Bellman's algorithm, expressing speed-up factors between 7 and 100 for instances of 90,000 to 100,000 data points (see Fig. 2). Overall, our results suggest that the PLA-GA is particularly well suited to compression by PLA for long time series.

2 Preliminaries

For $m, n \in \mathbb{N}$, let $[m..n]$ denote the set of all natural numbers in the interval $[m, n]$. A *time series* is a function $S\colon \mathbb{R}_{\geq 0} \to \mathbb{R}$ with a finite domain of $n \in \mathbb{N}^+$ elements, where n is the *length* of S. We call the domain of S, denoted by $\mathrm{dom}(S)$, the *time points of* S. Furthermore, for an index $i \in [1..n]$, let t_i^S denote the i-th smallest time point of S, and let v_i^S denote its function value $S(t_i^S)$. Given a time series S of length n, we call a time series C of length m a *compression of* S if and only if $\mathrm{dom}(C) \subseteq \mathrm{dom}(S)$, $m \leq n$, and if $t_1^S = t_1^C$ and $t_n^S = t_m^C$. Note that while we demand that the time points of C are a subset of those of S, we make no such claims for the function values of C and S. We view C as a piecewise linear approximation (PLA) of S by interpolating values for the time points in $\mathrm{dom}(S) \setminus \mathrm{dom}(C)$ using linear functions. More formally, we define the *PLA of S via C*, denoted as C^*, to be a time series of length n with $\mathrm{dom}(C^*) = \mathrm{dom}(S)$

such that, for all $i \in [1..m-1]$ and all $t \in \text{dom}(S) \cap [t_i^C, t_{i+1}^C]$, it holds that $C^*(t) = v_i^C + (t - t_i^C)(v_{i+1}^C - v_i^C)/(t_{i+1}^C - t_i^C)$.

Approximation Error by PLA. Given a time series S and a compression C, we quantify the approximation error via the *mean squared error* (MSE), which is a common measures [10,14,23,25]. We define $\text{MSE}(C, S) = \frac{1}{n}\sum_{i=1}^{n}\left(v_i^S - C^*(t_i^S)\right)^2$.

3 The Piecewise Linear Approximation Genetic Algorithm

We introduce the $(\mu+\lambda)$ *piecewise linear approximation genetic algorithm* (PLA-GA; Algorithm 1), a memetic genetic algorithm following a $(\mu+\lambda)$ GA outline and using PLA to compress a time series of length n down to length $m \leq n$, minimizing the MSE. The PLA-GA makes use of a *seeding* and a *local search* operator, both of which are tailored to improve the quality of compressed time series. Please find the source code as well as a detailed description of the PLA-GA on GitHub.[1]

Each individual represents a compressed time series C of length m for a given time series S of length n. Since we minimize the MSE of C^* and S, we only store the time points dom(C); we compute the values v_i^C optimally with respect to the MSE only if needed. That is, we represent a compressed time series C by a bit string x of length n such that, for all $i \in [1..n]$, $x_i = 1$ if $t_i^S \in \text{dom}(C)$, and $x_i = 0$ otherwise, as is common [11,12]. Note that x has exactly m 1s. Recall that we demand each compressed time series to contain the first and last time point of S. Thus, for each individual x, it holds that $x_1 = x_n = 1$. If a bit string fulfills these conditions, we call it *valid*, otherwise *invalid*.

Algorithm 1: The $(\mu+\lambda)$ PLA-GA with parameters $\alpha \in [0,1]$, $k \in \mathbb{N}$, $\mu \geq k$, $\lambda \in \mathbb{N}^+$, and $m \geq 2$, compressing a time series S of length n down to length $m \leq n$, minimizing the MSE

```
1  P1 ← μ − k individuals, each uniformly at random from {x ∈ {0,1}^n | x is valid};
2  P2 ← k individuals, each generated by one of the k seeding operators;
3  P ← P1 ∪ P2;
4  while termination criterion not met do
5      L ← ranking_selection_λ(P, S);
6      O ← recombine(L);
7      bitflip_repair(O);
8      bitswap_mutation_α(O);
9      foreach x ∈ O do local_search(x, S);
10     P ← out of P ∪ O, choose μ best individuals;
```

[1] https://github.com/arthurz0/ga-for-time-series-compression-by-pla/.

In order to determine the *fitness* of an individual for C, that is, its MSE to S, we first determine the values v_i^C optimally as proposed by Marmarelis [25]. This approach is based on solving a system of m linear equations, where each equation represents the error of a piecewise linear function of C to S. Solving these equations takes time in $\mathcal{O}(n)$. Afterward, we compute the MSE.

4 Experimental Evaluation

We evaluate the performance of the PLA-GA empirically in different settings. In Sect. 4.2, we analyze the impact of the memetic operators on the PLA-GA's performance by comparing different variants of hybridization. We find that both memetic operators improve the quality of the solutions and the time to find them.

Afterward, we compare the PLA-GA against other algorithms that compress a time series S of length n to a time series C of length m. These algorithms take a simplified approach to PLA by only choosing points of the original series, that is, for all $t \in \text{dom}(C)$, it holds $C(t) = S(t)$, called *sampling.*

In Sect. 4.3, we compare the PLA-GA against two memetic EAs of Duràn-Rosal et al. [10,11], which, to the best of our knowledge, are the currently best EAs for compression by PLA. Since the EAs require a predetermined budget N of evaluations, we give the PLA-GA a roughly equivalent computation time budget. We observe that the PLA-GA outperforms the EAs during the entire budget.

In Sect. 4.4, we analyze the quality of the MSE achieved by the PLA-GA when it is not bound to a fixed time budget. To this end, we use Bellman's algorithm [4] as a baseline, which is deterministic and optimal under the restriction of sampling. First, we compare the quality reached when giving both algorithms the same run time budget on data sets of up to 30,000 points. The results depend strongly on the data set, ranging from solutions of similar quality to significant improvements of up to 0.55 times the MSE of Bellman in the best experiment. Then, we analyze how much time the PLA-GA needs to find solutions of the same quality as Bellman on problem instances of increasing size. We observe that the PLA-GA scales far better, reaching speed-up factors between 7 and 100 for problem sizes of 90,000 or 100,000 points.

In all of our experiments, we run the PLA-GA with $\mu = 200$, $\lambda = 200$, and $\alpha = \frac{0.8}{m-2}$ (and $k = 3$). These parameters were determined in pilot experiments. The experiments were conducted on an Intel(R) Core(TM) i5-6200U CPU @ 2.30 GHz with 8 GB RAM, unless noted otherwise.

4.1 Data Sets

Most of our time series originate from the publicly accessible UCR Time Series Classification Archive [32], which contains sets of time series for classification purposes. We use the data sets *Rock*, *Ham*, *HandOutlines*, *Mallat*, *Phoneme*, and *StarLightCurves* from different application fields. Since we aim to conduct experiments on time series of up to 100,000 points but the classification data sets consist of instances with a length between 431 (Ham) and 2844 (Rock) points, we concatenate the instances of each data set until the desired length is reached. An exception to this is Mallat (8192 points), where we use the same version as Duràn-Rosal et al. in [10,11].

Further, we use the "PAMAP2 Physical Activity Monitoring Dataset" [28,29] from the UCI Machine Learning Repository [33], containing data from real-world activity tracking. In particular, we use the columns 5 (*Subject5*) and 6 (*Subject6*) from the protocol of *Subject103*, both of which contain 252,311 points of 3D-acceleration data. Since the time series are too long for our purposes, we only take the first n points, where the specific value of n is mentioned in each experiment.

4.2 Evaluation of Hybridization

We examine the benefit added by hybridization and evaluate its cost, demonstrating that both, seeding and local search, improve the PLA-GA.

Setup. We run the PLA-GA in the following four configurations: with no seeding and no local search (PLA-GA-B), with seeding but no local search (PLA-GA-S), with local search but no seeding (PLA-GA-L), and the main algorithm PLA-GA, which uses both seeding and local search. We measure their MSE per iteration and determine the run time cost of seeding and local search. We have 50 independent runs per algorithm and choose $m = 0.01n$.

Results. Fig. 1 shows the MSEs of each PLA-GA variant by iteration. Although it is hardly noticeable in the figure, note that, bar statistical inaccuracies, PLA-GA and PLA-GA-S have the same best MSE in the initial population (iteration 0), as both use seeding, which deterministically introduces high-quality solutions. Similarly, both PLA-GA-B and PLA-GA-L start with roughly the same best MSE, as neither uses seeding. Further, the MSEs of each algorithm are concentrated around the median, as the interquartile ranges are merely visible.

On all data sets, PLA-GA-B performs worst and the PLA-GA best. In between we have PLA-GA-L, which starts worse than PLA-GA-S but overtakes it after some time, on Phoneme even at iteration 1. This substantiates the intuition that seeding helps during the start of the algorithm (since it adds good individuals to the initial population), whereas the gain of local search is incremental and adds up over time. On some data sets, the three better configurations end up with very close MSEs, though it should be considered that small differences gain significance the closer the solutions get to the optimum. When looking at the number of iterations required to reach a specific level of quality, the speed boost granted by the hybridization operators becomes evident.

For the full picture, one must further consider the computational effort of the hybridization operators. The creation of the initial population without seeding costs roughly as much as 1.3 iterations without local search; adding seeding increases this cost by roughly 25 %, so the computational effort of seeding is

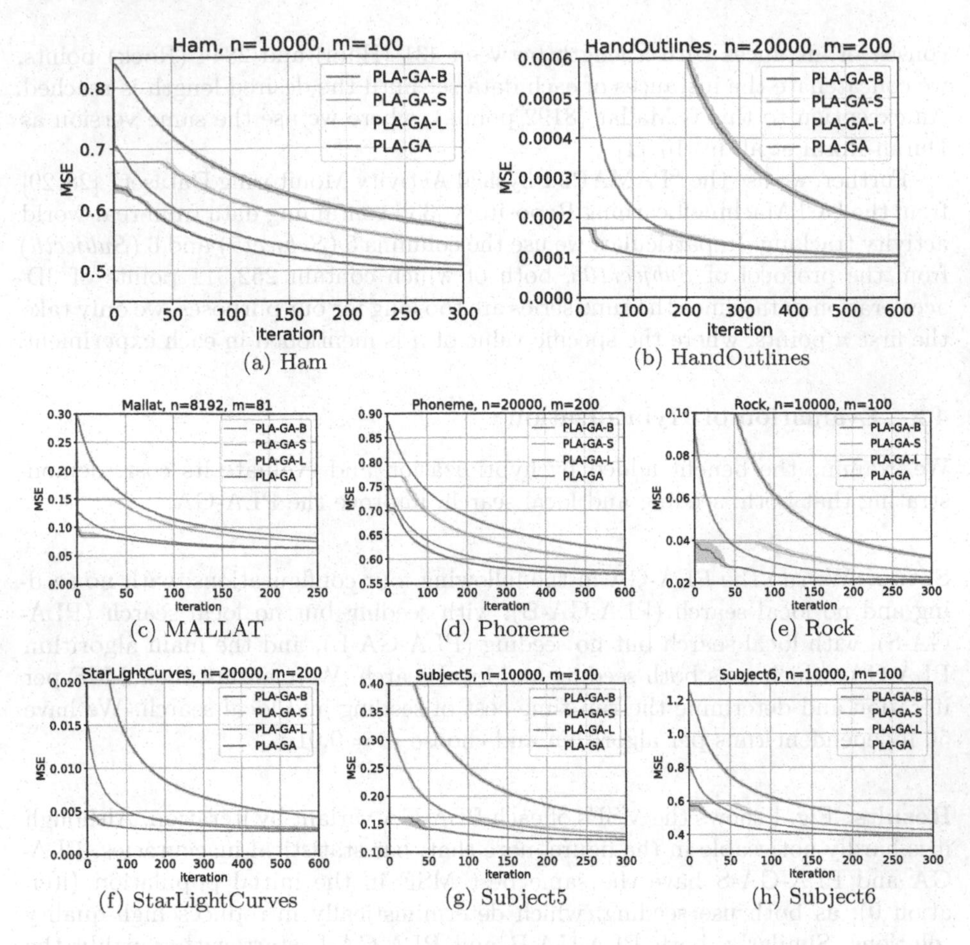

(a) Ham (b) HandOutlines

(c) MALLAT (d) Phoneme (e) Rock

(f) StarLightCurves (g) Subject5 (h) Subject6

Fig. 1. The MSE of the best solution for four different variants of the PLA-GA (Algorithm 1) plotted against the iteration, compressing a time series of length n down to length m. The solid lines represent the median of 100 runs per algorithm, and the surrounding shaded areas depict the mid 50%. The legend of each plot denotes the order of the curves in the last iteration. Please also refer to Sect. 4.2.

almost irrelevant for the total run time. The local search, on the other hand, is very expensive: an iteration with the local search takes roughly 2.9 times as long as an iteration without. Nonetheless, the quality benefit of local search makes it worthwhile, leading to an overall better performance despite its high cost.

4.3 Comparison with Other Evolutionary Algorithms

Duràn-Rosal et al. propose several memetic EAs for compressing time series by PLA under the sampling restriction [10–12]. Their best are the particle swarm algorithm *ACROTSS* [10] and the coral reef algorithm *DBBePTOSS* [11]. Both algorithms require a bound N on the number of evaluations in advance, as they

Table 1. Median best MSE of each of the 100 runs for the algorithms DBBePTOSS [11], ACROTSS [10], and the PLA-GA (Algorithm 1) when compressing a time series of length n. The results state the MSE after the first and the last iteration within the given budget. Scale denotes the factor that applies to all MSEs in the table. Please also refer to Sect. 4.3.

Data set	n	Scale	DBBePTOSS		ACROTSS		PLA-GA	
			Start	End	Start	End	Start	End
Mallat	8192	10^{-5}	15889	2207	1427	999	626	410
Ham	10000	10^{-3}	982	474	399	330	246	196
Rock	10000	10^{-5}	15264	1405	1122	918	906	682
Subject5	10000	10^{-4}	5813	1148	920	791	750	598
Subject6	10000	10^{-3}	1251	408	347	308	301	256
HandOutlines	20000	10^{-7}	45619	538	377	335	294	215
Phoneme	20000	10^{-3}	1122	588	534	496	484	393
StarLightCurves	20000	10^{-6}	30833	1115	789	684	319	185

perform local searches after certain iterations, relative to N. When comparing these algorithms to the PLA-GA, we provide it with a similar budget. However, since the PLA-GA's computational effort of a fitness evaluation is much higher due to calculating optimal function values for the compressed time series first, we instead give the PLA-GA a wall-clock-time budget matching the converted median run time of the fastest competing approach.

Setup. For the EAs of Duràn-Rosal et al., we choose the parameters as stated in the respective publications, including the budget of $N = 3.5n$ evaluations and compression length $m = 0.025n+1$. For each algorithm, we start 100 independent runs and log the MSE of the best individual in the population in each iteration. Unfortunately, the EAs of Duràn-Rosal et al. are implemented in Matlab, while the PLA-GA runs in Julia. Thus, we cannot directly take the wall-clock time of the EAs as budget for the PLA-GA. Instead, we divide the times of the EAs by 5, which is roughly the speed-up of our implementation of Bellman's algorithm [4] when run in Julia compared to Matlab. However, we acknowledge that no constant conversion factor exists [2,30].

Results. Table 1 shows the median best MSE of each algorithm for the first and last iteration. The PLA-GA clearly outperforms the competing approaches, as even its starting values are consistently better than the end values of DBBePTOSS and ACROTSS. This is most likely due to the less restrictive approach of the PLA-GA and its use of seeding.

Table 2. Median MSE of the best solution of 50 runs of the PLA-GA (Algorithm 1) after a computation time equivalent to Bellman's algorithm [4] when compressing a time series of length n. Ratio shows the median MSE of the PLA-GA divided by that of Bellman's. Best ratio is the best MSE of the PLA-GA divided with the MSE of Bellman. Scale denotes the factor that applies to all MSEs in the table. Please also refer to Sect. 4.4.

Data set	n	Scale	Bellman	PLA-GA	Ratio	Best ratio
Mallat	8192	10^{-4}	911	741	0.81	0.79
Ham	20000	10^{-3}	512	496	0.97	0.96
Rock	20000	10^{-3}	270	184	0.68	0.64
Subject5	30000	10^{-3}	131	121	0.92	0.91
Subject6	30000	10^{-3}	300	287	0.96	0.94
HandOutlines	20000	10^{-7}	1663	910	0.55	0.54
Phoneme	20000	10^{-3}	566	573	1.01	1.00
StarLightCurves	20000	10^{-5}	418	270	0.65	0.64

4.4 Comparison with Bellman

We compare the MSE and the wall-clock time of the PLA-GA against Bellman's algorithm [4], which deterministically computes an optimal solution in the restricted setting of sampling. First, we examine the MSE of the PLA-GA when giving it a budget equivalent to Bellman's wall-clock-time. Second, we analyze how the PLA-GA's run time scales with the input size n in comparison to Bellman's algorithm, which has a rather slow run time of $\mathcal{O}(n^2 m)$.

Setup. Both sets of experiments have 50 independent runs per algorithm and $m = 0.01n$. In the first set, the PLA-GA has a run time budget equivalent to the wall-clock-time of Bellman's algorithm, and the time series have up to 30,000 points. For the second set, we stop the PLA-GA once it finds a solution equal to Bellman's quality or better for the first time. The second set was conducted on an Intel(R) Core(TM) i5-6500 CPU @ 3.20 GHz and 8 GB RAM.[2]

Results. Table 2 compares the MSE of the solutions found by the PLA-GA and Bellman with the same amount of time; a small ratio is desirable. The results vary drastically between the experiments, ranging from similar MSEs on some data sets to huge improvements over Bellman on others. This variance is to be expected, as the difference between Bellman and the best ratio, which is an upper bound on the global optimum, also differs between each data set.

[2] Bellman's algorithm was modified to calculate the segment errors incrementally on the fly. This reduced the memory complexity from $\mathcal{O}(n^2)$ to $\mathcal{O}(mn)$, allowing us to run larger experiments. This modification does not affect the asymptotic run time.

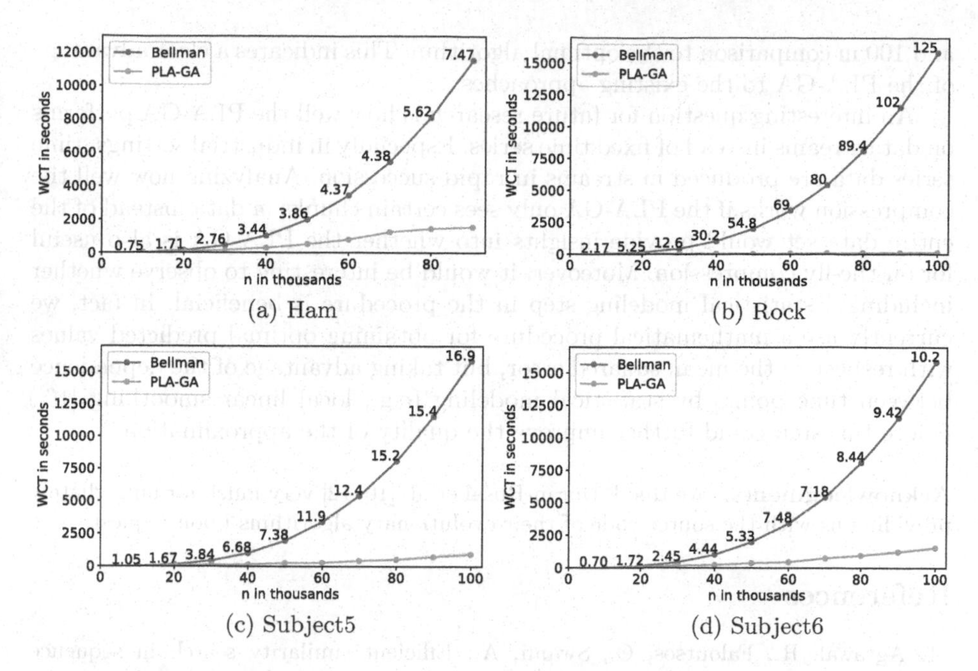

(a) Ham (b) Rock

(c) Subject5 (d) Subject6

Fig. 2. Median Wall-clock time it takes for the PLA-GA (Algorithm 1) and Bellman's algorithm [4] on increasing instance sizes. The PLA-GA is stopped once it finds a solution of at least Bellman's quality. In each figure, the lower curve shows the run time of the PLA-GA. The numbers over Bellman's curve denote the speed-up factor for that value of n, which is the quotient of Bellman's run time divided by that of the PLA-GA. Please also refer to Sect. 4.4.

Figure 2 compares the run times of Bellman and the PLA-GA with increasing problem size n and shows the resulting speed-up factors. The results differ strongly between the data sets, reaching a speed-up of 7 for Ham on 90,000 and 100 for Rock on 100,000. Thus, the PLA-GA seems to be more sensitive to the type of data than Bellman. Nonetheless, the speed-up factor clearly grows with n across all four data sets, indicating that the PLA-GA scales far better than Bellman.

5 Conclusions

We introduced the PLA-GA, which is a memetic $(\mu + \lambda)$ GA for compressing a time series by PLA. We showed that its two memetic operators – seeding and local search – have a positive impact on the algorithm's solution quality and run time. Further, we ran experiments comparing the PLA-GA to other EAs for time series compression and an optimal deterministic algorithm for a more restrictive setting. We observed that the PLA-GA outperforms all of these approaches with respect to solution quality and run time, leading to speed-up factors between 7 and 100 in comparison to the optimal algorithm. This indicates a clear advantage of the PLA-GA to the existing approaches.

An interesting question for future research is how well the PLA-GA performs on data streams instead of fixed time series. Especially in industrial settings, time series data are produced in streams in rapid succession. Analyzing how well the compression works if the PLA-GA only sees certain chunks of data instead of the entire data set would provide insights into whether the PLA-GA is also useful for on-the-fly compression. Moreover, it would be interesting to observe whether including a statistical modeling step in the procedure is beneficial. In fact, we currently use a mathematical procedure for obtaining optimal predicted values with respect to the mean squared error, but taking advantage of the dependence between time points by statistical modeling (e.g., local linear smoothing [16]) before this step could further improve the quality of the approximation.

Acknowledgments. We thank Durán-Rosal et al. [10–12] very much for immediately providing us with the source code of their evolutionary algorithms upon request.

References

1. Agrawal, R., Faloutsos, C., Swami, A.: Efficient similarity search in sequence databases. In: Proceedings of FODO 1993, pp. 69–84 (1993)
2. Aruoba, S.B., Fernández-Villaverde, J.: A comparison of programming languages in economics. Working Paper 20263, National Bureau of Economic Research (2014)
3. Baragona, R., Battaglia, F., Poli, I.: Evolutionary Statistical Procedures. An evolutionary computation approach to statistical procedures designs and applications. Springer (2011).https://doi.org/10.1007/978-3-6SD42-16218-3
4. Bellman, R., Kotkin, B.: On the approximation of curves by line segments using dynamic programming. RAND Corporation, II. Technical report (1962)
5. Cucina, D., Rizzo, M., Ursu, E.: Multiple changepoint detection for periodic autoregressive models with an application to river flow analysis. Stoch. Environ. Res. Risk Assess. **33**(4–6), 1137–1157 (2019)
6. Doerr, B., Doerr, C., Kötzing, T.: Static and self-adjusting mutation strengths for multi-valued decision variables. Algorithmica **80**(5), 1732–1768 (2018)
7. Doerr, B., Fischer, P., Hilbert, A., Witt, C.: Detecting structural breaks in time series via genetic algorithms. Soft. Comput. **21**(16), 4707–4720 (2017)
8. Douglas, D.H., Peucker, T.K.: Algorithms for the reduction of the number of points required to represent a digitized line or its caricature. Cartograph. Int. J. Geograph. Inf. Geovis. **10**(2), 112–122 (1973)
9. Dunning, T., Friedman, E.: Time Series Databases: New Ways to Store and Access Data. O'Reilly and Associates, Sebastopol (2014)
10. Durán-Rosal, A.M., Gutiérrez, P.A., Poyato, Á.C., Hervás-Martínez, C.: A hybrid dynamic exploitation barebones particle swarm optimisation algorithm for time series segmentation. Neurocomputing **353**, 45–55 (2019)
11. Durán-Rosal, A.M., Gutiérrez, P.A., Salcedo-Sanz, S., Hervás-Martínez, C.: Dynamical memetization in coral reef optimization algorithms for optimal time series approximation. Progress in AI **8**(2), 253–262 (2019)
12. Durán-Rosal, A.M., Peáa, P.A.G., Martínez-Estudillo, F.J., Hervás-Martínez, C.: Time series representation by a novel hybrid segmentation algorithm. In: Proceedings of HAIS 2016, vol. 9648, pp. 163–173 (2016)

13. Eiben, A.E., Smith, J.E.: Introduction to Evolutionary Computing. NCS. Springer, Heidelberg (2015). https://doi.org/10.1007/978-3-662-44874-8
14. Elmeleegy, H., Elmagarmid, A.K., Cecchet, E., Aref, W.G., Zwaenepoel, W.: Online piecewise linear approximation of numerical streams with precision guarantees. PVLDB **2**(1), 145–156 (2009)
15. Esling, P., Agon, C.: Time-series data mining. ACM Comput. Surveys (CSUR) **45**(1), 1–34 (2012)
16. Fan, J., Yao, Q.: Nonlinear Time Series. Nonparametric and parametric methods. Springer (2008). https://doi.org/10.1007/978-0-387-69395-8
17. Fu, T.C.: A review on time series data mining. Eng. Appl. Artif. Intell. **24**(1), 164–181 (2011)
18. Hollmig, G., et al.: An evaluation of combinations of lossy compression and change-detection approaches for time-series data. Inf. Syst. **65**, 65–77 (2017)
19. Hung, N.Q.V., Jeung, H., Aberer, K.: An evaluation of model-based approaches to sensor data compression. IEEE Trans. Knowl. Data Eng. **25**(11), 2434–2447 (2013)
20. Hurvitz, P.M.: GPS and accelerometer time stamps: proper data handling and avoiding pitfalls. In: Proceedings of ACM SIGSPATIAL 2015, pp. 94–100 (2015)
21. Keogh, E., Chu, S., Hart, D., Pazzani, M.: Segmenting time series: a survey and novel approach. In: Data Mining in Time Series Databases, pp. 1–21. World Scientific (2004)
22. Last, M., Horst, B., Abraham, K.: Data mining in time series databases. World Scientific (2004)
23. Lin, J.W., Liao, S.W., Leu, F.Y.: Sensor data compression using bounded error piecewise linear approximation with resolution reduction. Energies **12**(13), 2523 (2019)
24. Lovrić, M., Milanović, M., Stamenković, M.: Algoritmic methods for segmentation of time series: an overview. JCBI **1**(1), 31–53 (2014)
25. Marmarelis, M.G.: Efficient and robust polylinear analysis of noisy time series. arXiv preprint, CoRR abs/1704.02577 (2017)
26. Ramer, U.: An iterative procedure for the polygonal approximation of plane curves. Comput. Graph. Image Process. **1**(3), 244–256 (1972)
27. Ratanamahatana, C.A., Lin, J., Gunopulos, D., Keogh, E., Vlachos, M., Das, G.: Mining time series data. In: Data Mining and Knowledge Discovery Handbook, pp. 1069–1103. Springer, Boston (2005). https://doi.org/10.1007/0-387-25465-X_51
28. Reiss, A., Stricker, D.: Creating and benchmarking a new dataset for physical activity monitoring. In: Proceedings of PETRA 2012, pp. 1–8 (2012)
29. Reiss, A., Stricker, D.: Introducing a new benchmarked dataset for activity monitoring. In: Proceedings of ISWC 2012, pp. 108–109 (2012)
30. Sinaie, S., Nguyen, V.P., Nguyen, C.T., Bordas, S.: Programming the material point method in Julia. Adv. Eng. Softw. **105**, 17–29 (2017)
31. Struzik, Z.R., Siebes, A.: Wavelet transform in similarity paradigm. In: Proceedings of PAKDD, pp. 295–309 (1998)
32. The UCR time series classification archive. http://www.springer.com/lncs. Accessed 27 June 2020
33. UCI machine learning repository. http://archive.ics.uci.edu/ml. Accessed 27 June 2020
34. Visvalingam, M., Whyatt, J.D.: Line generalisation by repeated elimination of points. Cartograph. J. **30**(1), 46–51 (1993)
35. Welch, W.: Algorithmic complexity: three NP-hard problems in computational statistics. J. Stat. Comput. Simul. **15**, 17–25 (1982)

36. Zhao, H., Dong, Z., Li, T., Wang, X., Pang, C.: Segmenting time series with connected lines under maximum error bound. Inf. Sci. **345**, 1–8 (2016)
37. Zordan, D., Martínez, B., Vilajosana, I., Rossi, M.: On the performance of Lossy compression schemes for energy constrained sensor networking. ACM Trans. Sens. Netw. **11**(1), 15:1–15:34 (2014)

Sensor Drift Compensation Using Robust Classification Method

Guopei Wu[1], Junxiu Liu[1(✉)], Yuling Luo[1,2], and Senhui Qiu[1,3(✉)]

[1] School of Electronic Engineering, Guangxi Normal University, Guilin, China
j.liu@ieee.org, qiusenhui@gxnu.edu.cn
[2] Guangxi Key Lab of Multi-source Information Mining and Security, Guilin, China
[3] Guangxi Key Laboratory of Wireless Wideband Communication and Signal Processing, Guilin, China

Abstract. Gas sensor drift affects the performance of chemical sensing. In this paper, a Long Short Term Memory (LSTM) network and a Support Vector Machine (SVM) are used for gas sensor drift compensation to improve gas classification performance. An improved dynamic feature extraction method is developed to reduce feature dimensions. A public time series chemical sensing dataset is used for evaluation, which was collected by 16 metal-oxide gas sensors over three years. Results show that a high classfication accuracy can be achieved using the proposed method compared to other studies, which demonstrates the robustness of the proposed method for sensor drift compensation.

Keywords: Sensor drift · Support vector machine · Long short term memory network · Gas sensors

1 Introduction

Gas sensors are used for identifying different types of gases and monitoring the composition of gases such as food producing, environmental and biomedical monitoring [1], where reliability of sensors is an important indicator in gas classification systems. However, sensor drift affects the gas recognition performance. There are different forms of drifting including zero, span and concept drifts etc. [2]. Sensor drift occurs due to the interference of many factors such as atmospheric pressure, temperature, humidity variations and other uncontrollable changes from external environment [3]. Moreover, materials of sensors are unstable during the process of gas recognition. Physical and chemical interactions in micro-structure of gas sensors or deployments in polluted environment can lead to changes of sensor surface. For example, sensors' ageing affects the sensor signal collection process, which makes the output sensor signals unstable. These factors reduce recognition accuracy in gas classification systems.

Sensor drift should be considered in the data processing stage [4], and it can be addressed by using some compensation technologies. By using compensation, classification accuracy of gas can still be maintained. Compensation technologies

H. Yang et al. (Eds.): ICONIP 2020, LNCS 12534, pp. 605–615, 2020.
https://doi.org/10.1007/978-3-030-63836-8_50

can be divided into the hardware and software compensations. For the former, it mainly focuses on repairing gas sensors, which includes removing aged modules or using stable sensors. The disadvantage is the high cost of replacing sensors [2]. For the latter, software compensation includes univariate, multivariate and machine learning methods. Univariate method is sensitive to sampling rates and not suitable for solving sensor drift in severe environment. Multivariate method can simulate drift, but the process is complex. In recent studies, machine learning methods are used to resolve sensor drift. Support Vector Machine (SVM), K nearest neighbour, random forest are usually used in gas recognition [3,5]. It is necessary to reduce dimension of gas data as original gas data is complex, and accuracy is low if original data is used for gas recognition. Machine learning methods can classify features with low dimension and address sensor drift in gas classification systems. Principal components analysis is used for feature extraction and partial least squares regression is used for classification in [6]. By removing components of drift direction from collected data, a better performance can be achieved. A signal processing technique is used by [7] for sensor drift compensation. However, a chemically stable gas is necessary for reference in these studies, which is difficult to conduct. In recent studies, deep learning methods are used in gas recognition. A domain adaptive extreme learning machine [8] is used to mitigate sensor drift. Deep learning such as recurrent neural network is used in many studies [9–13]. Gas data belongs to long time-series data, and recurrent neural network is not suitable for classifying it. Therefore, the Long Short Term Memory (LSTM) network is used for classification, and Pearson product-moment correlation coefficient is used for feature extraction by [2]. However, classfication accuracy can be further improved.

In this paper, an improved feature extraction method is proposed by using ReliefF algorithm. Features' dimension is reduced by adjusting weights obtained from the ReliefF. LSTM and two fully-connected layers are used for gas recognition. For fair comparison with other studies, SVM is also trained as a baseline classifier for evaluating the proposed algorithm. A public gas dataset namely 'Gas Sensor Array Drift Dataset' in the machine learning repository is used for classifying. Contribution of this paper is summarized as follows: (1). An improved feature extraction algorithm is proposed in this paper. (2). Gas sensor drift can be mitigated using the proposed methods and performance is better in most batches compared to other works using the same dataset. The rest of the paper is organised as follows. Section 2 is dataset description. Section 3 introduces the proposed algorithm. Section 4 introduces structure of network. Section 5 presents experiments and analysis of results. Section 6 is the conclusion.

2 Dataset

The dataset [4] used in this work is a public gas dataset for gas recognition. It consists of six different gases collected during 36 months using 16 metal-oxide sensors. Gas sensors are put into different test chambers containing Ethanol, Ethylene, Acetone, Ammonia, Acetaldehyde and Toluene. Number of collected

Table 1. Gas dataset [4]

Batch ID	Month ID	Number of samples
1	1,2	445
2	3,4,8,9,10	1244
3	11,12,13	1586
4	14,15	161
5	16	197
6	17,18,19,20	2300
7	21	3613
8	22,23	294
9	24,30	470
10	36	3600

samples is 13,910. For each sample, there are eight kinds of features calculated using data collected from gas sensors. According to [4], two features are calculated by maximum ascent and decay in internal resistance of gas sensors in steady state, and other six features are calculated through exponential moving average filter, which is given by

$$y[k] = (1 - a)y[k-1] + a(r[k] - r[k-1]), \tag{1}$$

where $a = 0.1, 0.01, or\ 0.001$, $y[k]$ represents the real scalar and initial state of y is set to zero, r is time curve of sensor resistance. Thus, feature dimension of each sample is 128 (16×8).

Original features are pre-processed using normalization, and each feature is scaled to the range of -1 to $+1$, which is expressed by

$$y[n] = 2 \times \frac{x(n) - Min(x)}{Max(x) - Min(x)} - 1, \tag{2}$$

where x represents original value of features, $Max(x)$ and $Min(x)$ represent the maximum and minimum value of all features, respectively. Data in this dataset is organized into ten batches, where each batch consists of different amounts of samples. Data combination in gas dataset is shown by Table 1. The number of samples is different in each batch according to [4]. Features are extracted in the proposed algorithm for gas recognition using SVM and LSTM.

3 Feature Extraction Algorithm

As described in Sect. 2, feature dimension is 128 for each sample. According to [2], good accuracy can be achieved by reducing dimension of features in gas dataset. Moreover, computational complexity time can be reduced and classification time can be saved.

ReliefF is improved from Relief for multi-class tasks [14]. It can help remove unnecessary data from dataset and output the weights of different features in the dataset. Thus, dimension can be reduced compared to original data. Weights are calculated according to correlations among features of samples with different labels. After setting the thresholds weights, features with high weights are kept for classification.

The ReliefF is improved in this work. First, the gas dataset is divided into training data and testing data. Then the sample R is selected and the nearest neighborhood of the same and different classes are calculated to find the nearest neighbors by $diff$ function. $diff$ is defined by

$$diff\left(A, I_1, I_2\right) = \frac{\left|value\left(A, I_1\right) - value\left(A, I_2\right)\right|}{Max(A) - Min(A)} \tag{3}$$

where I_1 and I_2 are two samples selected from attribute A. After that, the nearest neighbor H in the same class and the nearest neighbor M in different classes are selected. Then R_i, M, H and weighting vector are used for calculating the distance of adjacent sample A between the same and different classes. Weight $W[A_f]$ is updated by

$$W\left[A_f\right] = W_0\left[A_f\right] + \frac{diff\left(A_P, R_i M\right)}{m} - \frac{diff\left(A_F, R_i, H\right)}{m} \tag{4}$$

where initial value of $W_0[A_f]$ is set to 0, m represents the operations are repeated for m times.

For gas dataset used in this work, features of each sample are extracted with ReliefF and weights of each sample are obtained. For each batch, weights of all features are shown by Fig. 1, where the distribution of high weights are mainly limited on position of 0–60 and 110–128. Features with high weights mean they are more important among all features. Thus, a threshold value should be set and features with weights higher than the threshold value can be kept. Thus, dimension of samples' features can be reduced. Threshold is selected dynamically in order to achieve the best accuracy in classifiers. Considering dimension of each sample, threshold value is set to 0–0.018.

4 LSTM

LSTM is effective in classifying long timing information, which can be used for analyzing gas samples. The model used in this paper consists of four layers, which includes the input layer, the LSTM layer, the hidden layer and the output layer. Structure of the model is shown by Fig. 2. Input sequence X is fed to LSTM layer, and output of LSTM layer is sent to the hidden layer. After that, output of hidden layer is sent to output layer for classification. C and H are two outputs from current state to the next state of LSTM cell.

As threshold of feature extraction algorithm is not fixed, dimension of input sequence X also varies in this work. According to threshold value, dimension

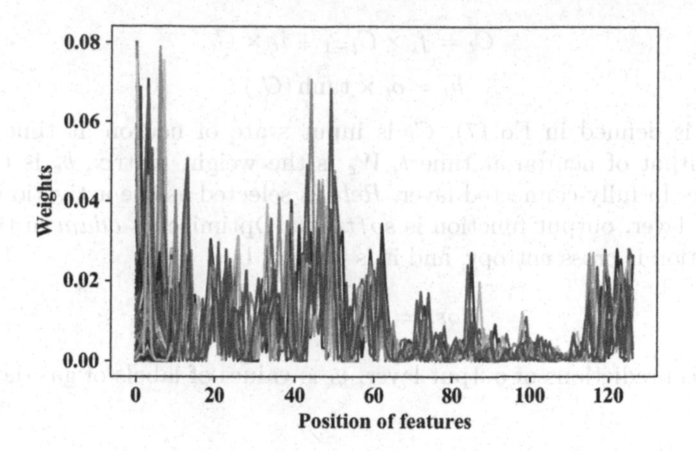

Fig. 1. Feature weight distribution

of X is from 29 to 128. LSTM cell consists of forget, input and output gates. Outputs of these three gates can be expressed by

$$f_t = \sigma\left(w_f \times h_{t-1} + w_f \times x_t + b_f\right), \tag{5}$$

$$i_t = \sigma\left(W_i \times [h_{t-1}, x_t] + b_i\right), \tag{6}$$

$$o_t = \sigma\left(w_o \times h_{t-1} + w_o \times x_t + b_o\right), \tag{7}$$

where f, i, o are results of the three gates, W_f, W_i, W_o are weights of three gates, which are initially set to 0. b_f, b_i and b_o are bias terms. States can be expressed by

$$\tilde{C}_t = \tanh\left(W_c \times [h_{t-1}, x_t] + b_c\right), \tag{8}$$

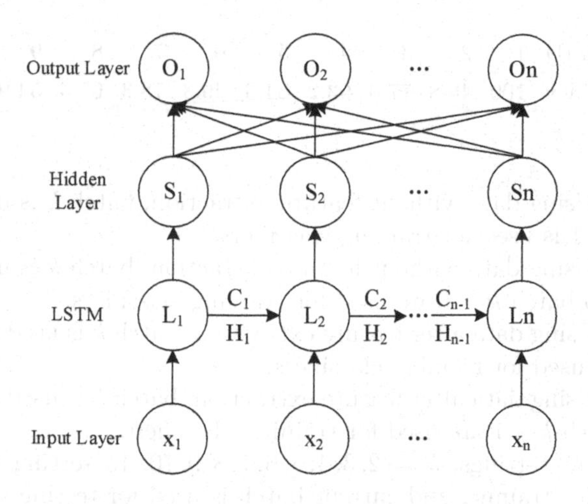

Fig. 2. Structure of LSTM

$$C_t = f_t \times C_{t-1} + i_t \times \tilde{C}_t, \tag{9}$$

$$h_t = o_t \times \tanh(C_t), \tag{10}$$

where o_t is defined in Eq. (7), C_t is input state of neuron at time $t+1$, h_t is final output of neuron at time t, W_c is the weight matrix, b_c is the bias of input gate. In fully-connected layer, *Relu* is selected as the activation function. In output layer, output function is $softmax$. Optimizer is *adam* in this model. Loss function is cross entropy, and it is defined by

$$loss = -\sum y_i \log \widehat{y}_i, \tag{11}$$

where y_i is predictions of output layer, y_i is values of labels of gas dataset.

5 Results

For gas recognition, sensor drift affects the performance of classifiers and accuracy of each batch is gradually decreasing with time. The SVM is used for benchmarking. In gas dataset, batch 1 is used as training set, and batch 2 to 10 are used as testing set. Threshold in this experiment is set to 0.018. Accuracy of each batch is shown by Table 2, where data used for training and testing is original data in dataset. Accuracy of batch 1 is 100% because training and testing data are same. Accuracy of each batch gradually decreases during the whole testing process. It indicates that gas sensor is drifting when sensor is used for monitoring the gas in external environment. Experiments in this paper mainly focus on mitigating gas sensor drift by using the proposed feature extraction algorithm and classifiers. In order to study senor drift of gas sensors, four settings are considered as follows:

Table 2. Accuracies (%) of SVM

Batch ID	1	2	3	4	5	6	7	8	9	10
Accuracy	100	46.8	77.0	93.2	84.3	49.3	78.3	67.3	54.9	32.4

Setting 1: Using data without feature extraction, batch k is used for testing, and batch $k-1$ is used for training classifiers.

Setting 2: Using data without feature extraction, batch k is used for testing, and batch 1 to batch $k-1$ are used for training classifiers.

Setting 3: Using data after feature extraction, batch k is used for testing, and batch $k-1$ is used for training classifiers.

Setting 4: Using data after feature extraction, batch k is used for testing, and batch 1 to batch $k-1$ are used for training classifiers.

In these four settings, $k = 2,3,4,5,6,7,8,9,10$. In setting 1, the previous batch is used for training and current batch is used for testing to minimize difference between training set and testing set due to drift of gas sensors. However, because of different amount of samples in each batch, classification results may be affected by training set due to lacking enough data. Thus, in setting 2, all previous batches are used for training and current batch is used for testing on classifiers. Data used in setting 3 and setting 4 are extracted features, and they are set for comparison with setting 1 and setting 2 in order to verify effectiveness of the proposed feature extraction algorithm.

Table 3. Accuracies (%) and thresholds of all batches by using SVM under 4 settings

Batch ID	Setting1	Setting 2	Setting 3	Setting 4
2	46.8(0.018)	46.8(0,018)	47.2(0,01)	47.2(0.01)
3	77.0(0.018)	53.5(0.018)	77.5(0.08)	81.3(0.018)
4	93.2(0.018)	91.9(0.018)	93.2(0.01)	92.5(0.018)
5	84.3(0.018)	98.5(0.018)	86.3(0.01)	99.0(0.018)
6	49.3(0.018)	75.5(0.018)	49.3(0)	75.7(0.01)
7	78.3(0.018)	85.6(0.018)	78.3(0)	86.6(0.005)
8	67.3(0.018)	94.9(0.018)	65.6(0)	94.9(0)
9	54.9(0.018)	77.2(0.018)	56.0(0.001)	77.2(0.01)
10	32.4(0.018)	67.2(0.018)	42.8(0.018)	67.2(0.018)

Table 4. Accuracies (%) and thresholds of all batches by using LSTM under 4 settings

Batch ID	Setting1	Setting 2	Setting 3	Setting 4
2	74.4(0.018)	74.4(0.018)	80.2(0.01)	80.2(0.01)
3	82.7(0.018)	78.0(0.018)	79.0(0.01)	97.2(0.018)
4	77.6(0.018)	80.1(0.018)	91.3(0.01)	99.4(0.018)
5	98.0(0.018)	99.0(0.018)	98.5(0.01)	98.5(0.018)
6	43.6(0.018)	77.6(0.018)	40.1(0.01)	76.1(0.01)
7	86.8(0.018)	87.0(0.018)	86.0(0.01)	86.5(0.01)
8	94.2(0.018)	92.2(0.018)	86.1(0.01)	86.4(0.01)
9	64.9(0.018)	76.8(0.018)	52.8(0.01)	76.2(0.01)
10	34.2(0.018)	74.4(0.018)	45.2(0.01)	72.3(0.01)

A linear kernel SVM with search space $2^{[-10:10]}$ is used for experiments under four settings. After training and testing on SVM, accuracies of all batches in four experiments are obtained, which are shown by Table 3. Numbers in brackets represent threshold values of all batches. Accuracies of setting 1 are selected as benchmarking results. Best accuracy achieves 99.4% at batch 4 among all settings. Best average accuracy achieves 80.2% under setting 4. It should be noticed that lowest accuracy is achieved at batch 10. According to Table 1, collection of batch 10 is at least six months later than batch 9, which causes serious sensor drift. Performance of SVM is also shown by Fig. 3. Performance of setting 4 is better than that of setting 1, which indicates that the algorithm is effective in mitigating sensor drift of gas sensors. Moreover, performance of setting 2 and setting 4 using all previous batches for training the classifier are better than that of settings using only one batch for training.

For LSTM, experiments are the same as SVM under 4 settings. For parameters of LSTM, epoch is set to 200, batch size is set to 100, and learning rate is set to 0.01. The number of neurons in input layer and step of LSTM cell is the same as dimension of input, which is decided by threshold value. Neurons in hidden and output layer are 100 and 6, respectively. The used neural network is set by using an open source platform named Tensorflow [15]. After training and testing, accuracy of each batch is obtained and shown in Table 4. Best accuracy achieves 99.4% at batch 4 among all settings. Best average accuracy achieves 85.9% under setting 4. Lowest accuracy is also obtained at batch 10. Performance of LSTM is also shown by Fig. 4. The best performance is setting 4. Performance of settings using extracted features is better than settings using original data, indicating the robustness of the proposed method.

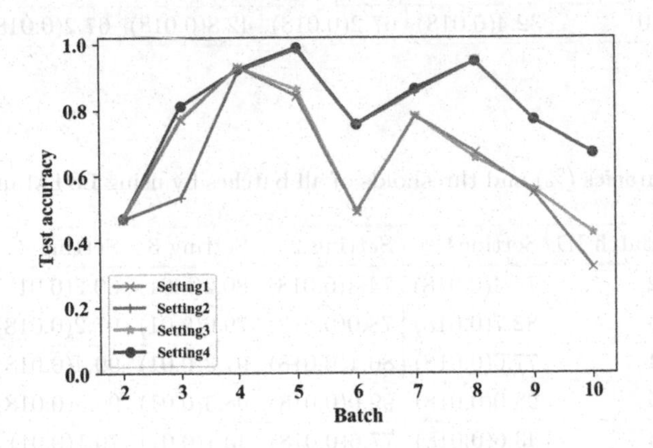

Fig. 3. Accuracy of each batch under 4 settings by using SVM

After collecting best accuracy of each setting on each classifier, performance of gas recognition in this paper is shown by Fig. 5. Performance of this paper is much better compared to the benchmarking method, which indicates that gas sensor drift can be resolved using the proposed feature extraction algorithm and classifiers. For complexity analysis, amount of parameters in LSTM is 22,467.

From experiments under four settings, it can be shown that using data processed by the proposed methods have better performance than other works. In Table 3, accuracies of setting 3 are the same as or higher than that of setting 4 in batch 2, 3, 4, 5, 6, 7, 9, 10. Accuracies of setting 4 are the same as or higher than that of setting 2 in all batches. In Table 4, accuracies of setting 3 are higher

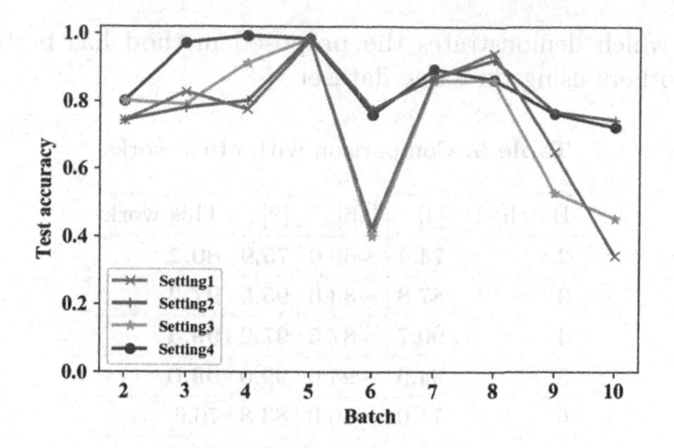

Fig. 4. Accuracy of each batch under 4 settings by using LSTM

than that of setting 1 in batch 2, 4, 5, 10. Accuracies of setting 4 are higher than that of setting 2 in batch 2, 3, 4. Moreover, accuracies of the proposed methods are better than that of benchmarking methods, which demonstrates the effectiveness and robustness of gas sensor drift compensation methods proposed in this paper.

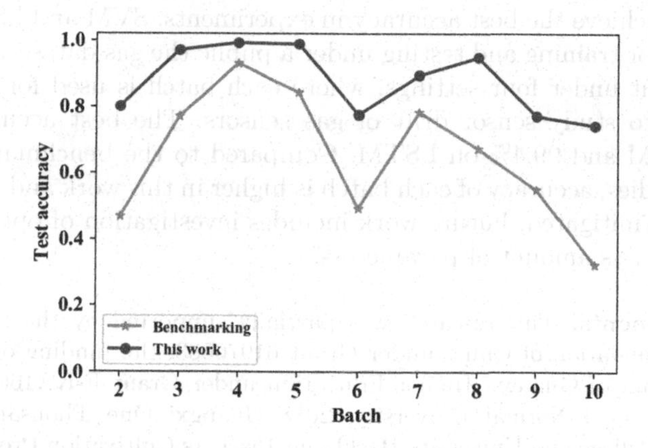

Fig. 5. Best accuracies in this work compared with the benchmarking method

Three works using the same dataset are used for comparison. Best accuracies of 9 batches in each study are shown by Table 5. In the approach of [4], component correction method and SVM are used for gas recognition, but accuracy is not high (54.3%~94.9%). Accuracies of each batch in the proposed work are higher than [16] except batch 6 and 10. Pearson product-moment correlation cofficient is used for feature extraction in the approach of [2], and SVM and LSTM are used as classifiers, where a higher accuracy is obtained compared to [4]. In this paper, the number of batches with higher accuracies is more than other works, which demonstrates the proposed method has better robustness compared to others using the same dataset.

Table 5. Comparison with other works

Batch ID	[4]	[16]	[2]	This work
2	74.4	~69.0	75.9	**80.2**
3	87.8	~83.0	95.5	**97.2**
4	90.7	~87.5	97.2	**99.4**
5	94.9	~94.6	99.0	**99.0**
6	71.0	~80.0	83.8	76.6
7	83.5	~79.0	88.5	**89.5**
8	91.8	~81.0	94.2	**94.9**
9	69.2	~76.0	85.9	77.2
10	54.3	~75.0	83.4	74.4

6 Conclusion

In this paper, a feature extraction algorithm is proposed and two classifiers are used for gas classification. Features with different dimensions are dynamically extracted to achieve the best accuracy in experiments. SVM and LSTM are used as classifiers for training and testing under a public the gas dataset. Experiments are carried out under four settings, where each batch is used for testing in all experiments to study sensor drift of gas sensors. The best accuracy achieves 99.0% on SVM and 99.4% on LSTM. Compared to the benchmarking method and other studies, accuracy of each batch is higher in this work and sensor drift in gas sensors is mitigated. Future work includes investigation of optimized neural network with less amount of parameters.

Acknowledgments. This research was partially supported by the National Natural Science Foundation of China under Grant 61976063, the funding of Overseas 100 Talents Program of Guangxi Higher Education under Grant F-KA16035, the Diecai Project of Guangxi Normal University, 2018 Guangxi One Thousand Young and Middle-Aged College and University Backbone Teachers Cultivation Program, research fund of Guangxi Key Lab of Multi-source Information Mining & Security (19-A-03-02), research fund of Guangxi Key Laboratory of Wireless Wideband Communication and Signal Processing, the Young and Middle-aged Teachers' Research Ability Improvement Project in Guangxi Universities under Grant 2020KY02030, and the Innovation Project of Guangxi Graduate Education under Grant YCSW2020102.

References

1. Llobet, E., Brezmes, J., Vilanova, X., Sueiras, J.E., Correig, X.: Qualitative and quantitative analysis of volatile organic compounds using transient and steady-state responses of a thick-film tin oxide gas sensor array. Sens. Actuators B: Chem. **41**(1), 13–21 (1997)

2. Zhao, L., Xiao, M., Yu, H.: Sensor drift compensation based on the improved LSTM and SVM Multi-Class ensemble learning models. Sensors **19**(18), 3844–3849 (2019)
3. Rehman, A.U., Bermak, A.: Heuristic Random Forests (HRF) for drift compensation in electronic nose applications. IEEE Sens. J. **19**(4), 1443–1453 (2019)
4. Vergara, A., Vembu, S., Ayhan, T., Ryan, M.A., Homer, M.L., Huerta, R.: Chemical gas sensor drift compensation using classifier ensembles. Sens. Actuators B: Chem. **166**(1), 320–329 (2012)
5. Adhikari, S., Saha, S.: Multiple classifier combination technique for sensor drift compensation using ANN and KNN. In: 2014 IEEE International Advance Computing Conference (IACC), pp. 1184–1189 (2014)
6. Artursson, T., Eklöv, T., Lundström, I., Mårtensson, P., Sjöström, M., Holmberg, M.: Drift correction for gas sensors using multivariate methods. J. Chemom. **14**(5), 711–723 (2000)
7. Llobet, E., Brezmes, J., Vilanova, X., Sueiras, J.E., Correig, X.: Drift compensation of gas sensor array data by orthogonal signal correction. Chemometr. Intell. Lab. Syst. **100**(1), 28–35 (2010)
8. Zhang, L., Zhang, D.: Domain adaptation extreme learning machines for drift compensation in E-Nose systems. IEEE Trans. Instrum. Meas. **64**(7), 1790–1801 (2015)
9. Liu, J.: Exploring self-repair in a coupled spiking astrocyte neural network. IEEE Trans. Neural Netw. Learn. Syst. **30**(3), 865–875 (2019)
10. Fu, Q., et al.: Improving learning algorithm performance for spiking neural networks. In:2017 IEEE 17th International Conference on Communication Technology (ICCT), pp. 1184–1189 (2017)
11. Luo, Y., et al.: Forest fire detection using spiking neural networks. In: Proceedings of the 15th ACM International Conference on Computing Frontiers, pp. 371–375 (2018)
12. Liu, J., et al.: Financial data forecasting using optimized echo state network. In: International Conference on Neural Information Processing (ICONIP), pp. 1–11 (2018)
13. Liu, J., Huang, X., Luo, Y., Yi, C.: An energy-aware hybrid particle swarm optimization algorithm for spiking neural network mapping. In: International Conference on Neural Information Processing (ICONIP), pp. 805–815 (2017)
14. Kira, K., Rendell, L.A.: A practical approach to feature selection. Mach. Learn. Proc. **1992**, 249–256 (1992)
15. Abadi, M., et al.: TensorFlow : a system for large-scale machine learning. In: 12th USENIX Symposium on Operating Systems Design and Implementation, pp. 265–283 (2016)
16. Verma, M., Asmita, S., Shukla, K.K.: A regularized ensemble of classifiers for sensor drift compensation. IEEE Trans. Neural Netw. Learn. Syst. **16**(5), 1310–1318 (2016)

SpringNet: Transformer and Spring DTW for Time Series Forecasting

Yang Lin[1(✉)], Irena Koprinska[1(✉)], and Mashud Rana[2]

[1] School of Computer Science, University of Sydney, Sydney, NSW, Australia
ylin4015@uni.sydney.edu.au, irena.koprinska@sydney.edu.au
[2] Data61, CSIRO, Sydney, Australia
mdmashud.rana@data61.csiro.au

Abstract. In this paper, we present SpringNet, a novel deep learning approach for time series forecasting, and demonstrate its performance in a case study for solar power forecasting. SpringNet is based on the Transformer architecture but uses a Spring DTW attention layer to consider the local context of the time series data. Firstly, it captures the local shape of the time series with Spring DTW attention layers, dealing with data fluctuations. Secondly, it uses a batch version of the Spring DTW algorithm for efficient computation on GPU, to facilitate applications to big time series data. We comprehensively evaluate the performance of SpringNet on two large solar power data sets, showing that SpringNet is an effective method, outperforming the state-of-the-art DeepAR and LogSparse Transformer methods.

Keywords: Time series forecasting · Solar power forecasting · Transformer · Dynamic Time Warping · Deep learning · Dynamic programming

1 Introduction

Time series forecasting is an important task in many domains, e.g. forecasting stock prices, sales and spending, traffic flow, electricity consumption and generated solar power. The traditional autoregressive and state-space models fit each of the related time series independently and require expertise in manually selecting trend and seasonality which limits their applicability [1].

Recently, deep learning methods have been investigated as an alternative. Salinas et al. [2] proposed DeepAR, a probabilistic forecasting model based on sequence-to-sequence Long Short Term Memory (LSTM) neural networks. However, the vanishing and exploding gradient problem of LSTM makes training difficult, especially when processing long sequences. The Transformer architecture [3] has been recently proposed to model sequential data with attention mechanism only, without any recurrent or convolutional layers. Its main advantage is the ability to access any part of the historical sequence regardless of distance. Li et al. [1] proposed the LogSparse Transformer, a modification of the Transformer

H. Yang et al. (Eds.): ICONIP 2020, LNCS 12534, pp. 616–628, 2020.
https://doi.org/10.1007/978-3-030-63836-8_51

for time series forecasting, aiming to overcome the problem of locality-agnostics and memory bottleneck by employing convolutional attention layers and sparse attention mechanism. However, the LogSparse Transformer may have limited ability to capture the time series shape information because the shape could be distorted after being projected into latent space with lower dimension after convolutions. For example, two series with similar shapes that are shifted or scaled over the time axis may have completely different results after convolutions.

On the other hand, Dynamic Time Warping (DTW) [4] is a classic trajectory similarity measure that can handle temporal distortions, such as shifting and scaling in the time axis. It has also been used in sequential modelling tasks, including time series analysis [5–7]. The main drawback of DTW is its high complexity, due to the non-parallelizable characteristics of dynamic programming, which limits its applicability. A variation of DTW is the Spring DTW algorithm [8], which identifies subsequences in a data stream, that are similar to a given template. We propose to use the String DTW algorithm as an attention mechanism for Transformer architectures.

In this paper, we present a new deep learning approach, SpringNet, for time series forecasting. SpringNet is based on the Transformer architecture but utilizes Spring DWT attention layers that measure the similarities of query-key pairs of sequences. We assume that attending to the shape of time series patterns directly would be beneficial to achieve accurate prediction. SpringNet is the first Transformer that attends to the shape of time series patterns directly with Spring attention layers. We also propose a batch version of the Spring DTW algorithm for GPU acceleration, by identifying a batch of matched subsequences concurrently.

The effectiveness of the proposed SpringNet approach is comprehensively evaluated for solar power forecasting using two big data sets. The results show that SpringNet outperformed the state-of-the-art deep learning models DeepAR and LogSparse Transformer and the persistence baseline, especially under random fluctuations of data. The batch version of the Spring DWT algorithm in SpringNet was also found to be significantly faster than the original Spring DWT.

2 Case Study: Solar Power Forecasting

Solar photovoltaic (PV) power is a cost-effective and sustainable electricity source. However, the power output is highly variable as it depends on the weather conditions. PV power forecasting is needed to quantify the uncertainty associated with power generation and ensure the successful integration of PV systems into the electricity grid. Previous work on solar power forecasting includes statistical methods such as autoregressive integrated moving average [9] and machine learning methods such as neural networks [10] and support vector regression [11].

2.1 Data Sets

We use two solar power data sets: Sanyo[1] and Hanergy.[2] They contain solar power generation data from two PV plants in Alice Springs, Northern Territory, Australia. The Sanyo dataset contains solar power data from 01/01/2011 to 31/12/2017, and the Hanergy dataset - from 01/01/2011 to 31/12/2016.

We also collected weather data from nearby weather stations. The weather data includes *temperature, humidity, global and diffuse radiation.* In addition, we also prepared weather forecast data based on historical weather data. Specifically, weather forecasts are formed by adding 20% Gaussian noise to the observed weather data since we did not have access to weather forecast data. Both solar power and weather data are aggregated to 30-min intervals by taking the average , and only the data between 7 am and 5 pm is considered [10]. The missing values in the raw data (1.25% in Sanyo and 3.24% in Hanergy) are filled using the multivariate imputation by chained equations algorithm [12]. Both solar power and weather data are normalized to have zero mean and unit variance.

In addition to solar power and weather data, we also consider the calendar information as inputs to the prediction models. The calendar information (time features) we consider include *month, hour-of-the-day, minute-of-the-hour* [1,2].

2.2 Problem Statement

We use the solar power for day d with associated covariate information for days d (weather and calendar features) and day $d+1$ (weather forecast and calendar features) to forecast the solar power for the next day $d+1$. Specifically, given is a set of N: 1) solar power time series $\{\mathbf{PV}_{i,1:T_l}\}_{i=1}^{N}$, where $\mathbf{PV}_{i,1:T_l} \triangleq [\mathrm{PV}_{i,1}, \mathrm{PV}_{i,2}, ..., \mathrm{PV}_{i,T_l}]$, T_l is the input sequence length, $T_l = 20$ (1 day), and $\mathrm{PV}_{i,t} \in \Re$ is the ith PV power generated at time t; 2) associated time-based multi-dimensional covariate vectors $\{\mathbf{X}_{i,1:T_l+T_h}\}_{i=1}^{N}$, where $T_h = 20$ (1 day) denotes the length of forecasting horizon. The covariates for our case study include: weather $\{\mathbf{W1}_{i,1:T_l}\}_{i=1}^{N}$, weather forecasts $\{\mathbf{WF}_{i,T_l+1:T_l+T_h}\}_{i=1}^{N}$ and calendar features $\{\mathbf{Z}_{i,1:T_l+T_h}\}_{i=1}^{N}$. Our goal is to predict the PV power for the the next T_h time steps after T_l, i.e. $\{\widehat{\mathbf{PV}}_{i,T_l+1:T_l+T_h}\}_{i=1}^{N}$.

The overall structure of SpringNet is illustrated in Fig. 1. The model's input and output arrangement is the same as that of DeepAR [2] and LogSparse Transformer [1]. At each time step, the inputs of the model are the **PV** values at its previous time step and the covariates **X** (weather **W1** and calendar **Z** features) at the current time step. At the first time step of the encoder, $\mathrm{PV}_{i,0}$ value is initialized as zero. The decoder uses weather forecasts **WF** features instead of weather features **W1**. The decoder is autoregressive, which takes the observation at the previous time step $\widehat{\mathrm{PV}}$ as an input at the current time step.

[1] http://dkasolarcentre.com.au/source/alice-springs/dka-m4-b-phase.
[2] http://dkasolarcentre.com.au/source/alice-springs/dka-m16-b-phase.

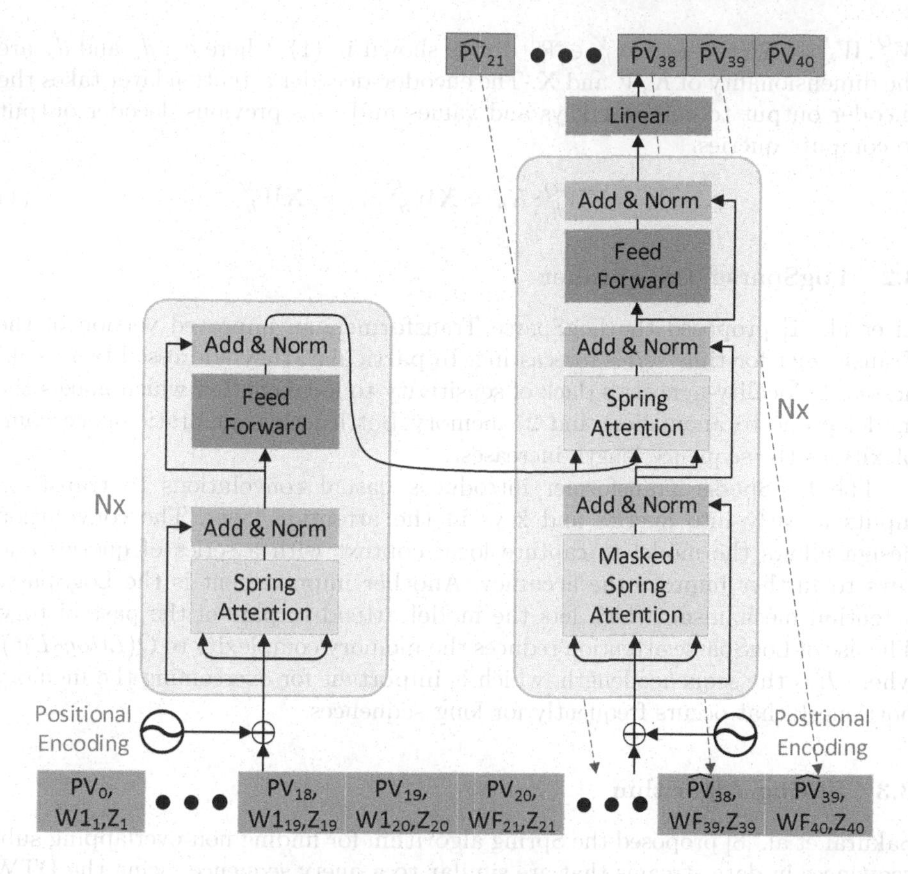

Fig. 1. Summary of SpringNet

3 Background

3.1 Transformer

The Transformer [3] is a new architecture which uses only attention mechanism for processing sequential data. Compared to the widely used sequence models, it does not use any recurrent or convolutional layers, but keeps the encoder-decoder design and uses stacked multi-head self-attention and fully connected layers, which could run in parallel.

Each layer of the encoder contains a multi-head self-attention layer followed by a feed-forward layer, while that of decoder contains an additional encoder-decoder attention layer between the self-attention layer and the feed-forward layer. The multi-head attention uses scaled dot product with the queries Q, keys K and values V. The queries, keys and values are obtained from previous layer output for self-attention and encoder output for encoder-decoder attention. Given input $\mathbf{X}$ of the attention layer, the hth query, key and value matrix can be computed through linear projections with the trainable weights:

$W_h^Q, W_h^K \in \Re^{d_x \times d_k}$ and $W_h^V \in \Re^{d_x \times d_v}$ as shown in (1), where d_k, d_v and d_x are the dimensionality of K, V and $\mathbf{X}$. The encoder-decoder attention layer takes the encoder output to compute keys and values and uses previous decoder output to compute queries.

$$Q_h = \mathbf{X}W_h^Q; K_h = \mathbf{X}W_h^K; V_h = \mathbf{X}W_h^V \quad (1)$$

3.2 LogSparse Transformer

Li et al. [1] proposed the LogSparse Transformer, an improved version of the Transformer for time series forecasting. In particular, they addressed two weaknesses: 1) locality-agnostics (lack of sensitivity to local context which makes the model prone to anomalies) and 2) memory bottleneck - quadratic space complexity as the sequence length increases.

The LogSparse Transformer introduces casual convolutions to transform inputs linearly into queries and keys in the attention layer. The convolution design allows the model to capture local context with a series of queries and keys to further improve the accuracy. Another improvement is the LogSparse attention mechanism, which lets the model attend to part of the past history. The use of LogSparse attention reduces the memory complexity to $O(L(log_2 L)^2)$, where L is the sequence length, which is important for overcoming the memory bottleneck that occurs frequently for long sequences.

3.3 Spring Algorithm

Sakurai et al. [8] proposed the Spring algorithm for finding non-overlapping subsequences in data streams that are similar to a query sequence, using the DTW distance measure. Compared to the naive DTW subsequence searching method, which has $O(n^3 m)$ time complexity (where n is the length of the sequence and m is the length of the query sequence), the Spring algorithm with star padding and Subsequence Time Warping Matrix (STWM) is significantly faster and requires $O(m)$ space and $O(m)$ time per time-tick.

The Spring algorithm has attracted significant interest due to its effectiveness and efficiency. Cai et al. [7] proposed DTWNet which uses the Spring algorithm as a feature extractor for time series classification.

4 Proposed Approach: SpringNet

4.1 Motivation and Novelty

The daily pattern in solar power data could vary significantly over time because it is highly sensitive to weather conditions. For examples, although the overall solar power pattern repeats everyday (an inverse U shape with a peak in the middle of the day), fluctuations caused by weather conditions could occur several times during the day, significantly changing this pattern. The time of occurrence and the magnitude of these fluctuations vary substantially.

Algorithm 1: SpringNet attention algorithm

input : $Q, K \in \Re^{n_{batch} \times n_{head} \times L \times d_k}$, $V \in \Re^{n_{batch} \times n_{head} \times L \times d_v}$, $L_{sub} \in \mathbb{Z}^+$, $\mathcal{F} : (\Re^{N \times L_{sub} \times d_k}, \Re^{N \times L \times d_k}, \mathbb{Z}^+) \rightarrow \Re^{N \times n_{top} \times 2}$, $n_{top} \in \mathbb{Z}^+$

output: $O \in \Re^{n_{batch} \times n_{head} \times L \times d_v}$

1 init: $Q_{sub} \triangleq \Re^{L \times n_{batch} \times n_{head} \times L_{sub} \times d_k}$; $K_{temp} \triangleq \Re^{L \times n_{batch} \times n_{head} \times L \times d_k}$; $V_{sub} \triangleq \Re^{n_{batch} \times n_{head} \times n_{top} \times d_v}$; $D \triangleq \Re^{n_{batch} \times n_{head} \times n_{top} \times 1}$; $N = L \times n_{batch} \times n_{head}$;

`// Preprocessing`

2 **for** $l \leftarrow 1$ **to** L **do**

3 $\quad Q_{sub}[l, :, :, :, :] = Q[:, :, l : l + L_{sub}, :]$;

4 $\quad K_{temp}[l] = K$;

5 **end**

6 reshape Q_{sub} to $\Re^{N \times L_{sub} \times d_k}$;

7 reshape K_{temp} to $\Re^{N \times L \times d_k}$;

`// Batch Spring DTW function`

8 $matrix \leftarrow \mathcal{F}(Q_{sub}, K, n_{top})$;

9 $V_{sub}, D \leftarrow V, matrix$;

10 $O = \text{softmax}(D) \times V_{sub}$;

The Transformer cannot capture the local context of such time series data with its canonical self-attention layers [1]. While the LogSparse Transformer is able to capture local context, it could miss the time series shape information during convolutions.

On the other hand, DTW was designed to measure time series similarity with temporal distortions. Motivated by the success of recent works that combine DTW with deep learning [5–7], we leverage DTW to compute attention scores.

The Spring DTW algorithm for subsequences matching is effective and efficient (compared to the naive DTW) but is not suitable to be implemented on GPU because it processes one pair of sequences at a time. To the best of our knowledge, recent applications still use the original version of the Spring algorithm and proceed on a sample-by-sample basis.

Below we present our proposed approach, SpringNet, which assumes that it is beneficial for forecasting models to capture series shape information, especially when the repeatable fluctuations occur frequently. SpringNet is a Transformer architecture that attends to the shape of time series patterns directly by using the SpringNet attention algorithm. The SpringNet attention algorithm is based on the Spring subsequence matching algorithm but allows to process a batch of query-key sequence pairs concurrently and is thus suitable for GPU computation.

4.2 Model Architecture

We adopt the general Transformer architecture but replace the multi-head attention layer of Transformer or the convolutional attention of LogSparse Transformer with our Spring attention layer, as shown in Fig. 1. The SpringNet attention algorithm is illustrated in Algorithm 1, where L_{sub} is the query subsequence

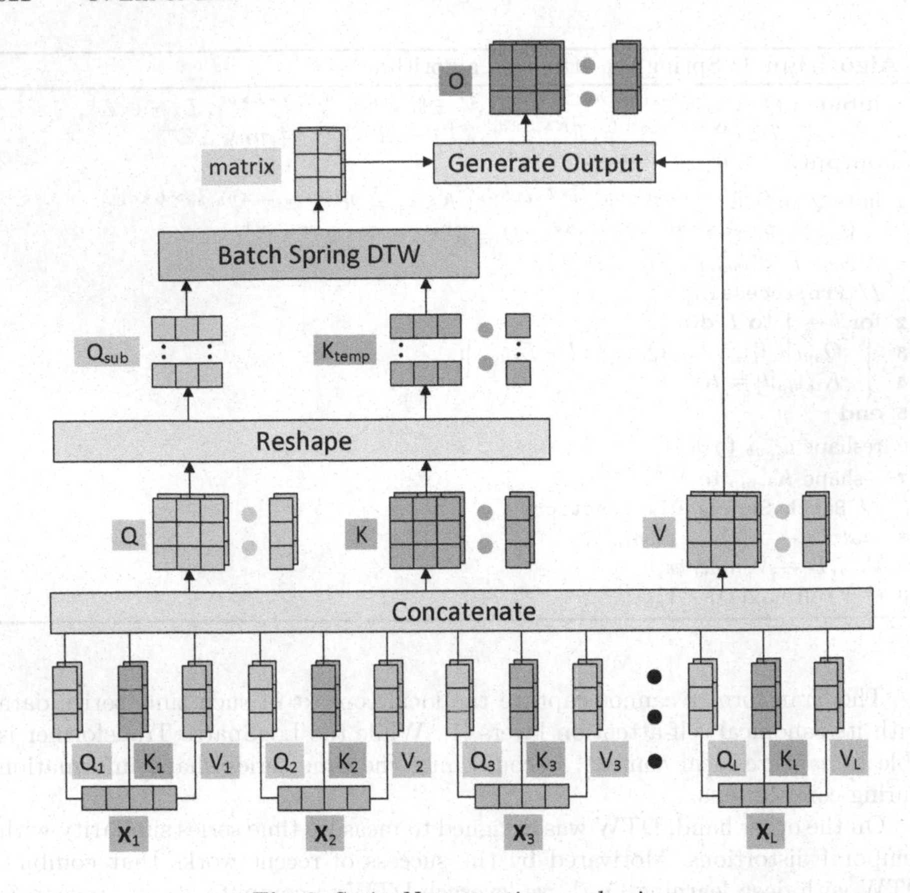

Fig. 2. SpringNet attention mechanism

length, n_{top} is the number of best-matched subsequences from the keys and $\mathcal{F}$ denotes the batch Spring DTW function as shown in Algorithm 2. Instead of mapping a piece of subsequence into a query and key like the convolutional attention, Spring attention transforms single time point into individual query and key and observes the shape of query and key series pattern. Spring attention identifies the subsequences of keys that match query series.

In lines 2 to 7 of Algorithm 1, we preprocess the queries and keys by extracting subsequences from queries, repeating keys and reshaping the tensors as the input of batch Spring DTW function $\mathcal{F}$. $Q[:,:,l : l + L_{sub},:]$ in line 3 indicates the extraction of a tensor from Q with all elements in the $1st$, $2nd$ and $4th$ axis and the elements from the lth to the $l+L_{sub}$ position in the $3rd$ axis of Q. $\mathcal{F}$ produces the *matrix* that stores the DTW distance D of n_{top} best-matched subsequences from the key series and their ending indexes. Then, values which indexes are stored in *matrix* are extracted as V_{sub}. Finally, we use DTW distance D and matched values V_{sub} to compute the Spring attention output O.

Whenever n_{top} is less than the number of keys, the Spring attention is sparse. The n_{top} controls the number of time steps that are attended to. Thus, the attention could be sparser and memory usage could be lower with smaller n_{top}. L_{sub} controls the Spring attention locality and SpringNet with short L_{sub} tends to capture short-term pattern.

Figure 2 shows the feedforward dataflow of the SpringNet attention mechanism. Similarly to the Transformer, SpringNet extracts queries, keys and values from individual inputs and concatenates them as tensors. Then, these queries and keys are reshaped to Q_{sub} and K_{temp} (see lines 2 to 7 of Algorithm 1) and passed to the Batch Spring DTW algorithm (function $\mathcal{F}$). Finally, the values and *matrix* computed by the Batch Spring DTW algorithm are used to generate the attention output O (see lines 9 to 10 of Algorithm 1).

4.3 Batch Spring Attention

We follow the design of the Spring algorithm for subsequence mining but propose a batch version in Algorithm 2. The batch Spring attention algorithm achieves the same functionality as the Spring algorithm (see Sect. 3.3). The advantage of our Spring attention algorithm is the ability to process multiple query-key pairs concurrently on GPU, in order to speed up the Spring attention layer's feedforward speed.

We create multiple matrices and arrays to store temporary variables: 1) D_{prev} and D_{now} store the previous and current DTW distance, S_{prev} and S_{now} store the previous and current starting position of all samples; the four matrices come from the STWM; 2) *Dis* stores the DTW distance of matched subsequences, J_e stores the ending position of matched subsequences; both arrays are used to update the *matrix* via function updateMatrix, which is the output of batch Spring DTW function. The function updateMatrix ensures *matrix* only keeps the *Dis* and J_e of n_{top} best-matched subsequences.

Our Algorithm 2 has two nested loops starting at line 4 and 5 to identify all subsequences in parallel, while the original Spring DTW algorithm would have a third outer loop to iterate through all subsequence templates. In each Spring attention layer, there are $n_{batch} \times n_{head} \times L$ subsequence templates (Q_{sub}). In lines 5 to 15 of Algorithm 2, we compute the DTW distance and subsequence starting position of the subsequence point at the *jth* time step. The candidate subsequences are identified and saved in lines 16 to 21. Finally, we update the *Dis*, J_e and STWM to proceed to the next time step in lines 22 to 29.

5 Experimental Setup

All prediction models were implemented using PyTorch 1.5 and CUDA 10.1. For both data sets, we use the last year as test set, the second last year as validation set for hyperparameter tuning, and the remaining data (5 years for Sanyo and 4 years for Hanergy) as training set.

Algorithm 2: Batch Spring DTW algorithm

input : $Q_{sub} \in \Re^{N \times L_{sub} \times d_k}$, $K_{temp} \in \Re^{N \times L \times d_k}$, $n_{top} \in \mathbb{Z}^+$
output: $matrix \in \Re^{N \times n_{top} \times 2}$

```
1  init: D_prev, D_now, S_prev, S_now ≜ ℜ^{N×L_sub}; J_e, Dis, check ≜ ℜ^N; k ∈ ℤ+;
       matrix ≜ ℜ^{N×n_top×2}.
2  D_prev[:, :], J_e[:], Dis[:], matrix[:, :, :] = ∞;
3  D_now[:, :], S_prev[:, :], S_now[:, :] = 0;
4  for j ← 1 to L do
       // Update subsequences DTW distance and starting position
5      for i ← 1 to L_sub do
6          if i == 1 then
7              D_now[:, i] = ||K_temp[:, j, :] − Q_sub[:, i, :]||;
8              S_now[:, i] = j;
9          else
10             D_now[:, i] = ||K_temp[:, j, :] − Q_sub[:, i, :]||+
                   min(D_now[:, i − 1], D_prev[:, i], D_prev[:, i − 1], axis = 2);
11             Distance ← concatenate(D_now[:, i − 1], D_prev[:, i],
                   D_prev[:, i − 1]) along the 3rd axis;
12             Start ← concatenate(S_now[:, i − 1], S_prev[:, i],
                   s_prev[:, i − 1]) along the 3rd axis;
13             S_now[:, i] = Start[:, :, argmin(Distance, axis = 3)];
14         end
15     end
       // Identify new matched subsequences
16     check[:] = 0;
17     for i ← 1 to L_sub do
18         check[D_now[:, i] >= Dis[:] ∩ S_now[:, i] > J_e[:]]+ = 1;
19     end
20     index = (check == L_sub);
       // Store the information of new matched subsequences
21     matrix[index] = updateMatrix(matrix[index], Dis[index], J_e[index]);
       // Reset for the incomig time point
22     Dis[index] = ∞;
23     index ← duplicate index along the 2nd dimension for L_sub times;
24     D_now[S_now <= J_e ∪ index] = ∞;
25     index = (D_now[:, L_sub] <= Dis[:]);
26     Dis[index] = D_now[index, L_sub];
27     J_e[index] = j;
28     D_prev = D_now;
29     S_prev = S_now;
30 end
```

We use three models for comparison: two state-of-the-art autoregressive deep learning models (DeepAR and LogSparse Transformer) and a persistence model. DeepAR [2] is a widely used sequence-to-sequence forecasting model, while the LogSparse Transformer [1] is a recently proposed variation of the Transformer architecture for time series forecasting; we denote it as "Transformer" in Tables 1 and 2. The persistence baseline is a typical baseline in forecasting and considers the PV power output of the previous day as the prediction for the next day.

Table 1. Hyperparameters for all models

Model	δ	d_{hid}	n_{layer}	$d_k\&d_v$	n_{head}	L_{sub}	n_{batch}
DeepAR (Sanyo)	0	8	3	–	–	–	256
Transformer (Sanyo)	0.2	12	3	6	3	–	256
SpringNet (Sanyo)	0	24	2	6	3	3	256
DeepAR (Hanergy)	0.2	16	4	–	–	–	512
Transformer (Hanergy)	0.2	12	3	4	2	–	512
SpringNet (Hanergy)	0	12	2	4	2	3	512

All deep learning models are optimized by mini-batch gradient descent with the Adam optimizer, with variable learning rate (initial $\lambda = 0.005$, decay factor $= 0.5$) and maximum number of epochs 100. We used Bayesian optimization for hyperparameter search with a maximum number of iterations 20.

For all models, the dropout rate δ is chosen from $\{0, 0.1, 0.2\}$, the hidden layer dimension size d_{hid} and number of layers n_{layer} are chosen from $\{8, 12, 16, 24, 32\}$ and $\{2, 3, 4, 5\}$. For LogSparse Transformer and SpringNet, the query and value's dimension size $d_k\&d_v$ and number of heads n_{head} are chosen from $\{4, 6, 8, 12\}$ and $\{2, 3, 4, 6, 8\}$. For LogSparse Transformer, we use restart attention range of 20 and local attention range of 3. For SpringNet, the number of best-matched subsequences n_{top} is set to 5 and the subsequence length L_{sub} is chosen from $\{2, 3, 4\}$.

The selected best hyperparameters for all models are listed in Table 1 and used for the evaluation on the test set.

6 Results and Discussion

Table 2 shows the Mean Absolute Error (MAE) and Root Mean Squared Error (RMSE) of all models for the two data sets. The best result for each metric and data set is highlighted in bold. On both data sets, all deep learning models outperform the baseline model. SpringNet is the most accurate model in terms of MAE on both data sets, followed by LogSparse Transformer and DeepAR. In terms of RMSE, SpringNet is the best performing model for Sanyo and the second best for Hanergy. Overall, SpringNet is the best performing model which shows the effectiveness of the proposed approach which attends to the shape of the time series patterns directly.

In addition, the Batch Spring DTW algorithm significantly speeds up the model training. On the same Tesla P100-16GB GPU, the feedforward process

Table 2. Accuracy of all models

Model	Sanyo		Hanergy	
	MAE	RMSE	MAE	RMSE
Persistence	0.522	0.985	0.703	1.174
DeepAR	0.276	0.381	0.370	0.505
Transformer	0.267	0.384	0.360	**0.478**
SpringNet	**0.258**	**0.380**	**0.358**	0.494

of a single Spring DTW layer (Algorithm 2) takes 2.081 s and 2.082 s per batch on Sanyo and Hanergy set respectively. In comparison, for the original Spring algorithm, the processing time is 12.057 h and 18.426 h per batch on the Sanyo and Hanergy data sets correspondingly. The running time of our Spring DTW layer is relatively stable with respect to the batch size, as long as it does not exceed the GPU memory constraint, while that of the original Spring algorithm without parallelism increases as the number of series pairs increases.

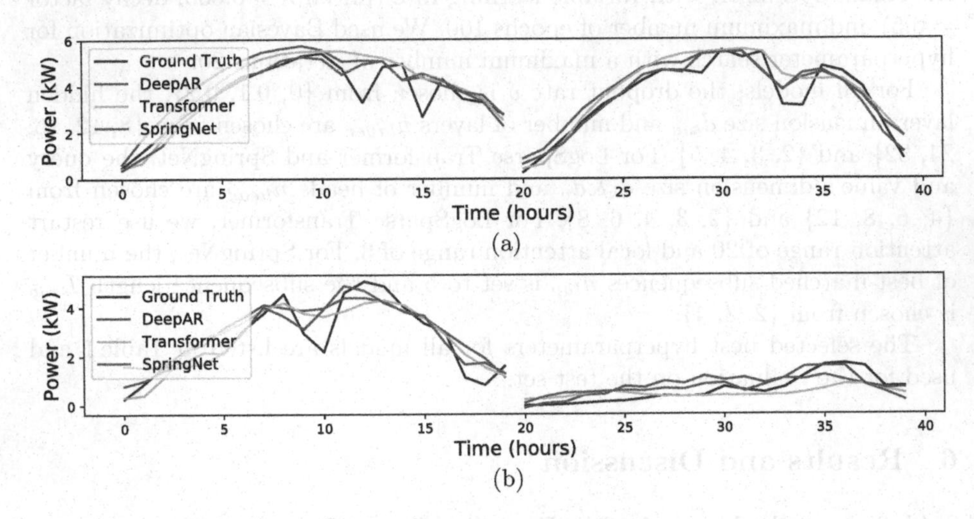

Fig. 3. Actual vs predicted data: (a) Sanyo dataset and (b) Hanergy dataset

Figure 3 plots the actual and predicted solar power data for two days, for both data sets. The actual data shows different levels of fluctuations, higher for Sanyo and lower for Hanergy. We can see that the predictions provided by SpringNet are the closest to the ground truth. SpringNet also forecasts the data fluctuations much better than DeepAR and LogSparse Transformer which tend to produce smooth curves. This shows that SpringNet is able to capture the time series pattern and deal well with repeated fluctuations.

Hence, based on the results, we conclude that SpringNet is a promising method for solar power forecasting - it outperforms all models used for comparison and is more robust to fluctuations. The attention layer in SpringNet helps to capture repeatable fluctuation patterns and provide accurate forecasts especially in the presence of fluctuations.

7 Conclusions

In this paper, we present SpringNet, a new deep learning based approach for time series forecasting, and demonstrate its performance in a case study for solar PV power forecasting. SpringNet is based on the LogSparse Transformer architecture but uses Spring DTW attention layers. We propose the use of Spring attention to overcome the weakness of LogSparse Transformer to effectively capture the time series shape information. In addition, we propose the Batch Spring DTW algorithm to speed up the feedforward operation of the Spring DTW attention algorithm. SpringNet is a generic time series forecasting approach and can be used in different domains. We present a case study for solar power forecasting, using two big data sets from solar plants located in Australia. The results show that SpringNet outperforms the state-of-the-art deep learning forecasting methods DeepAR and LogSparse Transformer, and also a persistent baseline used for comparison. Our experiments suggest that SpringNet can capture shape information from trajectories and make robust predictions. In summary, the results validate our hypothesis that attending to the shape of time series pattern directly is beneficial. We also found that the Batch Spring attention algorithm was significantly faster than the sequential version.

Hence, we conclude thatSpringNet with Batch Spring Attention mechanism is a promising method for time series forecasting.

References

1. Li, S., et al.: Enhancing the locality and breaking the memory bottleneck of Transformer on time series forecasting. In: Conference on Neural Information Processing Systems (NeurIPS) (2019)
2. Salinas, D., Flunkert, V., Gasthaus, J., Januschowski, T.: DeepAR: probabilistic forecasting with autoregressive recurrent networks. Int. J. Forecast. **36**, 1181–1191 (2020)
3. Vaswani, A., et al.: Attention is all you need. In: Conference on Neural Information Processing Systems (NeurIPS) (2017)
4. Sakoe, H., Chiba, S.: Dynamic programming algorithm optimization for spoken word recognition. IEEE Trans. Acoust. Speech Signal Process. **26**, 43–49 (1978)
5. Cuturi, M., Blondel, M.: Soft-DTW: a differentiable loss function for time-series. In: International Conference on Machine Learning (ICML) (2017)
6. Guen, V.L., Thome, N.: Shape and time distortion loss for training deep time series forecasting models. In: Conference on Neural Information Processing Systems (NeurIPS) (2019)

7. Cai, X., Xu, T., Yi, J., Huang, J., Rajasekaran, S.: DTWNet: a dynamic time warping network. In: Conference on Neural Information Processing Systems (NeurIPS) (2019)
8. Sakurai, Y., Faloutsos, C., Yamamuro, M.: Stream monitoring under the time warping distance. In: International Conference on Data Engineering (ICDE) (2007)
9. Pedro, H.T., Coimbra, C.F.: Assessment of forecasting techniques for solar power production with no exogenous inputs. Solar Energy **86**, 2017–2028 (2012)
10. Lin, Y., Koprinska, I., Rana, M., Troncoso, A.: Pattern sequence neural network for solar power forecasting. In: International Conference on Neural Information Processing (ICONIP) (2019)
11. Rana, M., Koprinska, I., Agelidis, V.G.: 2D-interval forecasts for solar power production. Solar Energy **122**, 191–203 (2015)
12. Azur, M., Stuart, E., Frangakis, C., Leaf, P.: Multiple imputation by chained equations: what is it and how does it work? Int. J. Methods Psychiatr. Res. **20**, 40–49 (2011)

U-Sleep: A Deep Neural Network for Automated Detection of Sleep Arousals Using Multiple PSGs

Shenglan Yang[1], Bijue Jia[1], Yao Chen[1], Zhan ao Huang[1], Xiaoming Huang[2], and Jiancheng Lv[1](✉)

[1] College of Computer Science, Sichuan University, Chengdu 610065, People's Republic of China
{yangshenglan,jiabijue,chenyaoscu,huangzhanao}@stu.scu.edu.cn, lvjiancheng@scu.edu.cn

[2] CETC Cyberspace Security Research Institute Co., Chengdu 610041, China
apride@gmail.com

Abstract. Sleep disorders can seriously affect human health. Most of the previous studies focused on sleep disorders of apnea, but few on non-apnea. This type of sleep disorder is complex and difficult to detect by traditional methods. In this paper, a physiological time series segmentation network U-Sleep based on deep learning is proposed to analyze these sleep disorders. U-Sleep is a time series convolution network based on U-Net architecture. U-Sleep uses the sequence to sequence input-output mode to map multiple complete original polysomnograms to a single tag sequence. This enables our model to automatically learn the variable interaction between different signals and any related time dependence, and automatically extract such arousal features from the rich physiological time series. We conducted three-fold cross-validation, and use the ensemble model strategy to get the final detection results. Experiments on the datasets of PhysioNet show that the average performance of the final model is: accuracy(0.886), F1(0.892), AUROC(0.916), AUPRC(0.797), SE(0.804), and AI(6.240).

Keywords: Neural networks · LSTM · Sleep arousal · Polysomnography

1 Introduction

Sleep quality is closely related to human health. Poor sleep quality can easily lead to poor mental state, which can lead to health problems, especially obesity, depression, liver disease, mental diseases, and some cardiovascular and cerebrovascular diseases with high risk of death. Sleep quality will decrease with the frequent occurrence of sleep arousal, that is, temporary interruptions of wakefulness into sleep or spontaneous increase of the vigilance level.

H. Yang et al. (Eds.): ICONIP 2020, LNCS 12534, pp. 629–640, 2020.
https://doi.org/10.1007/978-3-030-63836-8_52

Sleep disorder mainly refers to the arousals caused by the interruption of the sleep process. Apnea and hypoventilation are the most common sleep interruptions. This kind of sleep disorder will have a more direct impact on human health and is a more serious disease, so there are many studies. Sleep interruption of non-apnea includes pain, bruxism, insomnia, muscle twitches, vocalization, snoring, periodic leg movements, and respiratory-related diseases (such as Cheyne Stokes breathing, respiratory disorders, etc.). There are few studies on this kind of sleep disorder, and the main reasons are as follows: first, this kind of research is expensive and difficult to detect with traditional methods; second, compared with apnea, artificial detection of such sleep arousals has been shown to have a lower scoring reliability [3].

Polysomnography (PSG) is widely used in the sleep laboratory. The detection of sleep disorders is usually to mark the arousal caused by sleep disorders, which is usually done manually by sleep experts through several periods of PSG records. This method has obvious defects: first, the task is very tedious and time-consuming; second, the knowledge and experience of the rater will have an important impact on the scoring results. Therefore, the development of PSG automatic arousal detection systems in the form of efficient, fast, and reliable algorithms can provide strong help for clinicians.

The main purpose of this paper is to realize the automatic detection of non-apnea disorders and promote the relevant medical research. We propose a feed-forward network which uses forward segmentation to map multiple complete PSGs into a tag sequence. In this paper, we make the following contributions: First, a network with a full convolution encoder-decoder structure is proposed to realize multi-scale feature recognition. Second, a unified preprocessing method for biological signals is proposed, which greatly reduces the workload and solves the problem that it is difficult to analyze multiple biological signals at the same time. third, We input complete sleep records into the model to ensure that the model can capture any relevant time dependence and automatically extract arousal features from abundant physiological time series. Fourth, LSTM uses skip connection structure to add location information and strengthen the determination of arousal boundary points. More discussion about this study will be given in later sections.

The rest of this paper is arranged as follows. Section 2 briefly introduces the related work. The U-Sleep model is introduced in detail in Sect. 3. In Sect. 4, we describe and analyze the experiment in detail, including the data preprocessing and the performance evaluation. Section 5 concludes our work.

2 Related Work

The American Academy of Sleep Medicine (AASM) defines the arousal as an abrupt shift in electroencephalogram (EEG) signal frequency[1]. Besides, AASM mentions that arousal can be associated with an increase in chin electromyographic (EOG) signals [2]. It's hard to analyze events in different biological signals at the same time, Therefore, previous researchers mainly study sleep based on EEG signal and adopt the standard parameter methods for time and frequency domains. Cho combined Support Vector Machine (SVM) with previous work, proposed an automatic arousal algorithm based on time-frequency analysis and SVM classifier using a single channel EEG [6]. Pacheco and Vaz used a k-means classifier after obtaining the frequency and power of the EEG and EMG respectively [10]. Alvarez-Estevez and Moret-Bonillo compared different classification models after selecting intervals based on the frequency from two EEG derivations, and on the amplitude from one submental EMG [11].

Deep learning performs well in sequence problems [4]. Many people use it for sleep research to avoid extracting manual features. Sleep staging is an important preliminary examination in the diagnosis of sleep disorders. In [20], an end-to-end learning method is proposed, which uses the temporal context of each 30 seconds window of data to classify sleep stages without extracting manual features. Similarly, Perslev proposed the U-Time model which implicitly classifies each time point of the input signal and aggregates these classifications at a fixed time interval to achieve the segmentation of the sleep stage [12]. In the detection of sleep disorders, most of the work is conducive to the detection of apnea. Sinam filtered ECG to generate two-dimensional images, and finally used a convolutional neural network based pre-training model (AlexNet) to predict apnea disease [16]. In [13], Kim proposed a recursive neural network sleep arousal detection model with MFCC as feature vector Sanchis used an artificial neural network for pattern recognition of arousal [8].

The above methods have studied sleep from many aspects, but there are still some problems, which are mainly due to the following reasons: first, sleep disorders of non-apnea are difficult to detect because of their numerous types. Second, The changes of PSG are very subtle when sleep disorders occur, these methods do not fully mine the hidden features behind the data. Third, The influence of sleep on human body is multifaceted. Detection from a single or a few biological signals will have a greater impact on the detection results.

3 Methodology

In this work, we propose a sleep disorder detection model U-Sleep based on time series to solve the above three problems. Its inspiration comes from the popular U-Net [14] architecture originally proposed for image segmentation and the long short term memory networks. In this part, we first give some symbols used in this work and define the problem of sleep arousal detection. In the rest of this section, we will introduce the architecture of the model, then discuss the main components of U-Sleep.

3.1 Problem Definition

Let $x \in \mathbb{R}^{s \times \tau}$ be a physiological signal at sampling rate s for τ seconds, article i sleep record can be abstractly expressed as a physiological signal set $X_i = \{x_1, x_2, \cdots, x_C\}$. Where C represents the number of channels. Each channel is a one-dimensional signal with a length of $m = s \times \tau$. Then define e as the frequency we want to output. That is, each label is based on $n = e \times \tau$ sampled points. In other words, our goal is to map X_i to a label of n. Let $\hat{Y}_i$ be the label sequence. The formula of problem definition could be expressed as the following: $\hat{Y}_i = f(X_i, \theta)$, where $X_i \in \mathbb{R}^{C \times m}$, $\hat{Y}_i \in \mathbb{R}^n$, and $f(\cdot)$ is the model with parameter θ we are going to construct in this paper.

3.2 Model Architecture

In this work, we see some similarities between sleep disorder detection and image segmentation. We propose a U-Sleep model variant based on the concept of U-Net which has a good performance in image segmentation to detect sleep disorders. We use the method of sequence to sequence to input a complete sleep record into the network for learning and realize the segmentation of one-dimensional sleep data sequence. The input of U-Sleep is multiple fixed-length PSG signals, each length is m. The network is composed of encoder block, decoder block, and LSTM layer which are shown in Fig. 1.

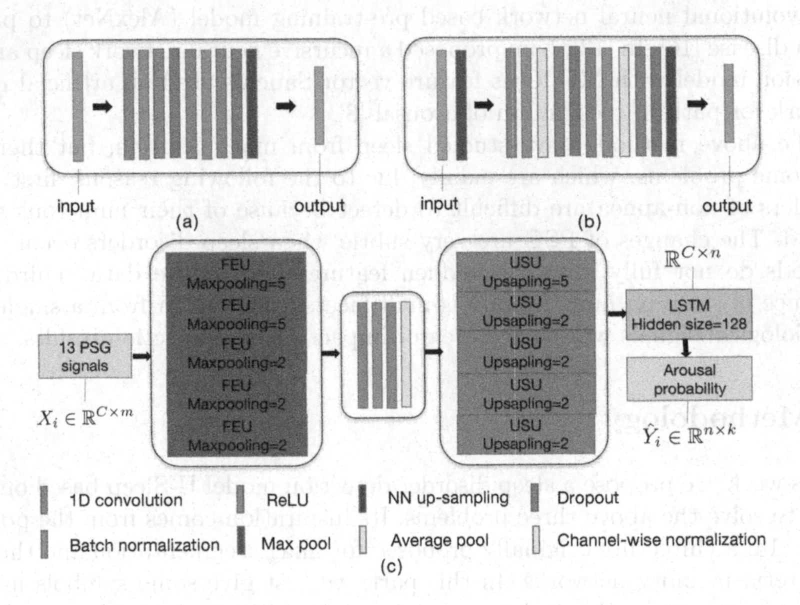

Fig. 1. (a): FEU(Feature Extraction Unit), with max-pooling and no channel-wise normalization. (b): USU(Up-sampling Unit), with up-sampling and channel-wise normalization. (c): Structural overview of the network architecture, including an encoder block(5 FEU), a decoder block(5 USU), and an LSTM block

Encoder Block. The encoder block consists of five similar FEUs, the architecture of each unit illustrated in Fig. 1(a). It takes input tensor $X \in \mathbb{R}^{B\times C\times m}$ where B is the batch size. The process at each unit l is defined as:

$$V^l = f_r(f_r(X^{l-1} * {W_0}^l + {b_0}^l) * {W_1}^l + {b_1}^l) \tag{1}$$

where $*$ denotes the convolutional operation and $f_r(\cdot)$ is Relu activation function. The $W_{0,1}^{1,\ldots,l}$ and the $b_{0,1}^{1,\ldots,l}$ are parameters. Each unit will produce a scale vector $V^l \in \mathbb{R}^{C\times d_l}$, where the width $d^l = m/\prod_1^l pw^l$, and the pw is the pooling window size. After five FEUs, there is an additional convolution unit that will enforce learning abstract features and make the d_l of the V^l correspond to the d^l of the P^l one by one. After this, we transform the out to a feature $V^L = f_r(V^5 * W^L + b^L)$. In the lowest layer, the aggressive down-sampling reduces the input dimensionality by a factor $5 \times 5 \times 2 \times 2 \times 2 = 200$. This can greatly reduce computing and memory requirements. Sleep data is usually recorded overnight, even if the sampling frequency is very low, the data length is very large, and the analysis of sleep arousal needs at least 10 seconds of data before and after observation. The input of complete sleep data is conducive to network learning. The input X can be a complete PSG record or a subset of it. Our model is based on convolution, the input length m is variable. Although the m is adjustable, it must be large enough to meet all the maximum pooling operations, that is, the $d^5 \geq 1$. In theory, this is equivalent to $m_{min} = 200$. Too small m prevents the model from exploiting long-term relationships, thereby reducing performance.

Decoder Block. The decoder block consists of five similar USUs, Each unit has a similar structure to FEU. Before the signal enters the convolution layer, it goes through the nearest-neighbor up-sampling processing [18]. This block receives a collection of tensors V^l. The resulting feature maps $P^l \in \mathbb{R}^{C\times d^l}$ are concatenated with the corresponding V^l at the same scale. The process at each unit l is defined as:

$$P^l = f_r(f_{pw}(f_r(f_{pw}((P^{l-1} \oplus V^{l-1}) * {W_0}^l + {b_0}^l)) * {W_1}^l + {b_1}^l)) \tag{2}$$

where $\oplus$ is the concatenation operation, f_{pw} is the position-wise normalization function. In particular, $V^0 = V^L$. Before using the ReLU activation function, position-wise normalization with a channel-specific affine transform is also applied to batch normalization outputs. With this, we transform output The dilation in the convolution layer is also used to expand the receptive field. The decoder block produces an output $P^L \in \mathbb{R}^{B\times C^L \times n}$.

After the decoder, there is the LSTM consists of one or more self-connected memory cells c^i: the input gate ig_t, forget gate fg_t, and output gate og_t. The multiplication gate of LSTM allows its storage unit to keep information for a long time. This process can be expressed as follows:

$$ig_t = \sigma(W_{ix} \cdot P_t^L + W_{ih} \cdot h_{t-1} + b_i) \tag{3}$$

$$fg_t = \sigma(W_{fx} \cdot P_t^L + W_{fh} \cdot h_{t-1} + b_f) \tag{4}$$

$$og_t = \sigma(W_{ox} \cdot P_t^L + W_{oh} \cdot h_{t-1} + b_o) \quad (5)$$

$$C_t = fg_t * C_{t-1} + ig_t * \widetilde{C_t} \quad (6)$$

$$h_t = og_t * \tanh(C_t) \quad (7)$$

where $\sigma(\cdot)$ is the sigmoid function, $tanh(\cdot)$ is the hyperbolic tangent function. We get the final feature vector $\widetilde{X} \in \mathbb{R}^{B\times C\times n}$. Then, we reduce the dimension of the out feature vector by two fully connected layers, which can be defined as:

$$\hat{X} = f_c(W_2^{fc} \cdot f_c(W_1^{fc} \cdot \widetilde{X} + b_1^{fc}) + b_2^{fc}) \quad (8)$$

Finally, a softmax function is applied to transforms the $\hat{X}$ to a probability vector $\hat{Y}$.

4 Experimental Results and Discussion

4.1 Data Description

We briefly review the characteristics of the dataset in our works. Data were contributed by the Massachusetts General Hospital's (MGH) Computational Clinical Neurophysiology Laboratory (CCNL), and the Clinical Data Animation Laboratory (CDAC), including PSG data from 994 subjects, adding up to the overall 135 GB. Each polysomnographic recording set contains 13 signals, include 6-channel EEG, single-channel EOG, 3-channel EMG, respiratory Airflow, single-lead ECG, and SaO2 signals. All signals are digitized at 200 Hz. According to the provided annotation file, we remarked the target arousal regions by 1 and the non-target arousal regions by 0. More details regarding the dataset and available annotations for different sleep analysis purposes are provided in [17].

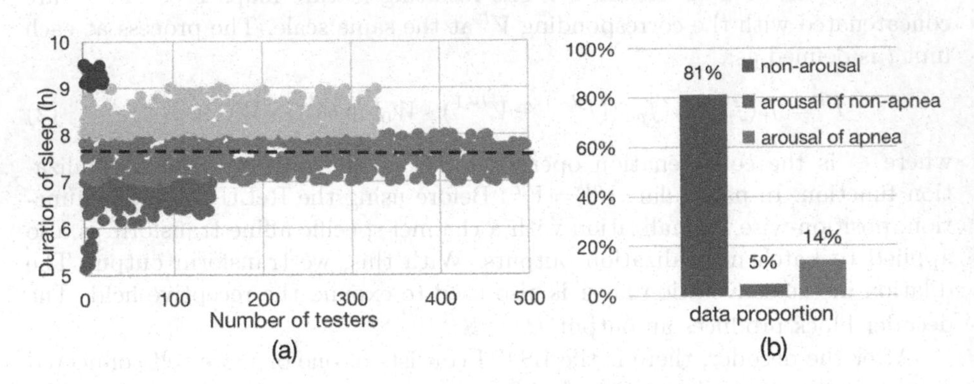

Fig. 2. (a): The sleep duration distribution of the dataset. (b): The data proportion of the dataset. (Color figure online)

According to our statistical analysis of the dataset, we found that most of the subjects slept time is between 7–8 h, and the average time is 7.7 h. The sleep duration distribution of the dataset is shown in Fig. 2. The red dotted line in Fig. 2.(a) represents the average sleep duration, and (b) shows the proportion of positive and negative samples, there is a data skew problem. Moreover, the duration of most arousal events (99.7%) is less than 2 min, and the average event duration is 30 ± 15 s. The arousal events are asymmetrically distributed in all sleep stages of the dataset. The 994 data were randomly split into training set (70.4%), validation set (10%), and test set (19.6%).

4.2 Pre-processing

In this work, the 13 channels PSGs are used. First, all channels are sampled down 50 Hz and removed the DC bias. Secondly, each channel is normalized separately by fast Fourier transform (FFT) convolution. Note that the normalization here is the standardization of the sliding window with 10-min size and 80% overlap, rather than the whole record. According to the guidelines, the baseline of breathing is established within 2 min, that is to say, the size of the baseline window is 2 min. A high percentage overlap ensures that any important variation in breathing is not normalized out. Finally, considering the stability and efficiency of the calculation memory, we unify the sampling time of each record to 7.5 h, that is, each record has $50 \times 7.5 \times 3600 = 1350000$ sampling points. For data longer than 7.5 h, discard the redundant data directly, and fill in zero value for data shorter than 7.5 h. We plot in Fig. 3 an example of 60 s of PSG recording with sleep arousal annotations.

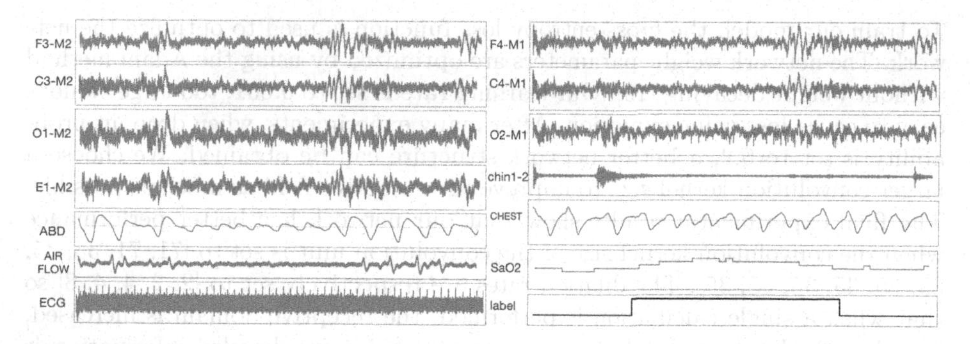

Fig. 3. Time series waveforms given as inputs to the model (blue), and labels determining whether the patient was in a target arousal phase (red). (Color figure online)

4.3 Evaluation Metric

In order to evaluate the effectiveness of the model, we use several performance metrics, includes accuracy, F1 score, area under the accurate recall curve (AUPRC), area under the receiver operating characteristic curve (AUROC), sleep efficiency (SE), and arousal index (AI). They are defined as follows:

$$Accuracy = \frac{TP + TN}{TP + TN + FP + FN}, \qquad F1 - score = \frac{2 \times P \times R}{P + R} \tag{9}$$

$$AUPRC = \sum_j P_j(R_j - R_{j+1}) \tag{10}$$

$$AUROC = \sum_{j \in (TN+FP)} \frac{TP_j}{(TP+FN) \times (TN+FP)} \tag{11}$$

$$SE = \frac{TST}{TRT}, \quad AI = \frac{N_a \times 60}{TST} \tag{12}$$

where

$$P = \frac{TP}{TP+FP}, \qquad R = \frac{TP}{TP+FN} \tag{13}$$

TP: Both the reference and the system detection indicate an arousal state. FN: The reference indicates an arousal state, but the system detects a non-arousal state. FP: The system detects an arousal state but the reference indicates it is not. FN: Both the reference and the system detection indicate a non-arousal state [7]. TST, TRT, N_a correspond to the total sleeping and recording times and number of arousals lasting more than 10s, respectively. These metrics are used to identify sleep disorders and estimate their severity. The higher the AI value, the more serious sleep disorder. Note that this is the gross AUPRC (i.e., for each possible value of j, the P and R are calculated for the entire test database), which is not the same as averaging the AUPRC for each record.

4.4 Hyperparameter Settings

To train our model, the cross-entropy loss function is used to optimize the network. The network weight parameters are optimized by using the Adam method without weight decaying [19]. Our learning rate is set to 0.001, too large value is easy to cause loss value explosion. After many experiments, when dropout probability is set to 0.2, a better network structure can be obtained. We choose a larger convolution kernel size to improve the receptive field as much as possible. The final experimental results show that our network has better performance when the convolution kernel size of the convolution unit is set to [71, 71, 35, 35, 35, 35, 35, 35, 35, 35]. The dilation rates are respectively set to [2, 2, 4, 4, 8] so that when a single calculation is performed, the receptive domain is increased, but the calculation amount is not increased, and more detailed information is retained, which can improve the accuracy of the upper sampling part.

4.5 Ablation Study

We use the AUPRC and AUROC scores as evaluation metrics of the validation set data in the process of network training, to determine whether the network is optimized with the increase of training time. In each epoch, we randomly select 200 complete nocturnal records and send them to the network for processing. If the result of the verification set does not improve or the loss of verification is not further reduced, the learning process is stopped. Note that in our work, the definition of the epoch is different from the general concept. It is the application of 200 whole night records to the network, rather than the whole training set. Moreover, a record may appear in multiple epochs. Figure 4 shows the two metrics evaluated and the loss value as the model trained. In order to improve the generalization ability of the model, the available data are divided into three folds. We repeat the whole training process three times in different folds data, and finally average the detection of the three models to obtain the final ensemble model.

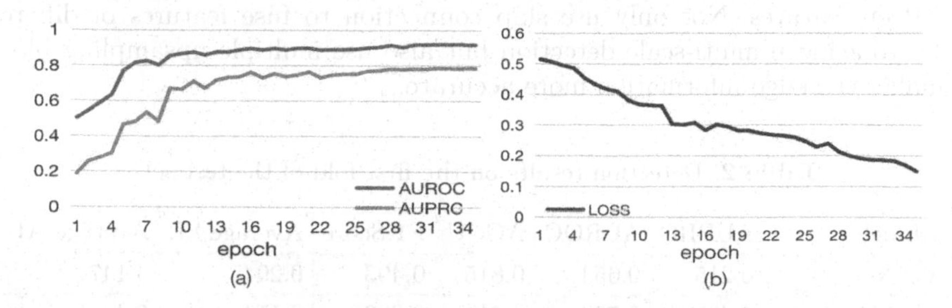

Fig. 4. (a): The AUPRC and AUROC for the training. (b): The loss value for the training. The training employed early stopping to prevent overfitting

Table 1. The performance of a single model and the ensemble model on the test set

Model	AUPRC	AUROC	ACCU	F1-Score	Average SE	Average AI
Model1	0.785	0.912	0.879	0.882	0.792	6.188
Model2	0.731	0.889	0.820	0.841	0.773	6.236
Model3	0.751	0.904	0.872	0.867	0.799	6.213
Ensemble model	**0.797**	**0.916**	**0.886**	**0.892**	**0.804**	**6.240**

According to the details given in Sects. 3.2, our network output frequencies are 10 Hz. The original data is 200 Hz, we do a up-sampling processing of the network output, and upgrade it to the 200 Hz to evaluate the performance of our network in the test data set. The average detection values are measured and shown in Table 1. The actual average SE and the actual SE on three different folds data are 0.838, 0.815, 0.829, 6.398, 6.469, 6.402, respectively. The ensemble model gives the best results.

4.6 Performance Comparison

In order to evaluate the effectiveness of the model, we compared U-Sleep with CNN [9], CNN-LSTM [15], BRNN-LSTM [5]. These models all trained for a fixed number of epochs without early stopping, we think that the direct application of the original implementation is not conducive to our comparison. Therefore, we re-implemented three models and plugged it into our U-Sleep training pipeline. This ensures that the models use the same early stopping mechanisms and PyTorch implementations. As shown in Table 2, U-Sleep shows the best performance in both accuracy and precision. We think the CNN's biggest problem is that it can't model longer sequence information. LSTM can memorize the information of longer steps, so the CNN-LSTM has achieved better results. In BRNN-LSTM, the input of each layer includes the information of all previous layers, a richer description and discrimination of features is formed. U-Sleep combines shallow and deep features. Not only use skip connection to fuse features of different scales to achieve multi-scale detection but also use multiple upsampling blocks to makes the edge information more accurate.

Table 2. Detection results on the first fold of the test set

Model	AUPRC	AUROC	ACCU	F1-Score	Average SE	Average AI
CNN	0.215	0.653	0.815	0.493	0.294	5.147
CNN-LSTM	0.467	0.788	0.811	0.803	0.510	5.468
BRNN-LSTM	0.745	0.881	0.847	0.845	0.782	6.103
U-Sleep	**0.797**	**0.916**	**0.886**	**0.892**	**0.804**	**6.240**

4.7 Discussion

U-Sleep is a new method of mapping multiple complete sleep map sequences to a single label sequence by using the ability of full convolutional encoder-decoder structures. We first preprocess the physiological signals in a unified way, which greatly reduces the workload and solves the problem that it is difficult to analyze multiple signals at the same time. Sleep has many effects on the human body. Detection of sleep disorders from a single or a few biological signals has a great impact on the detection results. The learning of multiple biological signals can effectively improve the accuracy of detection results. The codec structure of the network realizes the multi-scale feature recognition of sleep disorders. The low-resolution information of the encoder after multiple undersampling can provide the context information of the detection target in the whole sequence. This feature is helpful to judge the type of sleep disorder. The decoder combines the features of the encoder to form a thicker feature map. The high-resolution information transferred directly from the encoder to the decoder at the same height through concatenate operation can provide more fine features, such as gradient, for the detection of sleep disorders. This enables our model to automatically learn the variable interactions between different biological signals. At the same time, the fusion of multi-scale features enables the model to fully mine the small changes of PSG. At last, we use the sequence advantage of LSTM to aggregate the output. The input of the complete sequence ensures that the model can capture any relevant time dependence and automatically extract arousal features from the rich physiological time series. By analyzing multiple signals at the same time, our model has achieved good results in the detection of non-apnea.

5 Conclusions

In this study, we propose a feed-forward model to automatically detect the arousals in sleep. The performance and simplicity of the proposed model encourage us to further improve the model, make use of additional features, and build a new deep neural network model for different physiological signals.

The biggest limitation of this study is the single data set. All the records are collected in the same place using the same equipment. 13 PSG channels are interdependent, and the marker data only uses the judgment results of a professional, which is a great test for the generalization ability of the model.

If each record has multiple independent annotations, you can compare the performance measures of different annotators with each other and model outputs. If the model output and multiple manual annotations show great differences, we need to do further work to make this model reach the artificial level, or better understand how an imperfect model can be effectively integrated into the clinical workflow. This work shows the potential of clinical application. With the development of research, it is expected to become an automatic real-time detection system for sleep disorders.

Acknowledgement. This work is supported in part by the National Key Research and Development Program of China under Contract 2017YFB1002201, in part by the National Natural Science Fund for Distinguished Young Scholar under Grant 61625204, and in part by the State Key Program of the National Science Foundation of China under Grant 61836006.

References

1. Rosenberg, R.S., Van Hout, S.: The American academy of sleep medicine inter-scorer reliability program: respiratory events. J. Clin. Sleep Med. **10**(04), 447–454 (2014)
2. Berry, R.B., et al.: Rules for scoring respiratory events in sleep: update of the 2007 AASM manual for the scoring of sleep and associated events. J. Clin. Sleep Med. **8**(05), 597–619 (2012)
3. Engleman, H.M., Douglas, N.J.: Sleep 4: sleepiness, cognitive function, and quality of life in obstructive sleep apnoea/hypopnoea syndrome. Thorax **59**(7), 618–622 (2004)
4. Jia, B., Lv, J., Liu, D.: Deep learning-based automatic downbeat tracking: a brief review. Multimedia Syst. **25**(6), 617–638 (2019). https://doi.org/10.1007/s00530-019-00607-x
5. Ragnarsdóttir, H., Marinósson, B., Finnsson, E., Gunnlaugsson, E., Ágústsson, J. S., Helgadóttir, H.: Automatic detection of target regions of respiratory effort-related arousals using recurrent neural networks. In: 2018 Computing in Cardiology Conference (CinC), pp. 1–4. IEEE, Holland (2018)

6. Cho, S.P., Lee, J., Park, H. D., Lee, K.J.: Detection of arousals in patients with respiratory sleep disorders using a single channel EEG. In: 2005 IEEE Engineering in Medicine and Biology 27th Annual Conference, pp. 2733–2735. IEEE, Shanghai, China (2006)
7. Mesaros, A., Heittola, T., Virtanen, T.: Metrics for polyphonic sound event detection. Appl. Sci. **66**, 162 (2016)
8. Sanchis, J.R.S., Guerrero, J., Olivas, E.S., Lopez, A.J.S.: Neural networks for the detection of EEG arousal during sleep. In: Proceedings of the 6th Internet World Congress for Biomedical Sciences, Ciudad Real, Spain (2000)
9. Miller, D., Ward, A., Bambos, N.: Automatic sleep arousal identification from physiological waveforms using deep learning. In: 2018 Computing in Cardiology Conference (CinC), pp. 1–4. IEEE, Holland (2018)
10. Pacheco, O.R., Vaz, F.: Integrated system for analysis and automatic classification of sleep EEG. In: Proceedings of the 20th Annual International Conference of the IEEE Engineering in Medicine and Biology Society, Biomedical Engineering Towards the Year 2000 and Beyond, vol. 20, pp. 2062–2065. IEEE (1998)
11. Álvarez-Estévez, D., Moret-Bonillo, V.: Identification of electroencephalographic arousals in multichannel sleep recordings. IEEE Trans. Biomed. Eng. **58**(1), 54–63 (2010)
12. Perslev, M., Jensen, M., Darkner, S., Jennum, P.J., Igel, C.: U-Time: A fully convolutional network for time series segmentation applied to sleep staging. In: Advances in Neural Information Processing Systems, pp. 4417–4428. (2019)
13. Kim, H., Jun, T. J., Kim, D.: Recurrent neural networks-based sleep arousal detection model using MFCC as a feature vector. (2018)
14. Ronneberger, O., Fischer, P., Brox, T.: U-net: convolutional networks for biomedical image segmentation. In: International Conference on Medical image computing and computer-assisted intervention, LNCS, vol. 9351, pp. 234–241. Springer, Cham (2015). https://doi.org/10.1007/978331924574428
15. Shoeb, A., Sridhar, N.: Evaluating convolutional and recurrent neural network architectures for respiratory-effort related arousal detection during sleep. In: 2018 Computing in Cardiology Conference (CinC), pp. 1–4. IEEE, Holland (2018)
16. Singh, S.A., Majumder, S.: A novel approach OSA detection using single-lead ECG scalogram based on deep neural network. J. Mech. Med. Biol. **19**(04), 1950026 (2019)
17. Ghassemi, M.M., et al.: You snooze, you win: the physionet/computing in cardiology challenge 2018. In: 2018 Computing in Cardiology Conference (CinC), pp. 1–4. IEEE (2018)
18. Odena, A., Dumoulin, V., Olah, C.: Deconvolution and checkerboard artifacts. Distill **1**(10), e3 (2016)
19. Kingma, D. P., Ba, J.: Adam: A method for stochastic optimization. arXiv preprint arXiv:1412.6980 (2014)
20. Chambon, S., Galtier, M.N., Arnal, P.J., Wainrib, G., Gramfort, A.: A deep learning architecture for temporal sleep stage classification using multivariate and multimodal time series. IEEE Trans. Neural Syst. Rehabil. Eng. **26**(4), 758–769 (2018)

Author Index

Abbasi, Nida Itrat 79
Ahmed, Sheraz 38, 318, 544
Alshammari, Bandar 26
Amini, Massih-Reza 282
Anwar, Saeed 172
Aouada, Djamila 380
Asim, Muhammad Nabeel 38

Babbar, Rohit 282
Bezerianos, Anastasios 79
Brewer, Madeline 343

Cai, Weidong 185
Cai, Zhenting 366
Caldwell, Sabrina 172
Calisa, Rajanish 185
Catchpoole, Daniel 3
Chang, Wei-Cheng 555
Chaudhury, Anumita 79
Chen, Cheng 98
Chen, Dongkai 532
Chen, Tianyi 532
Chen, Wenxiao 366
Chen, Yao 629
Chen, Yufei 432
Chen, Zulong 479
Cheng, Luqi 98
Cheng, Wen Xin 306

Dabrowski, Joel Janek 579
Deng, Qiujun 432
Dengel, Andreas 38, 318, 544
Dodballapur, Veena 185
Dong, Tengran 532
Dong, Xuyuan 444
Dragomir, Andrei 79
Du, Jonathan 15
Dudek, Grzegorz 519

Eghan, Moses J. 353

Fan, Lingzhong 98
Figueiredo, Mário A. T. 258
Friedrich, Tobias 592

Gao, Min 455
Gao, Neng 467
Gedeon, Tom 172
Gheorghe, Lucian 149
Gu, Lingyun 110
Guo, Xudong 331

Hamano, Junji 79
Han, Hongfang 110, 122
Hao, Yu 555
Hong, Zhongxian 110
Hu, Jiahao 63
Huang, Xiaoming 629
Huang, Zhan ao 629

Ibrahim, Muhammad Ali 38
Imura, Jun-ichi 149

Jastrzębski, Stanisław 294
Jia, Bijue 629
Jiang, Zhi 479
Jin, Canghong 532
Juda, Mateusz 258
Jurj, Sorin Liviu 232

Katuwal, Rakesh 306
Kennedy, Paul 3
Khan, Anis 15
Khan, Shah Khalid 26
Khosravi, Abbas 197
Khushi, Matloob 3, 15, 26, 49
Kokai, Yuki 90
Kołomycki, Maciej 258
Koprinska, Irena 616
Kou, Yingshuai 467
Krejca, Martin S. 592

Lagodzinski, J. A. Gregor 592
Lai, Guokun 555
Leung, Chi-Sing 162
Li, Chen 63
Li, Suiwu 504
Li, Xuewei 444, 504
Li, Yu 479

Li, Yun 568
Lin, Shilu 395
Lin, Yang 616
Liu, Hongwei 444, 504
Liu, Junxiu 605
Liu, Wenda 366
Liu, Xianhui 432
Liu, Yang 172
Liu, Zheng 568
Lu, Wenhao 162
Lu, Xuesong 110
Luo, Lang 432
Luo, Yuling 605
Lv, Jiancheng 629

Ma, Chenxiang 208
Malik, Muhammad Imran 38
Maziarka, Łukasz 294
Mehrgardt, Philip 49
Mercier, Dominique 318
Ming, Zhong 407
Mir, Adnan 26
Mumtaz, Wajid 245

Nahavandi, Saeid 197
Nambu, Isao 90
Naseem, Usman 3, 15, 26
Neruda, Roman 270
Ni, Yongxin 407
Nowak, Aleksandra 294

Opritoiu, Flavius 232
Oyedotun, Oyebade K. 380

Pan, Weike 407
Panta, Adhish 3
Pauletto, Loïc 282
Pei, Dan 366
Pełka, Paweł 519
Peng, Keqin 63
Poon, Simon 49
Poon, Simon K. 15, 26
Prusinowski, Karol 135
Przewięźlikowski, Marcin 220

Qayyum, Abdul 245
Qazi, Atika 26
Qin, Zhenyue 172
Qiu, Senhui 605
Qu, Hong 353

Rahman, Ashfaqur 579
Rahman, Jessica Sharmin 343
Rana, Mashud 616
Razzak, Imran 245
Rizzo, Manuel 592
Rong, Wenge 63

Saint-Auret, Sony 79
Shan, Shajia 568
Shen, Ziyu 568
Siddiqui, Shoaib Ahmed 544
Śmieja, Marek 220, 258
Smyl, Slawek 519
Song, Shiming 208
Song, Yang 185
Spurek, Przemysław 294
Struski, Łukasz 220, 258
Suganthan, P. N. 306
Sum, John 162
Sysko-Romańczuk, Sylwia 135

Tabarisaadi, Pegah 197
Tabor, Jacek 294
Tang, Haoran 420
Thakor, Nitish V. 79
Tu, ChenYang 467
Turkson, Regina Esi 353

Vidnerová, Petra 270
Vladutiu, Mircea 232

Wada, Yasuhiro 90
Waheed, Nazar 26
Wang, Chen 395
Wang, Haixian 110, 122
Wang, Jiong 467
Wang, Runsheng 455
Wang, Xiaoyu 149
Wang, Yuchen 353
Wang, Ziyi 479
Wei, Mengting 122
Wen, Junhao 420
Winckler, Nicolas 282
Withana, Anusha 49
Wróblewska, Anna 135
Wu, Di 395
Wu, Guopei 605

Xia, Bin 568
Xiao, Wendong 479
Xiong, Xingliang 110, 122
Xiong, Zhang 63
Xiu, Li 491
Xu, Haowen 366
Xu, Junhai 98
Xu, Tianyi 444, 504

Yan, Jianfeng 122
Yan, Jiangpeng 491
Yang, Shenglan 629
Yang, Yiming 555
Yang, Yujiu 491
Yu, Donghan 555
Yu, Jian 444, 504
Yu, Mei 444, 504
Yu, Qiang 208
Yu, Ruiguo 444, 504

Zahn, Arthur 592
Zeng, Jun 420
Zhan, Zhuoxin 407
Zhang, Junwei 455
Zhang, Lingxiao 491
Zhang, Ruohong 555
Zhang, YiFan 579
Zhang, Yifei 467
Zhang, Zhen 98
Zhao, Mankun 444, 504
Zhao, Quanwu 455
Zhen, Chengchang 444
Zhong, Zhicong 395
Zhou, Yipeng 395

Printed in the United States
By Bookmasters